Analytical Ultracentrifugation
Techniques and Methods

Analytical Ultracentrifugation
Techniques and Methods

Edited by

David J. Scott, Stephen E. Harding and Arthur J. Rowe
National Center for Macromolecular Hydrodynamics,
University of Nottingham, UK

RSCPublishing

ISBN 0-85404-547-3

A catalogue record for this book is available from the British Library

Published by The Royal Society of Chemistry,
Thomas Graham House, Science Park, Milton Road,
Cambridge CB4 0WF, UK

Registered Charity Number 207890

For further information see our web site at www.rsc.org

Typeset by Macmillan India Ltd, London, UK
Printed by Biddles Ltd, Norfolk, UK

Preface

The state of the analytical ultracentrifugation (AUC) field is in very good health some 85 years after the invention of the technique. There has, since the publication in 1992 of the last Royal Society of Chemistry volume ("Analytical Ultracentrifugation in Biochemistry and Polymer Science"), been an explosion of interest in the field fuelled by the increase in the number of analytical ultracentrifuges and the wide range of new methods of data analysis. Allied to this has been the realisation that simply knowing the gene sequence of a protein is not enough, even in this post-genomic era, to describe a protein's solution state properties. This provides great optimism for the years ahead for the AUC field.

This collection of invited chapters evolved out of the 7th UK AUC meeting at Oxford in 2004, when it was recognised that there were many new users to the field who had not trained in the laboratories of the 'usual suspects' of the discipline. There was a great demand for this volume among these users; therefore we have endeavoured to write and edit a resource that not only provides a summary of the state of the field as we find it in 2005, but introduces the novice to the huge breadth and depth of applications of the technique.

It is hoped that this volume will be a resource to all users in the field. As always, users in the AUC field can keep in contact with the latest developments via the RASMB notice board (http://www.bbri.org/RASMB/rasmb.html), and new users are urged very strongly to join the debate.

David Scott
Stephen Harding
Arthur Rowe
Nottingham, 2005

Contents

Contributors

Z. Aziz, *University of Nottingham, Information Services, University Park, Nottingham NG7 2GG, UK.*

J. Behlke, *Max Delbrück Center for Molecular Medicine, 13092 Berlin, FRG; Tel./Fax: +49 30 9406 2802; E.-mail: behlke@mdc-berlin.de.*

G. Bernardi, *Stazione Zoologica Anton Dohrn, Villa Comunale, 80121 Naples, Italy.*

L. Bourdillon, *Institute of Chemical Sciences and Engineering, Faculty of Basic Sciences, Swiss Federal Institute of Technology Lausanne, CH-1015 Lausanne, Switzerland.*

P.J.G. Butler, *MRC Laboratory of Molecular Biology, Hills Road, Cambridge, CB2 2QH, UK; Tel:+44 1223 248011 (Switchboard) +44 1223 402296 (DDI for PJGB); Fax: +44 1223 213556; e-mail: pjgb@mrc-lmb.cam.ac.uk.*

N. Carels, *Stazione Zoologica Anton Dohrn, Villa Comunale, 80121 Naples, Italy.*

O. Clay, *Stazione Zoologica Anton Dohrn, Villa Comunale, 80121 Naples, Italy.*

H. Cölfen, *Max-Planck-Institute of Colloids and Interfaces, Colloid Chemistry, Am Mühlenberg, 14424 Potsdam, Germany.*

J.J. Correia, *Department of Biochemistry, University of Mississippi Medical Center, Jackson, MS 39216, USA.*

M.A. Daugherty, *Department of Biochemistry, Given B409, The University of Vermont College of Medicine, Burlington, VT 05405, USA.*

B. Demeler, *The University of Texas Health Science Center at San Antonio, Department of Biochemistry, MC 7760, 7703 Floyd Curl Drive, San Antonio, TX 78229-3901, USA; Tel.: 210 567 6592; Fax: 210 567 1136.*

C.J. Douady, *Stazione Zoologica Anton Dohrn, Villa Comunale, 80121 Naples, Italy and Equipe d'Hydrobiologie et Ecologie Souterraines & Plateforme d'Ecologie Moléculaire, Laboratoire d'Ecologie des Hydrosystèmes Fluviaux, Université Claude Bernard Lyon 1-UMR CNRS 5023, F-69622 Villeurbanne Cedex, France.*

H. Durchschlag, *Institute of Biophysics and Physical Biochemistry, University of Regensburg, D-93040 Regensburg, Germany.*

K.G. Fleming, *T.C. Jenkins Department of Biophysics, Johns Hopkins University, 3400 North Charles Street, Baltimore, MD 21218, USA; Tel.: 410 516 7256; Fax: 410 516 4118; e-mail: Karen.Fleming@jhu.edu.*

M.G. Fried, *Department of Molecular and Cellular Biochemistry, University of Kentucky College of Medicine, 800 Rose Street, Lexington, KY 40536, USA.*

P.B. Furtado, *Department of Biochemistry and Molecular Biology, Darwin Building, University College London, Gower Street, London WC1E 6BT, UK.*

J. Garcia de la Torre, *Departamento de Química Física, Universidad de Murcia, 30071 Murcia, Spain; e-mail: jgt@um.es.*

H.E. Gilbert, *Department of Biochemistry and Molecular Biology, Darwin Building, University College London, Gower Street, London WC1E 6BT, UK.*

R.J.C. Gilbert, *Division of Structural Biology, Henry Wellcome Building for Genomic Medicine, University of Oxford, Roosevelt Drive, Oxford OX3 7BN, UK and Oxford Centre for Molecular Sciences, Central Chemistry Laboratory, University of Oxford South Parks Road, Oxford. OX1 3QH, UK; Tel.: +44 1865 287535; Fax: +44 1865 287547; e-mail: gilbert@strubi.ox.ac.uk.*

S.E. Harding, *NCMH, Physical Biochemistry Laboratory, University of Nottingham, School of Biosciences, Sutton Bonington LE12 5RD, UK; Tel.: +44 9516148; Fax: +44 95161452; e-mail: Steve.Harding@nottingham.ac.uk.*

L.A. Holladay, *Molecular Interactions Resource, Division of Bioengineering and Physical Science, Office of Research Services, National Institutes of Health, Bethesda, MD 20893, USA.*

A. Krebs, *Structural and Computational Biology Programme, European Molecular Biology Laboratory, D-69117 Heidelberg, Germany.*

I. Laidlaw, *Analytik Ltd, Unit 4 The Acorn Centre, Chestnut Avenue, Biggleswade, Bedfordshire SG18 0RA, UK.*

Y.C. Lee, *Department of Biochemistry and Molecular Biology, Darwin Building, University College London, Gower Street, London WC1E 6BT, UK.*

M.S. Lewis, *Molecular Interactions Resource, Division of Bioengineering and Physical Science, Office of Research Services, National Institutes of Health, Bethesda, MD 20892, USA.*

A. Ortega, *Departamento de Química Física, Universidad de Murcia, 30071 Murcia, Spain; e-mail: jgt@um.es.*

H.E. Perez Sanchez, *Departamento de Química Física, Universidad de Murcia, 30071 Murcia, Spain; e-mail: jgt@um.es.*

S.J. Perkins, *Department of Biochemistry and Molecular Biology, Darwin Building, University College London, Gower Street, London WC1E 6BT, UK.*

M.M. Reily, *Molecular Interactions Resource, Division of Bioengineering and Physical Science, Office of Research Services, National Institutes of Health, Bethesda, MD 20892, USA.*

O. Ristau, *Max Delbrück Center for Molecular Medicine, 13092 Berlin, FRG; Tel./Fax: +49 30 9406 2802; e-mail: behlke@mdc-berlin.de.*

A.J. Rowe *NCMH, University of Nottingham, School of Biosciences, Sutton Bonington, LE12 5RD, UK.*

P. Schuck, *Protein Biophysics Resource, Division of Bioengineering & Physical Science, ORS, OD, National Institutes of Health, Bethesda, MD 20892, USA; e-mail: pschuck@helix.nih.gov.*

D.J. Scott, *National Centre for Macromolecular Hydrodynamics, School of Biosciences, University of Nottingham, Sutton Bonington, Leicestershire, LE12 5RD, UK; e-mail: dj.scott@nottingham.ac.uk.*

P.J. Sherwood, *Analytical Ultracentrifugation Research Laboratory, Boston, Biomedical Research Institute, Watertown, MA 02472, USA.*

C.A. Sontag, *Department of Biochemistry, University of Mississippi Medical Center, Jackson, MS 39216, USA.*

W.F. Stafford, *Analytical Ultracentrifugation Research Laboratory, Boston, Biomedical Research Institute, Watertown, MA 02472, USA.*

M. Steinmetz, *Analytik Ltd, Unit 4 The Acorn Centre, Chestnut Avenue, Biggleswade, Bedfordshire SG18 0RA, UK.*

Z. Sun, *Department of Biochemistry and Molecular Biology, Darwin Building, University College London, Gower Street, London WC1E 6BT, UK.*

C.G. Tate, *MRC Laboratory of Molecular Biology, Hills Road, Cambridge CB2 2QH, UK; Tel.: +44 1223 248011 (Switchboard) +44 1223 402296 (DDI for PJGB); Fax: +44 1223 213556; e-mail: pjgb@mrc-lmb.cam.ac.uk.*

C. Wandrey, *Institute of Chemical Sciences and Engineering, Faculty of Basic Sciences, Swiss Federal Institute of Technology Lausanne, CH-1015 Lausanne, Switzerland.*

P.R. Wills, *Department of Physics, University of Auckland, Auckland, New Zealand.*

D.J. Winzor, *Department of Biochemistry, University of Queensland, Brisbane, Australia.*

P. Zipper, *Physical Chemistry, Institute of Chemistry, University of Graz, A-8010 Graz, Austria.*

CHAPTER 1

A *Brief Introduction to the Analytical Ultracentrifugation of Proteins for Beginners*

DAVID J. SCOTT AND PETER SCHUCK

1 Introduction

Analytical ultracentrifugation (AUC) is one of the classical methods for the characterization of purified proteins in dilute solutions, and as such there is a large body of theoretical and practical studies in the literature going back eight decades, covering aspects including the technical implementation of AUC,[1] the theoretical foundation of ultracentrifugation in thermodynamics and physical chemistry of macromolecules,[2] and the mathematical analysis of ultracentrifugation experiments.[3] Developed in the 1920s by The Svedberg,[4] and redesigned into a commercial instrument by Edward Pickels,[5] it became a central technique in the development of biochemistry and molecular biology. While the technique was in decline in the 1970s and 1980s, new instrumentation and numerical analysis in the 1990s stimulated renewed interest in AUC, in particular for the study of protein interactions.[6]

For the novice, however, the highly developed technical aspects in conjunction with the multitude of analytical approaches can be daunting. Therefore, this introduction to the book is unashamedly aimed at the novice who has come into contact with the methodology of the analytical ultracentrifuge and wishes to use the instrument for the characterization of globular protein samples in solution. Such a novice can be a molecular biologist, a protein crystallographer or NMR spectroscopist, or working as part of a high-through-put proteomics project, with interest in the characterization of the aggregation state, heterogeneity and thermodynamic characterization of reversible interactions of proteins. The aim of the present introduction is to provide help in where to start, or how to design and analyse a successful AUC experiment. Necessarily, this requires a very selective presentation, and for more in-depth information and descriptions of selected sedimentation velocity (SV) and sedimentation equilibrium (SE) methods, the reader is referred to recent reviews, monographs, practical protocols, and websites,[1,7–25] and, of course, subsequent chapters in this book. Fortunately, with regard to the data interpretation, the mathematical details of advanced analysis methods are largely encapsulated in software. However, in addition to general knowledge in modelling of data and non-linear regression, their use requires understanding of the basic concepts behind them.

2 What Can I Do with My Protein Sample?

At its most basic level, AUC simply consists of the application of a centrifugal force with the simultaneous real-time observation of the resulting redistribution of the macromolecule. Analysis can be performed from first principles, given quantitative and rigorous data on a particular sample. This requires no label or other chemical modification of the proteins; such as would occur in cross-linking or many fluorescence experiments; and no interaction with any matrix or surface, as would be required for gel filtration or surface plasmon resonance. As proteins are studied in solution, the experimentalist has direct access to their solution properties: a key strength of the technique. A central feature of sedimentation experiments for the study of protein interactions is that faster sedimenting complexes are transported through a solution of the slower sedimenting components. Consequently, reversibly formed complexes that dissociate can readily re-associate during the experiment, thus permitting the hydrodynamic and thermodynamic characterization of even weak and transient interactions.

Analytical ultracentrifugation of non-interacting proteins can reveal the molar mass, gross shape, and the heterogeneity of the sample. The latter includes the detection of even trace quantities of oligomers and aggregates, which can be of interest in biotechnology applications or aid in the interpretation of biosensor experiments. For interacting systems, protein complexes can be characterized with regard to their stoichiometry and the thermodynamic and kinetic constants of complex formation. Importantly, sedimentation techniques can distinguish between multiple coexisting complexes of different stoichiometries and also provide information on self-association properties, on mixed self- and hetero-association. The latter can be a crucial information for the biophysical study of protein interactions with other techniques, such as isothermal titration calorimetry. Hydrodynamic separation also yields information on the low-resolution structure of protein complexes and can enable the detection of conformational changes.

Although a large number of specialized centrifugation techniques have been developed for a variety of studies, such as analytical zone sedimentation,[11] difference sedimentation,[26] synthetic boundary measurements,[27] density gradient[28,29] or fractionating[30,31] and short-column techniques,[32] the vast majority of ultracentrifugation experiments for the characterization of proteins are conducted by either conventional loading SV or long-column SE, which will be described in the following sections.

Ideally, both SV and SE techniques should be carried out on your sample, starting with SV to characterize the purity of the material. It is possible to characterise effectively a sample with SV using the latest analysis methods, however, both methods are highly complementary. The essential piece of information needed to plan an AUC experiment is to make sure at least three different concentrations of the sample are analysed, covering as wide a concentration range as possible. This is because concentration-dependent behaviour of the sample provides an information-rich data set that is highly effective in characterizing the solution properties of proteins, in particular, their interactions. This applies equally for SV and for SE. For equilibrium experiments there is the additional requirement that data should be obtained, where possible, in a sequence of experiments at three different rotor speeds.

The quantity of material required is typically of the order of a few hundred micrograms. Owing to the concentration gradients established during centrifugation, a 10- to 1000-fold concentration range is typically observed in a single cell, and a size range of three decades in molar mass can be covered in a single experiment. Interacting components under study may have sizes ranging from peptides to very large multiprotein complexes. In general, affinities in the range of 10^4–10^8 M^{-1} can be determined, and kinetic dissociation rate constants of the order of ~10^{-5}–10^{-2} s^{-1} can be distinguished.

3 General Principles

Conceptually, the analytical ultracentrifuge can be thought of as a conventional preparative centrifuge that is equipped with an optical system for the observation of the protein distribution in real time during the centrifugation. The acquired data report on the spatial gradients that result from the application of the centrifugal field, and their evolution with time. Analytical rotors accept specialized assemblies for containing, typically, 100–400 μL of sample between windows that are optically transparent and perpendicular to the plane of rotation. The optical detection system is synchronized with the rotor revolution, such that data are acquired only while the sample assembly is in the light path (Figure 1).

The most commonly used optical detection systems are a dual-beam UV/VIS spectrophotometer equipped with monochromator (ABS) and a highly sensitive laser interferometer which records the refractive index gradients (IF). The ABS system requires typical loading concentrations between 0.1 and 1.5 OD, dependent on the

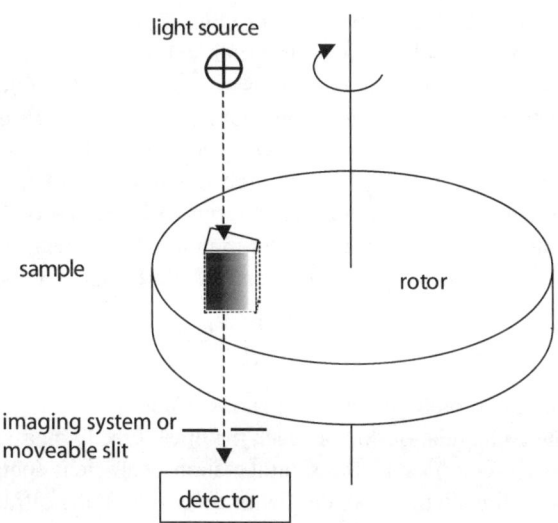

Figure 1 *Schematics of the analytical ultracentrifugation detection*

type of experiment, and has the advantage of being able to selectively detect the protein, for example at 280 nm (for the aromatic amino acids), 230 nm (peptide backbone), or for characteristic chromophores in the VIS, if present. The IF system is not selective, but completely linear in concentration and offers high signal-to-noise ratios and rapid data acquisition (which can be particularly valuable for observing the time course of sedimentation). Drawbacks of the IF system are the experimental requirement to have a precise match of the composition of the reference buffer, and systematic time-invariant signal offsets to be accounted for in the data analysis (see below). More practical information with regard to the selection of the optical system can be found in refs. 14, 16 and 33.

Two basic types of experiments are possible: (a) the application of a high centrifugal force and the analysis of the time course of the sedimentation process, termed sedimentation velocity (SV); and (b) the application of a low centrifugal force that permits the diffusion to balance the sedimentation such that a time-invariant equilibrium gradient can be observed, termed sedimentation equilibrium (SE). Both SV and SE approaches are uniquely suited for the study of protein interactions. First, as a basis for the analysis of protein interactions, it is necessary to become familiar with the principle of sedimentation for non-interacting proteins.

The sedimentation process is governed by three factors – the gravitational force, the buoyancy and the hydrodynamic friction. The gravitational force is $F_{sed} = m\omega^2 r$ (with m the protein mass, ω^2 the rotor angular velocity, and r the distance from the centre of rotation). Since it is proportional to the square of the rotor speed, adjusting the rotor speed permits the study of a wide range of particle sizes, ranging from small peptides to very large protein complexes (<1 kDa to >1 GDa). The buoyancy force $F_b = -m\bar{v}\rho\omega^2$ (with $\bar{v}$ the effective protein partial-specific volume and ρ the solvent density) opposes the sedimentation (following Archimedes principle) and is governed by the mass of the displaced solvent. Thus, protein partial-specific volumes are important, and the density effects of protein glycosylation, bound detergent and preferential hydration may be relevant considerations. Finally, the frictional force is governed by the hydrodynamic translational frictional coefficient as well as the migration velocity, and can be expressed as $F_f = s(kT/D)\omega^2$ (with k the Boltzmann constant, T the absolute temperature, and D the diffusion constant), where the sedimentation coefficient $s = v/\omega^2$ is a molecular constant (with v the absolute migration velocity). This permits the measurement of the low-resolution shape of the proteins and their complexes in terms of Stokes radii. A key parameter is the sedimentation coefficient s (measured in units of Svedberg, with $1\,S = 10^{-13}$ s). From the balance of these three forces, one can derive the Svedberg equation

$$\frac{s}{D} = \frac{M(1-\bar{v}\rho)}{RT} \tag{1}$$

(with M denoting the protein molar mass, and R denoting the gas constant),[4] which provides a fundamental relationship between the three directly measurable quantities for a single protein component: the sedimentation coefficient (obtained from the migration of the sedimentation boundary with time in SV), the diffusion coefficient (obtained from the spread of the sedimentation boundary with time in SV) and the molar mass (obtained from the exponential gradient in SE).

4 Sedimentation Velocity

The principle behind SV is quite straightforward: at a high rotor speed, typically 40 000 rpm or above, the vast majority of proteins will sediment to the bottom of the cell. This deceptively simple process is highly information rich, as the sedimentation depends both on the size and shape of the protein. Hence a 50-kDa protein may sediment slower than a 40-kDa protein due to the heavier protein being highly elongated. If a protein self-associates with concentration then its apparent rate of sedimentation will change when either raising or lowering the loading concentration. SV will also give information on solution heterogeneity.

4.1 Setting Up a Simple Sedimentation Velocity Experiment

When first characterizing a sample by AUC, a typical amount of protein required would be 1 mL at 1 OD_{280}, assuming the use of the ABS detection system, or at 0.5–1 mg mL^{-1}, assuming the IF detection system. This will be enough for SV with three cells at three-fold dilutions, and have some sample left to perform an SE run. The three velocity cells should be loaded with 400 μL of sample: this will be a sample volume that fills *ca.* 85% of the length of the cell. Typically, concentrations can be at 1:1, 1:3 and 1:10 of the stock concentration. Generally, the concentration range should be as wide as possible, with the constraint not to exceed 1 mg mL^{-1} (to prevent non-ideal sedimentation), and to remain well within the detection limits (*e.g.*, 0.05 mg mL^{-1} $< c$ for the IF system, and 0.05 OD $< c <$ 1.5 OD for the ABS system). The reason for the concentration series is simple: for a self-associating system, lowering the concentration sufficiently will cause oligomers to dissociate into smaller species, which will sediment slower. However, if the protein is present in a variety of oligomers that do not interconvert, then dilution will have no effect whatsoever on the proportion of oligomers. Hence, a dilution series is essential to characterizing any protein in solution. Indeed, one SV experiment (of three dilutions) will very quickly tell the experimentalist in a matter of hours whether their protein sample undergoes a complex set of self-associations of exquisite biological necessity, or that the protein is an aggregated mess, and another sample preparation is needed.

The protein purity should be >95%, and it is recommended to perform size-exclusion chromatography as the last preparative step. With regard to the buffer requirements, it is useful to include at least 20 mM salts to suppress electrostatic interactions. Obviously, no solvent component should interfere with the optical detection, and reference buffer precisely matched in volume and composition is needed for experiments with the IF system. For most proteins under common experimental conditions, the effective partial-specific volume can be predicted with sufficient precision from the amino acid sequence,[19,34] for example, using the software SEDNTERP, which also permits to calculate the solvent density and viscosity from tabulated data.[35] Greater care must be used with proteins containing non-amino acid components (*e.g.*, glycoproteins, proteins with prosthetic groups, detergent-solubilized proteins), or buffer conditions containing glycerol, sucrose or other components increasing the density and potentially leading to preferential solvation.[23,34,36]

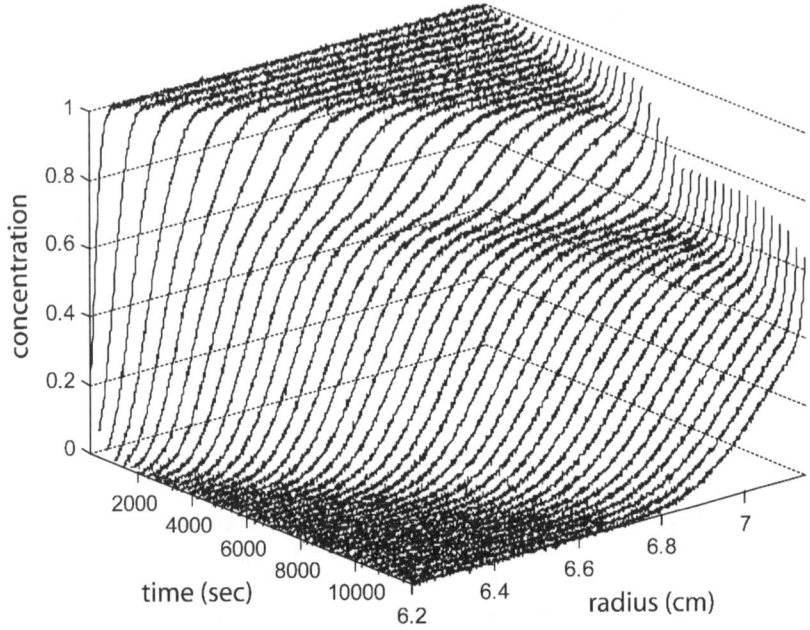

Figure 2 *Sedimentation profiles of two species of 50 kDa, 4 S, and 150 kDa, 8 S, sedimenting at 50 000 rpm*

In the absence of other considerations (and unless the protein complexes under study exceed several 100 kDa in size), the highest possible rotor speed (50 000 or 60 000 rpm, dependent on the rotor type used) is recommended for the highest resolution. A practical aspect crucial for the SV experiment is a thorough temperature equilibration to prevent convection in the solution column to occur. Very detailed practical instructions for planning and starting the SV run can be found in ref. 33.

Typical sedimentation data may look like as shown in Figure 2, showing initially the solution plateau, the evolution of the sedimentation boundaries of a non-interacting two-component mixture and the solvent plateau. Owing to the sector-shaped solution column, the solution plateau concentration decreases with time as the proteins move toward larger radial position with wider cross-sections. This amounts typically to 10–20% during the SV experiment.

To harness this powerful method of analysis, a suitable method of interpretation is needed. This is the concern of the next section.

4.2 Sedimentation Velocity Analysis

As indicated above, the major parameter derived from SV experiments is the sedimentation coefficient (s), which is the velocity per unit centrifugal force:

$$s = \frac{1}{\omega^2 r} \frac{dr}{dt} = \frac{m}{f} \qquad (2)$$

where s is the sedimentation coefficient, $\omega^2 r$ the centrifugal field, r the radius, t time of sedimentation, m the buoyant molecular mass and f the frictional coefficient. Although the sedimentation coefficient (or s value) is a molecular constant, the experimentally observed sedimentation coefficients will vary with temperature and solution density, and are always converted into a standard state of 20 °C in water at infinite dilution (denoted $s_{20,w}^0$). This can be achieved, for example, using the software SEDNTERP, written by Dr. John Philo.[35] A comparison of the $s_{20,w}^0$ value with that predicted theoretically for a smooth compact sphere of the same mass and density gives the frictional ratio, f/f_0, which reports on the shape of the molecule (by conversion into hydrodynamically equivalent ellipsoids with the software SEDNTERP, or by comparison with the theoretically predicted s value derived by hydrodynamic modelling from the crystal structure using the software HYDRO).[37] Because the $s_{20,w}^0$ value cannot exceed that of a smooth compact sphere of the same mass and density, the $s_{20,w}^0$ value can be used in some cases to deduce the oligomeric state of the protein.[19]

The simplest analytical method to derive the sedimentation coefficient from experimental data consists in reading the boundary displacement with time, and application of Equation (2). Although this illustrates conceptually the source of information, it does not provide the most precise estimate for the s value, and neglects a significant amount of information to be extracted from the sedimentation data. As can be seen in Figure 1, in addition to a displacement of the sedimentation boundary, diffusional broadening can be observed, which can be rich in information.

4.2.1 Modelling with Discrete Lamm Equation Solutions

In the other extreme, it is possible to base the analysis on the detailed sedimentation–diffusion equation, the Lamm equation[38]

$$\left(\frac{\partial c}{\partial t}\right)_r = -\frac{1}{r}\left\{\frac{\partial}{\partial r}\left[s\omega^2 r^2 c - Dr\left(\frac{\partial c}{\partial r}\right)_t\right]\right\}_t \tag{3}$$

which describes the evolution of the complete concentration profiles during the SV experiment in the sector-shaped ultracentrifugal solution column. Several software packages allow fitting Equation (3) to experimental data, in order to determine both s and D (and consequently the molar mass M; see Equation (1)) for one or several species. These include BPCFIT, LAMM, SEDANAL, SEDFIT, SEDPHAT, SVEDBERG and ULTRASCAN. Modern approximate analytical and numerical algorithms for solving Equation (3) are described in refs. 39–44, and adaptations to account for non-ideal solutions, solvent and pressure gradients have been developed.[45–47] Although rigorous in theory, in practice, this approach is not easily applicable in many cases, due to the need to know in advance the number of protein species. This can be difficult given the exquisite sensitivity of SV to heterogeneity, including trace amounts of degradation products or impurity. Further, as illustrated in ref. 48, if species are present with slightly different s value (such that the difference in migration due to the different s values is less than the diffusional broadening), unrecognized heterogeneity will lead to an overestimation of D, and consequently an underestimate of the molar mass. Therefore, in practice, direct modelling with

Equation (3) will frequently give only a lower limit of the molar mass. If, on the other hand, Equation (3) can be fitted to the experimental data with residuals within the noise of the data acquisition with the correct molar mass, this provides a stringent demonstration of sample homogeneity.

Several strategies have been developed to determine sedimentation coefficient distributions, which can report on the heterogeneity of the sample, and in some cases may be used to derive molar mass information. Some are outlined in the following in chronological order, making increasing use of the computational power of desktop computers to apply more mathematically complex approaches embedded in the software packages. They also differ in the ability and strategy used to account for the systematic radial-dependent baseline profiles that occur in data acquired with the IF system.

4.2.2 The van Holde–Weischet Distribution

The integral sedimentation coefficient distribution $G(s)$ introduced in 1978 by van Holde and Weischet[49] basically exploits the observation that the displacement from diffusion proceeds proportionally to the square root of time, while the sedimentation is linear with time. Technically, this is accomplished by dividing the sedimentation boundaries into fractions (horizontal slices), reflecting the radii R_i of certain intervals of the fractional plateau concentration, followed by a transformation of the radius coordinate into apparent sedimentation coefficients $s*$ with

$$s* = \frac{1}{\omega^2 t} \ln \frac{r}{r_m} \tag{4}$$

$s*$ simply reflects the velocity at which an ideal particle would have to migrate in order to travel from the meniscus to the position r within the time t. The s_i^* values from the corresponding boundary fractions of the different scans are then extrapolated to infinite time according to the formula

$$s_i^*(t) = s - \frac{2\sqrt{D}}{\omega^2 r_m} \Phi^{-1} \left(1 - \frac{2i}{N} \right) \times \frac{1}{\sqrt{t}} \tag{5}$$

with Φ^{-1} denoting the inverse error function, which is based on an Faxén-type approximate solution of the Lamm equation for a single species.[3,49] A drawback of this method is that it requires the selection of the subset of experimental scans for the analysis, such that the scans to be analysed exhibit clear solution and solvent plateaus permitting the boundary division. For peptides and small proteins, this is not always possible. Also, it is difficult to apply to data from the IF detection exhibiting systematic radial-dependent baseline offsets.[50]

The method can be found in the software ULTRASCAN[8,51] and SEDFIT.[50] The application is illustrated in Figure 3. For mixtures of sedimenting species, the $G(s)$ distribution ideally shows piecewise vertical segments corresponding in length to the relative amount of species with certain s values. Because the underlying Equation (5) is valid for a single species only, $G(s)$ cannot resolve species that do not show clearly separating boundaries (*i.e.* if the difference in migration due to the different s values is less than the diffusional broadening).[50] In this case, the $G(s)$ transformations will exhibit

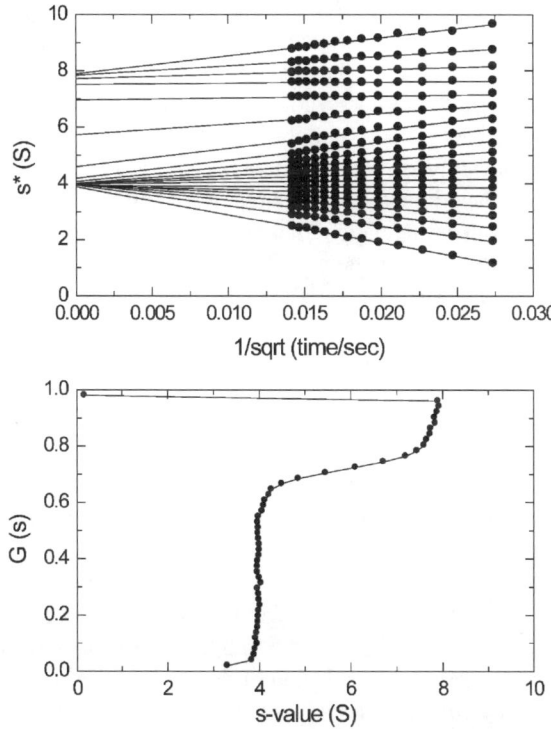

Figure 3 *Van Holde–Weischet analysis of the data shown in Figure 1. Upper panel: the s_i^* values from each boundary (vertical columns of symbols), and the extrapolation of the corresponding boundary fractions in an inverse square root of time-scale (lines). Lower panel: the extrapolated s values for each fraction, which compose the integral distribution $G(s)$*

diagonal lines with positive slopes within the range of s values of the sedimenting species. This highlights the particular strength of the method, which is the diagnosis if the sample is homogeneous, heterogeneous, or if it exhibits concentration-dependent repulsive interactions leading to non-ideal sedimentation. (The latter is characterized by a decreasing sedimentation coefficient with increasing concentrations.) In summary, for homogeneous samples, $G(s)$ will show a vertical line, for heterogeneous or associating systems it will be a diagonal line with positive slope, and for non-ideal sedimentation it will be a diagonal line with negative slope. Demeler discusses this methodology in greater depth in Chapter 11.

4.2.3 Apparent Sedimentation Coefficient Distributions g*(s)

A different strategy to analyse experimental SV data is the differential sedimentation coefficient distribution $g^*(s)$. The asterisk on the g emphasizes that it is a distribution of apparent sedimentation coefficients, meaning the distribution of s values of hypothetical non-diffusing particles. By design, the area under the $g^*(s)$ distribution is equal to the amount of sedimenting material with that sedimentation coefficient.

Two different techniques to derive $g^*(s)$ are currently in use. The first is a data transformation, first introduced in 1992 by Stafford[52] and further developed by Philo,[53] based on a time derivative dc/dt of the data and termed $g(s^*)$. The other is based on a direct least-squares fit of the data with an integral equation for the distribution, termed ls-$g^*(s)$, described in 2000 by Schuck and Rossmanith.[54]

The dc/dt-based method was very important because it provided the first approach, short of subtracting experimentally measured water blanks, that was specifically adapted to account for the systematic radial-dependent baseline offsets of the IF data acquisition system. The time-dependent offsets are reduced by a vertical alignment procedure prior to the data analysis, and the radial-dependent offsets vanish by taking pair-wise differences of scans. The approximation is made that the pair wise difference $\Delta c/\Delta t$ is taken for the time derivative dc/dt. This is used, in turn, in the transformation

$$g(s^*) = \left(\frac{\partial c}{\partial t}\right)\frac{1}{c_0}\left(\frac{\omega^2 t^2}{\ln(r/r_m)}\right)\left(\frac{r}{r_m}\right)^2 \tag{6}$$

(with c_0 denoting the initial loading concentration) which is based on the equations for the sedimentation of non-diffusing species, and combined in an iterative procedure to account for radial dilution effects.[52] Details on the computational approach can be found in ref. 55. The distribution from the dc/dt approach has been termed $g(s^*)$. A limitation of this approach is the validity of the approximation $\Delta c/\Delta t \sim dc/dt$, which can cause artificial broadening of the $g(s^*)$ curves and skewing at small s values,[53,54] to an extent that is increasing with larger boundary displacement between scans and with larger time interval covered by the set of scans considered in the analysis. The dc/dt method can be found in the software DCDT+, and in this implementation, the maximum number of scans is recommended for which the errors remain within an acceptable tolerance. An approach to quantitatively account for effects of the approximation $\Delta c/\Delta t \sim dc/dt$ in the modelling of $g(s^*)$ curves is described in ref. 53. Figure 4 illustrates the result of this method.

The second approach to calculate the apparent sedimentation coefficient distribution is based on the definition of $g^*(s)$ as a distribution of non-diffusing species, each sedimenting according to a step-function $U(s, r, t)$

$$U(s, r, t) = e^{-2\omega^2 st}\begin{cases}0 & \text{for } r < r_m e^{\omega^2 st} \\ 1 & \text{else}\end{cases} \tag{7}$$

which together produce a signal

$$a(r, t) \cong \int g^*(s)U(s, r, t)\,ds \tag{8}$$

Taking advantage of the increased computational power, it is possible to directly fit Equation (8) to the experimental data.[54] Adjustments accounting for solvent compressibility can be made in Equation (7).[46] Because this method is based on the least-squares fit of the data (hence the designation ls-$g^*(s)$), the usual goodness of fit criteria, such as r.m.s. deviation and residuals, are available. To avoid noise amplification in solving Equation (8), regularization is applied to calculate the simplest distribution consistent with the raw data.

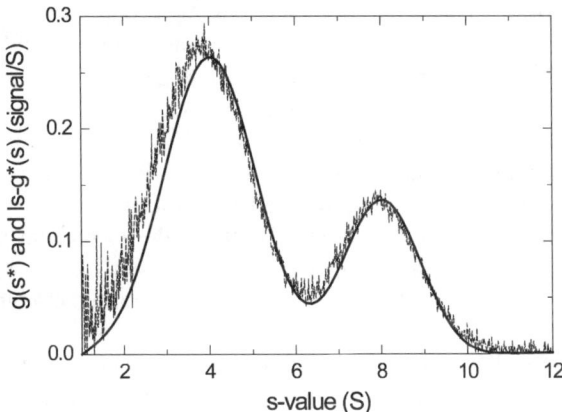

Figure 4 *Application of g(s*) from the dc/dt method (thin dotted line), and the ls-g*(s) method (bold solid line) to the data shown in Figure 2*

The systematic offsets of the IF systems are taken into account by explicitly calculating the best-fit estimates of the radial-dependent baseline profiles as well as the time-dependent offset from the data to be analysed, taking advantage of a new algebraic approach that can replace the alignment and differentiation procedure with minimal amplification of noise.[56] This can be used to remove the systematic noise contribution for inspection of the fit in the original data space. A reconstruction of the sedimentation model from the dc/dt-based $g(s^*)$ distribution has been described, which can allow to inspect the goodness-of-fit of the $g(s^*)$ representation.[20] An overview of computational details can be found in ref. 48. Although in theory the ls-$g^*(s)$ distribution is equivalent to the $g(s^*)$ distribution from dc/dt approach, because of the absence of the $\Delta c/\Delta t \sim dc/dt$ approximation, it does not have the stringent requirements for scan subset selection, and is therefore more widely applicable,[54] and more suitable, in particular, to data obtained at higher rotor speed or from the ABS optical detection system with the typical large boundary displacement between successive scans. This approach is implemented in SEDFIT, and results are also shown in Figure 4.

The limitation of its application is the underlying idea of an apparent sedimentation coefficient distribution, which is that the sedimentation–diffusion process under observation can be modelled by an apparent distribution of non-diffusing species, such that the leading and trailing parts of the diffusionally broadened boundary can be described by populations of species apparently sedimenting faster or slower, respectively. In the direct boundary model, the quality of the approximation of the sedimentation boundary as a superposition of non-diffusing species can be tested by inspecting the quality of fit. As in any data analysis problem, if the model does not adequately describe the data, no trust can be put in the parameters of the model, or in this case, the sedimentation coefficient distribution. For example, in the ls-$g^*(s)$ distribution, a failure of fitting the original data well may result in false peaks, and requires shortening the time interval of scans considered for analysis.

It can be shown in theory – at least for data that cover a narrow time interval – that the shape of the $g^*(s)$ distribution is approximately Gaussian, with a half-width related to the diffusion coefficient of the protein. As a consequence, it is possible in a second stage of the SV analysis to fit the $g(s^*)$ distributions with Gaussians, and to use the Svedberg equation (Equation (1)) to derive the molar mass.[53] For several distinct non-interacting species it is possible to fit multiple Gaussian peaks where the number of peaks correspond to the number of species. Although this technique has the advantage of not being biased by sedimenting species at s values outside the diffusion-broadened Gaussian peak, in theory it is not as accurate as the approach of solving the Lamm equation, Equation (3), due to the theoretical approximations made.[53] Further, if there is unrecognized heterogeneity of species within one $g^*(s)$ peak, as outlined above, it will only provide a lower limit to the molar mass.

A conceptual connection exist between the $g^*(s)$ and the integral $G(s)$ distributions by van Holde–Weischet, which can be obtained via extrapolation of $g^*(s)$ to infinite time,[50] and between $g^*(s)$ and $c(s)$, which is described in the following section and can be considered an extension of $g^*(s)$ from an apparent distribution of non-diffusing particles to the distribution of diffusing particles.

4.2.4 Sedimentation Coefficient Distributions $c(s)$

The goal of the $c(s)$ distribution is to calculate a distribution of sedimenting species taking into account their diffusion. This is accomplished by an extension of Equation (8), in which the full solution of the Lamm equation for sedimentation and diffusion is considered for each sedimenting species:

$$a(r, t) \cong \int c(s)\chi_1(s, D(s), r, t)\, \mathrm{d}s \qquad (9)$$

where $\chi_1(s, D, r, t)$ denotes the normalized Lamm equation solution (Equation (3)).[57] Essentially, analysing the experimental data with Equation (9) is equivalent to finding the best combination of a large set of Lamm equation solutions that fits the data best. The problem of estimating the extent of diffusion $D(s)$ for each species with a certain s value is approached with a hydrodynamic scaling law that assumes that all particles have a uniform frictional ratio, f/f_0, taking advantage of the weak size dependence of $D \sim s^{-1/2}$ and the weak dependence of the frictional ratio on macromolecular shape. The parameter f/f_0 is adjusted to a best-fit estimate in the fitting process. A comparison of this method with other approaches can be found in ref. 50, and the detailed procedure is described in ref. 48. Like the ls-$g^*(s)$ distribution, the systematic noise contributions from IF detection system are calculated explicitly and can be subtracted from the raw sedimentation profiles, and the $c(s)$ distribution produces a detailed model of the original sedimentation profiles which can be assessed with conventional measures of the goodness-of-fit. This approach is implemented in the software SEDFIT and SEDPHAT. The results of the application to the data in Figure 2 are shown in Figure 5.

Because diffusion is accounted for by using Lamm equation solutions as model functions in Equation (9), the resulting sedimentation coefficient distribution has sharp features that are not diffusionally broadened. The information on boundary spreading can be recovered by applying the Svedberg equation (Equation (1)) for

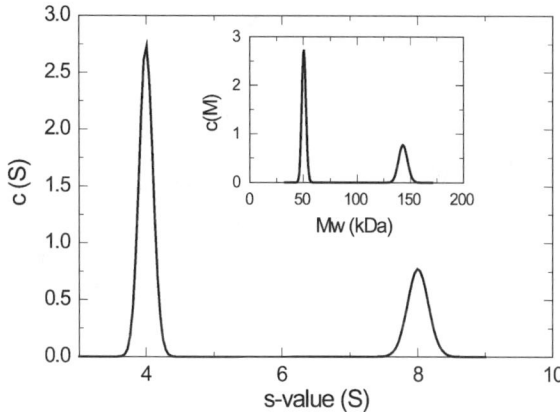

Figure 5 *Application of c(s) to the data shown in Figure 2. The inset shows the corresponding c(M) distribution from the same fit*

each pair of s and D values to transformation $c(s)$ into a molar mass distribution $c(M)$. Because the relationship between s and D for each species can be adjusted to specific problems, several variants of $c(s)$ are available for studies of protein conformational changes, for bimodal sedimentation coefficient distributions, for sedimentation of macromolecules in the presence of small molar mass species and for global analysis of multiple data sets with combinations of discrete and piecewise continuous distributions. A generalization to take advantage of multisignal detection for the study of heterogeneous protein–protein interactions has been described.[58]

Beyond the high resolution and the sensitivity to trace components, the $c(s)$ method has the property of a being able to be employed over the whole data set of SV profiles from the very beginning of sedimentation to the complete depletion of visible material, a distinct advantage over the $g^*(s)$ and van Holde–Weischet methods. The $c(s)$ method can also be used to analyse data from very small peptides, where the plateau region has collapsed before the meniscus has cleared. Such samples are dominated by back-diffusion from the bottom of the cell, and have in the past been very difficult to analyse, however, the $c(s)$ method copes as well with these as with fast sedimenting samples. A disadvantage of the method is introduced by the possible ambiguity of solving Equation (9) with slightly different distributions that may fit the data equally well. Maximum entropy regularization is embedded in the algorithm in order to produce always the simplest distribution that fits the data with a quality statistically indistinguishably from the best-fit, but this can have a tendency to merge neighbouring peaks. Further, caution must be used in the interpretation of the $c(M)$ distribution, which is valid only if there is a single major peak, otherwise it is limited to the case where all species have the same frictional ratio f/f_0, which may not be known *a priori*, or models for bimodal distributions must be used. Further, $c(s)$ can provide a significant amount of detail on the sample, but a pre-requisite of a reliable interpretation is that a very good fit of the raw sedimentation data can be achieved. This method is covered in detail in Chapter 2.

4.3   Analysis of Interacting Systems by Sedimentation Velocity

The experimental data basis for the study of protein interactions should include SV data obtained over a large range of different loading concentrations of the individual protein components. Then, for hetero-associations it is necessary to obtain data at different reactant concentrations and molar ratios, ideally covering a range from far below to far above the K_D of the interaction. A good starting point for the analysis of interacting systems with SV is to derive the $c(s)$ or $g^*(s)$ distributions. While the van Holde–Weischet method is excellent to diagnose qualitatively the presence of an interaction, by displaying a concentration-dependent increase in the sedimentation coefficients and positive sloped $G(s)$ distributions, no theory is currently known to allow rigorous thermodynamic or hydrodynamic parameters of reversibly interacting proteins to be derived from the quantitative analysis of $G(s)$. For clarity in the following sections, the diagnosis and analysis of interaction is illustrated as it appears in the $c(s)$ method, which is described in more detail in Chapter 2. However, Chapter 3 discusses the implementation of the time-difference modelling in some depth, while more detailed information on the van Holde–Weischet method can be found in Chapter 11.

The presence of an interaction changes the sedimentation profiles in SV in three different ways: (1) owing to the molar mass of the complex being higher than that of the individual components, the average sedimentation coefficient is increased in a concentration-dependent manner, reflecting the concentration dependent population of the complex; (2) because of the altered sedimentation process, which is now influenced also by the chemical reaction between the sedimenting species, the boundary shapes have different features; and (3) owing to the (even transient) formation of complexes, boundary components can appear with co-sedimentation of the individual protein components and therefore the boundary features may not necessarily reflect the properties of the seperate species, but rather reflect the sedimenting system. Correspondingly, different analysis strategies can be applied that exploit these different aspects.

First, the weight-average sedimentation coefficient s_w can be obtained by integration of $g^*(s)$ or $c(s)$ distributions and plotted as a function of loading concentration.[20] Although it may seem counter-intuitive at first due to the hydrodynamic and kinetic processes observed in SV, this isotherm $s_w(c)$ permits a rigorous thermodynamic analysis of the interaction, because s_w can be strictly obtained from applying mass balance consideration to the sedimentation boundaries.[1,59] Irrespective of the boundary shape or the peak structure in $c(s)$ or $g^*(s)$, from inspecting the concentration dependence of s_w the presence or absence of an interaction can be discerned. The quantitative analysis of $s_w(c)$ can proceed by standard techniques to derive s values of the individual species and complexes and the equilibrium binding constant.[59] For a 1:1 hetero-association, the isotherm takes the form

$$s_w(c_A^{tot}, c_B^{tot}) = \frac{\varepsilon_A c_A^{free} s_A + \varepsilon_B c_B^{free} s_B + (\varepsilon_A + \varepsilon_B) K c_A^{free} c_B^{free} s_{AB}}{\varepsilon_A c_A^{tot} + \varepsilon_B c_B^{tot}}$$

(10)

$$c_A^{tot} = c_A^{free} + K c_A^{free} c_B^{free} \qquad c_B^{tot} = c_B^{free} + K c_A^{free} c_B^{free}$$

(with ε denoting the extinction coefficients and K the equilibrium association constant). This approach was reviewed recently in refs. 20 and 60. For convenience, this and other isotherm models described below for sedimentation analysis have been implemented for global modelling in the software SEDPHAT.

The effect of the boundary shapes can be observed in different ways. With the $c(s)$ methodology, interactions are detected either by the emergence of new peaks or by shifts in the peak position as the concentration of one or both reactant is altered. Peaks that remain constant in position but change in relative area are indicative of complex formation that is slow on the time-scale of sedimentation. A shift of peak positions indicates more rapid chemical interconversion of species during sedimentation, such that the sedimenting system assumes an average sedimentation rate between those of the reacting species. This is illustrated in Figure 6 for hetero-associating systems (analogous illustrations for self-association can be found in refs. 23 and 33). For slow interactions, the relative areas of the $c(s)$ peaks as a function of loading concentration can be analysed in the form of isotherms of species populations to determine K_D. For rapid interactions, Gilbert and Jenkins have theoretically analysed the sedimentation of the system. For the case of a heterogeneous interactions of a two-component mixture, they predicted the formation of two boundaries – one reflecting the undisturbed sedimentation of one free species, and the second reflecting the co-sedimentation of free species of both components together with the complex. These boundaries occur in distinct proportions and at well-defined sedimentation coefficients in between the s values of the free component and the complex.[61,62] Accordingly, integration of $c(s)$ peaks of fast reactions can give isotherms of the amplitude and s value of the reaction boundary that can be modelled on the basis of Gilbert–Jenkins theory,[63] in combination with the isotherms of weight-average s-values, to derive the binding constant of the interaction.

The analyses of isotherms derived from the weight-average s values, the species populations, and the reaction boundary can also give valuable insights to define the reaction scheme and the stoichiometry. Such an analysis is discussed in more detail by Schuck in Chapter 2, and in the refs. 20, 63 and 64.

Finally, a method to obtain more direct insights in the stoichiometry of the complexes formed is the multisignal sedimentation coefficient distribution $c_k(s)$, available in the software SEDPHAT and described in ref. 58. Conceptually, it exploits spectral differences of proteins to analyse the composition of the different $c(s)$ peaks, which can be advantageous, in particular, in the presence of extended associations forming multiple complexes.

4.4 Reaction Kinetics in Sedimentation Velocity

While AUC is not normally a technique associated with the determination of reaction rate information, as indicated in Figure 6, it is sensitive to the rate of chemical conversion relative to the time-scale of sedimentation. For SV experiments with currently accessible rotor speeds, reactions can be considered fast, or essentially instantaneous, on the time-scale of sedimentation if the chemical off-rate constant k_{off} is larger than 10^{-2} s^{-1}. On the other extreme, complexes can be considered essentially stable species if the off-rate constants are lower than 10^{-5} s^{-1}. Therefore, inspection of $c(s)$ curves can permit the qualitative diagnostics of the reaction kinetics. The range of kinetic rate constants where the complex formation is comparable in time scale to the sedimentation, in particular reactions with k_{off} between 10^{-3} and 10^{-4} s^{-1}, falls in between those indicated for slow and fast reactions in Figure 6.

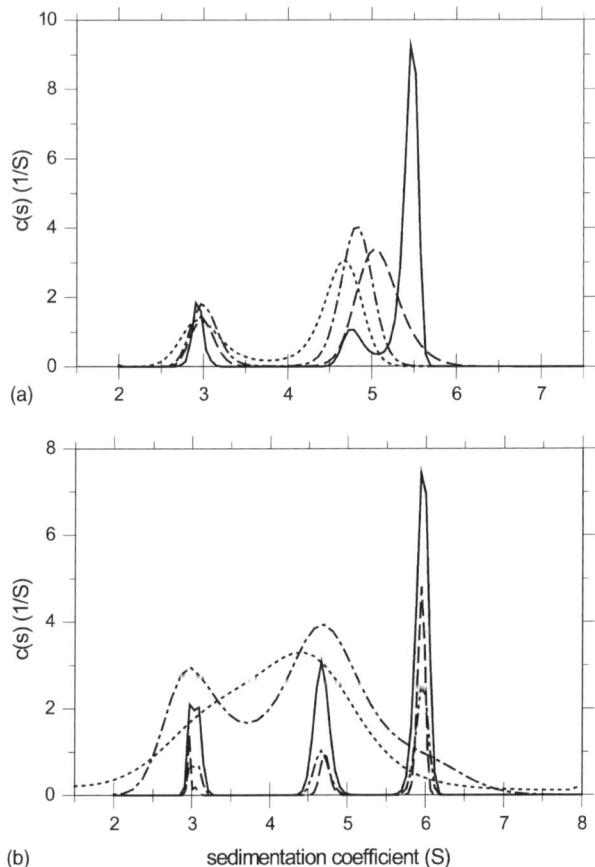

Figure 6 *Examples of c(s) distributions for interacting systems. Shown are (scaled) distributions for the hetero-association of a 30-kDa protein and 3 S with a second species of 50 kDa and 4.5 S forming a 6 S complex of 1:1 stoichiometry. The influence of the reaction kinetics can be seen in comparison of the fast interaction (a) and a slow interaction (b) on the time-scale of sedimentation. Data were simulated mimicking interference optical detection at a rotor speed of 50 000 rpm, with the usual signal-to-noise ratio. While for rapid reactions the peak position of the reaction boundary changes with concentration, slow reactions result in peak positions that are independent of concentrations. However, it must also be considered that the distributions at the lowest concentrations are broadened because of the limiting signal to noise ratio, whereas the distributions obtained from high concentrations are sharper. For rapid reactions even at concentrations 10-fold above K_D the highest peak does not correspond to the complex s value. For slow reactions, the peaks reflect approximately the populations of the sedimenting species, while at fast reactions the peaks reflect the effective sedimentation properties of the reacting system. (a) Rapid interaction with $k_{off} = 1 \times 10^{-2}\ s^{-1}$ and an equilibrium constant of $K_D = 10\ \mu M$, with equimolar loading concentrations of 1 (dotted line), 3 (dash-dotted line), 10 (dashed line), and 30 μM (solid line). (b) Slow interaction with $k_{off} = 3.2 \times 10^{-5}\ s^{-1}$, $K_D = 3.2\ \mu M$, and equimolar loading concentrations at 0.3 (dotted line), 1 (dash-dotted line), 3 (dash-dot-dotted line), 10 (dashed line), and 30 μM (solid line)*

Although the isotherm approach from integration of $c(s)$ peaks, in particular the isotherm of weight-average s values $s_w(c)$ can still provide a valid and highly useful approach to characterize the interaction, the direct modelling of the data with Lamm equations incorporating reaction terms can provide a more complete description. An example of such a Lamm equation is that for a 1:1 hetero-association:

$$\frac{\partial c_i}{\partial t} = \frac{1}{r}\frac{\partial}{\partial r}\left[rD_i\frac{\partial c_i}{\partial r} - s_i\omega^2 r^2 c_i\right] + \sigma_i q, \quad i = 1, 2, 3 \tag{11}$$

where

$$q = \frac{k_{off}}{K_d}c_1 c_2 - k_{off} c_3 \quad \text{and} \quad \sigma_1 = \sigma_2 = -\sigma_3 = -1 \tag{12}$$

q represents the chemical flux, and the indices 1 and 2 describe the individual free reactants, and 3 the complex. Algorithms for solving Equation (11) have been reviewed in 1981 by Cox and Dale,[7] but it has become practical only recently to perform these calculations efficiently enough to model globally this equation to data obtained from different loading concentrations and molar ratios, and thereby estimate the equilibrium and chemical rate constant. This exploits the specific influence of the reaction kinetics on the boundary shape, as illustrated in Figure 7. This

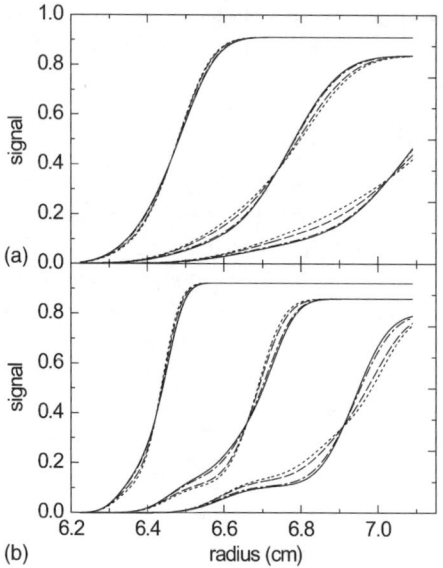

Figure 7 *Boundary shapes differ for systems with different reaction kinetics. Characteristic shapes are more pronounced for larger species with correspondingly smaller diffusion coefficient. (a) Concentration profiles at 3 000, 9 000, and 12 000 s for the sedimentation of the system shown in Figure 6, at a concentration three-fold above K_D. Kinetic rate constants are $k_{off} = 10^{-2}$ (dotted line), 10^{-3} (dashed line), 10^{-4} (dash-dotted line), and 10^{-5} (solid line). (b) Sedimentation under similar conditions but with three-fold larger protein species, at times of 500, 1000, and 1500 s*

approach has been implemented for modelling time-difference data in SEDANAL,[65] as outlined in more detail in Chapter 3, and for direct modelling of the raw data with explicit models for the systematic noise components in the software SEDPHAT.[64,66] The advantage of this approach is that it permits to fully exploit the details of the boundary shapes, as opposed to the weight-average sedimentation coefficient and the amplitudes and s values of the $c(s)$ peaks of the reaction boundaries only. The disadvantage is that these analyses are significantly more difficult and time-consuming in practice, have a far higher requirement for sample purity and a relatively narrow range of kinetic rate constants for which the analysis is well determined.

It should be strongly emphasized that it is not always possible to assign unambiguously the correct model using SV alone, as there can be a similarity in the boundary shapes of a rapid higher-order complex formation with those from a slow multi-stage reactions. Ambiguity of the boundary interpretation is more problematic for small proteins, where the hydrodynamic separation and the chemical reaction is masked by a higher diffusion coefficient. In this case, the combination with information from other biophysical techniques, in particular those reporting on the time-scale of reaction or on the potential association scheme, can be highly valuable and provide the necessary information required for correct model assignment.

Whenever possible the interaction should also be studied by SE, which is described in the following. SE is independent of reaction kinetics and hydrodynamic parameters, but has intrinsically a lower size-dependent resolution and therefore benefits very significantly from a prior characterization of sample purity by SV. SV and SE are highly complementary, and both approaches should give consistent results.[67,68]

5 Sedimentation Equilibrium

At lower rotor speeds, the transport of sample down the centrifuge cell is balanced by its desire to diffuse back up the cell due to the creation of a concentration gradient. SE is established when no change in the concentration distribution of any component is detectable (Figure 8). The time required to attain SE is proportional to the square of the solution column height, and, for proteins <200 kDa, is usually on the order of 1–2 days for sample volumes of 100–180 μL.

For a single species under ideal conditions, *i.e.* in the absence of repulsive interactions due to volume exclusion or charge interactions, the concentration distribution can be described by

$$a(r, t) = c(r_0)\varepsilon d \exp\left[M(1 - \bar{v}\rho)\frac{\omega^2}{2RT}(r^2 - r_0^2) \right] \tag{13}$$

where $c(r_0)$ is the concentration at a reference radius, ε the molar extinction coefficient and d the optical pathlength (usually 1.2 cm).[4] Ideally, the observed concentration range can easily be varied over 100-fold for one sample; using several absorbance wavelengths and interference optics means that 1 to 2000-fold concentration range can easily be probed. This should be combined with SE experiments conducted at several different loading concentrations and molar ratios of mixtures, as well as the acquisition of SE data consecutively at two to three different rotor speeds.

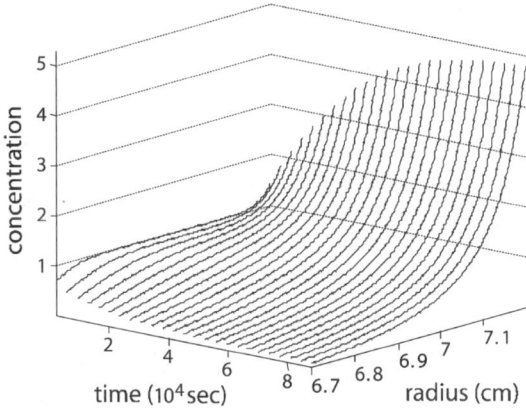

Figure 8 *Concentration distributions during the approach to equilibrium. Sedimentation of the protein mixture from Figure 2 is shown, at a shorter solution column and at a rotor speed of 12 000 rpm. Only the final equilibrium scans will be subjected to data analysis with thermodynamic models of sedimentation equilibrium*

Therefore, SE is extraordinarily powerful at assaying self-associating systems, where different oligomeric states are populated at different concentration ranges.

5.1 Setting Up a Sedimentation Equilibrium Experiment

When performing a SE experiment, the one parameter most likely to cause the novice the most difficulty is the correct selection of rotor speed. Too slow a rotor speed and there is not enough curvature in the data for adequate curve fitting; too fast a rotor speed and the majority of the sample will pellet to the bottom of the cell, making measurement impossible. As a rough guide, when working with 180 μL samples, for an average molar mass M of the protein species of interest, typically one can select a set of three rotor speeds with

$$\text{rpm} = \sqrt{\frac{100}{M(\text{kDa})}} \times 8000, \quad \sqrt{\frac{100}{M(\text{kDa})}} \times 12\,000, \quad \text{and} \quad \sqrt{\frac{100}{M(\text{kDa})}} \times 15\,000 \qquad (14)$$

Typical SE profiles are shown in Figure 9, showing a typical 'low-speed' profile[69] at the lowest rotor speed with concentration ratios of ~5:1 from meniscus to bottom, and the so-called 'meniscus-depletion' condition[70] at the highest rotor speed. The SE experiment will start with the approach to equilibrium at the lowest rotor speed. When SE has been attained and the data acquired, the rotor speed is increased to the second, and, again after attainment of SE and data acquisition to the third (the highest).

The time required for attaining equilibrium with the assumed 180 μL samples (~5-mm columns) at the first rotor speed is typically of the order of 24–48 h, and equilibria at the higher rotor speed are approached in slightly less time. For 3 mm column length, 16–20 h are typical, but higher rotor speeds than those in Equation (14) have to be chosen to achieve sufficient curvature, and still shorter times are possible

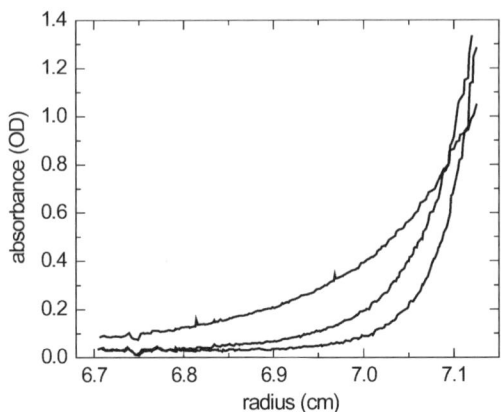

Figure 9 *Typical shapes of absorbance distributions in sedimentation equilibrium at different rotor speeds*

with short-column techniques.[18,71] It should be noted, however, that the information content of the concentration profiles significantly increases with larger column height.[16] Longer times are required for proteins with higher molar mass; as well as for reversible proteins complexes where the dissociation rate constant is very small, and for very elongated molecules which have unusually low diffusion coefficients. Additionally when working with buffers of high viscosity or density, such as 5–10% glycerol, additional time to reach equilibrium is required. Typically, in order to minimize protein degradation during the extended time of the experiments, in the absence of other considerations, SE is frequently conducted at low temperature.

It is very important to check that equilibrium has truly been attained. This cannot be assessed from the shape of the concentration profiles, but is done, typically, by subtracting successive SE traces taken in intervals of at least 4 h apart. When no systematic deviation is seen across the cell, then the sample is regarded to have attained equilibrium. The software WINMATCH[72] greatly facilitates this procedure, and permits to visualize the asymptotic approach to equilibrium by displaying the r.m.s. difference of all previous scans relative to the last scan.

SE experiments are *always* conducted over a range of initial loading concentrations. This should ideally be combined with data obtained at multiple wavelengths of the ABS detection and ideally in combination with interference optics in order to increase the dynamic range of the concentration gradient detected in the SE experiment. For the latter, however, a mechanical stabilization process for the centrifuge cell assemblies has to be applied,[22–24,73] and the systematic radial-dependent signal offsets of the IF detection have to be removed by subtraction of a water blank or by computational analysis.[22] The use of multiple rotor speeds and loading concentrations is particularly important for the analysis of protein–protein and protein–DNA interactions. The data acquired over a range of signals, concentrations, and rotor speeds can then be analysed by global analysis, and from these appropriate errors in the parameters can be defined. More details on the planning SE experiments can be found in the protocol in ref. 33.

As mentioned above, it is highly recommended to perform a characterization of the proteins by SV prior to the SE experiment, in order to establish the interaction model and the purity of the samples. As in SV, the study of heterogeneous protein–protein interactions requires the individual components to be characterized separately.

For the data analysis, beyond the global analysis of data from different rotor speeds and loading concentration, strategies to improve the modelling and the resulting parameter estimates include the application of a constraint for conservation of mass, which relates the total observed mass in equilibrium with the loading concentration[17] or among the data from the different rotor speeds,[22,74] as well as the algebraic decomposition of systematic signal offsets.[22] An analysis of optimal statistical weights for the data points is provided in Chapter 8. Several software packages are available for the analysis of SE, with varying degree of flexibility and capabilities for the analysis of self-association and hetero-association.

5.2 Self-Association

Sedimentation equilibrium data can be fitted with specific models for self-association and from this, binding constants can be determined. For any interacting system, SE is established if the sample is both in 'mechanical' sedimentation–diffusion equilibrium and in chemical equilibrium, and therefore, all interacting species distribute so that they follow mass action law at all places in the centrifuge cell. Taking the simplest example, a monomer–dimer self-association, mass action means that the amount of dimer can be expressed as $c_2 = K_{12}c_1^2$ (with concentrations and binding constants expressed in molar units), and for the SE profile then follows the relationship

$$a(r) = c_1(r_0)\varepsilon_1 d \exp\left[M_b \frac{\omega^2}{2RT}(r^2 - r_0^2) \right]$$
$$+ K_{12}c_1(r_0)^2 2\varepsilon_1 d \exp\left[2M_b \frac{\omega^2}{2RT}(r^2 - r_0^2) \right]$$

(15)

where M_b denotes the monomer buoyant molar mass $M = (1 - \bar{v}\rho)$. This assumes an ideally sedimenting system in the absence of repulsive interactions from charge or steric repulsion, which is fulfilled in most cases for dilute protein solutions (<1 mg mL^{-1}) in buffers containing sufficient ionic strength. Wills and Winzor discuss non-ideality at some length in SE in Chapter 4; Minton and co-workers[75] also assess the effect of non-ideality for highly concentrated solutions rigorously.

As the sedimentation creates a concentration gradient, a SE experiment over several speeds and loading concentrations has the potential to contain information over a large portion of the binding isotherm. Several other biophysical techniques require fitting of exponentials to experimental data. However, in contrast to most techniques, the exponents in SE are usually known (*i.e.*, M_b should be known), which makes the problem more tractable. As outlined above, the effective partial-specific volume, and hence M_b, for most proteins under common experimental conditions can be predicted with sufficient accuracy from amino acid composition or mass spectrometry[19,34] (and for glycoproteins the carbohydrate composition).[76,77] For proteins under conditions of high co-solvent concentration leading to preferential solvation,[23,34,36] or proteins containing other non-amino acid components, such as protein–detergent or protein–lipid complexes, the density increment (or effective partial-specific volume) can be determined, for example, by conducting SE[78] or SV[79] experiments in H_2O and D_2O, or by densitometry.[80] This topic is addressed in Chapters 7 and 19. Fortunately, SE itself can often be used to determine directly experimentally M_b, if conditions can be established where the proteins are monomeric. If M_b is included as a fitting parameter, correlations of M_b and K_{12} can occur that increase the uncertainty of both, in particular for experiments where concentrations are achieved that only partially span the binding isotherm.

5.3 Heterogeneous Interactions

The simplest heterogeneous interaction is for two components A and B to form a 1:1 complex. Here, we can specify the mass action law $c_{AB} = K_a c_A c_B$. The total signal seen at sedimentation equilibrium then becomes:

$$
a(r) = c_A(r_0)\varepsilon_A d \exp\left[M_{b,A}\frac{\omega^2}{2RT}(r^2 - r_0^2)\right] + c_B(r_0)\varepsilon_B d \exp\left[M_{b,B}\frac{\omega^2}{2RT}(r^2 - r_0^2)\right]
$$
$$
+ K_a c_A(r_0)c_B(r_0)(\varepsilon_A + \varepsilon_B)d \exp\left[(M_{b,A} + M_{b,B})\frac{\omega^2}{2RT}(r^2 - r_0^2)\right]
$$

(16)

It is possible to combine Equations (15) and (16) for more extended association equilibria, and to include mixed self-association and further hetero-associations. More in-depth thermodynamic theory of SE of multicomponent mixtures can be found in the chapter by Wills and Winzor (Chapter 4), and in the work by Minton and co-workers.[75] In the analysis of heterogeneous associations, the buoyant molar mass values $M_{b,A}$ and $M_{b,B}$ can be determined directly by studying the individual protein components separately.

One difficulty in the analysis of hetero-associations arises if the molar mass values of the components differ by less than 20% or more than a factor 5, because in this case the unbound or the bound components cannot be directly distinguished from the shape of the sedimentation profiles due to mathematical correlation of their exponential signal contributions. This makes a direct analysis of the binding very difficult. This problem can be overcome in two ways. First, multiwavelength analysis techniques can be applied if the protein components exhibit significantly different absorption properties (for example, extinction coefficients at 280 and 250 nm), or if one of the components has extrinsic chromophoric labels.[22,76,81,82] This is particularly useful for protein–DNA interactions (see Chapter 10). Similar, the global multisignal analysis with ABS and IF detection system is possible, exploiting differences in the weight-based extinction coefficients of proteins.[22] It is also possible to intrinsically label proteins by incorporating fluorescent analogues of tryptophan, such as 7-azatryptophan or 5-hydroxytryptophan.[83,84,85] Second, such interactions can also be studied by imposing a mathematical constraint whereby the total mass of each component in the solution column is related either to the loading concentration[10,17,86] or conserved between the equilibria at different rotor speeds.[22,74] Further improvement in the data analysis can be possible if the experiment is conducted in titration series with serial dilutions or if one component is kept constant – designs that enable to establish mathematical relationships between the total mass of material implied in the mathematical models for the different cells.[22,23]

6 Conclusions

This review is by no means exhaustive. Further important topics covered in this book include methods to determine large-scale properties of genomes (Chapter 5), synthetic boundary formation (Chapter 6), and polysaccharide and polymer characterization (Chapters 12, 147 and 23). One particular fertile set of methodologies is relating sedimentation studies to those structures derived from X-ray crystallography (Chapters 20 and 21) and small-angle scattering (Chapters 13, 15, 16 and 18). It is hoped that this review has given some help to the novice in understanding the principles behind an analytical ultracentrifugation experiment and that it facilitated access to the in-depth discussion of the techniques found in subsequent chapters in this book, and the references therein.

References

1. H. K. Schachman, *Ultracentrifugation in Biochemistry*, Academic Press, New York, 1959.
2. C. Tanford, *Physical Chemistry of Macromolecules*, Wiley, New York, 1961.
3. H. Fujita, *Foundations of Ultracentrifugal Analysis*, Wiley, New York, 1975.
4. T. Svedberg and K. O. Pedersen, *The Ultracentrifuge*, Oxford University Press, London, 1940.
5. B. Elzen, Scientists and rotors. The development of biochemical ultracentrifuges, Dissertation, University Twente, Enschede, 1988.
6. H. K. Schachman, in *Analytical Ultracentrifugation in Biochemistry and Polymer Science*, S. E. Harding, A. J. Rowe and J. C. Horton (eds), Royal Society of Chemistry, Cambridge, 1992, 3–15.
7. D. J. Cox and R. S. Dale, in *Protein-Protein Interactions*, C. Frieden and L. W. Nichol (eds), Wiley, New York, 1981.
8. B. Demeler, H. Saber and J. C. Hansen, *Biophys. J.*, 1997, **72**, 397–407.
9. J. C. Hansen, J. Lebowitz and B. Demeler, *Biochemistry*, 1994, **33**, 13155–13163.
10. J. S. Philo, *Proceedings of Ultrasensitive Biochemical Diagnostics II*, SPIE, San Jose, CA, 1997, 170–177.
11. J. Lebowitz, M. Teale and P. Schuck, *Biochem. Soc. Transact.*, 1998, **26**, 745–749.
12. G. Rivas, W. Stafford and A. P. Minton, *Methods: A Companion to Methods in Enzymology*, 1999, **19**, 194–212.
13. F. Arisaka, *Tanpakushitsu Kakusan Koso*, 1999, **44**, 82–91.
14. T. M. Laue, *Current Protocols Protein Sci.*, 1999, 20.3.1–20.3.13.
15. T. M. Laue and W. F. I. Stafford, *Annu. Rev. Biophys. Biomol. Struct.*, 1999, **28**, 75–100.
16. P. Schuck and E. H. Braswell, in *Current Protocols in Immunology*, J. E. Coligan, A. M. Kruisbeek, D. H. Margulies, E. M. Shevach and W. Strober (eds), John Wiley & Sons, New York, 2000, 18.8.1–18.8.22.
17. J. S. Philo, *Method. Enzymol.*, 2000, **321**, 100–120.
18. J. J. Correia and D. A. Yphantis, in *Analytical Ultracentrifugation in Biochemistry and Polymer Science*, S. E. Harding, A. J. Rowe and J. C. Horton (eds), The Royal Society of Chemistry, Cambridge, U.K., 1992, 231–252.

19. J. Lebowitz, M. S. Lewis and P. Schuck, *Protein Sci.*, 2002, **11**, 2067–2079.
20. P. Schuck, *Anal. Biochem.*, 2003, **320**, 104–124.
21. J. L. Cole, *Method. Enzymol.*, 2004, **384**, 212–232.
22. J. Vistica, J. Dam, A. Balbo, E. Yikilmaz, R. A. Mariuzza, T. A. Rouault and P. Schuck, *Anal. Biochem.*, 2004, **326**, 234–256.
23. A. Balbo and P. Schuck, in *Protein–Protein Interactions*, E. Golemis and P. D. Adams (eds), Cold Spring Harbor Laboratory Press, Cold Spring Harbor, New York, in print.
24. www.nih.gov/od/ors/dbeps/PBR/AUC.htm
25. www.analyticalultracentrifugation.com
26. M. S. Springer and H. K. Schachman, *Biochemistry*, 1974, **13**, 3726–3733.
27. D. J. Cox, *Science*, 1966, **152**, 359–361.
28. G. J. Howlett and P. D. Jeffrey, *J. Phys. Chem.*, 1973, **77**, 1250–1258.
29. J. B. Ifft, *Biophys. Chem.*, 1976, **5**, 137–157.
30. M. A. Bothwell, G. J. Howlett and H. K. Schachman, *J. Biol. Chem.*, 1978, **253**, 2073–2077.
31. S. Darawshe, G. Rivas and A.P. Minton, *Anal. Biochem.*, 1993, **209**, 130–135.
32. D. A. Yphantis and T. Arakawa, *Biochemistry*, 1987, **26**, 5422–5427.
33. http://www.nih.gov/od/ors/dbeps/PBR/AUCProtocols.htm
34. J. Lebowitz, M. S. Lewis and P. Schuck, *Protein Sci.*, 2003, **12**, 2649–2650.
35. T. M. Laue, B.D. Shah, T. M. Ridgeway and S. L. Pelletier, in *Analytical Ultracentrifugation in Biochemistry and Polymer Science*, S. E. Harding, A. J. Rowe and J. C. Horton (eds), The Royal Society of Chemistry, Cambridge, 1992, 90–125.
36. H. Eisenberg, *Biological Macromolecules and Polyelectrolytes in Solution*, Clarendon Press, Oxford, 1976.
37. J. Garcia De La Torre, M. L. Huertas and B. Carrasco, *Biophys. J.*, 2000, **78**, 719–730.
38. O. Lamm, *Ark. Mat. Astr. Fys.*, 1929, **21B**(2), 1–4.
39. J.-M. Claverie, H. Dreux and R. Cohen, *Biopolymers*, 1975, **14**, 1685–1700.
40. L. A. Holladay, *Biophys. Chem.*, 1979, **10**, 187–190.
41. J. S. Philo, *Biophys. J.*, 1997, **72**, 435–444.
42. P. Schuck, *Biophys. J.*, 1998, **75**, 1503–1512.
43. P. Schuck, C. E. MacPhee and G. J. Howlett, *Biophys. J.*, 1998, **74**, 466–474.
44. J. Behlke and O. Ristau, *Biophys. Chem.*, 2002, **95**, 59–68.
45. P. Schuck, *Biophys. Chem.*, 2004, **108**, 187–200.
46. P. Schuck, *Biophys. Chem.*, 2004, **187**, 201–214.
47. A. Solovyova, P. Schuck, L. Costenaro and C. Ebel, *Biophys. J.*, 2001, **81**, 1868–1880.
48. J. Dam and P. Schuck, *Method. Enzymol.*, 2004, **384**, 185–212.
49. K. E. van Holde and W. O. Weischet, *Biopolymers*, 1978, **17**, 1387–1403.
50. P. Schuck, M. A. Perugini, N. R. Gonzales, G. J. Howlett and D. Schubert, *Biophys. J.*, 2002, **82**, 1096–1111.
51. B. Demeler and van K. E. Holde, *Anal. Biochem.*, 2004, **335**, 279–288.
52. W. F. Stafford, *Anal. Biochem.*, 1992, **203**, 295–301.
53. J. S. Philo, *Anal. Biochem.*, 2000, **279**, 151–163.
54. P. Schuck and P. Rossmanith, *Biopolymers*, 2000, **54**, 328–341.
55. W. F. Stafford, *Method. Enzymol.*, 1994, **240**, 478–501.
56. P. Schuck and B. Demeler, *Biophys. J.*, 1999, **76**, 2288–2296.
57. P. Schuck, *Biophys. J.*, 2000, **78**, 1606–1619.
58. A. Balbo, K. H. Minor, C. A. Velikovsky, R. Mariuzza, C. B. Peterson and P. Schuck, *Proc. Natl. Acad. Sci. USA*, 2005, **102**, 81–86.
59. R. F. Steiner, *Arch. Biochem. Biophys.*, 1954, **49**, 400–416.
60. J. J. Correia, *Method. Enzymol.*, 2000, **321**, 81–100.

61. G. A. Gilbert and R. C. Jenkins, *Nature*, 1956, **177**, 853–854.
62. G. A. Gilbert, *Proc. Royal Soc. London A*, 1959, **250**, 377–388.
63. J. Dam and P. Schuck, *Biophys. J.*, 2005, **89**, 651–666.
64. J. Dam, C. A. Velikovsky, R. Mariuzza, C. Urbanke and P. Schuck, *Biophys. J.*, 2005, **89**, 619–634.
65. W. F. Stafford and P. J. Sherwood, *Biophys. Chem.*, 2004, **108**, 231–243.
66. www.analyticalultracentrifugation.com/sedphat/sedphat.htm
67. S. Tarabykina, D. J. Scott, P. Heizyle, T. J. Hill, J. R. Tame, M. Kriajevska, I. Lafitte, P. J. Denick. G. G. Dodson, N. J. Maitland, E. R. Lukanidia, I. B. Bronstein, *J. Biol. Chem.*, 2001, **276**, 24212–24222.
68. J. Dam, R. Guan, K. Natarajan, N. Dimasi, L.K. Chlewicki, D.M. Kranz, P. Schuck, D.H. Margulies and R. A. Mariuzza, *Nature Immunol.*, 2003, **4**, 1213–1222.
69. E. G. Richards, D. C. Teller and H.K. Schachman, *Biochemistry*, 1968, **7**, 1054–1076.
70. D. A. Yphantis, *Biochemistry*, 1964, **3**, 297–317.
71. D. L. Sackett and R. E. Lippoldt, *Biochemistry*, 1991, **30**, 3511–3517.
72. www.biotech.uconn.edu/auf
73. A. T. Ansevin, D. E. Roark and D. A. Yphantis, *Anal. Biochem.*, 1970, **34**, 237–261.
74. D. E. Roark, *Biophys. Chem.*, 1976, **5**, 185–196.
75. S. Zorrilla, M. Jimenez, P. Lillo, G. Rivas and A. P. Minton, *Biophys. Chem.*, 2004, **108**, 89–100.
76. M. S. Lewis, R. I. Shrager and S.-J. Kim, in *Modern Analytical Ultracentrifugation*, T.M. Schuster and T. M. Laue (eds), Birkhäuser, Boston, 1993, 94–115.
77. S. Shire, *Determination of Molecular Weight of Glycoproteins by Analytical Ultracentrifugation*, Beckman Instruments, Palo Alto, CA, 1992.
78. J. Edelstein and H. Schachman, *J. Biol. Chem.*, 1967, **242**, 306–311.
79. Y. Gohon, G. Pavlov, P. Timmins, C. Tribet, J. L. Popot, and C. Ebel, *Anal. Biochem.*, 2004, **334**, 318–334.
80. C. Ebel, H. Eisenberg and R. Ghirlando, *Biophys. J.*, 2000, **78**, 385–393.
81. P. Schuck, *Progr. Colloid. Polym. Sci.*, 1994, **94**, 1–13.
82. F. Dölle and D. Schubert, *Progr. Colloid Polym. Sci.*, 1997, **107**, 77–81.
83. T. M. Laue, D. F. Senear, S. Eaton and J. B. Ross, *Biochemistry*, 1993, **32**, 2469–2472.
84. D. J. Scott, S. Leejeerajumnean, J. A. Brannigan, R. J. Lewis, A. J. Wilkinson and J. G. Hoggett, *J. Mol. Biol.*, 1999, **293**, 997–1004.
85. D. J. Scott, A. L. Ferguson, M. T. Gallegos, M. Pitt, M. Buck and J. G. Hoggett, *Biochem. J.*, 2000, **352**, 539–547.
86. M. S. Lewis and R. J. Youle, *J. Biol. Chem.*, 1986, **261**, 11572–11577.

CHAPTER 2

Diffusion-Deconvoluted Sedimentation Coefficient Distributions for the Analysis of Interacting and Non-Interacting Protein Mixtures

PETER SCHUCK

Abstract

By modeling the evolution of concentration profiles measured in a sedimentation velocity experiment with a distribution of Lamm equation solutions it is possible to calculate diffusion-free sedimentation coefficient distributions, $c(s)$. For non-interacting or slowly interacting mixtures of sedimenting macromolecules, these report on the number and s value of the species present in solution. For rapidly interacting mixtures, the sedimentation boundaries and peaks in $c(s)$ can be viewed as approximations of the asymptotic reaction boundary from Gilbert–Jenkins theory, which describe characteristic features of the reacting system, and provide a rigorous approach to the weight-average sedimentation coefficient. The extension to sedimentation coefficient distributions of the individual components in a mixture, $c_k(s)$, calculated by global modeling of multiple signals from the sedimentation process, can reveal the stoichiometry of protein complexes. This chapter reviews the theory, describes some aspects related to the experimental setup, and illustrates the practical use of $c(s)$ and $c_k(s)$ in several examples.

1 Introduction

The analysis of the macromolecular redistribution following the application of a high gravitational field provides very rich information on macromolecules and macromolecular mixtures.[1,2] In the past, sedimentation velocity (SV) analytical ultracentrifugation has played an important role in several fields, including physical chemistry, polymer chemistry, biochemistry, and molecular biology, and it is continuing to provide unique information for the study of both synthetic and biological macromolecules in solution. It can permit the determination of homogeneity, molar mass, and gross shape of macromolecules, the thermodynamic and kinetic characterization of protein–protein interactions, as well as the detection of ligand-induced conformational changes. For both self-association and heterogeneous associations of proteins, frequently the number and type of complexes formed and the timescale of the interaction can be discerned and the equilibrium constants can be determined.

The theoretical description of the sedimentation process has long been established, starting in Svedberg's laboratory in 1929 with Lamm's derivation of the partial differential equation for macromolecular sedimentation and diffusion in the centrifugal field.[3] Several generations of physical chemists have continued this work and contributed to the theoretical and experimental development of this technique. SV is still a modern tool and active area of research, which has rapidly developed in the last decade as the Lamm equation has become tractable in routine data analysis,[4-11] and increasingly more comprehensive models for the sedimentation process can be used. In conjunction with modern data acquisition systems of the commercial analytical ultracentrifuges, this has permitted the analysis of significantly increased data sets comprising one or more signals from the complete sedimentation process and across the whole solution column.

In the present communication, we review how Lamm equation solutions can be used in an approach of direct least-squares modeling of the complete sedimentation data in order to describe distributions of sedimenting particles. For samples that exhibit a continuous, more or less broad distribution of sizes, such as carbohydrates, synthetic polymers, or even populations of single-cell organisms, this can lead to a characterization of the shape and polydispersity of the distribution of sedimentation parameters. For samples that consist of a few discrete macromolecular species this description can help to identify the number, size, and s value of the sedimenting species, as well as molecular interactions. The latter is particularly important as analytical ultracentrifugation can provide unique information on protein–protein interactions and the formation of multi protein complexes. Key features of the modeling with distributions of Lamm equation solutions, termed $c(s)$, are that both diffusion and the characteristic noise structure of the sedimentation data can be deconvoluted, leading to high resolution and sensitivity. It also lends itself to the incorporation of multiple signals from the sedimentation process into a global analysis, resulting in component sedimentation coefficient distributions, for example, revealing stoichiometry of mixed protein complexes. The present chapter gives an overview of the theory, practice, and limitations of this approach.

2 Theory

An unknown macromolecular mixture in a SV experiment can be characterized via a differential distribution of sedimentation coefficients, $c(s)$, where the concentration of species with sedimentation coefficients between s_1 and s_2 is given by the integral $\int_{s_1}^{s_2} c(s)\,ds$. In order to calculate the distribution $c(s)$ consistent with least-squares data analysis, we require that if we sum over the sedimentation signals of all species the resulting total signal matches the experimental data as close as possible. This can be expressed as[12]

$$a(r, t) \cong \int_{s_{\min}}^{s_{\max}} c(s) \chi_1(s, D(s), r, t) \, \mathrm{d}s \qquad (1)$$

with $a(r, t)$ denoting the experimental signal at radius r and time t, and $s_{\min}$ and $s_{\max}$ the minimum and maximum s value of the distribution. $\chi_1(s, D, r, t)$ represents the characteristic signal pattern that a single macromolecular species at unit concentration would produce (Figure 1). The latter was given by Lamm in the form of a differential equation for sedimentation and diffusion[3] as

$$\frac{\partial \chi}{\partial t} = \frac{1}{r} \frac{\partial}{\partial r} \left[rD \frac{\partial \chi}{\partial r} - s\omega^2 r^2 \chi \right] \qquad (2)$$

where r denotes the distance from the center of rotation, ω the rotor angular velocity, and s and D the macromolecular sedimentation and diffusion coefficient, respectively. Adjustments can be made to account for compressible solvents, or sedimentation in dynamic density gradients.[13,14] It is apparent from Figure 1 that the sedimentation profiles $\chi_1(s, D, r, t)$ exhibit large qualitative differences and are strongly size-dependent. (An exception to the latter is very elongated structures, which can reach a limiting s-value.)[15] Accordingly, the modeling problem in Equation (1) is relatively well posed – it consists essentially in unraveling which combination of sedimentation patterns will fit the observed data best.[12] Numerical details how Equations (1) and (2) can be solved are described in ref. 16. This method and those described in the following are implemented in the programs SEDFIT and SEDPHAT. They can be obtained from the website www.analyticalultracentrifugation.com, where tutorials and detailed information on the use of the software can also be found.

It should be noted that SV is conceptually a comparatively simple experiment, consisting only of the application of a gravitational force to the macromolecules in free solution, and, as a consequence, the sedimentation/diffusion process is very accurately described by Equation (2). This can make the analysis very quantitative. It should comprise the entire sedimentation process, starting from the first signs of non-uniform distributions close to the meniscus and bottom, to the complete depletion of all material within the observable solution column, or the attainment of sedimentation equilibrium [Figure 2(a)]. Systematic noise components in the data acquisition can be taken into account easily using algebraic methods. They lead to two additional terms in Equation (1) that explicitly express the systematic noise as radial-dependent baseline profiles and time-dependent but radially uniform offsets.[17] Least-squares optimal estimates of these noise components can be subtracted from the data without introducing bias (as long as the noise parameters are maintained as unknowns in the following analyses), which permits a better visual inspection of the sedimentation profiles and the details of the fit. This is illustrated in Figure 2. Also, a best-fit meniscus position can be calculated and is usually well-determined by the data. This can lead to a higher degree of precision than permitted by the visual inspection of the optical artifact at the meniscus, with the latter being used only for estimating limits for minimum and maximum values of the meniscus. In general, the distribution model Equation (1) leads to fits within the noise of the data acquisition, as judged by the root-mean-square (rms) deviation of

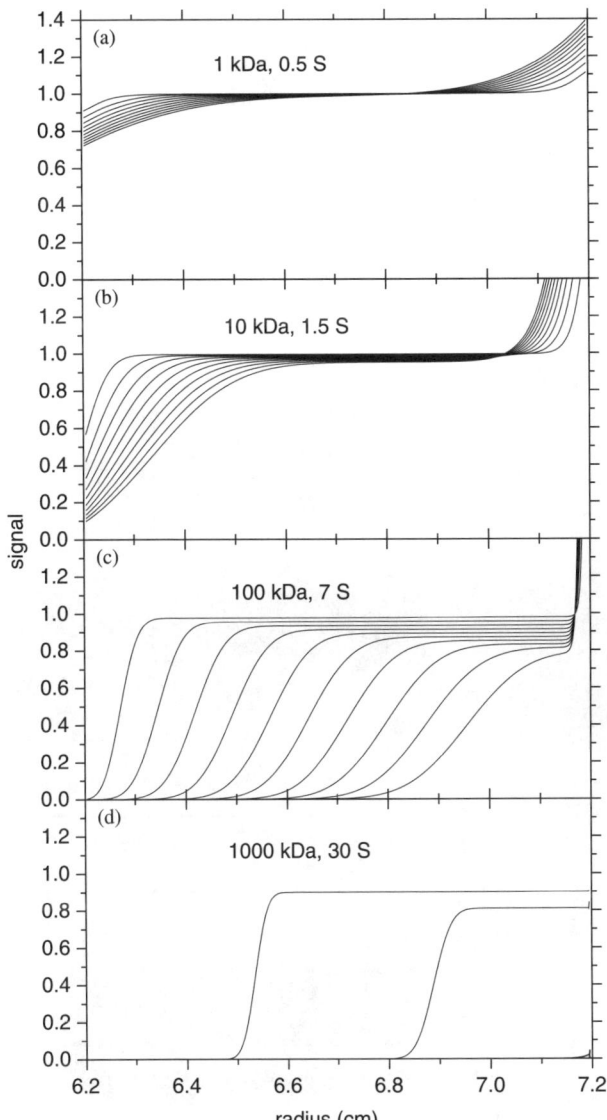

Figure 1 *Characteristic shapes of sedimentation profiles simulated for globular species of 1 kDa (a), 10 kDa (b), 100 kDa (c), and 1 GDa (d) at 50 000 rpm and unit loading concentration. Scans are depicted in time-intervals of 10 min after the start of centrifugation. A larger size range can be accommodated easily by adjusting the rotor speed and the time interval of data acquisition*

the fit and the absence of systematic misfit of the sedimentation boundary in the residuals bitmap. This is a pre-requisite for the detailed interpretation of the results.

The numerical solution of the Fredholm integral equation Equation (1) requires discretization of the range of s values of the distribution. With a spacing of s values

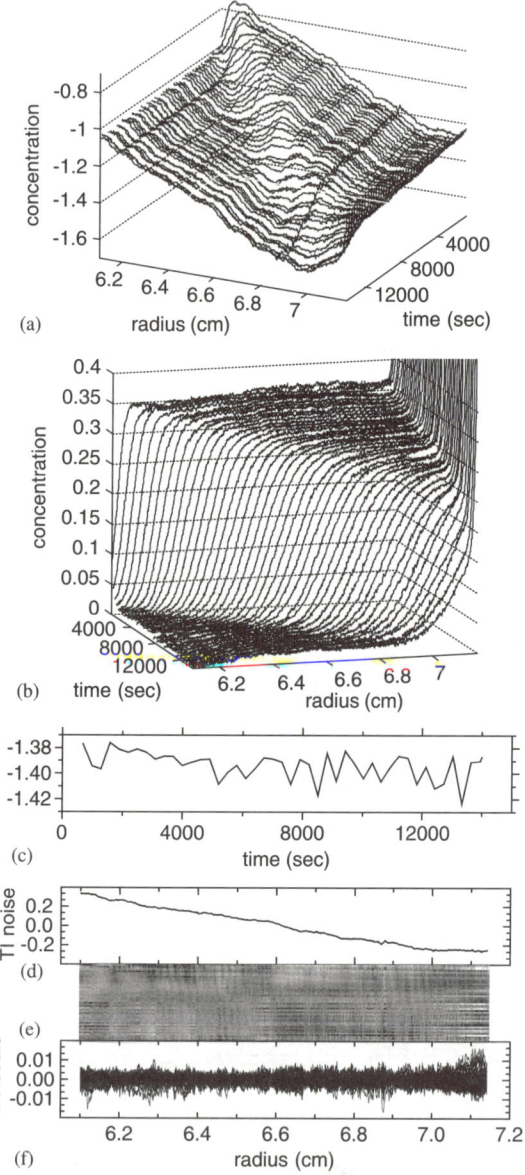

Figure 2 *Illustration of the data analysis of an IgG sample at 40 000 rpm. (a) Raw data from interference optical data acquisition. (b) Same data after subtraction of the systematic noise contributions, which are the radial-invariant time-dependent off-sets (c), and the time-invariant radial-dependent baseline profile (c). The systematic noise was calculated as part of the c(s) analysis shown in Figure 3(a). The rmsd of the fit was 0.0027 fringes, with the residuals in (f). The residual bitmap in (e) is scaled to ±0.01 fringes from black to white. It shows that the remaining noise is dominated by vertical and horizontal lines, which are likely caused by higher-order vibrations of the optical system. Importantly, no diagonal pattern is visible which would indicate the presence of a systematic misfit of the sedimentation boundary*

of the order of 0.1 S, the inversion of Equation (1) is generally well-conditioned and numerically stable. This is due to the characteristic shapes of $\chi_1(s, D, r, t)$, but also aided by the positivity requirement for the distribution $c(s)$. However, it can be shown mathematically that in the limit of very fine discretization of s the calculated distribution $c(s)$ can become very noisy and many different curves $c(s)$ may fit the data equally well within the noise of the data acquisition.[12] A well-established technique to suppress noise amplification in $c(s)$ and to avoid detail in the distribution not warranted by the data is regularization. It is based on predefined measures of likelihood or parsimony of the distributions, such as maximum entropy $-\int c(s)\log(c(s))\,ds$ for discrete mixtures, or minimal total curvature $\int |c''(s)|^2\,ds$ for broader distributions. Regularization ensures that if there is any statistical ambiguity in the resulting distribution, the most likely distributions will be selected among all those that fit the data within a predefined confidence limit.[18,19] The rationale is that, following Occam's razor, the solution with the highest parsimony is to be preferred. While this can introduce a small bias, by design it is certain that the selected result produces fits of the raw data that are of a quality statistically indistinguishable from the overall best fit. Regularization also provides a tool to explore alternate interpretations of the data, both by switching the regularization method and by changing the predefined confidence limit to achieve weaker or stronger regularization (*e.g.* going from a P value of 0.7 to 0.95). If detailed features of the calculated distribution disappear on a higher confidence limit, they should be interpreted with care, while features that appear independent over a large range of P values are details that are an integral part of the observed data.

Complicating the analysis outlined above is the fact that two parameters, s and D, are required to specify the shape of the single macromolecular species sedimentation boundaries, but we calculate only a one-dimensional sedimentation coefficient distribution. Several different options are available to arrive at an estimate for D. A very straightforward approach can be taken if all species in solution have the same diffusion coefficient, which can be measured separately, for example, distributions of apo-ferritin and ferritin. If D is known, the Svedberg relationship[1]

$$M(1 - \bar{v}\rho) = \frac{sRT}{D} \tag{3}$$

(with the protein partial-specific volume $\bar{v}$, the solution density ρ, the gas constant R and the absolute temperature T) can be used to transform the measured sedimentation coefficient distribution $c(s)$ directly into a molar mass distribution $c(M)$.[12,20]

Another special case is that of negligible diffusion during the time of the sedimentation experiment, for example with very large particles. Here, the approximation $D=0$ can be used and the resulting sedimentation coefficient distribution is termed $ls\text{-}g^*(s)$.[21] The Lamm equation solutions simplify to step functions for incompressible solvents, and to step functions with sloping plateaus if solvent compressibility is taken into account.[14] Conceptually, the $ls\text{-}g^*(s)$ distribution is related to $g(s^*)$ by dc/dt or dc/dr, although there are large differences in the practical application, among them the applicability to data with steep boundaries and large time intervals.[21] If the $ls\text{-}g^*(s)$ analysis is applied to the sedimentation analysis of particles that do exhibit significant diffusion, the result is termed an apparent sedimentation coefficient distribution. Such distributions are convoluted by functions that are in first-approximation Gaussians,[21] from which for single species D and M may be estimated.[22] However, more direct approaches to combine single species analyses with distributions will be described below.

A general approach to arrive at realistic estimates for D in unknown distributions is based on the description of D as a monotonous single-valued function of s. The second assumption is made that, in a first approximation, all sedimenting species have similar frictional ratios f/f_0, which can be described with the weight-average frictional ratio $(f/f_0)_w$. For this case, we can derive the hydrodynamic scaling law

$$D(s) = \frac{\sqrt{2}}{18\pi} kTs^{-1/2}(\eta(f/f_0)_w)^{-3/2}((1 - \bar{v}\rho)/\bar{v})^{1/2} \qquad (4)$$

(with k denoting the Boltzmann constant and η the solvent viscosity).[16] Several observations are in favor of this approach: (1) It is well-known that the values of f/f_0 are only very weakly dependent on the macromolecular shape[23] (examples for commonly observed values of the hydrated frictional ratio are 1.2–1.3 for relatively globular proteins, 1.5–1.8 for asymmetric or glycosylated proteins, and larger values for very asymmetric or unfolded proteins or linear chains). (2) Diffusion is much less size-dependent than sedimentation (for a particle of radius R, it is $D{\sim}R^{-1}$, while $s{\sim}R^2$). (3) The peak positions of $c(s)$ are determined substantially by the measured displacement with time of the sedimentation boundary midpoint, and are largely independent of the precise value of $(f/f_0)_w$, which mainly affects the peak shape and the resolution of species with different sedimentation coefficients. In fact, a family of $c(s)$ with different values of $(f/f_0)_w$ – going continuously from $c(s)$ with $(f/f_0)_w = \infty$ (which is identical to the case of $D(s) = 0$ in the ls-$g*(s)$ method and shows Gaussian-broadened peaks) to $c(s)$ with correct weight-average $(f/f_0)_w$ (which shows sharp diffusion deconvoluted peaks) – will exhibit a constant peak position. (4) Since the boundary spreading from differential migration and from diffusion is different, $(f/f_0)_w$ can be extracted by non-linear regression from the experimental data, in addition to the $c(s)$ distribution.[20] (5) The extent of diffusion relative to sedimentation decreases with higher rotor speed. As a consequence, any errors in the approximations involved in the description of diffusion can be minimized at the highest rotor speeds, which also optimize the hydrodynamic resolution. (Although $c(s)$ is very well suited for sedimentation at high rotor speeds, it can also be used very effectively for data that exhibit a large boundary spread, such as from small molecules or from macromolecules at low rotor speed.) This is the standard $c(s)$ approach used in most of the applications so far.[24] The results of the analysis for the data of Figure 2 is shown in Figure 3(a).

If the weight-average frictional ratio $(f/f_0)_w$ is well determined by the data (which can be tested by comparing the rms deviation of $c(s)$ models with non-optimal f/f_0 values), and if the distribution shows a single major peak, $c(s)$ can be transformed using Equation (3) to a molar mass distribution $c(M)$. Ordinarily, this will give estimates within 10% of the true molar mass (provided the correct partial-specific volume is known), which can give a reliable answer, for example, to the question whether the protein represented in a major $c(s)$ peak is monomeric or dimeric. Caution should be applied, however, since, in contrast to $c(s)$, the peak locations of $c(M)$ will be strongly dependent on that the $(f/f_0)_w$ is correct for the individual species, which may not be true, in particular, for trace components.

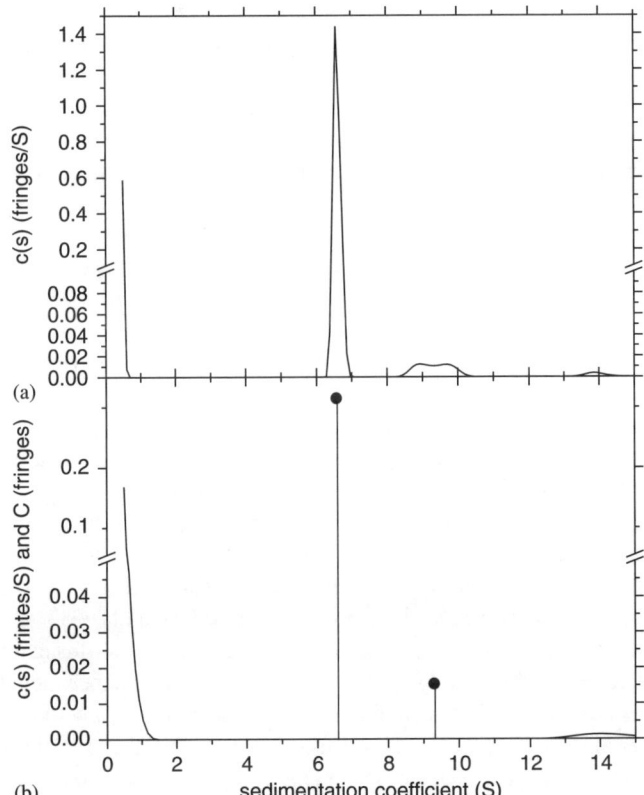

Figure 3 *(a) c(s) distribution from the analysis of the data shown in Figure 2. (b) Hybrid continuous/discrete distribution [Equation (5)] with two discrete species of best-fit apparent molar mass of 159 kDa and 6.59 S, and a second species constrained to be of twice the molar mass and with best-fit s value of 9.32 S. The continuous sections cover the range from 0.5 to 5 S and 12 to 20 S, with a fix f/f₀ value of 1.3. It should be noted that the molar mass values of the discrete species were calculated independently of the f/f₀ value of the continuous sections*

A more precise approach to extract molar mass information from the sedimentation profiles is the combination of the continuous distribution $c(s)$ with discrete species, as in the hybrid discrete/continuous distribution model of SEDPHAT. Here, several continuous segments can surround intervals of s values that contain only one or more discrete species:

$$a(r, t) \cong \sum_i C^{(i)} \chi_1(s_i, M_i, r, t) + \sum_j \int_{s_{\min,j}}^{s_{\max,j}} c^{(j)}(s) \chi_1(s, (f/f_0)_{w,j}, r, t) \, ds \qquad (5)$$

with $c^j(s)$ denoting the continuous distribution of species with sedimentation coefficients in different non-overlapping intervals $I_j = [s_{\min,j}, s_{\max,j}]$ which can be characterized by separate frictional ratios, in combination with discrete species at loading concentrations $C^{(i)}$ (in signal units) with s values s_i outside the intervals I_j.[25] The discrete species are described by Lamm equation solutions, each with the two parameters

s and M, and are not connected to the determination of $(f/f_0)_{w,j}$ for the continuous segments. In the implementation of SEDPHAT, up to 10 discrete species can be fitted, and their molar mass values can be constrained to be multiples, for example, of a monomer molar mass. Such an analysis can make use of the prior knowledge of the molar mass relationship and discrete nature of the sedimenting species, while at the same time providing enough flexibility to describe in the continuous $c(s)$ segments possibly present larger species, such as aggregates, as well as contaminations with smaller molar mass or degradation products. An example is shown in Figure 3(b) for an antibody sample. If multiple SV data sets are available from the same sample, they can be combined in a global analysis with the sedimentation coefficient and molar mass values of the discrete species treated as global parameters.

For the practical application of the $c(s)$ analysis, it is important to select the limits s_{min} and s_{max} of the distribution wide enough to allow for the description of all sedimenting species. However, a range much wider than necessary will significantly increase the computational time. A good range can be ensured by observing the calculated distribution at the extreme values. A non-zero value of $c(s_{max})$ indicates the possible presence of material sedimenting faster than s_{max}, and suggests that this value should be increased. A non-zero value of the distribution at the minimum value s_{min}, $c(s_{min})$, can indicate either the presence of material sedimenting slower than s_{min}, in which case s_{min} should be reduced, or it can be caused by a correlation of the theoretical sedimentation pattern at s_{min} with the baseline parameters (this can be the case, in particular, for experiments at low rotor speed and with short experiment times). The two cases can be distinguished by comparing the distribution $c(s)$ with the result after lowering the value of s_{min}: if $c(s_{min})$ increases, s_{min} is in the regime of baseline correlation; alternatively, if $c(s_{min})$ decreases, a smaller species was probably present and of s_{min} should be further reduced until $c(s_{min}) = 0$, or until s_{min} is in the regime of the baseline correlation. In case of a baseline correlation, s_{min} can be kept well below the species of interest, but the extreme value at $c(s_{min})$ should be ignored. In this case, the correlation at s_{min} only indicates the overparameterization of the description of small molecule sedimentation, which is without consequence and does not introduce a bias into the results of the remaining $c(s)$ distribution. Only the reverse situation – a constraint in the description of sedimentation – would create a bias and must be avoided.

A related practical question is that of the discretization of the s range and the spacing of the values. Frequently, for studies on medium-sized proteins, a density of grid points of 10 per S unit can be a good compromise between 'smoothness' of the calculated distribution and excessive computational time. It should be verified that the numerical resolution of s values is sufficiently high not to constrain the quality of the fit. In some cases, logarithmic spacing can be an effective way to span a large range of s-values.

For the characterization of sedimentation coefficient distributions of multicomponent mixtures of spectrally dissimilar macromolecules, it is useful to follow the sedimentation process simultaneously at different characteristic signals and to acquire multiple data sets $a_\lambda(r, t)$.[26] For the analysis, it is possible to generalize Equation (1) and formulate component sedimentation coefficient distributions $c_k(s)$ as

$$a_\lambda(r, t) \cong \sum_{k=1}^{K} \varepsilon_{k\lambda} \int_{s_{min}}^{s_{max}} c_k(s)\chi_1(s, (f/f_0)_{k,w}, r, t)\, ds \quad \lambda=1 \dots \Lambda, K \leq \Lambda, \det(\varepsilon_{k\lambda})\neq 0 \quad (6)$$

where each macromolecular component k is described as contributing in a characteristic way to the signal λ according to a molar extinction coefficient (or molar signal increment) matrix $\varepsilon_{k\lambda}$. This matrix can be determined from separate experiments of the individual sedimenting components. An example is the analysis of protein–protein interactions between proteins of different weight-based extinction coefficients: here, the acquisition of absorbance data simultaneous to the interference data allows distinguishing the sedimentation behavior of each protein component in the mixture. The component sedimentation coefficient distributions $c_k(s)$ are in molar units, and their ratio allows unraveling protein stoichiometries in complexes.

Refinements are possible, for example, by constraining different intervals of sedimentation coefficients to certain ranges of possible complex stoichiometries.[26] Another extension is the combination with discrete species analogous to the hybrid continuous/discrete model of Equation (5). This can take the form

$$a_\lambda(r, t) \cong \sum_i \sum_{k=1}^{K} \varepsilon_{k\lambda} C_k^{(i)} \chi_1(s_i, M_i, r, t) + \sum_j \sum_{k=1}^{K} \varepsilon_{k\lambda} \int_{s_{\min,j}}^{s_{\max,j}} c_k^{(j)}(s) \chi_1(s, (f/f_0)k_{j,\mathrm{w}}, r, t) \, \mathrm{d}s \quad (8)$$

with $C_k^{(i)}$ representing the concentration of protein component k sedimenting as a discrete species i (characterized by the sedimentation coefficient s_i and the molar mass M_i), and $C_k^{(j)}(s)$ the sedimentation coefficient distribution of component k in the continuous segment between $s_{\min,j}$ and $s_{\max,j}$.

3 Experimental

The practical techniques for conducting SV experiments have long been worked out in great detail at the highest level of sophistication.[1,2] The direct least-squares modeling of the sedimentation profiles comprising the entire sedimentation process to within the noise of the data acquisition requires a certain experimental design, and introduces particular sensitivity toward certain experimental aspects which I will comment on in the following.

As is well-known, macroscopic convection can be caused by temperature gradients, and therefore thorough temperature equilibration of the rotor is necessary. In our experience with current instrumentation, after the temperature set point and the temperature reading are within 0.2 °C, extending the equilibration time to additional 60 min has given best results. During this equilibration period, the rotor should be at rest, since rotation – even at a low rotor speed – will not only cause some macromolecular sedimentation but also lead to misleading entries in the scan files for the elapsed time after start of centrifugation. (This is in contrast to some other sedimentation and analysis techniques, which may be less sensitive toward initial rotation.) Rotor acceleration can be accounted for in SEDFIT and SEDPHAT in the finite element solution of the Lamm equation. If initial convection does occur, it may result in larger residuals close to the meniscus predominantly at the beginning of sedimentation. Unfortunately, because the further sedimentation will depend on the entire history of the experiment, later scans will also be affected in their information and the detailed analysis is problematic even if the initial scans are removed from the analysis and an apparently high quality of fit is obtained.

Regarding the loading composition of the sample, the $c(s)$ method currently requires the absence of repulsive non-ideal sedimentation, and accordingly, low macromolecular concentrations are required. Fortunately, the optical systems are highly sensitive, and concentration < 0.05 mg mL^{-1} can routinely be studied with both the absorbance and the interference optical detection system. For the interference optical system, the solution in the sample and reference sector should be precisely matched in volume and chemical composition, in order to cancel the potentially large signal originating from buffer salts. It is possible to analyze unmatched samples by explicitly modeling the redistribution of the buffer salts (see below), but frequently at the cost of the level of detail that can be reliably extracted from the analysis.

The highest possible rotor speed will give the best hydrodynamic resolution, and in our laboratory, we routinely use Epon centerpieces at rotor speeds of 50 000 and 60 000 rpm after verifying the absence of leaks.[27] Intermediate rotor speeds contain more molar mass information, and the method can also be applied – at lower resolution – to data at low rotor speeds and the approach to equilibrium. Regarding the period of time for data acquisition, if possible it should extend from the beginning of the sedimentation to the complete depletion or the absence of further observable migration. The Lamm equation modeling approach is not restricted in the consideration of early scans or large time intervals, and these can carry valuable information on the absence or presence of very small and large particles, as well as the shape of the baseline profiles.

As indicated above, the meniscus position should be included as a fitting parameter to be optimized by non-linear regression, because it generally cannot be determined experimentally with sufficient precision. It may not coincide with the peak maximum of the characteristic optical artifact, but the latter can be used as a guide for minimal and maximal values for this parameter. Fitting for the meniscus position conceptually amounts to extracting s-values from the evolution of the visible boundary positions, rather than their distance from the meniscus. Similarly, if the sedimentation data to be analyzed include regions of backdiffusion from the bottom of the solution column, or if a small molar mass species is included in the model, the bottom position of the solution column should also be determined by non-linear regression.

A step-by-step experimental protocol for novices in analytical ultracentrifugation, which also includes the basic steps for data analysis in SEDFIT and SEDPHAT for SV and sedimentation equilibrium can be found in refs. 27 and 28.

4 Results

Figure 2 illustrates the results of the $c(s)$ model in the original data space. The interference optical raw data, shown in Figure 2(a), are clearly superimposed by well-known systematic radial-dependent baseline offsets ('TI noise'). Additionally superimposed are time-dependent offsets constant in radius ('RI noise') which are less well visible in the presentation of Figure 2(a). As a consequence, the $c(s)$ boundary model included parameters for TI and RI offsets. It was fitted to the raw data, including meniscus position and the weight-average frictional value as unknown parameters. After the $c(s)$ fit has converged at an rms deviation of 0.0027 fringes, the raw data was decomposed into the different contributions. Figure 2(b) shows the calculated signal after subtraction of the systematic signal offsets. The time-dependent offsets are shown in Figure 2(c), and the radial-dependent baseline is shown in Figure 2(d). The residuals of the fit are encoded in a grayscale and shown as a bitmap in Figure 2(e), with lines of pixels corresponding to different scans (first scan at the top, last at the bottom), and rows of pixels corresponding to different radial positions. In this representation, systematic misfits of the boundary would be visualized as diagonal patterns. These appear to be absent in Figure 2(e), and the horizontal and vertical stripes suggests that part of the residuals likely originate from vibrations or thermally induced noise. The magnitude of the maximal residuals is shown in Figure 2(f). In general, we found a similar procedure useful in the analysis of absorbance optical data, which appeared to differ mainly in the absence of time-dependent offsets ('RI noise') and in the radial-dependent baseline patterns exhibiting smaller slopes (data not shown). The high quality of the fit shows that the calculated $c(s)$ distribution represents the data well, which justifies a detailed interpretation.

The calculated $c(s)$ distribution is shown in Figure 3(a). It exhibits a major peak for the monomeric species at ~6.5 S (S-values are given without buffer correction), a baseline-separated broader peak at 9–10 S, comprising of 4.8% of the material, and traces of higher oligomers at 14 S. Next, we made the assumption for the 9 S species to be dimers, enabling a more refined analysis with the hybrid discrete/continuous model of Equation (5). We inserted two discrete species at 6.5 and 9 S, constraining their molar mass to be in the ratio 2:1, and treated the monomer molar mass and the s values of the monomer and dimer as unknowns. In order to avoid trace contamination of other species to bias these estimates for the discrete species, they were combined with two sections of continuous sedimentation coefficient distributions, from 0.5 to 5 S and 12 to 20 S. The resulting distribution is shown in Figure 3(b).

This example indicates the sensitivity of the $c(s)$ analysis. Simulations show that from a set of profiles covering the complete sedimentation process at a signal-to-noise ratio of 200:1 (which can be readily achieved, *e.g.* at a loading concentration of 0.3–0.4 mg mL^{-1} of protein in the interference optics), minor peaks consisting of 0.2% of the total protein concentration should in theory be detectable with statistical confidence. The statistical accuracy of the calculated $c(s)$ distribution can be assessed by Monte–Carlo simulations.[29] For some applications, it may be advantageous to apply the statistical tests to the integral of $c(s)$ over a certain s range, rather than calculating the error contours of the $c(s)$ distribution.

In order to test the resolution of $c(s)$ for sedimentation profiles that are governed by diffusional spread, SV experiments were simulated for a mixture of species with sedimentation coefficients of 1, 2, and 3 S at equal concentration. Even at rotor speeds of 50 000 and 60 000 rpm, diffusion broadening does not allow clearly sedimentation boundaries to establish (Figure 4). The resulting $c(s)$ distributions show clearly defined peaks. Limitations in the deconvolution of diffusion are encountered at the lower rotor speed, which shows a degrading resolution and a small shift in the 1 S peak.

A second example is the interaction between D1.3 monoclonal antibody and hen egg lysozyme (Figure 5). Panels (a)–(c) show the fringe displacement data of D1.3 (a), lysozyme (b), and a mixture (c) after subtraction of the systematic noise components. A sloping signal offset can be discerned in the data from the sedimentation of the antibody (a), which are signals from the redistribution of unmatched buffer salts superimposed to the macromolecular sedimentation. These are modeled as an extra discrete

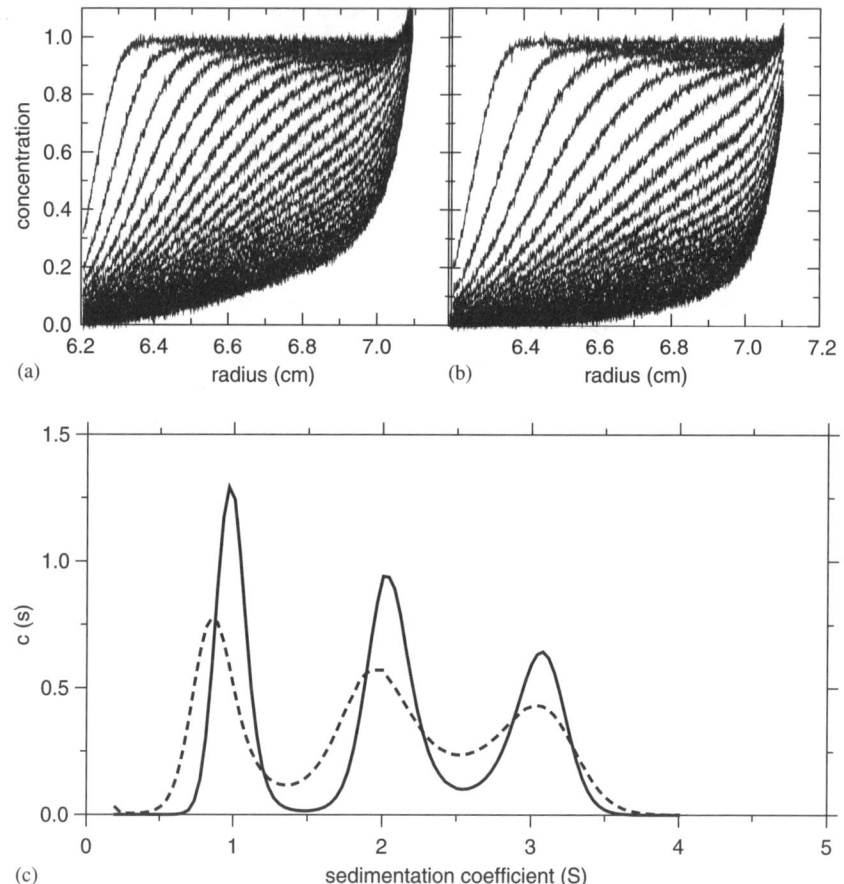

Figure 4 *Resolution of the c(s) distribution for mixtures of species with low molar mass. Sedimentation profiles were simulated for three species with 6 kDa and 1 S, 10 kDa at 2 S, and 20 kDa and 3 S, respectively, at a loading concentration of 0.33 each, with 0.01 simulated noise. Sedimentation was simulated for data acquisition in 10 min intervals (every second scan shown) for rotor speeds of 50 000 rpm (a) and 60 000 rpm (b). (c) shows the c(s) distributions calculated assuming the presence of TI noise, an unknown meniscus position and optimizing the weight-average frictional ratio. c(s) from the 50 000 rpm data (dotted line) and from the 60 000 rpm data (solid line)*

species at 0.12 S and buoyant molar mass of 55.4 Da (best-fit estimates from the $c(s)$ analysis). Compared to the $c(s)$ peak of the antibody alone (dashed line in panel d), the mixture (solid line) shows a peak shifted by 8% (0.5 S) toward a larger sedimentation coefficient, indicating binding of lysozyme to the antibody. Because the concentrations used are >100-fold the equilibrium dissociation constant and lysozyme is in large excess, it can be assumed that all available sites on D1.3 in this antibody preparation are saturated and the 7.1 S peak in the mixture represents the antibody–antigen complex, possibly in addition to contaminations with unreacted antibody.

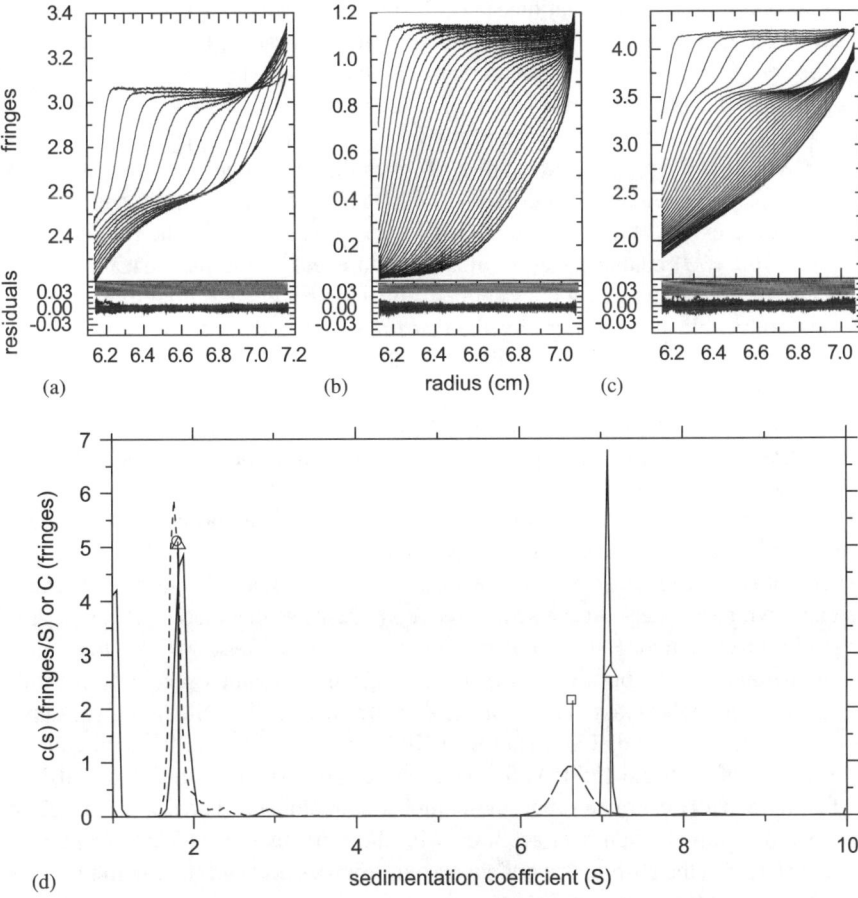

Figure 5 *Analysis of the interaction between D1.3 monoclonal antibody and hen egg lysozyme, using interference optical fringe displacement data acquired in SV experiments at 50 000 rpm, 21 °C. (a)–(c) Fringe displacement profiles (every third scan shown) of D1.3 (a), hen egg lysozyme (b), and a mixture of both (c). The residuals are from the c(s) analysis shown in (d). It leads to an rms deviation of 0.0028 (a), 0.0028 (b), and 0.0043 (c) fringes, respectively. The residuals bit-map was scaled to a maximum dynamic range of 0.02. (d) c(s) distribution of D1.3 (dashed line), hen egg lysozyme (dotted line), and the mixture (solid line). For comparison, the results of the hybrid discrete continuous model are indicated as open symbols, showing only the discrete species for the hen egg lysozyme (open circles, M* = 15.6 kDa), the D1.3 antibody (open squares, M* = 150 kDa), and the two macromolecular species observed in the mixture (open triangles, M* = 15.1 and 175 kDa). The discrete species are in units of fringes, scaled 5-fold for clarity*

It is interesting to examine how information on the stoichiometry of the complex could be extracted from the sedimentation data. Three different approaches are possible. First, the relative shift in the *s* value may be taken to estimate the increase in mass with the 2/3-power rule, assuming the constancy of shapes. This would suggest an increase in mass by 12%, or ~17 kDa, which is 1.2-fold the molar mass of the lysozyme.

A second source of information can be the measurement of the species molar mass from boundary spreading. Using the best-fit frictional ratios, the $c(s)$ distributions can be transformed into molar mass distributions, which yields a peak of 163 kDa (apparent molar mass assuming $\bar{v}\rho=0.73$) for the antibody $((f/f_0)_w=1.6)$, and 15.8 kDa $((f/f_0)_w=1.3)$ for lysozyme, respectively. For the individual proteins, these values are within the typical precision of $c(M)$ from the true molar mass values. For the mixture, however, unrealistic values of 19.4 kDa for the free lysozyme and 141 kDa for the complex are obtained, with a best-fit $(f/f_0)_w$ value of 1.4. The reason for the unsuccessful $c(M)$ characterization of the mixture lies in the fact that the weight-average frictional ratio of 1.4 does neither reflect the frictional ratio of lysozyme nor the antibody well. It is too high for the lysozyme and leads to an overestimation of its mass, and *vice versa* for the antibody. Such a situation can be encountered if the $c(s)$ distribution does not exhibit a single major peak. This problem may be addressed by calculating $c(s)$ with two $(f/f_0)_w$ values for different ranges of s values. The prerequisite for this technique is that the sedimenting species are well separated in s value, which is fulfilled in the present case.

A better approach for extracting the molar mass information from the data in Figures 5(a)–(c) is the hybrid discrete/continuous model [Equation (5)], if the peak regions of $c(s)$ are replaced by discrete species for which s, and M are treated directly as unknown parameters. At the same time, continuous segments of $c(s)$ can be used to model trace components at higher and lower s values. These will use the scaling method Equation (4), but the frictional ratio of the continuous segments are independent of the molar mass values for the discrete species. This method gives apparent molar mass values of 15.6 kDa for the lysozyme, 150 kDa for the antibody, and two species of 15.1 and 175 kDa for the mixture [open symbols in Figure 5(d)].

The third method to determine the complex stoichiometry consists in the global analysis of sedimentation data acquired with the refractive index sensitive interference system and the absorbance optical system [Figures 6(a) and (b)]. With the wavelength set to 280 nm, the global analysis can exploit differences in the fraction of aromatic amino acids of the antibody and lysozyme. In general, this requires differences in the mg/ml-based extinction coefficient of the two proteins. In the present case, although the mg/mL-molar extinction coefficients are significantly different (1.58 OD_{280} /(mg mL^{-1}×cm) for IgG and 2.57 OD_{280} /(mg mL^{-1}×cm) for lysozyme, respectively), a 2:1 complex and free antibody differ only by 10% due to the large difference in molar mass. The component sedimentation coefficient distributions $c_k(s)$ according to Equation (6) are shown in ref. 26, indicating a molar ratio of 1.8:1 and the presence of unreacted antibody. The result of the multisignal hybrid discrete/continuous analysis following Equation (8) is shown in Figure 6(c). The molar ratio of the content of antibody and antigen in the 7.1 S component is 1.74:1, consistent with the previously reported results for the continuous multisignal $c_k(s)$ analysis.[26] The best-fit estimate for the molar mass of the complex species is 171 kDa.

$c(s)$ is based on equations describing superpositions of non-interacting species. This is a good approximation for the application to mixtures of reversibly interacting proteins if the complexes formed in the loading mixture are maintained during the sedimentation experiment. Sufficient stability of species can originate from a low off-rate constant or can be achieved with loading concentrations far above or

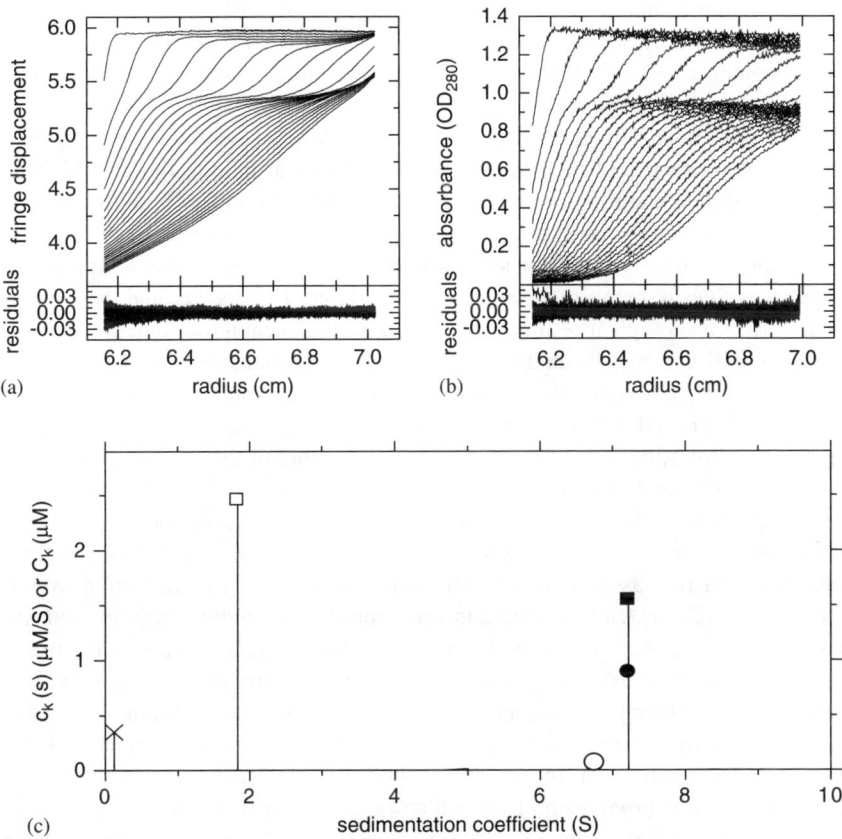

Figure 6 *Multisignal analysis of the antibody–antigen mixture is shown in Figure 5(c). Shown are the interference optical fringe displacement data (a) and absorbance optical data at 280 nm (b) acquired simultaneously during sedimentation. For clarity, only every third scan is shown. In a preliminary analysis, similar multisignal data pairs from the antibody and the antigen were modeled, based on a theoretical molar interference optical signal increment ($2.75 \times Mw$ fringes/M cm), and the value for the protein molar extinction coefficient at 280 nm was determined. Values of 2.22×10^5 and 3.42×10^4 OD_{280}/M cm were sobtained for the D1.3 antibody and lysozyme, respectively. This was used as a constraint in the analysis of the mixture. The model consisted of the following discrete species: of a small buffer component at 0.12 S contributing only to the interference data (cross), the free lysozyme (open square), free antibody (open circle), and the bound lysozyme (solid square) and antibody (solid circle). The bound species were constrained to be co-sedimenting by establishing a link in molar mass and sedimentation coefficient. This was combined with two continuous sections for both the lysozyme and the antigen component (not visible on this scale)*

below the equilibrium dissociation constant. If these conditions are not fulfilled, the sedimentation coefficient distribution will be governed by the features of the sedimentation boundary caused by the chemical reaction. It is of interest to study how these features will be represented in the $c(s)$ curves. For this purpose, using the finite element Lamm equation solution with reaction terms described in ref. 30, the sedimentation of an interacting system of the type $A+B \gtrless AB$ was simulated for interacting proteins of 30 and 50 kDa, at different reaction rate constants, and at different concentrations of the components.

For fast reactions, Gilbert and Jenkins have developed a theoretical description for the asymptotic shapes of the sedimentation boundary at infinite time,[31] which predicts the sedimentation to proceed with an undisturbed boundary with the sedimentation rate of one of the free species, in combination with a reaction boundary that is composed of a mixture of both free species and the complex, and sediments at a range of s values in between the s-values of the larger free species and the complex. Figure 7(a) shows typical results obtained for a fast reaction (dissociation rate constant $k_{off}=0.01$ s^{-1}, which can usually be considered instantaneous on the time-scale of sedimentation) at equimolar loading concentrations from 0.1–3-fold K_d. Typical features of the $c(s)$ distribution are a bimodal boundary shape, with a peak appearing at a concentration-independent s value close to that of the smaller protein component, and a second peak at a concentration-dependent position. This is consistent with the predictions by Gilbert for the migration of the reaction boundary of such a system in the limit of no diffusion.[31] A quantitative analysis of the concentration dependence of the faster boundary component (calculated by integration over the faster $c(s)$ peak) can yield valuable information on the binding constant and the s value of the complex [32] [insets in Figure 7(a)]. Similarly, isotherms of the relative signals of the two boundary components can be analyzed on the basis of Gilbert–Jenkins theory.[32] This can significantly improve the traditional analysis of the concentration dependence of the overall weight-average sedimentation coefficient.[32–34] For comparison, the $g^*(s)$ distributions calculated as ls-$g^*(s)$ in Figure 7(b) show a single broad peak, from which no finer structure can be discerned and only the overall weight-average s value be determined. Similarly, the integral sedimentation coefficient distributions $G(s)$ from the van Holde–Weischet extrapolation method [35] does not allow to discern information on the timescale of reaction, or any further quantitative analysis.[32]

Figure 8(a) shows the $c(s)$ distributions for a system with slower complex dissociation, $k_{off}=3.2 \times 10^{-5}$ s^{-1} (for equimolar loading concentrations from 0.1–10-fold K_d). Peaks can be discerned at constant positions (the $c(s)$ curves at the lowest concentrations are broadened due to a limited signal-to-noise ratio), resolving the sedimenting species. The reaction is reflected here in the changes of the relative heights of the peaks, reflecting redistribution according to mass action law. The isotherms of the weight-average s value and the signal amplitudes of the three boundary components, which can be modeled jointly in a global analysis, are shown in the insets. Interestingly, the ls-$g^*(s)$ distribution in Figure 8(b) again show a single broad peak, qualitatively similar to those in Figure 7(b). Similarly, also the integral sedimentation coefficient distributions $G(s)$ do not resolve the species and resemble those from Figure 7(b).

In comparison, Figures 7(a) and 8(a) illustrate the qualitatively different behavior of the $c(s)$ distribution for slow ($k_{off} < 10^{-5}$ s^{-1}) and fast ($k_{off} > 10^{-3}$ s^{-1}) reactions, which can be diagnosed from the concentration dependence of the $c(s)$ distributions (going from concentrations significantly lower than K_d to concentrations significantly higher

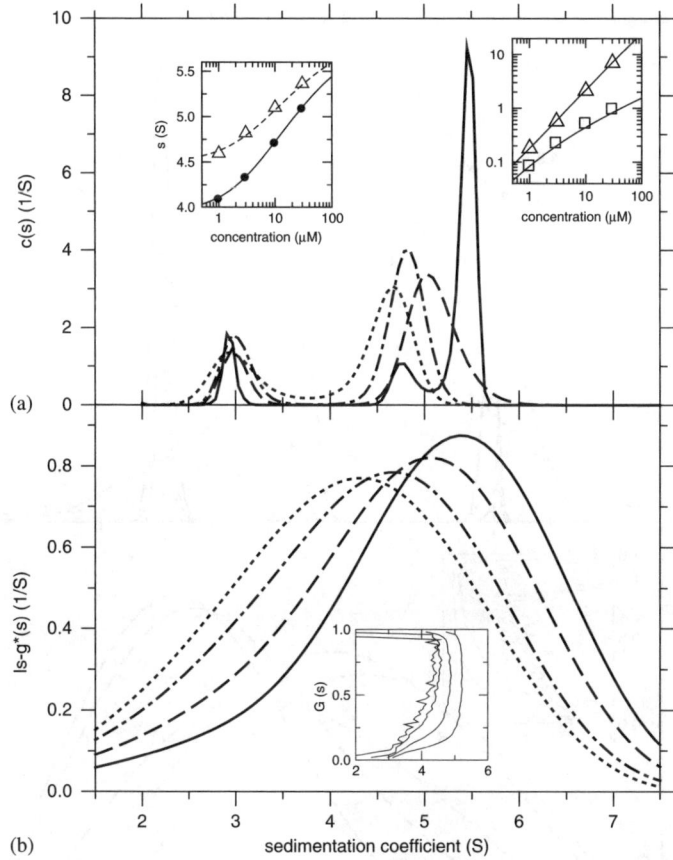

Figure 7 *Results of sedimentation coefficient distribution analysis applied to the rapidly inter-acting proteins. Calculations are based on a species with 30 kDa and 3 S binding to a second species with 50 kDa and 4.5 S to form a 6 S complex with an equilibrium dissociation constant of $K_d = 10$ μM, and an off-rate constant $k_{off}=1\times10^{-2}$ s^{-1}. Sedimentation was simulated for a rotor speed of 50 000 rpm, mimicking signal and noise of interference optical data acquisition, with 50 scans in intervals of 5 min sub-jected to the c(s) analysis. Component concentrations were equimolar at 1 (dotted line), 3 (dash-dotted line), 10 (dashed line), and 30 μM (solid line). (a) The result of the c(s) analysis, using meniscus, time-invariant noise, and weight-average frictional ratio as unknown parameters. The left inset shows the isotherms of the weight-aver-age s value (solid circles) and the s value of the fast boundary component (open tri-angles), both determined by integration of the c(s) distribution, and the theoretical isotherms predicted from mass action law (solid line) and Gilbert–Jenkins theory (dotted line), respectively. The right insets shows the amplitudes in signal units of the undisturbed (squares) and reaction boundary component (triangles), and the theo-retical isotherms predicted from Gilbert–Jenkins theory (solid lines). (b) The appar-ent sedimentation coefficient distribution ls-g*(s) without deconvolution of diffusion, applied to the scans from 50–100 min only to ensure a reasonable fit of the data. All distributions are scaled to unit area. The inset shows the integral sedimentation coef-ficient distributions from the van Holde–Weischet method, calculated with the least-squares method described in ref. 20*

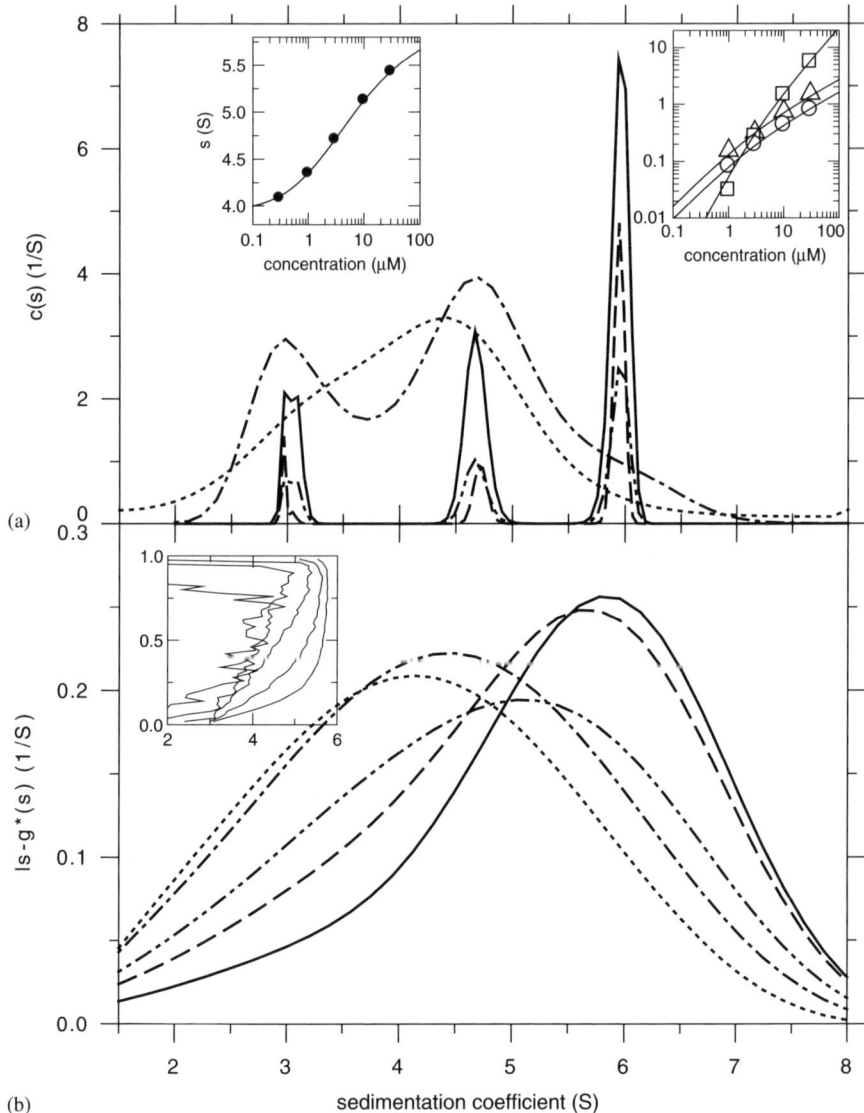

Figure 8 *Results of the sedimentation coefficient distribution analysis applied to slowly inter-*
acting proteins. Calculations are based on the same system as in Figure 7, but with
K_d=3.2 μM and k_{off}=3.2×10^{-5} s^{-1}. Component concentrations were equimolar at
0.3 (dotted line), 1 (dash-dotted line), 3 (dash-dot-dotted line), 10 (dashed line),
and 30 μM (solid line). All other parameters of the simulated sedimentation and the
analysis were identical to those in Figure 7. (a) The c(s) distributions, with the
isotherm of the weight-average s value (left inset, solid circles), and the isotherms
of the concentrations of the individual species (right inset, in signal units) with free
A (circles), free B (triangles), and complex (squares), and the corresponding theo-
retical isotherms based on mass action law (solid lines). (b) The ls-g*(s) distribu-
tions. All distributions are scaled to unit area, except curves for 1 and 3 μM in (a),
which are scaled 10-fold for clarity. The inset shows the integral sedimentation
coefficient distributions from the van Holde–Weischet method

than K_d). If the peak positions are changing, the reaction is fast on the timescale of sedimentation, and the distribution reflects the features of the reaction boundary. If the peak positions are constant but changing only in heights, this suggests stability of the complex on the timescale of sedimentation, and supports the interpretation of peaks as reflecting parameters of the sedimenting species. For fast reactions, the interaction can be characterized by integrating the two characteristic peaks and modeling the isotherm of s values and boundary heights on the basis of Gilbert–Jenkins theory,[32] while for slow reactions the isotherms of populations of the individual species can be determined by integration of $c(s)$, which can be modeled by applying mass action law and mass conservation to the loading mixture. In either case, as shown in ref. 34, weight-average sedimentation coefficients can be obtained rigorously from the integration of the complete $c(s)$ profiles. For any of the isotherm analyses, the integration can be conducted such that peaks from contaminating species can be excluded if they sediment outside the range of s values of the interacting species.

5   Discussion

The examples above illustrate that a detailed model for the sedimenting macromolecular mixture can be extracted from the least-squares modeling of the concentration profiles observed in SV analytical ultracentrifugation. This is consistent with the results reported previously on this methodology,[12,16,20] and with the results of applications to the study of many macromolecular mixtures.[24] Because the $c(s)$ distributions are superpositions of Lamm equation solutions, the hydrodynamic resolution is not confined to clearly visible sedimentation boundaries, but extends to the discrimination of macromolecular species within broad boundaries, where the rms displacement from diffusion during the sedimentation process exceeds the separation from differential sedimentation.[20]

Data from the entire sedimentation experiment can be modeled, and the analysis can be combined with the algebraic methods to calculate best-fit estimates of the systematic noise contributions. This method provides the least bias and noise amplification, and allows to subtract the estimated systematic noise contributions from the raw data. Although these estimates are model-dependent (and therefore the consideration systematic noise offsets must be maintained during the subsequent analyses of the corrected data), it should be noted that the degrees of freedom in the data interpretation resulting from the unknown offsets are inherent in the data and cannot be eliminated by any data pretreatment without bias. However, we found that the visualization of the fit in the raw data space can be very instructive, and it can be very helpful for the design and interpretation of sedimentation experiments to verify the relationship of different boundary components, or other data aspects, to the peaks in the $c(s)$ distribution. In our experience, details of the sedimentation coefficient distributions should be interpreted only after a critical inspection of the raw data and the quality of the fit, ensuring that all features of the sedimentation process in the sample have been described in the model. The latter also includes the sedimentation of unmatched buffer salt.

SV can be a very sensitive tool for the detection of trace components, a topic of high importance in biotechnology applications.[36] In principle, if a sedimenting oligomeric species sediments faster than the leading edge of the diffusion envelope of a monomer, its presence will manifest itself in a slope (or migrating step) in the plateau region of the monomer. (Theoretically, the highest rotor speeds should promote this configuration.) Considering the large number of data points reporting on the shape of the plateau during the course of the experiment, it can easily be seen how the detection sensitivity can be far below the noise for a single data point. The $c(s)$ and ls-$g^*(s)$ distributions can extract this information as they permit the detailed modeling of data sets – including the plateau region – and since they can describe fast-moving species with large boundary displacement between scans. The characterization of species sedimenting within the limits of the diffusion broadened boundary of the monomer is difficult. The use of prior knowledge, such as the existence of discrete species with molar mass constraints combined with segments of continuous distributions [Figure 3(b)] is a possibility to improve the analysis. In this approach, the continuous segments may only describe the species of secondary interest outside the s range of the discrete species, but they ensure that those are properly modeled and not bias the estimate of the species of interest.

The description of diffusion in the sedimentation profiles constituting the basis functions of the distribution (Figure 1) provides for the deconvolution of diffusion in the sedimentation coefficient distributions, which is the source of the relatively high resolution. The extent of diffusion for each species is estimated by the assumption that, in first approximation, all species have a very similar frictional ratio. Since the numerical value of the frictional ratio usually does not vary greatly, this will accommodate a large range of shapes. (It should be noted that there is no assumption of an actual specific shape or frictional ratio, since the value of $(f/f_0)_w$ is determined by regression from the data.) By design, this approach will work best at the highest rotor speeds, when the extent of diffusion is small compared to sedimentation. Nevertheless, as illustrated in Figure 4, it can also work very well for resolving species with a large diffusional spread, caused either by low rotor speeds or by small macromolecule size.

If prior knowledge is available regarding the molar mass or dissimilar frictional shape of certain species (or if the course of the analysis leads to such conclusion), more flexibility of the $c(s)$ analysis is possible by dividing the range of s values into different regimes, and by combination with discrete species with or without molar mass constraints. Possible pitfalls of such more complex parameterizations of the distribution are correlations, but these can be avoided by considering the characteristic shapes of the sedimentation profiles provided by the different segments and species assembled in the distribution. Further, we have shown previously that the discrete species in the hybrid distributions can be described by global parameters in the framework of global analysis of several SV data sets.[25] This can be the global analysis of different samples with common discrete species[25] or the global analysis of the same sample at different rotor speeds.[37]

Although in the discussion above the purpose of estimating the extent of diffusion was essentially a means to arrive at higher resolution sedimentation coefficient distributions, of course, the diffusion itself can provide valuable information on the nature of the sample. As illustrated in the example of Figures 5(a) and (b) [corresponding to the dotted and dashed line in Figure 5(d)] the $c(s)$ distribution may be converted into a $c(M)$ molar mass distribution using the Svedberg relationship to calculate the molar mass from each s value in the distribution and its estimated diffusion coefficient. In our experience this works well if a single major peak is present in the distribution (usually within 10% of the correct molar mass value, provided the correct density and partial-specific volume is known). This will ensure that the weight-average frictional ratio is representing well the major species of interest. In many cases, such a determination of molar mass by $c(M)$ analysis can be superior to the modeling with single or few discrete species.[16] In contrast, the peaks of less abundant fractions of the sample may be at an incorrect molar mass position if they exhibit significantly different frictional ratio from the calculated weight average. Such a dependence of peak positions on the value of $(f/f_0)_w$ is usually not observed in the $c(s)$ distribution.[20] The example of the mixture in Figure 5(c) shows how $c(M)$ can lead to incorrect results for mixtures of similarly abundant species with dissimilar frictional ratio. Better results are obtained when the peak regions of $c(s)$ are substituted by discrete species, combined with continuous sections in between [using again the hybrid discrete/continuous model in SEDPHAT, Equation (5)], such that the hydrodynamic scaling law, Equation (4), and assumptions of shape similarity are abandoned for the species of interest (see above).

A third source of information from the sedimentation process specifically for the characterization of heterogeneous macromolecular mixtures is the different signal contributions to the interference optical and absorbance optical detection system, and/or to the absorbance at different wavelengths. This was demonstrated by Steinberg and Schachman for the simultaneous analysis of constituent sedimentation coefficients. (This has been implemented in SEDPHAT for the global analysis of isotherms of signal-average sedimentation coefficients obtained at different loading compositions and observed at different optical signals.) It has been shown recently that characteristic distinguishable optical signals can be exploited to unravel the sedimentation coefficient distributions of the individual components in mixtures, $c_k(s)$.[26] Owing to the large data basis in SV, differences in the number of aromatic amino acids in proteins can be sufficient for distinguishing their sedimentation for mixtures of two or three different proteins, if data are acquired by refractive index sensitive interference optics and absorbance optics at 280 and 250 nm.[26] This can be enhanced by chromophoric labeling of selected protein components. First applications have shown the potential for the combined spectral and hydrodynamic resolution to unravel the stoichiometry of multiple co-existing protein complexes, such as in the multistep assembly of vitronecting and plasminogen activator inhibitor-1,[38] and in the assembly of the multiprotein complex of adaptor proteins after T-cell activation.[39]

In the present context, we have shown how the global multisignal $c_k(s)$ analysis can reveal the stoichiometry of the protein complex (Figure 6), extended to the combination of discrete species and continuous segments of the distribution [Equation (8)]. Owing to the large molar mass ratio, the spectral differences of complexes with different stoichiometry was relatively small, posing a very stringent test for the spectral decomposition. Nevertheless, the results clearly indicate the formation of 2:1 complexes, consistent with the previous results from the continuous $c_k(s)$ analysis.[26] It should be noted that this does neither rely on the particular s value of the complex, nor the boundary spreading, and therefore appears to be more robust against the onset of the influence of chemical reactions on the time-scale of sedimentation in reaction boundaries.[26,32]

It is a problem inherent to all known sedimentation coefficient distributions that they are based on equations describing the non-interacting superposition of species, which is strictly not suitable for the characterization of reversibly interacting macromolecules. Nevertheless, it can be a good approximation if the protein complexes are relatively stable on the timescale of sedimentation, which could be either due to a low dissociation rate constant of the complex, or if the complex can be maintained in virtual saturation through use of concentrations far above the equilibrium dissociation constant. If these conditions are not fulfilled, the sedimentation is governed by the simultaneous migration and chemical conversion. Gilbert has described in detail the shapes of the resulting reaction boundaries for the case of instantaneous reactions in the absence of diffusion, and shown that they exhibit bimodal shape for two-component mixtures, with a leading boundary component that migrates at a rate intermediate between that expected of a stable complex and stable free species.[31] Independently, Krauss *et al.* have shown that the reaction boundary for small reactants can be described approximately by a single diffusion coefficient.[5,30,40] Using computer simulations for the sedimentation of reacting systems with different complex stability and at different loading concentrations,[30,34] we have shown that whether or not $c(s)$ peaks represent sedimenting species or features of the reaction boundary can be diagnosed from a series of experiments at different loading concentrations covering a range from significantly lower to significantly higher than K_d. If the peak positions of $c(s)$ remain at a constant position but change in the ratio of peak area, a relatively slow interaction is indicated for which the species interpretation is a good approximation. On the other hand, dependence of the $c(s)$ peak positions on the loading concentration indicates the presence of a fast reaction with $c(s)$ reflecting the reaction boundary. However, it is also possible that the sedimentation data alone may not have sufficient information to permit the assessment of the interaction kinetics.[30]

In both cases, the integration of $c(s)$ distribution over all peaks caused by species participating in the reaction will provide a correct measure of the weight-average s value (or signal-average s value, respectively). The isotherm of the dependence of the signal-average s values on the loading concentration can be modeled to determine equilibrium constants.[34,41] This can be combined in a global model with isotherms utilizing the resolution of undisturbed and reaction boundary or the population of the different species, respectively, dependent on the timescale of reaction kinetics.

As an alternative to the isotherm modeling approach following $c(s)$ analysis for interacting systems, the raw sedimentation data may be modeled directly with Lamm equation solutions incorporating reaction terms.[42] Efficient numerical solutions of these Lamm equations have been recently introduced and implemented in SED-PHAT.[30] In this context, our results suggest that the application of the $c(s)$ analysis can be helpful initially to determine the correct interaction model, in particular, in the form of the multisignal component sedimentation coefficient distributions $c_k(s)$. More details on the behavior of $c(s)$ and multisignal $c_k(s)$ applied to reacting systems, and the comparison with Lamm equation modeling and isotherm analyses can be found in refs. 27, 30, 32 and 34.

References

1. T. Svedberg and K. O. Pedersen, *The Ultracentrifuge*, Oxford University Press, London, 1940.
2. H. K. Schachman, *Ultracentrifugation in Biochemistry*, Academic Press, New York, 1959.
3. O. Lamm, *Ark. Mat. Astr. Fys.*, 1929, **21B(2)**, 1–4.
4. L. A. Holladay, *Biophys. Chem.*, 1979, **10**, 187–190.
5. C. Urbanke, B. Ziegler and K. Stieglitz, *Fresenius Z. Anal. Chem.*, 1980, **301**, 139–140.
6. J. S. Philo, *Biophys. J.*, 1997, **72**, 435–444.
7. P. Schuck, *Biophys. J.*, 1998, **75**, 1503–1512.
8. P. Schuck, C. E. MacPhee and G. J. Howlett, *Biophys. J.*, 1998, **74**, 466–474.
9. B. Demeler, J. Behlke and O. Ristau, *Method. Enzymol.*, 2000, **321**, 36–66.
10. J. Behlke and O. Ristau, *Biophys. Chem.*, 2002, **95**, 59–68.
11. W. F. Stafford and P. J. Sherwood, *Biophys. Chem.*, 2004, **108**, 231–243.
12. P. Schuck, *Biophys. J.*, 2000, **78**, 1606–1619.
13. P. Schuck, *Biophys. Chem.*, 2004, **108**, 187–200.
14. P. Schuck, *Biophys. Chem.*, 2004, **187**, 201–214.
15. C. A. MacRaild, D. M. Hatters, L. J. Lawrence and G. J. Howlett, *Biophys. J.*, 2003, **84**, 2562–2569.
16. J. Dam and P. Schuck, *Method. Enzymol.*, 2004, **384**, 185–212.
17. P. Schuck and B. Demeler, *Biophys. J.*, 1999, **76**, 2288–2296.
18. S. W. Provencher, *Comp. Phys. Comm.*, 1982, **27**, 213–227.
19. P. C. Hansen, *Rank-Deficient and Discrete Ill-Posed Problems: Numerical Aspects of Linear Inversion*, SIAM, Philadelphia, 1998.
20. P. Schuck, M. A. Perugini, N. R. Gonzales, G. J. Howlett and D. Schubert, *Biophys. J.*, 2002, **82**, 1096–1111.
21. P. Schuck and P. Rossmanith, *Biopolymers*, 2000, **54**, 328–341.
22. J. S. Philo, *Anal. Biochem.*, 2000, **279**, 151–163.
23. C. R. Cantor and P. R. Schimmel, *Biophysical Chemistry. II. Techniques for the Study of Biological Structure and Function*, W.H. Freeman, New York, 1980.
24. www.analyticalultracentrifugation.com/references.htm
25. H. Boukari, R. Nossal, D. L. Sackett and P. Schuck, *Phy. Rev. Lett.*, 2004, **93**, 098106.
26. A. Balbo, K. H. Minor, C. A. Velikovsky, R. Mariuzza, C. B. Peterson and P. Schuck, *Proc Natl. Acad. Sci. USA*, 2005, **102**, 81–86.
27. A. Balbo and P. Schuck, in *Protein–Protein Interactions*, E. Golemis and P. D. Adams, (eds), Cold Spring Harbor Laboratory Press, Cold Spring Harbor, New York, in press.
28. www.n.h.gov/od/ors/dbeps/PBR/AUCProtocols.htm
29. W. H. Press, S. A. Teukolsky, W. T. Vetterling and B. P. Flannery, *Numerical Recipes in C*, University Press, Cambridge, 1992.
30. J. Dam, C. A. Velikovsky, R. Mariuza, C. Urbanke and P. Schuck, *Biophys. J.*, 2005, **89**, 619–634.
31. G. A. Gilbert and R. C. Jenkins, *Nature*, 1956, **177**, 853–854.
32. J. Dam and P. Schuck, *Biophys. J.*, 2005, **89**, 651–666.
33. R. F. Steiner, *Arch. Biochem. Biophys.*, 1954, **49**, 400–416.
34. P. Schuck, *Anal. Biochem.*, 2003, **320**, 104–124.
35. K. E. van Holde and W. O. Weischet, *Biopolymers*, 1978, **17**, 1387–1403.
36. S. J. Shire, Z. Shahrokh and J. Liu, *J. Pharm. Sci.*, 2004, **93**, 1390–1402.
37. Z. F. Taraporewala, P. Schuck, R. F. Ramig, L. Silvestri and J. T. Patton, *J. Virol.*, 2002, **76**, 7082–7093.

38. K. H. Minor, C. R. Schar, G. E. Blouse, J. D. Shore, D. A. Lawrence, P. Schuck, and C. B. Peterson, *J. Biol. Chem.*, 2005, in press.

39. J. C. Houtman, H. Yamaguchi, M. Barda-Saad, B. Bowden, R. Titterence, E. Appella, P. Schuck and L. E. Samelson, submitted.

40. G. Krauss, A. Pingoud, D. Boehme, D. Riesner, F. Peters and G. Maass, *Eur. J. Biochem.*, 1975, **55**, 517–529.

41. I. Z. Steinberg and H. K. Schachman, *Biochemistry*, 1966, **5**, 3728–3747.

42. D. J. Cox and R. S. Dale, in *Protein-Protein Interactions*, C. Frieden, L. W. Nichol (eds), Wiley, New York, 1981.

CHAPTER 3

Models for Direct Boundary Fitting of Indefinite Ligand-Linked Self-Association

J. J. CORREIA, C. A. SONTAG, W. F. STAFFORD AND P. J. SHERWOOD

Abstract

Analysis of sedimentation velocity data for indefinite self-associating systems is historically achieved by fitting of weight-average sedimentation coefficients ($\hat{s}_w$). There are two main methods for the estimation of weight-average s values from sedimentation velocity data: $g(s)$ from DCDT, and $c(s)$ from Sedfit. By analyzing a set of experiments performed with vinblastine-induced tubulin spirals, we demonstrate that both approaches are appropriate and complementary. In addition, a more robust curve-fitting method that directly fits the shape of the indefinite sedimenting boundary has been incorporated into SEDANAL. This approach utilizes a set of fitting routines that perform global non-linear least-squares (NLLS) fits by a combination of finite-element simulations and a fitting algorithm that uses a simplex method to search parameter space. An overview of appropriate use, limitations and current ongoing developments in these areas is provided.

1 Introduction

This chapter is concerned with the analysis of sedimentation velocity data from systems that undergo indefinite self-association. The traditional method involves analysis of the concentration dependence of the weight-average sedimentation coefficient ($\hat{s}_{20,w}$).[1] Weight-average s is derived from the second moment of the boundary or a discrete distribution calculation equal to $\Sigma\, s_i\, c_i\, /\, \Sigma\, c_i$, where the subscript i is summed over all species, and c_i is expressed in weight-concentration units. Numerous ultracentrifuge studies have been performed on the ligand-induced self-association of tubulin, and those ligands include Mg^{2+},[2,3] drugs like vinca alkaloids[4–9] and microtubule-associated proteins like MAP2[10] and stathmin.[11,12] Additional examples of indefinite systems include insulin self-association in the absence of Zn^{2+},[13] spectrin,[14] glutamate dehydrogenase,[15] myelin basic protein in the presence of detergents,[16] FtsZ[17,18] and Rev.[19] The advantage of sedimentation velocity analysis is that the entire boundary, and thus the entire reactive species distribution, is measured, and contributes to the $\hat{s}_w$ behavior and to the overall shape of the boundary. Deviations from reversible behavior, like inactive monomers or irreversible aggregates, also contribute to the weight average of the boundary and can thus bias the analysis. Extensive introductions to this method have been published[1,20] and should be referred to for many of the earlier details. In this chapter, we first compare the use of $g(s)$ and $c(s)$ for estimation of the weight-average s values for indefinite systems. In general, the two methods are equivalent and provide complementary information for use in fitting the concentration dependence of $\hat{s}_w$. Second, direct boundary fitting by a combination of finite-element simulations of the reaction boundary[21,22] and an NLLS-fitting algorithm that uses a simplex method to search parameter space[23] to fit to time difference (ΔC) data have now been incorporated into SEDANAL.[24] This approach allows for convenient inclusion of models derived from HYDRO[25] (s_N vs. N) that correspond to linear and helical or curved polymers.[20] In addition, indefinite or isodesmic models can be combined with other discrete models where only one or some of the reactants can participate in both reaction pathways. Isodesmic refers to an indefinite model where all the K_is for successive polymer growth are equal. Here, we describe our initial applications of this novel approach using SEDANAL, outlining how one can validate the results by comparison with weight-average methods, and conclude with a summary of ongoing developments and challenges for quantitative analysis of complex systems of this type.

2 Methods

All methods for this chapter have been described elsewhere[1,9,20,24] and only specific details are given. The antimitotic and tubulin-binding drug vinblastine, the buffer Pipes (piperazine-1,4-bis(2-ethanesulfonic acid)) and the nucleotide GDP (guanosine 5′-diphosphate sodium salt) were purchased from Sigma.

3 Results

Figure 1 presents the $g(s)$ and $c(s)$ analysis of data from a set of 15 velocity runs performed with 0–43.6 μM vinblastine and 2 μM GDP-tubulin in typical buffer conditions (80 mM Pipes, 50 μM GDP, 0.5 mM MgCl$_2$, pH 6.8, at 19.7 °C). The speed is adjusted as the oligomer size increases to maintain a fairly constant number of scans for the analysis. As expected, the $g(s)$ distributions appear to be noisier and broader than the $c(s)$ distributions, reflecting experimental diffusion. Also, as expected for $c(s)$ analysis done with regularization, the $c(s)$ distributions have a tendency to be peaky and exhibit shoulders or peaks that move more slowly than the faster peaks. One has a tendency with $c(s)$ distributions to assign molecular meaning to the shape of peaks and shoulders. For example, are the trailing shoulders in the $c(s)$ intermediates in the reaction, or, since this region of the distribution seems to move with the peak position, is it a feature of the kinetics of the reaction boundary?

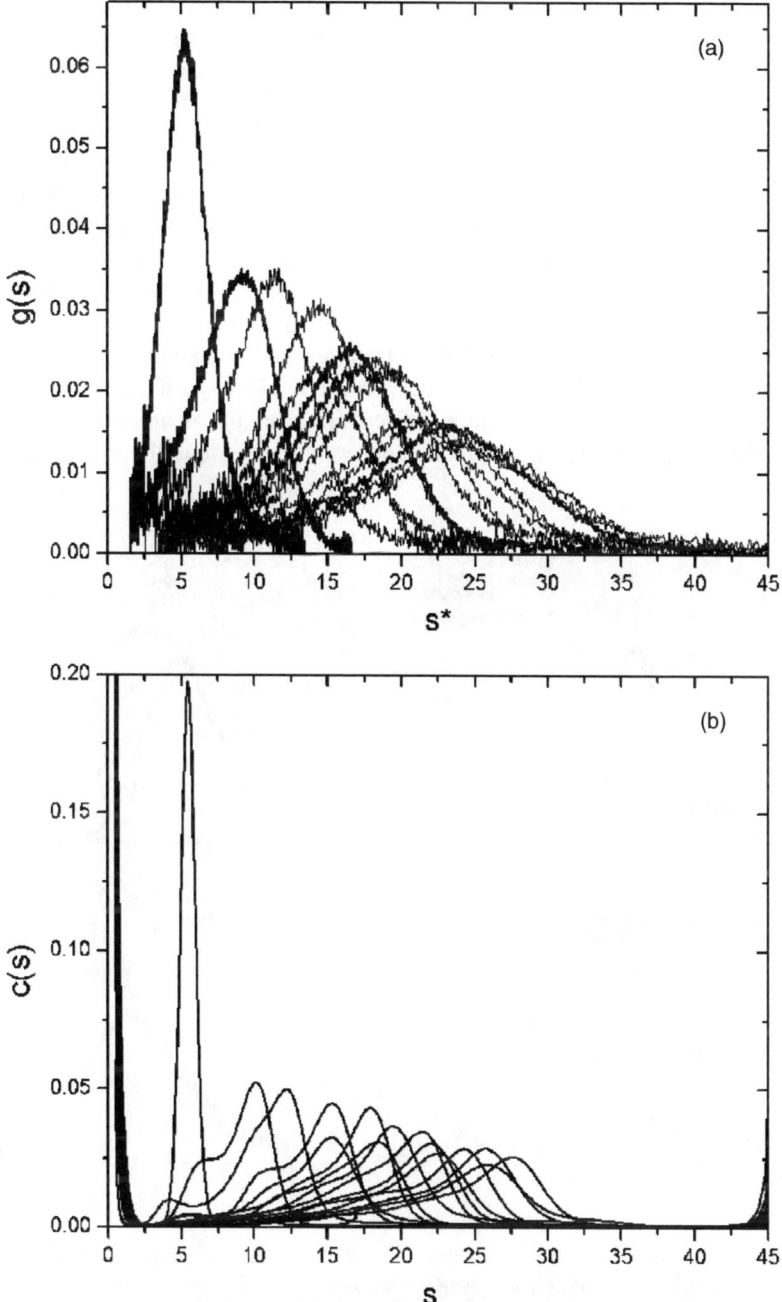

Figure 1   *Data were collected for 0–43.6 μM vinblastine and 2 μM GDP-tubulin (in 80 mM Pipes, 0.5 mm MgCl₂, 50 μM GDP, pH 6.8) as described in Sontag et al.,[20] and then analyzed with DCDT+ and Sedfit to produce (a) g(s) distributions, and (b) c(s) distributions*

To investigate this, a number of these curves have been superimposed to show how the two analysis methods compare (Figure 2). For example, at 2.3 μM vinblastine, the $c(s)$ suggests a shoulder and a slow peak, while the corresponding $g(s)$ reveals no evidence of these features. At the two intermediate drug concentrations (12.1 and 25.3 μM), the centripetal or slower side of $g(s)$ tracks the $c(s)$ curves very well, suggesting a very broad sedimenting boundary, but again without revealing inflections or shoulders. At the highest drug concentration, $g(s)$ has a break that corresponds to a shoulder in the $c(s)$. These observations are consistent with the typical features and signal-to-noise expected for $c(s)$ and $g(s)$ analysis of XLA data. However, it is our preference to be cautious about interpreting shape features in a $c(s)$ distribution, unless it appears in a family of curves and is verified by shapes in the corresponding $g(s)$ distributions. This caution is derived from our understanding of $c(s)$ as a mathematical distribution that fits the boundary shape, but is only a hypothesis about the molecular distribution in the solution. In general, given the assumptions $c(s)$ analysis makes,[26] it appears to be most appropriate for non-interacting systems.

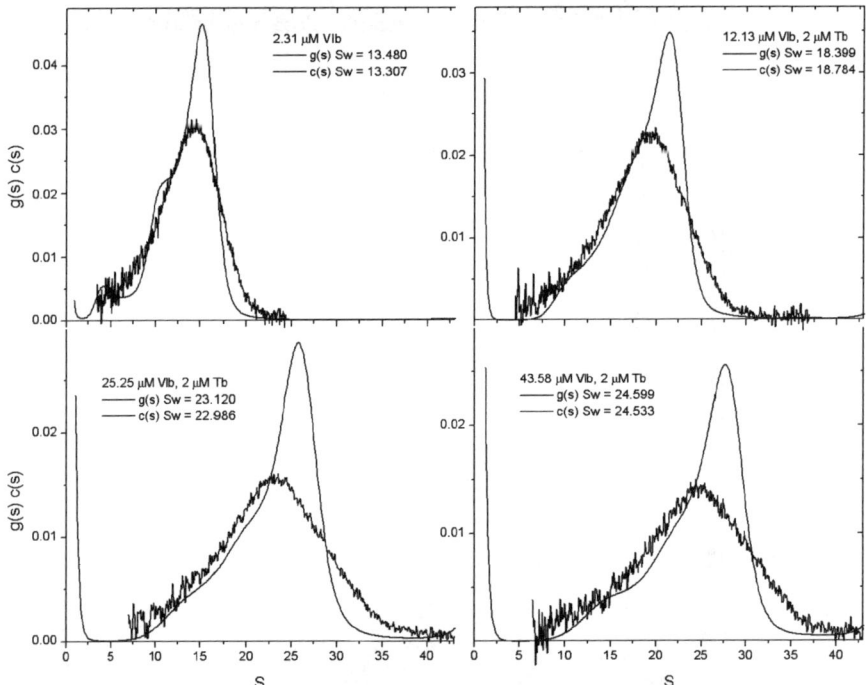

Figure 2   *Selective distribution plots from Figure 1 are superimposed to demonstrate the similarity and differences between the $g(s)$ and $c(s)$ patterns. The tendency of $c(s)$ generated with regularization to be peaky is typically offset with either a longer trailing tail or a sharp drop at the leading edge. Nonetheless, since both distributions fit the experimental boundary shape they necessarily have similar weight-average values, as demonstrated at 2.31 μM (13.480, 13.307 s), 12.13 μM (18.399, 18.784 s), 25.25 μM (23.120, 22.986 s) and 43.58 μM (24.599, 24.533 s) vinblastine concentration*

Verification from multiple loading concentrations, $g(s)$ analysis and direct boundary fitting are required to test any hypothesis about the molecular distribution, including peak positions, concentration dependence, reaction mechanism and energetics. To demonstrate this, the $g(s)$ and $c(s)$ distributions were integrated to generate $\hat{s}_w$ estimates, and those are plotted in Figure 3 *vs.* drug concentration. These data were then fit to an isodesmic model using a ligand-induced indefinite polymerization scheme.[9] In this model K_1 corresponds to drug binding to tubulin heterodimer, and K_2 corresponds to the polymerization of liganded tubulin–drug complex to the growing spiral. For isodesmic models described here K_2 applies to all steps in spiral growth. It is dramatically clear that the $g(s)$- and $c(s)$-derived $\hat{s}_w$ data (Figure 3) nearly superimpose, and the best fits give identical results, within error. This proves that $g(s)$ and $c(s)$ analyses, when properly applied, are equivalent for estimation of weight-average s values and quantitative measurement of indefinite equilibrium constants. This had been previously verified by Schuck,[27] where he demonstrated that $g(s)$ by back-transformation correctly fits the boundary, although his Figures 1 and 5 suggest that the $g(s)$-derived $\hat{s}_w$ values fell on a different curve than the $c(s)$-derived values, something we have never seen in our numerous comparisons of $c(s)$ and $g(s)$. They both work equally well. [The caveat to this is that $c(s)$ may be more useful at very low signal-to-noise. Furthermore, given the ability to analyze all the scans in a run,

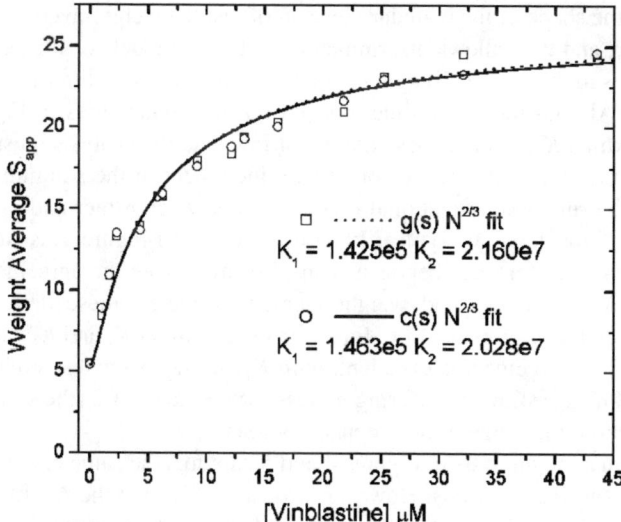

Figure 3 *A plot of weight-average s values, derived from integration of the $g(s)$ ($\square$) and $c(s)$ (o) distribution data in Figure 1, vs. free vinblastine concentration, and fit with a ligand-mediated indefinite isodesmic $N^{2/3}$ model as described in Sontag et al.[20] Note the excellent agreement between the individual points (the error bars on these $\hat{s}_w$ measurements are within the symbol sizes) and the best-fit values. The derived K_1, K_2 values are identical within error ($K_1 = 1.425 \times 10^5$ vs. $1.463 \times 10^5 M^{-1}$; $K_2 = 2.16 \times 10^7$ vs. $2.028 \times 10^7 M^{-1}$). Thus, in spite of the differences in the shapes of the $g(s)$ and $c(s)$ distributions, the weight-average values are identical because they both fit the shape of the sedimenting boundary*

$c(s)$ provides the ability to integrate selected regions of the distribution from very broad multiple-peaked distributions. Since $g(s)$ analysis involves a narrower span of scans, selective integration, while possible, is not as versatile, although a wide distribution analysis (DCDT–WDA),[28] also incorporated into SEDANAL, provides expanded multi-speed functionality.]

In the weight-average approach, a complete titration of either drug[9] or protein concentration[4] is required to extract K_1 and K_2. The K_1K_2 values extracted by this ligand-mediated analysis are related to the indefinite association constant K_{iso} or $K_{2,app}$ by the formula,[5,6]

$$K_{2,app} = \frac{K_2}{\left[1 + \left(\dfrac{1}{(K_1[\text{Drug}]_{\text{free}})}\right)\right]^2} \tag{1}$$

We use K_{iso} or $K_{2,app}$ interchangeably, but make the distinction between $K_{2,app}$ analysis, where K_1 and K_2 are determined directly, and SEDANAL K_{iso} fits of individual samples at different drug concentrations, where an isodesmic model is applied independent of a model involving drug concentration. Thus, at a fixed free-drug concentration, SEDANAL direct boundary fitting can determine K_{iso}. Alternatively, fitting of these K_{iso} values *vs.* drug concentration can be used to determine K_1 and K_2 values and estimates of $K_{2,app}$. The advantage of direct boundary fitting is that it also investigates the shape of the boundary instead of just the weight-average behavior of the boundary, and thus allows discrimination between models or inclusion of additional features to the model. To demonstrate this, an isodesmic fitter has been coded into SEDANAL and the vinblastine-induced reaction boundaries in Figure 1 have been fit to extract K_{iso}. [Since these are fits of raw data the analysis must be done in s_{app} units, using the actual $(1-v\rho)$ or density increment of the solution at the temperature of the run, and a functional relationship between n-mer and s_n values (see Sontag *et al.*[20] for details)]. Typical fits are presented in Figure 4 as ΔC *vs.* radius plots for pairs of differences between scans. To investigate the agreement between SEDANAL K_{iso} estimates and weight-average estimates, these data are plotted against drug concentration and fit to Equation (1) to extract K_1 and K_2 values, shown in Figure 5. The agreement is excellent, with K_1 and K_2 estimates being essentially identical and $K_{2,app}$ estimates differing by less than a factor of 2 (the solid *vs.* dotted lines) throughout the range of drug concentrations.

It is clear that the analysis in Figures 4 and 5 captures the same energetic features as the weight-average analysis. However, it is also clear that the ΔC fits have some systematic, albeit small, deviations from the data. The advantage of direct boundary fitting with SEDANAL is that one can test different models to uncover the origin of the deviations. In this instance, we used an $N^{2/3}$ model, which assumes constant axial ratio. As described elsewhere,[20] a more correct model for vinblastine-induced spirals is a helical 42-bead model. Figure 6 presents typical isodesmic SEDANAL fits using this helical model. In general, the rms of the fits are worse relative to the $N^{2/3}$ fits (Figure 4), and the systematic deviations demonstrate less agreement in the centripetal or trailing regions of the boundary, with a more pronounced peak at the maximum (possibly the very features captured in the $c(s)$ distributions). These data were

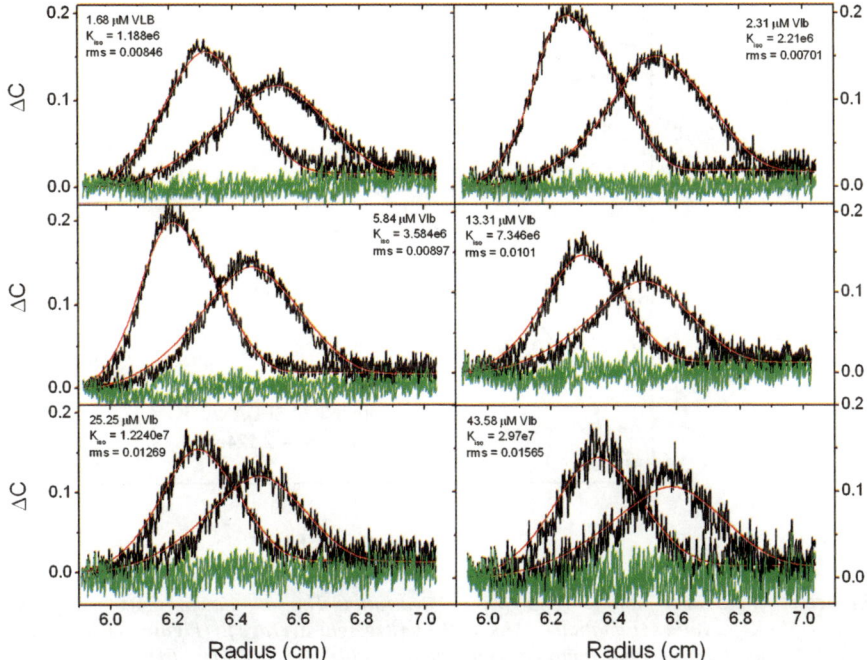

Figure 4 *SEDANAL isodesmic analysis of selected data sets from Figure 1. Twenty scans of data from each drug concentration are fit to an indefinite association model (K_{iso}) according to Equation (1) using an $N^{2/3}$ model for s_n vs. N, the degree of association (see refs. 20 and 24 for details). The data are presented as ΔC vs. radius with only the first and last difference pair shown. The data are in black, the fit in red, and the residuals are in green*

also compared to the weight-average 42-bead helical estimates of $K_{2,app}$ by fitting K_{iso} vs. drug concentration (Figure 7). The agreement is again excellent, being a factor of 1.5 in K_2 and in $K_1 K_2$. Thus, the helical fitting again captures similar energetic features to that of the weight-average fitting; however, the systematic deviations in ΔC (Figure 6) suggest that there are missing or incorrect features in the model.

To investigate this further, Figure 8(a) presents the distribution of species implied by the $N^{2/3}$ and the 42-bead models at the highest drug concentrations. For the $N^{2/3}$ model the distribution peaks at 5-mer and becomes insignificantly populated above 50-mer. Alternatively, the 42-bead helical model peaks near 35-mer and is significantly populated beyond 200-mer. [These simulations were done with a Turbo Pascal DOS program (JJC), but they can now also be done in SEDANAL using its equilibrium calculator.] The wide discrepancy in *n*-mer distribution is due to the functional dependence of s_n on *n*-mer.[20] To show this, we plot the weight concentration of *n*-mer *vs.* the corresponding s_n (Figure 8(b)). Using an $N^{2/3}$ model, the s_n values change rapidly with *n*-mer value and, thus, only a few polymers (<50) in the distribution are required to describe the boundary shape. Alternatively, the helical 42-bead model has a relatively flat dependence of s_n upon *n*-mer value,[20] and thus even a very broad distribution up to 200-mer (Figure 8(a)) has difficulty describing the shape of the

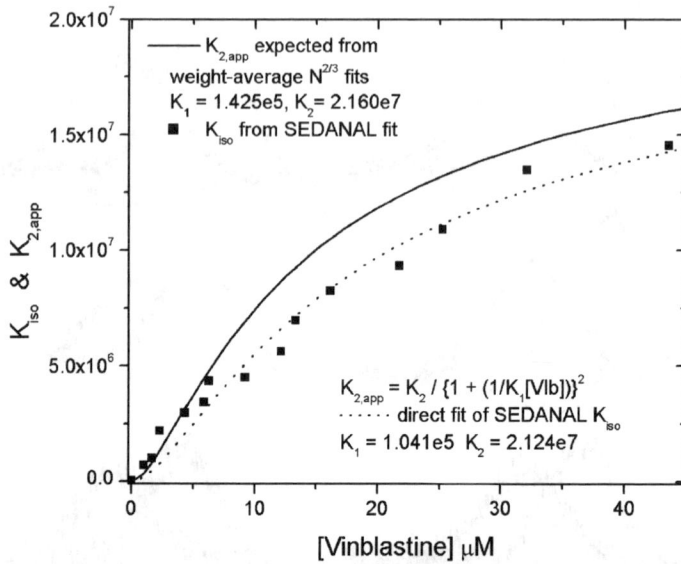

Figure 5   *A plot of K_{iso} and $K_{2,app}$ vs. drug concentration. The solid line presents the expected $K_{2,app}$ values (Equation (1)) using the best weight-average fit (Figure 3) with the $N^{2/3}$ model. The solid squares present the K_{iso} values derived from SEDANAL fits of the same data, also using the $N^{2/3}$ model. These K_{iso} data are then fit according to Equation (1) (the dotted line) and demonstrate excellent agreement with the weight-average estimates ($K_1 = 1.425 \times 10^5$ vs. 1.041×10^5 M^{-1}; $K_2 = 2.16 \times 10^7$ vs. 2.124×10^7 M^{-1}). This method validates the accuracy of the SEDANAL approach.*

broad experimental boundaries. This highlights why the $N^{2/3}$ model has been useful as a fitting function of weight-average data.[1–9] Thermodynamic or Wyman analysis (see Sontag et al.[20] for details), a plot of log K_{iso} vs. log[vinblastine$_{free}$], can be conducted as an additional test (Figure 9). The slope of this line reflects the number of ligands bound during the association process. For the $N^{2/3}$ model the slope is 0.79, consistent with previous results (see Sontag et al.[20] for details) and the fact that the binding of one drug molecule per tubulin heterodimer is linked to spiral formation. For the helical model the slope is 1.58, twice the $N^{2/3}$ value, and consistent with the difficulty in using the helical model to fit the boundary shape. This again indicates the advantage of using SEDANAL to fit these data. Since the more physically reasonable helical model fits less well, we must conclude that additional factors are required in the model. Both hydrodynamic (K_s) and thermodynamic non-ideality (BM1) can easily be included using the model editor provided with SEDANAL[24] (also described in the SEDANAL_users_manual.pdf). For example,

$$s = s_0/(1 + K_S c) \qquad (2)$$

Inclusion of these parameters in the fitting has a small effect on a Wyman analysis of the data (Figure 9), bringing the slopes in both instances only slightly closer to 1 (0.79–0.82 for $N^{2/3}$, and 1.59–1.51 for helical). Nonetheless, this is still not sufficient

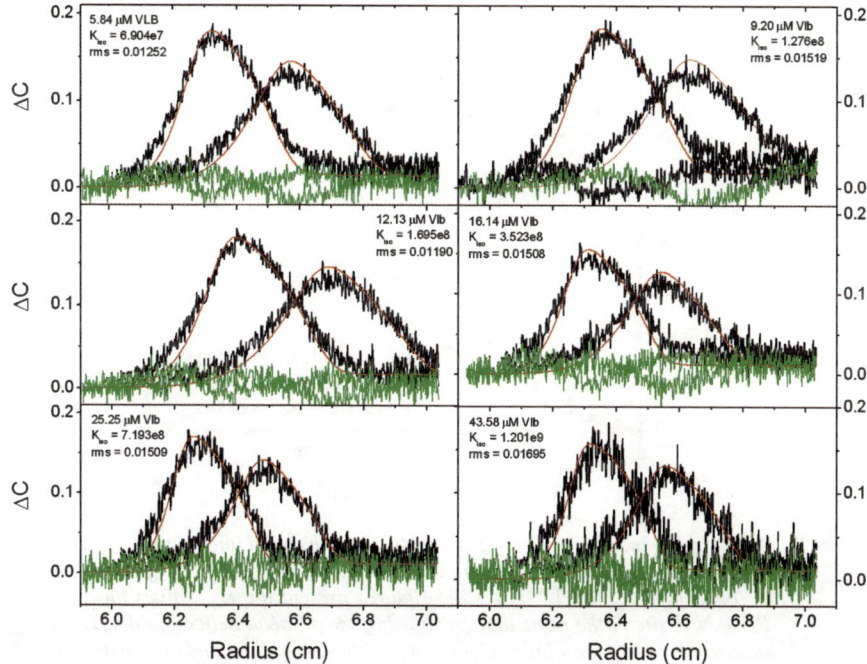

Figure 6 *SEDANAL isodesmic analysis of selected data sets from Figure 1. Twenty scans of data for each drug concentration are fit to an indefinite association model (K_{iso}) according to Equation (1), using a helical 42-bead HYDRO model for s_n vs. N, the degree of association (see refs. 20 and 24 for details). The data are presented as ΔC vs. radius with only the first and last difference pair shown. The data are in black, the fit in red, and the residuals are in green*

to declare the fits good enough to describe the data (the residuals are still not random, not shown), implying that additional factors must be included.

4 New Developments

We are currently exploring additional alterations to these isodesmic models.

1. One significant experimental factor is that tubulin is heterogeneous and composed of multiple gene products and numerous post-translational modifications.The data in Figure 6 exhibit a sharper peak and more pronounced tailing than the fit. This could be due to heterogeneity in tubulin isotype affinity for the spirals (although this was not observed in an earlier $\hat{s}_w$ analysis).[29] Re-analysis of data from purified isotypes[29] is currently underway to explore this possibility.
2. The indefinite model used in these fits assumes an isodesmic mechanism where all the elongation K's are equal. An alternate model, isoenthalpic, assumes constant enthalpy, but variable polymer-size-dependent entropy for polymer growth.[30] This model is anticipated to decrease the affinity as a function of polymer length, but has not yet been tested.

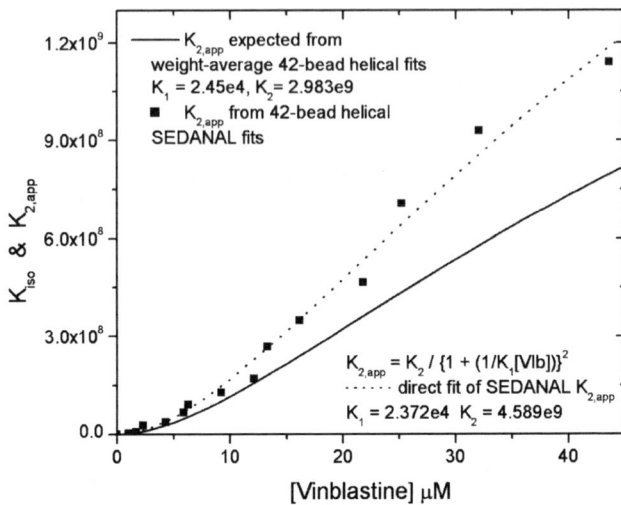

Figure 7 *A plot of K_{iso} and $K_{2,app}$ vs. drug concentration. The solid line presents the expected $K_{2,app}$ values (Equation (1)) using the best weight-average fit (Figure 3) with the 42-bead helical model.[20] The solid squares present the K_{iso} values derived from SEDANAL fits of the same data, also using the 42-bead helical model. These K_{iso} data are then fit according to Equation (1) (the dotted line) and demonstrate excellent agreement (a factor of 1.5 for K_2 and $K_1 K_2$) with the weight-average estimates. As discussed in the text, in spite of this, the systematic deviations in the ΔC fits suggest additional modification to the model is required*

3. The version of SEDANAL used to analyze the data for this chapter used a constant radial point density, typically 1600 points across the cell. We have found numerous numerical problems, especially at the cell base, where the concentration builds to extremely high levels and numerical solutions begin to fail, often producing negative concentrations. In the Claverie method, used here to solve the Lamm equation, each numerical iteration must proceed throughout the cell, from meniscus to base, to account for conservation of mass. To deal with pelleting at the base as well as sharp boundaries early in the run, variable radial spacing has now been introduced into SEDANAL. We typically use a 10-fold higher point density at the meniscus and a 100-fold higher point density at the base. Typical isodesmic fits now require 2400–4800 points (with 10 000 being the current maximum number of points).

4. An additional possibility to account for the systematic deviations observed in these fits is that the boundaries are broadened by slow kinetics in the re-equilibration of these broad distributions.[29,31] Incorporation of kinetics is programmed into SEDANAL, and we are currently working on adding it to indefinite models.

5. The current isodesmic models being tested ignore the explicit step of drug binding, but rather fit according to Equation (1). This is the basis of the distinction between $K_{2,app}$ and K_{iso}. The slow sedimentation of free drug coupled to the ligand-linked association may contribute to the observed boundary trailing. The direct inclusion of a ligand-mediated model into SEDANAL has yet to be tested.

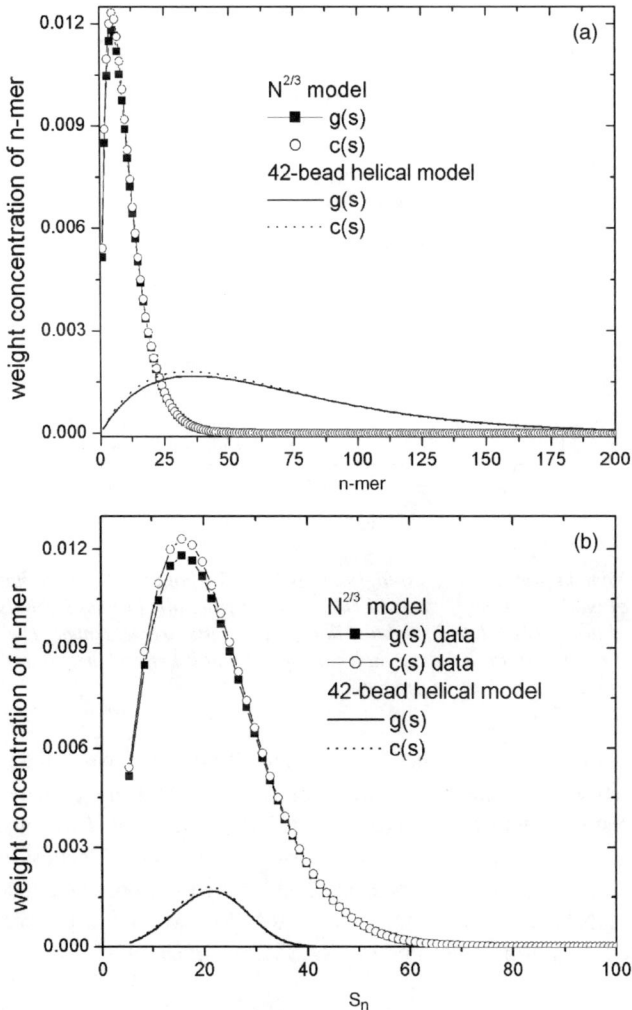

Figure 8 *(a) A plot of weight fraction of n-mer vs. n-mer for both the $N^{2/3}$ and the 42-bead helical models. The data are simulated at 42.6 μM vinblastine with the g(s)- and c(s)-derived best fits (Figure 3) to demonstrate the similarity of the two weight-average estimates. (b) A plot of weight fraction of n-mer vs. s_n values for that n-mer for both the $N^{2/3}$ and the 42-bead helical models. The broad distribution of the $N^{2/3}$ simulation explains the ability to fit the broad experimental distribution. The narrow distribution for the 42-bead helical model explains, in part, the difficulty in fitting the data*

5   Conclusions

In this chapter we have compared the use of $g(s)$ and $c(s)$ for estimation of weight-average s values and the energetics for indefinite associating systems, verifying that they produce identical results. A direct boundary-fitting approach for indefinite

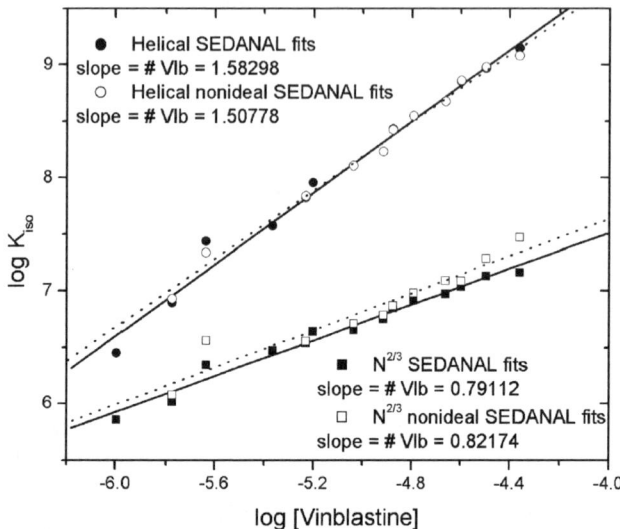

Figure 9   *A Wyman plot (log K_{iso} vs. log[vinblastine]) for both the $N^{2/3}$ and the 42-bead hel-*
ical model.[20] In both cases SEDANAL data from ideal (open symbols, solid lines)
and non-ideal (closed symbols, dotted lines) fits are presented. The slope equals
the number of bound drug molecules per tubulin heterodimer that are coupled to
the overall association process

polymerization has now been incorporated into SEDANAL. We validate the use of
SEDANAL isodesmic models by comparison with weight-average methods, and further demonstrate the steps required to estimate ligand-mediated (K_1K_2) and thermodynamic parameters. Numerous extensions to these models (kinetics, isoenthalpic,
ligand-mediated) are being developed and tested and will be presented in a future publication. SEDANAL is available through the RASMB Analytical Ultracentrifugation
Software Archive at http://www.bbri.org/RASMB/rasmb.html.

Acknowledgement

We thank Sharon Lobert and Holland Alday for careful reading of the manuscript.
This is publication #39 from the UMMC Analytical Ultracentrifuge Facility.

References

1. J. J. Correia, *Method. Enzymol.*, 2000, **321**, 81.
2. R. P. Frigon and S. N. Timasheff, *Biochemistry,* 1975, **14**, 4559.
3. R. P. Frigon and S. N. Timasheff, *Biochemistry*, 1975, **14**, 4567.
4. G. C. Na and S. N. Timasheff, *Method. Enzymol.*, 1985, **117**, 459.
5. G. C. Na and S. N. Timasheff, *Method. Enzymol.*, 1985, **117**, 496.
6. G. C. Na and S. N. Timasheff, *Biochemistry*, 1986, **25**, 6214.
7. G. C. Na and S. N. Timasheff, *Biochemistry*, 1986, **25**, 6222.
8. S. Lobert, A. Frankfurter and J. J. Correia, *Biochemistry,* 1995, **34**, 8050.

9. S. Lobert and J. J. Correia, *Method. Enzymol.*, 2000, **323**, 77.
10. J. C. Lee, N. Tweedy and S. N. Timasheff, *Biochemistry*, 1978, **17**, 2783.
11. P. A. Curmi, S. S. L. Anderson, S. Lachkar, O. Gavet, E. Karsenti, M. Knossow and A. J. Sobel, *Biol. Chem.*, 1997, **272**, 25029.
12. L. Jourdain, P. Curmi, A. Sobel, D. Panataloni and M. F. Carlier, *Biochemistry*, 1997, **36**, 10817.
13. P. D. Jeffrey, B. K Milthorpe and L. W. Nichol, *Biochemistry*, 1976, **15**, 4660.
14. A. Henniker and G. B. Ralston, *Biophys. Chem.*, 1994, **52**, 251.
15. R. Josephs, H. Eisenberg and E. Reisler, in *Protein–Protein Interactions*, R. Jaenicke and E. Helmreich (eds), Springer-Verlag, New York, 1972, 57.
16. R. Smith, *Biochemistry*, 1982, **21**, 2697.
17. G. Rivas, A. Lopez, J. Mingorancei, M. J. Ferrandizi, S. Zorrilla, A. P. Minton, M. Vicentei and J. M. Andreu, *J. Boil. Chem*, 2000, **275**, 11740.
18. G. Rivas, J. A. Fernandez and A. P. Minton, *PNAS*, 2001, **98**, 3150.
19. J. L. Cole, J. D. Gehman, J. A. Shafer and L. C. Kuo, *Biochemistry*, 1993, **32**, 11769.
20. C. A. Sontag, W. F. Stafford and J. J. Correia, *Biophys. Chem.*, 2004, **108**, 215.
21. J. M. Claverie, H. Dreux and R. Cohen, *Biopolymers*, 1975, **14**, 1685.
22. G. P. Todd and R. H. Haschemeyer, *PNAS*, 1981, **78**, 6739.
23. J. A. Nelder and R. Mead, *Comput. J.*, 1965, **7**, 308.
24. W. F. Stafford and P. J. Sherwood, *Biophys. Chem.*, 2004, **108**, 231.
25. J. G. de la Torre, S. Navarro, M. C. Lopez Martinez, F. G. Diaz and J. J. Lopez Cascales, *Biophys. J.*, 1994, **67**, 530.
26. P. Schuck, *Biophys. J.*, 2000, **78**, 1606.
27. P. Schuck, *Anal. Biochem.*, 2003, **320**, 104.
28. W. F. Stafford and E. H. Braswell, *Biophys. Chem.*, 2004, **108**, 273.
29. S. Lobert, A. Frankfurter and J. J. Correia, *Cell Motil. Cytoskel.*, 1998, **39**, 107.
30. R. C. Chatelier, *Biophys. Chem.*, 1987, **28**, 121.
31. S. Lobert, J. W. Ingram, B. T. Hill and J. J. Correia, *Mol. Pharmacol.*, 1998, **53**, 908.

CHAPTER 4

Allowance for Thermodynamic Nonideality in Sedimentation Equilibrium

PETER R. WILLS AND DONALD J. WINZOR

1 Introduction

Although the expressions for sedimentation equilibrium were derived initially from the viewpoint of a balance between centrifugal and diffusional fluxes,[1] current theory stems from a realization by Goldberg[2] that solute distribution is amenable to rigorous thermodynamic description. This development led to analysis of the concentration dependence of apparent molecular weight for a single, nonassociating solute in terms of its true molar mass (the value extrapolated to zero solute concentration) and an activity coefficient that quantified the concentration dependence.[3-6] The exact nature of that empirically defined activity coefficient has been recognized only recently.[7]

The next theoretical breakthrough was the demonstration that the sedimentation equilibrium procedure had the potential to accurately define the thermodynamic activity of a solute in incompressible solutions.[8-11] This potential was strengthened further by the realization that statistical mechanics provided a means of interpreting the experimentally derived thermodynamic activities and activity coefficients.[10,12-15]

Originally, the characterization of protein self-association by direct analysis of the experimental concentration distribution entailed, for ideal systems, evaluation of the equilibrium concentration of monomers throughout the distribution, and hence deduction of the equilibrium constant(s) by combining that quantity with the total concentration at each radial distance.[9,15,16] Despite its simplicity, this procedure has usually been overlooked in favour of an alternative direct analysis that is based on a nonlinear least-squares minimization procedure to obtain the best-fit description of the total concentration distribution as the sum of exponential expressions for each putative oligomeric species.[17,18] Although this simulation-based procedure (readily available in software packages such as ASSOC) is suitable for quantifying ideal solute self-association, the intricacies of the algorithms used to achieve the best-fit solution[19] have deterred any attempt to incorporate realistic allowance for effects of thermodynamic nonideality.

The preparation of our chapter for the previous edition of this treatise on analytical ultracentrifugation[20] has provided the impetus for a reappraisal of the theory of sedimentation equilibrium that incorporates the advantages that stem from the consideration of thermodynamic nonideality on the statistical-mechanical basis of excluded volume.[14,21] A logical starting point for the present deliberations is, therefore, the quantitative description of thermodynamic nonideality.

2 Description of Thermodynamic Nonideality

The chemical potential, μ_A, of solute in a two-component solution (solute A and solvent S) is a function of temperature, T; pressure, P; and composition. Changes in μ_A arising from alterations in composition at fixed temperature may be expressed in terms of thermodynamic activity provided that either the pressure or the solvent chemical potential (μ_S) is held constant as the concentration of solute is varied. In thermodynamic studies, the chemical potential of solute is considered to be the sum of a standard-state value (μ_A°) and a part that, under ideal conditions, depends logarithmically upon solute concentration. The concentration of ideal solute can be measured on the mole fraction, molal or molar scale (as well as the corresponding weight-based scales), and the choice of concentration scale can dictate the magnitude of the standard-state chemical potential.

For very dilute solutions all concentration scales bear approximately linear relationships to each other, whereupon the variations in μ_A can be taken into account through the logarithmic term, irrespective of the concentration variable chosen.[22,23] At higher concentrations, however, the approximate proportionality between concentration scales can no longer be relied on. Consequently, the chemical potential of solute is best written in terms of the concentration scale that is pertinent to the thermodynamic activity defined for the particular conditions of measurement.

2.1 Definition of Thermodynamic Activity

In partition studies, such as equilibrium dialysis, osmometry, and gel chromatography, the chemical potential of solute is monitored under the condition of constant solvent chemical potential, a constraint under which μ_A is most simply defined[24] as

$$(\mu_A)_{T,\mu_S} = (\mu_A^\circ)_{T,\mu_S} + RT \ln z_A \tag{1a}$$

$$z_A = \gamma_A C_A \tag{1b}$$

where the thermodynamic activity of solute, z_A, is a molar quantity and therefore most naturally expressed as the product of its molar concentration, C_A, and the corresponding activity coefficient, γ_A.

On the other hand, the corresponding expressions for the chemical potential of solute in experiments conducted under the usual laboratory constraints of constant temperature and pressure, P, are

$$(\mu_A)_{T,P} = (\mu_A^\circ)_{T,P} + RT \ln a_A \tag{2a}$$

$$a_A = y_A m_A \tag{2b}$$

in which the molal thermodynamic activity, a_A, is the product of the molal solute concentration, m_A, and the corresponding activity coefficient, y_A. In this regard, the molality of solute is defined as

$$m_A = n_A/(n_S M_S) \tag{3}$$

where the dimensionless ratio n_A/n_S (moles of solute per moles of solvent) is divided by the molar mass of solvent, M_S, to retain the customary unit (mol kg^{-1}) for m_A.

2.2   Virial Expansions

Another means of incorporating the consequences of thermodynamic nonideality is to employ polynomial expansions – a practice well embedded in osmometry theory, where the osmotic pressure, Π, is written as a virial expansion in molar concentration, C_A.

$$\left(\frac{\Pi}{RT}\right)_{T,\mu_S} = C_A + B_2 C_A^2 + B_3 C_A^3 + \cdots \tag{4}$$

Furthermore, by a purely thermodynamic argument, it can be shown[24] that these virial coefficients (B_2, B_3, ...) also describe the standard molar activity coefficient.

$$\ln \gamma_A = 2B_2 C_A + \frac{3}{2} B_3 C_A^2 + \cdots \tag{5}$$

A consequence of this interrelationship between γ_A and the virial coefficients is that B_2, B_3, *etc.*, find simple statistical-mechanical interpretation in terms of the physical interactions between pairs, triplets and larger clusters of solute molecules.[21,24]

In situations where the chemical potential of solute is defined under the constraints of fixed temperature and pressure, the counterparts of Equations (4) and (5) become

$$-\frac{(\mu_S - \mu_S^0)_{T,P}}{RT\rho_S} = m_A + C_2 m_A^2 + C_3 m_A^3 + \cdots \tag{6}$$

$$\ln y_A = 2C_2 m_A + \frac{3}{2} C_3 m_A^2 + \cdots \tag{7}$$

where ρ_S is the solvent density, and C_2 and C_3 denote the second and third molal virial coefficients, respectively. Unlike their molar counterparts, C_2 and C_3 are not interpretable on the statistical-mechanical basis of excluded volume. For incompressible solutions, however, the following relationships, expressed in terms of the solute partial specific volume, $\bar{v}_A$, exist between the two sets of virial coefficients:

$$B_2 = C_2/\rho_S + M_A \bar{v}_A \tag{8a}$$

$$B_3 = C_3/\rho_S^2 + 2C_2 M_A \bar{v}_A/\rho_S + (M_A \bar{v}_A)^2 \tag{8b}$$

Although the essential incompressibility of aqueous solutions thus renders possible the expression of z_A in terms of either molar or molal solute concentration, this thermodynamic activity, defined in Equation (1), remains a molar quantity. In other words, the use of an inappropriate solute concentration scale is countered by a change in the magnitude of the activity coefficient ($z_A/m_A \neq \gamma_A$; $a_A/C_A \neq y_A$). It is therefore important, in any experimental situation, to select the concentration scale appropriate for the most direct analysis of data so that advantage can be taken of statistical-mechanical concepts for the quantitative interpretation of the experimentally determined osmotic virial coefficients.

2.3 Statistical-Mechanical Interpretation of Osmotic Virial Coefficients

As noted by McMillan and Mayer,[21] the osmotic virial coefficients relate to molecular properties of the solute. In statistical-mechanical terms, B_2 and B_3 are essentially cluster integrals over Mayer f-functions,[21,25] $f_{ij}(x_{ij})$, that are specified in terms of the potential of mean force, $u_{ij}(x_{ij})$, between pairs of molecules (i and j) that are separated by distance x_{ij}. Specifically,

$$f_{ij}(x_{ij}) = \exp[- u_{ij}(x_{ij})/(kT)] - 1 \tag{9}$$

where k is the Boltzmann constant. For a spherically symmetrical potential of mean force the second virial coefficient is given by the expression

$$B_2 = - 2\pi L \int_0^\infty f_{12}(x_{12}) x_{12}^2 \, dx_{12} \tag{10}$$

in which subscripts 1 and 2 refer to any two molecules of solute, A. The presence of Avogadro's number, L, in Equation (10) is required to convert the virial coefficient from a molecular into a molar quantity. Subject to the assumption that the triplet potential can be represented in a pairwise additive fashion, the counterpart of Equation (10) for the third virial coefficient has the form

$$B_3 = - \frac{8\pi^2 L^2}{3} \int_0^\infty f(x_{12}) x_{12} \, dx_{12} \int_0^\infty f(x_{13}) x_{13} \, dx_{13} \int_{|x_{12}-x_{13}|}^{x_{12}+x_{13}} f(x_{23}) x_{23} \, dx_{23} \tag{11}$$

which reflects the interactions between pairs in a cluster of three molecules.

The calculation of virial coefficients for proteins is rendered tractable by assuming the adequacy of their description in terms of a hard sphere model with radius R_A and a net charge Z_A spread uniformly over its surface. This simplified model allows magnitudes for B_2 and B_3 to be calculated at about the level of precision to which they can be measured experimentally. By dividing the relevant potential of mean force into hard-sphere (excluded volume) and electrostatic contributions, it follows that $u_{ij}(x_{ij})$ is given by the relationships

$$u_{ij}(x_{ij}) = \begin{cases} \infty & x < 2R_A \tag{12} \\ \dfrac{Z_A^2 e^2}{\varepsilon(1 + \kappa R_A)^2} \dfrac{\exp[-\kappa(x_{ij}-2R_A)]}{x_{ij}}, & x \geq 2R_A \tag{13} \end{cases}$$

in which κ is the Debye–Hückel inverse screening length, e the electronic charge and ε the dielectric constant of the solvent medium. Integration of Equations (10) and (11) is facilitated by expanding the exponential term as

$$f_{ij}(x_{ij}) = -\frac{u_{ij}(x_{ij})}{kT} + \frac{1}{2!}\left[\frac{u_{ij}(x_{ij})}{kT}\right]^2 - \cdots \tag{14}$$

After incorporation of Equations (12)–(14), the required integrations of Equations (10) and (11) give rise to the following expressions[26] for the second and third virial coefficients of a single nonassociating solute:

$$B_2 = \frac{16\pi LR_A^3}{3} + \frac{(1+2\kappa R_A)}{4\kappa^2}\left(\frac{Z_A}{1+\kappa R}\right)^2\left(\frac{8\pi e^2}{\varepsilon kT}\right) - \frac{1}{128\pi\kappa}\left(\frac{Z_A}{1+\kappa R}\right)^4\left(\frac{8\pi e^2}{\varepsilon kT}\right)^2 + \cdots \tag{15}$$

$$B_3 = \frac{160\pi^2 L^2 R_A^6}{9} + 4\pi R_A^5\left(\frac{Z_A}{1+\kappa R_A}\right)^2\left(\frac{8\pi e^2}{\varepsilon kT}\right)\left(\frac{11}{15} - \frac{38}{90}\kappa R_A + \cdots\right)$$

$$- \frac{R_A^3\exp(4\kappa R_A)}{3\kappa}\left(\frac{Z_A}{1+\kappa R_A}\right)^4\left(\frac{8\pi e^2}{\varepsilon kT}\right)^2(1+\cdots) \tag{16}$$

$$+ \frac{\exp(-2\kappa R_A)}{768\pi\kappa^3}\left(\frac{Z_A}{1+\kappa R_A}\right)^6\left(\frac{8\pi e^2}{\varepsilon kT}\right)^3 + \cdots$$

Since κ is related to molar ionic strength, I, by the expression

$$\kappa = \sqrt{\frac{8\pi e^2 I}{\varepsilon kT}} \tag{17}$$

the relatively simple relationship

$$B_2 \approx \frac{16}{3}\pi LR_A^3 + \frac{Z_A^2(1+2\kappa R_A)}{4I(1+\kappa R_A)^2} \tag{18}$$

is obtained by truncating Equation (15) at the Z^2_A term – a result generated by considering the Mayer f-function to be approximated by the first term on the right-hand side of Equation (14). Although this simplified form of Equation (15) has been used for the evaluation of B_2 on several occasions,[7,20,27,28] a subsequent investigation[29] has revealed its significant overestimation by this means for a system with $R_A=2.44$ nm, $Z_A=10$, and $I=0.05$ M. In other words, Equation (18) only provides a reasonable estimate of the second virial coefficient for self-interaction in situations where the combination of a relatively low charge density and a moderately high ionic strength improves the validity of the inherent assumption that $u_{ij}(x_{ij}) \ll 1$. It is therefore preferable to include the next term of Equation (15) or to calculate virial coefficents (B_2, B_3) by numerical integration of Equations (10) and (11).

3   Theory of Sedimentation Equilibrium

As indicated in our contribution to the previous edition of this book,[20] there were ambiguities in the original thermodynamic derivation of sedimentation equilibrium theory.[4] Since then we have developed a more direct means of casting the basic sedimentation equilibrium expression in a form that establishes the nature of the thermodynamic activity being monitored.[7,26]

3.1 Sedimentation Equilibrium Equation for a Single Solute

For a solution of single solute with molecular mass M_A, the criterion for sedimentation equilibrium at constant temperature T can be written in terms of the solute chemical potential as

$$d\mu_A - M_A\omega^2r\ dr = 0 \tag{19}$$

where ω is the angular velocity of the rotor and r the radial distance from its centre. In the original derivation[4] $d\mu_A$ was first expanded in terms of pressure, P, and molal solute concentration, m_A, as independent variables to give the expression

$$d\mu_A = (\partial\mu_A/\partial P)_{T,m_A}\ dP + (\partial\mu_A/\partial m_A)_{T,P}\ dm_A \tag{20}$$

For incompressible solutions the pressure term in Equation (20) could be eliminated by making the substitutions $(\partial\mu_A/\partial P)_{T,m_A}=M_A\bar{v}_A$ and $dP=\omega^2r\rho(r)\ dr$ to obtain

$$M_A[1 - \bar{v}_A\rho(r)]\omega^2r\ dr = (\partial\mu_A/\partial m_A)_{T,P}\ dm_A \tag{21}$$

as the description of the sedimentation equilibrium distribution in terms of the molal concentration. $\rho(r)$ is the solution density at radial distance r. However, a transition from the molal to the molar concentration scale is required to render this expression compatible with optical recording of the sedimentation equilibrium distribution.

The simplest way to obtain an expression in terms of molar concentration was to express the solute chemical potential as,[4]

$$d\mu_A - (\partial\mu_A/\partial P)_{T,C_A}\ dP + (\partial\mu_A/\partial C_A)_{T,P}\ dC_A \tag{22}$$

leading to

$$M_A[1 - \bar{v}_A\rho(r)]\omega^2\ rdr = (\partial\mu_A/\partial C_A)_{T,P}\ dC_A \tag{23}$$

However, two difficulties arise in attempts to employ this statement of the condition for sedimentation equilibrium. First, the composition dependence of the solution density, $\rho(r)$, has to be taken into account to achieve the separation of variables required for integration of the left-hand side of either Equation (21) or (23).[5,20,30,31] Second, the form of the differential $(\partial\mu_A/\partial C_A)_{T,P}$ in Equation (23) is not straightforward, in that it essentially entails differentiation of the logarithm of a molal activity with respect to molar concentration. The significance of this subtlety has been recognized only within the past decade.[7,20,21,31]

The first difficulty is readily accommodated by noting that the composition-dependent buoyancy factor for an incompressible solute may be written as[5,7,20,31]

$$[1 - \bar{v}_A\rho(r)] = (1 - \bar{v}_A\rho_s)\ [1 - M_A\bar{v}_AC_A(r)] \tag{24}$$

where ρ_S is the solvent density. The separation of variables required for integrating is thus achieved by expressing Equation (23) as

$$M_A(1 - \bar{v}_A\rho_S)\omega^2r\,dr = \frac{(\partial\mu_A/\partial C_A)_{T,P}}{(1 - M_A\bar{v}_AC_A)}\,dC_A \tag{25}$$

It now remains to obtain a more tractable form of the $(\partial\mu_A/\partial C_A)_{T,P}$ term – preferably as a function of $(\partial\mu_A/\partial C_A)_{T,\mu_S}$ because of its relevance to the statistical-mechanical interpretation of virial coefficients.

As a starting point, we consider the differentiation of Equation (22) with respect to molar concentration under conditions of constant temperature, T, and chemical potential of solvent, μ_S. To facilitate that differentiation we eliminate the dP term in Equation (22) by noting that for an incompressible solution,

$$dP = C_A(\partial\mu_A/\partial C_A)_{T,\mu_S}dC_A \tag{26}$$

is a condition of the general requirement[14] that $\Sigma\mu_i\,dn_i = V\,dP + S\,dT$, where V denotes volume and S entropy. It then follows directly that

$$(\partial\mu_A/\partial C_A)_{T,P} = (\partial\mu_A/\partial C_A)_{T,\mu_S}[1 - C_A(\partial\mu_A/\partial P)_{T,C_A}] \tag{27}$$

or, on making the substitution $(\partial\mu_A/\partial P)_{T,C_A} = M_A\bar{v}_A$ for incompressible solutions

$$(\partial\mu_A/\partial C_A)_{T,P} = (\partial\mu_A/\partial C_A)_{T,\mu_S}(1 - M_A\bar{v}_AC_A) \tag{28}$$

Upon incorporating this expression for $(\partial\mu_A/\partial C_A)_{T,P}$ into Equation (25), we obtain the relationship

$$M_A(1 - \bar{v}_A\rho_S)\omega^2r\,dr = (\partial\mu_A/\partial C_A)_{T,\mu_S} \tag{29}$$

where, from Equation (1), the right-hand side may be written as $RT\,d\ln z_A$. The differential form of the sedimentation equilibrium equation for an incompressible solution comprising solvent S and a single solute A is, therefore,

$$M_A(1 - \bar{v}_A\rho_S)\omega^2r\,dr = RT\,d\ln z_A \tag{30}$$

which is readily integrated between limits r_F and r to yield

$$z_A(r) = z_A(r_F)\exp\left[\frac{M_A(1 - \bar{v}_A\rho_S)\omega^2(r^2 - r_F^2)}{2RT}\right] \tag{31}$$

where $z_A(r_F)$, the constant of integration, is the thermodynamic activity of solute at the chosen reference radial distance r_F. Although most experimental studies have employed a reference radial position within the limits of the solution column subjected to sedimentation equilibrium, the rotor centre ($r_F = 0$) has also been used on occasion.[26,32] Inasmuch as the exponential term is a renormalization of the variable r when the buoyant mass, angular velocity, and temperature are already specified, Equation (31) will be written as

$$z_A(r) = z_A(r_F)\,\psi_A(r) \tag{32a}$$

$$\psi_A(r) = \exp\left[\frac{M_A(1 - \bar{v}_A\rho_S)\omega^2(r^2 - r_F^2)}{2RT}\right] \tag{32b}$$

Several important features emerge from Equations (30) and (31). First, in any sedimentation equilibrium experiment, the measured distribution reflects the molar thermodynamic activity, z_A (or its weight concentration equivalent, $M_A z_A$). Second, on the grounds that the differential of Equation (31) with respect to r^2 gives

$$\frac{\mathrm{d}\ln z_A(r)}{\mathrm{d}r^2} = \frac{M_A(1 - \bar{v}_A\rho_S)\omega^2}{2RT} \tag{33}$$

the apparent molar mass, M_A^{app}, obtained on the basis of assumed ideal behaviour ($C_A = z_A$, $\gamma_A = 1$) is related to its true value by the expression

$$M_A^{\mathrm{app}} = \frac{M_A}{1 + C_A(\partial\ln\gamma_A/\partial C_A)_{T,\mu_S}} = \frac{M_A}{1 + 2B_2C_A + (3/2)B_3C_A^2 + \cdots} \tag{34}$$

Consequently, the activity coefficient deduced from the dependence of $1/M_A^{\mathrm{app}}$ upon solute concentration is the molar activity coefficient, γ_A, the parameter amenable to statistical-mechanical interpretation. Third, derivation of the basic equation in this manner has clearly confirmed earlier deductions, either implicit[33-35] or explicit,[7,20,26,28,30] that use of the solvent density in the buoyancy term is not an approximation, as implicated in the original derivation.[4] Finally, the original formulation of the condition for sedimentation equilibrium as Equation (23) is now seen to be basically correct but incomplete because of failure to identify the constraints imposed on the defined standard state.

Although Equation (34) clearly allows characterization of the thermodynamic nonideality of a single-solute solution from the concentration dependence of apparent molecular mass, more direct procedures are available. For example, combination of Equations (1b), (5) and (32) yields the relationship[7,36]

$$\psi_A(r) = \frac{C_A(r)}{z_A(r_F)}\exp\{2B_2C_A(r) + (3/2)B_3[C_A(r)]^2 + \cdots\} \tag{35}$$

which, in principle, allows the second and third virial coefficients, as well as the constant $z_A(r_F)$, to be obtained by nonlinear curve-fitting of $[\psi_A(r), C_A(r)]$ data from the entire sedimentation equilibrium distribution to this expression.

3.2 Incorporation of the Effects of Solute Self-Association

In Section 2, the thermodynamic nonideality of a single species was accommodated by expressing its activity coefficient as a virial expansion in solute concentration

[Equation (5)]. Application of the same approach to a mixture of species leads to the relationship[26]

$$\ln \gamma_i = 2B_{ii}C_i + \sum_{j \neq i}B_{ij}C_j + \frac{3}{2}B_{iii}C_i^2 + \sum_{j \neq i}B_{iij}C_iC_j$$

$$+ \frac{1}{2}\sum_{j \neq i}B_{ijj}C_j^2 + \frac{1}{2}\sum_{j \neq i}\sum_{\substack{k \neq j \\ k \neq i}}B_{ijk}C_jC_k + \cdots$$

(36)

where B_{ii} and B_{iii} are the respective second and third virial coefficients for self-interaction. B_{ij} is the second virial coefficient for nonideality of species i that arises from its coexistence with a concentration C_j of another species, whereas B_{iij}, B_{ijj} and B_{ijk} are third virial coefficients encompassing physical interactions between dissimilar species in a mixture. Allowance for effects of thermodynamic nonideality by this means clearly poses a daunting task because of its requirement for knowledge of the exact composition of the mixture – the information sought from any experiment. Admittedly, the situation is simplified considerably for a self-associating solute, for which the law of mass action relates the concentrations of the various species to that of the monomer. Specifically,

$$C_i = K_iC_1^i\left(\frac{\gamma_1^i}{\gamma_i}\right)$$

(37)

where C_i denotes the molar concentration of oligomeric species i ($i=1$, monomer; $i=2$, dimer; and so on), and K_i (with $K_1 \equiv 1$) the molar association constant relating the thermodynamic activity of oligomer i to that of the monomer. Inasmuch as the assignment of a magnitude to the activity coefficient ratio in Equation (37) is conditional upon specification of the oligomeric composition of the solution [Equation (36)], this approach is necessarily iterative. Examples of its application are to be found in sedimentation equilibrium studies of the self-association of lysozyme[15] and α-chymotrypsin.[37] This work was prompted by the theoretical study[12] that highlighted the inadequacy of allowing for effects of thermodynamic nonideality by means of the Adams–Fujita approximation[6]

$$\ln \gamma_i = i\overline{B}M_1\Sigma C_iM_i$$

(38)

where $\overline{B}$ is an empirical constant, and ΣC_iM_i the total weight concentration of self-associating solute. The popularity of that approach, which still persists in current computer software for sedimentation equilibrium analysis, is surprising in view of its inherent assumption[37] that thermodynamic nonideality has no material effect on solute self-association [because in Equation (37), $\gamma_1^i/\gamma_i=1$ for all species].

Whereas the above approach leads to the expression of activity coefficients as a series expansion that requires specification of the concentrations of all oligomeric solute states present, Hill and Chen[38] adopt the thermodynamic stance that a self-associating system is still a single-solute component, whereupon the effects of self-association must be incorporated into the virial coefficients describing the thermodynamic nonideality of the solute constituent. Such action leads to the thermodynamic activity of monomer, z_1, being written as the following series expansion in terms of the total base-molar concentration of solute: $\overline{C}_A = \Sigma C_i M_i / M_1$,

$$z_1(r) = \overline{C}_A(r) \exp \left\{ 2B_2 \overline{C}_A(r) + \left(\frac{3}{2}\right) B_3 [\overline{C}_A(r)]^2 + \cdots \right. \tag{39}$$

Of necessity, this expression is formally identical with that obtained by combining Equations (1) and (5) for a nonassociating single solute, except that the virial coefficients contain additional terms to account for the effects of self-association. Specifically,

$$B_2 = B_{11}^* - K_2 \tag{40a}$$

$$B_3 = \left(\frac{2}{3}\right)(B_{111}^* - K_3) - K_2(2B_{11}^* - B_{12}^*) + K_2^2 \tag{40b}$$

where K_2 and K_3 are the association constants for dimer and trimer formation [Equation (37)], respectively, and the star superscripts on virial coefficients are used to denote that these refer only to the operation of nonassociative forces among the relevant cluster of molecules. Thus, B_{11}^* and B_{111}^* are the respective second and third virial coefficients describing interactions due to nonassociative forces between monomers [Equations (15) and (16)], whereas B_{12}^* is the corresponding second virial coefficient for nonassociative interaction between monomer and dimer. For dissimilar species i and j with spherical geometry, this parameter is given by the expression[26]

$$B_{ij}^* = \frac{4\pi L}{3}(R_i + R_j)^3 + \frac{Z_i Z_j(1 + \kappa R_i + \kappa R_j)}{2\kappa^2(1 + \kappa R_i)(1 + \kappa R_j)}\left(\frac{8\pi e^2}{\varepsilon kT}\right)$$
$$- \frac{1}{64\pi\kappa}\left(\frac{Z_i Z_j}{(1 + \kappa R_i)(1 + \kappa R_j)}\right)^2 \left(\frac{8\pi e^2}{\varepsilon kT}\right)^2 + \cdots, \quad i \neq j \tag{41}$$

or its approximate counterpart [*cf.* Equation (18)]

$$B_{ij}^* \approx \frac{4\pi L}{3}(R_i + R_j)^3 + \frac{Z_i Z_j(1 + \kappa R_i + \kappa R_j)}{2I(1 + \kappa R_i)(1 + \kappa R_j)}, \quad i \neq j \tag{42}$$

Substitution of Equation (32) for $z_1(r)$ in Equation (39) leads to the relationship [*cf.* Equation(35)]

$$\psi_1(r) = \frac{\overline{C}_A(r)}{z_1(r_F)} \exp\left\{2B_2\overline{C}_A(r) + \left(\frac{3}{2}\right)B_3[\overline{C}_A(r)]^2 + \cdots\right\} \tag{43}$$

from which magnitudes of the two virial coefficients [as well as the constant parameter $z_1(r_F)$] can, in principle, be obtained by nonlinear curve-fitting of the $[\psi_1(r), \overline{C}_A(r)]$ data set to Equation (43). The association constants may then be calculated from the resulting values of B_2 and B_3 via Equations (40a) and (40b) after the assignment of values to the various virial coefficients for nonassociative interactions

$(B_{11}^*, B_{12}^*$ and $B_{111}^*)$ on the basis of the effective radii and net charges of monomeric and oligomeric solute species.[27]

A problem with the above approach is the requirement for rapid convergence of the virial expansion series. Although this condition is likely to be met for weak self-associations, larger values of K_2 and K_3 can lead to situations in which the series converges too slowly, or even diverges.[27,39] However, that limitation is readily overcome[27,29] by retaining the Hill–Chen approach[38] but replacing Equation (39) by the relationship that expresses $\overline{C}_A(r)$ as a function of $z_1(r)$, namely,

$$\overline{C}_A(r) = z_1(r)\{1 + 2b_2z_1(r) + 3b_3[z_1(r)]^2 + \cdots\} \tag{44}$$

where the coefficients b_2, b_3, ..., come from the osmotic pressure relationships[24]

$$\frac{\Pi}{RT} = z_1 + b_2z_1^2 + b_3z_1^3 + \cdots \tag{45a}$$

$$\overline{C}_A(r) = z_1\frac{\partial(\Pi/RT)}{\partial z_1} \tag{45b}$$

The coefficients b_i are related to their more familiar osmotic counterparts B_i [Equation (4)] through the expressions

$$b_2 = -B_2 \tag{46a}$$

$$b_3 = -\frac{B_3}{2} + 2B_2^2 \tag{46b}$$

Upon substitution of Equations (40a) and (40b) for the conventional osmotic virial coefficients and introduction of the ψ function [Equation (32)] to express the variation of $z_1(r)$ in terms of the invariant parameter $z_1(r_F)$, Equation (44) becomes

$$\overline{C}_A(r) = z_1(r_F)\psi_1(r) + 2(K_2 - B_{11}^*)[z_1(r_F)\psi_1(r)]^2$$
$$+ 3\left(K_3 - K_2B_{12}^* + B_{11}^{*2} - \frac{B_{111}^*}{2}\right)[z_1(r_F)\psi_1(r)]^3 + \cdots \tag{47}$$

The base-molar concentration of solute throughout the sedimentation equilibrium distribution is thus a polynomial expression with $\psi_1(r)$ as the independent variable.[27,29] As in the previous procedure [Equation (43)], the evaluation of K_2 and K_3 is conditional upon specification of magnitudes for B_{11}^*, the excluded volume contribution to the quadratic coefficient, as well as B_{12}^* and B_{111}^*, the corresponding contributions of monomer–dimer and monomer–monomer–monomer interactions to the term in $[\psi_1(r)]^3$. However, the advantage of employing Equation (47) for the characterization of solute self-association is its general applicability to systems for which truncation of the power series at the cubic term suffices to describe the relationship between $\overline{C}_A(r)$ and $\psi_1(r)$. In this regard, the current unavailability of a theoretical expression for the quartic term is not restrictive inasmuch as the precision of experimentally determined $[\overline{C}_A(r), \psi_1(r)]$ data does not suffice to warrant the inclusion of a term in $[\psi_1(r)]^4$.

3.3 Interactions between Dissimilar Macromolecular Reactants

Despite the greater biological prevalence of interactions between dissimilar macromolecular reactants, protein self-association has been the predominant phenomenon studied by sedimentation equilibrium. Studies of heterogeneous association reactions have usually entailed the evaluation of binding constants by iterative simulation of sedimentation equilibrium distributions to obtain the best-fit description of experimental results for the two constituents[40–42] in a protocol adapted from its counterpart for the characterization of solute self-association.[17] However, the problem with this analysis is its inability to accommodate realistic allowance for the effects of thermodynamic nonideality. Discussion is therefore confined to an adaptation of the above Hill–Chen approach for self-association that incorporates rigorous statistical-mechanical treatment of thermodynamic nonideality into the analysis of interactions between dissimilar reactants.[32]

For illustrative purposes we consider a mixture comprising an acceptor A that interacts with macromolecular ligand B to form complexes C ($\equiv$AB) and D ($\equiv$AB$_2$). In terms of virial coefficients for excluded volume interactions between species, the expansion of osmotic pressure as a function of thermodynamic activity is

$$\frac{\Pi}{RT} = z_A + z_B + z_C + z_D + b_{AA}z_A^2 + b_{AB}z_Az_B + b_{BB}z_B^2 + b_{AAA}z_A^3 + b_{BBB}z_B^3$$
$$+ b_{AAB}z_A^2z_B + b_{ABB}z_Az_B^2 + b_{AC}z_Az_C + b_{BC}z_Bz_C + \cdots \tag{48}$$

where the subscripts to the b_i denote the molecular species interacting in a cluster. However, the thermodynamic activities of the two complexes, C and D, may be written in terms of stoichiometric equilibrium constants (K_{AB}, K_{ABB}), and the two reactant activities raised to the appropriate power as

$$z_C = K_{AB}z_Az_B \tag{49a}$$

$$z_D = K_{ABB}z_Az_B^2 \tag{49b}$$

Incorporation of these two equations into Equation (48) gives

$$\frac{\Pi}{RT} = z_A + z_B + b_{AA}z_A^2 + (b_{AB}^* + K_{AB})z_Az_B + b_{BB}z_B^2 + b_{AAA}z_A^3 + b_{BBB}z_B^3$$
$$+ (b_{AAB}^* + b_{AC}^*K_{AB})z_A^2z_B + (b_{ABB}^* + b_{BC}^*K_{AB} + K_{ABB})z_Az_B^2 + \cdots \tag{50}$$

as the dependence of osmotic pressure upon thermodynamic activity correct to terms of third order; once again, the star superscript is used to denote a virial coefficient that accounts for only non-associative forces in a molecular cluster. On the basis that $\overline{C}_i = z_i\,\partial(\Pi/RT)/\partial z_i$, the expressions for $\overline{C}_A$ and $\overline{C}_B$ in terms of thermodynamic activity become

$$\overline{C}_A(r) = z_A + 2b_{AA}z_A^2 + b_{AB}z_Az_B + 3b_{AAA}z_A^3 + 2b_{AAB}z_A^2z_B + b_{ABB}z_Az_B^2 + \cdots \tag{51a}$$

$$\overline{C}_B(r) = z_B + 2b_{BB}z_B^2 + b_{AB}z_Az_B + 3b_{BBB}z_B^3 + b_{AAB}z_A^2z_B + 2b_{ABB}z_Az_B^2 + \cdots \tag{51b}$$

in which the various coefficients are described in terms of their traditional osmotic counterparts by the relationships[32]

$$b_{AA} = -B_{AA}; \quad b_{BB} = -B_{BB} \tag{52a}$$

$$b_{AB} = b^*_{AB} + K_{AB} = K_{AB} - B^*_{AB} \tag{52b}$$

$$b_{AAA} = -B_{AAA}/2 + 2B^2_{AA}; \quad b_{BBB} = -B_{BBB}/2 + B^2_{BB} \tag{52c}$$

$$b_{AAB} = b^*_{AAB} + b^*_{AC}K_{AB} = -B^*_{AAB}/2 + 2B^*_{AB}B_{AA} + B^{*2}_{AB} - B^*_{AC}K_{AB} \tag{52d}$$

$$b_{ABB} = b^*_{ABB} + K_{ABB} + b^*_{BC}K_{AB} = -\frac{B^*_{ABB}}{2} + 2B^*_{AB}B_{BB}$$
$$+ B^{*2}_{AB} + K_{ABB} - B^*_{BC}K_{AB} \tag{52e}$$

Experimental estimation of the coefficients b_{AB}, b_{AAB}, and b_{ABB} appearing in Equation (51) thus opens up a possible means of determining the two equilibrium constants (K_{AB}, K_{ABB}), provided that magnitudes can be assigned to the virial coefficients describing excluded volume interactions. Expressions have already been given for estimating the three virial coefficients, B_{AA}, B_{BB}, and B^*_{AB} [Equations (15) and (41)] as well as the two third virial coefficients for self-interaction, B_{AAA} and B_{BBB} [Equation (16)]. Unfortunately, there are currently no corresponding expressions for B^*_{AAB} and B^*_{ABB}, without which the full potential of Equations (51a) and (51b) for characterizing heterogeneous associations cannot be realized. Consequently, consideration has to be restricted to the characterization of a 1:1 interaction from results over a concentration range where nonideality is described satisfactorily by Equations (51a) and (51b) truncated at second virial coefficient terms.

For a mixture of two components there are clearly two ψ functions, $\psi_A(r)$ and $\psi_B(r)$, but that for acceptor (A) can be expressed in terms of $\psi_B(r)$ because of the relationship[32,43]

$$\psi_A(r) = [\psi_B(r)]^p \tag{53a}$$

$$p = \frac{M_A(1 - \bar{v}_A\rho_S)}{M_B(1 - \bar{v}_B\rho_S)} \tag{53b}$$

From Equations (32) and (51)–(53) it therefore follows that the radial dependencies of the constituent concentrations $\bar{C}_A(r)$ and $\bar{C}_B(r)$ are described by the relationships

$$\bar{C}_A(r) = z_A(r_F)[\psi_B(r)]^p + 2B_{AA}[z_A(r_F)]^2[\psi_B(r)]^{2p}$$
$$+ (K_{AB} - B^*_{AB})z_A(r_F)z_B(r_F)[\psi_B(r)]^{p+1} + \cdots \tag{54a}$$

$$\bar{C}_B(r) = z_B(r_F)\psi_B(r) + 2B_{BB}[z_A(r_F)]^2[\psi_B(r)]^2$$
$$+ (K_{AB} - B^*_{AB})z_A(r_F)z_B(r_F)[\psi_B(r)]^{p+1} + \cdots \tag{54b}$$

Because $z_A(r_F)$ and $z_B(r_F)$ are both constants within a given sedimentation equilibrium experiment, $\overline{C}_A(r)$ and $\overline{C}_B(r)$ are given by specific multinomial functions of the single independent variable $\psi_B(r)$.

A model-independent approach to the evaluation of K_{AB} entails the separate elucidation of $z_A(r_F)$ and $z_B(r_F)$ as the ordinate intercepts of the respective dependencies of $\overline{C}_A(r)/[\psi_B(r)]^p$ upon $\psi_B(r)$, and of $\overline{C}_B(r)/\psi_B(r)$ upon $[\psi_B(r)]^p$, whereupon K_{AB} remains the only parameter of unknown magnitude to be determined from the two constituent concentration distributions (after specification of values for B_{AA}, B_{BB}, and B_{AB}^*). As noted previously,[32] the calculation of $z_A(r_F)$ and $z_B(r_F)$ throughout the distributions on the basis [Equation (32)] that $z_i(r)=z_i(r_F)\psi_i(r)$ allows amalgamation of the results into a single determination of K_{AB} by their analysis in terms of the expression

$$\overline{C}_A(r)+\overline{C}_B(r)=z_A(r)+z_B(r)-2B_{AA}[z_A(r)]^2 \tag{55}$$

$$-2B_{BB}[z_B(r)]^2+2(K_{AB}-B_{AB}^*)z_A(r)z_B(r)+\cdots$$

Although that procedure has the merit of being model-independent, the values of $z_A(r)$ and $z_B(r)$ used for analysis of the data according to Equation (55) rely heavily upon the respective estimates of $z_A(r_F)$ and $z_B(r_F)$ that are obtained without the advantage of any input from the sedimentation equilibrium distribution for the other constituent. The alternative is to introduce model dependence from the outset in a procedure that accommodates concomitant analysis of both distributions in terms of Equations (54a) and (54b) to find the best-fit values for $z_A(r_F)$, $z_B(r_F)$, and K_{AB} in a given sedimentation equilibrium experiment.

4 Analysis of Distributions for a Nonassociating Solute

In sedimentation equilibrium studies of proteins the major emphasis centred initially on use of the technique for molecular mass determination, whereupon thermodynamic nonideality tended to be regarded as an annoying phenomenon that necessitated the extrapolation of apparent molecular mass to zero protein concentration in order to obtain the true value. Consequently, the reported virial coefficients B (usually expressed in units of mL mol g^{-2}) were merely operational values deduced from the slope of the concentration dependence of $1/M^{app}$ written as

$$1/M^{app} = 1/M_A + 2Bc_A \tag{56}$$

where c_A is the weight concentration of protein (measured in g mL^{-1}). From the reciprocal form of Equation (34) it is evident that $B=1000B_2/(M_A)^2$ when the standard units of L mol^{-1} are chosen for B_2 and g mol^{-1} for M_A. Despite a tradition for reporting virial coefficients in this strange manner, we shall depart from convention by using B_2, expressed on a molar basis, as the primary quantity, which is logical from the statistical-mechanical viewpoint.

Characterization of the thermodynamic nonideality of a protein has become a matter of prime concern only recently. This development reflects the requirement of magnitudes for virial coefficients, B_{ii}, of the two reactants in sedimentation equilibrium studies of heterogeneous associations [Equation (54)]. Furthermore, determination of the second virial coefficients for the two reactants affords a way of assessing their effective radius, R_i, and, on occasion, net charge, Z_i – parameters needed for the calculation of the cross-term virial coefficient B_{ij} [Equations (41) and (42)] as well as B_{iii} [Equation (16)] in the event that consideration needs to be accorded cubic terms in thermodynamic activity. We therefore begin this section by outlining the use of sedimentation equilibrium for determining the second virial coefficient of a nonassociating protein.

4.1 Measurement of the Second Virial Coefficient

The fundamental molecular parameter in sedimentation equilibrium is the buoyant molecular mass of the protein, $M_A(1-\bar{v}_A\rho_S)$, which means that the assignment of magnitudes to the protein's partial specific volume ($\bar{v}_A$) and the solvent density (ρ_S) is a prerequisite for the determination of molar mass (M_A). Although only the buoyant molar mass is required for the calculation of $\psi_A(r)$ [Equation (32b)], expression of the protein concentration on a molar basis, $C_A(r)$, is clearly reliant upon M_A. These two quantities are most readily obtained by conducting a low-speed sedimentation equilibrium experiment[3] on a sufficiently dilute protein solution to justify the assumption that $C_A(r)$ is a satisfactory approximation of $z_A(r)$ in Equation (31). Once $M_A(1-\bar{v}_A\rho_S)$ and M_A have been obtained in this manner, B_2 ($\equiv R_{AA}$) can be determined from a sedimentation equilibrium distribution that covers a wider protein concentration range.

Procedural aspects of the evaluation of the second virial coefficient are illustrated by a sedimentation equilibrium study of isoelectric carbonic anhydrase.[36] Figure 1(a) presents the results of a sedimentation equilibrium experiment at 15 000 rpm in which the protein concentration ranged between 0.03 and 0.22 mg mL^{-1} across the distribution. The solid line, which represents the best-fit description in terms of Equation (31) with $z_A(r)$ and $z_A(r_F)$ replaced by the respective absorbances at 280 nm [$r_F=7.0000$ cm, $A(r_F)=0.201$], signifies a buoyant molar mass of 7420 (± 80) g mol^{-1}, and a molar mass of 28 500 (± 300) g mol^{-1}.[36] Figure 1(b) presents the Rayleigh interference record for a solution of isoelectric carbonic anhydrase subjected to sedimentation equilibrium at 32 000 rpm. In this experiment of meniscus-depletion design[44] the range of protein concentration (3.33 fringes$\equiv 1$ mg mL^{-1})[45] suffices for determination of the second virial coefficient. Analysis of the results in terms of Equation (35) with $r_F=6.9998$ cm is summarized in Figure 2(a), where the best-fit description (————) corresponds to a reference thermodynamic activity, $z_A(r_F)$ of 5.36 (± 0.01) μM and a second virial coefficient, B_2, of 163 (± 9) L mol^{-1}. An alternative analysis in terms of the inverse relationship [*cf.* Equation (54)],

$$C_A(r) = z_A(r_F)\psi_A(r) - 2B_2[z_A(r_F)\psi_A(r)]^2 + \cdots \tag{57}$$

is shown in Figure 2(b), from which the respective estimates of $z_A(r_F)$ and B_2 are 5.35 (± 0.01) μM and 147 (± 6) L mol^{-1}.[36] In neither case does the accuracy of the [$C_A(r)$,

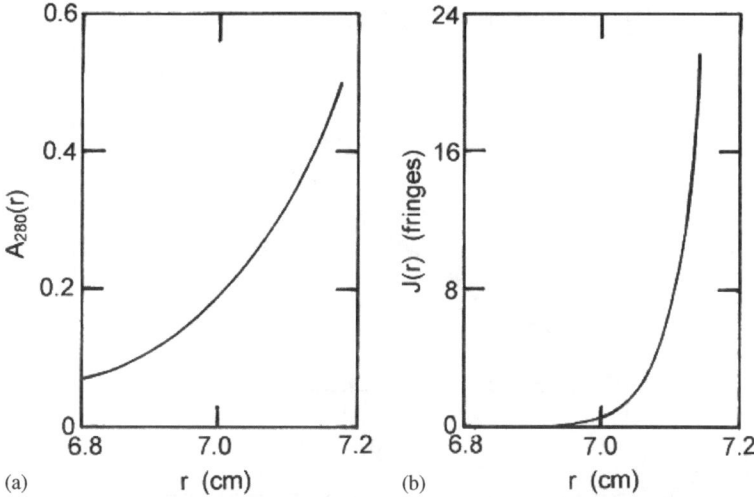

Figure 1 *Determination of the second virial coefficient, B_2, for isoelectric carbonic anhydrase (pH 5.3, $I=0.1$ M). (a) Low-speed (15 000 rpm) sedimentation equilibrium distribution for the determination of buoyant molecular mass and hence the ψ function. (b) Rayleigh distribution from a high-speed run (32 000 rpm) for the evaluation of B_2 [data taken from Winzor et al.[36]]*

$\psi_A(r)$] data suffice for extension of the virial expansion beyond the second virial coefficient term.

Selection of the isoelectric point for determination of the second virial coefficent certainly simplifies the estimation of effective radius, R_A, from an expression such as Equation (15) or its truncated version [Equation (18)] because of the absence of any charge contribution to B_2. However, the second virial coefficient required to allow for nonideality in the characterization of a heterogeneous association [Equation (54)] is the value of B_{AA} (or B_{BB}) that reflects the net charge under the conditions (pH, ionic strength) of the interaction study. In that regard, determination of the second virial coefficient at the isoelectric point as well as under the conditions of interest renders possible the evaluation of Z_A (or Z_B) – a parameter required for estimation of B_{AB} via Equation (41) or Equation (42). For example, interpretation of the second virial coefficient of 417 L mol^{-1} found for ovalbumin (pH 8.5, $I=0.11$ M) in terms of Equation (18) with the estimate of 2.80 nm for R_A deduced from the B_2 of 217 L mol^{-1} for isolectric ovalbumin (pH 4.6) signifies a net protein charge (Z_A) of -14 under slightly alkaline conditions.[7] Furthermore, the feasibility of using thermodynamic nonideality for simultaneous evaluation of the effective radius and net charge of a nonassociating protein is illustrated in Figure 3, which summarizes the ionic strength dependence of the second virial coefficient reported[31] for lysozyme in acetate–chloride buffer (pH 4.5). Analysis of these results in terms of Equation (18) yields respective estimates of 14.1 (±0.02) and 1.69 (±0.06) nm for Z_A and the effective radius,[7] the latter being in excellent agreement with its accepted Stokes counterpart (1.72 nm) for the enzyme.[31]

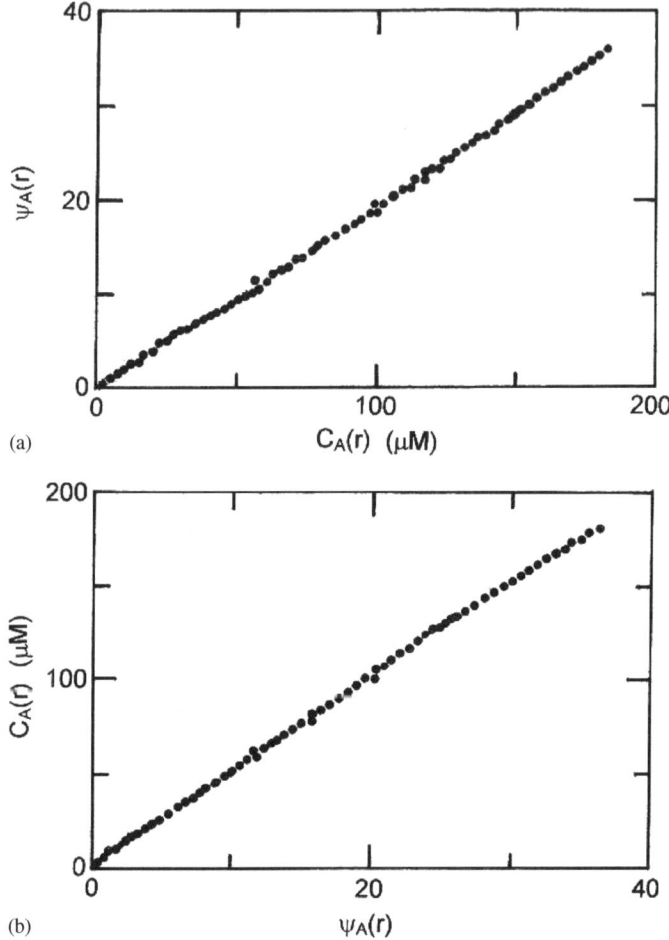

Figure 2 *Determination of the second virial coefficient, B_2, for isoelectric carbonic anhydrase (pH 5.3, $I=0.1$ M) by sedimentation equilibrium. (a) Analysis of the distribution from Figure 1(b) according to Equation (35). (b) Corresponding analysis according to Equation (57) [data taken from Winzor et al.[36]]*

4.2 Determination of Protein–Polymer Virial Coefficients

Theoretical considerations have led to the prediction and demonstration that the strength of protein interactions can be enhanced by high concentrations of macromolecular solutes such as dextran and poly (ethylene glycol),[46–52] a phenomenon that is attributed to the consequences of polymer concentration on the activity coefficients of the interacting species. Because an extended polymer chain bears little resemblance to the rigid impenetrable sphere on which theoretical calculation of virial coefficients is usually based, the experimental evaluation of virial coefficients for the excluded-volume effect of polymers on proteins becomes a matter of importance. It transpires that sedimentation equilibrium provides a very simple procedure for their determination.[53]

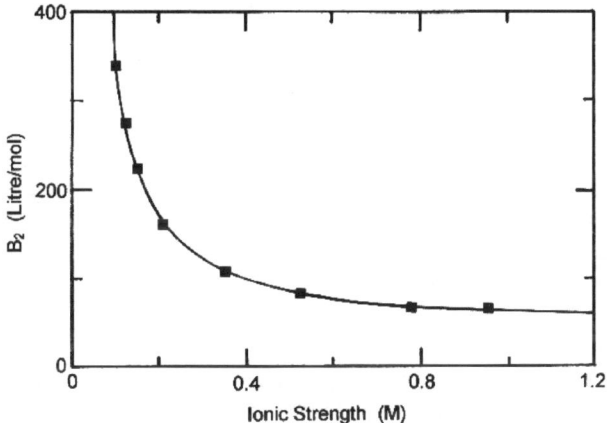

Figure 3 *Evaluation of the effective radius, R_A, and net charge, Z_A, of lysozyme (pH 4.5) by curve-fitting the ionic strength dependence of the second virial coefficient, B_2, to Equation (18). [data taken from Behlke and Ristau[31]]*

Consider a situation in which parallel sedimentation equilibrium runs are conducted on a solution of protein alone and on a protein–polymer mixture for which the absorption optical system records only the protein distribution. By employing a low protein concentration and a relatively high concentration of polymer, it is possible to obtain separate sedimentation equilibrium distributions for the two components, that for the polymer being deduced from the Rayleigh interference record of the distribution. By selecting a common reference radial position, r_F, in the two distributions, it is possible to compare directly the radial dependence of the ratio of protein thermodynamic activities. From Equation (32) those thermodynamic actvities are given by the relationships

$$z_A(r) = z_A(r_F)\psi_A(r) \tag{58a}$$

$$[z_A(r)]_M = [z_A(r_F)]_M\psi_A(r) \tag{58b}$$

where subscript M denotes thermodynamic activities in the cell containing polymer (M). After replacement of z_A by the product of γ_A and C_A [Equation (1b)], the ratio of protein concentrations at a given radial distance becomes

$$\frac{C_A(r)}{[C_A(r)]_M} = \frac{C_A(r_F)}{[C_A(r_F)]_M}\frac{[\gamma_A(r)]_M}{[\gamma_A(r_F)]_M} \tag{59}$$

on the grounds that assumed ideal behaviour in the run on protein alone $[\gamma_A(r)=\gamma_A(r_F)=1]$ should be a reasonable approximation because of the low protein concentration being used. Upon noting that the counterpart of Equation (5) for the logarithm of the protein activity coefficient in the mixture, namely,

$$\ln \gamma_A = 2B_{AA}C_A + B_{AM}C_M + \cdots \tag{60}$$

can be approximated by the protein–polymer term because $C_A \ll C_M$, Equation (59) may be written as

$$\ln \frac{C_A(r)}{[C_A(r)]_M} = \ln \frac{C_A(r_F)}{[C_A(r_F)]_M} + \left(\frac{B_{AM}}{M_M}\right)[c_M(r) - c_M(r_F)] \tag{61}$$

in which the ratios of concentrations may be replaced by the corresponding ratios of absorbances A_λ. The second virial coefficient (in L g^{-1} rather than L mol^{-1}) can thus be obtained from the slope of the dependence of $\ln\{A_\lambda(r)/[A_\lambda(r)]_M\}$ upon the difference in weight concentration c_M of polymer at r and r_F, a quantity determined readily from the Rayleigh distribution.

The application of this approach to the evaluation of B_{AM}/M_M for a system with carbonic anhydrase as the protein and poly (ethylene glycol) 20 000 as the polymer is illustrated in Figure 4, where the slight curvature of the data set is considered to reflect polydispersity of the polymer rather than an influence of higher-order virial coefficient terms. Linear regression analysis of the data set signifies a second virial coefficient (B_{AM}/M_M) of 31 ($\pm$3) mL g^{-1}, a value that can be rationalized satisfactorily in terms of the Jansons and Phillips[54] model of the system as an excluded volume interaction between a rigid, impenetrable spherical protein and a polymer modelled as an effective Brownian walk of specified root-mean-square end-to-end length.[53]

An interesting aspect of the determination of B_{AM}/M_M by this means is the absence of terms in Equation (61) for angular velocity or molecular mass. These factors do, of course, play an important role in the design of the experiment because of their influence on (i) the accuracy with which $A_\lambda(r)/[A_\lambda(r)]_M$ may be estimated, and (ii) the range of polymer concentration for estimation of the slope. A rotor speed of 14 000 rpm was used to generate the sedimentation equilibrium distributions in the

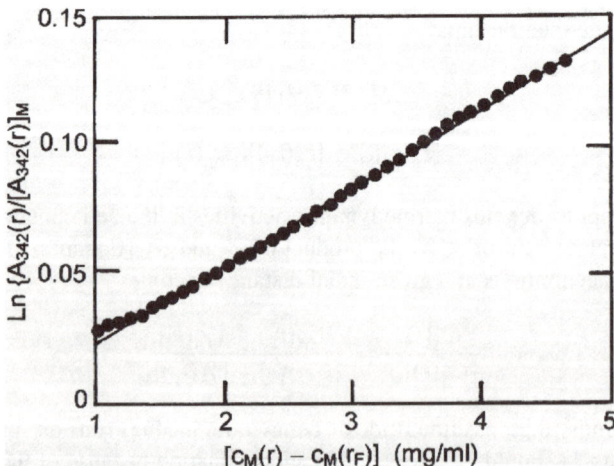

Figure 4   *Application of Equation (61) to obtain the second virial coefficient, B_{AM}/M_M, for the excluded volume interaction between carbonic anhydrase and poly(ethylene glycol) 20 000 in phosphate–chloride buffer (pH 7.0, $I=0.39$ M). [data taken from Wills et al.[53]]*

above experiments on carbonic anhydrase (0.28 mg mL^{-1}) in the absence and presence of poly (ethylene glycol) 20 000 (5.2 mg mL^{-1}).

4.3 Molecular Crowding by Small Cosolutes

An ability to occupy space and hence to exert excluded volume effects is not a property that is confined to macromolecular solutes. In this regard the thermodynamic nonideality effected by small cosolutes such as glycerol and sucrose have certainly tended to be considered in terms of preferential solvation[55–59] rather than on the statistical-mechanical basis of excluded volume, but the formal equivalence of the two treatments has now been established.[60,61] Indeed, the excluded volume and preferential solvation treatments merely afford alternative ways of accounting for the magnitude of B_{AM}, the second virial coefficient for the protein–cosolute interaction.

In keeping with the Casassa and Eisenberg[34] approach, we consider the situation in which a solution of nonassociating protein (A) is dialysed exhaustively against buffer containing a high concentration C_M of small inert cosolute, M, before being subjected to ultracentrifugation at angular velocity ω and temperature T. At sedimentation equilibrium the distributions of the protein and cosolute are described in terms of thermodynamic activity by the expressions

$$z_A(r) = z_A(r_F) \exp\left[\frac{M_A(1 - \bar{v}_A \rho_S)\omega^2(r^2 - r_F^2)}{2RT}\right] \tag{62a}$$

$$z_M(r) = z_M(r_F) \exp\left[\frac{M_M(1 - \bar{v}_M \rho_S)\omega^2(r^2 - r_F^2)}{2RT}\right] \tag{62b}$$

where Equation (31) has been used in preference to the short-hand form [Equation (32)] to emphasize that the density in the buoyancy factor is unequivocally that of the unsupplemented buffer (ρ_S). Provided that the magnitude of $M_M(1 - \bar{v}_M \rho_S)$ is sufficiently small for essentially no redistribution of cosolute at sedimentation equilibrium, it can be helpful to regard M as part of the solvent. Under those circumstances the protein concentration distribution can be described by the relationship [Equation (5.27) of Eisenberg[35]]

$$C_A(r) = C_A(r_F) \exp\left[\frac{M_A(1 - \phi'_A \rho_d)\omega^2(r^2 - r_F^2)}{2RT}\right] \tag{63}$$

where ϕ'_A is the apparent partial specific volume of the protein defined under the constraints of constant chemical potentials of solvent, μ_S, and cosolute, μ_M;[55,56] and where ρ_d is the density of the cosolute-supplemented diffusate. Although the predialysis step has thus simplified the analysis to that for an ideal single-solute system, an encumbrance to the use of Equation (63) is the necessity for experimental measurement of the apparent partial specific volume ϕ'_A, an endeavour that requires a large amount of protein.[57–59] Further simplification of the approach was afforded by the demonstration[62] that the buoyant molecular mass in Equation (63) is defined correct to first order in C_M as

$$M_A(1 - \phi'_A \rho_d) = M_A(1 - \bar{v}_A \rho_S) - (1 - \bar{v}_M \rho_S)B_{AM}M_M C_M(r_F) \tag{64}$$

where the first two terms are the buoyant molecular masses, $(2RT/\omega^2)(\mathrm{d}\ln A_\chi/\mathrm{d}r^2)$, that emanate from absorption scans for low protein concentrations in the presence and absence, respectively, of a nonabsorbing cosolute. The molecular mass of the protein may thus be obtained from the ordinate intercept of the dependence of buoyant molecular mass upon cosolute concentration, thereby avoiding the necessity of measuring ϕ'_A.

As well as providing a way of obtaining the molecular mass of a protein from sedimentation equilibrium studies in concentrated cosolute environments, Equation (64) also offers a simple means of measuring the second virial coefficient for the protein–cosolute interaction. On the grounds that C_M, the concentration of cosolute in the diffusate, represents a reasonable estimate of $C_M(r_F)$ because of the relatively minor redistribution of a small solute during the sedimentation equilibrium experiment, the magnitude of B_{AM} may be deduced from the slope of the dependence of the protein buoyant molecular mass upon C_M. This procedure is illustrated in Figure 5(a), which shows the effect of sucrose concentration on the buoyant molecular mass of thyroglobulin.[62] Interpretation of these results on the basis [Equation (64)] that the ordinate intercept is $M_A(1-\bar{v}_A\rho_S)$ leads to a molecular mass of 660 000 (±33 000) g mol^{-1} for thyroglobulin, an estimate consistent with the accepted value of 670 000 g mol^{-1}.[63] Furthermore, the estimate of 1500 (±300) L mol^{-1} for B_{AM} that is inferred from the slope of Figure 5(a) essentially matches the effective hydrodynamic volume of 1490 L mol^{-1} that is deduced from the Stokes radius of 8.4 nm. Such similarity of the excluded and hydrodynamic volumes is reasonable inasmuch as $B_{AM}=4\pi L(R_A+R_M)^3/3\approx4\pi LR_A^3/3$ for a small uncharged cosolute ($R_A\gg R_M$).

Another application of Equation (64) is its use for determining the subunit molecular mass of a protein from sedimentation equilibrium studies with a denaturant as

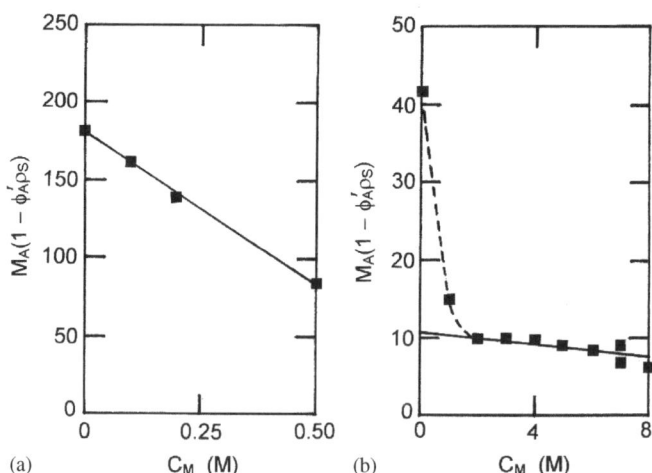

Figure 5 *Effect of the concentration of small cosolute, M, on the buoyant molar mass of a protein. (a) Bovine thyroglobulin (0.2–0.3 mg mL^{-1}) in Tris–chloride buffer (pH 7.5, I=0.1 M) supplemented with sucrose. (b) Rabbit muscle aldolase (0.25–0.35 mg mL^{-1}) in acetate–chloride buffer (pH 5.0, I=0.24 M) supplemented with urea [data taken from Jacobsen et al.[62]]*

cosolute M. That possibility is illustrated in Figure 5(b), which presents the dependence of buoyant molecular mass of aldolase as a function of urea concentration.[62] At low urea concentrations the steep slope of the dependence can be attributed to progressive disruption of quaternary structure. However, at urea concentrations beyond that required for dissociation of aldolase into subunits the dependence should reflect nonideality arising from excluded volume interactions between denatured subunits and cosolute (urea). Linear regression analysis of results at the higher concentrations signifies an ordinate intercept of 10 900 ($\pm$1400) g mol^{-1}. Combination of this estimate of buoyant molar mass with the partial specific volume of 0.742 mL g^{-1} [64] and the measured buffer density signifies a molar mass of 43 000 ($\pm$5000)g mol^{-1} for aldolase subunits,[62] a value in keeping with that of 160 000 g mol^{-1} for the tetrameric native enzyme.

A major outcome of these considerations of molecular crowding by small cosolutes is the demonstration of a simple and economic means for estimating the molecular mass of protein subunits by sedimentation equilibrium in the presence of high concentrations of a denaturant. Casassa and Eisenberg[34] had provided the initial lead by showing that the analysis of sedimentation equilibrium distributions for proteins in the presence of denaturant could be simplified to that for the standard single-solute system by prior dialysis of the protein solutions against buffer supplemented with denaturant. Subsequently, the analysis was simplified further by the derivation[62] of Equation (64), the application of which merely requires knowledge of the partial specific volume of the protein and the density of the unsupplemented buffer in order to determine the subunit molecular mass of a protein.

5 Experimental Studies of Protein Self-Association

The characterization of protein self-association by sedimentation equilibrium was traditionally based on assessment of the concentration dependence of weight-average molecular mass[65–67] by a procedure devised initially for the analysis of light-scattering data.[68] Although there are still proponents of that approach,[52,69,70] direct analysis of the equilibrium concentration distributions in terms of the thermodynamic activity of monomer [Equation (47)] affords a simpler procedure – particularly in instances where rigorous allowance is to be made for effects of thermodynamic nonideality on the statistical-mechanical basis of excluded volume. Thirty years have elapsed since the initial demonstration[9] of a procedure for determining the monomer activity, $z_1(r)$, as a function of total protein concentration, $\overline{C}_A(r)$. However, the major developments in optimizing the use of that information have mainly occurred within the past decade. A brief outline of the steps leading to replacement of the original omega analysis[9] by the current procedure based on the ψ function[27] seems in order.

5.1 The Progression from Omega to Psi

Delineation of the thermodynamic activity of monomer throughout a sedimentation equilibrium distribution was approached initially via the omega function, $\Omega(r)$, defined as

$$\Omega(r) = \frac{\overline{C}_A(r)}{\overline{C}_A(r_F)} \exp\left[\frac{M_A(1 - \bar{v}_A \rho_S)\omega^2(r^2 - r_F^2)}{2RT} \right] \quad (65)$$

where $\overline{C}_A(r_F)$ is the total (base-molar) concentration of solute at a selected reference radial distance, r_F, within the limits of the solution column. This function was then plotted against $\overline{C}_A(r_F)$ to determine $z_1(r_F)$ from the ordinate intercept on the grounds that

$$\lim_{\overline{C}_A(r) \to 0} \Omega(r) = \Omega_0 = \frac{z_1(r_F)}{\overline{C}_A(r_F)} \tag{66}$$

Knowledge of the thermodynamic activity of monomer at the reference radial distance then allowed its calculation throughout the distribution via Equation (31), which could be written in the form

$$z_1(r) = \frac{\Omega_0 \overline{C}_A(r_F)}{\Omega(r)} \tag{67}$$

It was this access to the relationship between $z_1(r)$ and $\overline{C}_A(r)$ that prompted the development of strategies for the incorporation of nonideality into sedimentation equilibrium analysis by means of species activity coefficients calculated on the statistical-mechanical basis of excluded volume.[15,37,71]

As emphasized by Morris and Ralston,[72,73] a drawback of the above omega analysis[9] is the extent of reliance that needs to be placed on the accuracy of a curvilinear extrapolation to obtain Ω_0 as the limiting value of $\Omega(r)$. This undesirable feature of the procedure is clearly evident in Figure 6(a), which shows the extrapolation involved in elucidating Ω_0 from results[16,37] for α-chymotrypsin (pH 3.9, $I=0.20$ M) over a concentration range (0–2 mg mL^{-1}) for which ideal behaviour may reasonably be assumed [$\gamma_i(r) \approx 1$; $i=1, 2, \ldots$]. In an attempt to increase the confidence level of Ω_0, the relationship for total protein concentration, $\overline{C}_A = C_1 + 2K_2C_1^2 + \cdots$, was combined with the experimental definition of $\Omega(r)$ to give the expression

$$\Omega(r) = \frac{\overline{C}_A(r)}{\overline{C}_A(r_F)\psi_1(r)} = \Omega_0 + 2K_2\Omega_0^2\overline{C}_A(r_F)\psi_1(r) + \cdots \tag{68}$$

The essentially linear dependence of $\Omega(r)$ upon $\overline{C}_A(r_F)\psi_1(r)$ for the same data set [Figure 6(b)] certainly improves the certainty with which the ordinate intercept may be determined. However, it was soon realized[27] that upon substitution of $z_1(r_F)/\overline{C}_A(r_F)$ for Ω_0, Equation (68) may be written

$$\overline{C}_A(r_F)/\psi_1(r) = z_1(r_F) + 2K_2[z_1(r_F)]^2\psi_1(r) + \cdots \tag{69}$$

which suggests the evaluation of $z_1(r_F)$ and K_2 from the dependence of $\overline{C}_A(r_F)/\psi_1(r)$ upon $\psi_1(r)$. Such a plot [Figure 6(c)] is effectively Figure 6(b) with the transposition of $\overline{C}_A(r_F)$ from the abscissa to the ordinate variable.

Comparison with Equation (47) shows that Equation (69) describes the recommended psi analysis of sedimentation equilibrium distributions reflecting solute self-association in instances where chemical reaction predominates over thermodynamic nonideality ($K_2 \gg B_{11}^*$). This analysis based on the ψ function is thus seen to be an evolutionary development of the omega analysis that allows separation of the

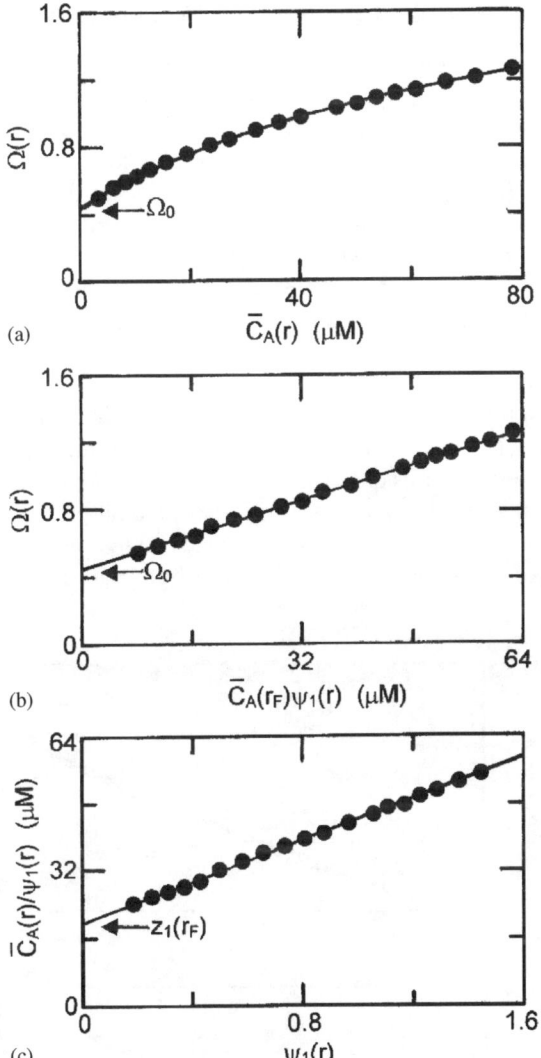

Figure 6 *Analysis of sedimentation equilibrium distributions for α-chymotrypsin dimerization (pH 3.9, I=0.20 M). showing the progression from use of (a) the Ω function[9] to (c) the ψ function[27] for the characterization of solute self-association. The plots are derived from that shown in (b), which was a modification of the omega analysis devised to improve the extrapolation to obtain $\Omega_0 = C_1(r_F)/\bar{C}_A(r_F)$ as the ordinate intercept [data in (b) taken from Jacobsen and Winzor[37]]*

dependent $[C_A(r)]$ and independent $[\psi_1(r)]$ variables – a feature lacking in its predecessor.[9] The consequent elimination of $\Omega(r)$ in favour of $\psi_1(r)$ for subsequent analyses of sedimentation equilibrium distributions for interacting systems should therefore not be regarded as the demise of the Ω function but rather its reincarnation in an even better form.

5.2   Studies of Nonideal Protein Self-Association

The initial application of Equation (47) to the characterization of nonideal solute self-association by sedimentation equilibrium[27] is summarized in Figure 7(a), which presents results for α-chymotrypsin in acetate–chloride buffer under conditions of lower ionic strength (pH 4.1, $I=0.08$ M) than those pertaining to Figure 6. This change was made to decrease the magnitude of K_2 and thereby enhance the relative influence of thermodynamic nonideality on the equilibrium concentration distributions. Results from a high-speed sedimentation equilibrium experiment with an upper concentration limit of 2.2 mg mL^{-1}(O) have been combined with those (●) from a low-speed run ($1.8<\bar{c}_A(r)<7.0$ mg mL^{-1}) by selecting reference radial positions (r_F) corresponding to a common $\bar{c}_A(r_F)$ of 1.93 mg mL^{-1} [$\bar{C}_A(r_F)=77.2$ μM)]. Attempts to include the cubic term of Equation (47) in the nonlinear curve-fitting procedure lead to non-reconcilable magnitudes of the quadratic and cubic coefficients, the latter being enormously positive rather than necessarily negative for this monomer–dimer system. Analysis of the results in terms of Equation (47) truncated at the quadratic term yields values of 59.2 (±0.5) μM for $z_1(r_F)$ and 2510 (±80) L mol^{-1} for $(K_2-B^*_{11})$. A dimerization constant of 2820 (±80) M^{-1} is thus deduced on the basis of the second virial coefficient, B^*_{11}, of 309 L mol^{-1} that is calculated from Equation (18) for an α-chy-

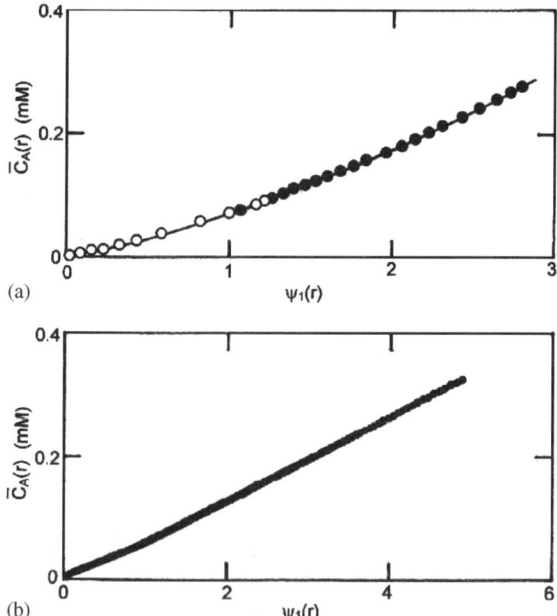

(a)

(b)

Figure 7   *Use of the ψ function for the characterization of α-chymotrypsin dimerization. (a) Analysis of high-speed (O) and low-speed (●) sedimentation equilibrium distributions (pH 4.1, $I=0.08$ M) in terms of Equation (47) with a common value (77.2 μM) for $C_A(r_F)$ [data taken from Wills et al.[27]] (b) Corresponding analysis of a high-speed (30 000 rpm) sedimentation equilibrium distribution (pH 4.1, $I=0.05$ M) with $r_F=7.1000$ cm [data taken from Wills and Winzor[29]]*

motrypsin monomer with $R_1 = 2.44$ nm and $Z_1 = +10$. In that regard R_1 is taken as the Stokes radius, whereas the net charge (valence) is the experimental value determined by measurement of the Donnan distribution of ions in equilibrium dialysis.[74]

Analysis of the two sedimentation equilibrium distributions on the basis of a common $\overline{C}_A(r_F)$ has the advantage that the same magnitude of $z_1(r_F)$ must apply to both distributions. An alternative curve-fitting procedure could entail the selection of a common r_F: K_2 would then be an invariant parameter to be evaluated by global fitting of the two dependencies of $\overline{C}_A(r)$ upon $\psi_1(r)$, whereas $z_1(r_F)$ would be a local curve-fitting parameter for each sedimentation equilibrium distribution. However, in the only application of such global curve-fitting of four sedimentation equilibrium distributions for lysozyme,[75] the returned magnitudes of K_2 (504 M^{-1}) and K_3 (43 400 M^{-2}) differed significantly from the respective estimates of 585 ($\pm$37) M^{-1} and 204 000 ($\pm$24 000) M^{-2} deduced[27] by selecting r_F to effect the return of a single $z_1(r_F)$ value. The latter should provide the more reliable analysis because of the fewer parameters being evaluated by the curve-fitting process.

Improved resolution of Rayleigh fringe distributions in the Beckman XL-I ultracentrifuge renders possible the examination of a wider concentration span in a single experiment. Whereas the results shown in Figure 7(a) required the conduct of two sedimentation equilibrium runs in a Beckman model E ultracentrifuge, those presented in Figure 7(b)[29] emanate from a single high-speed experiment on α-chymotrypsin (pH 4.0, $I = 0.05$ M) in its successor. For this system the even lower ionic strength necessitates inclusion of the cubic term in the expansion of total concentration as a function of monomer activity. Specifically, the results are fitted to the truncated form of Equation (47) with $K_3 = 0$,

$$\overline{C}_A(r) = z_1(r_F)\psi_1(r) + 2(K_2 - B_{11}^*)[z_1(r_F)\psi_1(r)]^2$$
$$+ 3\left(B_{11}^{*2} - \frac{B_{111}^*}{2} - K_2 B_{12}^*\right)[z_1(r_F)\psi_1(r)]^3 \tag{70}$$

using calculated values of all virial coefficients (B_{11}^*, B_{12}^* and B_{111}^*) obtained[29] by numerical integration of the Mayer f-function [Equations (10) and (11)] on the basis of spherical geometry for the dimer ($R_2 = 2^{1/3} R_1$). Respective estimates of 1571 ($\pm$ 8) M^{-1} and 53.7 ($\pm$0.1) μM are returned for K_2 and $z_1(r_F)$, the monomer activity at the selected reference radial position (r_F) of 7.1000 cm. Consideration of the dimer to be a dumbell of monomers decreases the value of B_{12}^* from 1140 to 1057 L mol^{-1}, and thereby decreases the estimate of K_2 to 1452 ($\pm$8) M^{-1}. However, from the viewpoint of reaction energetics the two ΔG^0 estimates of -17.9 and -17.7 kJ mol^{-1} differ trivially.

5.3   Self-Association in the Presence of Cosolutes

From a physiological viewpoint the consequences of thermodynamic nonideality arising from self-interaction may well be minor compared with those reflecting excluded-volume interactions with other species in the cellular environment.[76-78] For example, the extent of self-association of a minor protein component such as a glycolytic enzyme within the red blood cell would be subject to the influence of molecular crowding by the very high haemoglobin concentration (~5 mM) therein. In situations where the concentration of cosolute, C_M, greatly exceeds that ($\overline{C}_A$) of a reversibly dimerizing protein it follows from Equations (36) and (37) that

$$(K_2^{app})_M = C_2/C_1^2 \simeq K_2 \exp[(2B_{1M} - B_{2M})C_M] \qquad (71)$$

where $(K_2^{app})_M$ is the apparent dimerization constant (ratio of species concentrations raised to the appropriate powers) in the presence of cosolute; and where K_2 is effectively the corresponding parameter obtained for the dilute solute solution in the absence of cosolute. On the grounds that $2B_{1M} > B_{2M}$ an enhanced extent of solute self-association is the predicted outcome of excluded volume interactions with cosolutes. Experimental confirmation of that prediction is provided by tracer sedimentation equilibrium studies[79,80] of the effect of bovine serum albumin on the buoyant molecular mass of fibrinogen [Figure 8(a)]; and of the effect of Dextran T10 on that of tubulin [Figure 8(b)]. Because each sedimentation equilibrium distribution for labelled solute has been analysed in terms of a single, weight-average buoyant molecular mass, the concentrations in Figure 8 refer to the composition of the mixture subjected to sedimentation equilibrium. Inasmuch as separate distributions were measured for the solute and cosolute, a far greater amount of experimental information on the systems could have been extracted by determining point-average values of buoyant molecular mass $[\overline{M}_A(r)]$ as a function of $\overline{C}_A(r)$ and $\overline{C}_M(r)$; or, more directly, by using the ψ function to evaluate $z_1(r)$ as a function of $\overline{C}_A(r)$ and $\overline{C}_M(r)$.

An enhanced extent of protein self-association is also likely to result from the high concentrations of small cosolutes (osmolytes) such as sucrose, glycine betaine, and

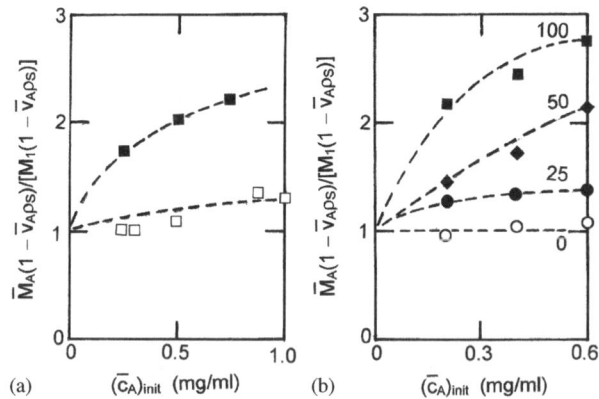

(a) $(\overline{C}_A)_{init}$ (mg/ml)   (b) $(\overline{C}_A)_{init}$ (mg/ml)

Figure 8   *Applications of the tracer sedimentation equilibrium technique[79,80] for detection of the enhancement of protein self-association by high concentrations of macromolecular cosolutes. (a) Concentration dependence of the weight-average buoyant molecular mass of radiolabelled fibrinogen in Hepes–chloride buffer (pH 7.4), supplemented with Ca^{2+} and Mg^{2+} (both 1 mM) in the absence (□) and presence (■) of bovine serum albumin (40 mg mL^{-1}). (b) Corresponding dependencies for rhodamine-labelled fibrinogen in the presence of the indicated concentrations (mg mL^{-1}) of Dextran T10 [data taken from Rivas et al.[52]]*

proline that are expressed by plants (and other organisms) in response to biological stress.[81] Although a decreased magnitude of the covolume difference $(2B_{1M} - B_{2M})$ in Equation (71) for a small cosolute seemingly diminishes the likelihood of observing enhanced solute dimerization, the effect of this decrease in covolume difference is more than offset by the large molar concentration of cosolute (>0.1 M) that is generated in response to biological stress. The plausibility of this reasoning has been substantiated by sedimentation equilibrium studies of α-chymotrypsin dimerization (pH 3.9, $I=0.20$ M) in the presence of monosaccharides, disaccharides and tetrasaccharides.[82,83]

At the rotor speed used to effect a satisfactory sedimentation equilibrium distribution for the enzyme (20 000 rpm) there is effectively no redistribution of the small cosolute, and hence its concentration throughout the distribution may be approximated by that, C_M, in the solution subjected to centrifugation $[C_M(r) \approx C_M]$. Consequently, the same apparent dimerization constant should prevail throughout the protein concentration distribution [Equation (71)]. Application of the psi approach to the equilibrium absorbance scans (280 nm) shown in Figure 9(a) for α-chymotrypsin dimerization (pH 3.9, $I=0.20$ M) alone (O) and in the presence of 0.2 M raffinose (●) leads to estimates of 49 000 M^{-1} and 91 000 M^{-1} for K_2 and $(K_2^{app})_M$, respectively.[83] Furthermore, the prediction [Equation (71)] of a linear dependence of $\ln(K_2^{app})_M$ upon C_M is observed [Figure 9(b)] in sedimentation equilibrium studies with the cosolutes glucose (■) and sucrose (◆) as well as raffinose (●), as is the predicted increase in slope with size of the sugar. However, the slopes signify second virial coefficients that are only about one-third of B_{AM} values based on excluded volume interactions between rigid impenetrable spheres. Those disparities could well reflect some penetration of the protein domain by the small cosolutes, or the existence of preferential chemical interactions of cosolute with monomeric enzyme. In that regard the high cosolute concentrations being used to effect displacement of the monomer–dimer equilibrium position could well render significant a very weak interaction between cosolute and protein, and hence give rise to nonconformity with the requirement for cosolute inertness that is inherent in calculations of excluded volume effects. Since a potential for weak chemical reactivity clearly also applies to macromolecular cosolutes, this study of α-chymotrypsin dimerization[83] serves to illustrate a potential quantitative limitation of thermodynamic noinideality predictions based on the statistical-mechanical basis of excluded volume.

6 Characterization of Heterogeneous Associations

Despite their greater prevalence in biology, interactions between dissimilar macromolecular reactants have until recently been accorded very little attention by the ultracentrifugation community. Apart from an isolated study of an ovalbumin–lysozyme interaction 25 years ago by means of the Ω function,[84] the use of sedimentation equilibrium for the characterization of acceptor–ligand interactions has been restricted to post-1990, by which time the advantages of direct analysis of the recorded distribution(s) had been realized. However, use of the weight-average molecular mass of one constituent ($\overline{M}_i$; $i=$ A or B) has been employed to good effect

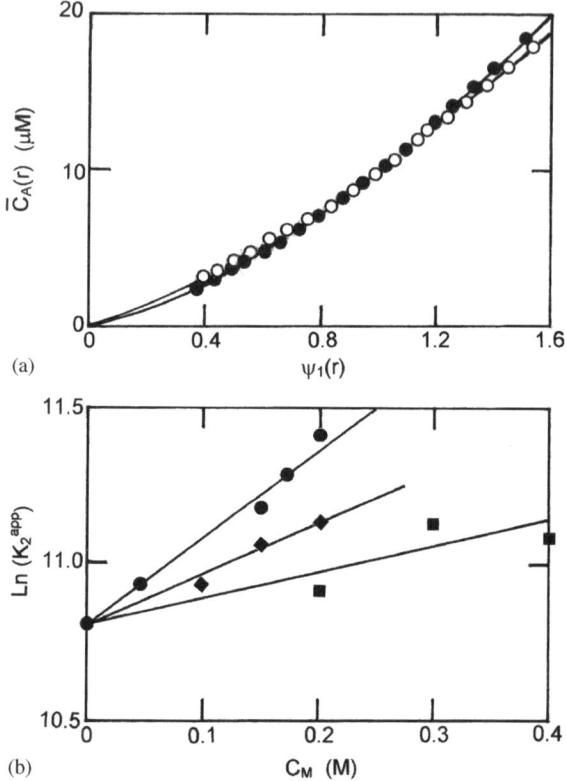

(a)

(b)

Figure 9 *Effect of sugars on the dimerization of α-chymotrypsin (pH 3.9, I=0.2 M). (a) Psi analysis of sedimentation equilibrium distributions for enzyme alone (O) and in the presence of 0.2 M raffinose (●), the r_F values being selected for a common $\overline{C}_A(r_F)$ of 9.9 μM. (b) Dependence of the apparent dimerization constant upon cosolute concentration for glucose (■), sucrose (◆) and raffinose (●) [data taken from Patel et al.[83]]*

in studies of heterogeneous association equilibria by the tracer sedimentation equilibrium technique.[85–87] This approach is illustrated in Figure 10, which refers to the interaction between complement subcomponents C1r (component A) and C1s (component B) in mixtures containing a fixed concentration $(\overline{C}_A)$ of radiolabelled C1r and a wide range of of C1s concentrations.[86] An increase in ligand concentration $(\overline{C}_B)$ gives rise to a progressive change in $\overline{M}_A$ from that of C1r (175 to 321 kDa), which corresponds to the species AB_2. The reaction stoichiometry is thereby established. Provided that a single intrinsic binding constant, k_{AB}, suffices to describe the interaction of C1s with two sites on homodimeric C1r, the results shown in Figure 10 require characterization in terms of the following relationship for the weight-average molecular mass of C1r constituent, $\overline{M}_A$,

$$\overline{M}_A = \frac{M_A + 2k_{AB}C_BM_{AB} + k_{AB}^2C_B^2M_{AB_2}}{1 + 2k_{AB}C_B + k_{AB}^2C_B^2} \tag{72}$$

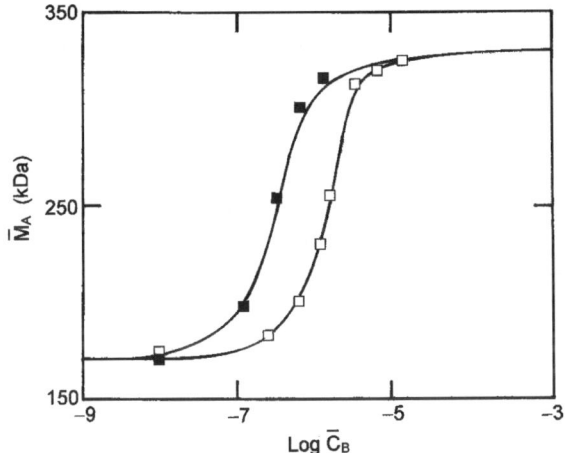

Figure 10   *Tracer sedimentation equilibrium studies of the interaction between complement subcomponents C1r (A) and C1s (B) in Hepes–chloride buffer (pH 7.4) supplemented with 1 mM Ca^{2+}. Data points denote the weight-average molecular mass of C1r component deduced from experiments with a range of loading concentrations, $\overline{C}_B$, of C1s and fixed loading concentrations, $\overline{C}_A$, of 0.6 μM (■) and 3.0 μM (□) ^{3}H-C1r. [data taken from Rivas et al.[86]]*

The concentration of free C1s (C_B) is the solution to the two simultaneous equations expressing mass conservation of C1r (A) and C1s (B), namely,

$$\overline{C}_A = C_A + 2k_{AB}C_AC_B + k_{AB}^2C_AC_B^2 \tag{73a}$$

$$\overline{C}_B = C_B + 2K_2C_B^2 + 2k_{AB}C_AC_B + 2k_{AB}^2C_AC_B^2 \tag{73b}$$

where the consituent ligand concentration ($\overline{C}_B$) contains an additional term to cover the dimerization of free C1s governed by association constant K_2. Nonlinear curve-fitting of the results (Figure 10) to Equations (72) and (73) leads to respective estimates of 3.2×10^7 and 6.3×10^6 M^{-1} for k_{AB} and K_2 as the quantitative description of the C1r–C1s interacting system.[86]

As noted in relation to the corresponding approach for characterizing protein self-association in the presence of cosolutes (Figure 8), the compaction of an entire sedimentation equilibrium distribution into a cell-average molecular mass represents inefficient use of the large amount of quantitative information contained therein for an interacting system. We therefore return to the earlier theme of extracting concentration distributions of individual species by direct analysis of experimental sedimentation equilibrium patterns reflecting interaction between dissimilar macromolecular species.

6.1   Analysis of An Equilibrium Distribution Reflecting both Constituents

To date there have been only three studies in which the characterization of an acceptor–ligand interaction has been based on the extraction of model-independent

distributions for free reactants from sedimentation equilibrium experiments on reacting mixtures; and in none of those studies have the effects of thermodynamic non-
ideality been taken into account. Ironically, the inaugural study[84] employed the
Rayleigh interference optical system and hence encountered the most unfavourable
situation – that in which the sole experimental record is a distribution reflecting all
reacting species. This situation also prevailed in an XL-A study of the interaction
between an electron-transferring flavoprotein and trimethylamine dehydrogenase,[88]
because both reactants were flavoproteins with comparable spectral characteristics
throughout the range of accessible wavelengths.

In the characterization of the ovalbumin–lysozyme interaction from the Rayleigh
interference record of total concentration distribution advantage was taken of the fact
that the ratio of thermodynamic activity of the smallest species (lysozyme, species
B) to total concentration at a reference radial position can be obtained by extrapolating the Ω function for that species,[84]

$$\Omega_B(r) = \frac{C_B(r)}{C_B(r_F)} \exp\left[\frac{M_B(1 - \bar{v}_B \rho_S)\omega^2(r^2 - r_F^2)}{2RT}\right] \tag{74}$$

to zero total concentration [Figure 11(a)]. On the basis of assumed thermodynamic
ideality $[z_B(r_F)=C_B(r_F)]$, the concentration distribution for free lysozyme was then
calculated and subtracted from the total concentration distribution to obtain a residual distribution with ovalbumin (A) the smallest species. Repetition of the omega
analysis for species A on that distribution then yielded $M_A z_A(r_F)/[\bar{c}(r_F)-c_B(r_F)]$
[Figure 11(b)]. An obvious limitation of this conceptually elegant approach to the
determination of concentration distributions for the free reactants is its reliance upon
decidedly curvilinear extrapolations to obtain the estimates of $M_B z_B(r_F)$ and
$M_A z_A(r_F)$. Fortunately, this problem is circumvented to some extent in the comparable approach based on the ψ function,[88] which has been illustrated with a study of
the absorbance distribution (A_{280}) for a mixture of an electron-transferring flavoprotein (B) and trimethylamine dehydrogenase (A).

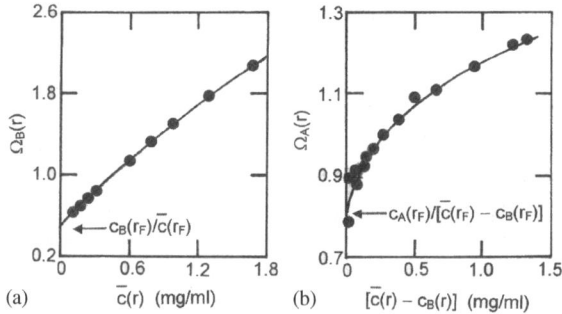

Figure 11   *Determination of the free concentrations of lysozyme (B) and ovalbumin (A) by*
omega analysis of a Rayleigh sedimentation equilibrium distribution for a mix-
ture of the two reactants. (a) Extraction of the free lysozyme concentration,
$c_B(r_F)$, in a mixture with a total protein concentration, $\bar{c}(r_F)$, of 0.52 mg mL^{-1}.
(b) Corresponding determination of $c_A(r_F)$. [data taken from Jeffrey et al.[84]]

Because the recorded total absorbance is the sum of contributions from different species, each characterized by its own absorption coefficient (ε_i), the distribution cannot be converted into one in concentration except in the special circumstance that all of those absorption coefficients are identical. Consequently, the analysis must be in terms of total absorbance, $A_{tot}(r)$, expressed as

$$A_{tot}(r) = \varepsilon_A l C_A(r) + \varepsilon_B l C_B(r) + \varepsilon_{AB} l C_{AB}(r) + \varepsilon_{AB_2} l C_{AB_2}(r) + \cdots \quad (75)$$

where l denotes the cell length (1.2 cm); and where the concentrations of complex species (AB, AB$_2$, ...) at each radial distance are related to those of free A and free B by the law of mass action. Introduction of the ψ function for each species allows Equation (75) to be written in terms of species concentrations at the reference radial position. Specifically,

$$A_{tot}(r) = \varepsilon_A l C_A(r_F)\psi_A(r) + \varepsilon_B l C_B(r_F)\psi_B(r) + \varepsilon_{AB} l C_{AB}(r_F)\psi_{AB}(r)$$
$$+ \varepsilon_{AB_2} l C_{AB_2}(r_F)\psi_{AB_2}(r) + \cdots \quad (76)$$

Upon noting [Equation (53)] that $\psi_A(r)=[\psi_B(r)]^p$, $\psi_{AB}(r)=[\psi_B(r)]^{p+1}$, ..., where p is the buoyant molecular mass of A relative to that of B, Equation (76) can be rearranged to the form

$$A_{tot}(r)/\psi_B(r) = \varepsilon_B l C_B(r_F) + \varepsilon_A l C_A(r_F)[\psi_B(r)]^{p-1} + \varepsilon_{AB} l C_{AB}(r_F)[\psi_B(r)]^p$$
$$+ \varepsilon_{AB_2} l C_{AB_2}(r_F)[\psi_B(r)]^{p+1} + \cdots \quad (77)$$

The dependence of $A_{tot}(r)/\psi_B(r)$ upon $[\psi_B(r)]^{p-1}$ thus has an ordinate intercept of $\varepsilon_B l C_B(r_F)$, i.e., $A_B(r_F)$). The application of Equation (77) to obtain $C_B(r_F)$ from the absorbance distribution for a mixture of electron transferring flavoprotein (B) and trimethylamine dehydrogenase (A) is summarized in Figure 12(a), from which an estimate of $A_B(r_F)$ is obtained as the ordinate intercept.[88] Evaluation of $A_B(r_F)$ by such means allows delineation of $A_B(r)=A_B(r_F)\psi_B(r)$ and hence $C_B(r)=A_B(r)/[\varepsilon_B l]$ throughout the sedimentation equilibrium distribution.

Subtraction of $A_B(r)$ from $A_{tot}(r)$ yields a revised absorbance distribution in which A is now the smallest macromolecular species. This revised distribution may therefore be analyzed in terms of the relationship

$$[A_{tot}(r) - A_B(r)]/[\psi_B(r)]^p = \varepsilon_A l C_A(r_F) + \varepsilon_{AB} l C_{AB}(r_F)\psi_B(r)$$
$$+ \varepsilon_{AB_2} l C_{AB_2}(r_F)[\psi_B(r)]^2 + \cdots \quad (78)$$

in which $\psi_B(r)$ is retained as the independent variable. $A_A(r_F)$ may thus be estimated as the ordinate intercept of the dependence of $[A_{tot}(r)-A_B(r)]/[\psi_B(r)]^p$ upon $\psi_B(r)$ [Figure 12(b)], $A_A(r)$ calculated as $A_A(r)[\psi_B(r)]^p$ throughout the distribution, and $C_A(r)$ as $C_A(r)=A_A(r)/[\varepsilon_A l]$.

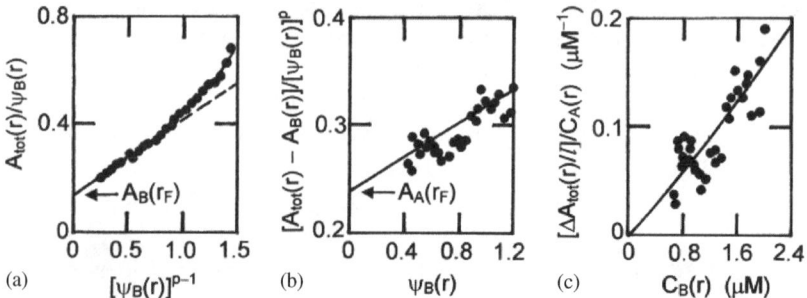

Figure 12 *Characterization of the interaction between an electron-transferring flavoprotein
(B) and trimethylamine dehydrogenase (A) by psi analysis of the absorbance,
A_{280}, distribution for a mixture of the reactants. (a) Determination of $A_B(r_F)$ by
means of Equation (77). (b) Corresponding determination [Equation (78)] of
$A_A(r_F)$. (c) Evaluation of the intrinsic binding constant k_{AB} via Equation (80) with
$K_{AB}=2k_{AB}$ and $K_{AB_2}=2k_{AB}^2$ [data taken from Wilson et al.[88]]*

The remaining task is to rationalize the residual absorbance in terms of stoichiometric equilibrium constants (K_{AB}, K_{AB_2}) and the relationship

$$A_{tot}(r)/l = [A_{tot}(r) - A_A(r) - A_B(r)]/l = K_{AB}\varepsilon_{AB}C_A(r)C_B(r)$$
$$+ K_{AB_2}\varepsilon_{AB_2}C_A(r)[C_B(r)]^2 + \cdots \tag{79}$$

where it is assumed that $\varepsilon_{AB}=(\varepsilon_A+\varepsilon_B)/2$ and $\varepsilon_{AB_2}=(\varepsilon_A+2\varepsilon_B)/3,....$ In keeping with
Equations (77) and (78) the analysis entails a rearranged form of this expression,
namely

$$[\Delta A_{tot}(r)/l]/C_A(r) = K_{AB}\varepsilon_{AB}C_B(r) + K_{AB_2}\varepsilon_{AB_2}[C_B(r)]^2 + \cdots \tag{80}$$

which allows the association constants to be obtained from the best-fit description of
$[\Delta A_{tot}(r)/l]/C_A(r)$ as a polynomial in $C_B(r)$. This final stage of the analysis is shown
in Figure 12(c), where the results were subjected to nonlinear regression analysis on
the basis of Equation (80) with $K_{AB}=2k_{AB}$ and $K_{AB_2}=k_{AB_2}$ to obtain an intrinsic binding constant k_{AB} of 6.7 $(\pm0.4)\times10^6$ M^{-1} for the interaction of the electron-transferring flavoprotein on two equivalent and independent sites on trimethylamine
dehydrogenase.[88]

6.2 Analysis of Separate Distributions for the Two Constituents

More precise characterization of an interaction should emanate from sedimentation
equilibrium experiments under circumstances where separate equilibrium distributions for the two constituents can be deduced. This advantageous situation can be
achieved by selecting constituent-specific assays in tracer sedimentation equilibrium[80] or, under favourable circumstances, by combining information from different
optical records of the same equilibrium distribution.[43,89] This approach is illustrated

by means of a study of an electrostatic interaction between cytochrome c (B) and ovalbumin (A) at low ionic strength (pH 6.3, I=0.03 M) in which sedimentation equilibrium distributions at 410 and 280 nm were used to resolve the separate radial dependencies of $\overline{C}_A(r)$ and $\overline{C}_B(r)$.[43] Because effects of thermodynamic nonideality were deemed to be negligible on the grounds that the total protein concentration did not exceed 1.3 mg mL^{-1} in any distribution, the constituent concentrations of cytochrome c were analysed in terms of the expression [see Equation (54b)]

$$\overline{C}_B(r)/\psi_B(r) = C_B(r_F) + K_{AB}C_A(r_F)C_B(r_F)[\psi_B(r)]^p \qquad (81)$$

The essentially linear dependence of $\overline{C}_B(r)/\psi_B(r)$ upon $[\psi_B(r)]^p$ [Figure 13(a)] renders fairly accurate the extrapolation to obtain $C_B(r_F)$ as the ordinate intercept. In that regard a value (±2 SD) of 5.83 (±0.02) μM is obtained by polynomial curve-fitting, irrespective of the order at which the polynomial is truncated. The consequent equilibrium distribution of free cytochrome c, calculated as $C_B(r)\psi_B(r)$, was then combined with the two constituent concentrations $[\overline{C}_B(r),\ \overline{C}_A(r)]$ to determine a magnitude of the binding function, $v(r)=[\overline{C}_B(r)-C_B(r)]/\overline{C}_A(r)$, at each radial distance. Results from that sedimentation equilibrium experiment and fourteen others are summarized as a binding curve in Figure 13(b), from which an equilibrium constant of 63 000 (±2000) M^{-1} is obtained for 1:1 interaction between the two reactants[43] – a stoichiometry also signified by the essentially linear dependence observed in Figure 13(a) [see Equation (81)].

6.3 Global Curve-Fitting of Sedimentation Equilibrium Distributions

A more popular approach to the characterization of heterogeneous associations by sedimentation equilibrium entails nonlinear curve-fitting of absorbance distributions

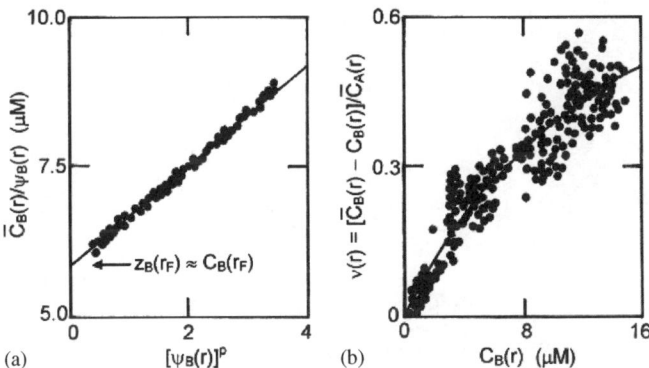

Figure 13 *Characterization of the interaction between cytochrome* c *(B) and ovalbumin (A) from separate sedimentation equilibrium distributions for the two constituents. (a) Evaluation of $C_B(r_F)$ at the reference radial position by psi analysis Equation (81). (b) Binding curve (every tenth point), together with the best-fit description in terms of 1:1 stoichiometry for the interaction [data taken from Winzor et al.[43]]*

at several wavelengths to a specific model of the interaction.[40–42] Indeed, this appears to provide the only feasible means of evaluating K_{AB} for a nonideal 1:1 interaction, because nonfulfilment of the approximation that $z_B(r_F) \approx C_B(r_F)$ in the above procedure negates the determination of $C_{AB}(r)$ as $[\bar{C}_B(r) - C_B(r)]$. Basically, nonlinear regression procedures need to be used for simultaneous analysis of the separate distributions for $\bar{C}_A(r)$ and $\bar{C}_B(r)$ in terms of Equations (54a) and (54b) to obtain the three fitted parameters $z_A(r_F)$, $z_B(r_F)$, and K_{AB}. In practice, the relationship

$$A_\lambda/l = \varepsilon_{\lambda,A}\bar{C}_A(r) + \varepsilon_{\lambda,B}\bar{C}_B(r) \tag{82}$$

is incorporated into the joint regression process to allow direct analysis of the two (or more) absorbance distributions. The results of one sedimentation equilibrium experiment from the ovalbumin–cytochrome c study[43] are shown in Figure 14, together with the best-fit descriptions $[z_A(r_F) = 8.625 \ (\pm 0.016) \ \mu M; \ z_B(r_F) = 1.622 \ (\pm 0.005) \ \mu M; \ K_{AB} = 76\,900–78\,400 \ M^{-1}]$ obtained by such analysis.[32] An asymmetric range of values for K_{AB} reflects use of the standard free energy, $\Delta G^0 = -RT \ln K_{AB}$, as the fitting parameter – a procedure recommended[17] for such joint regression analysis by the Levenberg – Marquardt method.

Disconcerting features of the two analyses of essentially the same experimental dataset are not only the disparity between the two global estimates of K_{AB} (63 000 M^{-1} from Figure 13(b), *cf.* 83 900 M^{-1}) but also the inference that both estimates contain little uncertainty (95% confidence levels of 2000 and 300 M^{-1}, respectively): the latter uncertainty estimate clearly bears no relation to the experimental situation. Although the later treatment[32] differed from the earlier analysis (Figure 13) in that allowance was included for effects of thermodynamic nonideality, the low protein concentrations (<1.3 mg mL^{-1}) in these experiments should have precluded any sizeable influence of the virial coefficient terms [Equation (54)] on the result.

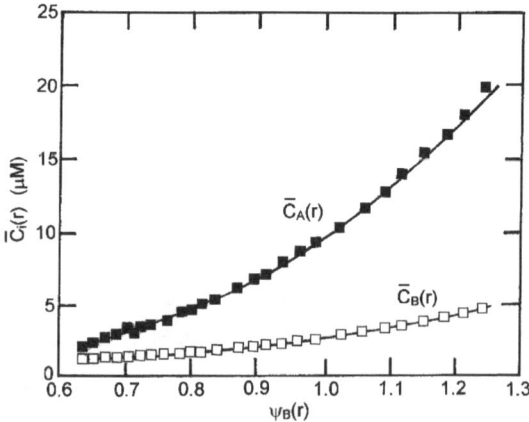

Figure 14   *Best-fit descriptions of the constituent distributions from a sedimentation equilibrium experiment on a mixture of cytochrome c (B) and ovalbumin (A) obtained by joint regression analysis in terms of Equations (54a), (54b), and (82) [data taken from Wills et al.[32]]*

The disparity between the estimates of K_{AB} must therefore reflect differences in the weighting that is given to each experiment. In that regard it is pertinent to examine an alternative means of examining the results from Figure 13(a) and its counterparts for the other experiments in the light of Equation (81) for a 1:1 interaction.

Whereas only the ordinate intercept $[C_B(r_F)]$ was used for delineation of the binding curve [Figure 13(b)], the slope of the dependence of $\bar{C}_B(r)/\psi_B(r)$ upon $[\psi_B(r)]^p$ provides the corresponding magnitude of $C_{AB}(r_F)$ in each experiment. As required, that estimate of $C_{AB}(r_F)$ coincided with the value inferred from $[\bar{C}_B(r)-C_B(r)]$. On the grounds that $C_A(r_F)$ may be calculated as $\bar{C}_A(r_F)-\bar{C}_B(r_F)+C_B(r_F)$, a value of K_{AB} may be obtained from the species concentrations at the reference radial position in each experiment. Alternatively, linear least-squares curve-fitting of those species concentrations from all experiments to the expression

$$C_{AB}(r_F)/C_A(r_F) = K_{AB}C_B(r_F) \tag{83}$$

provides another global estimate of the binding constant (Figure 15). Whereas the mean of the separate K_{AB} values is 81 000 M^{-1}, the estimate deduced from the slope of Figure 15 is 60 000 ($\pm$12 000) M^{-1} because of the smaller weighting that is accorded the data for small values of $C_B(r)$ – a region where the small magnitude of $C_B(r_F)$ is difficult to define. This region is also accorded less weight in analysis of the binding curve [Figure 13(b)]. Indeed, the close agreement between the best-fit K_{AB} values obtained by these procedures signifies a comparable weighting of data from different experiments. Thus, despite its lack of sophistication, the analysis summarized in Figure 15 almost certainly provides a more realistic estimate of K_{AB} and its inherent experimental uncertainty than the description emanating from the more elaborate global curve-fitting procedure.[32]

Although no quantitative study of a heterogeneous association has yet been reported under conditions requiring account to be taken of the effects of thermodynamic

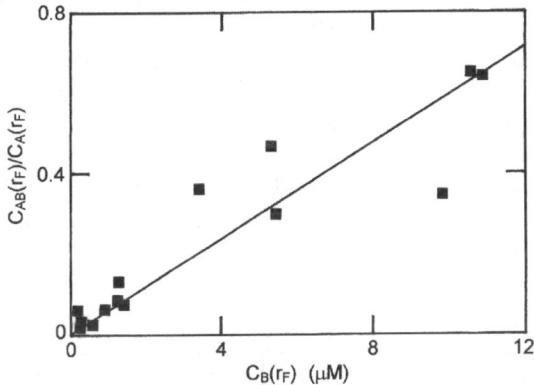

Figure 15 *Alternative, model-dependent characterization of the interaction between cytochrome c (B) and ovalbumin (A) by combining the magnitudes of species concentrations at the reference radial distance in Equation (83) [data taken from Winzor et al.[43]]*

nonideality, the later treatment[32] of the sedimentation equilibrium results for the oval-bumin–cytochrome c system[43] has illustrated all of the steps needed to analyze the individual constituent distributions. Assignment of magnitudes to the various virial coefficients renders possible the joint application of Equations (54a), (54b), and (82) to obtain estimates of $z_A(r_F)$, $z_B(r_F)$, and K_{AB} from each experiment; and thereby $z_{AB}(r_F)$ as $K_{AB}z_A(r_F)z_B(r_F)$. In view of the above demonstration (Figure 15) that analysis of thermodynamically ideal data in terms of Equation (83) yields an estimate of K_{AB} commensurate with that from a binding curve for a 1:1 interaction, the dependence of $z_{AB}(r_F)/z_A(r_F)$ upon $z_B(r_F)$ is suggested as a means of obtaining a global estimate of the binding constant and its inherent confidence limits.

7   Concluding Remarks

This review has summarized developments in the treatment of thermodynamic non-ideality in sedimentation equilibrium that have followed from the unequivocal demonstration that the distributions require theoretical description in terms of the molar thermodynamic activity (z_i) of the sedimenting species. Since that break-through, which was the highlight of our discussion of the present topic in the first edition of this treatise,[20] those advances have included:

(i) a more direct derivation of the basic sedimentation equilibrium equation for a single solute that considers specifically the constrictions imposed on the standard state definition;

(ii) replacement of the Ω function by its ψ counterpart for the extraction of the thermodynamic activity of a free reactant from sedimentation equilibrium distributions;

(iii) further consideration of the Hill–Chen concept that chemical interaction should be regarded as a form of thermodynamic nonideality – an approach that simplifies considerably the analysis of nonideal associations;

(iv) illustration of a modified Hill–Chen approach with studies of α-chy-motrypsin dimerization under conditions where second virial coefficients suffice to account for the effects of thermodynamic nonideality, as well as the procedure to be adopted when nonideality is not restricted to the consequences of nearest-neighbour interactions;

(v) the measurement of second virial coefficients describing excluded-volume interactions between proteins and small uncharged cosolutes, and the interpretation of those virial coefficients in such terms rather than in terms of protein solvation;

(vi) illustration of a rigorous means by which effects of thermodynamic nonideality may be incorporated into the analysis of sedimentation equilibrium distributions that allow resolution of the two constituent concentrations in a mixture of two dissimilar reactants.

The theoretical barriers to the quantitative allowance for effects of thermodynamic nonideality in protein interactions have thus been largely overcome. However, there remain practical barriers that are likely to limit the applications of the theoretical developments to experimental situations. Inability to accommodate the consequences of reactant heterogeneity has always been a practical impediment to quantitative studies of interactions by sedimentation equilibrium, and hence the requirement for homogeneous reactants for such studies is an inevitable restriction to widespread application of the technique. However, the greatest barrier to rigorous allowance for the effects of thermodynamic nonideality in sedimentation equilibrium studies is undoubtedly the requirement for assignment of a net charge (valence) to each interacting species. Despite attempts spread over 70 years, there is no simple procedure for unequivocal determination of the net charge of a macromolecule.[90–92] A solution to that problem is thus likely to be the next advance in the quest to make rigorous allowance for effects of thermodynamic nonideality in sedimentation equilibrium studies.

References

1. T. Svedberg and K. O. Pedersen, *The Ultracentrifuge*, Clarendon Press, Oxford, 1940.
2. R. J. Goldberg, *J. Phys. Chem.*, 1953, **57**, 194.
3. K. E. Van Holde and R. L. Baldwin, *J. Phys. Chem.*, 1958, **62**, 734.
4. J. W. Williams, K. E. Van Holde, R. L. Baldwin and H. Fujita, *Chem. Rev.*, 1958, **58**, 715.
5. H. Fujita, *Mathematical Theory of Sedimentation Analysis*, Academic Press, New York, 1962.
6. E. T. Adams, Jr. and H. Fujita, in *Ultracentrifugal Analysis in Theory and Experiment*, J. W. Williams (ed), Academic Press, New York, 1963, 119.
7. P. R. Wills, D. R. Hall and D. J. Winzor, *Biophys. Chem.*, 2000, **84**, 217.
8. R. H. Haschemeyer and W. F. Bower, *Biochemistry*, 1970, **9**, 435.
9. B. K. Milthorpe, P. D. Jeffrey and L. W. Nichol, *Biophys. Chem.*, 1975, **3**, 169.
10. P. D. Jeffrey, L. W. Nichol, D. R. Turner and D. J. Winzor, *J. Phys. Chem.*, 1977, **81**, 776.
11. P. D. Jeffrey in *Protein–Protein Interactions*, C. Frieden and L.W. Nichol (eds), Wiley, New York, 1981, 213.
12. A. G. Ogston and D. J. Winzor, *J. Phys. Chem.*, 1975, **79**, 2496.
13. P. D. Ross and A. P. Minton, *J. Mol. Biol.*, 1977, **112**, 437.
14. T. L. Hill, *Thermodynamics for Chemists and Biologists*, Addison–Wesley, Reading, MA, 1968.
15. P. R. Wills, L. W. Nichol and R. J. Siezen, *Biophys. Chem.*, 1980, **11**, 71.
16. R. Tellam, J. de Jersey and D. J. Winzor, *Biochemistry*, 1979, **24**, 5316.
17. M. L. Johnson, J. J. Correia, D. A. Yphantis and H.R. Halvorson, *Biophys. J.*, 1981, **36**, 575.
18. J. J. Correia and D. A. Yphantis, in *Analytical Ultracentrifugation in Biochemistry and Polymer Science*, S. E. Harding, A. J. Rowe and J. C. Horton (eds), Royal Society of Chemistry, Cambridge, 1992, 231.
19. M. L. Johnson and M. Straume, in *Modern Analytical Ultracentrifugation*, T. M. Schuster and T. M. Laue (eds), Birkhäuser, Boston, MA, 1994, 37.
20. P. R. Wills and D. J. Winzor, in *Analytical Ultracentrifugation in Biochemistry and Polymer Science*, S.E. Harding, A. J. Rowe and J.C. Horton (eds), Royal Society of Chemistry, Cambridge, 1992, 311.
21. W. G. McMillan and J. E. Mayer, *J. Chem. Phys.*, 1945, **13**, 276.
22. C. Tanford, *Physical Chemistry of Macromolecules*, Wiley, New York, 1961.
23. M. P. Tombs and A. R. Peacocke, *The Osmotic Pressure of Biological Macromolecules*, Clarendon Press, Oxford, 1974.

24. T. L. Hill, *J. Chem. Phys.*, 1959, **30**, 93.
25. J. E. Mayer, *J. Chem. Phys.*, 1950, **18**, 1426.
26. P. R. Wills and D. J. Winzor, *Prog. Colloid Polym. Sci.*, 2002, **119**, 113.
27. P. R. Wills, M. P. Jacobsen and D. J. Winzor, *Biopolymers*, 1996, **38**, 119.
28. P. R. Wills, M. P. Jacobsen and D. J. Winzor, *Prog. Colloid Polym. Sci.*, 1997, **107**, 1.
29. P. R. Wills and D. J. Winzor, *Biophys. Chem.*, 2001, **91**, 253.
30. P. R. Wills, W. D. Comper and D. J. Winzor, *Arch. Biochem. Biophys.*, 1993, **300**, 206.
31. J. Behlke and O. Ristau, *Biophys. Chem.*, 1999, **76**, 13.
32. P. R. Wills, M. P. Jacobsen and D. J. Winzor, *Biophys. J.*, 2000, **79**, 2178.
33. M. Fixman, *J. Phys. Chem.*, 1958, **62**, 374.
34. E. F. Casassa and H. Eisenberg, *Adv. Protein Chem.*, 1964, **19**, 287.
35. H. Eisenberg, *Biological Macromolecules and Polyelectrolytes in Solution*, Clarendon Press, Oxford, 1976.
36. D. J. Winzor, L. E. Carrington and S. E. Harding, *Biophys. Chem.*, 2001, **93**, 231.
37. M. P. Jacobsen and D. J. Winzor, *Biophys. Chem.*, 1992, **45**, 119.
38. T. L. Hill and Y. D. Chen, *Biopolymers*, 1973, **12**, 1285.
39. W. F. Stafford and D. A. Yphantis, *Biophys. J.*, 1972, **12**, 1359.
40. M. S. Lewis, R. I. Shrager and S. J. Kim, in *Modern Analytical Ultracentrifugation: Acquisition and Interpretation of Data for Biological and Synthetic Polymer Systems*, T. M. Schuster and T. M. Laue, Birkhäuser, Boston, MA, 1994, 94.
41. T. Kim, T. Tsukiyama, M. S. Lewis and C. Wu, *Protein Sci.*, 1994, **3**, 1040.
42. M. F. Bailey, B. E. Davidson, A. P. Minton, W. H. Sawyer and G. J. Howlett, *J. Mol. Biol.*, 1996, **263**, 671.
43. D. J. Winzor, M. P. Jacobsen and P. R. Wills, *Biochemistry*, 1998, **37**, 2226.
44. D. A. Yphantis, *Biochemistry*, 1964, **3**, 297.
45. P. Voelker, *Prog. Colloid Polym. Sci.*, 1995, **99**, 162.
46. T. C. Laurent, *Eur. J. Biochem.*, 1971, **21**, 498.
47. H. J. Bosma, G. Voordow, A. De Kok and C. Veeger, *FEBS Lett.*, 1980, **120**, 179.
48. L. W. Nichol, A.G. Ogston and P.R. Wills, *FEBS Lett.*, 1981, **126**, 18.
49. S. B. Zimmerman and B. Harrison, *Proc. Natl. Acad. Sci. USA*, 1987, **84**, 1871.
50. T. C. Jarvis, D. M. Ring, S. S. Daube and P. H. von Hippel, *J. Biol. Chem.*, 1990, **265**, 15160.
51. M. K. Reddy, S. E. Weitzel and P. H. von Hippel, *Proc. Natl. Acad. Sci. USA*, 1993, **90**, 3211.
52. G. Rivas, J. A. Fernandez and A. P. Minton, *Biochemistry*, 1999, **38**, 9379.
53. P. R. Wills, Y. Georgalis, J. Dijk and D. J. Winzor, *Biophys. Chem.*, 1995, **57**, 37.
54. K. M. Jansons and C. G. Phillips, *J. Colloid Interface Sci.*, 1990, **137**, 75.
55. G. Cohen and H. Eisenberg, *Biopolymers*, 1968, **6**, 1077.
56. E. Reisler and H. Eisenberg, *Biochemistry*, 1969, **8**, 4572.
57. J. C. Lee and S.N. Timasheff, *Arch. Biochem. Biophys.*, 1974, **165**, 268.
58. J. C. Lee and S. N. Timasheff, *J. Biol. Chem.*, 1981, **256**, 7193.
59. V. Prakash and S. N. Timasheff, *Anal. Biochem.*, 1981, **117**, 330.
60. P. R. Wills and D. J. Winzor, *Biopolymers*, 1993, **33**, 1627.
61. D. J. Winzor and P. R. Wills in *Protein–Solvent Interactions* R. B. Gregory (ed), Marcel Dekker, New York, 1995, 483.
62. M. P. Jacobsen, P. R. Wills and D. J. Winzor, *Biochemistry*, 1996, **35**, 13173.
63. R. F. Steiner and H. Edelhoch, *J. Am. Chem. Soc.*, 1961, **83**, 1435.
64. J. F. Taylor and C. Lowry, *Biochim. Biophys. Acta*, 1956, **20**, 109.
65. E. T. Adams, Jr. and J.W. Williams, *J. Am. Chem. Soc.*, 1964, **86**, 3454.
66. D. E. Roark and D. A. Yphantis, *Ann. N.Y. Acad. Sci.*, 1969, **164**, 245.
67. D. C. Teller, *Method. Enzymol.*, 1973, **27**, 346.
68. R. F. Steiner, *Arch. Biochem. Biophys.*, 1952, **39**, 333.

69. R. C. Chatelier and A. P. Minton, *Biopolymers*, 1987, **26**, 507.

70. A. P. Minton, *Anal. Biochem.*, 1990, **190**, 1.

71. P. D. Jeffrey, B. K. Milthorpe and L. W. Nichol, *Biochim. Biophys. Acta*, 1977, **495**, 195.

72. M. Morris and G. B. Ralston, *Biophys. Chem.*, 1985, **23**, 49.

73. M. Morris and G. B. Ralston, *Biochemistry*, 1989, **28**, 8561.

74. C. L. Ford and D. J. Winzor, *Biochim. Biophys. Acta*, 1983, **756**, 49.

75. D. J. Winzor, M. P. Jacobsen and P. R. Wills, *Prog. Colloid Polym. Sci.*, 1999, **113**, 69.

76. A. P. Minton, *Mol. Cell. Biochem.*, 1983, **55**, 119.

77. S. B. Zimmerman and A. P. Minton, *Ann. Rev. Biophys. Biomol. Struct.*, 1993, **22**, 27.

78. A. P. Minton, *J. Biol. Chem.*, 2001, **276**, 10577.

79. G. Rivas, K. C. Ingham. and A. P. Minton, *Biochemistry*, 1994, **33**, 2341.

80. G. Rivas, K. Tangemann, A. P. Minton and J. Engel, *J. Mol. Recognit.*, 1996, **9**, 31.

81. L. G. Paleg and D. Aspinall, *Physiology and Biochemistry of Drought Resistance in Plants*, Academic Press, Sydney, 1981.

82. K. E. Shearwin and D. J. Winzor, *Biophys. Chem.*, 1988, **31**, 287.

83. C. N. Patel, S. M. Noble, G. J. Weatherly, A. Tripathy, D. J. Winzor and G. J. Pielak, *Protein Sci.*, 2002, **11**, 997.

84. P. D. Jeffrey, L. W. Nichol and R. D. Teasdale, *Biophys. Chem.*, 1979, **10**, 379.

85. T. M. Laue, D. F. Senear, S. Eaton and A. J. B. Ross, *Biochemistry*, 1993, **32**, 2469.

86. G. Rivas, K. C. Ingham and A. P. Minton, *Biochemistry*, 1994, **33**, 2341.

87. G. Rivas and A. P. Minton, *Biochem. Soc. Trans.*, 2003, **31**, 1015.

88. E. K. Wilson, N. S. Scrutton, H. Cölfen, S. E. Harding and D. J. Winzor, *Eur. J. Biochem.*, 1997, **243**, 393.

89. H. Cölfen, S. E. Harding, K. M. Värum and D. J. Winzor, *Carbohydr. Res.*, 1996, **30**, 45.

90. D. J. Winzor, *Anal. Biochem.*, 2004, **325**, 1.

91. D. J. Winzor, L. E. Carrington and S. E. Harding, *Anal. Biochem.*, 2004, **333**, 114.

92. D. J. Winzor, S. Jones and S. E. Harding, *Anal. Biochem.*, 2004, **333**, 225.

CHAPTER 5

Using Analytical Ultracentrifugation of DNA in CsCl Gradients to Explore Large-Scale Properties of Genomes

OLIVER CLAY, NICOLAS CARELS, CHRISTOPHE J. DOUADY
AND GIORGIO BERNARDI

Introductory Remark

Analyses of absorbance profiles of DNA in CsCl density gradients at sedimentation equilibrium, obtained by analytical ultracentrifugation (AUC), played an important role in molecular genetics and genomics during almost half a century (1957–2004). They allowed accurate calculations of GC (base composition) distributions, GC mosaicism and gene densities in vertebrate genomes that have now been amply confirmed via completed sequences, including those of human, mouse, and pufferfish. We here review general principles guiding past and present uses of salt gradient AUC for exploring genomic DNA, and discuss open problems of AUC/CsCl inference that should become tractable with the aid of a few more entirely sequenced vertebrate genomes.

1 Historical Introduction

1.1 Measuring GC in an Ultracentrifuge

One of the most elementary properties of a DNA sequence is its base composition. If the orientations of the base pairs are neglected, the base composition is simply the sequence's GC: the molar fraction of guanine and cytosine or, equivalently, the proportion of the base pairs in the sequence that are GC rather than AT.

An early, elegant discovery opened the way to rapid experimental measurement of GC. In 1957, Meselson et al.[1] had introduced the methodology of CsCl gradient ultracentrifugation: a salt (CsCl) gradient is built up during ultracentrifugation, and a biomolecule such as DNA eventually finds its equilibrium position in that gradient (see ref. 2 for a quantitative introduction to density gradient ultracentrifugation). Two years later, it was found that the GC levels of essentially all DNAs, except for heavily methylated or otherwise modified DNAs, were linearly related to their positions in the CsCl gradient at sedimentation equilibrium, *i.e.*, to the DNAs' buoyant densities in CsCl.[3–6] The GC of an organism's DNA could therefore be routinely determined by AUC at sedimentation equilibrium, which is usually attained in less than 24 h.

1.2 Measuring GC Distributions in an Ultracentrifuge

The CsCl/AUC method allows far more information to be gathered than just the total GC% of a genome, however. This important point went almost unnoticed at first. Indeed, one can fragment a genome in a number of ways (yielding different molar mass or molecular weight distributions, *i.e.*, different fragment length distributions), and then obtain, via AUC, the distribution or histogram of the GC levels of the fragments. Comparisons of such GC distributions within species (at different molar masses) and between species (preferably at similar molar masses) can lead to novel deductions about the large-scale structures of the species' genomes, their functional correlates, and their interspecific differences. The power of this approach is most obvious in the case of warm-blooded vertebrates, which have the widest GC distributions of all taxa, at fragment lengths above 10–30 kb (kilobase pairs; 30 kb of DNA correspond to about 20 MDa.)

Having said this, there were a few technical problems that needed to be overcome before a CsCl absorbance profile, *i.e.*, an AUC scan of a band of DNA at sedimentation equilibrium, could be routinely converted into its underlying GC distribution. At a first approximation, a CsCl absorbance profile is a convolution of the true GC distribution and a Gaussian-shaped diffusion broadening: when the molar mass of the DNA molecules or fragments is below about 50 kb, they will move appreciably around their equilibrium positions, thus widening the CsCl profile. The shorter the DNA fragments, the more the profile is widened. If one can reliably estimate this Gaussian diffusion broadening, one can recover the GC distribution from a DNA sample whose molar mass is as low as 10 kb (see ref. 7 and references therein).

A detailed theoretical treatment of the CsCl equilibrium profile was provided in a series of articles by Vinograd, Hearst and Schmid in the 1960s and early 1970s. If one demands the highest possible precision, one is faced by a multitude of factors that should, in principle, influence the CsCl profile. Such factors include pressure effects, charges on the DNA molecules, DNA methylation (typically present in small amounts), and light bending (see ref. 7 and its online supplement for references). When one is just trying to reconstruct a reasonably accurate GC distribution of a genome, many of these corrections become almost negligible, or almost cancel (*e.g.*, in the case of pressure corrections, if one uses a marker to calculate GC). One of these published corrections that can notably influence the calculations is the correction for virial effects. Another effect arises because DNA, CsCl salt, and water do not form an ideal solution. The solvation (hydration) of the

DNA/CsCl solute can be estimated, and this leads to a corrected expression for the broadening of a CsCl profile as a function of molecular weight and DNA concentration. As a check, the resulting expression was used to estimate the molar masses of intact phages from their CsCl profiles:[8,9] the estimates agreed well with independent calculations of these phages' genomic lengths. For vertebrates, the dependence of the broader CsCl profiles' widths on DNA concentration is more complicated (see Section 6).

1.3 Using the Ultracentrifuge to Probe Vertebrate Genomes: Discounting Diffusion and Repetitive DNA

In the mid-1970s began the first dedicated quantitative analyses of vertebrate profiles. The relative compositional heterogeneity of vertebrates (compared, for example, with bacteria or phages, at scales >10 kb) implies that their CsCl profiles encode non-trivial information, on the large-scale structure of the vertebrates' chromosomes, which can be resolved via AUC.[10–13] A few raw CsCl profiles of calf and some other mammals had been published earlier, but without the diffusion discounting that would have been necessary, given their relatively low molar masses; furthermore, in one or two cases so much DNA was loaded that the maximal absorbance was no longer proportional to the amount of DNA present (saturation). Such conditions did not yet allow quantitative interspecies comparisons.

Another condition made it difficult to progress: the presence of large amounts of highly repetitive DNA, or 'satellite' DNA, in some well-studied species. The species included calf, whose thymus DNA had been a paradigm for mammalian DNA since the early 1950s. The name 'satellite' was motivated by the ultracentrifugation metaphor: such repetitive DNA bands are often (but not always) found outside the main part of a species' CsCl profile (see refs. 11 and 14 for example). This satellite DNA is prone to rapidly contract or expand, on an evolutionary timescale. Indeed, tracts of repetitive DNA can be shrunk or extended either by slippage during replication, in which the template strand and its copy become shifted relative to each other (so that a sequence can afterwards be present in tandem duplicate or not at all), and/or by ectopic homologous recombination between sequences that happen to be similar but are not at corresponding chromosomal positions (so that there is a looping out of intervening DNA, leading to different lengths for the two homologous chromosomes). By eliminating satellite DNA from a GC distribution, we are left with a desired picture of evolutionarily stable DNA, which can be used to characterize large-scale genomic differences between different mammalian or vertebrate orders.

Filipski *et al.*[10] first quantified the contributions of satellite DNA to the bovine genome: it turned out that one-quarter of the genome consisted of such highly repetitive DNA. The large proportion explained why the CsCl profile of calf was broader than those of many other mammals, such as human. It later turned out that calf, other ruminants, and some geomyoid rodents are not typical mammals in this respect: most mammals have far less satellite DNA. The highly repetitive satellite DNA also sometimes bands anomalously, *i.e.*, it can form peaks at positions corresponding to 'wrong' GC levels.[15,16] The satellite peaks are typically narrow compared to the total profile (see Figure 1 for example), because their DNA has almost constant base composition over long tracts, and because satellite DNA tends to aggregate[12] (see also Section 6). The satellite peaks can differ markedly among closely related species, or even among populations of the same species, because of the instability (rapid contraction, expansion and/or extinction) of the tracts. In a genome that has only few, isolated satellite peaks, the differences will affect the relative heights of the peaks, whereas in a genome that has many or closely overlapping satellite peaks the overall shape of the CsCl profile can change, so that it becomes difficult to resolve the peaks.

The method for experimentally quantifying satellite contributions, used with success for the bovine, mouse and guinea pig genomes,[10,15] was then also applied to other vertebrate genomes. The GC distributions of different taxa could now be compared after satellites had been taken into account.[11,12]

GC distributions measured at different molecular weights (fragment lengths) yield information on the organization of base composition along the chromosomes of a

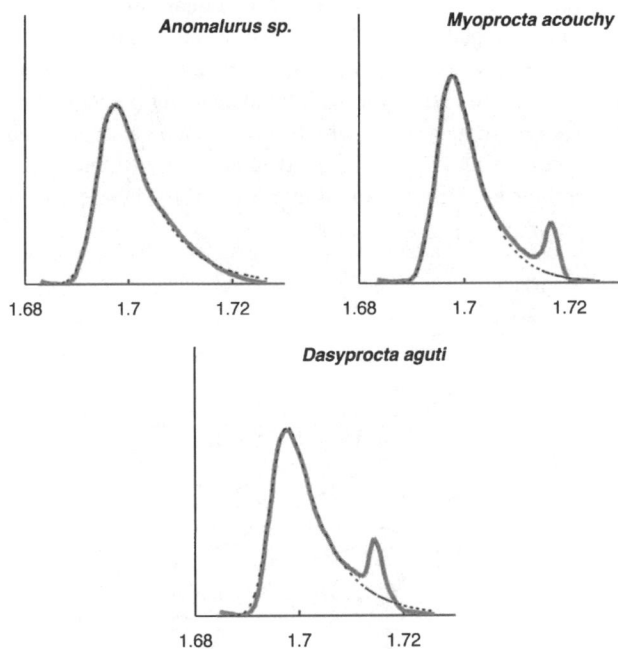

Figure 1 *Equilibrium CsCl absorbance profiles of DNA from three rodent species, Anomalurus sp. (scaly-tailed flying squirrel), Myoprocta acouchy (green acouchi), and Dasyprocta aguti (agouti). All three samples had molar masses above 50 kb, so that the CsCl profiles are essentially the GC distributions of the species' genomes. The horizontal axis is calibrated in buoyant density (g cm^{-3}; see Section 2); the vertical axis shows area-normalized UV absorbance readings at 260 nm, i.e., relative amounts of DNA. Buoyant densities of 1.68, 1.70 and 1.72 correspond to GC levels of 20.4, 40.8, and 61.2%, respectively. The narrow peaks are due to highly repetitive satellite DNA. Dashed lines show fits to a unimodal, asymmetric curve (Gaussian-broadened truncated exponential). Data are from ref. 14*

species. For example, a random sequence of GCs and ATs would yield a GC distribution that narrows rapidly as the molecular weight increases (binomial distribution). In contrast, the GC distributions of mammals such as human and mouse remained invariant – with no narrowing – for all molecular weights above 70 kb that could be analyzed at that time, which extended to well over 300 kb. This was an altogether unexpected result: it implied that mammalian chromosomes are mosaics of long (typically ⩾300 kb) regions of fairly homogeneous base composition, *i.e.*, of long GC-rich regions alternating with long GC-poor regions.[12] It should be pointed out that this result, which concerns properties of the genome sequence, was derived via AUC alone, well before DNA sequences were available. The AUC/CsCl analysis of DNA, at different molecular weights, and its interpretation in terms of a mosaic genome structure, were however motivated by the results after fractionating numerous eukaryotic genomes in cesium sulfate gradients in the presence of sequence-specific ligands, and decomposing the fractions' CsCl profiles into Gaussian components.[11,12] The organization of mammalian chromosomes into the long regions, later called isochores,[13] as well as the invariance of the GC distribution in the 100–300 kb range, have now been confirmed, a quarter of a century later, by the full human and mouse sequences.[17,7] Figure 2 shows the molecular weight dependence of human, bacterial, and random DNA, calculated from sequence data; similar curves were obtained much earlier via AUC.[12,13] The lower bound of ⩾300 kb for the average isochore length in mammals (which corresponded to experimental limits for measuring molar masses at the time of the original study) remains correct. In fact, many isochores are far longer and extend over several megabases.

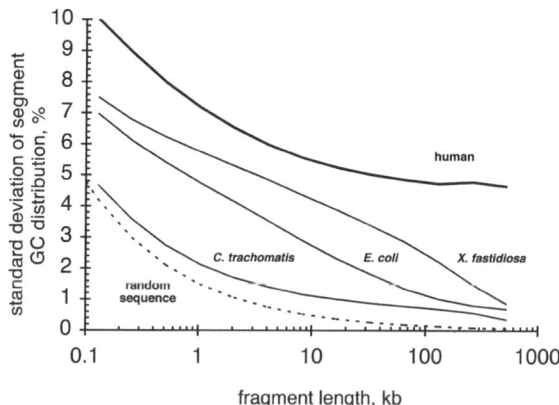

Figure 2   *Standard deviations (widths) of the GC distributions of different genomes, plotted against fragment length on a logarithmic scale. Solid curves, from top to bottom: human, with plateau due to isochore structure; an unusually heterogeneous bacterium, Xylella fastidiosa; a bacterium with intermediate heterogeneity, Escherichia coli; and a very homogeneous bacterium, Chlamydia trachomatis. The dashed curve at bottom left illustrates that no known species is as homogeneous as a random DNA sequence consisting of independent, identically distributed nucleotides. For large fragment sizes, human and other vertebrates have much wider GC distributions than bacteria. Data are from ref. 61*

The invariance of a typical mammal's GC distribution, in the 70–300 kb range, has also a practical implication. One then needs only one CsCl profile in this range, in order to predict the mammal's profiles at all other molar masses in the range (invariance zone): they are essentially identical to the observed profile.

On the topic of repetitive DNA, a final point should be mentioned. In addition to the highly repetitive satellite DNA, there is also another class of repetitive DNA, the so-called interspersed repeats or middle-repetitive DNA. These repeats include the shorter SINES (*e.g.* Alu elements in human) and the longer LINES (*e.g.* L1). Studies of the entire genome sequences have now shown that repeats account for as much as 45% of the human DNA and 35% of the mouse DNA. Since the repeats are interspersed, they have often coevolved with their flanking DNA over quite long evolutionary time spans, they are relatively stable, and they show GC levels approaching those of their local environment. One would therefore not feel authorized to simply 'remove' such repeats from GC distributions, for phylogenetic or genome comparative studies. Instead, it seems most appropriate to regard the interspersed repeats as an integral part of the genomic DNA that coevolves with the unique DNA, and to retain the GC distribution of satellite-free DNA (*i.e.* excluding highly repetitive DNA but including interspersed repeats) as the trait to be analyzed and compared among species. The AUC alone then gives us all the information we need in order to extract the relevant GC distributions.

2   Equations for CsCl Gradients

We now give equations to quantify some of the points made above.

The buoyant density ρ of a molecule in CsCl is defined as the density of the salt solution where that molecule is found, at sedimentation equilibrium. This buoyant density is essentially a linear function of the molecule's radial distance r from the ultracentrifuge axis,[18]

$$\rho = \rho_m + \omega^2 (r^2 - r_m^2)/(2\beta_B) \tag{1}$$

where ρ_m and r_m are the buoyant density and radial position of a suitable marker DNA (such as phage 2C of *Bacillus subtilis*, which has $\rho_m = 1.7420$ g cm^{-3}), ω is the angular speed (usually we work at 44 000 rpm) and β_B is $1.190–1.195 \times 10^9$ cgs units for Beckman models E and XL-A under standard conditions; note that $r^2 - r_m^2 \approx 2r_m(r - r_m)$, an expression that is linear in r. In turn, the GC level of a DNA molecule is linearly related to the molecule's buoyant density in CsCl,[6]

$$GC\% = \frac{100\% \times (\rho - 1.660 \text{ g cm}^{-3})}{0.098} \tag{2}$$

so GC is again a linear function of the radial distance r.

Diffusion broadening contributes to the total variance of a CsCl absorbance profile. We formally re-express distances r or buoyant densities ρ as equivalent GC percentages x, specified by Equation (2) above. We can then write the absorbance profile, *i.e.*, the frequency distribution (probability density) of the DNA's radial position in the AUC cell, as the convolution

$$f(x) = h*\varphi(x) \tag{3}$$

where $h(x)$ is the GC distribution, $\varphi(x)$ is a Gaussian point spread function describing the diffusion broadening, and $*$ denotes the convolution integral. Thus the variances σ^2 add

$$\sigma_f^2 = \sigma_h^2 + \sigma_\varphi^2 \tag{4}$$

The standard deviation σ_h of the GC distribution is often denoted H, and is called the compositional heterogeneity of the DNA. A reasonable estimate of the diffusion contribution to the profile variance is given, at 25 °C and other standard conditions, by

$$\sigma_\varphi^2 \approx 44.5 \text{ kb}/l \tag{5}$$

where l is the average fragment length in kilobases (see ref. 7, and for the full formulae ref. 8). There is actually a slight GC dependence, the formula (5) being intended for use at around 50% GC; at 30% GC the right-hand side should be 44.0 kb/l, at 70% GC, it should be 45.0 kb/l. There is also a concentration dependence, which is however difficult to estimate (see Section 6). We see that when l is 50 kb or higher, the diffusion broadening (standard deviation of the spread function) corresponds to less than 1% GC. Thus, for a vertebrate profile with a heterogeneity of 4.3% GC, the proportion of the total profile variance explained by diffusion is about 0.04.

Vertebrate GC distributions, or raw absorbance profiles, can be modeled by superpositions of Gaussians, although typically this means fitting many parameters. Alternately, after removing narrow satellite peaks one can often model the underlying GC distribution $h(x)$ or absorbance profile $f(x)$ by a convolution of a left-truncated exponential $g(x)$ and a Gaussian,

$$h(x) = g*\psi(x)$$

$$f(x) = h*\varphi(x) = (g*\psi)*\varphi(x) = g*(\psi*\varphi)(x) \tag{6}$$

Note that the convolution of two Gaussians is again a Gaussian; as usual the variances add. This means, for example, that we can estimate the GC distribution $h(x)$ from a fit of the raw CsCl profile: we convolve the fitted $g(x)$ with a Gaussian whose variance is the variance of the fitted Gaussian $\psi*\varphi$ minus the diffusion variance (5). Explicit formulae for g, h, and f in terms of the truncation or cutoff point x_c, exponential rate k, and Gaussian widths can be found in refs. 14 and 7.

3   Anatomy of a CsCl Absorbance Profile

Figure 1 shows, schematically, three CsCl absorbance profiles from a rodent study.[14] The molar mass was above 50 kb for all samples: the diffusion can therefore be practically neglected and the absorbance profiles faithfully represent the GC distributions. As we have seen, the GC distributions are obtained by a simple linear recalibration of the horizontal axis; only the GC of the narrow peaks, formed by satellites, may be unreliable.

When the satellite peaks are discounted, the underlying profiles of different species of mammals, birds, and many cold-blooded vertebrates have a similar broad, positively asymmetric shape, or functional form. The conserved shape can be characterized to high accuracy by an exponential decrease broadened by a Gaussian. In other words, this functional form is a convolution of a left-truncated exponential function and a Gaussian point spread function (dashed curve in Figure 1). The Gaussian spread function is, in turn, the convolution (Equation 6) of two narrower Gaussians: one of them represents the diffusion, while the other is inherent to the GC distribution. In summary, what differs among mammalian species – generally more so among distantly related species, and less so among closely related species, as one would expect for a phylogenetically informative trait – is not the functional form of the profile, but for example its position along the GC axis and its width.

Given the conservation of the basic profile shape, two or three essential parameters suffice to capture most of the compositional variation among genomes of different species (or higher-level taxonomic groups such as families, suborders, or orders). For many purposes, one position parameter (mode or mean) and one width parameter (standard deviation or asymmetry, *i.e.*, mode–mean difference) are enough. The functional form mentioned above gives two different width parameters, one characterizing an asymmetric (exponential-like) broadening and the other the symmetric (Gaussian) broadening of the GC distribution; the total variance is the sum of these two variances.[14,7]

How does one rid a GC distribution of its satellite DNA components? At the species level and above, such components or peaks typically just represent annoying phylogenetic noise. One approach, which is still the best but quite laborious, is to characterize the satellites experimentally. This involves several steps. In one method,[11,12,15,19] a mercury- or silver-based, oligonucleotide-specific DNA ligand is first added to the DNA. After preparative ultracentrifugation in a cesium sulfate gradient one fractionates the DNA, removes the ligand from the fractions and re-centrifuges them in analytical CsCl gradients, so that one can then observe the behavior of the peaks as one passes from one fraction to the next. Usually this method detects and quantifies all but the most stubbornly cryptic satellites. Subtraction of the satellite components leaves us the GC distribution of the satellite-free DNA.

A simpler expedient, which often gives very satisfactory results, is to simply postulate that the usual, basic shape of vertebrates' satellite-free GC distributions, or CsCl profiles, will be conserved for the taxa that are being examined. The automatic fitting of this special shape to the raw CsCl profile (or, alternatively, to a GC distribution of an entirely sequenced vertebrate) yields estimates of three fit parameters: an exponential decline rate, a position parameter, and a Gaussian spreading parameter. (Details are given in ref. 7 and its online supplement.) Satellite peaks are bypassed by the fitting program (see Figure 1), except in very rare cases (geomyoid rodents, some ruminants), where satellites are so abundant that they form a continuum rather than isolated peaks, and thus distort the profile beyond recognition.

Standard deviation, mean, mode, and so on of the satellite-free GC distribution can be calculated directly from the fit parameters.

4   The Biological Meaning of GC Distributions

4.1   GC-Rich DNA and GC-Poor DNA are Functionally Different

After the technical points discussed so far, we now make an excursion to illustrate why extracting and comparing genomic GC distributions is a biologically relevant pursuit.

Of prime interest in AUC-assisted genomics is the understanding of nuclear genomes, how they are organized, how their organization evolved or changed in different taxa, and how this organization is related to the functional activities of the genomes and their genes. The analysis of GC distributions alone can give deep insights into the workings of vertebrate genomes and the differences among them.

The mosaic structure of mammalian and avian genomes, *i.e.*, the alternating of GC-rich and GC-poor isochores along their chromosomes, has a number of important biological correlates. GC-rich DNA above, say, 50% GC is rare (accounting for not more than about 10% of human DNA, for example). This is seen in the positively asymmetric CsCl profiles of mammals and birds, which have their modes near 40% GC.

In contrast to the scarcity of GC-rich DNA is the abundance of GC-rich genes. For example, the histogram of GC levels of human genes (rather than of human DNA) is almost flat-topped, with even a slight negative asymmetry. Consequently, the gene density in GC-rich regions is much higher than in GC-poor regions. Proper alignment of the two distributions can be achieved using the orthogonal regression line (major axis) of a bivariate landscape or scatterplot, showing GC levels of genes' third codon positions along the vertical axis and GC levels of the surrounding DNA along the horizontal axis. For human genes, this line has a slope close to 3: GC levels of genes residing in GC-rich isochores stand out, compositionally, against their environment (see refs. 20 and 21 and the figures and references therein). GC of third codon positions is a particularly sensitive monitor of the GC level of the DNA surrounding the gene. Indeed, third positions are only mildly constrained by the protein that the gene must encode: in almost all codons the third nucleotide can be changed from G to C, or from A to T, without changing the encoded amino acid.

In human, the gene density was estimated earlier, by aligning the distributions for genes and DNA, to be about 17 times higher in the GC-richest regions of the genome than in the GC-poorest regions.[20,22] This ratio was confirmed by Lander *et al.*,[23] a decade later, using the entire human genome sequence. The GC-rich isochores could thus be viewed as a genome core, and the GC-poorest isochores as gene deserts.

Apart from gene density there are numerous other, functionally important differences between GC-rich and GC-poor isochores. For example, genes residing in GC-rich DNA have shorter introns than genes in GC-poor DNA;[24] repetitive DNA categorically remains at low levels in the GC-richest DNA, but not in the GC-poorest DNA;[25,26] GC-rich DNA has preferentially an open chromatin structure whereas GC-poor DNA is typically in closed, compact chromatin; in GC-rich DNA, CpG islands (regions having unusually high densities of unmethylated CpG dinucleotides) are frequent and often cover the entire gene in GC-rich DNA, while in GC-poor DNA they are much rarer and typically do not cover much more than the promoter of the gene;[27,28] GC-rich DNA apparently recombines more frequently than GC-poor DNA, in eukaryotes from yeast to human ($R \approx 0.4$),[29,30] so that impressively long haplotype blocks[31] would be unlikely to occur in gene-dense regions; GC-rich isochores replicate early in S phase whereas GC-poor isochores replicate late;[32] the GC-richest DNA tends to loop far into the nucleus during interphase, while the GC-poorest DNA typically stays close to the periphery of the nucleus.[33]

4.2   The Territories of the GC-Richest Isochores are in the Interior of the Interphase Nucleus

This last correlate of GC is a recent discovery that deserves some comment.

The general topic of chromosomal or sub-chromosomal territories – preferred regions within the cell's nucleus, around which the DNA diffuses – has been discussed in a number of articles in the last few years; we cite here, with their references, Saccone *et al.*,[33] Zink *et al.*,[34] and Boyle *et al.*[35] The latter group have calculated the average position of each human chromosome (centroid) and its distance from the edge of the nucleus, during interphase. We have plotted these distances against the GC of the chromosomes and obtain an impressively linear relation (Figure 3(a); $R=0.91$). At sub-chromosomal scales, the same tendency has been shown directly, by fluorescent *in situ* hybridization (FISH) experiments: mammalian and avian sub-chromosomal regions or bands that hybridize with the GC-richest DNA tend to localize in the interior of the nucleus, while those that hybridize with the GC-poorest DNA localize at the periphery[33] [see Figure 3(b)]. As a caricature, an ultracentrifuge cell could thus be viewed as a direct physical model of the biological cell: the distance into the AUC cell where one finds a large region's DNA at sedimentation equilibrium tells us the average distance that region extends into the interior of the nucleus at interphase.

We still need, however, to clarify how large a sub-chromosomal region must be for its position to depend on GC, a problem that is still largely open. For example, if a 100 kb region having 30% GC were immediately adjacent to a 100 kb region having 60% GC, one would not expect to find the GC-poor region at the nuclear edge, and all of the immediately adjacent GC-rich region far away in the nuclear interior. More generally, what region around a gene best predicts that gene's position (territory) in the nucleus: is it the isochore, a chromosomal band, or a symmetric window of 50, 100, or 1000 kb, for example, that should represent the local GC?

Chromosomal and sub-chromosomal territories have recently received attention in the context of translocations that can lead to medical syndromes or disease, notably cancer. One would intuitively expect that two regions, located on two different chromosomes, are more likely to experience an accidental translocation if they are in close physical proximity during relevant phase(s) of the cell cycle.[36] The data obtained so far for interphase are, however, slightly disappointing. Although known translocation-prone pairs of sub-chromosomal regions in human (such as *MYC:IGH*, involved in Burkitt's lymphoma) are significantly more often close to each other in interphase

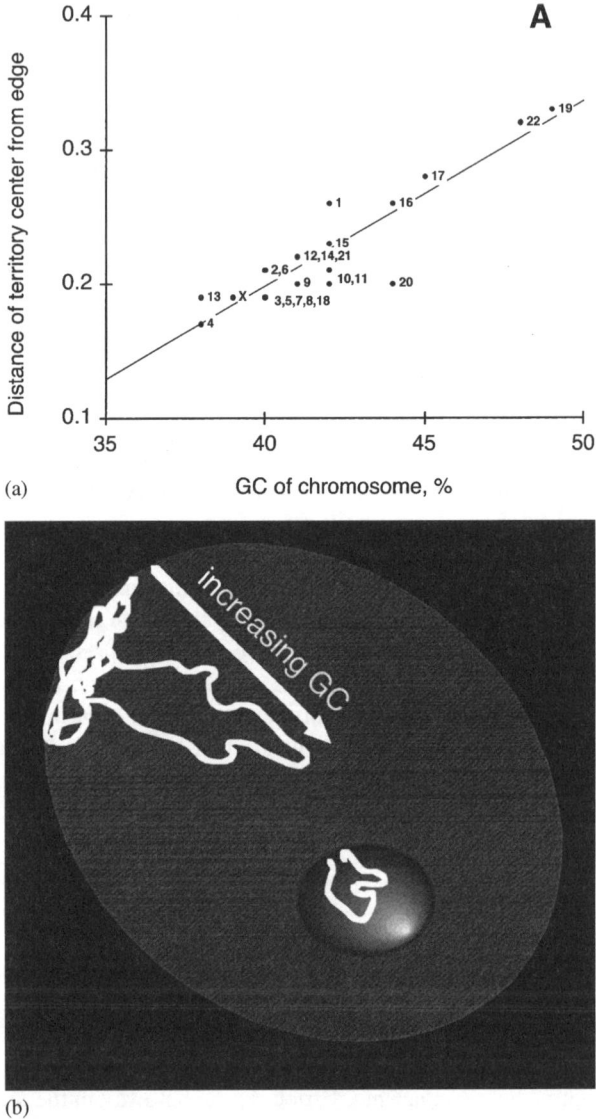

(a)

(b)

Figure 3   *(a) Distance of human chromosomes' centroids (territory centers) from the edge of the nucleus at interphase correlates with the chromosomes' average GC level (R=0.91). We have taken the distances from Boyle et al.,[35] where they are normalized for nuclear size, and plotted them against the GC of the chromosomes (labeled; here the data of ref. 62 are used, more recent sequence releases give essentially the same plot). (b) Caricature of GC-rich mammalian or avian DNA looping into the interior of a cell's nucleus during interphase, while the GC-poor DNA remains close to the periphery. The GC-rich DNA for ribosomal RNA genes (rDNA) is an exception, since it loops into separate nucleoli. Adapted from ref. 33*

than pairs that do not undergo translocations (such as *MYC:TGFBR2*), when one looks at the numbers the difference is small (32.7 *vs.* 22.2%, in the observations made by Roix *et al.*[37]). This suggests that physical proximity during interphase is likely to be only one of the several factors determining translocation probabilities. Modified criteria, and/or the inclusion of additional factors, may improve the predictive power of interphase distances – and thus possibly of GC differences – for translocations or other functionally relevant interactions. Such interactions might also include contacts in *trans* that could facilitate transcriptional activation during interphase (*cf.* ref. 38).

4.3 Changes in GC Distributions During Vertebrate Evolution

Homeothermy evolved twice, independently, in the lineages leading to the present-day mammals and birds. Similarly, the GC-richest DNA is found only in mammals and birds. In cold-blooded vertebrates (except some reptiles[39]), the corresponding isochores are clearly GC-poorer, and the contrast between GC-rich and GC-poor isochores did not evolve to the dramatic differences seen in mammals and birds [Figure 4(a)]. In other words, the GC-enrichment in the gene-densest regions of the genome evolved twice, independently, in precisely the two lineages leading to the present-day warm-blooded vertebrates. This and other evidence (see ref. 40 for detailed discussions) suggests that the GC-richest regions of the ancestral genomes became GC-richer in response to a need for higher thermal stability of the DNA (and possibly also RNA or proteins) as the body temperature increased. It therefore appears that primarily the GC-richest DNA loops into the interior of the nucleus at interphase (see above) and is thus 'exposed', while the GC-poorest DNA remains in a more compact form, and in a more closed chromatin structure, near the edge of the nucleus and therefore does not have the same need to be stabilized.

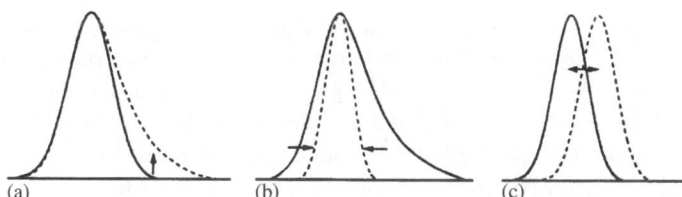

(a) (b) (c)

Figure 4 *Schematic portrayal of three elementary compositional shifts that characterize vertebrate evolution. A compositional shift is a relatively rapid, concerted change in GC during evolution, affecting part or all of a taxon's GC distribution. (a) The increase in GC affects mainly the genome's GC-richest regions; such a shift occurred during the transition from cold- to warm-blooded vertebrates (major shift; upward arrow). (b) Both GC-rich and GC-poor DNA is 'eroded' from the genome; such a shift occurred in a group of myomorph rodents that include mouse and rat (minor shift; inward arrows). (c) Genome-wide increase or decrease in GC occurred, for example, during fish evolution (horizontal shift; lateral arrows). Shifts are often composed of two or more of the elementary events depicted here. Conversely, many mammalian orders have witnessed no substantial shifts (conservative mode). Modified from ref. 63*

Mouse, and a number of other myomorph rodents, underwent a 'reverse' shift, well after the GC-rich isochores were already established in mammals (see ref. 41 and the publications cited therein). The mouse has a narrower GC distribution than other eutherian mammals: the GC-poorest and especially the GC-richest DNA is missing in mouse [Figure 4(b)].[41,14]

The murid example illustrates how biological findings can be deduced or predicted by studying and comparing AUC profiles. In view of the biological correlates listed above, the loss of the GC-rich tail in murid profiles suggests that CpG islands should be very different in mouse than in human, and one does indeed find striking differences when one compares the (orthologous) CpG islands of these two species. In mouse the CpG islands are severely 'eroded', covering less of the gene, exhibiting less contrast in CpG density with the surrounding DNA, and in some cases being absent altogether.[27,28,42,43] Such differences tie in well with other observations that have been made for mouse, compared to human and other mammals: its DNA repair is less meticulous, its inactive X chromosome is more often accidentally activated, its methylation-free or hypomethylated genes are more often accidentally methylated, and its substitution rate is higher.[44,45]

Fishes present another type of change. Not only are their GC distributions narrower and more symmetric than, for example, mammals, birds, and some reptiles, but the positions (modes or means) of these distributions on the GC axis are often far from those of most terrestrial vertebrates, ranging from about 37 to 49% GC.[46,47] Fishes have therefore apparently undergone several 'horizontal' shifts: the shape and width of the CsCl profile is often well conserved while its position shifts substantially [Figure 4(c)].

In the context of gene density calculations in human, we discussed the linear relation between the GC of human genes, or their third positions, and the GC of their surrounding DNA. How well are such relations conserved across different species, or how quickly do they change during evolution? If such relations and their equations were perfectly conserved throughout a taxonomic group, the GC level of a sequenced GC-rich gene, *i.e.*, of a 'compositional marker' sequence, would be predictable from the species' AUC profile and would add little new information: a profile with a GC-rich tail would predict a correspondingly high GC for the sequenced marker gene, and *vice versa*. A preliminary analysis,[7] of 24 species/genera from seven eutherian orders for which both the AUC profile and a GC-rich marker sequence were available (exon 28 of the gene for von Willebrand factor, vWF; see ref. 48), suggests that the linear relation linking their GC levels does change, but slowly, in the sense that the relation is typically well conserved within orders but can differ among orders.

5   Comments on Molar Mass Dependence

Molar mass distributions of vertebrate DNA samples are often quite narrow. For our purposes, we can then treat the samples as if they were monodisperse, *i.e.*, as if all the fragments had the same length (the mean). Here, as elsewhere, we assume that the mean molar mass of each sample is at least 10 kb. Extracted DNA with lower mean molar masses will often be polydisperse, may have been degraded and/or partly denatured, and can be difficult to use for comparative AUC studies.

For many species, particularly if they are rare or endangered, we will not have the luxury of high (>50 kb) molecular weight DNA. Our sample may instead have fragments of, say, 15 kb. For another species, such as human or mouse, we may be able to prepare (and inject into the AUC cell using the usual protocol) DNA having a much higher molecular weight, *e.g.*, 90 kb. Such high molecular weight DNA provides more information on the large-scale properties of the genome. Can we then compare the two species? At 15 kb a mammalian GC distribution still broadens as one decreases the molecular weight, whereas at 90 kb it is almost invariant (see Figure 2). From 90 kb fragments, we could obtain the profile for 15 kb fragments by chopping or shearing the 90 kb fragments, but we could not reconstruct 90 kb fragments from 15 kb fragments even in principle, since we would need adjacency information that we do not have. So what can we do? More generally, how can we compare 50 or 200 species, each one represented by a sample that has a different molar mass?

As an expedient, we can simply plot the raw data for the estimated GC distributions, for example with standard deviation on the vertical axis (after using Equation (5) to correct for diffusion) and mean fragment length on the horizontal axis (see ref. 7 for an example). Although we cannot extrapolate through individual points, the general shape of the scatterplot does give a rough idea of how a given GC distribution's width should change as a function of fragment length.

A more satisfactory solution may be possible in future, after studying individual genomes from different orders and comparing, for each species, its CsCl profile widths at different molecular weights. Whereas earlier studies, such as the mouse study of Macaya *et al.*,[12] could only be done experimentally, the sequencing of entire vertebrate genomes now allows accurate studies to be conducted quickly on the computer (see, *e.g.*, ref. 49 for an in-depth study of the human chromosomes). One spin-off of such studies is an indication of conserved properties, shapes, or regions in the standard deviation *vs.* molar mass plots or their equivalent representations, namely correlograms and Fourier spectra (discussed in ref. 49). Progress along such lines should allow results from a few sequenced model genomes to be extrapolated to the many vertebrate taxa for which we have no genome sequence. Indeed, for many known species we will have at most an AUC profile and possibly one or two sequenced nuclear genes, for example from gene-specific sequencing projects that were designed to cover a wide range of taxa (see, *e.g.*, refs. 48 and 50). At the time of writing, entire genome sequences of vertebrates are available for human, chimpanzee, mouse, rat, and two species of pufferfish, and for parts of the chicken, dog, and zebrafish genomes; a few sequenced contigs exist even for platypus. The systematic study of such long genomic sequences may soon tell us where and how we could normalize GC distributions for molecular weight.

6 Comments on Concentration Dependence

Concentration dependence of vertebrate CsCl profiles is a more complicated problem than molecular weight dependence, for several reasons. Luckily its effects are less pronounced, and much of what follows should be unnecessary if one just wants to extract a reasonable estimate of a species' GC distribution.

If we load twice as much DNA into the ultracentrifuge cell, from the same sample, we might expect the absorbance profile to be twice as high, at all parts of the band that forms at sedimentation equilibrium.

Instead, we observe slight, systematic distortions. In other words, the area-normalized CsCl profile changes as a function of the total DNA loaded or, correspondingly, as a function of the peak concentration or maximum absorbance. This is true even when maximum concentrations remain within the recommended range, *i.e.*, giving a maximum absorbance (optical density) between about 0.3 and 1.0. One would like to predict how the profile will change, *i.e.*, if and how one can normalize for concentration in general, but this is still a largely open problem. Unlike the molar mass problem, this problem is entirely an AUC problem, since obviously the true GC distribution will not depend on the amount of the same DNA that one analyzes.

We begin by considering a sample of identical DNA molecules, so that its GC distribution is, in principle, an infinitely sharp peak (delta function). What will happen to the sample's CsCl profile as we increase the amount of DNA loaded, beginning at zero (infinite dilution)? When the amount loaded is low, we are presumably in the domain of classical virial effects. The DNA and CsCl in water will not be an ideal solution, and there will be solvation (hydration). Increasing the total concentration will broaden the profile; typically the width will increase exponentially with increasing concentration.[51,9] As the amount of DNA loaded increases, however, there is a possibility that some of the molecules will aggregate, which would increase the effective molar mass and thus narrow the profile. The conditions for aggregation are still poorly understood, but it is likely to occur, for example, in satellite DNA (or when molecules/fragments have 'sticky' ends).[12]

Vertebrate profiles have relatively broad, and often asymmetric GC distributions, which compound our problem. A peak region (high concentration) may then be narrowed by aggregation at the same time as the GC-rich tail (low concentration) is broadened by virial effects, to different extents at different parts of the tail. In technical terms: if the two opposing effects of concentration could be modeled, we would afterwards face not just a simple convolution problem, but a complicated, general folding problem. Indeed, whereas molar mass does not depend on GC in a typical preparation of DNA (no correlation between molar mass and GC),[52] local concentration obviously does. In summary, changing the total concentration can lead to local, non-monotonous horizontal contractions and expansions of the profile. These may, in turn, slightly shift the mode in seemingly unpredictable ways. One way to approach the problem is by experimentally obtaining plots of profile parameters (mode, standard deviation) *vs.* maximum absorbance (profile height at the mode). When this is done for bacteriophages, the plots of log(standard deviation) *vs.* maximum absorbance have slopes that differ from one phage species to another,[51,9,8] suggesting that the slope can depend on the genomes' sequences. Even for these essentially infinitely narrow GC distributions, the only way to gauge the effect of concentration was to obtain profiles for several concentrations and then extrapolate to infinite dilution. Comparing the GC distributions of entirely sequenced vertebrate or other genomes with their AUC profiles may now give better insight and show how concentration-related profile distortions are related to GC and/or sequence features other than GC.

7 Conclusion and Perspectives

We have discussed some of the applications of CsCl gradient AUC in genomics. In particular, we have described large-scale GC variations and other GC-related genomic features that can be deduced from CsCl profiles, either alone or in conjunction with sequence data (where sparse sequence data often suffice).

CsCl gradient ultracentrifugation also serves other uses. Early investigations employed the labeling of DNA and/or RNA. Two well-known experiments of this type are that of Meselson and Stahl,[53] who showed that DNA replication is semi-conservative, and that of Hall and Spiegelman[54] (discussed in ref. 2), who showed the existence *in vivo* of RNA molecules, now called messenger RNAs (mRNAs), which are complements of one strand of a DNA genome. In the first experiment, DNA that had incorporated heavy nitrogen (N^{15}) had a shifted AUC peak; in the second experiment DNA was labeled with H^3, RNA with P^{32}, and preparative ultracentrifugation in CsCl was followed by fractionation and assessment of the levels of labeled DNA and RNA in each fraction.

CsCl profiles and their parameters (*e.g.*, mode, standard deviation) can also be used to fine-tune phylogenies, or to highlight regions of accepted phylogenies that may benefit from closer attention. We have found that AUC profile data often confirm established phylogenies, but differ from them in some regions that are widely regarded as tentative or controversial (see, *e.g.*, ref. 14). While it might be tempting to propose profile parameters as phylogenetically faithful traits, such reliance can mislead. If only CsCl data are available for a group of species, compositional shifts (see above) could be confounded with drifts that would indicate large distances. A more prudent course is to use AUC profile data together with phylogenies obtained from independently gathered sequence or other data. One should however keep in mind that nuclear protein-coding genes will be subject to the same GC shifts as the species' CsCl profiles, so third and often first codon positions must be checked for interspecific stationarity before they are used for phylogenetic purposes.[55]

Another application of CsCl gradient AUC is in analyzing mixtures of genomes of naturally associating (symbiotic or parasitic) species. By locating the peaks of the resulting CsCl landscape and inferring their GC levels, one can screen possible candidate species, unambiguously identify the species by follow-up studies, and quantify their relative abundances in the mixture or community to which they belong *in vivo*.[56,57] The extrapolation of this principle to diagnostics, *e.g.*, for quick GC-based identification of the microbial species in a sample, seems straightforward and promising. Indeed, the GC levels of prokaryotes span a wide range, their profiles are typically narrow compared to this range (see Figure 2), and different species are therefore likely to be resolved by AUC.

A use of the CsCl method that has recently been revived is for detecting, quantifying, and monitoring protein–DNA clamp formations or other salt-stable complexes or interactions (*e.g.*, refs. 58–60). The application has so far been largely limited to studies of topoisomerase and salt-stable clamps that are associated with it, but the principle is general: well-spaced peaks tell the relative abundances of salt-stable complexes between DNA and 1, 2, 3, … protein molecules.

It seems fair to conclude that CsCl gradient AUC of DNA remains a powerful tool in functional, comparative, and evolutionary genomics, as well as in molecular biology. It can be hoped that increasingly varied and creative applications will be devised in the near future, as more researchers reconsider this method when designing experiments.

Acknowledgements

We thank Salvatore Bocchetti for the technical AUC work that made our recent studies possible. We also thank Kamel Jabbari and Salvatore Saccone for helpful discussions, Maria Costantini for sharing data, and Laura Giangiacomo for literature references.

References

1. M. Meselson, F. W. Stahl and J. Vinograd, *Proc. Natl. Acad. Sci. USA*, 1957, **43**, 581.
2. K. E. van Holde, W. C. Johnson and P. S. Ho, *Principles of Physical Biochemistry*, Prentice-Hall, Upper Saddle River, NJ, 1998.
3. N. Sueoka, J. Marmur and P. Doty, *Nature*, 1959, **183**, 1429.
4. J. Marmur and P. Doty, *Nature*, 1959, **183**, 1427.
5. R. Rolfe and M. Meselson, *Proc. Nat. Acad. Sci. USA*, 1959, **45**, 1039.
6. C. L. Schildkraut, J. Marmur and P. Doty, *J. Mol. Biol.*, 1962, **4**, 430.
7. O. Clay, C. Douady, N. Carels, S. Hughes, G. Bucciarelli and G. Bernardi, *Eur. Biophys. J.*, 2003, **32**, 418.
8. C. W. Schmid and J. E. Hearst, *Biopolymers*, 1972, **11**, 1913.
9. J. E. Hearst and C. W. Schmid, *Meth. Enzymol.*, 1973, **27**, 111.
10. J. Filipski, J. P. Thiery and G. Bernardi, *J. Mol. Biol.*, 1973, **80**, 177.
11. J. P. Thiery, G. Macaya and G. Bernardi, *J. Mol. Biol.*, 1976, **108**, 219.
12. G. Macaya, J. P. Thiery and G. Bernardi, *J. Mol. Biol.*, 1976, **108**, 237.
13. G. Cuny, P. Soriano, G. Macaya, and G. Bernardi, *Eur. J. Biochem.*, 1981, **115**, 227.
14. C. Douady, N. Carels, O. Clay, F. Catzeflis and G. Bernardi, *Mol. Phylogenet. Evol.*, 2000, **17**, 219.
15. G. Corneo, E. Ginelli, C. Soave and G. Bernardi, *Biochemistry*, 1968, **7**, 4373.
16. R. D. Wells and J. E. Blair, *J. Mol. Biol.*, 1967, **27**, 273.
17. A. Pavlícek, J. Paces, O. Clay and G. Bernardi, *FEBS Lett.*, 2002, **511**, 165.
18. J. B. Ifft, D. M. Voet and J. Vinograd, *J. Phys. Chem.*, 1961, **65**, 1138.
19. J. Cortadas, G. Macaya and G. Bernardi, *Eur. J. Biochem.*, 1977, **76**, 13.
20. S. Zoubak, O. Clay and G. Bernardi, *Gene*, 1996, **174**, 95.
21. S. Cruveiller, K. Jabbari, O. Clay and G. Bernardi, *Brief. Bioinform.*, 2003, **4**, 43.
22. D. Mouchiroud, G. D'Onofrio, B. Aïssani, G. Macaya, C. Gautier and G. Bernardi, *Gene*, 1991, **100**, 181.
23. E. S. Lander *et al.*, *Nature*, 2001, **409**, 860.
24. L. Duret, D. Mouchiroud and C. Gautier, *J. Mol. Evol.*, 1995, **40**, 308.
25. J. Paces, R. Zíka, V. Paces, A. Pavlícek, O. Clay and G. Bernardi, *Gene*, 2004, **333**, 135.
26. A. Pavlícek, K. Jabbari, J. Paces, V. Paces, J. Hejnar and G. Bernardi, *Gene*, 2001, **276**, 39.
27. B. Aïssani and G. Bernardi, *Gene*, 1991, **106**, 173.
28. B. Aïssani and G. Bernardi, *Gene*, 1991, **106**, 185.
29. J. L. Gerton, J. DeRisi, R. Schroff, M. Lichten, P. O. Brown and T. D. Petes, *Proc. Natl. Acad. Sci. USA*, 2000, **97**, 11383.

30. S. M. Fullerton, A. Bernardo Carvalho and A. G. Clark, *Mol. Biol. Evol.*, 2001, **18**, 1139.
31. M. J. Daly, J. D. Rioux, S. F. Schaffner, T. J. Hudson and E. S. Lander, *Nat. Genet.*, 2001, **29**, 229.
32. C. Federico, S. Saccone, and G. Bernardi, *Cytogenet. Cell Genet.*, 1998, **80**, 83; C. Federico, L. Andreozzi, S. Saccone and G. Bernardi, *Chromosome Res.*, 2000, **8**, 737.
33. S. Saccone, C. Federico and G. Bernardi, *Gene*, 2002, **300**, 169.
34. D. Zink, A. Bolzer, C. Mayr, W. Hofmann, N. Sadoni and K. Überla, *Gene Ther. Mol. Biol.*, 2001, **6**, 1.
35. S. Boyle, S. Gilchrist, J. M. Bridger, N. L. Mahy, J. A. Ellis and W. A. Bickmore, *Hum. Mol. Genet.*, 2001, **10**, 211.
36. T. Pederson, *Nat. Genet.*, 2003, **34**, 242.
37. J. J. Roix, P. G. McQueen, P. J. Munson, L. A. Parada and T. Misteli, *Nat. Genet.*, 2003, **34**, 287.
38. H.-P. Müller and W. Schaffner, *Trends Genet.*, 1990, **6**, 300.
39. S. Hughes, O. Clay and G. Bernardi, *Gene*, 2002, **295**, 323.
40. G. Bernardi, *Structural and Evolutionary Genomics: Natural Selection in Genome Evolution*, Elsevier Science, Amsterdam, 2004.
41. G. Bernardi, *Gene*, 2000, **259**, 31.
42. K. Matsuo, O. Clay, T. Takahashi, J. Silke and W. Schaffner, *Somat. Cell Mol. Genet.*, 1993, **19**, 543.
43. F. Antequera and A. Bird, *Proc. Natl. Acad. Sci. USA*, 1993, **90**, 11995.
44. R. Holliday, *Understanding Ageing*, Cambridge University Press, Cambridge, UK, 1995.
45. W. H. Li, *Molecular Evolution*, Sinauer, Sunderland, MA, 1997.
46. G. Bernardi and G. Bernardi, *J. Mol. Evol.*, 1990, **31**, 282.
47. G. Bucciarelli, G. Bernardi and G. Bernardi, *Gene*, 2002, **295**, 513.
48. O. Madsen, M. Scally, C. J. Douady, D. J. Kao, R. W. DeBryk, R. Adkins, H. M. Amrine, M. J. Stanhope, W. W. de Jong and M. S. Springer, *Nature*, 2001, **409**, 610.
49. W. Li and D. Holste, *Phys. Rev. E*, 2005, **71**, 041910.
50. W. J. Murphy, E. Eizirik, S. J. O'Brien, O. Madsen, M. Scally, C. J. Douady, E. Teeling, O. A. Ryder, M. J. Stanhope, W. W. De Jong, and M. S. Springer, *Science*, 2001, **294**, 2348.
51. C. W. Schmid and J.E. Hearst, *J. Mol. Biol.*, 1969, **44**, 143.
52. N. Sueoka, *Proc. Natl. Acad. Sci. USA*, 1959, **45**, 1480.
53. M. Meselson and F.W. Stahl, *Proc. Natl. Acad. Sci. USA*, 1958, **44**, 671.
54. B. D. Hall and S. Spiegelman, *Proc. Natl. Acad. Sci. USA*, 1961, **47**, 137.
55. C. Saccone, C. Lanave, G. Pesole and G. Preparata, *Meth. Enzymol.*, 1990, **183**, 570.
56. M. Costantini, B. Lafay and G. Matassi, *Boll. Mus. Ist. Biol. Univ. Genova*, 2002, **66-67**, 48.
57. M. Costantini, *Gene*, 2004, **342**, 321.
58. R. E. Depew, L. F. Liu and J. C. Wang, *J. Biol. Chem.*, 1978, **253**, 511.
59. T. Hu, S. Chang and T. Hsieh, *J. Biol. Chem.*, 1998, **273**, 9586.
60. V. H. Oestergaard, L. Bjergbaek, C. Skouboe, L. Giangiacomo, B.R. Knudsen and A.H. Andersen, *J. Biol. Chem.*, 2004, **279**, 1684.
61. P. Bernaola-Galvàn, J. L. Oliver, P. Carpena, O. Clay and G. Bernardi, *Gene*, 2004, **333**, 121.
62. J. C. Venter et al., *Science*, 2001, **291**, 1304.
63. G. Bernardi, S. Hughes and D. Mouchiroud, *J. Mol. Evol.*, 1997, **44** (Suppl 1), S44.

CHAPTER 6

LAMMNUM: A Program to Study Self-Associating Macromolecules in Sedimentation Velocity Experiments

JOACHIM BEHLKE AND OTTO RISTAU

1 Introduction

Analytical ultracentrifugation is used to analyze protein self-association. It is employed to compare enzymatic activity and quaternary structure of proteins and to ascertain whether the crystal structure derived from X-ray scattering is identical with the solution structure. Furthermore, structure analysis by high-resolution NMR requires knowledge of the state of association and the type of equilibria. The classical approaches are the equilibrium sedimentation and the concentration-dependent analysis of weight-average sedimentation coefficients, the isotherm method.[1] Another possibility of the shape analysis of sedimentation velocity boundaries has been developed.[2-4] These approaches provide association constants by fitting a set of sedimentation velocity profiles, which can be quickly obtained as an important aspect for less stable proteins. Furthermore, it is possible to omit the region near the cell base, which occasionally contain small quantities of aggregates. This is an advantage compared with the thermodynamically well-described method of sedimentation equilibrium. Our program allows the reliable determination of binding constants for self-associating macromolecules from sedimentation velocity runs.

2 Theoretical Background

The program LAMMNUM with a numerical solution of the Lamm equation was written according to refs. 2, 3 and 5. It allows estimating equilibrium constants (K^w). The program works with numerically calculated partial derivatives for each parameters s, D, and K^w according to ref. 6. Additionally the meniscus and bottom radii can be estimated. This is done by enhancing the bottom radius or diminishing the meniscus radius in the finite element scheme by 1/10th of the grid point distance. The difference quotient calculated with the solution vector of the changed and the unchanged finite element grid is not exactly located at the basic grid (experimental points as by the other to estimating parameters (s, D)), therefore the difference quotients is calculated only one-sided not symmetrical spaced around the basic points to spare computing time. Normally we estimate these two radii in each whole boundary fit procedure. The mathematical method of the nonlinear fit is the same as that described in ref. 7. The additional iteration of the Gaussian normal equations described in this paper is further increased to overcome the poor numerical condition of the information matrix.[8] The grid point distance in the interference approach is the same as in the data files. In the absorption approach the nearest radius distance found in the data files is used (mostly 0.001 cm). The time step standard is 1 s but this value can be reduced. The fit region can be independently chosen in comparison with the meniscus or bottom radius. This is important to exclude impurities (aggregates) in the bottom region. It is sufficient to expand the fit region until the crossing point of the concentration traces in order to estimate the bottom radius. The concentration (absorbance) at this point does not markedly exceed the loading value. A clearly to low estimated bottom radius compared to the physical one is a distinct reference to concentration depending on sedimentation coefficient. For the analysis of radius- and time-independent noise, the algebraic method developed by Schuck and Demeler[9] was used. The association constant (K^w) was introduced into the Claverie scheme according to ref. 3. This scheme is based on the constituent sedimentation and diffusion coefficients defined in refs. 10 and 11 and describes the dependence of the sedimentation and diffusion coefficients on concentration and radial distance by a finite element function such as the moving boundary. To get more numerical stability we used the relations for spherical-shaped complexes for the dependence of sedimentation and diffusion coefficients on the degree n of association as:

$$s_n = n^{2/3} s_M$$
$$D_n = n^{-1/3} D_M$$

(1)

The program SEDFIT of Schuck[3] also applicable for estimation of association constants does not use this relation. To enhance the variability of the program LAMMNUM, other exponents as in Equation (1) can also be chosen, but cannot be estimated.

The sedimentation coefficient of solutions with self-associating components is

$$s(c_w) = \frac{\sum_{i=1}^{i=n} c_i s_i}{\sum_{i=1}^{i=n} c_i}$$

(2)

The corresponding diffusion coefficient is expressed as follows:

$$D(c_w) = \frac{\sum_{i=1}^{i=n} D_i \frac{\partial c_i}{\partial r}}{\sum_{i=1}^{i=n} \frac{\partial c_i}{\partial r}}$$

(3)

Here c_i is the weight concentration of the i-mer. Taking the law of mass action into account

$$c_i = K_i^w c_M^i, \quad K_1^w = 1 \tag{4}$$

the constituent sedimentation or diffusion coefficients are

$$s(c_w) = s_M \frac{1 + \sum_{i=2}^{i=n} i^{2/3} K_i^w c_M^{i-1}}{1 + \sum_{i=2}^{i=n} K_i^w c_M^{i-1}} \tag{5}$$

$$D(c_w) = D_M \frac{1 + \sum_{i=2}^{i=n} i^{2/3} K_i^w c_M^{i-1}}{1 + \sum_{i=2}^{i=n} i K_i^w c_M^{i-1}} \tag{6}$$

In Equations (5) and (6), s_M or D_M are the sedimentation or diffusion coefficient, respectively, of the monomers at the same solution conditions. An additional dependence of the sedimentation parameter on concentration can also take into account. For s_M then holds

$$s_{M,C} = \frac{s_M}{1 + k_s c_w} \tag{7}$$

The concentration of the monomer c_M can be calculated with the help of mass conservation and the equilibrium constants K^w from the total weight concentration c_w at each grid point (without virial effects). For the monomer concentration c_M, the following equation must be solved:

$$c_w = c_M + \sum_{i=2}^{i=n} k_i^w c_M^i \tag{8}$$

Up to five instant equilibria of arbitrary degree can be fitted. However the fit procedure is also possible without considering an equilibrium. Our program LAMMNUM works internally with reduced dimensionless concentrations c/c_{load}. Therefore the binding constants are implemented as dimensionless quantities. The real values are

$$K_n^{w,real} = K_n^w c_{load}^{-(n-1)} \tag{9}$$

with c_{load} given in weight concentration (g L^{-1}). The corresponding constant in molar concentrations is

$$K_n^M = K_n^{w,real} \frac{M^{(n-1)}}{n} \tag{10}$$

with M the molecular mass of monomers. The program works with the quantity $\ln(1 + K^w)$ as the fit parameter rather than with the dimensionless association constant itself. This is of advantage for the expected wide range of data. The real concentration vectors necessary for the least-squares method are arrived by multiplication of the finite element solution vectors with the loading concentration.

3 Estimation of Initial Values for K^w

The most frequently occurring protein interactions are monomer–dimer equilibria. For this case, we have empirically found a method to obtain initial values for the binding constant K^w. The concentration dependence of sedimentation coefficients according to Equation (7) with negative k_s value suggests an instantaneous equilibrium.

As an alternative to Equation (7), for small k_s values, another relation

$$s_c = s_0(1 + k_s c) \tag{11}$$

is applicable. A useful estimation of the binding constant K^w from the parameter k_s is not possible. As shown in Figure 1, the differentiated concentration profiles of the moving boundary for an instantaneous equilibrium are asymmetric.

The asymmetry arises with growing distance of the moving boundary from the meniscus. We have empirically found that Equation (11) with some modification as given by

$$s_c = s_0[1 + k_s c_w(1 - 0.545 c_w + 0.035 c_w^2)] \tag{12}$$

can be used to get a rough approximation of the frictional ratio when only a few of the first concentration profiles with small asymmetry were fitted. For this procedure, only Faxén-type data sets are used. This means that the bottom region is excluded

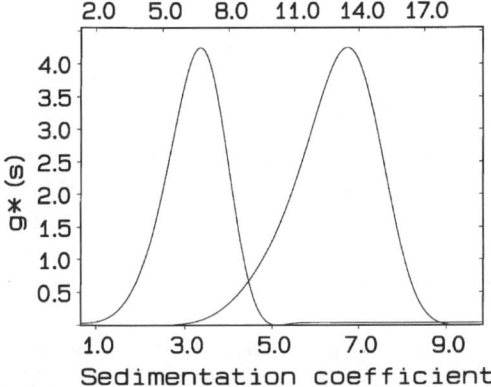

Figure 1 *Differentiated synthetic concentration profiles for a monomer–dimmer equilibrium ($M_{monomer}$=69 kDa; c=1 mg mL^{-1}; s=5 S; D=6.54×10^{-7}cm^2 s^{-1}; 60 000 rpm; K_a=6.9×10^4 M^{-1}). The left curve generated after 2250 s belongs to the upper abscissa, the right one after 4500 s to the lower abscissa*

from the fit and the bottom radius must be held constant during the fit. In order to get satisfactory results, one control parameter is included in the program. The product $s\omega^2 t(r_f - r_m)/(r_b - r_f)$ is a measure for the travel of the files used in the fit that does not exceed ≈ 0.4. Here, r_f describes the end point of the fit region. A further criterion is a quick convergence of the fit (iteration number < 10).

From the estimated frictional ratio (monomer = dimer) and the known monomer molecular mass, the sedimentation coefficient s_c of the monomer can be calculated. The resulting s_c value can be compared with the s_E value estimated together with the frictional ratio. In the case of a monomer–dimer equilibrium, the s_E value can have a maximum of $2^{2/3} = 1.5874$ times of the calculated s_c value for the monomer. The quotient

$$q = \frac{s_E - s_C}{1.5874 s_c - s_c} = \frac{s_E/s_c - 1}{0.5874} \tag{13}$$

is a measure for the strength of dimerization. As expected, the quotient reaches the value 1 for complete association ($K^w = \infty$) and the value 0 for no interaction ($K^w = 0$). Thus the binding constant itself can be roughly estimated by the formula

$$K_w = 4\tan(q\pi/2) + 4\tan^2(q\pi/2) \tag{14}$$

where the quotient q is changed to an angle in radians.

Normally the maximal error of the frictional ratio is smaller than $\pm 3\%$. The deviation of the binding constant K^w is up to $\pm 50\%$, but this considerable error is not a problem for our program LAMMNUM that has excellent convergence properties typical for fit procedures working with derivatives instead of the simplex technique. The somewhat longer computing time of the derivative technique is in most cases compensated for the quicker convergence. The extinction for the centrifuge cell (1.2 cm path length) can be derived from the 3000 rpm trace with the help of the known loading concentration. When using other exponents than that in Equations (5) and (6), the frictional ratio of the dimer deviates from that of the monomer.

4   Data Acquisition

In order to test the efficiency of our program LAMMNUM, synthetic concentration profiles were generated by using an alternative variant according to ref. 5. This approach uses two Claverie schemes, one for monomers and the other for dimers. The equilibrium between the two concentrations localized in both schemes is recalculated for each grid point and after each time step. The accurate second estimate of the association constant using our program mean that the two approaches[3,5] yield identical results (not shown). To test the dependence of the estimated bottom radius on the amplitude of k_s in Equation (7), we simulate some synthetic file sets. The sets were generated using the approach in ref. 3 using Equations (5) and (7). Experimental concentration profiles were from sedimentation velocity experiments obtained in a type XL-A analytical ultracentrifuge (Beckman, Palo Alto, CA). The data from previous experiments were partially re-evaluated.[12–14]

5 Results

5.1 Estimation of Initial K^w Values

Since the equilibrium constant of an instantaneous equilibrium analyzed from sedimentation velocity runs depends on several parameters such as molecular mass, and sedimentation and diffusion coefficients, it is advisable to estimate an approximate K^w from the early profiles as proposed in Section 3. As seen in Figure 1, the asymmetry of the differentiated concentration profiles becomes clearer with increasing running times. Therefore, only the first profiles were used to estimate K^w. In Table 1, the data derived from experiments with the protein Be3 that forms a monomer–dimer equilibrium are presented and compared with the final calculated values using the program LAMMNUN. The deviation between estimated and calculated data is moderate; this procedure is helpful to estimate K^w for the reliable calculation of binding constants, especially when limited information of the protein is available.

5.2 Analysis of Synthetic Data Files

The program LAMMNUM can estimate up to five equilibria of arbitrary degree in a self-associating system. Fits of monomer–dimer equilibria have been demonstrated above in data acquisition. To demonstrate the excellent convergence for more complicated situations, we have simulated a data set containing two equilibria, a monomer–dimer and a monomer–tetramer, which has dimensionless association constants $K_2^w c_{load}=1$ and $K_4^w (c_{load})^3=2$. In spite of wrong initial values ($K_2^w c_{load}=0$ and $K_4^w (c_{load})^3=20$), the program finds the correct results after seven iterations using only knowledge of the molecular weight (Figure 2). When considering a loading concentration of 1 g L^{-1} and a molecular mass of 49.42 kDa for the monomer, a real molar dimer constant of $K_2^M=2.47 \times 10^4$ L mol^{-1} and a real tetramer constant $K_4^M=6.03\times10^{13}$(L mol^{-1})^3 are obtained. For the initial value of the sedimentation coefficient, a frictional ratio of 1.4 was used which provides a reasonable starting value together with the inputted molecular weight.

To demonstrate the fit of a monomer–dimer equilibrium that also contains a nonassociating component, we have simulated such a data set by the addition of two appropriate data files. The nonassociating component was assumed to be an impurity of

Table 1 *Comparison of estimated initial $K^w_{c_{load}}$ values with the final values obtained from sedimentation velocity profiles of the transport protein Be3 at different temperatures*

$T(^{\circ}C)$	Initial values	Final values obtained with LAMMNUM	
		Whole boundary	Faxén type
10	8.5	13.2	12.8
15	12.0	9.7	13.5
20	12.5	11.4	12.9
25	3.3	5.6	6.2
30	2.9	3.1	4.6

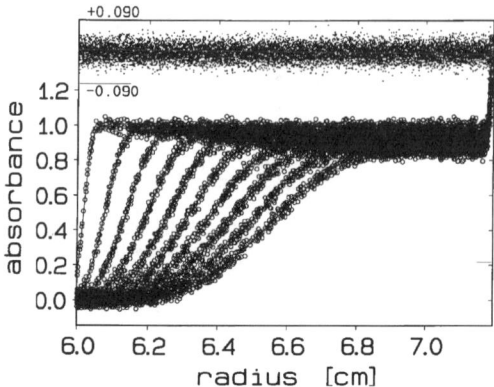

Figure 2 *Fit of synthetic data set with a monomer–dimer and a monomer–tetramer equilibrium; 50 000 rpm; noise, 2%; time step, 1 s; radial step, 0.001 cm; $M_{monomer}$ = 49.42 kDa, held constant*

Input data	Initial values	Estimated parameter
$s = 4$ S	$s = 3.69$ S	$s = 3.997$ S
$D = 7.308$ F	$D = 6.265$ F	$D = 7.303$ F
$r_m = 6.0$ cm	$r_m = 6.0$ cm	$r_m = 6.0$ cm
$r_b = 7.20$ cm	$r_b = 7.19$ cm	$r_b = 7.200$ cm
$K^w_2 = 1$	$K^w_2 = 0$	$K^w_2 = 0.992$
$K^w_4 = 2$	$K^w_4 = 20$	$K^w_4 = 1.990$
$c_{load} = 1$	$c_{load} = 0.972$	$c_{load} = 0.9998$

10% related to the main component (Figure 3). By entering the known molecular weight of the monomer the program shows excellent convergence. After seven iterations, the fit was completed.

The bottom radius is estimated as evidence for dependence of sedimentation velocity on concentration. The fit program works without consideration of Equation (7). The theoretical bottom radius is 7.2 cm. The fit region was limited at the crossing point of the concentration traces. Surprisingly, the binding constant depends also very strongly on the amplitude of k_s. The somewhat worse fit without considering the concentration dependence of the sedimentation velocity is scarcely discernible in the presence of noise. The results are shown in Table 2.

5.3 Analysis of Experimental Sedimentation Profiles

In the following section, the program LAMMNUM is applied to study the equilibrium constants of some proteins.

5.3.1 Cro Repressor

The Cro repressor of λ phage is a small DNA binding protein of 66 amino acids and is only active as a dimeric molecule. The dimeric interphase consists of a β strand from the C-terminal part of each subunit. When substituting the amino acid Val in

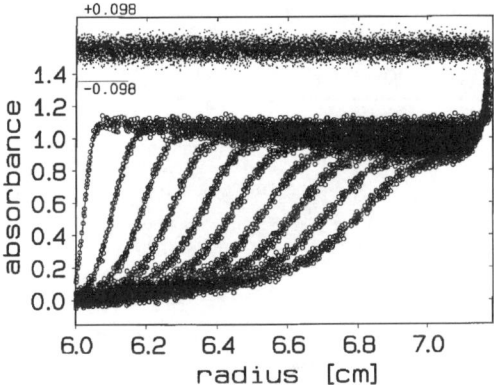

Figure 3   *Fit of synthetic data set with a monomer–dimer equilibrium and one independent component without equilibrium; 50 000 rpm; noise, 2%; time step, 1 s; radial step, 0.001 cm; equilibrium component, $M_{monomer}$ = 55.728 kDa, held constant*

Input data	Initial values	Estimated parameter
s = 4 S	s = 4.16 S	s = 4.000 S
D = 6.482 F	D = 6.74 F	D = 6.482 F
r_m = 6.0 cm	r_m = 6.0 cm	r_m = 6.000 cm
r_b = 7.2 cm	r_b = 7.187 cm	r_b = 7.200 cm
K_2^w = 3	K_2^w = 1	K_2^w = 2.977
c_{load} = 1	c_{load} = 0.902	c_{load} = 1.0002
Independent component		
s = 2 S	s = 1.5 S	s = 2.002 S
D = 10.336 F	D = 10 F	D = 10.36 F
c_{load} = 0.1	c_{load} = 0.15	c_{load} = 0.09925

position 55 by Cys, the mutant spontaneously dimerizes.[15] This leads to an enhanced stability; however, the covalently linked dimers retain their capacity to dimerize to a tetramer. The association process can be treated as a monomer–dimer equilibrium and was analyzed from sedimentation velocity experiments (Figure 4). To get a good fit, it was necessary to consider the decrease in sedimentation velocity depending on concentrations in Equation (7). Without this, the bottom radius was estimated to a low value (~7.15 cm), which was caused by the very steep ascent of the concentration profiles near the bottom. The association constant obtained in this way was K_a = (3.7 ± 0.2)×10³ M⁻¹, in satisfactory agreement with K_a = (2.9 ± 0.2)×10³ M⁻¹ obtained with the sedimentation equilibrium technique with our program POLYMOLE.[16]

5.3.2   Hexokinase

The solution structure of yeast hexokinase has been recently studied in detail with respect to the monomer–dimer equilibrium.[12,13] The equilibrium is influenced by the addition of substrate and effectors; it is more strongly affected by substituting Ser14 with other amino acids. Figure 5 demonstrates the fitted sedimentation profiles of

Table 2 *The dependence of bottom radius and other sedimentation parameter on the amplitude of k_s in Equation (7), a quantity for the dependence of sedimentation velocity on concentration*

k_s (L g^{-1})	0.00	0.05	0.1	0.15	0.2	0.25
Bottom radius (cm)	7.20	7.1882	7.1794	7.1715	7.1648	7.1585
Meniscus radius (cm)	6.20	6.1997	6.1995	6.1993	6.1991	6.1989
Sedimentation constant (S)	4.0	3.679	3.423	3.209	3.027	2.869
Diffusion constant (F)	7.309	6.723	6.254	5.864	5.532	5.243
Binding constant (L g^{-1})	3.0	5.42	9.31	15.76	26.78	46.49
Friction ratio	1.2	1.305	1.402	1.486	1.586	1.673

The fit region reaches from meniscus 6.2 cm to the crossing point of the concentration traces. This point varies from 7.141 cm at k_s=0.05 until 7.113 cm at k_s =0.25. The molecular weight of 49.416 kDa was held constant during the fit. Loading concentration was 1g L^{-1}; rotor speed, 40 000 rpm; partial-specific volume, 0.73 mL g^{-1}; solution density, 1 g mL^{-1}.

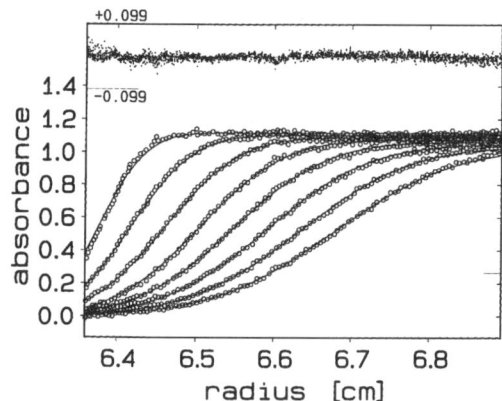

Figure 4 *Sedimentation profiles of the λ Cro repressor (Cro-V55C) dissolved in 25 mM Gly-HCl (pH 2.9) containing 200 mM KCl. The loading concentration was 1.59 g L^{-1}, T = 20 °C, speed = 50 000 rpm. Considering a molecular mass of 14.226 kDa, the following parameters were obtained: $s_{20,w}$ = 1.82 ± 0.02 S; $D_{20,w}$ = (12.1 ± 0.02)10^{-7} cm^2 s^{-1}; f/f$_0$ = 1.1; k_s = (0.14 ± 0.02) mL mg^{-1} (see Equation (7)); K_a = (3.7 ± 0.2)10^3 M^{-1}*

yeast hexokinase Ala14 resulting in a K_a=(1.4 ± 0.10)×10^5 M^{-1}. This is in excellent agreement with K_a=(1.5 ± 0.1)×10^5 M^{-1} obtained with the sedimentation equilibrium technique.

6   Discussion

Many proteins are capable of forming association–dissociation equilibria, which are of biological and drug development importance. For the analysis of such equilibria by analytical ultracentrifugation, two methodological variants can be used: the thermodynamically well-defined sedimentation equilibrium method or the hydronamic sedimentation velocity technique. In addition to the weight-average sedimentation

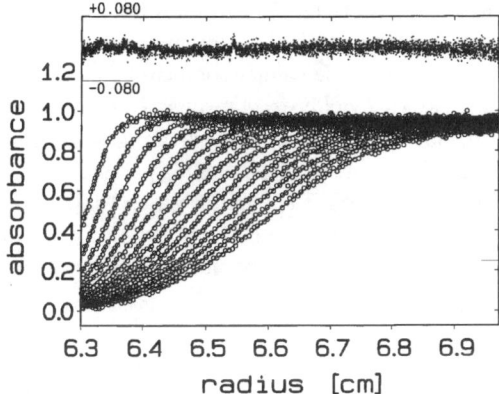

Figure 5 *Sedimentation profiles of a yeast hexokinase mutant (S14A) in 50 mM potassium-phosphate buffer (pH 7.4), containing 1 mM PMSF and 1 mM EDTA, protein concentration: 0.62 g L^{-1}, T=20 °C. Assuming a molecular mass of 53.948 kDa for the monomer the following data were estimated: $s_{20,w}$ = 3.828 ± 0.009 S; $D_{20,w}$ = (6.65 ± 0.02)10^{-7} cm^2 s^{-1}; f/f$_0$=1.27; K_a=(1.52 ± 0.10)10^5 M^{-1}*

velocity approach that is generally used, shape analysis of the moving boundary described[5] is suitable to analyze association equilibria. The latter approach allows one to determine association constants by fitting one set of sedimentation velocity profiles and thus results can be quickly obtained, an important factor for less stable proteins. In contrast to any differential sedimentation coefficient distribution approach, shape analysis uses the original data which means that the signal-to-noise ratio is not changed or made worse.

The first program for shape analysis was developed by Schuck.[3] Recently, Stafford and Sherwood[17] described a program SEDANAL for homologous and heterologous association that is not based on the approach of Schuck.[3] In both programs the well-known relation (Equation (1)) that describes the dependence of sedimentation parameters on the degree of association is not implemented. Therefore, one more parameter needs to be determined for each component. To overcome this disadvantage we have implemented this relation in our program LAMMNUM. Furthermore, LAMMNUM allows the introduction of the molecular mass and calculates the diffusion constant from the current s value during the fit by the Svedberg relation. Additionally it is possible to enter the frictional ratio for calculating the sedimentation coefficient. If the sedimentation coefficient is also held constant, the number of parameters is reduced by 2 (s, D). Otherwise, the input of an approximate frictional ratio yields a good initial value for s. These possibilities can give suitable starting values for difficult fit problems. After some successful iterations, the desirable parameters can also be floated. When molecular weight is not known, the input of an approximate frictional ratio alone can be used as a numerical constraint.

The additional dependence of the sedimentation parameter on concentration according to Equation (7) can also be taken into account. Because of the employment of a constant exponent for the dependence of sedimentation or diffusion coefficient on the degree of association (Equation (1)) and the application of numerical derivatives, our program LAMMNUM works very reliably. It is desirable to use high rotor speed and great column height to get long paths of the moving boundary and minimal plateau region. Extended running time alone cannot compensate high rotor speed. In spite of wrong initial values, LAMMNUM yields exact results in contrast to the well-known program SEDFIT, which show no convergence in such cases. This is caused by the calculation of the partial derivatives for each estimating parameter, an essential advantage compared with the simplex method.[18] The initial values for the estimating parameters can be distinctly inaccurate for this reason. The more expensive calculation is mostly compensated in time by the quicker convergence of the derivative method.

Acknowledgement

We are grateful to Drs. V.V. Rogov (Poustchino, Russia) and T. Kriegel (Dresden, FRG) for providing the study on their proteins (Cro repressor and hexokinase).

References

1. R. P. Frigon and S. N. Timasheff, *Biochemistry*, 1975, **14**, 4559.
2. J. -M. Claverie, H. Dreux and R. Cohen, *Biopolymers*, 1975, **14**, 1685.
3. P. Schuck, *Biophys. J.,* 1998, **75**, 1503.
4. P. Schuck, *Anal. Biochem.*, 2003, **320**, 104.
5. D. J. Cox and R. S. Dale, in *Protein–Protein Interaction*, C. Frieden and L. W. Nichol (eds), Wiley, New York, 1981, 173.
6. G. P. Todd and R. H. Haschemeyer, *Proc. Natl. Acad. Sci. USA*, 1981, **78**, 6739.
7. B. Demeler, J. Behlke and O. Ristau, *Method. Enzymol.*, 2000, **321**, 38.
8. H. R. Schwarz, H. Rutishauser and E. Stiefel, in *Numerik Symmetrischer Matrizen*, B. G. Teubnered (ed), Stuttgart, 1968.
9. P. Schuck and B. Demeler, *Biophys. J.*, 1999, **76**, 2288.
10. R. F. Steiner, *Arch. Biochem. Biophys.*,1954, **49**, 400.
11. D. J. Cox, *Arch. Biochem. Biophys.*, 1969, **129**, 106.
12. J. Behlke, K. Heidrich, M. Naumann, E. -C. Müller, A. Otto, R. Reuter and T. Kriegel, *Biochemistry*, 1998, **37**, 11989.
13. R. Golbik, M. Naumann, A. Otto, E. -C. Müller, J. Behlke, R. Reuter, G. Hübner and T. M. Kriegel, *Biochemistry*, 2001, **40**, 1083.
14. H. Fabian, K. Fälber, K. Gast, D. Reinstädler, V. V. Rogov, D. Naumann, D. F. Zamyatkin and V. V. Filimonov, *Biochemistry*, 1999, **38**, 5633.
15. A. J. Hubbard, L. P. Bracco, S. J. Eisenbeis, R. B. Gayle, G. Beaton and M. H. Caruthers, *Biochemistry*, 1990, **29**, 9241.
16. J. Behlke and O. Ristau, *Biochemistry*, 1997, **36**, 5149.
17. W. F. Stafford and P.J. Sherwood, *Biophys. Chem.*, 2004, **108**, 231.
18. M. L. Johnson and M. L. Faunt, *Method. Enzymol.*, 1992, **210**, 1.

CHAPTER 7

Correcting for the Buoyancy of Macromolecules: Density Increments and Apparent Partial Specific Volumes with Particular Reference to the Study of Membrane Proteins

P. J. G. BUTLER AND C. G. TATE

1 Introduction

An inherent problem with measuring the molecular mass of any macromolecule by sedimentation methods is the Archimedian buoyancy of the molecule, so that the buoyant mass is less than the macromolecular mass. In their classic introduction to ultracentrifugation, Svedberg and Pedersen[1] introduced the term $(1-\bar{v}\rho)$, where $\bar{v}$ is the partial specific volume of the macromolecule and ρ is the solvent density. This term acts as a correction factor between the macromolecular mass (M) and the buoyant mass, with $\bar{v}\rho$ as the buoyancy. It is particularly convenient since both $\bar{v}$ and ρ can be calculated by summation of terms and normalisation,[2–5] due to their being extensive properties, and this is very conveniently done using the program Sednterp.[6] Although there are problems with this approach, in that $\bar{v}$ is theoretically the volume occupied by a unit mass of the macromolecule *in vacuo* and ρ, as used, is the *solvent* density, rather than that of the solution which is displaced by the macromolecule, for simple systems it can give good measurements of the mass.

The situation is, however, much more complex with multicomponent systems, where significant solvation or co-solvation of the macromolecule may occur, and it becomes important to consider the apparent partial specific volume, which allows for selective binding of solvent components.[7] This becomes particularly important for membrane proteins, which are usually only soluble when released from the lipid membrane with detergent, resulting in replacement of many of the lipid molecules with detergent molecules, so that the protein exists in solution in a protein–detergent micelle.

The importance of solvation and other effects on the partial specific volume was recognised early in the development of the field,[2,8] yet too often, little was done about it. However, the increasing interest not only in membrane proteins, but also in other macromolecular systems where solvation or co-solvation may be significant, means that this effect is becoming more widely important in the analysis of data from analytical ultracentrifugation. Various techniques have been devised, either for obtaining suitable parameters to allow the determination of the macromolecular mass or for attempting to eliminate the effects of co-solvation, and we discuss them here, with a particular emphasis on membrane proteins as these often present the most extreme examples.

2   Density Increments and Apparent Partial Specific Volumes

2.1   Measurement and Use of Density Increments

2.1.1   Background

Casassa and Eisenberg[7,9,10] made a thermodynamic analysis of multicomponent systems, concentrating in particular on osmotic pressure, the refractive index increment and the partial volume of macromolecular solutes. These latter properties are directly relevant to Rayleigh light scattering and analytical ultracentrifugation. In this analysis, they considered three components: the pure solvent (component 1), the non-diffusible macromolecule (component 2), and diffusible solutes (components 3, *etc.*); which they considered in a condition of constant chemical potential of all diffusible components (*i.e.* effectively at dialysis equilibrium). In this discussion, we will confine ourselves to considerations relevant to the ultracentrifuge, but it is interesting and important to note the close relationship of these to osmotic pressure. This leads to the fact that the virial coefficients describing non-ideality are the osmotic virial coefficients, and also to the need for constant chemical potentials of all diffusible components, for the equations to apply strictly.

For analytical ultracentrifugation, the parameter is the density increment of the macromolecule $[(\partial\rho/\partial c_2)_\mu]$, which is defined (and measured) from the equation

$$\left(\frac{\partial\rho}{\partial c_2}\right)_\mu = \frac{(\rho_{\text{solution}} - \rho_{\text{solvent}})}{c_2} \qquad (1)$$

where ρ is the density, c_2 the concentration and numerical (and word) subscripts indicate the component being considered, while the partial derivatives (and subscript μ) indicate the condition of constant chemical potential. In principle, Equation (1) should be evaluated as $c_2 \rightarrow 0$, but in practice c_2 is often small enough that any error is negligible – in other cases $(\partial\rho/\partial c_2)_\mu$ can be measured at various concentrations and extrapolated to $c_2 = 0$.

Importantly, Casassa and Eisenberg[7] also showed that

$$\left(\frac{\partial\rho}{\partial c_2}\right)_\mu = (1 - \phi'\rho_0) \qquad (2)$$

where ϕ' is the 'apparent' partial specific volume of component 2, under these specific solvent conditions, and ρ_0 the solvent density.

Equation (2) shows that the classic term $(1 - \bar{v}\rho)$ in the sedimentation equations can be substituted by the fundamental parameter $(\partial\rho/\partial c_2)_\mu$ and that this will take into account all solvation effects.

2.1.2 Sedimentation Equilibrium

The equation describing the apparent, weight-average molecular mass of the macromolecule $(\overline{M}_{w,app})$ can be reformulated as

$$\overline{M}_{w,app} = \frac{\mathrm{d\ln}\, c_2}{\mathrm{d}r^2} \frac{2RT}{(\partial\rho/\partial c_2)_\mu \omega^2} \tag{3}$$

where r is the radius, ω the angular velocity (in radians^{-1}) and R and T are the gas constant and temperature (in K), respectively. Since $\overline{M}_{w,app}$ is the apparent mass (allowing for the effects of non-ideality and aggregation), this formulation can also be applied in all the other equations relating the concentration distribution to radius and weight-average mass, allowing one to use similar forms to fit the original data with different models,[11] whether these are for non-ideality or aggregation.

The importance of this formulation is that the molecular mass obtained is strictly that of the component whose concentration is measured as c_2 when determining the density increment.[7] This is perhaps most easily understood by remembering the close relationship between analytical ultracentrifugation and osmotic pressure, with the latter depending upon the number of molecules (or moles) per millilitre.[10] Once the calculation is thought of in these terms, it is obvious that the mass obtained will be related directly to the way the concentration is entered into the equation, and concentrations of protein alone, *e.g.* determined by amino acid analysis, will give the mass of the protein irrespective of the solvation or co-solvation state.

2.1.3 Sedimentation Velocity

By a similar process, the Svedberg equation[1] becomes[10]

$$M_2 = \frac{s}{D} \frac{RT}{(\partial\rho/\partial c_2)_\mu} \tag{4}$$

where M_2 is the molecular mass of a single component 2, s its sedimentation coefficient and D its diffusion coefficient. A similar formulation can again be extended to all the related equations, including those used in applications such as the DCDT+[12] and SEDANAL[13] (indeed SEDANAL v3.45 now takes $(\partial\rho/\partial c_2)_\mu$ as a parameter).

2.2 Alternatives to Measurement of Density Increments

2.2.1 Calculation of Apparent Partial Specific Volumes and Density Increments

An alternative approach to the measurement of density increments, which had an apparent appeal in allowing the traditional formulation of the equations in terms of partial specific volumes, was developed by Tanford and Reynolds.[14,15] They showed, from Casassa and Eisenberg,[7] that the apparent partial specific volume could also be calculated from those of the separate components in a micelle and derived the equation

$$(1 - \phi'\rho_0) = (1 - \bar{v}_P\rho_0) + \delta_D(1 - \bar{v}_D\rho_0) + \delta_L(1 - \bar{v}_L\rho_0) \tag{5}$$

where the generic term $\bar{v}_C$ is the partial specific volume of component C (either protein (P), detergent (D), or lipid (L)) and δ_C is weight fraction of component C in the micelle, compared to the protein.

Eisenberg has since pointed out[10] that this is equivalent to taking the density increment as a proportional sum of those of the components, although for convenience he only expressed this for a single co-solvent species. However, his equation can be readily generalised as:

$$\left(\frac{\partial\rho}{\partial c_2}\right)_\mu = (1 - \bar{v}_2\rho_0) + \sum_{i=3}^{n+2}\xi_i(1 - \bar{v}_i\rho_0) \tag{6}$$

where n is the number of co-solvent components and ξ_i is given by

$$\xi_i = \left(\frac{w'_i - w_i}{w_2}\right) = \left(\frac{\partial w_i}{\partial w_2}\right)_\mu \tag{7}$$

i.e. it is an interaction parameter indicating the change in gram molality (w_i) of component i with the change in gram molality of component 2 at constant chemical potential of solvent and co-solvent components. To a reasonable approximation this is equal to the weight ratio of component i to component 2.

2.2.2 Density Matching of Co-solvent Component

Before the development of modern instruments for measuring the density of small volumes of liquid with high precision, such as the Paar density-meter,[16] it could be difficult to determine $(\partial\rho/\partial c_2)_\mu$ directly, and there still remains the problem of establishing a constant chemical potential (*i.e.* 'dialysis equilibrium') for all 'diffusible' solvent and co-solvent components between the solvent and the macromolecular solution when many detergents are used to solubilise a membrane protein (see discussion below in Section 4). Even greater difficulties occur in measuring values for ξ_i, since this not only still requires the condition of 'dialysis equilibrium' so that the amount of co-solvent components binding to the macromolecule can be determined from differences in concentration in the solution and solvent, but also requires a method for measuring the concentration of molecules like detergent or lipid.

One means of avoiding these problems, for detergent binding, was suggested by Reynolds and Tanford,[15] who pointed out that if the solvent density is adjusted to meet the condition

$$\rho_0 = \frac{1}{\bar{v}_D} \tag{8}$$

the term for detergent in Equation (5) becomes zero and, provided the amount of lipid (or any other additional co-solvent component) is negligible, the partial specific volume of the protein alone ($\bar{v}_p$) can be substituted for ϕ' and there is no need to know the amount of detergent bound to the protein. This technique is known as 'density matching' and it has been widely employed when investigating the aggregation of detergent-solubilised membrane proteins, although often without any apparent consideration of the possible effects of bound lipid, even though Reynolds and Tanford[15] had pointed out the importance of this when they originally discussed the method. Two main methods of density matching are available, together with a combination of them. The solvent density can be increased, by addition of $H_2^{18}O$, D_2O, or $D_2^{18}O$ to the normal H_2O in the solvent, to obtain a density match to the detergent. One additional complication of this technique is the need, for accuracy, to include the effects of partial deuteration of the macromolecule on its partial specific volume. Alternatively, one of a limited number of rather specific detergents can be chosen, which have densities very close to water and therefore display neutral buoyancy. (Possible problems with these detergents are discussed in Section 3 below.)

As an alternative to modifying the water density (*i.e.* component 1), small molecules, such as glycerol or sucrose, have been used to increase the solvent density. However, such components have to be considered as further co-solvent components and possible binding/exclusion in the solvation of the macromolecule considered. Such effects have been established for a long time, *e.g.*, the differential binding of H_2O to the coat protein of tobacco mosaic virus, in some aggregation states, in a sucrose solution was measured by weight differences at dialysis equilibrium and shown to be ~ 26 mol/mol protein,[17] while, unsurprisingly, the converse effect was seen when DNA was compared in Na^+ and Cs^+ solutions.[18] In the first case, solvation is occurring to the exclusion of the co-solvent, while in the second, co-solvent is binding. When the water density is varied, it is generally assumed that exclusion will not occur, since even $D_2^{18}O$ is not very dissimilar to H_2O, and the need for correction for partial deuteration has already been mentioned. While the magnitude of these effects is a matter of continuing discussion,[19,20] it is clear that this will increase with increasing concentrations of density medium and so may well become significant during density matching.

3 The Problem with Membrane Proteins and the Choice of Detergent

3.1 General Considerations

Integral membrane proteins exist in a bipartite world; part of the protein is hydrophobic and lies in the centre of the membrane and this transmembrane portion is sandwiched between two hydrophilic regions that extend into the aqueous environment on each side of the membrane. Purification of the protein therefore requires the disruption of the membrane by detergents and the presence of detergents throughout all the purification steps to maintain the membrane protein in a native conformation. Thus the detergent's role is to mimic the lipid bilayer immediately around the membrane protein, but, instead of forming large molecular bilayers, the detergent forms a discrete micelle around the hydrophobic portion of the protein. It is this protein–detergent complex that sediments during centrifugation or migrates in size-exclusion chromatography, and this complex is invariably much larger than the membrane protein alone. For example, the *Escherichia coli* small multidrug transporter EmrE was purified in the detergent *n*-dodecyl-β-D-maltoside (DDM);[21] the purified protein migrated as a 137 kDa complex on size-exclusion chromatography which was due to the presence of 218 molecules of DDM and five lipid molecules associated with the EmrE dimer (M_r: 30.4 kDa). Lipid molecules will invariably be associated with any purified membrane protein, unless the purification was performed under denaturing conditions, *e.g.* in organic solvents, or using harsh detergents like sodium dodecyl-sulphate (SDS) (see below). In fact, many membrane proteins will become inactive if these lipid molecules are removed,[22] which can happen during purification or if the wrong detergent is used.

The choice of detergent for purification and analytical ultracentrifugation has to be determined empirically and it is entirely possible that different detergents may be required at different steps in the purification process and analysis. The over-riding concern is to maintain the membrane protein in a functional state, but this can be difficult to ascertain once the protein is in detergent. Many membrane proteins that are involved in photosynthesis or the mitochondrial electron transport chain contain chromophores that are essential for function and have characteristic absorbance spectra that change if the protein becomes inactive or loses some of its subunits. Many receptors have a well-defined pharmacological profile of inhibitors and ligands that bind tightly to the protein, and this can be used as a diagnostic for functionality during purification (see, for example, refs. 23 and 24); it is reasonable to assume that if the three-dimensional structure of a binding site is maintained, then the rest of the protein's structure that forms the scaffold for the binding site is also maintained. However, for many membrane proteins there is no simple, direct assay to determine whether the protein is in a native conformation in detergent and caution must be exercised in interpreting any biophysical data from such preparations.

3.2   Choice of Detergent

There are now hundreds of detergents commercially available and they span the complete spectrum from 'harsh' (*i.e.* likely to denature or inactivate a membrane protein) to 'mild' (*i.e.* likely to maintain the native membrane protein structure). The general rule of thumb is that harsh detergents have a small hydrophilic head group and a short aliphatic chain, whereas mild detergents have large hydrophilic head groups and large hydrophobic tails. Charged head groups make a detergent more denaturing than uncharged head groups, and hydrophobic tails that are polyaromatic hydrocarbons make a detergent milder than if the hydrophobic region is a simple alkyl chain. SDS is regarded as a very harsh detergent and there are only a very few membrane proteins whose structures are unaffected by it, *e.g.* the *E. coli* outer membrane protein OmpA.[25] In contrast, DDM, which differs from SDS only in that the sulphate head group is changed to the disaccharide maltose, is regarded as a mild detergent and has been used extensively for the purification of receptors and transporters (see, for example, refs. 26–28). Octylglucoside (OG) could be regarded as a detergent with intermediate properties; in comparison to DDM, OG has a shorter alkyl chain (8 carbon *vs.* 12 carbon) and the headgroup is half the size (glucose as opposed to maltose) and OG is therefore harsher than DDM. Some of the mildest detergents such as digitonin and *N,N*-bis(3-D-gluconamidopropyl)cholamide (bigCHAP) have large head groups composed of sugars and polyaromatic hydrocarbon tails. Unfortunately, the choice of a detergent is usually not as simple as choosing the mildest detergent available. Some transporters prefer polyoxyethylene detergents as opposed to the alkylmaltoside series, and some detergents will not solubilise the protein from the membrane, but they may be good for steps later in the purification. In addition, the right buffer composition and salt concentration are vital for the stability of many membrane proteins. Finally, the degree of purification is also important; for example, the use of too many ion-exchange chromatography steps can inactivate a membrane protein,[29] possibly due to the removal of too many lipid molecules from the hydrophobic domain. As can be appreciated, this multi-factorial problem makes the purification of membrane proteins challenging.

3.3 Preparation of Sample

Once the membrane protein is purified, the sample has to be made suitable for analytical ultracentrifugation. This is again made problematic due to the presence of detergents. If after purification a protein is too dilute for analysis, the normal procedure is to concentrate the protein using any one of a number of commercial concentrators containing a membrane of defined molecular weight cut-off that retains the protein upon centrifugation while allowing the passage of buffers and salts. For soluble proteins, there is no difficulty with this technique because the protein is the only macromolecule in solution. In contrast, concentration of a membrane protein invariably results in the concentration of the detergent. When a detergent is above its critical micellar concentration (CMC), which is usual during membrane protein purification, the detergent molecules self-assemble to form micelles (reviewed in ref. 30) that can often contain 40–150 molecules per micelle, which translates to an apparent M_r of 20–100 kDa. Thus, detergent micelles are often concentrated along with the membrane protein and, consequently, this results in a change of density of the solution, which of course must be accurately defined for the analysis of analytical centrifugation data. The situation is particularly difficult if a density measurement is to be performed on the purified membrane protein to determine its density increment. For these measurements, the membrane protein needs to be at a concentration of at least 1 mg mL^{-1}, and preferably much higher for accurate measurements. A requirement for this measurement is that the protein solution is effectively at dialysis equilibrium with all the solvent components, including the detergent. Unfortunately, the rate of dialysis of many mild detergents is extremely slow (days to weeks), because the CMC of many detergents is low. The CMC approximates to the effective concentration of detergent monomer in solution, and it is only the monomeric detergent (if the micelle is large) that can freely diffuse across a dialysis membrane, thus, the lower the CMC, the slower the rate of dialysis. This problem can be avoided by eluting the protein from, for example, an ion-exchange column using a step gradient of high salt so that the protein is eluted in the minimum volume.[21,26] Alternatively, a detergent that forms a small micelle such as deoxycholate could be used; the protein–detergent complex can then be separated from detergent micelles by size-exclusion chromatography.[31] If the detergent has a high CMC and the protein is sufficiently stable, then dialysis over a period of days will be possible.

4 Examples of Specific Applications for Proteins in Multi-Component Systems or with Unusual Compositions

No attempt will be made to give a comprehensive survey of this field, as it is clearly not possible within the space of a single chapter. However, a number of examples have been chosen, to illustrate the use of each of the methods listed above for detergent-solubilised membrane proteins, and also to show how direct measurement of the density increment enables proteins with unusual compositions, or in high concentrations of urea, to be characterised.

4.1 Density Matching

4.1.1 Proteins in Protein–Detergent Micelles

In their initial description of density matching, Reynolds and Tanford[15] used the technique to determine the molecular mass of the AI apoprotein from high-density serum lipoprotein, solubilised in either $C_{12}E_8$ [$C_{12}H_{25}(OCH_2CH_2)_8OH$] or Lubrol WX. In this case, they did not simply use single solvent densities to match those of the detergents, but rather looked at the dependence of the apparent $M_P(1 - \phi'\rho_0)$ upon solvent density and, from Equation (5), took $\phi' = \bar{v}_P$ when the density matched the detergent. They found that the protein was monomeric under these conditions.

In the other examples which we give, the detergent was chosen to match the solvent density. The oligomerisation state of cytochrome c oxidase, was determined after solubilisation in octyltetra/pentaoxyethylene (C_8E_4), a detergent chosen because its density is very close to that of water.[32] This gave the result of a heterodimer, with a single large and small subunit, under these conditions.

A more detailed study was of the dimerisation of the transmembrane α-helix from glycophorin A, as a fusion protein with staphylococcal nuclease, solubilised with octyloxyethylene dodecyl ether (C_8E_5).[33,34] In this study, the authors were able to obtain data on the monomer–dimer equilibrium and to estimate the dissociation constant (K_d) as 240 ($\pm$ 50) nM, while point mutants showed lower affinity, with higher dissociation constants.

4.1.2 Measuring Detergent or Lipid Densities

Another slant on the use of density matching has been to measure the partial specific volume of detergents or lipids. The density of DDM was estimated by sedimentation equilibrium of the detergent in solvents with varying D_2O:H_2O ratios and extrapolation of the solvent density to the point where the detergent did not sediment, giving a partial specific volume for DDM of 0.814 mL g^{-1}.[26] Similarly the partial specific volume of phosphatidylcholine in vesicles was measured by the use of density matching, with varying ratios of $D_2O:H_2O$, and interpolation to the density at which there was neither sedimentation nor flotation.[35] This gave $\phi'=0.9814$ mL g^{-1} in water.

4.2 Calculation of Apparent Partial Specific Volume

Calculation of the apparent partial specific volume by weighted summation of those of the components of the complex has been used in a number of cases, to determine the molecular mass of transport proteins solubilised from the membrane in detergent.

The lactose transport protein of *Streptococcus thermophilus* (LacS) was studied, solubilised with DDM, and shown to act essentially as a stable heterodimer of the 51 kDa integral membrane domain and the 19 kDa regulatory domain, with this ~70 kDa dimer dimerising further, with a monomer–dimer association constant (K_a) of 4.4 mL mg^{-1}.[26,36] In the latter paper, the authors also looked at the xyloside transporter from *Lactobacillus pentosus* (XylP), again in DDM, and again concluded that it was present in a monomer–dimer equilibrium, with K_a of 62 mL mg^{-1} and therefore largely dimeric under the experimental conditions.[36]

A study of the ammonium transporter from *E. coli* (AmtB), again solubilised in DDM, showed that it is trimeric, with no evidence of dissociation under the conditions employed.[37]

4.3 Obtaining the Condition of 'Dialysis Equilibrium'

A major technical problem, with many detergents, is obtaining the condition of 'dialysis equilibrium' or, more strictly, constant chemical potential of all solvent and co-solvent components since, unlike small co-solvent molecules, neither detergent molecules nor lipids pass readily through dialysis membranes, due to their tendency to form micelles with high molecular mass and dimensions. Equilibration will only be rapid if the CMC of the detergent is high (>20 mM), but even then this may take days to complete (see Section 3).

In some other cases, when the protein forms a micelle with the detergent which is sufficiently larger than that of the detergent alone, these can be separated by size-exclusion chromatography, with the column equilibrated with buffer containing the desired detergent concentration. Under these circumstances, the protein–detergent micelles elute ahead of any disturbed detergent concentration 'peak' due to excess (or too low) detergent concentration in the applied sample from solubilisation, and in the equilibrating buffer. This was the case with the major intrinsic protein from bovine eye lens (MIP).[31] The main problem even in such cases is avoiding too much dilution of the applied sample, so that the final concentration is sufficiently high for accurate measurement of differences in either density or component concentrations, compared with the solvent.

Another method is binding of the solubilised protein to an affinity matrix, *e.g.,* by having a His-tag and binding this to a Ni^{2+}-nitriloacetic acid (Ni-NTA) resin, and then washing into the desired detergent concentration. The protein–detergent micelles can then be eluted with a high concentration of a small eluting agent dissolved in buffer with the correct free detergent concentration, and the eluting agent can be removed by dialysis, again keeping the detergent concentration constant. This method has a further advantage in allowing sample concentration as well as detergent and lipid equilibration. Examples of matrix:eluant combinations are ion-exchange resins with high salt concentrations, for any sufficiently charged protein with a neutral detergent, and Ni-NTA matrices with high imidazole concentrations, for proteins which have a His-tag. Both methods were successfully used with the multidrug transporter from *E. coli*, EmrE.[21]

4.4   Measurement of Density Increment

4.4.1   C-reactive Protein from Limulus polyphemus

The C-reactive protein (CRP) from *Limulus polyphemus* (the horseshoe crab) is a member of the 'pentraxin' family of proteins, so called because of their five-fold symmetry. However, despite showing every other property which matched other pentraxins, electron micrographs of CRP from *L. polyphemus* were unclear and could not be readily interpreted. We therefore carried out a study of the subunit composition of this protein,[38] since the early evolutionary state of *L. polyphemus* made it of particular interest.

A major complication in a study of CRP is that the protein is not only highly glycosylated, with variable sugars at the various sites, but it is also polygenic and so does not have a unique amino acid composition when prepared from the hemolymph of the horseshoe crab. However, the monomeric masses of all the protein species could be measured by electrospray mass spectroscopy, and they were found to be within the range of 25–25.5 kDa, containing ~ 1 kDa of carbohydrate. In view of the uncertainty of the exact composition of the purified protein, it was decided that the most reliable method of obtaining the mass of the aggregate was sedimentation equilibrium, with an experimentally determined density increment (since this would inherently be a weight average of those for individual components). Measurement of the protein concentration was also complicated by the heterogeneity, but this was done by interferometry since all species would be expected to have a refractive increment of 0.19 mL g^{-1}.[39] The density increment was found to be 0.273 (± 0.003) (giving $\phi' = 0.722$ mL g^{-1}). The molecular mass of the aggregates was found to vary between ~ 150 and ~ 300 kDa, corresponding to a mixture of 6-mer and 12-mer aggregates. In view of the heterogeneous protein, no attempt was made to fit the data more accurately, but in the light of the structures of other pentraxins it was possible to conclude that this CRP is based on a hexameric assembly, rather than the pentameric one found in vertebrates.[38]

4.4.2   Yeast Prion Protein Determinant Ure2

In bakers' yeast (*Saccharomyces cerevisiae*) there is a non-Mendelian inheritance of control of the ability to utilise ureidosuccinate in the presence of ammonia,[40] which has been shown to be due to a prion form of the chromosomally inherited protein Ure2.[41,42] Genetic studies showed that the first 65 residues of the N-terminus of Ure2 protein are required for expression of the prion phenotype[43] and this has been shown to correspond to aggregation of the Ure2 protein.[44] We were therefore interested in analysing the aggregation states of Ure2 protein and various truncation fragments containing N-terminal sequences.[45]

Size-exclusion chromatography and sedimentation velocity experiments showed that both full-length His-tagged Ure2, and various deletion mutants, exist as a monodisperse species, with $s_{20,w} \approx 4.3$–4.7, and apparently dimeric. (Full-length Ure2, with no His-tag, was highly aggregated.)

We therefore chose the most soluble deletion mutant, Δ15-42Ure2-His, for more careful analysis by sedimentation equilibrium. Because of the unusual properties of prion proteins, we could not safely calculate the partial specific volume, and therefore measured the density increment. This was found to be 0.223, corresponding to $\phi' = 0.771$ mL g^{-1} which is high for a typical protein. Using the measured value, Δ15-42Ure2-His was found to be a non-ideal dimer.

4.4.3 Bovine F$_1$-ATPase Inhibitor Protein IF$_1$

The activity of the proton-pumping ATP synthase (F$_1$F$_0$-ATPase) of mitochondria is regulated by ADP, the proton motive force (ΔμH$^+$) and a natural inhibitor, IF$_1$.[46] Loss of ΔμH$^+$, for example, following ischaemia, can reverse the action of ATP synthase, allowing it to hydrolyse ATP and concomitantly pump protons out of the mitochondria, and this activity is inhibited by IF$_1$. IF$_1$ binds to the F$_1$ component of the ATP synthase at pH values below neutrality but not above (p$K_a \sim 6.5$).[47] We were therefore interested in investigating the oligomerisation states of IF$_1$ at both acid and alkaline pH values, and of a mutant (IF$_1$-H49K) which eliminates the pH sensitivity and is active over a wide range of pH.[48]

One potential complication was the presence of a series of heptad repeats, which made it likely that IF$_1$ might fold into a coiled-coil and therefore have an unusual density increment. We therefore measured the density increment for IF$_1$ at the pH values used for analysis and the results are shown in Table 1. In practice, the values obtained at each pH are very similar, despite a range of protein concentrations, confirming an absence of concentration effect, and correspond to $\phi' \approx 0.760$ mL g^{-1}, which is higher than might be expected for a typical protein.

Applying these measured density increments to sedimentation equilibrium data obtained with IF$_1$ at pH values 5.0, 6.5 and 8.0 gave the results shown in Figure 1.[48] At all pH values, IF$_1$ appears as a single aggregate, with the smallest size at pH 5.0 where the aggregate corresponds to a dimer of the chemical monomer ($M_1 \approx 9580$ Da).[49] The data at pH 5.0 were well fitted to the model for a non-ideal 'monomer', where this was taken as an actual dimer, with the residuals small and random. At higher pH values, this dimer dimerises further, to give tetramers, which we hypothesised might lead to

Table 1 *Measurement of density increment for bovine IF$_1$ at various pH values*

pH	ρ_0 (g mL^{-1})	$\Delta\rho$ (mg mL^{-1})	c_2 (mg mL^{-1})	$\left(\dfrac{\partial\rho}{\partial c_2}\right)_\mu$	ϕ' (mL g^{-1})
5.0	1.002128	3.422	14.32	0.239	0.759
6.5	1.001527	2.510	10.80	0.232	0.766
8.0	1.000957	6.179	25.80	0.239	0.760

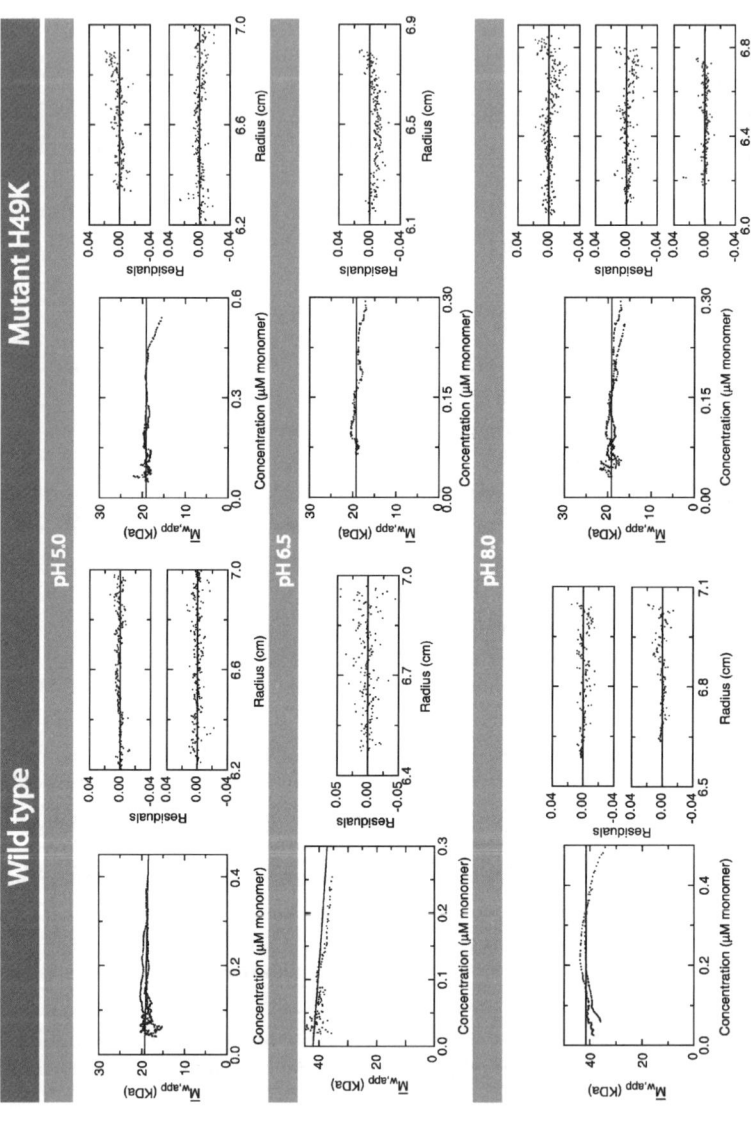

Figure 1 *Sedimentation equilibrium analysis of bovine IF$_1$ (wild-type and mutant H49K). Analyses were performed at pH 5.0, 6.5 and 8.0 and ionic strength 100 mM. Plots show $\overline{M}_{w,app}$ against concentration (as molarity of monomer) and the residuals (against radius) from fitting a model for a non-ideal 'monomer' to the scan data. The 'monomer' employed was a dimer (wild type pH 5.0, H49K at all pH values) or a tetramer (wild type pH 6.5 and 8.0) (Adapted from ref. 48, with permission from the American Society for Biochemistry and Molecular Biology.)*

inactivation of IF_1. To test this hypothesis, we looked at the mutant IF_1-H49K, which remains fully inhibitory over the whole pH range. This was found to remain dimeric at all pH values (Figure 1), compatible with the hypothesis that the IF_1 dimer is active and the tetramer inactive.

Subsequently it was shown that IF_1 will dimerise the F_1-ATPase[50] and also, when its structure was solved, that it exists as a coiled-coil dimer,[51] explaining the complete lack of dissociation once the newly synthesised monomers have dimerised.

4.4.4   Escherichia coli *Multidrug Transporter EmrE*

The *E. coli* multidrug transporter EmrE is a member of the family of small multidrug transporters found in many bacteria, which transport a range of bacteriocidal drugs out of the bacterium, enhancing its drug resistance – which is a major clinical problem. EmrE is only 110 amino acids long, and analysis suggests that it comprises four transmembrane α-helices, with very small domains, at the termini or in the loops, on either side of the membrane. This is in contrast to most other bacterial solute transporters, which usually have $\geqslant 10$ transmembrane helices, and leads to the question of whether it acts as an obligate oligomer and, if so, how large.

EmrE extracted from membranes of *E. coli*, overexpressing EmrE, with DDM was shown to crystallise, with a minimal functional unit of dimer, and, importantly, to bind substrate (tetraphenylphosphonium; TPP^+) both in solution and in the crystal.[24,52,53] However, there is also a reported X-ray structure of EmrE in which it appears as a dimer of dimers, suggesting that it is a tetramer in detergent,[54] although there is no information about substrate binding for these preparations. This uncertainty about the oligomerisation state made its determination in solution of particular interest and this was done on both 'delipidated EmrE' and a so-called 'high-molecular weight (MW) EmrE' preparations, where this latter had been extracted as gently as possible from the membranes, to minimise loss of any firmly bound lipid.[21]

The small size of EmrE ($M_1 = 15.192$ kDa for the His_6-tagged protein), together with the minimal domains outside the membrane, results in a very high ratio of bound detergent and so it is important to determine the density increment for the detergent-solubilised protein. The results of direct determinations of this for both delipidated and high-MW EmrE are shown in Table 2.

The measured density increments are 3–4 times those which would be expected for EmrE alone, based on a partial specific volume for the protein ($\bar{v}_2$) calculated with Sednterp[5] from the amino acid composition, and so we decided to calculate the expected values, using Equations (6) and (7).[10] This requires knowledge of the partial specific volumes of, and interaction parameters (ξ_n) between, all relevant components, *i.e.* EmrE (component 2), DDM (component 3) and the lipid (component 4) – in practice, analysis of the lipid showed that this was almost entirely phophatidylethanolamine (PE). The interaction parameters (ξ_3 and ξ_4; see Table 3) were measured from the weight ratios of DDM and PE to EmrE. The partial specific volumes of DDM and PE were taken from the literature; 0.814 $mL^{-1}g$ for DDM[26, 55] and the temperature-dependent values shown in Table 3 for PE.[56] Using these values, we calculated density increments (Table 3) of 0.885 for delipidated EmrE and 1.002 for high-MW EmrE. The

Table 2 *Measurement of density increments for delipidated and high-MW EmrE*

	Delipidated EmrE (20 °C)	*High-MW EmrE (5 °C)*
Solvent density (ρ_0)	1.003936 ± 0.00002 g mL^{-1}	1.005514 ± 0.00002 g mL^{-1}
Solution density (ρ)	1.005119 ± 0.0004 g mL^{-1}	1.007033 ± 0.00003 g mL^{-1}
Density difference ($\Delta\rho$)	1.18 ± 0.06 mg mL^{-1}	1.52 ± 0.05 mg mL^{-1}
Protein concentration (Δc)	1.23 ± 0.05 mg mL^{-1}	1.33 ± 0.05 mg mL^{-1}
Density increment ($\partial\rho/\partial c_2)_\mu$	0.96 ± 0.09	1.14 ± 0.08
Apparent partial specific volume (ϕ')	0.040 ± 0.004 mL g^{-1}	-0.14 ± 0.08 mL g^{-1}

Table 3 *Calculation of density increments for delipidated and high-MW EmrE*

	Delipidated EmrE (20 °C)	*High-MW EmrE (5 °C)*
Weight ratio DDM/EmrE (ξ_3)	3.52	3.8
Weight ratio PE/EmrE (ξ_4)	0.13	1.25
Partial specific volume of PE ($\overline{v}_4$)[56]	0.946 mL g^{-1}	0.932 mL g^{-1}
Solvent density (ρ_0)	1.003936 g mL^{-1}	1.005514 g mL^{-1}
Density increment ($\partial\rho/\partial c_2)_\mu$	0.885	1.002

calculated value for delipidated EmrE is similar to the measured value, and the small difference may be due to small errors in the mass of detergent and lipid bound to EmrE. The calculated density increment of the high-MW EmrE preparation is also similar to the measured value, although the discrepancy is slightly bigger than for the delipidated EmrE preparation. This difference may have arisen because the high lipid content of this preparation resulted in anomalous behaviour which led to a lower estimate of the density increment, perhaps because it was not truly at dialysis equilibrium; certainly this preparation behaved anomalously during velocity ultracentrifugation (see below) for uncertain reasons. The measured density increments and solvent densities were used for analysis of the analytical sedimentation results.[21]

The results of sedimentation equilibrium runs for the two preparations are shown in Figure 2.[21] In each case, the data are best fitted by an ideal monomer–dimer equilibrium, with acceptable residuals from the fitting and plots of $\overline{M}_{w,app}$ against concentration overlying for cells loaded at different starting concentrations. The best estimates for the monomer masses, using the measured density increments, are 15.2 ± 0.15 kDa for delipidated EmrE and 19.25 ± 0.8 kDa for high-MW EmrE, at 20.0 and 5.0 °C, respectively, which are in reasonable agreement with the true monomer mass of 15.192 kDa (from the amino acid composition). It is noticeable that the least-accurate value is that for the more highly lipid-containing EmrE, showing that even the unusually high value measured for the density increment is still rather low. The fitting also gives best estimates for K_d of 2.5 ± 0.5 and 10 ± 2 μM, respectively. These estimates were used for plotting the curves shown in Figure 2.

The interest in characterising these materials arises from the observation that both will bind TPP$^+$, with a molar ratio of 1:2 TPP$^+$:EmrE[21] – which is of course fully compatible with the structure which has been shown to be an asymmetric dimer.[53] There is, however, the question whether substrate binding might cause higher

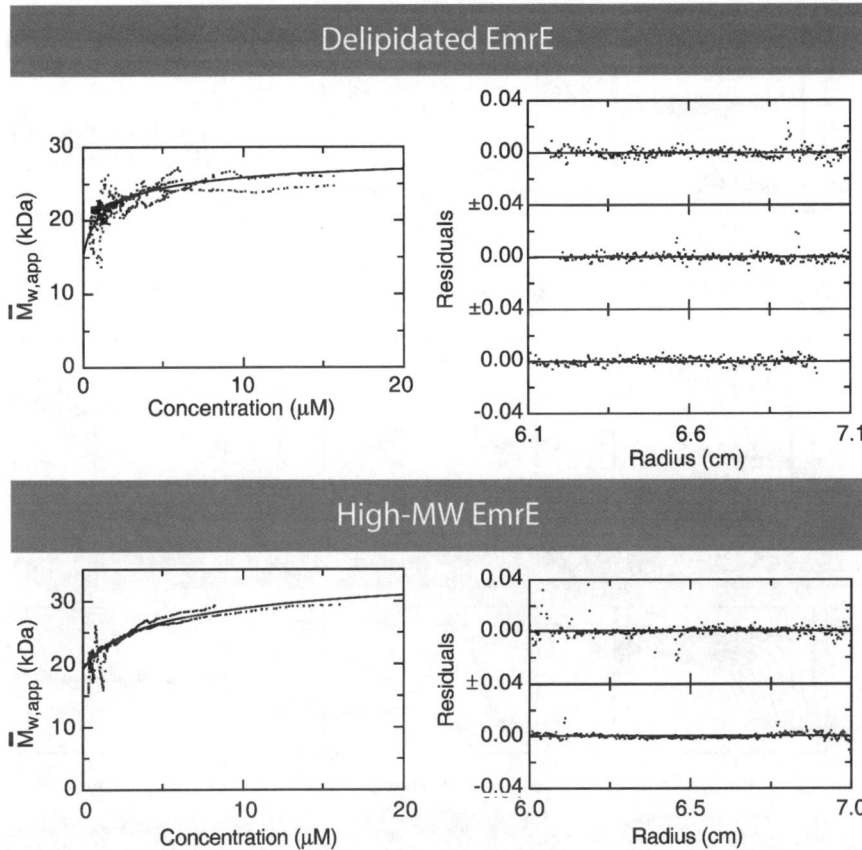

Figure 2 *Sedimentation equilibrium analysis of delipidated EmrE (at 20 °C) and high-MW EmrE (at 5 °C). Left panels: $\overline{M}_{w,app}$ against total monomer concentration for points from separate cells plotted together, together with lines corresponding to the curve for an ideal monomer–dimer equilibrium. Right panels: The residuals from fitting the raw data with this model, against radius for individual cells (Reprinted from ref. 21, with permission from Elsevier.)*

oligomerisation and so we compared both preparations with and without TPP$^+$, using sedimentation velocity. The results are shown in Figure 3,[21] with original scans and also the results of analysis with the application DCDT+.[12] Delipidated EmrE was found to sediment as a single component, with or without TPP$^+$, with $s_{20,w}=4.87\pm0.444$ and 4.90 ± 0.003 S and molecular masses of 26.5 ± 0.3 and 25.8 ± 0.3 kDa, respectively. High-MW EmrE shows a less symmetric peak of $g(s^*)$ and, agreeing with this loss of symmetry, it was necessary to fit the data with a model for two components. With or without TPP$^+$, the main component (>75% of the optical density) had $s_{20,w}=5.25\pm0.02$ and 5.25 ± 0.02 S and molecular masses of 21 ± 2 and 22 ± 2 kDa, respectively. The components corresponding to ~25% of the absorbance have $s_{20,w}=6.9\pm0.5$ and 7.3 ± 0.3 S and molecular masses of 9 ± 5 and 9.5 ± 5 kDa – both imprecise and highly unlikely values which are almost

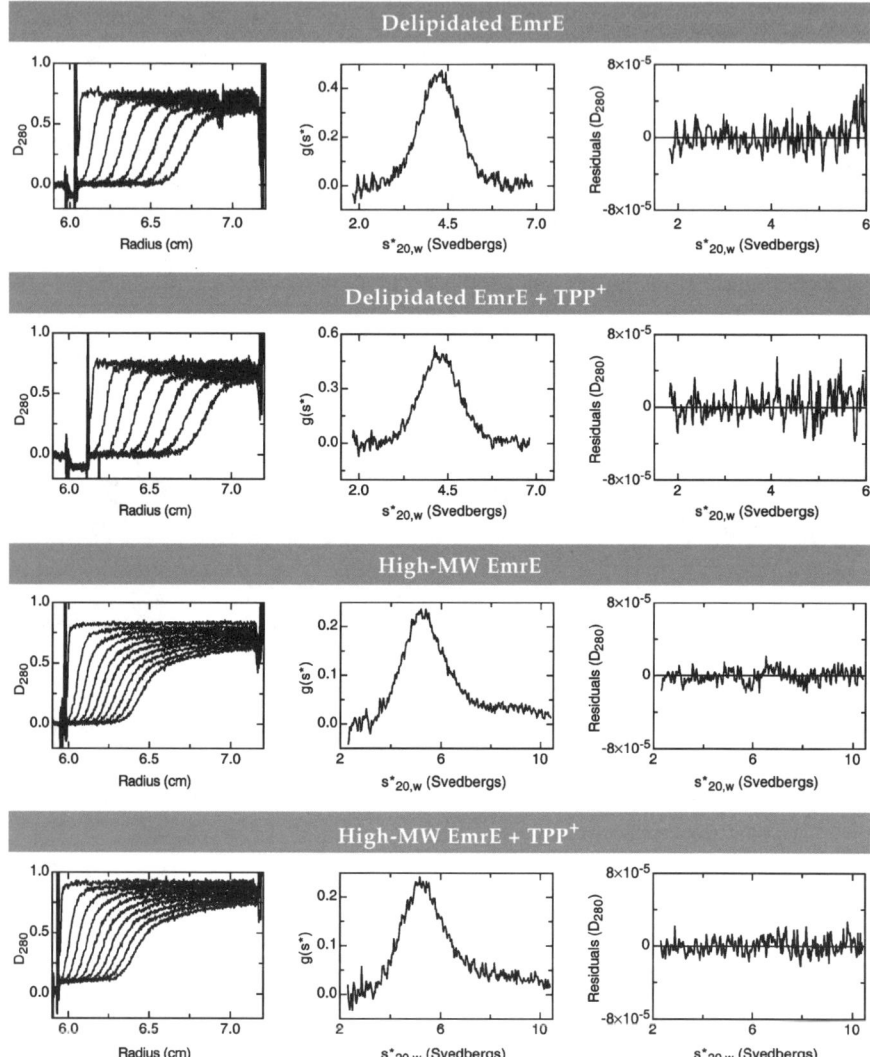

Figure 3 *Sedimentation velocity analysis of delipidated EmrE (at 20 °C) and high-MW EmrE (at 5 °C), with and without TPP⁺. Left panels: sample scans at equal time intervals (selected to be spaced apart); middle panels: Plots of g(s*) against s*; right panels: residuals for fitting Gaussian functions to dc/dt vs. s curves (these latter two pairs of panels correspond to the same 10 scans for each sample) (Reprinted from ref. 21 with permission from Elsevier.)*

certainly an artefact of the calculation in which the diffusion coefficient is estimated from the spreading of the boundary with time, and then used in an s/D calculation to obtain the molecular mass. If the boundary spreads from any cause other than diffusion (*e.g.*, due to some polydispersity), D will be overestimated and therefore too low a mass will be calculated. This has probably occurred in this case, and the polydispersity could be due to some slight aggregation and/or further lipid binding. Such boundary spreading, due to polydispersity, probably also explains the slightly lower molecular mass determined from velocity rather than equilibrium sedimentation.

The absence of evidence for monomeric EmrE in the velocity runs is not surprising, in view of the loading concentrations for the samples and the K_d values found in the equilibrium runs. Moreover it is not clear how rapid the association/dissociation rates will be under the conditions of detergent solubilisation – aggregation equilibration may occur in an equilibrium but not a shorter velocity run. Similarly, the nature of the minor component in the preparation of high-MW EmrE is unclear and it could be some kinetically trapped material.

The overall conclusion is clear that, under both these conditions of preparation, EmrE shows substrate binding and is dimeric, with no further association brought about by the substrate. The high-density increments show that without proper determination of the values to use any calculated molecular mass, and hence aggregation state, would have been in significant error.

5 Overall Conclusions

Analytical ultracentrifugation is often the system of choice for studying the oligomeric state of membrane proteins and three different solutions are described here to account for the detergent bound to membrane proteins during data analysis (measurement of the density increment, density matching, calculation of the density increment by summation). Each of the methods when applied appropriately gives an adequate answer to the question 'What is the oligomeric state of my purified membrane protein?' However, density matching may not be possible if the membrane protein is not stable in, for example, C_8E_5. The use of small osmolytes for density matching other detergents may also adequately determine the oligomeric state, although the assumptions made may reduce the accuracy of the final value calculated (see Section 2), which could make this method less appropriate for small membrane proteins.

Calculation of the density increment by summation can be extremely useful, as it provides independent checks on all the data from both analytical ultracentrifugation and size-exclusion chromatography. However, for accuracy and simplicity, measurement of the density increment is our preferred choice, but the preparation of suitable samples can be problematic and may not be possible for some membrane proteins that are unstable and cannot be purified in milligram quantities. However, in these cases, the determination of bound detergent will also be difficult, and it may be that density matching is the only appropriate method. Despite the difficulties of working with membrane proteins, due to their low abundance and instability after purification, there now exist a variety of techniques to allow the determination of their oligomeric state.

References

1. T. Svedberg and K. O. Pedersen, *The Ultracentrifuge*, Oxford University Press, London, 1940.
2. J. T. Edsall, in *Proteins, Amino Acids and Peptides as Ions and Dipolar Ions*, E. J. Cohn and J. T. Edsall (eds), Reinhold Publishing Corp., New York, 1943, 155.

3. H. Durschlag, in *Thermodynamic Data for Biochemistry and Biotechnology*, H. -J. Hinz (ed), Springer, New York, 1986, 45.

4. S. J. Perkins, *Eur. J. Biochem.*, 1986, **157**, 169.

5. T. M. Laue, B. D. Shah, T. M. Ridgeway and S. L. Pelletier, in *Analytical Ultracentrifugation in Biochemistry and Polymer Science*, S. E. Harding, A. J. Rowe and J. C. Horton (eds), Royal Society of Chemistry, Cambridge, UK, 1992, 90.

6. D. B. Hayes, T. Laue and J. Philo, *Sednterp*, available on *RASMB* website, 1995. (http://www.bbri.org/RASMB/rasmb.html)

7. E. F. Casassa and H. Eisenberg, *Adv. Prot. Chem.*, 1964, **19**, 287.

8. H. K. Schachman, *Ultracentrifugation in Biochemistry*, Academic Press Inc., New York, 1959.

9. H. Eisenberg, *Biological Macromolecules and Polyelectrolytes in Solution*, Clarendon Press, Oxford, UK, 1976.

10. H. Eisenberg, *Biophys. Chem.*, 2000, **88**, 1.

11. M. L. Johnson and M. Straume, in *Modern Analytical Ultracentrifugation*, T. M. Schuster and T. M. Laue (eds), Birkhäuser, Boston, USA, 1994, 37.

12. J. S. Philo, *Anal. Biochem.*, 2000, **279**, 151.

13. W. F. Stafford and P. J. Sherwood, *Biophys. Chem.*, 2004, **108**, 231.

14. C. Tanford and J. A. Reynolds, *Biochim. Biophys. Acta*, 1976, **457**, 133.

15. J. A. Reynolds and C. Tanford, *Proc. Natl. Acad. Sci. USA,* 1976, **73**, 4467.

16. O. Kratky, H. Leopold and H. Stabinger, *Method. Enzymol.*, 1973, **27**, 98.

17. C. L. Stevens and M. A. Lauffer, *Biochemistry*, 1965, **4**, 31.

18. G. Cohen and H. Eisenberg, *Biopolymers,* 1968, **6**, 1077.

19. H. Eisenberg, *Protein Sci.*, 2003, **12**, 2647.

20. J. Lebowitz, M. S. Lewis and P. Schuck, *Protein Sci.*, 2003, **12**, 2649.

21. P. J. G. Butler, I. Ubarretxena-Belandia, T. Warne and C. G. Tate, *J. Mol. Biol.*, 2004, **340**, 797.

22. P. K. Fyfe, K. E. McAuley, A. W. Roszak, N. W. Isaacs, R. J. Cogdell and M. R. Jones, *Trends Biochem. Sci.*, 2001, **26**, 106.

23. J. Tucker and R. Grisshammer, *Biochem. J.*, 1996, **317**, 891.

24. C. G. Tate, I. Ubarretxena-Belandia and J. M. Baldwin, *J. Mol. Biol.*, 2003, **332**, 229.

25. K. Dornmair, H. Kiefer and F. Jahnig, *J. Biol. Chem.*, 1990, **265**, 18907.

26. R. H. E. Frieson, J. Knol and B. Poolman, *J. Biol. Chem.*, 2000, **275**, 33527.

27. A. Ward, N. M. Sanderson, J. O'Reilly, N.G. Rutherford, B. Poolman and P. J. F. Henderson, in *Membrane Transport - A Practical Approach*, S. A. Baldwin (ed), Blackwell Press, Oxford, 2000, 141.

28. H. M. Weiss and R. Grisshammer, *Eur. J. Biochem.*, 2002, **269**, 82.

29. M. J. Lemieux, R. A. F. Reithmeier and D. N. Wang, *J. Struct. Biol.*, 2002, **137**, 322.

30. R. M. Garavito and S. Ferguson-Miller, *J. Biol. Chem.*, 2001, **276**, 32403.

31. N. König, G. A. Zampighi and P. J. G. Butler, *J. Mol. Biol.*, 1997, **265**, 590.

32. B. Ludwig, M. Grabo, I. Gregor, A. Lustig, M. Regenass and J. P. Rosenbusch, *J. Biol. Chem.*, 1982, **257**, 5576.

33. K. G. Fleming, A. L. Ackerman and D. M. Engleman, *J. Mol. Biol.*, 1997, **272**, 266.

34. K. G. Fleming and D. M. Engleman, *Proc. Natl. Acad. Sci. USA,* 2001, **98**, 14340.

35. C. -H. Huang and J. P. Charlton, *J. Biol. Chem.*, 1971, **246**, 2555.

36. E. J. M. L. Heuberger, L. M. Veenhoff, R. H. Duurkens, R. H. E. Frieson and B. Poolman, *J. Mol. Biol.*, 2002, **317**, 591.

37. D. Blakey, A. Leech, G. H. Thomas, G. Coutts, K. Findlay and M. Merrick, *Biochem. J.*, 2002, **364**, 527.

38. G. A. Tennent, P. J. G. Butler, T. Hutton, A. R. Woolfitt, D. J. Harvey, T. W. Rademacher and M. B. Pepys, *Eur. J. Biochem.*, 1993, **214**, 91.
39. G. E. Perlman and L. G. Longsworth, *J. Am. Chem. Soc.*, 1948, **70**, 2719.
40. F. Lacroute, *J. Bacteriol.*, 1971, **106**, 519.
41. R. B. Wickner, *Science*, 1994, **264**, 566.
42. R. B. Wickner, H. K. Edskes, M. L. Maddelein, K. L. Taylor and H. Moriyama, *J. Biol. Chem.*, 1999, **274**, 555.
43. D. C. Masison, M. L. Maddelein and R. B. Wickner, *Science*, 1995, **270**, 93.
44. H. K. Edskes, V. T. Gray and R. B. Wickner, *Proc. Natl. Acad. Sci. USA*, 1999, **96**, 1498.
45. S. Perrett, S. J. Freeman, P. J. G. Butler and A. R. Fersht, *J. Mol. Biol.*, 1999, **290**, 331.
46. J. E. Walker, *Curr. Opin. Struct. Biol.*, 1994, **4**, 912.
47. M. V. Panchenko and A. D. Vinogradov, *FEBS Lett.*, 1985, **184**, 226.
48. E. Cabezon, P. J. G. Butler, M. J. Runswick and J. E. Walker, *J. Biol. Chem.*, 2000, **275**, 25460.
49. M. J. van Raaij, G. L. Orriss, M. G. Montgomery, M. J. Runswick, I. M. Fearnley, J. M. Skehel and J.E. Walker, *Biochemistry*, 1996, **35**, 15618.
50. E. Cabezon, I. Arechaga, P .J. G. Butler and J. E. Walker, *J. Biol. Chem.*, 2000, **275**, 28353.
51. E. Cabezon, M. J. Runswick, A. G. Leslie and J. E. Walker, *EMBO J.*, 2001, **20**, 6990.
52. C. G. Tate, E. R. Kunji, M. Lebendiker and S. Schuldiner, *EMBO J.*, 2001, **20**, 77.
53. I. Ubarretxena-Belandia, J. M. Baldwin, S. Schuldiner and C. G. Tate, *EMBO J.*, 2003, **22**, 6175.
54. C. Ma and G. Chang, *Proc. Natl. Acad. Sci. USA*, 2004, **101**, 2852.
55. D. Schubert, C. Tziatzios, J. A. van den Broek, P. Schuck, L. Germeroth and H. Michel, *Prog. Colloid Polym. Sci.*, 1994, **94**, 14.
56. B. Tenchov, R. Koynova, M. Rappolt and G. Rapp, *Biochim. Biophys. Acta*, 1999, **1417**, 183.

CHAPTER 8

Application of Weighted Robust Regression to Equilibrium Ultracentrifugation

MARC S. LEWIS AND MICHAEL M. REILY

1 Introduction

The modern era of data analysis for analytical ultracentrifugation began approximately 30 years ago when investigators began directly fitting ultracentrifugal data with appropriate mathematical models using digital computers. Since that time, the analysis of absorbance equilibrium data has utilized non-linear least-squares curve-fitting using appropriate algorithms, of which the Levenberg–Marquardt is possibly the most frequently used. Beginning in 1997, Dimitriadis and Lewis[1,2] demonstrated that due to the non-Gaussian characteristics of the noise in the absorbance data, least-squares is not an optimal method and can introduce a systematic bias to the estimated parameters. They described methods for analyzing intensity data, which have Gaussian noise, and is thus appropriately fit using non-linear least-squares regression. Following this, Lewis and Reily[3] described means of determining appropriate weights for the fitting of intensity data, but also demonstrated that fitting intensity data could be unpredictably unreliable in terms of obtaining optimal fits. They demonstrated that, since the absorbance data had noise which was a logarithmically skewed Cauchy-type distribution, robust regression using the L-1 norm, which utilizes minimization of the sum of the absolute values of the residuals, appears to be an optimal method for the analysis of absorbance data. They also suggested that the balanced bootstrap appeared to be a very reliable method for obtaining optimal parameter error estimates. We present here a more detailed discussion of this approach to the analysis of absorbance equilibrium data.

2 Noise Distribution in Absorbance Data

Dimitriadis and Lewis[1,2] demonstrated that the noise distribution in intensity data is Gaussian and while the magnitude of the error in the intensity in the reference channel is essentially independent of radial position, the magnitude of the comparable error in the solution channel is attenuated proportionately to the ratio of the intensity of the solution channel to that of the reference channel. Since the absorbance, as a function of radius, is defined by

$$A(r) = \log_{10}\left(\frac{I_{\text{Incident}}}{I_{\text{Transmitted}}}\right)_r \tag{1}$$

understanding the nature of the noise distribution in absorbance data requires evaluation of the noise distribution of the ratio of the intensities and the effect of taking $\log_{10}$ of that ratio.

The ratio of two zero-mean Gaussian signals gives a signal whose noise distribution is a Cauchy distribution, which can be written as:[1]

$$P_z(z) = \frac{\sqrt{1-r^2}\sigma_1\sigma_2/\pi}{\sigma_2^2(z-r\sigma_1/\sigma_2)^2+\sigma_1^2(1-r^2)} \tag{2}$$

where $z = x/y$, σ_1 and σ_2 are the standard deviations of x and y, respectively, and r is the correlation coefficient between x and y for which intensity data will be zero since the intensities in the two sectors are independent. Another variable transformation is defined by $w=\log_{10}(z)$ which will give a signal which, except for the fact that we are still considering zero-mean values, will be similar to that of the absorbance. Using this transformation leads to the probability distribution of w:

$$P_w(w) = \frac{2\sqrt{1-r^2}\sigma_1}{\pi\sigma_2} \frac{e^w}{e^{2w}-2r(\sigma_1/\sigma_2)e^w+(\sigma_1/\sigma_2)^2} \tag{3}$$

Since $r=0$, this equation simplifies to

$$P_w(w) = \frac{1}{\pi \cosh(w-\overline{w})} \tag{4}$$

with a distribution mean $\overline{w}=\ln(\sigma_1/\sigma_2)$. The shift in the mean of the transformed variable indicates that the use of non-linear least-squares will introduce a systematic bias in the estimation process.

We must now take into consideration the fact that the intensity signals are not zero-mean. It is demonstrated in refs. 1 and 2 that the probability distribution is described by

$$P_z(z) = \frac{\ln(10)10^z}{2\beta(z)\pi\sigma_1\sigma_2} \exp(-\delta)\left[1 + \gamma(z)\sqrt{\frac{\pi}{\beta(z)}} \exp\left(\frac{\gamma(z)^2}{4\beta(z)}\right)\left(1 + \phi\left(\frac{\gamma(z)}{2\sqrt{\beta(z)}}\right)\right)\right] \tag{5}$$

where

$$\beta(z) = \frac{10^{2z}\sigma_2^2+\sigma_1^2}{2\sigma_1^2\sigma_2^2}, \quad \gamma(z) = \frac{10^z\overline{x}\sigma_2^2+\overline{y}\sigma_1^2}{\sigma_1^2\sigma_2^2}, \quad \delta = \frac{\overline{x}\sigma_2^2+\overline{y}\sigma_1^2}{2\sigma_1^2\sigma_2^2} \tag{6}$$

and the function $\phi(x)$ is the well-known probability integral, or error function defined by

$$\phi(x) = \frac{2}{\sqrt{\pi}}\int_0^x \exp(-t^2)\, dt \tag{7}$$

Computations using these equations indicate that the shift of the mean is a function of the ratio of the squares of the values of σ, that this value is independent of the mean values of x and y, and that these values lead to a sharpening of the distribution of z. Since, at an absorbency value of 1.0, the intensities will have a ratio of 10:1 and the ratio of values of σ will have a comparable ratio, there is an increasingly non-trivial shift of the distribution mean as the absorbance increases. The nature of the non-Gaussian noise distribution in absorbance data clearly indicates that the use of non-linear least-squares regression is not optimal and is not mathematically appropriate.

3   Robust Regression

When the error distribution in data is non-Gaussian, and where least-squares analysis is accordingly inappropriate, least-absolute value or L-1 norm regression is generally recognized as a highly acceptable alternative.[4] For a data set where the values of y_i, the dependent variable, are functions of x_i, the independent variable, the ith residual is defined as $y_i - f(x_i)$. While least-squares regression (L-2 norm) involves minimization of the sum of the squares of the residuals with the reciprocals of variance (σ^2) as weights, as in

$$\min \sum_{i=1}^{n} \frac{(y_i - f(x_i))^2}{\sigma_i^2} \tag{8}$$

L-1 norm regression minimizes the sum of the absolute values of the residuals with the reciprocals of σ as weights, as in

$$\min \sum_{i=1}^{n} \frac{\mathrm{abs}(y_i - f(x_i))}{\sigma_i} \tag{9}$$

It is readily apparent that, aside from the fact that L-1 regression is more appropriate for fitting data with non-Gaussian error, this form of robust regression is much less sensitive to outliers than least-squares analysis. This is because the sum of absolute values of the residuals is minimized with L-1 regression in comparison to the minimization of the sum of the squares of the reciprocals in L-2 regression. Thus, in L-2 regression, the effects of outliers are magnified because their deviations from the fitting line are squared and thus they have more weight in the fitting then do the absolute values of these same deviations. Outliers in absorbance data can be generally described as falling into two categories: those which are readily apparent to visual detection, such as the spikes which result from isolated flash lamp failures, and thus can be edited out, and those which are of a more subtle nature, arising from a variety of causes such as window flaws, smudges, fingerprints, *etc.*, and which may be quite difficult to detect visually. L-1 regression is a very effective tool for minimizing the impact of such outliers on the results obtained in the analysis.

There are different algorithmic approaches for performing L-1 regression. MLAB, for example, uses a modification of the Levenberg–Marquardt algorithm. For the software which we have developed, we have chosen to use the Nelder–Mead simplex method because it is a particularly robust direct search method for multidimensional unconstrained minimization.[5] While the simplex method usually requires significantly more iterations to obtain convergence than Levenberg–Marquardt or similar methods, contemporary personal computers are so fast that this is not a significant issue. Since none of the algorithms that are appropriate for performing L-1 regression is designed to yield estimates of fitting parameter error, this becomes a separate issue entirely and has no bearing on the choice of algorithm used. While there are a variety of ways of estimating fitting parameter errors, we have chosen to use the bootstrap, and in particular the balanced bootstrap, as our method for this purpose.

4 Balanced Bootstrap

The bootstrap has been described as a computer implementation of non-parametric or parametric maximum likelihood that has the advantage over the maximum likelihood method in that it allows the computation of maximum likelihood estimates of standard errors in conditions where no formulas are available.[6] The bootstrap simulation is thus a Monte–Carlo technique using only information from the original data set where samples are randomly selected from this data set to give a subset which is then analyzed using the mathematical model which was used to fit the original data set. This procedure is repeated a large number of times and then the stored returned values of the parameters are used to estimate the standard errors in the parameters obtained, fitting the original data set. There are a variety of possible approaches in the utilization of the bootstrap. Rather than describe all of these possible variations, we will only describe the procedure which we have used and the reasons for doing so.

One of the basic approaches to the bootstrap is to randomly delete data points from a data set and replace them with other data points which have been randomly selected from those points remaining in that set. The balanced bootstrap is a variation on this procedure, differing only in the fact that in the grand ensemble of all the data sets which are analyzed, each data point appears the same number of times. The result of this is that the balanced bootstrap ensures that the grand mean of the bootstrap resamples is identical to the original sample mean, thus removing any bias from the estimation of the parameter errors.[7] While there is some evidence that the balanced bootstrap returns the most optimal parameter error estimates since the variance of the mean is minimized, in general, the balanced bootstrap is considered to have an average performance about the same as the simple bootstrap estimate.[8] However, when applied to the analysis of equilibrium ultracentrifuge data, the use of the balanced bootstrap ensures that no region of the data is either overrepresented or underrepresented with concomitant distortion in the error analysis. This is critical because, in many cases, it appears that the standard deviations of the data are radially, and thus concentration, dependent.

The concept of the performance of the balanced bootstrap is quite readily visualized. Let us consider a data set with n entries so that we have an $n \times 3$ matrix with radii in column 1, absorbencies in column 2, and standard errors in column 3. If we wish to perform m bootstrap iterations on these data, we first concatenate m such matrices so that we have an $m*n \times 3$ matrix, which is considered to be the grand ensemble of the data. Each row is then assigned a random integer between 1 and $m*n$ with no repeats, and the matrix is sorted by these integers. The first new subset

Radius	Absorbance	Error	Random Int.
r1	ab1	σ1	5
r2	ab2	σ2	3
r3	ab3	σ3	6
r4	ab4	σ4	1
r1	ab1	σ1	13
r2	ab2	σ2	16
r3	ab3	σ3	9
r4	ab4	σ4	2
r1	ab1	σ1	11
r2	ab2	σ2	12
r3	ab3	σ3	10
r4	ab4	σ4	7
r1	ab1	σ1	15
r2	ab2	σ2	4
r3	ab3	σ3	8
r4	ab4	σ4	14

Original Data Set
Concatenated 4 times

Sort By Random Integer →

Radius	Absorbance	Error	Random Int.
r4	ab4	σ4	1
r4	ab4	σ4	2
r2	ab2	σ2	3
r2	ab2	σ2	4
r1	ab1	σ1	5
r3	ab3	σ3	6
r4	ab4	σ4	7
r3	ab3	σ3	8
r3	ab3	σ3	9
r3	ab3	σ3	10
r1	ab1	σ1	11
r2	ab2	σ2	12
r1	ab1	σ1	13
r4	ab4	σ4	14
r1	ab1	σ1	15
r2	ab2	σ2	16

4 New Data Sets Produced
For Balanced Bootstrap
Analysis

New data sets are boxed

Figure 1 *This flow chart illustrating the conceptual principles of the balanced bootstrap used to produce 4 new data sets using a 4-row data set*

matrix to be analyzed is rows 1 through n, the next is rows $n+1$ through $2n$, and so on until row $m*n$ is reached. The results of each analysis are stored in an appropriate matrix and when the analysis is complete, the standard errors of these stored values are calculated. This is illustrated in Figure 1. While facilitating the visualization of this procedure, this is not a computationally effective technique due to the enormous memory requirements of this configuration if an appropriate number of iterations is used.

For computational purposes, a new matrix is created, analyzed, and deleted before the production of the next new matrix. This requires far less memory than the procedure described above and can therefore be used on larger data sets with a large number of iterations. A new matrix is chosen by first creating an n-row 'selecting' vector that contains random integers between 1 and n. Each integer may be used multiple times or not at all within the vector. A separate 'counting' vector is used to keep track of how many times each integer has been used throughout the process. Once an integer has been used m times it cannot be used anymore. The selecting vector is then used as a template for a new matrix with each integer in the vector corresponding to a row in the original data matrix. For example, if the value of row 1 of the selecting vector were 5, the first row of the new matrix would be row 5 of the original matrix. Once a new matrix is created, it is analyzed and the parameter values are stored in appropriate matrices. This process is repeated m times, each time creating a new selecting vector subject to the counting rules. Finally, the stored parameter values are used to calculate the standard errors. This is illustrated in Figure 2.

In both cases, the column 3 values are to be used to calculate the weights for each analysis where the weights are normalized to unity using

$$w_i = \sigma_i/\overline{\sigma} = \sigma_i n/\sum_{i=1}^{n} \sigma_i \tag{10}$$

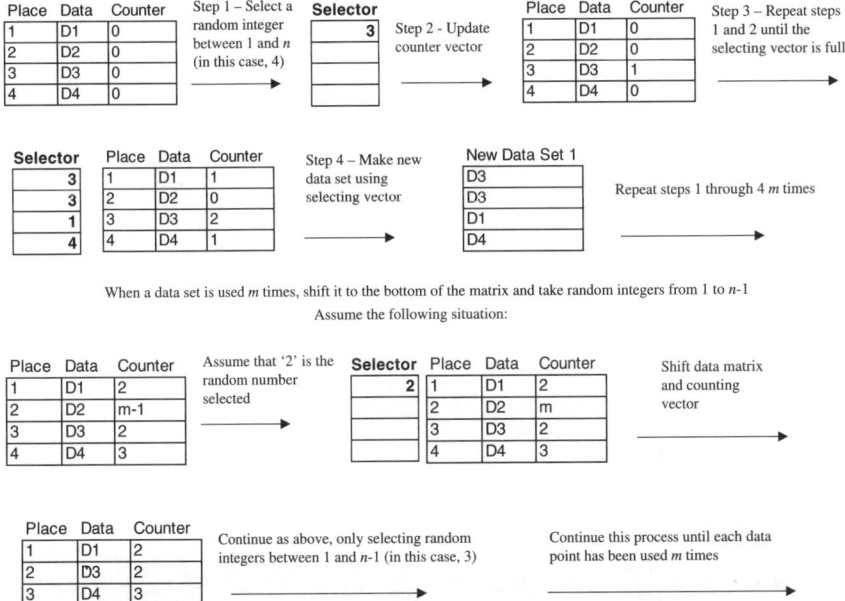

Figure 2 *This flow chart illustrating the practical method of producing m new data sets using the balanced bootstrap*

This permits direct comparison of weighted fits with unweighted fits where every value has a default weight of 1.

5 Examples

We will illustrate the application of the methodology described here with three examples: an ideal monomer (Figure 3; Table 1), a monomer–dimer association (Figure 4; Table 2), and a simple heterogeneous, two-component association (Figure 5; Table 3). In order to ensure proper noise distribution, the incident and transmitted intensities were generated with 5% RMS normally distributed noise and an absorbance was found by taking the base 10 logarithm. This was repeated 10 times and the mean value of the absorbancy was taken as its value and the calculated standard deviations were used for obtaining the weights.

We have chosen to present simulated data here because, unlike experimental data, we know that the generated parameter values are the optimal values for each data set. This permits the best possible comparison not only of the returned parameter values, but also of their standard errors and in particular, a comparison of the standard errors for least-squares fitting as calculated by the bootstrap and by the Levenberg–Marquardt algorithm utilized in MATLAB. The only disadvantage that this approach has is that it does not present the additional advantages of robust regression in dealing with problems arising from subtle outliers. Summarizing these results, two significant facts are apparent. The first of these is that, in every case,

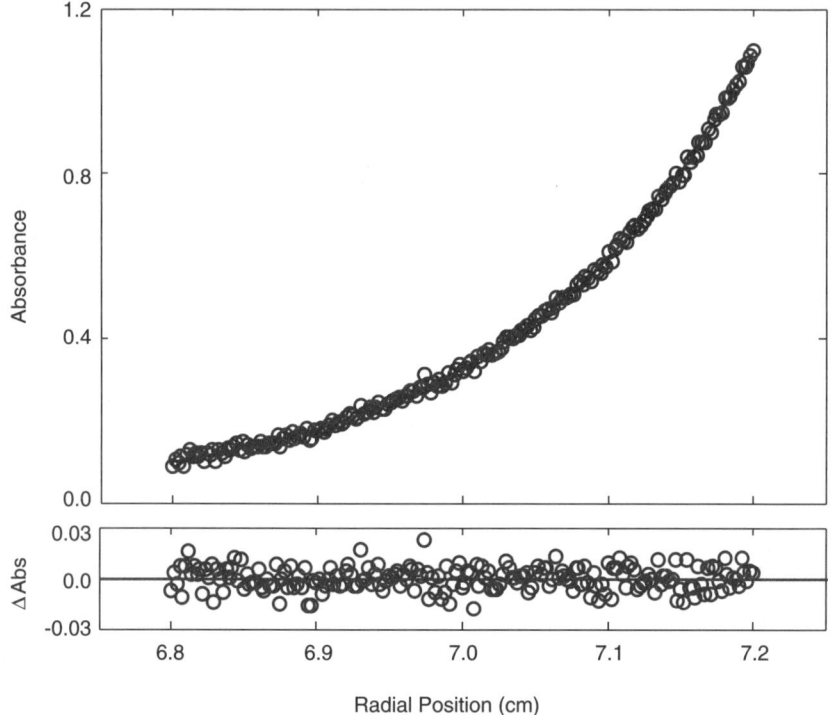

Figure 3   *Upper panel: simulated equilibrium ultracentrifuge data for an ideal monomer system. Lower panel: the distribution of residuals. Data were fit using an L-1 norm regression. The parameter values are given in Table 1*

Table 1   *Parameter values and errors for ideal monomer (Example 1)*

	Generating parameters	Robust method		Least-squares method		Levenberg– Marquardt method	
		Fitted value	*Error value[a]*	*Fitted value*	*Error value[a]*	*Fitted value*	*Error value[a]*
C_b	1	0.9987	0.003	1.0044	0.002	1.01	0.004
Mass	20000	19951	98.47	20254	68.42	19880	185.10

[a] ±1 S.E.; 67% confidence limit.

robust regression returned parameter estimates which were the closest to the generating parameters. The second of these is that for least-squares regression, the bootstrap returned significantly better parameter error estimates than did the Levenberg–Marquardt algorithm. This is not surprising since this algorithm assumes that the mathematical model is linear in the parameters, a condition which none of these models meet. Furthermore, we have observed that with experimental data, as

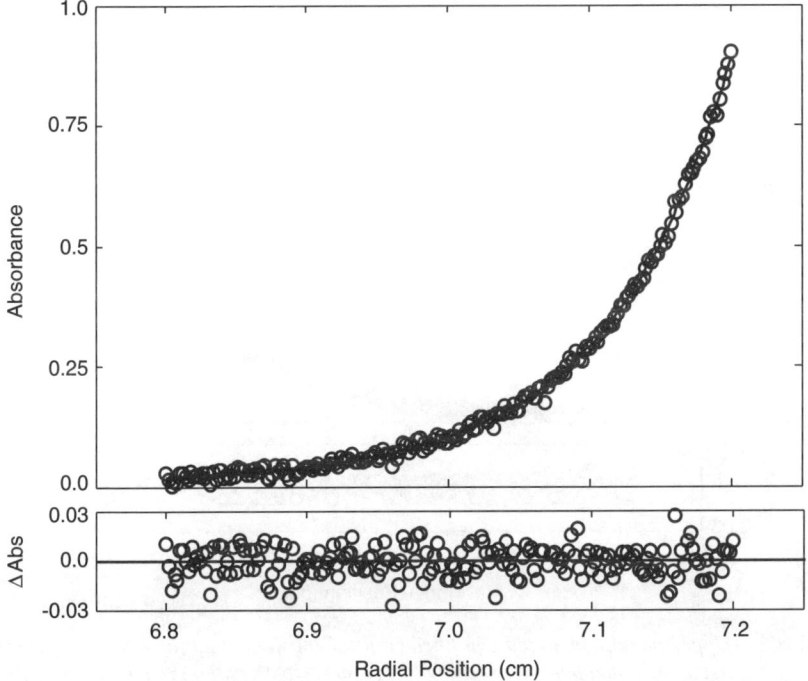

Figure 4 *Upper panel: simulated equilibrium ultracentrifuge data for a monomer–dimer association system. Lower panel: the distribution of residuals. Data were fit using an L-1 norm regression. The parameter values are given in Table 2*

Table 2 *Parameter values and errors for monomer–dimer association (Example 2)*

	Generating parameters	Robust method		Least-squares method		Levenberg–Marquardt method	
		Fitted value	Error value[a]	Fitted value	Error value[a]	Fitted value	Error value[a]
C_b	0.1	0.0971	0.009	0.1073	0.007	0.0102	0.012
$\ln k_{1,2}$	12	12.057	0.206	11.832	0.138	11.94	0.235

[a] ± 1 S.E.; 67% confidence limit.

the model becomes more complex and the data becomes noisier, the advantages of L-1 regression becomes more substantial.

6 Software

The software which we have developed and which was utilized for the preceding examples was originally written using MATLAB and then compiled as a stand-alone program running in the Windows environment for distribution. It is available upon

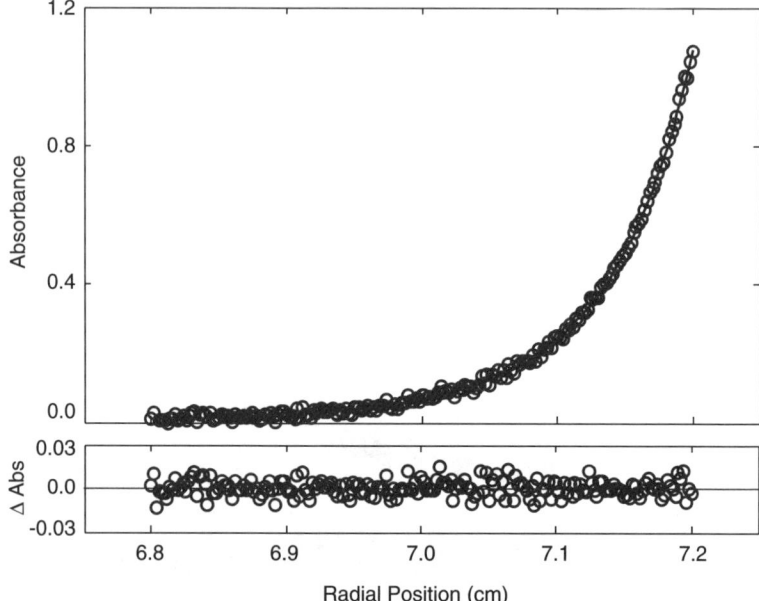

Figure 5 *Upper panel: simulated equilibrium ultracentrifuge data for a simple, heterogeneous, two-component system with 1:1 stoichiometry. Lower panel: the distribution of residuals. Data were fit using an L-1 norm regression. The parameter
values are given in Table 3*

Table 3 *Parameter values and errors for simple, heterogeneous, two-component
association (Example 3)*

	Generating parameters	Robust method		Least-squares method		Levenberg–Marquardt method	
		Fitted value	Error value[a]	Fitted value	Error value[a]	Fitted value	Error value[a]
$C_{b,A}$	0.1	0.102	0.0055	0.102	0.0038	0.104	0.111
$C_{b,B}$	0.1	0.0991	0.0054	0.101	0.0034	0.119	0.141
$\ln k_{A,B}$	12	11.989	0.063	11.96	0.054	11.74	0.201

[a] ± 1 S.E.; 67% confidence limit.

request from the authors. This program can generate and analyze data using the following models:

1. *Homogeneous association*: This model has the ability to fit up to seven predefined species simultaneously as well as two user-defined species, and can
   also be used for the analysis of monomer alone to determine values of
   $M(1-\bar{v}\rho)$ needed for the study of heterogeneous associations.

2. *Indefinite association*: This model can fit isodesmic associations, indefinite associations with exponential decay of the association constants, and either type of association model with ring closure.

3. *Heterogeneous association*: This model can fit associations of the type $mA + nB \leftrightarrow A_m B_n$ where m and n are user defined. This model can also be fit using a conservation of mass constraint.

4. *Species analysis*: This model is used to assist the investigator in determining the nature of the association being studied by determining relative amounts of species possibly present.

5. *Custom fit*: This model allows the user to enter their own fitting function for associations and fit using the simplex method.

6. *$\bar{v}$ calculator*: Permits calculation of the compositional partial specific volume from the amino acid composition using the consensus values of Perkins.[9]

Additionally, this software permits the use of both L-1 and L-2 norm regressions; the latter is appropriate for use in the analysis of interference data and has the advantage that parameter errors can be computed using the balanced bootstrap. Comparable software is available from Dr. Leslie Holladay (HOLLADAYL@aol.com) P.O. Box 244, Townsend, TN 37882, USA.

References

1. E. K. Dimitriadis and M. S. Lewis, *Progr. Colloid Polym. Sci.*, 1997, **107**, 20.
2. E. K. Dimitriadis and M. S. Lewis, *Method. Enzymol.*, 2000, **321**, 121.
3. M. S. Lewis and M. M. Reily, *Method. Enzymol.*, 2004, **384**, 232.
4. R. D. Armstrong and P. O. Bock, in *Robust Regression*, K. D. Lawrence and J. L. Arthur (eds), Marcel Decker, Inc., New York, 1990, 89.
5. J. C. Lagarias, J. A. Reeds, M. H. Wright and P. E. Wright, *SAIM J. Optim.*, 1998, **9**, 112.
6. T. Hastie, R. Tibshirani and J. Friedman, *The Elements of Statistical Learning*, Springer-Verlag, New York, 2003.
7. P. Hall, *The Bootstrap and Edgeworth Expansion*, Springer-Verlag, New York, 1992, 293.
8. B. Efron and R. J. Tibshirani, *An Introduction to the Bootstrap*, Chapman & Hall/CRC, Washington, DC, 1998.
9. S. J. Perkins, *Eur. J. Biochem.*, 1986, **157**, 169.

CHAPTER 9

The Synthetic Boundary Technique of Analytical Ultracentrifugation: A Valuable Tool to Study Structure Formation Processes Online

CHRISTINE WANDREY AND LAURENT BOURDILLON

1 Introduction

1.1 The Synthetic Boundary Technique

The synthetic boundary experiment is one of the basic experimental techniques in analytical ultracentrifugation.[1] Different types of so-called synthetic boundary cells have been constructed for various applications.[2–5] Some examples are presented in Figure 1. In 1952, for the first time, Kegels[2] and Pickels *et al.*[3] described ultracentrifugation cells in which sharp boundaries could be formed between solutions of different densities. The availability of the new cell types allowed investigations that were earlier feasible only with some difficulty. In the majority of applications, a sharp, stable boundary was obtained by layering one solution over another possessing a higher density while the ultracentrifuge operated at low rotational speed.

Schachman and Harrington[6] presented various ultracentrifugation studies with a synthetic boundary cell soon after. The early applications included

- extension of the sedimentation velocity method to substances with molar masses as low as 350 g mol^{-1};
- formation of a boundary between solutions of two different concentrations of the same macromolecule in order to obtain differential sedimentation constants;
- determination of hydrodynamic volumes of sedimenting components;
- measurement of the backward flow of solvent by the use of an indicator boundary formed by layering a virus solution over another virus solution containing an additional component; and

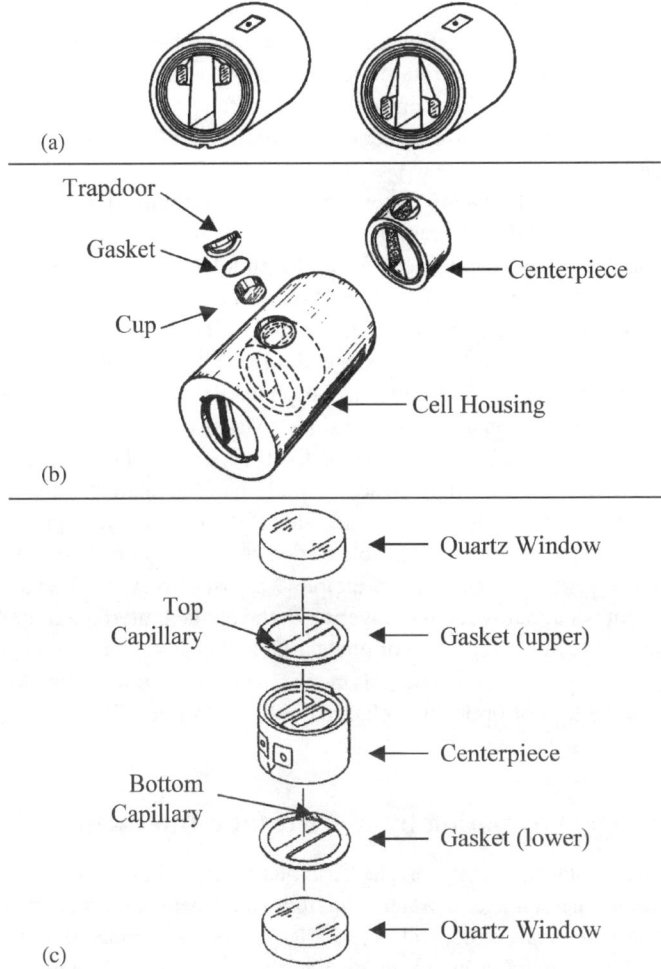

Figure 1 *Synthetic boundary cells. (a) Reprinted with permission from G. Kegels.[2] Copyright (2004), American Chemical Society. (b) Adapted from Pickels et al.[3] (c) Reprinted from Okajima et al.[4] with permission from Elsevier. Copyright (2004)*

- preliminary investigation of some fundamental problems in ultracentrifugation such as the sedimentation of slowly sedimenting materials in the presence of more rapidly sedimenting components, the analysis of equilibrium systems, or the role of convection.

In a modified experiment, a synthetic boundary cell allowed the formation of a synthetic boundary at the bottom level of the cell with self-adjusting meniscus and baseline [Figure 1(c)]. During centrifugation, each component of the solution of interest was separated according to its flotation characteristics. It was possible to calculate the flotation coefficient for each component.[4]

The technique was later used and reported by other authors.[7–10] They applied it, for example, to examine the sedimentation of small molecules (molar mass below 12 000 g mol^{-1}) for which the rate of sedimentation was insufficient to produce a sharp boundary that clears the meniscus,[7] to monitor products from enzyme reactions,[8] to measure differential sedimentation coefficients,[9] and to estimate extinction coefficients.[10]

Nowadays, synthetic boundary experiments are most frequently performed with a commercially available double-capillary-type sector centerpiece where the two sectors are connected by two capillaries, one at the middle on the upper side of the centerpiece to permit the solvent to pass, the other at the top of the same side allowing the backflow of air.[5] The experiment helps to determine the diffusion coefficient by monitoring the spreading of the boundary with time. The calculation of the diffusion coefficient requires the concentration in the plateau region, and the concentration gradient at the midpoint, as a function of time.[7] Moreover, portions of associates or aggregates can be identified by investigating the time-dependent concentration of the plateau region at low speed, where no sedimentation of the main fraction takes place. Furthermore, the concentration can be calibrated if the interference optics of the analytical ultracentrifuge is solely applicable to detect the concentration profile.

Recently, the synthetic boundary principle was used to study the reactions and interactions that occur subsequent to layering at the contact interface of the two solutions. Examples are the observation of nucleation and growth of inorganic colloids,[11] and the online study of electrostatic polymer network formation[12–14] yielding hydrogels by the interaction of oppositely charged polyelectrolytes. The latter application is the subject of this chapter.

1.2 Hydrogel Formation by Electrostatic Interaction

Hydrogels are either covalently or electrostatically cross-linked polymer networks containing a high percentage of water. The high water content renders them suitable for a variety of medical, biological, and biotechnological applications such as, for example, tissue engineering or immunoisolation.[15,16] Covalent linking by chemical reaction, in general, yields permanent and stable networks. Their degradation requires bond breakage. To the contrary, the stability and durability resulting from electrostatic cross-linking in aqueous media strongly depend on both the molecular characteristics of the polyelectrolytes involved and the medium conditions. Appropriate molecular characteristics are the type of the ionic group, the charge density, and the molar mass. Relevant medium conditions are the pH, the ionic strength, the ratio of the reacting partners, and the temperature. Electrostatic networks are considered to be temporary.

Hydrogel and hydrogel membrane formation through electrostatic interaction of oppositely charged polymers has attracted scientific and commercial interest owing to the simplicity of the cross-linking process and the broad application potential of the resulting materials.[17,18] The principle of polyelectrolyte network formation is shown in Figure 2. The two oppositely charged polymer chains connect by long-ranging Coulomb interaction. The small counterions are liberated. Hydrogel polyelectrolyte complex materials may be employed as components of flat devices or as self-containing hollow fibers, macrocapsules, or microcapsules.[17–20]

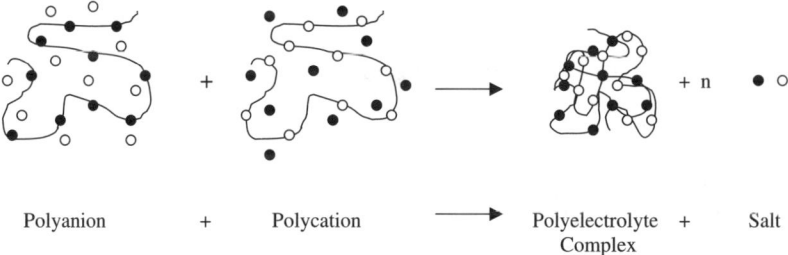

Polyanion   +   Polycation   ⟶   Polyelectrolyte   +   Salt
                                                                                Complex

Figure 2   *Polyelectrolyte complex formation*

Numerous combinations of anionic and cationic polymers have been examined with regard to their ability to form stable networks and membranes, also called polyelectrolyte complex membranes, with well-defined characteristics. However, only a few pairs of polyelectrolytes have been identified for this purpose.[17,21] The majority of combinations yield amorphous precipitates owing to the hydrophobicity of the polymer backbone, without the ability to swell in an aqueous medium. Some polyelectrolytes form hydrogels with small, divalent or multivalent ions. The key parameters of polyelectrolyte complex networks/membranes that are relevant for their practical applications are permeability, mechanical stability, elasticity, and durability. Certain applications require, in addition, biocompatibility.[22]

Relatively minor changes in the reaction conditions can alter the final network properties considerably, and the adjustment of the desired parameters is time-consuming. Furthermore, polyelectrolyte complexes are not static materials. All properties are equilibrium characteristics influenced by the environment. The prediction of network properties as a function of macromolecular, physicochemical, structural, and environmental conditions is yet to be elaborated. An approach that tries to establish general correlations between the physicochemical parameters and the network properties does not exist. Such screening suffers from the fact that an experimental technique that can follow the network/membrane formation online and monitor the various stages of the process, including formation, equilibration, and stabilization/destabilization, has not yet been used. The synthetic boundary technique combined with the high precision of the optical system of the Analytical Ultracentrifuge (AUC) Optima XL-I offers one possibility for such studies.

This chapter will focus on methodological aspects while presenting and discussing application examples only to the extent necessary to demonstrate the performance, advantages, and limitations of the synthetic boundary technique. A discussion of specific materials optimization is outside the scope of this chapter.*

---

*Different terms are used in this chapter to name the polymer network built by electrostatic interaction of polyelectrolytes. Despite differences in the specification all are applicable here synonymously. These are polymer network, polyelectrolyte complex, hydrogel, polyelectrolyte hydrogel, and simplex. As an exception, the terms membrane and polyelectrolyte complex membrane are restricted to a layer of limited dimension.

2   Hydrogel Formation in a Synthetic Boundary Cell

2.1   General Remarks and Requirements

Employing the synthetic boundary technique hydrogels can be produced by layering a polyanion on a polycation or vice versa. With regard to practical applications the latter procedure is more interesting owing to the known cell toxicity of cationic components, which are not recommended for mixing with cells and building the core, for example, of microcapsules. Therefore, the experimental procedure was developed by layering various polycations on a polyanion although the technique can be used without modification for the opposite case. As a basic requirement, the layering solution must have a lower density than the stationary solution. Otherwise, the polymer network formation will take place somewhere in the solution and will not be well defined in the interface region. It is also required that the position and the time-dependent formation of the network/membrane be without doubt detectable by the optical system of the AUC.

2.2   Description of the Experimental Principle

For all experiments a standard, synthetic, boundary capillary-type centerpiece (Beckman)[5] was used to investigate the spontaneous interfacial reaction between oppositely charged polyelectrolytes.

The geometry of the synthetic boundary centerpiece determines two dimensions of the three-dimensional (3D) network, the surface area. This is an area of $12 \times \sim 3$ mm fixed by the height of the centerpiece and the sector diameter at the position of the solvent meniscus. The height is the measure of the thickness of the liquid column that the light of the optics has to pass through. The third dimension of the polymer network is variable and depends on the reaction. The standard sector length is 14 mm in radial direction, from sector top to sector bottom, with the capillary 6 mm from the bottom. The maximum sector volume is 450 μL; however, for synthetic boundary experiments a solution volume of 150 μL is recommended for the solution sector.[5] The polyelectrolyte solution possessing a lower density is layered over the polyelectrolyte solution of higher density, which in most cases has, in addition, a higher viscosity. The horizontal projection (view from above) of a synthetic boundary centerpiece and the principle of the experiment are illustrated in Figure 3. Below the centerpieces, the appropriate absorbance and interference scans are schematically presented.

Each absorbance signal results from the difference of the absorbance measured in the solution sector (upper sector) and that measured in the solvent sector (lower sector). In principle, five ranges of such differences are expected from the interfacial reaction experiment. Indeed, some of the radial ranges will appear differently, depending on the progress of the experiment. Table 1 summarizes the ranges detected during different periods of the experiment.

The three radial ranges 1, 2, and 5 are present before the layering process has started [Figure 3(a)], whereas all five differences of the signal are visible during the layering process [Figure 3(b)]. Finally, when the layering has terminated, the four ranges 1, 3, 4, and 5 [Figure 3(c)] are expected, although range 5 may disappear when the network extends to the bottom of the sector.

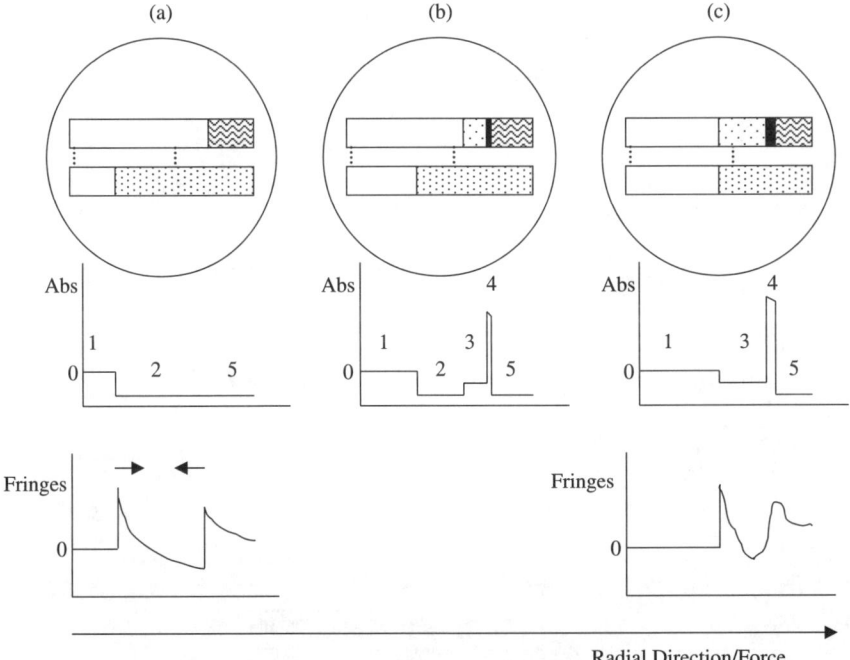

Figure 3 *Principle of electrostatic polymer network formation in a standard capillary type synthetic boundary cell. Component positions, absorption scans, and interference scans, (a) before the experiment, (b) during the layering process, (c) at the end of the experiment. Ranges 1–5 correspond to Table 1*

Table 1 *Absorbance detection during the synthetic boundary experiment*

Radial range	Signal source	Expected value		Layering process		
		PEL1		*Before*	*During*	*Completed*
		Absorbing	*Non-absorbing*			
1	Air–air	0	0	+	+	+
2	Air–PEL1	<0	0	+	+	−
3	PEL1–PEL1	≤0	0	−	+	+
4	Mem–PEL1	>0	>0	−	+	+
5	PEL2–PEL1	<0	0	+	+	+/(−)

PEL1, polyelectrolyte in the solvent sector, which passes the capillary; PEL2, polyelectrolyte in the solution sector; Mem, membrane; +, visible; −, not visible.

The interference scans, despite being less specific, are well suited to record the positions and movement of the menisci in both sectors, as it is schematized in Figure 3 below the schematic absorbance scans.

Two experimental examples are presented in Figures 4 and 5. The first is of a slightly UV-absorbing polycation, oligochitosan (OC), and the second is of a

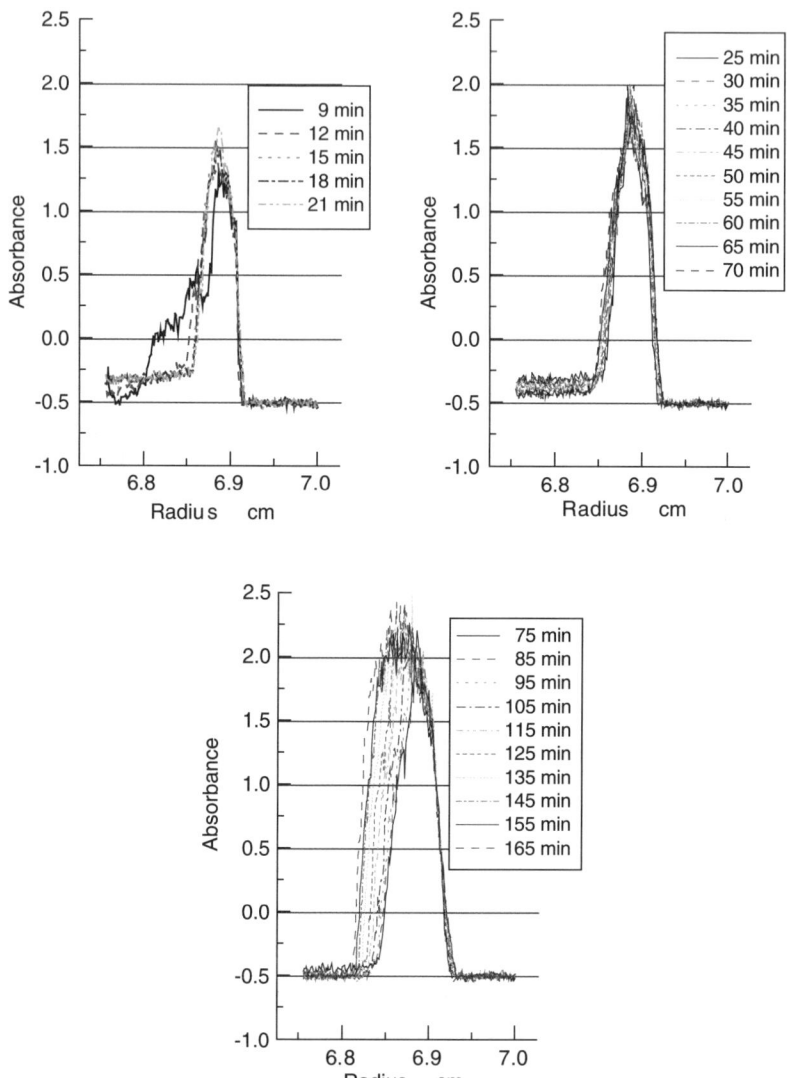

Figure 4 *Membrane formation of SA with a slightly UV-absorbing polycation, oligochi-*
tosan (OC2). 5000 rpm, 20 °C, 370 nm. From left to right: initial phase/membrane
growth, membrane growth/polycation consumption, polycation consumption/equi-
libration. Reprinted from Wandrey and Bartkowiak[29] with permission from
Elsevier. Copyright (2004)

non-UV-absorbing polycation, poly(L-lysine hydrobromide) (PLL). Three graphs are
presented for the experiment with OC, each one representing different time periods.

As soon as the polycation solution passes the capillary and is layered on the surface
of the polyanion solution, spontaneous structure formation begins in the boundary
zone, radial range 4. An absorbance signal >0 at the position of the boundary (range

4) detects a higher chitosan concentration in the upper sector than in the chitosan solution, indicating membrane formation [Figures 3(b) and 4]. To know to what extent the turbidity of the membrane contributes to the absorbance signal when absorbing polyelectrolytes react requires further detailed studies, but that the contribution of the turbidity can be remarkable becomes clearly visible from [Figure 5(a)]. The lack of a negative absorbance signal in radial ranges 2 and 3, to the left side of the membrane peak, confirms that PLL is not absorbing. Consequently, the membrane peak results exclusively from the network structure, *i.e.*, its turbidity. The photomicrograph of the membrane confirms the network structure concluded from the absorbance scans. Layering is completed when the solution meniscus in both sectors has reached the

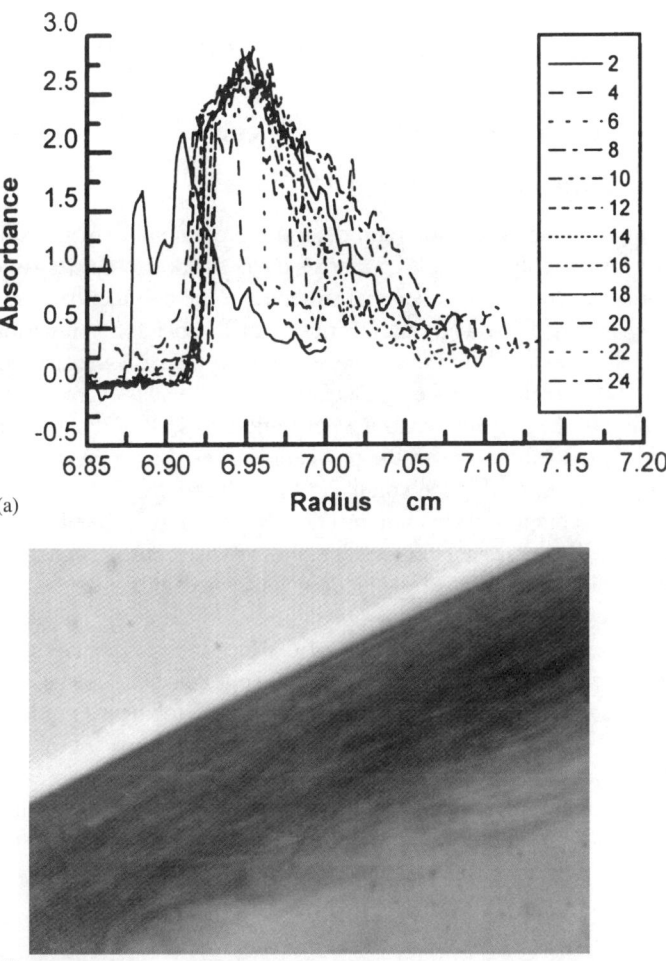

(a)

(b)

Figure 5 *Membrane formation of SA with a non-UV-absorbing polycation, poly(L-lysine hydrobromide), experiment 10 in Table 3. (a) Absorbance scans, (b) photomicrograph of the membrane. 5000 rpm, 20 °C, 320 nm. Reprinted from Wandrey et al.[30] with permission from Springer. Copyright (2004)*

same radial position [Figure 3(c)]. Time-dependent decrease of the absorbance in radial range 3, *i.e.*, the polycation concentration difference of both sectors, reflects time-dependent consumption of chitosan (range 3 in Figures 3(c) and 4). As already mentioned above, in some cases, dependent on the reaction conditions, range 5 can partly or completely disappear, particularly in experiments of long duration.

The simultaneous detection by the interference optics monitors refractive index differences between the two centerpiece sectors. For polyelectrolyte reactions the refractive index signal is composed of contributions of the polyanion, polycation, water, and low molar mass salt, which was already added to the polymer solutions or was liberated according to the reaction presented in Figure 2. Although the interference optics is not specific to one of the components, it will be shown in Section 4 that the information thereof can support the evaluation of the absorption scans and contribute to the general understanding of the hydrogel formation process and its kinetics.

2.3 Specific Methodological Aspects

2.3.1 Ratio of the Reacting Molecules

The polyanion to polycation ratio determines the number of possible network links in the polyelectrolyte complex, whereas the network structure (swollen or collapsed) primarily depends on the chemical structure of the polymer backbone as well as the local charge density. The latter varies with the pH for non-permanently-charged polyelectrolytes.[17,23,24] In the synthetic boundary experiment, the membrane's thickness and its structure can be regulated by the molecular characteristics and the concentration of the polyanion and the polycation, as well as the medium conditions. In addition, the total amount of the polycation layered onto the polyanion plays a significant role. By varying the volume of the polycation solution as well as its concentration, the complex formation reaction time can be limited. Less polycation, in general, yields thinner membranes under otherwise constant conditions. On the other hand, the size of the synthetic boundary centerpiece limits the range of manipulation possibilities of the polyanion to polycation ratio. Both the solvent and the solution volume can be used to regulate the ratio. Since the sedimentation force depends on the radial position, the filling height was kept constant for the polyanion solution (upper sector in Figure 3) during the method development in order to minimize the number of variables, although it has later been varied for specific experiments. The total volume of the sectors limits the solution volume, which can be layered. For example, for 100 μL polyanion solution the maximum volume for layering is 170 μL when 440 μL polycation solution is placed in the solvent sector. If the polyanion solution is increased to 150 μL then the maximum volume for layering will be 145 μL.

2.3.2 Surface Meniscus

Absorption scans can exhibit a meniscus peak. This would probably influence the detection of the initial membrane growth at the meniscus interface. Therefore, the occurrence of such a meniscus peak dependent on the solution concentration, the wavelength and the spin velocity was studied.[12] More in detail, standard double-sector cells were filled on the solution side with sodium alginate (SA) solutions of different concentrations. The cells were placed into a 4-hole rotor for simultaneous measurements of three concentrations. All solvent sectors were filled with the same polycation solution. Absorption scans resulting from the difference between the polyanion and the polycation solution are summarized in Figure 6 for various run conditions. The differences of the radial positions for the three concentrations are due to slightly different volumes in the solution sectors. The high viscosity of the alginate solutions limits, to a certain extent, very precise dosage. An error of 150 ± 5 μL can be calculated for the extreme radial positions in Figures 6(a) and (b).[25] Hence, this experiment simultaneously served to identify the precision of the solution deposition.

As expected, the height of the meniscus peak reduces with increasing velocity [Figures 6(a)–(c)] and modified wavelength, eventually disappearing.[12] The signal can become noisier at other wavelengths. Therefore, an optimum with respect to the signal-to-noise ratio has to be defined. Additionally, sedimentation force must be applied where the polycation solution passes the capillary between the two sectors at a sufficiently high speed while avoiding the sedimentation of any component.

As an example, for the investigation of the membrane formation of OC with SA, an optimum was assigned to a velocity range of 3000–8000 rpm and a wavelength range of 360–380 nm.[12] Interestingly, in all cases, when a stable membrane was formed, it did not sediment even when run velocities of 20 000 rpm were applied after membrane

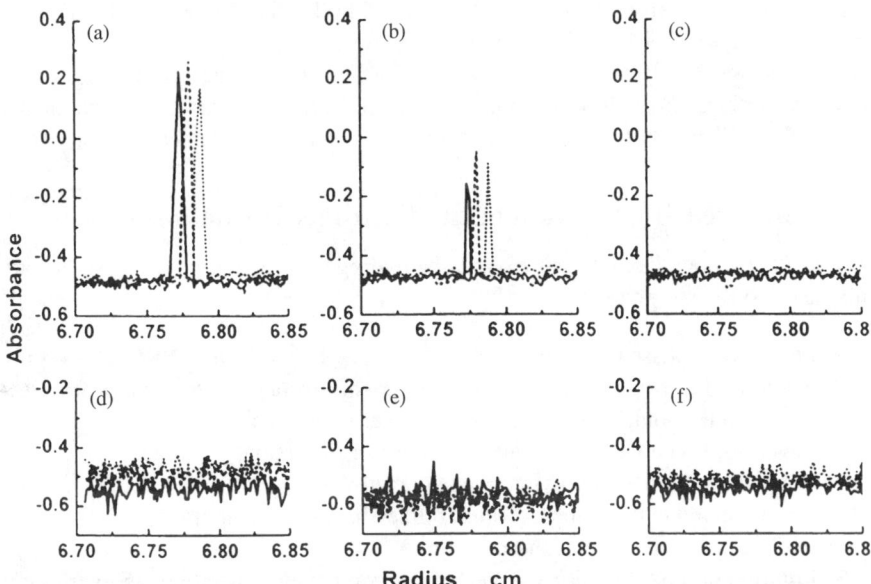

Figure 6 *Meniscus detection as a function of SA concentration, run velocity, and wavelength. Sodium alginate concentration: — 2%, --- 1.5%, ⋯⋯ 1.0%; (a) 3000 rpm, 380 nm; (b) 5000 rpm, 380 nm, (c) 12 000 rpm, 380 nm, (d) 3000 rpm, 360 nm, (e) 5000 rpm, 360 nm, (f) 10 000 rpm, 360 nm. Reprinted from Wandrey and Bartkowiak[29] with permission from Elsevier. Copyright (2004)*

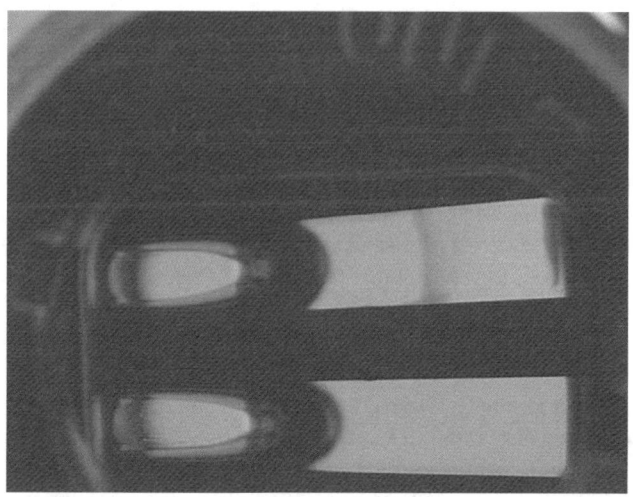

Figure 7   *Membrane location in the right centerpiece of the synthetic boundary cell. Reprinted from Wandrey and Bartkowiak[29] with permission from Elsevier. Copyright (2004)*

formation (Figure 7). Influence of the angular velocity on the membrane growth rate and direction was identified at high speed. Details will be discussed in Section 5.

Figure 7 was taken after approximately 2 h at 20 000 rpm. Even deceleration, removal of the cell from the rotor, and cell handling for the photograph did not alter the vertical position of the flat membrane in the centerpiece sector. Moreover, the membrane was stable over days, permitting further off-line characterization subsequent to removal from the cell. In contrast to the solution meniscus, which clearly shows the influences of the surface tension, the membrane remained flat.

2.4   Detailed Observation of the Hydrogel Formation Process

The synthetic boundary experiment and the hydrogel formation process can be subdivided into several phases:

- Preparation phase – running the AUC at very low velocity (1000–2000 rpm), where no layering occurs, to flatten the meniscus and to remove all polyelectrolyte in the solution sector from the lower optical lens.
- Layering, first contact, and initial membrane formation.
- Consumption of both polyions, network growth, until one component is completely bound in the complex, immobilized inside the membrane, or is excluded from contact with the second polyion.
- Equilibration by diffusion of one or both components, swelling, and shrinking
- Under certain conditions destabilization occurs with time.[13]

Figures 4 and 8 show examples of different phases.

Figure 8 provides one absorbance and one interference scan taken immediately after increasing the velocity to 5000 rpm (2 min), and subsequent to the first contact.

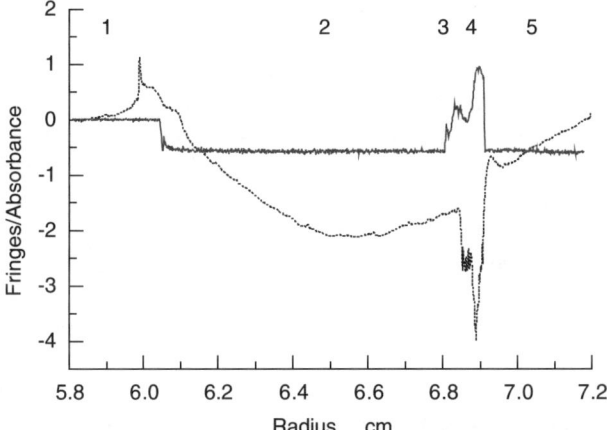

Figure 8 *Absorption (—) and interference scan (·····) taken at the onset of the experiment at 5000 rpm, 370 nm, and 20 °C. The scans were taken with a time shift of approximately 6 min due to the slowness of the UV/vis scanning. The scan ranges correspond to ranges 1–5 in Figure 3b. Reprinted from Wandrey and Bartkowiak[29] with permission from Elsevier. Copyright (2004)*

The interference scan appears very quickly (within seconds) and always detects the onset of the experiment. However, since minutes are required to monitor the UV scan for the total radial range of the centerpiece sector (radial position 5.8–7.2 cm) with a step width of 10 μm, the absorbance scan appears with a time delay of approximately 6 min. During the scan time, the meniscus in the solvent sector (polycation) moved from a radial position 5.989 to 6.047 cm. Hence, it is recommended for systematic investigations to scan only a limited range around the meniscus where the network is formed. At this position, rapid replication is desirable, and even necessary, in order to follow the membrane growth. To record a radial range of 2.5 mm with a step width of 10 μm requires less than 1 min.

Furthermore, Figure 8 serves to practically justify the experimental principle of Figure 3. The five radial ranges schematically presented in Figure 3 were proved experimentally for absorption and interference measurements. From the absorption scan the same intensity is detected for ranges 2 and 5 representing the difference between the non-absorbing SA solution in the solution sector and the slightly absorbing OC solution in the solvent sector. Ranges 3 and 4 correspond to the layered OC solution and the polyelectrolyte complex, respectively.

Figure 4(a) demonstrates the progress of the membrane formation process during a period of 21 min subsequent to the polycation–polyanion contact. It is obvious that the polycation consumption at the membrane surface (on the left side of the membrane peak) is very fast during the initial membrane formation. After 12 min the membrane growth slows down and is probably strongly diffusion-controlled for the components, which have to pass the initial membrane.

As shown in Figures 4(b) and (c), the membrane grows as long as the polycation is available. The membrane widens (membrane thickness equals the width of the scan peak at a given time) in both the top (left) and bottom (right) directions for the selected experimental conditions. Both the absorbance increases at the peak maximum and decreases in radial range 3 indicate polycation consumption. The preferred growth direction was found to depend on experimental conditions.

When all polycation has reacted some membranes start to equilibrate. This partly occurred in Figure 4(c). Equilibration depends on the molecular characteristics of the reacting polyelectrolytes and the particular polyanion and polycation concentrations. At higher run velocities the sedimentation force may influence the direction. Under certain conditions equilibration yields structures consisting of two phases, one of which is very compact whereas the other is like a weak gel.[12] It has to be emphasized that Figures 4 and 8 are selected as examples to demonstrate the various phases of the membrane formation process. They do not correspond to a desired optimal membrane. The occurrence and the extent of the specific phases strongly depend on the experimental conditions. This holds in particular for the growth direction, the membrane thickness, its equilibration, and phase separation.

Comprehensive investigation of all phases is of academic interest. For the majority of practical applications, however, the production process can be stopped after the defined time by interrupting the polyanion/polycation contact. As an example, microcapsules are removed from the receiving bath and washed to remove non-reacted polyelectrolyte molecules.[26] Contrarily, if the production of tubular homogeneous material is intended longer reaction and equilibration become indispensable.

2.5   Conclusions from Scans

Various levels of information can be obtained from absorption (A) and interference (I) scans. These include the velocity of layering (I), the polyelectrolyte concentration in the case of absorbing molecules (A), the membrane and solution positions (A, I), the membrane thickness (A, *I*), the membrane growth direction (A, I), the membrane growth velocity (A, *I*), and the membrane structure (*A*, *I*). Italic letters denote scans whose information is only qualitative, not yet clear, or needs more experimental data as well as the comparison of data from several polyanion/polycation systems.

2.6   Reproducibility and Sensitivity

Repeated runs under equivalent experimental conditions have been performed to identify the reproducibility and sensitivity of the modified synthetic boundary experiment.

Figure 9 shows the reproducibility comparing scans of two runs performed under identical conditions. It is visible that the scan of the first run recorded after 25 min reaction time [Figure 9(a)] is well reproduced by the repeat. The small differences between the chitosan concentrations on the left side of the membrane peak (radius <6.85 cm) probably result from fluctuations during layering. The differences disappeared with OC consumption after 70 min [Figure 9(b)]. The experiments demonstrated good scan reproducibility when the variation of the layering velocity was less than 20%. The smoothness of the curves is limited by the step width of the detection, which has its minimum at $10\,\mu m$.

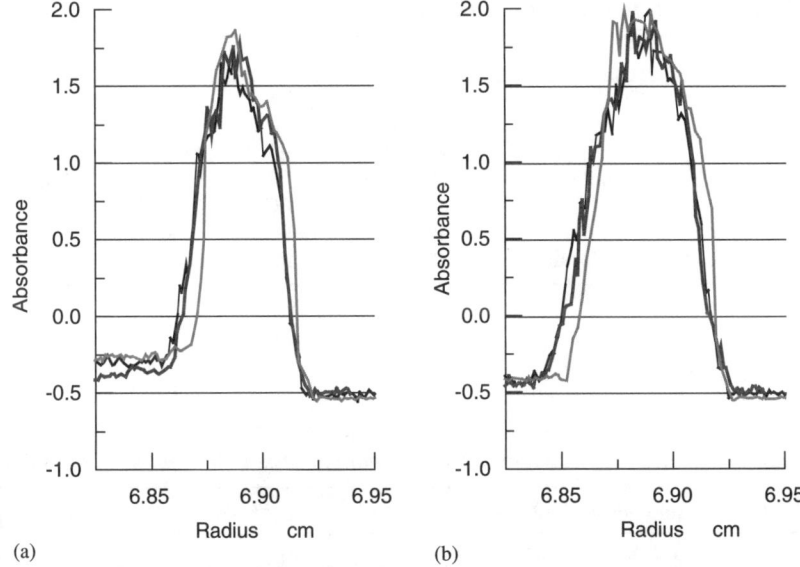

Figure 9 *Reproducibility and sensitivity of membrane formation experiments. 1.5% SA and 1.0% chitosan in 0.9% NaCl solution at 5000 rpm and 20 °C. Black scans run reproducibility with chitosan $M_n=2800\,g\,mol^{-1}$ and $M_w/M_n=1.7$; gray scan run performed with slightly different polycation chitosan $M_n=3000\,g\,mol^{-1}$ and $M_w/M_n=1.8$. Reaction time (a) 25 min, (b) 70 min. For better comparison the scans of the three separate experiments have been shifted to compensate the different radial positions of the menisci due to volume fluctuations. Reprinted from Wandrey and Bartkowiak[29] with permission from Elsevier. Copyright (2004)*

The third peak, grey in Figure 9, demonstrates the experimental sensitivity. In this experiment the solution of another polycation was layered on the same SA solution. Slight differences in the macromolecular characteristics of the OC ($M_n=3000$ g mol^{-1} and $M_n/M_w=1.8$ instead of $M_n=2800$ g mol^{-1} and $M_n/M_w=1.7$) caused changes in the scan width and shape.

3 Specific Studies – Application Examples

The method was evaluated on the basis of its the sensitivity in monitoring differences in hydrogel formation when the reacting components and experimental conditions vary. For this purpose the complex formation of SA with three polycations, poly(vinylamine hydrochloride) (PVAm), poly(L-lysine hydrochloride), and chitosan oligomers, was examined by varying specific experimental conditions such as concentration, pH, and molar mass.

3.1 Experimental Conditions

Table 2 summarizes the characteristics of the polyelectrolyte, and Table 3 summarizes the solution compositions and experimental conditions. The presence of CaCl$_2$

Table 2 *Polyelectrolyte characteristics*

Polyelectrolyte	Abbreviation	type	Molar mass (g mol^{-1})
Sodium alginate	SA	PA	204 000[a]
Oligochitosan	OC1	PC	2600[b]
	OC2	PC	2800[b]
Poly(L-lysine hydrobromide)	PLL1	PC	27 400–30 200[c]
	PLL2	PC	2.300–2.900[c]
	PLL3	PC	18 000[c]
Poly(vinylamine hydrochloride)	PVAm	PC	17 000[d]

[a] AUC.
[b] GPC.
[c] Provided by the supplier.
[d] Intrinsic viscosity.

Table 3 *Summary of selected experimental data necessary to support the discussion*

Experiment	SA concentration (%)	PC type	PC concentration (%)	pH
1	1.0	OC1	1.0	6.5
2	1.5	OC1	1.0	6.5
3	2.0	OC1	1.0	6.5
4	1.5	OC1	1.0	6.0
5	1.5	OC1	1.0	7.0
6	1.0	OC1	1.0	6.0
7	2.0	OC1	1.0	6.0
8	2.0	PLL1[a]	0.5	5.0
9	2.0	PLL2[a]	0.5	5.0
10	2.0	PLL3[b]	0.5	5.0
11	1.2	PLL3[b]	0.5	5.0
12	2.0	PVAm[c]	0.5	5.0

SA, sodium alginate, solution prepared in 0.9% NaCl; PC, polycation, solution prepared in 0.9% NaCl if not differently indicated; OC, oligochitosan; PLL, poly (L-lysine hydrobromide); PVAm, poly(vinylamine hydrochloride).
[a] In 0.45% NaCl+50 mM CaCl$_2$.
[b] In 1.22% CaCl$_2$.
[c] In 0.45% NaCl+0.5% CaCl$_2$.

in some of the polycation solutions causes a rapid gelation of SA. The calcium alginate gel is virtually clear and only minimally contributes to the absorbance signal. The presence of CaCl$_2$ is of interest for some practical applications, and also supports very fragile polyelectrolyte complex networks. Polycations replace CaCl$_2$ from the initially formed ionotropic gel.

It should be noted that the hydrogel components that originate from the polyion solution in the same centerpiece sector participating in network formation are denoted by a plus (+). Components derived from different sectors are denoted by a slash (/). For example, A+B/C+D represents hydrogel formation by layering C+D, the components in the solvent sector, onto A+B, the components of the solution placed into the solution sector of the synthetic boundary centerpiece. Low molar mass salts or other moieties added to the solutions that do not directly participate in the polyelectrolyte complex formation, such as, for example, sodium chloride, will be considered as part of the solvent not appearing in this form.

All experiments have been performed at 5000 rpm after a short period at 1000 rpm. The scans were taken at wavelengths between 320 and 380 nm with a step width of 10 μm limited to the radial range where the complex formation occurred. The filling volumes were 100 μL in the solution sector and 400 μL (experiments 1–9) or 420 μL (experiments 10–12) in the solvent sector. To ensure reproducibility, each experiment was carried out in duplicate. The layering velocity was controlled by the interference scans and could be reproduced within ±20%.

The experiments were designed to reveal the influence of

- the chemical structure of the cationic component,
- the molar mass of the cationic component,
- the concentration of the anionic component, and
- the pH

on the formation process and the hydrogel structure identified by the time-dependent scan shape. This section presents and analyzes the influences.

3.2   Discussion of the Experiments

3.2.1   Influence of the Chemical Structure of the Polycation

Figures 5 and 10 demonstrate the influence of the chemical structure of the cationic polyelectrolyte for otherwise constant conditions and similar molar mass of the two polycations compared, PLL and PVAm.

A thinner and more uniform membrane was obtained with SA/PVAm+CaCl$_2$ than with SA/PLL+CaCl$_2$. The two polycations differ in their charge density, with PVAm possessing the higher one. The last absorption scans, each recorded after 22 min for PVAm and after 24 min for PLL, correspond to photomicrographs showing a homogeneous gel for PVAm and a heterogeneous structure in the cell bottom direction for PLL, whereas the surface is well defined [Figure 5(b)]. The existence of such a smooth and compact surface can be concluded from the steep increase in the absorbance intensity on the left side of the scans in both Figures 5 and 10.

3.2.2   Influence of the Molar Mass of the Polycation

Results of a series of experiments where PLL of different molar masses has been employed as the polycation are presented in Figure 11. The networks obtained from the three-component system SA/PLL+CaCl$_2$ are shown. The calcium ions diffuse very fast into the SA solution but only partially replace the sodium counterions in the alginate type used in this study. The scans reveal different membrane dimensions and formation characteristics.

In this system, neither the polyanion (SA) nor the polycation (PLL) absorb at a wavelength of 370 nm. Therefore, the baseline appears at 0 absorbance on the left side of the membrane peak. The signal on the right side is caused by the slight turbidity of

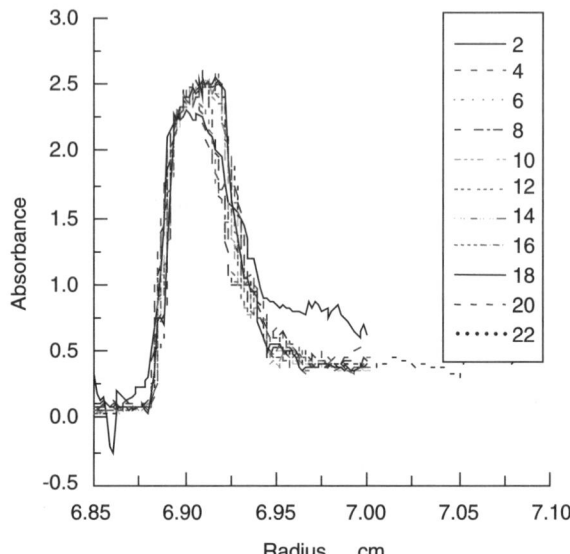

Figure 10 *Membrane formation of SA with poly(vinyl amine hydrochloride), experiment 12 in Table 3. 5000 rpm, 20 °C, 320 nm. Reprinted from Wandrey et al.[30] with permission from Springer. Copyright (2004)*

the supporting Ca-alginate gel, which is built immediately and more rapidly than the polyelectrolyte network. Nevertheless, the polyelectrolyte complex corresponding to range 4 in Figure 3 is clearly separated from the Ca-alginate gel, which occupies range 5. Figure 11 clearly reveals the influence of the molar mass of PLL on the scan width, with the lower molar mass forming broader peaks and thicker membranes.

The polyelectrolyte complex network SA/PLL is suggested to be denser in the case of the lower molar mass PLL2, for which a higher absorbance is detected in Figure 11. The suggestion has been verified by permeability measurements, clearly indicating a cut-off decrease (denser network) if the network SA/PLL was formed with lower molar mass PLL.[27] Furthermore, for the thinner and less compact membrane formed with PLL1 (higher molar mass), it can be concluded that less alginate has reacted with PLL1 than with PLL2 (lower molar mass). Consequently, more alginate remained available to react with calcium, yielding a higher calcium alginate concentration in range 5. The higher Ca-alginate concentration seems to initiate precipitation visualized by sedimentation profiles in the upper graph of Figure 11 on the right side of the membrane peak. The Ca-alginate concentration in the lower graph of Figure 11 is expected to be lower since more alginate is bound to PLL2. Consequently, the tendency to precipitate is diminished here.

3.2.3   Influence of the Concentration of the Polyanion

The plots in Figure 12 represent a superposition of the scans collected during the network formation when OC was layered on SA solutions of different concentrations,

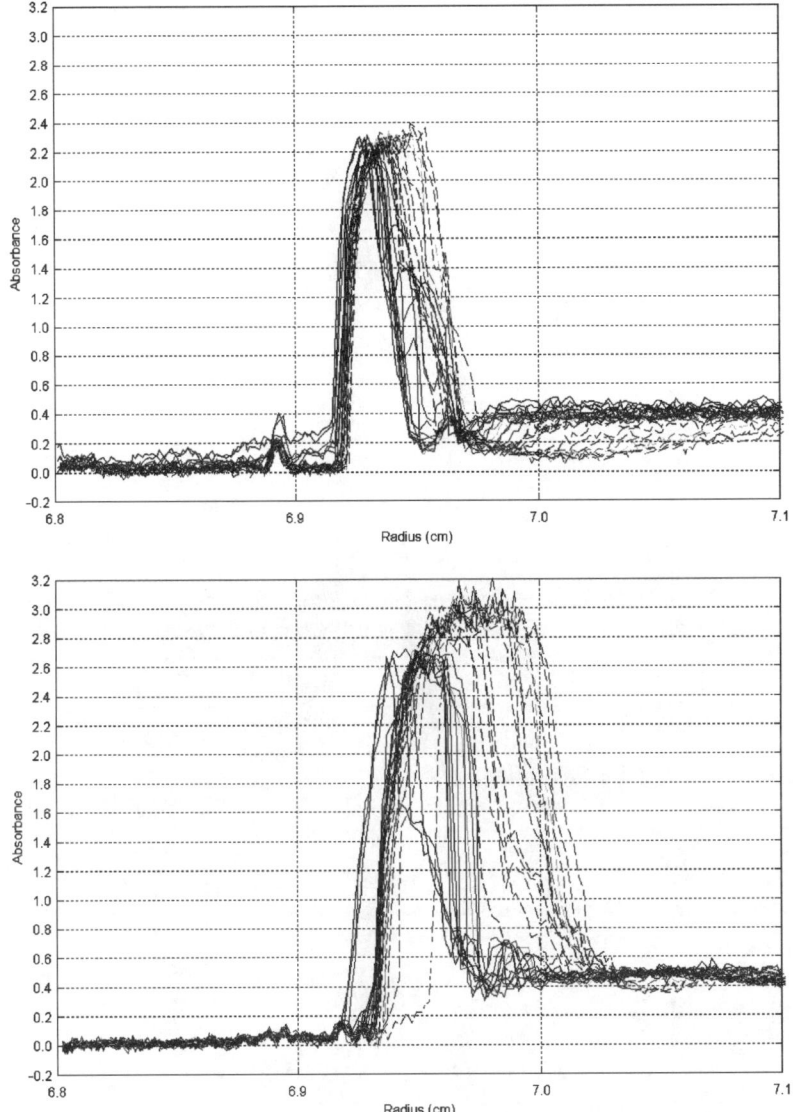

Figure 11   *Influence of the molar mass on the membrane formation of SA with poly(L-lysine hydrobromide), top: experiment 8 in Table 3, bottom: experiment 9 in Table 3. 5000 rpm, 20 °C, 370 nm. 30 scans, scan delay 2 min for the first 20 scans, 10 min for the last 10 scans. Reprinted from Bourdillon et al.[31] with permission from Springer. Copyright (2004)*

1.0, 1.5, and 2.0%, at constant pH=6.5. The influence of the concentration is clearly visible for all plots. The differences concern the peak area, the peak width, and the peak symmetry.

The negative intensity of the baselines on both sides of the membrane peak results from the absorbance of OC. The time-dependent, decreasing intensity on the left side

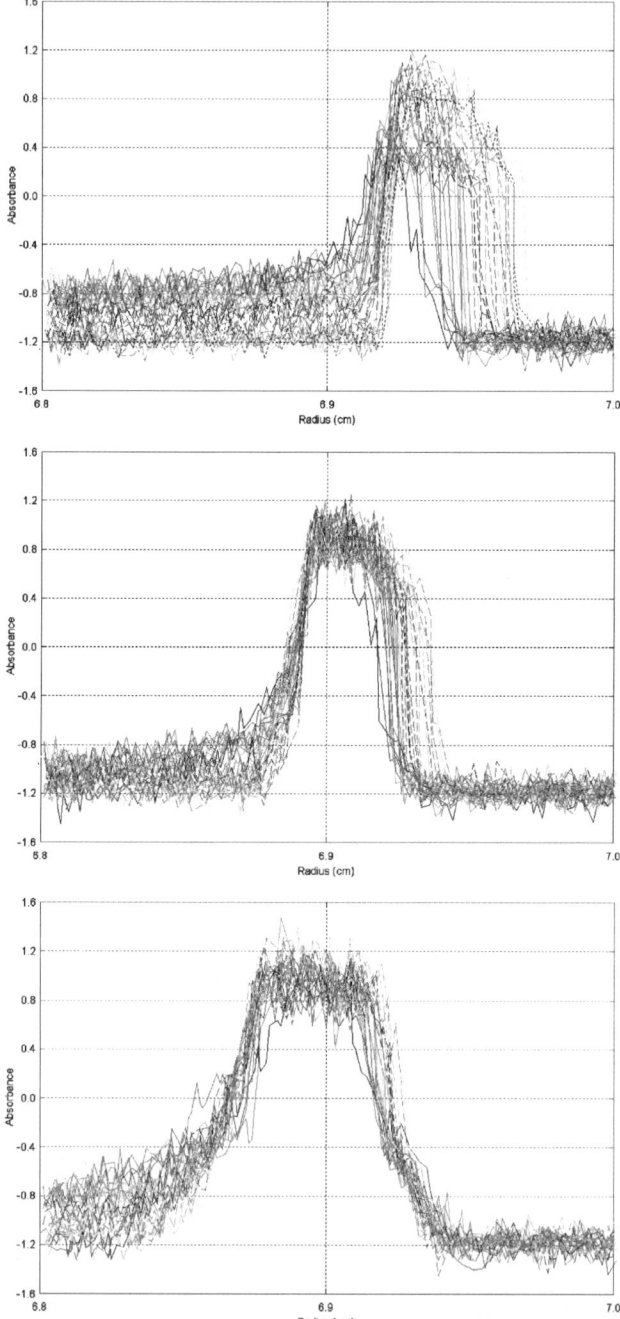

Figure 12 *Influence of the SA concentration on the membrane formation of SA with OC. From top to the bottom: experiments 1–3 in Table 3. 5000 rpm, 20 °C, 370 nm. 30 scans, scan delay 2 min for the first 20 scans, 10 min for the last 10 scans. Reprinted from Bourdillon et al.[31] with permission from Springer. Copyright (2004)*

of the membrane peak corresponds to the chitosan consumption with time. The constant absorbance on the right side confirms that SA is not absorbing, no OC is permeating without complex formation, and no sedimentation takes place. It is obvious that the shape of the membrane peak strongly depends on the initial concentration of the polyanion solution, although all SA/OC membranes exhibit their maximum at approximately 2.2 absorbance units.

The membrane growth in the SA solution is most strongly pronounced for the lowest SA concentration. A less dense network can be concluded for this case allowing faster diffusion and subsequent reaction on this site. The effect is less for 1.5% and almost missing for 2%. On the other hand, peak broadening occurs in its upper part, at absorbance >-0.4 on the right side indicating homogenization of the network. A well-defined surface can be concluded for the lowest concentration. Working at low SA concentration, the SA/OC membrane in Figure 12 exhibits a significant absorbance difference on both sides of the membrane peak, with a lower one in the direction of the SA solution direction.

3.2.4 Influence of the pH

For the three graphs in Figure 13 the pH has been varied in the range 6.0–7.0, but the SA concentration of 1.5% was kept constant. Here the strongest influence is visible on the peak symmetry. Less influence can be concluded for the scan width, the peak area, and height.

For the experiment at pH=6.0 (upper plot in Figure 13) the absorbance has a slightly different basic level, -1.1 rather than -1.2. It is known that the solubility of chitosan improves with decreasing pH as a consequence of a higher degree of protonation. An altered solution state can be suggested at pH=6.0 and, consequently, different absorption.

4 Scan Evaluation and Membrane Classification

Absorbance scans and interference scans were simultaneously collected as raw data during all synthetic boundary experiments performed, and have been partly presented in Section 3. The software package delivered with the AUC does not provide any software suitable for specific evaluation of scans recording interfacial reactions. Therefore, a procedure has been developed based on MATLAB software, Version 6.1.0, to extract scan data for subsequent evaluation of the absorbance scans. A 3D plot as shown in Figure 14 illustrates the time-dependent network formation and supports the evaluation. Turning the plot on the computer screen improves the visualization.

A number of criteria have been selected and defined for an objective evaluation of the absorbance scans, and the conclusions thereof have been found applicable to different types of polyelectrolyte networks formed at the flat interface in the synthetic boundary centerpiece. These are membrane thickness (MT_0), membrane compactness (MC), membrane homogeneity (MH), membrane symmetry (MS), and membrane formation kinetics. The component amounts, which are instantaneously available for the reaction, result from interference scans. The criteria for the absorption scans will be derived and explained in this section prior to the interference scan analysis.

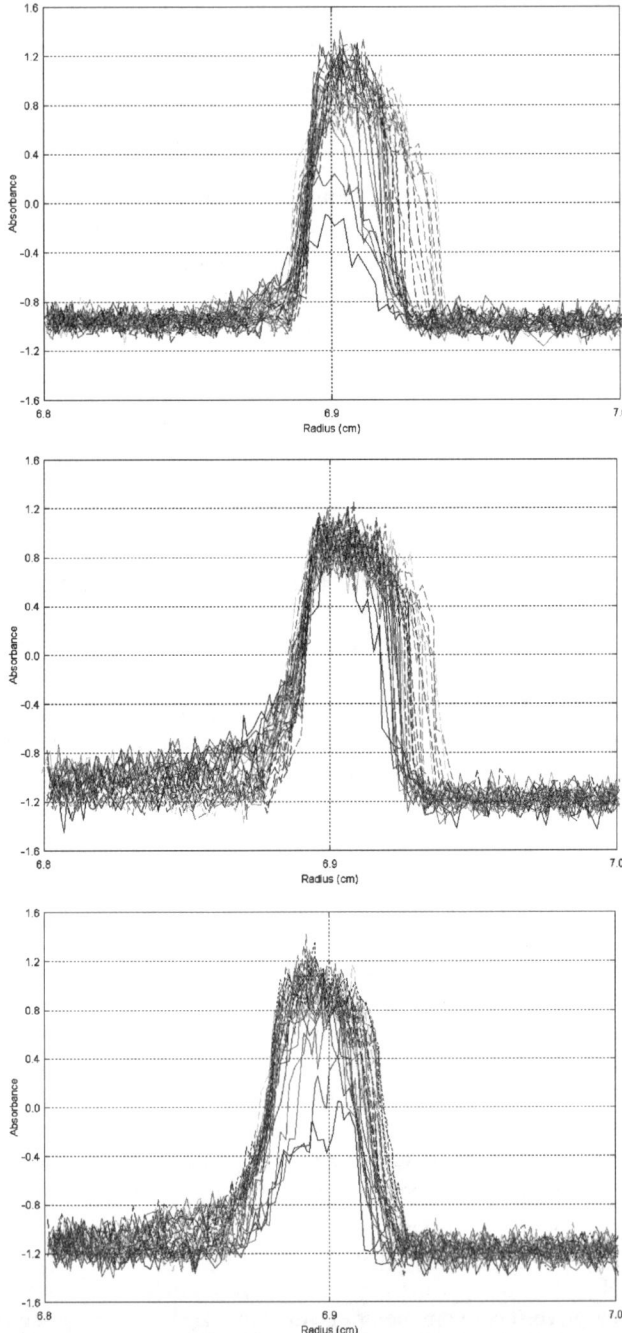

Figure 13   *Influence of the pH on the membrane formation of SA with OC. From top to the bottom: experiments 4, 2, and 5, in Table 3. 5000 rpm, 20 °C, 370 nm. 30 scans, scan delay 2 min for the first 20 scans, 10 min for the last 10 scans. Reprinted from Bourdillon et al.[31] with permission from Springer. Copyright (2004)*

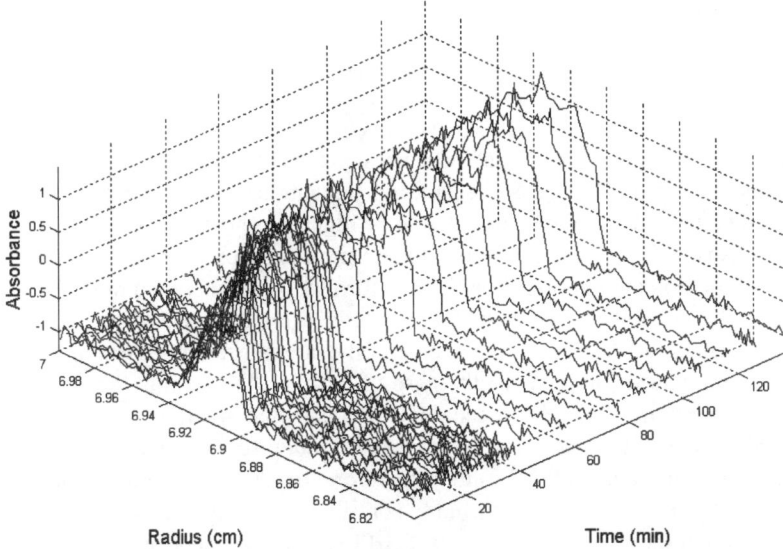

Figure 14 *3D presentation of a membrane formation experiment. 1.0% SA, 1.0% OCl, 0.9% NaCl, pH=7. 5000 rpm, 20 °C, 370 nm. 30 scans, scan delay 2 min for the first 20 scans, 10 min for the last 10 scans.*

Afterward, the applicability to different experimental findings will be examined and discussed for each. Here, the term 'membrane' stands for the polymer network composed of two oppositely charged polyelectrolytes, which can be considered as hydrogel. The term 'membrane' was selected for the classification to distinguish between the polyelectrolyte complex and the ionotropic gel formed by interaction of the polyelectrolyte with divalent counterions, here calcium ions, which were added in some experiments.

These criteria represent a current approach to a more detailed scan evaluation. However, they can be only qualitative parameters owing to the dependence on the turbidity via absorption optics.

4.1 Absorbance Scans

4.1.1 Membrane Thickness

Basically, the scan width should be the measure of the membrane thickness. However, at what minimum absorbance intensity can the network be considered a membrane if the absorbance varies with the radial position? For the evaluation, a minimum value of 0.5 units was arbitrarily assumed. In the case of SA/OC membranes in Figures 12 and 13, the membrane thickness was taken at -0.5 absorbance after 140-min reaction time (last scan). Performing the comparison at a fixed intensity seems to be more reliable than to take the values at the baseline, which is, in some cases, relatively noisy. The procedure is based on the assumption that the same chemistry yields the same network density.

4.1.2   Membrane Compactness

The membrane compactness parameter provides information on the structural density of the polyelectrolyte complex network. This compactness is determined on the basis of the maximum value of the absorbance scan.

Of course, the impact of several influences on the absorbance signal must be carefully distinguished. Clearly, self-absorbing polyanions and polycations will contribute. Their concentration is expected to be higher in the polyelectrolyte complex. The intensity will increase at the membrane peak position. Furthermore, since the ability of the polyanion/polycation pairs to form polyelectrolyte complex hydrogels and not solid precipitates is determined, to a large extent, by the hydrophilicity of the polymer backbone, the degree of swelling of the complex affects the signal by more or less pronounced turbidity. In the case of SA/OC, OC became absorbing after chain degradation.[28] The absorbance of a 1% chitosan solution is 1.2 at 370 nm detected by a -1.2 signal in the solution difference scans in the scan plots. The higher membrane signal (2.0–2.2) may, therefore, result from a higher chitosan concentration in the membrane or a turbidity contribution. As the second system, a combination of non-absorbing components, SA/PLL+$CaCl_2$, has been selected with the membrane peak only related to the turbidity. Considering all these aspects, a comparison of the compactness between different membrane chemistries needs further investigation.

4.1.3   Membrane Homogeneity

Equation (1) defines the homogeneity of the membrane. For the calculation, the difference between the membrane thickness (MT_0) and the scan width at a second different absorbance value (MT_1) is related to the absorbance difference, A_1-A_0, where the thickness has been measured, and the total membrane thickness, MT_0:

$$MH = 1 - \frac{MT_0 - MT_1}{(A_1 - A_0)MT_0} \tag{1}$$

The homogeneity becomes 1 according to Equation (1) for a rectangular peak where $MT_0 = MT_1$. The evaluation is schematically presented in Figure 15. The calculation considers the slope of the scan by relating the scan width difference ($MT_0 - MT_1$) at two absorbance values, to the appropriate absorbance difference ($A_1 - A_0$). This is assumed to correlate to a network density gradient on both sides of the membrane. Since the same slope will be more significant for a thinner membrane than for a thicker one, the ratio has been, in addition, related to the total membrane thickness. Such an evaluation provides more information than simply relating the peak area to the width at half of the maximum of the absorbance value.

Because the homogeneity parameter does not provide any information on which side of the membrane the heterogeneity occurs, a further parameter, the symmetry parameter, was introduced.

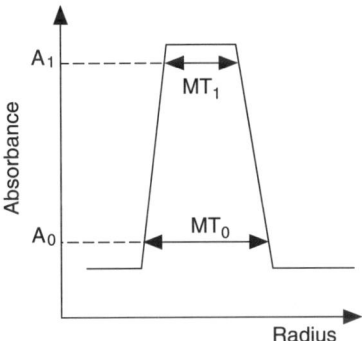

Figure 15 *Principle of the homogeneity calculation according to Equation (1). Membrane thickness at A_0 (MT_0), membrane thickness at a second absorbance value A_1 (MT_1). Reprinted from Bourdillon et al.[31] with permission from Springer. Copyright (2004)*

4.1.4 Membrane Symmetry

Figure 16 schematizes five extreme membrane types, including subforms, that can in general be expected from experiments and have been partly observed. Table 4 supports the classification.

Types I–III have peaks of continuous slopes on both sides, whereas types IV and V present discontinuity on one or both sides. More specifically, the latter can be subdivided into two geometric areas, with a rectangular one either on the bottom (types IVa, Va, Vb) or on the top (types IVb, Vc, Vd). The subdivision is marked by dotted lines in Figure 16.

As a most simple evaluation, the difference of the absorbance units on both peak sides, S_1 and S_2, related to the higher absorbance is proposed:

$$MS = 1 - \frac{S_1 - S_2}{S_1} \tag{2}$$

with $S_1 = A_1 - A_0$ and $S_2 = A_2 - A_0$; compare the schematic presentation in Figure 17. $MS = 1$ for $S_1 = S_2$, a fully symmetric membrane. Equation (2) tries to quantify the evaluation for a selected example, types Va and b. Type Va corresponds to peaks as obtained in Figures 12 and 13 at low SA concentration and low pH, respectively.

The molar mass of PLL has a significant impact on the membrane symmetry. The two membranes in Figure 11, however, possess a different shape if compared with the SA/OC membranes in Figures 12 and 13. The shape of the SA/OC membrane reflects an absorbance decrease at the top of the peak, whereas the peak shape for SA/PLL+CaCl$_2$ is characterized by the onset of an absorbance decrease in the middle of the peak. Equation (2) is not applicable in this case. More forms can be derived combining, for example, types 3 and 5. One extreme experimental example is presented in Figure 18. It shows a network that is compact in the center. The potential of the absorbance scans to monitor such a specific network structure is confirmed by the photomicrograph, which visualizes the structure concluded from the scans.

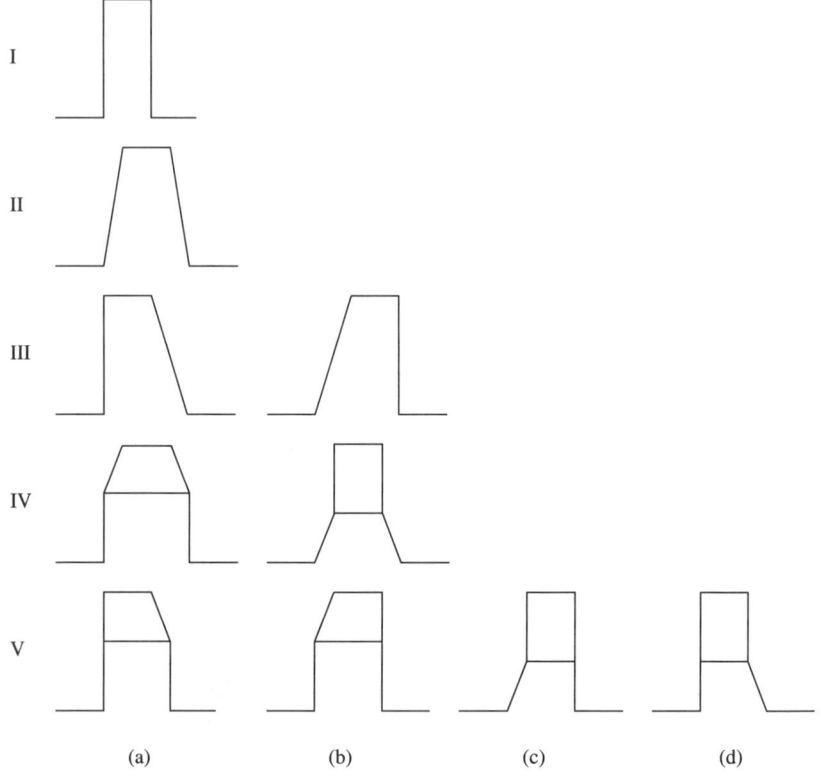

Figure 16   *Schematized membrane types I–V sub-groups a–d as described in Table 4*

Table 4   *Membrane classification*

Type	Homogeneous	Symmetric	Monotonous
I	+	+	+
II	−	+	+
III	−	−	+
IV	−	+	−
V	−	−	−

4.1.5   Kinetic Aspects

Different membrane formation processes are obvious from Figures 5 and 10–13. This concerns both the formation of a certain shape and the kinetics. For kinetic considerations, two approaches have been developed and applied to evaluate the membrane formation process and membrane quality on the basis of the experimentally obtained radial absorption scans. The first approach generalizes the membrane growth and its radial direction, resulting in a time-dependent description of the membrane geometry. The second one considers the membrane thickness as a function of time and might,

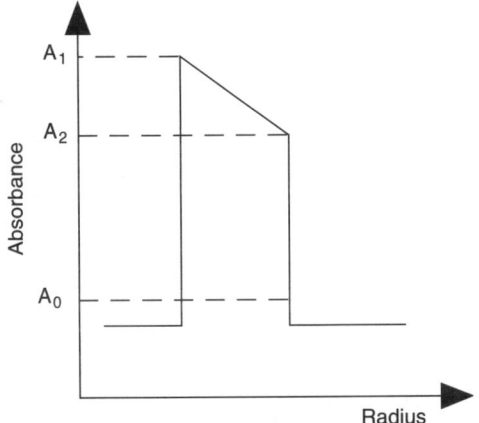

Figure 17 *Principle of the symmetry calculation according to Equation (2). Reprinted from Bourdillon et al.[31] with permission from Springer. Copyright (2004)*

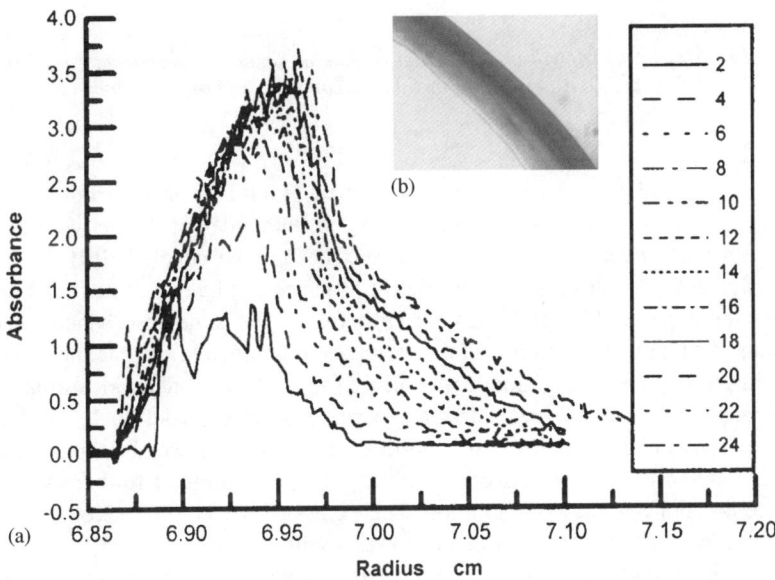

Figure 18 *Example of a heterogeneous, non-symmetric membrane, experiment 11 in Table 3. (a) absorbance scans, (b) photomicrograph of the membrane. 5000 rpm, 20 °C, 320 nm. Reprinted from Wandrey et al.[30] with permission from Springer. Copyright (2004)*

therefore, serve to calculate the growth velocity or shrinking of the membrane, but not the directions of growth or shrinking.

The first approach is presented in Figure 19 for SA/PLL+CaCl$_2$. The graph analyzes the movement of the contact interface in the initial phase, the growth direction, and may serve to calculate the growth velocity from the difference of the slopes

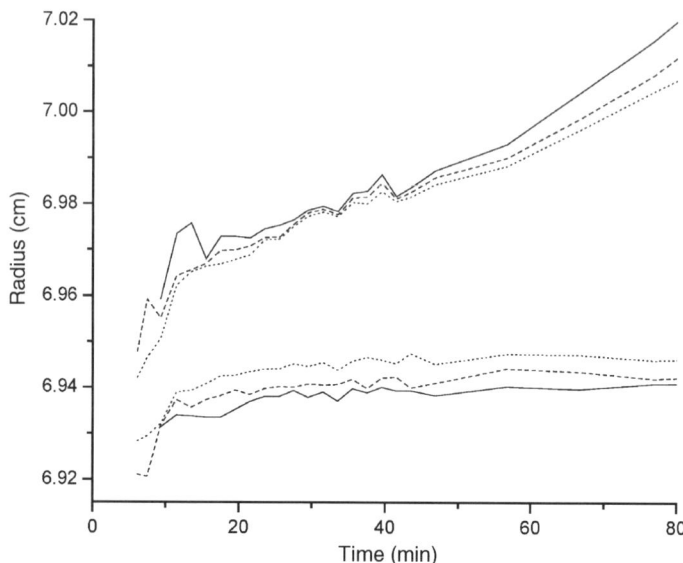

Figure 19 *Membrane growth as a function of time at various absorbance; (—) 1.0; (---)*
1.5; (······) 2.0; experiment 9 in Table 3, compare Figure 11, bottom

taken for both radial positions. Moreover, if there is a minor or no difference in the
radial positions at various absorbance values on both sides of the network a more
homogeneous network with well-defined surfaces can be concluded. If this is the
case only for the lower radius, a well-defined surface can be interpreted. The difference of the upper and lower curves corresponds, for the same absorbance value, to
the scan width at this position, and the extremes to the membrane thickness, MT_0.

Owing to the use of the same components for the SA/OC network formation in
experiments 1–7, the differences in the experimental findings definitely result from
the variations in the experimental conditions. Probably, a lower SA concentration
yields, initially, less dense networks, allowing the penetration of further OC chains
into the less concentrated SA solution. The excess of SA in the case of higher SA
concentrations will push the SA chains to react with the OC.

Using the second approach of time-dependent scan evaluation, the growth rate or
equilibration rate may be calculated. The formation process is described in Figure 20,
where the increase of the membrane thickness with time is compared for 1 and 2%
SA at pH=6.5. At 1% SA, the membrane thickness continuously increases, whereas
at 2% SA, a final thickness seems to be achieved after approximately 50 min. Clearly,
this process is also influenced by the availability of the OC, which was kept constant
for the two experiments.

Comparing the two approaches reveals that if instead of the increase of the membrane thickness as a function of time, the change in the radial positions at selected
absorbance with time is plotted, a more comprehensive description of the growth
process becomes accessible.

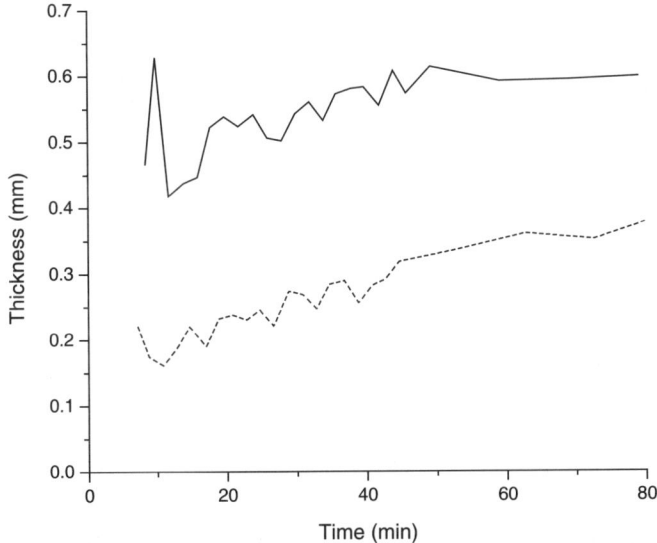

Figure 20 *Membrane thickness as a function of time. (—) experiment 3, (---) experiment 1 in Table 3, compare Figure 12. Reprinted from Bourdillon et al.[31] with permission from Springer. Copyright (2004)*

4.1.6 Non-specific Information

The total membrane thickness, MT_0, was estimated somewhat far from the baseline. In particular, in the case of SA/OC at 2% SA, a peak broadening at low absorbance appears. This corresponds to a less defined and rougher membrane surface. A well-defined and steep absorbance jump suggests a flat, sharp surface. In general, the surfaces of the PLL membranes are better defined than the SA/OC surfaces.

4.2   Interference Scans

The interference scans serve to calculate the solvent amounts in the two sectors. The movement of the meniscus in the polycation sector reflects the volume decrease, and the movement of the meniscus in the polyanion sector reflects the volume increase there. One example of the detection of the layering velocity is presented in Figure 21.

The inset of Figure 21 shows the movement of the time-dependent meniscus. The meniscus air/chitosan in the solvent sector (lower curve in the inset of Figure 21) is preferably recommended for the velocity evaluation, because of its more precise detection. In addition, no influence of mixing, diffusion, or complex formation can happen in this sector. Because of the trapezoidal cross-section of the sectors, the meniscus movement is not expected to be linear. From the known sector geometry (2.5° sector angle) the volume can be calculated.[26] For the example provided in Figure 21, a layering velocity of 4.9 μL min^{-1} was calculated.

In general, the layering velocity depends on the run velocity, the capillary diameter, and the solution properties of the solution to be layered. For the membrane formation

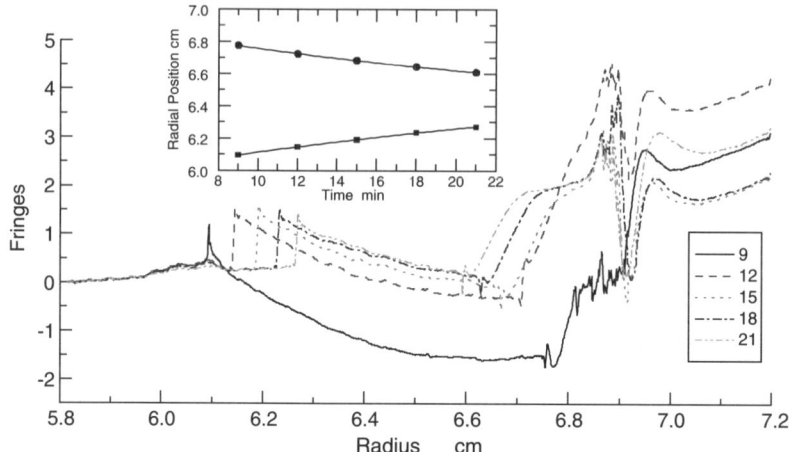

Figure 21   *Interference scan for the calculation of the layering velocity, scan time in minutes. Inset: time-dependent change of the radial position of the menisci in both cell sectors;* ■ *solvent sector;* ● *solution sector. Reprinted from Wandrey and Bartkowiak[29] with permission from Elsevier. Copyright (2004)*

experiments, a variation of the capillary diameter is recommended in order to realize similar velocities for solutions of different concentration and viscosity.

Information about the layering velocity can also be obtained from the absorption scans if they are recorded for the necessary broad radial range. Since absorption scans are more time-consuming, fast repetitions cannot be performed.

From the layering velocity, the amount of available polycation at a given time can be calculated. Together with the detection of the non-reacted polycation, conclusions about the total amount of complexed chitosan can also be drawn. The determination of the layering velocity is recommended for each experiment.

Both the decrease of the volume in the polycation sector and its increase in the polyanion sector should be identical, but this was not the case for most experiments. In Figure 22, the differences between the expected volume increase and the real increase are plotted, for three experiments. They decrease with time. More precisely, the volume in the polyanion sector increases less than expected from additivity. The volume loss probably identifies shrinking during the complex formation, a process that is difficult to follow experimentally but becomes, at least qualitatively, accessible.

5   Influence of the Sedimentation Force

All experiments and results reported so far correspond to conditions where no or a negligible influence of the sedimentation force was intended and realized. Figure 23 presents, for two different SA concentrations, 1.0 and 1.5%, the influence of the force on the membrane formation with OC when experiments were performed at 20 000 rpm. Both examples exhibit a strong network extension into the SA solution, but the membrane itself rests at the interface. Obviously, the lower SA concentration is

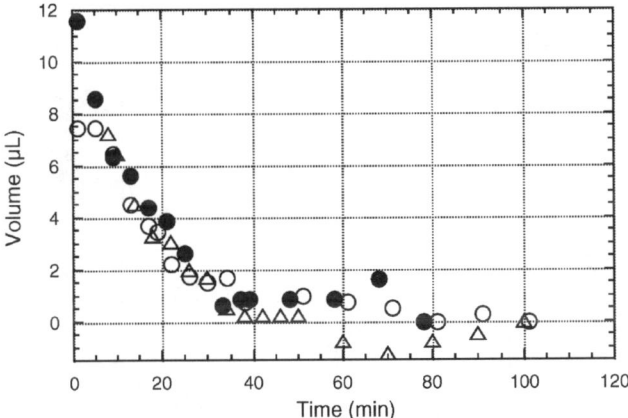

Figure 22 *Volume difference during the complex formation as a function of time, calculated as the difference between volume decrease in the polycation sector and increase in the polyanion sector. ● experiment 6, ○ experiment 4, △ experiment 7, all in Table. 3. Reprinted from Bourdillon et al.[31] with permission from Springer. Copyright (2004)*

advantageous for the growth process, resulting in a more homogeneous but also thicker network structure per unit time. Moreover, all of the chitosan is incorporated into the membrane after 65 min, which is monitored by identical low absorbance on both sides of the membrane. In contrast, the same level of absorbance was not reached for the higher concentration. Comparison with the experimental findings without force application (Figure 12) reveals different effects of the concentration on the final membrane structure. Higher mass transport due to force yields more homogeneity for lower concentrations. The growth is accelerated. Moreover, at high force the membrane surface becomes better defined at high concentration, which is demonstrated by the vertical absorbance jump.

6 Concluding Remarks

The synthetic boundary technique of analytical ultracentrifugation combined with the sensitivity and precision of the optical system of the OPTIMA XL-I was employed to investigate interfacial reactions at the flat interface of solutions containing oppositely charged macromolecules. The recently developed experimental procedure provides a multitude of information about the polymer network formation process and the network characteristics of the electrostatically linked network. Although some of the data provide qualitative information, as they depend on the turbidity reading via absorption optics, all characteristics together can contribute to establish qualitative and partly quantitative relationships to describe the influence of the polyelectrolyte characteristics and experimental conditions on the membrane-formation process and the membrane properties. Such relationships, to be used to support the development of membranes for specific applications such as cell immobilization or hydrogel delivery systems, are in progress.

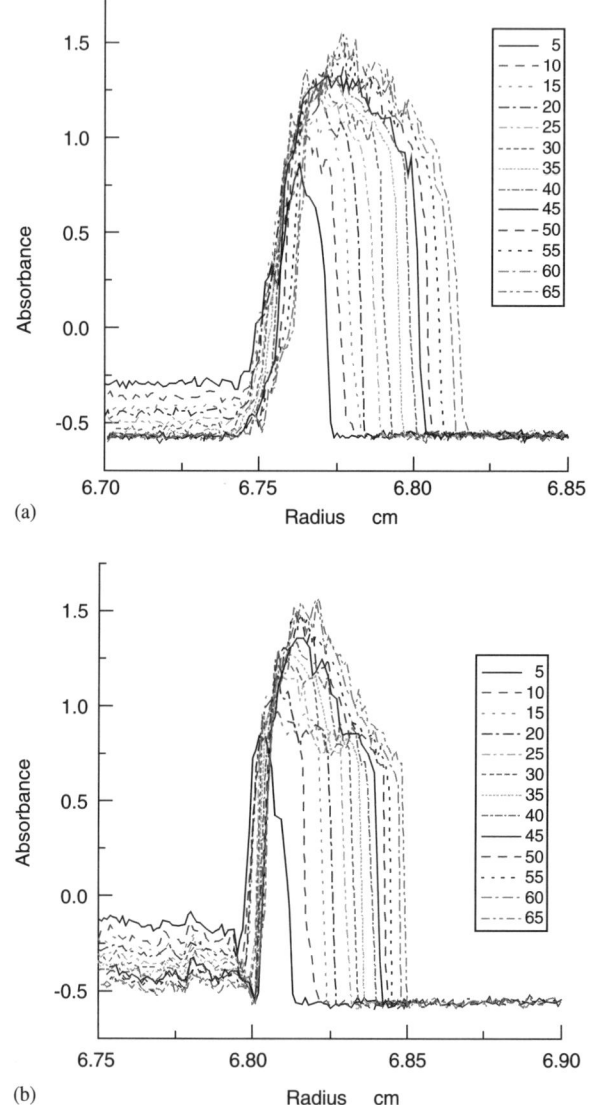

Figure 23 *Influence of the sedimentation force on the membrane formation process for two different SA concentrations with OC2 1.0% in 0.9% NaCl. (a) 1.0% SA , (b) 1.5% SA. 20 000 rpm, 20 °C, 380 nm*

In this chapter emphasis was laid on the description of the experimental principle. Experiments with various polyanions and polycations were selected to illustrate and evaluate the applicability and performance of the technique.

Future development of a fast-scanning UV optics is expected to facilitate the extension of the experimental principle to more, and in particular, faster, interfacial reactions, and will permit the use of up to seven cells at the same time. Furthermore, the feasibility of simultaneous diffusion studies to determine the 'membrane cut-off' should be evaluated.

Acknowledgements

The work is supported by the Swiss National Science Foundation, grants 2100-064996 and 200020-103615/1. The authors also thank P. Schuck, National Institutes of Health, Bethesda, USA, for his support in developing the evaluation software.

References

1. H. Cölfen, *Polym. News,* 2004, **29**, 101.
2. G. Kegels, *J. Am. Chem. Soc.,* 1952, **74**, 5532.
3. E. G. Pickels, W. F. Harrington and H. K. Schachman, *Proc. Natl. Acad. Sci.,* USA, 1952, **38**, 943.
4. Y. Okajima, Y. Homma and K. Seiki, *Biochim. Biophys. Acta,* 1993, **1169**, 103.
5. *An-60 Ti Analytical Rotor, Cells, and Counterbalance,* Beckman Instruments, Palo Alto, CA, USA, 1996.
6. H. K. Schachman and W. F. Harrington, *J. Polym. Sci.,* 1954, **12**, 379.
7. G. Ralston, *Introduction to Analytical Ultracentrifugation,* Beckman Instruments, Palo Alto, CA, USA, 1993.
8. D. L. Kemper and J. Evers, in *Methods in Enzymology XXVII. Enzyme Structure,* Part D, Academic Press, New York, 1973, 67.
9. R. Hersh and H. K. Schachman, *J. Am. Chem. Soc.,* 1955, **77**, 5228.
10 P. Voelker, *Progr. Colloid Polym. Sci.,* 1995, **99**, 162.
11. L. Börger, H. Cölfen and M. Antonietti, *Colloid Surface A,* 2000, **163**, 29.
12. C. Wandrey and A. Bartkowiak, *Colloid Surface A,* 2001, **180**, 141.
13. C. Wandrey, G. Grigorescu and D. Hunkeler, *Progr. Colloid Polym. Sci.,* 2002, **119**, 84.
14. L. Bourdillon and C. Wandrey, *Colloid Polym. Sci.,* 2004, **282**, 1247.
15. A. S. Hoffmann, *Adv. Drug Deliv. Rev.,* 2002, **43**, 3.
16. A. Kishida and Y. Ikada, in *Polymeric Biomaterials,* 2nd edn, S. Dumitriu (ed), Marcel Dekker, New York, 2002, 133.
17. H. Dautzenberg, W. Jaeger, J. Kötz, B. Philipp, C. Seidel and D. Stscherbina, *Polyelectrolytes: Formation Characterization and Application,* Hanser, Munich, 1994.
18. D. Hunkeler, *Trends Polym. Sci.,* 1997, **5**, 286.
19. D. E. Awrey, M. Tse, G. Hortelano and P. L. Chang, *Biotechnol. Bioeng.,* 1996, **52**, 472.
20. P. L. Chang, *Transfus. Sci.,* 1996, **17**, 35.
21. B. Gander, M. J. Blanco-Prieto, C. Thomasin, C. Wandrey and D. Hunkeler, in *Encyclopedia of Pharmaceutical Technology,* 2nd edn, J. Swarbrick and J. C. Boylan (eds), Dekker, New York, 2001, 481.
22. N. Angelova and D. Hunkeler, *Trends Biotechnol.,* 1999, **17**, 409.
23. B. Philipp, H. Dautzenberg, K.-J. Linow, J. Kötz and W. Dawydoff, *Progr. Polym. Sci.,* 1989, **14**, 91.
24. J. Mikulik, Z. Vinklárek and M. Vondruska, *Collect. Czech. Chem. Commun.,* 1993, **58**, 713.
25. C. H. Chervenka, *A Manual of Methods for the Analytical Ultracentrifuge,* Spinco Division Beckman Instruments, Palo Alto, CA, USA, 1969.
26. M. Lekka, D. Sainz-Serp, A. J. Kulik and C. Wandrey, *Langmuir,* Web publ. Oct. 9, 2004.

27. R. Gugerli, Doctoral Thesis no 2720, Swiss Federal Institute, Lausanne, Switzerland, 2003.
28. A. Bartkowiak and D. Hunkeler, *Chem. Mater.,* 2000, **12**, 206.
29. C. Wandrey and A. Bartkowiak, *Colloid Surface A,* 2001, **180**, 141–153.
30. C. Wandrey, G. Grigorescu and D. Hunkeler, *Progr. Colloid Polym. Sci.,* 2002, **119**, 84–91.
31. L. Bourdillon and C. Wandrey, *Colloid Polym. Sci.,* 2004, **282**, 1247–1257.

CHAPTER 10

Protein–DNA Interactions Studied at Sedimentation Equilibrium

MARGARET A. DAUGHERTY AND MICHAEL G. FRIED

1 Overview

The sedimentation equilibrium (SE) technique has been in use for nearly 80 years.[1] As a result, the literature describing applications of the technique is vast. Excellent reviews have been published during the past decade[2-6] and earlier work is also a valuable resource.[7-10] The analysis of macromolecule–ligand interactions developed over the same period and is the subject of a similarly large literature. Here, we will present results representative of recent work in our laboratories that demonstrate the promise and the challenges inherent in SE analysis of protein–nucleic acid interactions and will briefly discuss advances that are likely to prove valuable in future studies. We have used the methods described here in studies of sequence-specific and non-specific DNA interactions of transcription factors, DNA repair proteins and components of the DNA-replication machinery. In addition, these approaches are likely to be useful for the analysis of interactions that modulate chromatin structure and the protein–RNA interactions of ribosome assembly, tRNA charging and mRNA maturation. We hope that this review will be useful to workers in all these fields.

At SE a system of macromolecules will have a radial distribution of concentrations described by Equation (1).

$$c(r) = \sum_i c_{i,0} \exp[\sigma_i(r^2 - r_0^2)] \qquad (1)$$

Here, the summation is over all species, i; $c(r)$ is the observed concentration at radial position r, $c_{i,0}$ the concentration of the ith species at the reference position r_0 (typically close to the meniscus), the reduced molecular weight, σ_i is equal to $M_i(1-\overline{v}_i\rho)\omega^2/2RT$, where M_i is the molecular weight, $\overline{v}_i$ the partial specific volume (in mL g^{-1}), ρ the *solvent* density (g mL^{-1}), ω the rotor angular velocity, equal to rpm·π/30, R the gas constant (8.314×10^7 erg mol^{-1} K^{-1}) and T the absolute temperature.[*] For solutions with a small number[†] of components, direct fitting of Equation (1) to experimental data can yield relative concentrations and buoyant molecular weights of each component. Often the stoichiometry of an assembly can be deduced from its molecular weight, while the reference concentrations of individual species ($c_{i,0}$) can provide data allowing accurate and rigorous measurement of the equilibrium constant(s) governing the behavior of the system (see below). However, the $c_{i,0}$ and σ_i terms must be evaluated by curve fitting, and the difficulty of obtaining meaningful values increases rapidly with the number of species. This is the major challenge of multicomponent SE analysis; examples of such analyses will be presented that demonstrate both the strengths and the limitations of this approach. Tracer methods have been developed that reduce the number of species that contribute to the amplitude of the signal used as a proxy for concentration. Examples of this partial solution to the multicomponent problem using optical tracers will be discussed.

2   General Considerations

2.1   Simple Binding

A solution containing a homogeneous reaction represented by $nP+D \leftrightarrows P_nD$ has three macromolecular species that may be discriminated at SE. Its radial distribution of concentration is described by

$$c(r) = c_{P,0} \exp[\sigma_P(r^2-r_0^2)] + c_{D,0} \exp[\sigma_D(r^2-r_0^2)]$$

$$+ c_{P_nD,0} \exp[\sigma_{P_nD}(r^2-r_0^2)] \tag{2}$$

Here terms are defined as in Equation (1), subscripts P, D and P_nD indicate free protein, DNA and protein–DNA complex, respectively, and the reduced molecular weight terms (σ) are defined as described in Equation (1). Since for this reaction, $K_{obs}=[P_nD]/[P]^n[D]$, the term $c_{P_nD,0}$ is equal to $K_{obs}(c_{P,0})^n(c_{D,0})$. Under favorable conditions this allows the extraction of association constants from concentration distributions at SE. However, evaluation of the partial specific volume terms ($\bar{v}$) can be a

---

[*]Some authors prefer to define the reduced molecular weight as $\sigma_i \equiv M_i(1-\bar{v}_i\rho)\omega^2/RT$. This is larger by a factor of 2 than the σ_i defined in the text, and requires inclusion of an additional factor of 1/2 in Equation (1). Often data will contain a concentration-independent component due to a constant absorbance or refractive index difference between reference and sample sectors. For this reason, expressions describing absorbance or refractive index gradients commonly contain an additional constant 'baseline offset' term.

[†]Typically ≤3, although under favorable conditions, systems containing more species are amenable to analysis.

significant challenge.[‡] For most proteins, the value of $\bar{v}_p$ is adequately approximated by the mass-weighted average of amino acid residue partial specific volumes:

$$\bar{v}_P = \sum_A \frac{nM_A\bar{v}_A}{nM_A} \qquad (3)$$

where $\bar{v}_A$ is the residue partial specific volume (tabulated by Durschlag[11]), n the number of residues per monomer and M_A the mass per residue. The partial specific volumes of DNA and RNA depend sensitively on the ionic composition of the solution, the presence of secondary structure and the concentration of counterions present in the solution;[11] it may depend more weakly on base composition or sequence.[12] When the molecular weight of the DNA is known accurately from its sequence, its partial specific volume can be calculated from an independently measured value of σ_D.[13] Finally, the partial specific volume of the protein–DNA complex ($\bar{v}_{PD}$) may be approximated by the mass-weighted residue contributions:

$$\bar{v}_{PD} = \sum_{A,B} \frac{nM_A\bar{v}_A + mM_B\bar{v}_B}{nM_A + mM_B} \qquad (4)$$

where the sum is over all amino acid residues (A) and DNA bases (B) and n and m are the numbers of each residue in the complex. This is equivalent to the assumption that complex formation is not accompanied by significant change in the partial specific volumes of the components. This is often a good assumption[14] although there are notable exceptions.[15,16] Thus, if highly accurate measurement of the molecular weight, stoichiometry or association constant of a complex is required, experimental determination of $\bar{v}_{PD}$ is worthwhile. This can be obtained by performing parallel experiments in H_2O- and D_2O-containing buffers.[17] Simultaneous solution of Equations (5) and (6) gives $\bar{v}_{PD}$:

$$M_{\text{buoyant,}H_2O} = M(1 - \bar{v}\rho_{H_2O}) \qquad (5)$$

$$M_{\text{buoyant,}D_2O} = M(1 - \bar{v}\rho_{D_2O}) \qquad (6)$$

Here, $M_{\text{buoyant,}H_2O}$ and $M_{\text{buoyant,}D_2O}$ are the observed buoyant molecular weights and ρ_{H_2O} and ρ_{D_2O} are the measured densities of the H_2O- and D_2O-containing buffers.[§]

[‡]For proteins the term $(1 - \bar{v}\rho) \sim 0.27$, so a 1% error in $\bar{v}$ results in $\sim 3\%$ error in M_r.

[§]This calculation ignores changes in molecular weight that result from H–D exchange, which will depend on the identity of amino acids and non-protein components of the molecule(s) in question. The density of a solution can be estimated by summing density increments for each component.[4] Extensive published data allow accurate calculation of densities for dilute aqueous buffers with common components.[10,11,18] Alternatively, oscillating densimeters allow solvent densities to be measured at high precision. Since these values represent the actual densities of the buffer preparations, they should be used wherever possible.

2.2 Electrostatic Non-Ideality

Because DNA is highly charged, electrostatic repulsion can influence its radial concentration distribution at SE. Since it acts against the concentration of charged molecules, lower-than-expected molecular weights are a hallmark of this non-ideality. For DNA and RNA, electrostatic repulsion can be minimized by the inclusion of suitable counterions in the sample buffer. In most cases, 100–200 mM of a 1:1 salt such as KCl is enough to reduce this problem to an acceptable level. Fortunately, these concentrations are similar to the normal range in many organisms, allowing analyses to be carried out under physiologically appropriate conditions.

2.3 Components with Partial Activity

It is unlikely that all molecules of a protein preparation will be active in DNA binding. Translational truncation, incomplete posttranslational modification and proteolytic and oxidative damage are only a few of the potential causes of low specific activity in a protein sample. It is less-widely appreciated that nucleic acids can also be inactive in protein binding. Synthetic DNAs may be incompletely deprotected, contain sequence errors or have more than one favorable secondary structure. Biosynthesized DNAs may contain modified bases and/or backbone nicks. Often, inactive species have physical properties similar to those of the active forms, and contribute to the concentration gradient at SE. While the presence of a modest concentration of an inactive component will not prevent the determination of binding stoichiometry, it can significantly affect the estimation of K_{obs}. Thus, if an accurate value of K_{obs} is needed, one should test models that include an inactive component[¶] with $M_r=M_P$ or $M_r=M_D$ and only conclude that a component is fully active if they return concentration values that are equal, within error, to zero.

3   Detection of Sedimenting Species

3.1   Single Wavelength Analysis

Both proteins and nucleic acids absorb light strongly between 200 and 300 nm, with protein absorbance maxima near 215 and 280 nm (corresponding to peptide and aromatic absorbance bands) while the near-UV aromatic band of nucleic acid bases has a λ_{max} near 260 nm.[19] In spite of the partial overlap of protein and nucleic acid spectra, it is often possible to obtain useful results from data obtained at a single wavelength. The average value of ε_{260} (per nucleoside)[20] is $\sim 1\times10^4$ M^{-1} cm^{-1}. Ignoring possible hyper- and hypochromic effects, the molar extinction coefficient of an

---

[¶]Equation (7) is similar to Equation (2), except that it contains an additional term for an inactive protein or DNA component, designated as X*:

$$c(r) = c_{P,0}\exp[\sigma_P(r^2-r_0^2)] + c_{D,0}\exp[\sigma_D(r^2-r_0^2)] + c_{X*,0}\exp[\sigma_{X*}(r^2-r_0^2)] +$$

$$+ K_{obs}(c_{P,0})^n c_{D,0}\exp[\sigma_{P_nD}(r^2-r_0^2)] \tag{7}$$

oligonucleotide increases with length. This allows very sensitive detection of large nucleic acids. In contrast, the molar extinction coefficient of tryptophan is ~4000 M^{-1} cm^{-1}, and those of other amino acids are significantly less.[19] Since only ~1% of residues in a typical protein is tryptophan,[21] it is not uncommon for the A_{260} of a protein–DNA mixture to be dominated by the nucleic acid component.**

When the absorbance contribution of nucleic acid greatly exceeds that of protein, Equation (1) can be approximated by

$$A(r) = A_{D,0} \exp[\sigma_D(r^2-r_0^2)] + A_{PD,0} \exp[\sigma_{PD}(r^2-r_0^2)] \tag{8}$$

Here, $A(r)$, $A_{D,0}$ and $A_{PD,0}$ are the total absorbance at radial position r, the absorbance of DNA at the reference position and the absorbance of the protein–DNA complex at the reference position, respectively. Many studies start with this model.[13,22,23] If the molecular weights of protein and DNA are known independently, the stoichiometry of the complex can be inferred from the molecular weight of the complex. An example of such an analysis is shown in Figure 1.

More information is available when the protein's absorbance contribution is not negligible. Then, Equation (8) becomes

$$A(r) = A_{D,0} \exp[\sigma_D(r^2-r_0^2)] + A_{P,0} \exp[\sigma_P(r^2-r_0^2)] + A_{PD,0} \exp[\sigma_{PD}(r^2-r_0^2)] \tag{9}$$

with the additional term representing the absorbance distribution of the free protein. Estimation of the equilibrium constant of the binding reaction is straightforward when the concentrations of all three species are known accurately. Since $K_{obs} = [P_nD]/[P]^n[D]$, the reference absorbance of protein–DNA complex can be rewritten in terms of the corresponding absorbances of free protein and DNA:

$$A(r) = A_{D,0} \exp[\sigma_D(r^2-r_0^2)] + A_{P,0} \exp[\sigma_P(r^2-r_0^2)]$$

$$+ K'(A_{P,0})^n A_{D,0} \exp[\sigma_{PD}(r^2-r_0^2)] \tag{10}$$

Here, K' is the apparent equilibrium constant in absorbance units. The relationship to the molar-scale K_{obs} is given by

$$K_{obs} = K'\left(\frac{\varepsilon_P^n \varepsilon_D}{\varepsilon_{P_nD}}\right) l^{n-1} \tag{11}$$

This approach has two important caveats. First, the concentration gradients of the free protein component must be measured accurately. This can be difficult, since the

---

**For example, assuming the residue extinction coefficients cited above, a typical protein of 20 kDa might be characterized by ε_{260}~8000 M^{-1} cm^{-1}, while the corresponding value for an oligonucleotide of 20 residues will be ~1×10^5 M^{-1} cm^{-1}.

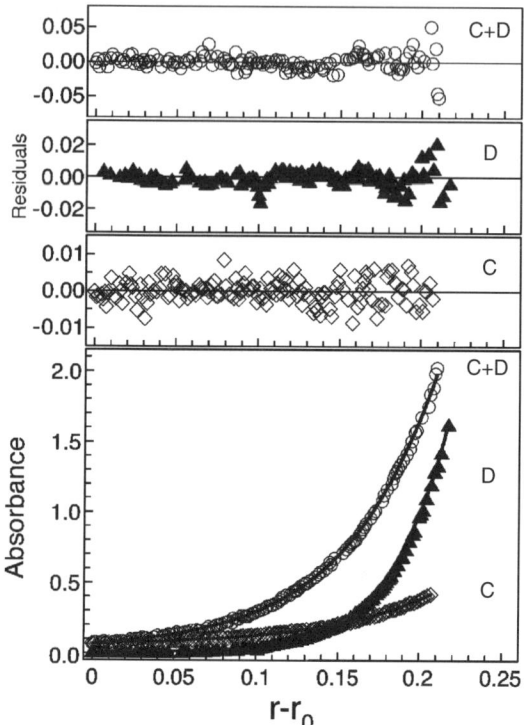

Figure 1 *SE of Escherichia coli CAP protein, a 214 bp lac promoter–operator restriction fragment and a nearly equimolar mixture of CAP protein and DNA. The samples were brought to equilibrium at 4 °C in a buffer consisting of 20 mM Tris (pH 8.0 at 4 °C), 1 mM EDTA, 200 mM KCl and 20 μM cAMP. The data for protein alone (curve C), obtained at $\lambda_{obs}=280$ nm, are consistent with a single-species model (Equation (1) with only one term), with $M_r=48\ 250\pm2100$ as expected for a homogeneous dimer (the sequence-based M_r(monomer)=23 619).[56] The data for DNA alone (Curve D; $\lambda_{obs}=260$ nm) are consistent with a single species of $M_r=143\ 500\pm3000$. The close agreement of this value with that expected for this DNA sequence ($M_r=141\ 668$) indicates the absence of significant electro-static non-ideality in the chosen buffer. The data for the protein–DNA complex (C+D; $\lambda_{obs}=260$ nm) are consistent with a two-species mixture, with $M_{r,1}=145\ 500\pm5500$, corresponding to free DNA and $M_{r,2}=190\ 300\pm4500$ corresponding to a 1:1 CAP–DNA complex. In all cases, the smooth curves represent global fits of the indicated models to data obtained at three rotor speeds. For protein alone these were 15 000, 18 000 (the data shown) and 21 000 rpm; for DNA alone, these were 9000, 12 000 (the data shown) and 15 000 rpm; for the CAP–DNA mixture, these were 7000, 9000 (the data shown) and 12 000 rpm*

absorbance of protein is unlikely to be large compared to DNA. Multiple wavelength analyses and the use of chromophoric or radioisotopic tracers offer solutions to this problem. These options are briefly examined below. Second, any change in nucleic acid absorbance due to complex formation can lead to misestimation of K'. Nucleic acid hyper- or hypochromicity can be detected by UV spectrophotometry; the absence of changes in extinction coefficient with complex formation should be verified whenever a change in nucleic acid secondary structure is a possibility. One approach that avoids this problem is to collect data at wavelengths in which protein absorbance is dominant. Attempts to do this with unmodified proteins have used wavelengths between 230 and 240 nm[24] or greater than 275 nm,[25] where nucleic acid absorbance is relatively small.

3.2 Multiple Wavelength Analysis

When both proteins and DNA contribute significantly to the absorbance, it can be advantageous to analyze the data at several wavelengths. Three variants on this theme have been described to date.[26–29] Behlke and co-workers have developed analysis programs for determining the stoichiometry and affinity of heterogeneously associating systems based upon global analysis of datasets obtained at multiple wavelengths.[25,29–32] Two constraints are imposed upon the system. The concentration for each component at each radial position is calculated using the assumption that mass is conserved in the system. This requires accurate knowledge of the loading concentration of each component. Second, in order to simplify binding constants, they assume that all binding sites are identical and independent. Lewis and co-workers approached the problem by scanning equilibrium data at a series of wavelengths chosen so that one macromolecule dominates the absorbance at a given wavelength. A matrix comprised of molar concentrations of each species at each radial position is constructed and used to obtain global fits to association models.[24,26,33] Finally, Ucci *et al.* and Bailey *et al.* have developed approaches in which they globally fit multiple SE datasets directly to specific association models.[27,28] An advantage of this approach is that the data are not transformed or rescaled before analysis. A caveat that all of these approaches share with single-wavelength analyses is that the extinction coefficients of complexed protein and complexed DNA are often unknown. The assumption that they are equal to those of the free components can lead to incorrect values of association constants. When correct values of association constants are needed, these assumptions should be tested by difference spectroscopy.

3.3 Extrinsic Chromophores

Often, protein or DNA can be covalently modified with an extrinsic chromophore, permitting selective monitoring in a wavelength range outside that of other molecules in the system. The use of such tracers can greatly simplify data analysis in systems containing many species with similar intrinsic spectra. Chromophores that have been found useful for labeling proteins include dansyl, $\lambda_{max}=335$ nm;[34] fluorescein, $\lambda_{max}=495$ nm;[35] coumarin derivatives, $\lambda_{max}\sim437$ nm;[36] FITC derivatives, $\lambda_{max}=490$ nm;[37] while useful DNA labels include fluorescein (available as a base-phosphoramidite with $\lambda_{max}\sim495$ nm[38]) and TAMRA, $\lambda_{max}=547$ nm.[39] An example of a sedimentation analysis in which the DNA was labeled with fluorescein at a single 3'-end is shown in Figure 2.

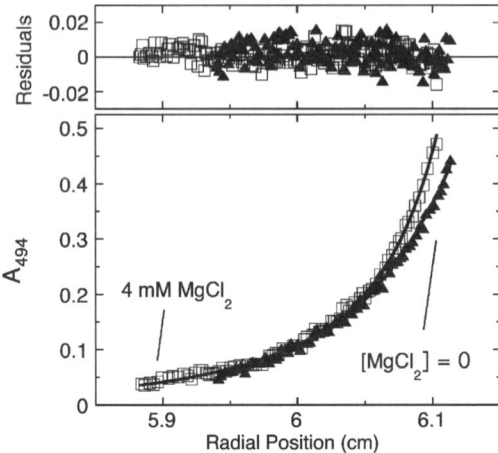

Figure 2 *Magnesium chloride-dependent interaction of yeast TATA-binding protein (TBP) with a fluorescein-labeled DNA duplex. Oligonucleotides of sequence 3'-F CGC TAT AAA AGG GC5' and 3'- GCC CTT TTA TAG CG-5', in which F denotes fluorescein, were obtained commercially. Duplex DNA was prepared as described.[57] TBP–DNA mixtures were prepared in buffer containing 20 mM Hepes (pH 7.9), 120 mM KCl with or without 4 mM $MgCl_2$, as indicated. Samples were brought to SE at 4 °C and 30 000 rpm ($[MgCl_2]=0$) or 27 000 rpm ($[MgCl_2]=4$ mM). Data were acquired using $\lambda_{obs}=494$ nm. The data obtained at $[MgCl_2]=0$ are fit by a single-species model (Equation (1) with one term); with $\overline{v}=0.55$ (typical of DNA). The fit returns $M_r=9994\pm200$, in good agreement with the value predicted for DNA alone ($M_r=9849$). The small, symmetrically distributed residuals demonstrate the compatibility of this model with the data. In contrast, data obtained at 4 mM $MgCl_2$ were compatible with a two-species model. Setting $M_{r,1}=9849$ and $\overline{v}_1=0.55$ returned a buoyant molecular weight of $27\,550\pm600$ for the second species. This suggests the presence of a complex of $M_{r,2}=39\,980\pm1920$, consistent with a sequence-based prediction for the 1:1 complex ($M_r=36\,867$)*

The high extinction coefficients of these chromophores allow the detection of relatively low concentrations of labeled molecules. In addition, the strong fluorescence of these chromophores makes them particularly well suited for use in fluorescence-detected sedimentation analyses.[40] Although the use of an extrinsic chromophore can avoid complications due to changes in the protein or DNA absorbance that may accompany complex formation, it should not be assumed that there is no change in extinction coefficient with binding. As with intrinsic chromophores, the absence of significant absorbance change should be verified by difference spectroscopy. A second caveat is that interactions with the extrinsic chromophore may perturb the protein–DNA interaction. A parallel analysis in which the chromophore is attached to a different macromolecular surface is often a suitable test for such perturbation.

3.4 Biosynthetic Incorporation of Chromophores

Using a bacterial strain that is auxotrophic for tryptophan, Ross and co-workers were able to substitute tryptophan residues with 5-hydroxytryptophan (5-OHTrp) in the bacteriophage λ_{c1} repressor protein.[41,42] This substitution shifted the protein's absorbance λ_{max} from 280 to 310 nm, allowing the detection of labeled protein in the presence of significant concentrations of DNA. The substitution with 5-OHTrp was without significant effect on several biological properties of the λ_{c1} repressor, and since the substitution is a conservative one, it is likely to be applicable with minimal perturbation in other proteins. This strategy has the potential to allow analysis of the mass distribution of an individual 5-OHTrp protein in the presence of unlabeled proteins, making possible the analysis of complicated protein–DNA interactions involving more than one protein species. Other strategies for the biosynthetic or partially biosynthetic incorporation of amino acid analogues into proteins are under development[43-45] and we anticipate that these will expand the repertoire of chromophoric residues available as optical probes in SE.

A second approach to the biosynthetic incorporation of chromophores is the production of chimeras in which the protein of interest is fused to another small protein (such as green fluorescent protein, GFP) that carries an easily detectable chromophore.[46,47] This strategy has seen wide use in molecular and cell biology, but none, to our knowledge, in analytical ultracentrifugation. However, the availability of affinity techniques for purification[48] and the facile detection of the GFP moiety can make such chimeras easier to acquire than the wild-type protein, while the presence of a unique chromophore can allow the study of chimeric proteins in very complex mixtures. In addition, the availability of GFP analogs with a wide range of chromophoric properties may allow the simultaneous, independent analysis of several differently labeled proteins in a single mixture. Finally, these proteins are strongly fluorescent, allowing their characterization by fluorescence-detected sedimentation.[40] These advantages are obtained at the cost of changes in the mass and hydrodynamic properties of the protein(s) and the potential for changes in their interactions with other molecules.[47] However, with proper care, chromophoric fusion proteins have the potential to provide a powerful and flexible route to the analysis of very complex protein–DNA mixtures.

4   Non-Specific Binding

Virtually all sequence-specific DNA binding proteins are also capable of binding non-cognate DNA sequences. Such sequences are present in vast molar excess over individual target sequences in a cell's genome. In addition, there are great many proteins such as histones, DNA replication and repair factors, and chromatin remodeling factors that bind DNA with little or no sequence preference. The characterization of such non-specific interactions can be an important step in the understanding of the functions of the systems in which they play a part.[49,50] Often, non-specific interactions are characterized by association constants in the range $10^4 \text{ M}^{-1} \leqslant K_a \leqslant 10^8 \text{ M}^{-1}$ that is particularly suitable for analysis by current SE techniques.

4.1   Long DNA Molecules

The most widely used analysis of non-specific binding follows the statistical approach of McGhee and von Hippel, who derived a modified Scatchard expression

(Equation (12)) for very long (effectively endless) binding lattices.[51]

$$\frac{v}{[P]} = K(1-nv)\left(\frac{(2\omega-1)(1-nv)+v-R}{2(\omega-1)(1-nv)}\right)^{n-1}\left(\frac{1-(n+1)v+R}{2(1-nv)}\right)^2 \qquad (12a)$$

with

$$R = ((1-(n+1)v)^2+4\omega v(1-nv))^{1/2} \qquad (12b)$$

Here, v is the binding density, equal to the average number of protein molecules bound to each DNA divided by the number of base-pairs per DNA molecule; [P] the free protein concentration, n the statistical binding site size in base-pairs, K the population-averaged association constant and ω the cooperativity parameter.[*†] If the molecular weights of free protein and DNA are known, values of v are easily extracted from the weight-average buoyant molecular weight of the DNA population, obtained by fitting a single species sedimentation model (Equation (1) with one term) to the DNA-absorbance data (Figure 3). Estimation of the concentration of free protein [P] is more problematic. If it cannot be measured by taking advantage of regions of the absorbance spectrum that are not dominated by DNA, it may be possible to approximate it using Equation (13).

$$[P] = [P]_{input}-m[DNA] \qquad (13)$$

Here $[P]_{input}$ is the concentration of protein molecules in the sample, ignoring binding; m is the average number of protein molecules bound to DNA (estimated from the molecular weight of the complex) and [DNA] is the concentration of DNA molecules in the sample.

An advantage of this approach is its simplicity. Since only the weight-average buoyant molecular weight need be measured, data suitable for the analysis can be obtained using very small sample volumes. On the other hand, it also means that a single sample provides only one measure of the v, [P] pair at each rotor speed. Typically, several samples are needed to accurately characterize binding by this approach (Figure 4). This contrasts with the analysis of homogeneous site-specific binding, in which values of stoichiometry and association constants are inferred at every point in the radial concentration profile (*cf.* Equation (2)).

4.2   Short DNAs

The method of McGhee and von Hippel provides an accurate analysis of binding affinity, cooperativity and site size when the DNA lattice contains hundreds of

---

[*†]The cooperativity parameter ω can be thought of as the equilibrium constant for protein–protein interaction, when both protein molecules are bound to adjacent sites on the DNA. Values of $\omega<1$ indicate that protein- interactions are unfavourable; when $\omega=1$ protein–protein interactions have no effect on the stability of the complex and when $\omega>1$, protein interactions stabilize the complex.

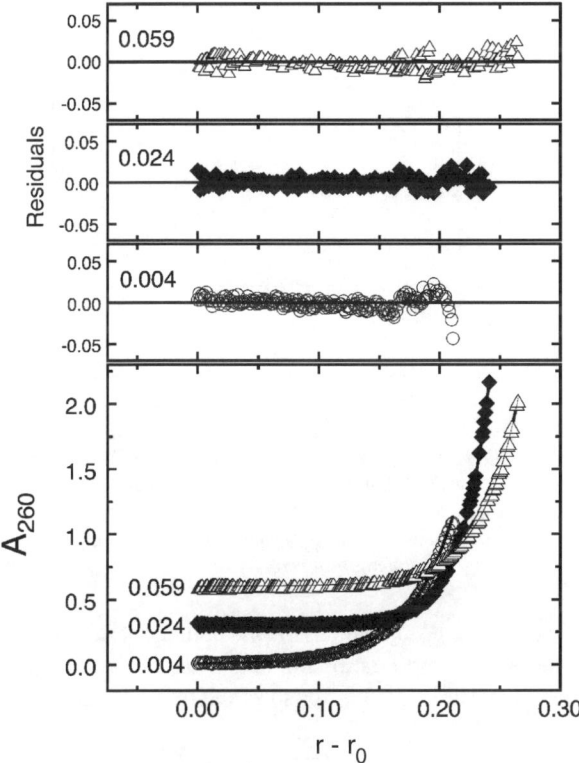

Figure 3 *Non-specific interaction of E. coli CAP protein with a large DNA fragment. CAP (M_r(dimer)=47 238) and DNA (a 2472 bp fragment from plasmid pMM02;[58,59] M_r ~1.636×10[6]). Samples were brought to equilibrium in 20 mM Tris (pH 8.0), 150 mM KCl. The absence of cyclic AMP ensures that CAP binding is non-specific.[60] Data were taken at 2000, 3000 and 4000 rpm at 4 °C; the solid curves are global fits to the combined data sets. For low CAP/DNA ratios (≤50), data were fit using the single-species model (Equation (1) with one term). When CAP/DNA >50 the absorbance of CAP becomes significant compared to that of DNA. For these samples, a two-species model was used with the one species fixed at the buoyant molecular weight of CAP and the second, corresponding to the CAP–DNA complex, allowed to float. Buffer density was measured with a Mettler DA-300 density meter. A spread-sheet calculation was used to predict the buoyant molecular weights of CAP–DNA complexes using input molecular weights and previously determined values of the partial specific volumes of CAP (0.737)[61] and DNA (0.55).[62] Binding densities (values shown to left of curves) were calculated from the CAP/DNA ratios that gave observed buoyant molecular weights. The data sets shown for samples of n=0. 004 and n=0.024 were obtained at 3000 rpm, that shown for n=0.059 was obtained at 2000 rpm*

binding sites and when the statistical binding site size is small compared to the length of the DNA. When these conditions are not met, the presence of partial binding sites at DNA ends can bias the binding statistics. Recently, Record and colleagues have verified the original estimate of McGhee and von Hippel that the correction term for

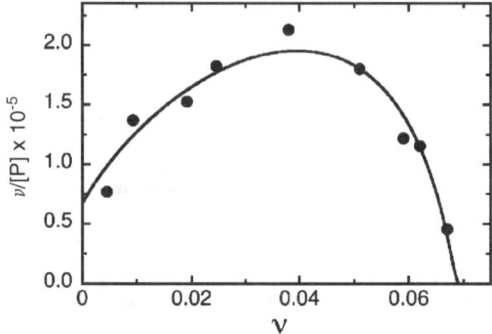

Figure 4 *Scatchard plot of the non-specific binding of CAP to a large DNA fragment. SE analyses were performed and binding densities calculated as described in the caption to Figure 3. Free CAP concentrations were calculated using $[P]=[P]_{input}-m[DNA]$, as described in the text. The smooth curve represents the fit of the McGhee–von Hippel equation (Equations (12a) and (12b)) to the data. This fit returned $K=66\,900 \pm 9900$ M^{-1}, cooperativity factor $\omega=77 \pm 12$ and statistical binding site size $n=14.4 \pm 0.2$ bp. All values are in substantial agreement with ones originally measured by Saxe and Revzin, using a sedimentation velocity technique[63]*

this 'end effect' is $(N-n+1)/N$, in which N is the number of monomer units in the lattice (bp in a duplex DNA) and n is the binding site size in monomer units.[51,52] The resulting modified Scatchard equation for short DNA lattices is as follows:

$$\frac{v}{[P]} = K(1-nv)\left(\frac{(2\omega-1)(1-nv)+v-R}{2(\omega-1)(1-nv)}\right)^{n-1}\left(\frac{1-(n+1)v+R}{2(1-nv)}\right)^{2}\left(\frac{N-n+1}{N}\right) \quad (14)$$

Here, K, n, v, ω and R are defined as in Equation (12). Given the easy availability of short, synthetic DNAs, we anticipate that this expression will be widely used in the future.*‡

5 Prospects

The past decade has seen a renaissance in SE analysis, catalyzed by the availability of modern instrumentation and new software for data analysis, and the increasing availability of biologically important macromolecules. We anticipate that this trend will accelerate. Fluorescence detection,[40] has lowered the concentration requirements of the technique, allowing study of hard-to-acquire macromolecules and equilibria with large association constants. Multiwavelength analysis[26–29] and methods for labeling proteins with chromophoric amino acid analogues (*e.g.* 5-hydroxytryp-

---

*‡These authors have also derived expressions for mixed specific and non-specific binding on lattices of limited length. Such expressions are useful when affinities for cognate and non-cognate sequences are close enough in magnitude that binding of cognate sequences is always accompanied by some occupancy of non-cognate sites.

tophan[41,42]) allow the simultaneous measurement of several proteins in complex mixtures, as well as the analysis of labeled proteins in the presence of several unlabeled species. Postcentrifugation analysis methods[53–55] allow the detection of components by tracer methods including direct counting of radioisotopes, radioimmunoassay and reverse transcriptase–polymerase chain reaction (RT–PCR). With these techniques, SE analyses can be carried out on target species at very low concentrations, in complex mixtures, and in the presence of high background concentrations of other macromolecules. These developments will make possible the analysis of systems of biological complexity, in the macromolecular concentration ranges that occur *in vivo*.

Acknowledgements

We wish to thank Drs. Gang Liu and Karen Smirnakis for DNA and CAP samples and Dr. Michael Brenowitz for the yeast TBP used in these experiments. This work was funded by NIH grant GM 070662.

References

1. T. Svedburg and K. O. Pedersen, *The Ultracentrifuge*, Johnson Reprint Corp., New York, 1956.
2. J. C. Hansen, J. Lebowitz and B. Demeler, *Biochemistry*, 1994, **33**, 13155–13163.
3. T. M. Laue, *Method Enzymol.*, 1995, **259**, 427–452.
4. G. Rivas, W. Stafford and A. P. Minton, *Methods*, 1999, **19**, 194–212.
5. T. M. Laue and W. F. Stafford, III, *Annu. Rev. Biophys. Biomol. Struct.*, 1999, **28**, 75–100.
6. J. Lebowitz, M. S. Lewis and P. Schuck, *Protein Sci.*, 2002, **11**, 2067–2079.
7. H. K. Schachman, *Ultracentrifugation in Biochemistry*, Academic Press, New York, 1959.
8. J. W. Williams, *Ultracentrifugation of Macromolecules*, Academic Press, New York, 1972.
9. K. E. Van Holde, in *The Proteins*, Vol. 1, H. Neurath and R. L. Hill (eds), Academic Press, New York, 1975, 225–291.
10. D. K. McRorie and P. J. Voelker, *Self-Associating Systems in the Analytical Ultracentrifuge*, Beckman Instruments, Inc., Palo Alto, 1993.
11. H. Durschlag, in *Thermodynamic Data for Biochemistry and Biotechnology*, H. -J. Hinz (ed), Springer-Verlag, Berlin, 1986, 45–128.
12. G. F. Bonifacio, T. Brown, G. L. Conn and A. N. Lane, *Biophys. J.*, 1997, **73**, 1532–1538.
13. K. Rippe, N. Mucke and A. Schulz, *J. Mol. Biol.*, 1998, **278**, 915–933.
14. M. A. Daugherty and M. G. Fried, *Method. Enzymol.*, 2003, **370**, 349–369.
15. A. A. Infante and R. Baierlein, *Proc. Natl. Acad. Sci. USA*, 1971, **68**, 1780–1785.
16. A. Gow, D.J. Winzor and R. Smith, *Biochim. Biophys. Acta*, 1985, **828**, 383–386.
17. S. J. Edelstein and H. K. Schachman, *Method. Enzymol.*, 1973, **27**, 82–98.
18. T. M. Laue, B. D. Shah, T. M. Ridgeway and S. L. Pelletier, in *Analytical Ultracentrifugation in Biochemistry and Polymer Science*, S. E. Harding, A. J. Rowe and J.C. Horton (eds), Royal Society of Chemistry, Cambridge, 1992, 90–125.
19. C. R. Cantor and P. S. Schimmel, *Biophysical Chemistry Part II: Techniques for the Study of Biological Structure and Function*, W. H. Freeman and Company, New York, 1980.
20. V. A. Bloomfield, D. M. Crothers and I. Tinoco, Jr., *Physical Chemistry of Nucleic Acids*, Harper & Row, New York, 1974, 7–21.

21. K. E. Van Holde, W. C. Johnson and P. S. Ho, *Principles of Physical Biochemistry*, Prentice-Hall, Upper Saddle River, NJ, 1998, 28–29.

22. L. A. Cassiday, L. L. Lebruska, L. M. Benson, S. Naylor, W. G. Owen and L. J. Maher, *Anal. Biochem.*, 2002, **306**, 290–297.

23. K. Wojtuszweski, M. E. Hawkins, J. L. Cole and I. Mukerji, *Biochemistry*, 2001, **40**, 2588–2598.

24. S. -J. Kim, T. Tsukiyama, M. S. Lewis and C. Wu, *Protein Sci.*, 1994, **3**, 1040–1051.

25. H. Xu, J. Frank, J. F. Holzwarth, W. Saenger and J. Behlke, *FEBS Lett.* 2000, **482**, 180–184.

26. M. S. Lewis, R. I. Shrager and S. -J. Kim, in *Modern Analytical Ultracentrifugation*, T. M. Schuster and T. M. Laue (eds), Birkhauser, Boston, MA, 1994, 81–93.

27. J. W. Ucci and J. L. Cole, *Biophys. Chem.*, 2004, **108**, 127–140.

28. M. F. Bailey, B. E. Davidson, A. P. Minton, W. H. Sawyer and G. J. Howlett, *J. Mol. Biol.*, 1996, **263**, 671–684.

29. J. Behlke and O. Ristau, *Eur. Biophys. J.* 1997, **25**, 325–332.

30. J. Behlke, O. Ristau and H. -J. Schonfeld, *Biochemistry*, 1997, **36**, 5149–5156.

31. J. Behlke and O. Ristau *Eur. Biophys. J.*, 2003, **32**, 427–431.

32. M. Schade, J. Behlke, K. Lowenhaupt, A. Hervert, A. Rich and H. Oschikinat, *FEBS Lett.*, 1999, **458**, 27–31.

33. N. K. Maluf and T. M. Lohman, *J. Mol. Biol.*, 2003, **325**, 889–912.

34. Molecular Probes, Inc., Online Handbook, Table 1.12, 2003. http://www.probes.com /handbook.

35. R. L. Weinberg, D. B. Veprintsev and A. R. Fersht, *J. Mol. Biol.*, 2004, **341**, 1145–1159.

36. M. Jezewska, S. Rajendran and W. Bujalowski, *Biochemistry*, 1998, **37**, 3116–3136.

37. S. P. Lee, E. Fuior, M. S. Lewis and M. K. Han, *Biochemistry*, 2001, **40**, 14081–14088.

38. S. Rajendran, M. J. Jezewska and W. Bujalowski, *J. Mol. Biol.*, 2001, **308**, 477–500.

39. L. E. Daniels, K. M. Wood, D. J. Scott and S. E. Halford, *J. Mol. Biol.*, 2003, **327**, 579–591.

40. I. K. MacGregor, A. L. Anderson and T. M. Laue, *Biophys. Chem.*, 2004, **108**, 165–185.

41. J. B. Ross, D. F. Senear, E. Waxman, B. B. Kombo, E. Rusinova, Y. T. Huang, W. R. Laws and C. A. Hasselbacher, *Proc. Natl. Acad. Sci. USA*, 1992, **89**, 12023–12027.

42. T. M. Laue, D. F. Senear, S. Eaton and J. B. A. Ross, *Biochemistry*, 1993, **32**, 2469–2472.

43. L. Wang, A. Brock, B. Herberich and P. G. Schultz, *Science*, 2001, **292**, 498–500.

44. I. Hirao, T. Ohtsuki, T. Fujiwara, T. Mitsui, T. Yokogawa, T. Okuni, H. Nakayama, K. Takio, T. Yabuki, T. Kigawa, K. Kodama, T. Yokogawa, K. Nishikawa and S. Yokoyama, *Nat. Biotechnol.*, 2002, **20**, 177–182.

45. T. Hohsaka, N. Muranaka, C. Komiyama, K. Matsui, S. Takaura, R. Abe, H. Murakami and M. Sisido, *FEBS Lett.*, 2004, **560**, 173–177.

46. R. Heim, D. C. Prasher and R. Y. Tsien, *Proc. Natl. Acad. Sci. USA*, 1994, **91**, 12501–12504.

47. O. V. Stepanenko, V. V. Verkhusha, V. I. Kazakov, M. M. Shavlovsky, I. M. Kuznetsova, V. N. Uversky and K. K. Turoverov, *Biochemistry*, 2004, **43**, 14913–14923.

48. R. I. Paramban, R. C. Bugos and W. W. Su, *Biotechnol. Bioeng.*, 2004, **86**, 687–697.

49. P. H. von Hippel, A. Revzin, C. A. Gross and A. C. Wang, *Proc. Natl. Acad. Sci. USA*, 1974, **71**, 4808–4812.

50. D. F. Stickle, K. M. Vossen, D. A. Riley and M. G. Fried, *J. Theor. Biol.*, 1994, **168**, 1–12.

51. J. McGhee and P. H. von Hippel, *J. Mol. Biol.*, 1974, **86**, 469–489.

52. O. V. Tsodikov, J. A. Holbrook, I. A. Shkel and M. T. Record, Jr., *Biophys. J.*, 2001, **81**, 1960–1969.

53. S. Darawshe and A. P. Minton, *Anal. Biochem.*, 1994, **220**, 1–4.

54. G. Rivas and A. P. Minton, *Biochem. Soc. Trans.*, 2003, **31**, 1015–1019.

55. G. Rivas and A. P. Minton, *J. Mol. Recognit.*, 2004, **17**, 362–367.
56. H. Aiba, S. Fugimoto and N. Ozaki, *Nucl. Acids Res.*, 1982, **10**, 1345–1361.
57. S. Khrapunov, N. Pastor and M. Brenowitz, *Biochemistry*, 2002, **41**, 9559–9571.
58. D. F. Stickle, G. Liu and M. G. Fried, *Eur. J. Biochem.*, 1994, **226**, 869–876.
59. M. G. Fried and J. M. Hudson, *Science*, 1996, **274**, 1930–1931.
60. M. G. Fried and D. M. Crothers, *J. Mol. Biol.*, 1984, **172**, 241–262.
61. D. Dyckman and M. G. Fried, *J. Biol. Chem.*, 2002, **277**, 19064–19070.
62. G. Cohen and H. Eisenberg, *Biopolymers*, 1968, **6**, 1077–1100.
63. S. A. Saxe and A. Revzin, *Biochemistry*, 1979, **18**, 255–263.

CHAPTER 11

UltraScan – A Comprehensive Data Analysis Software Package for Analytical Ultracentrifugation Experiments

BORRIES DEMELER

Abstract

With the introduction of the Beckman XL-I/A in the early 1990s, digital data acquisition from the analytical ultracentrifuge has made it possible to readily analyze sedimentation data with a PC. UltraScan,[1] one of the several data analysis software packages available for such analysis, is a comprehensive, multi-platform software package designed to not only address the tasks associated with the interpretation of analytical ultracentrifugation (AUC) experiments, but also to provide guidance with the design of sedimentation experiments and to address data management challenges arising in multi-user facilities. The availability of data in digital format has led to a wealth of AUC data, which demand new approaches for dealing with the large amounts of data generated. To address this challenge, UltraScan now includes a laboratory information management system (LIMS) which is based on a relational database. In addition, UltraScan integrates many routines for designing, analyzing, interpreting, and displaying sedimentation equilibrium and velocity experiments in a user-friendly graphical interface to make sophisticated analysis methods approachable for a wide audience, including new and less-experienced users. In this publication, an overview of the design philosophy of the software and its algorithms is presented, and the various modules, methods, and their applications are discussed. Examples for each method are shown, and a guide for the experimental design and implementation is given.

1 Overview

The UltraScan software package is the result of a collaborative effort extending over more than a decade now with many contributors.[2] Over the years, many modules have been added to the software, increasing functionality and generality. This publication offers an overview of the capabilities of the current state of the software. The software has been created with the intent to provide a convenient and high-performance data analysis environment for AUC experiments. The software is programmed in the platform independent C++/Qt language, and graphical versions for Linux, Unix, Windows, and Mac OS-X are available for free download.

The software addresses multiple issues related to the analysis: starting with the experimental design, the software offers modules that aid in the optimal design of the experiment. Once the data have been acquired, the data are edited with editing modules designed for different optical systems and centerpiece geometries and converted into a binary format suitable for rapid loading and analysis by a series of experimental analysis procedures. Sedimentation velocity analysis can be performed with several methods: determination of model-independent $G(s)$ distributions and partial concentrations can be accomplished with the van Holde–Weischet method, for samples with only a few discrete species, direct boundary fitting using finite element solutions of Lamm equation models can be used to determine s, D, molecular weight, frictional coefficients, and partial concentrations. The second moment analysis provides weight-average sedimentation coefficients and offers useful diagnostics for sedimentation experiments. Sedimentation equilibrium experiments can be analyzed with a global nonlinear least-squares regression analysis. Fitting statistics can be ascertained through either a bootstrap or Monte Carlo analysis. UltraScan includes a Beowulf interface, which allows this analysis to also be performed on a parallel Linux cluster to reduce computing time. Multi-wavelength analysis is complemented by a global extinction fitter suitable for determining intrinsic extinction coefficient distributions. Data analysis results can be directly exported in html format to a webserver, such that experimental results are available on the Internet.

To assist with data management tasks, an issue of concern especially for facilities with multiple machines and multiple investigators, an external relational database is incorporated into UltraScan. The database component of UltraScan has both a C++ implementation for direct database access from the software as well as a PHP interface for web-based access. In the relational database it is possible to link experimental data with associated information and not only store the data itself, but also the links to associated data. Examples for information that can be logically connected to the experimental data include: information describing details and the context of the experiment, the optical system used, the investigator, the cell type, the rotor, and the centerpiece geometry, sequence information of biopolymers (for purposes of estimating partial specific volume, molecular weight, extinction coefficients), buffer composition (for predicting hydrodynamic corrections) as well as data analysis results. PHP modules regenerate analysis reports on the fly by retrieving the relevant information through the Internet. An array of utility functions facilitates archiving of experimental data, calculation of ancillary constants from buffer composition and primary protein and nucleic acid sequences, molecular modeling, and common file operations.

2 Organization

The UltraScan software is organized into multiple modules that can be executed separately or simultaneously. Each module is a separate binary file linked against the main UltraScan library, which is dynamically loaded. This organization optimizes stability and memory needs, especially in a multi-user environment. Where appropriate, the code is multi-threaded to afford additional performance in a multi-processor environment. The software modules interface directly with several external programs. Database functionality is provided by MySQL, an Internet-enabled relational database. Context-specific help files, documentation, and data reports have been written in html and are accessed with an external browser such as Explorer, Netscape, or Mozilla. Data archives are generated with the public domain *tar* and *gzip* utilities to maintain cross-platform compatibility.[3] Web-based access is provided through the *Apache* webserver,[4] and a PHP[5] interface to the MySQL database.[6] Below, each module is discussed separately.

3 Modules

3.1 Experimental Design and Simulation Modules

An important part of AUC is the appropriate design of the run conditions, in particular the loading concentration, the rotor speed and the duration of the experiment, as well as wavelength selection. Suboptimal rotor speeds, incorrect run times, and scan times can lead to reduced information content of the resulting data. Incorrect sample concentration may lead to underrepresentation of a reversibly self-associating oligomeric species, while an incorrect wavelength selection can lead a noisy or non-linear signal. Often, some information is known about the protein from Sodium dodecyl sulfate (SDS) gel electrophoresis, mass spectrometry, gel filtration, or sequencing before analytical ultracentrifuge experiments are performed. Such information can be exploited to optimize the design of the experiment. A good estimate of the monomer molecular weight as well as an approximation of the molecular shape can then be used to predict the sedimentation and diffusion coefficient for the molecule. A modeling module facilitates the prediction of these coefficients from molecular weight, partial specific volume, buffer conditions, and shape model. Coefficients can be predicted for a sphere as well as by specifying the axial ratio for an oblate or prolate ellipsoids, or a long rod model (Figure 1).

Once sedimentation and diffusion coefficients are available, finite element solutions of the Lamm equation[7] are used to simulate experimental conditions. From these simulations, optimized experimental parameters can be derived. For velocity experiments, the simulations are used to predict the highest possible speed compatible with the scan speed, which is dependent on the number of cells to be scanned, the optical system, and the number of desired scans. For equilibrium experiments, the program will predict appropriate speeds based on the reduced molecular weight σ of the samples. Speeds appropriate for equilibrium experiments can be obtained by substituting the appropriate values for σ in Equation (1):

$$\text{rpm} = \frac{30}{\pi} \sqrt{\frac{2RT\sigma}{M(1-\bar{v}\rho)}} \tag{1}$$

where R is the gas constant, T the temperature in K, M the molecular weight, ρ the density of the buffer, $\bar{v}$ the partial specific volume and σ is defined by

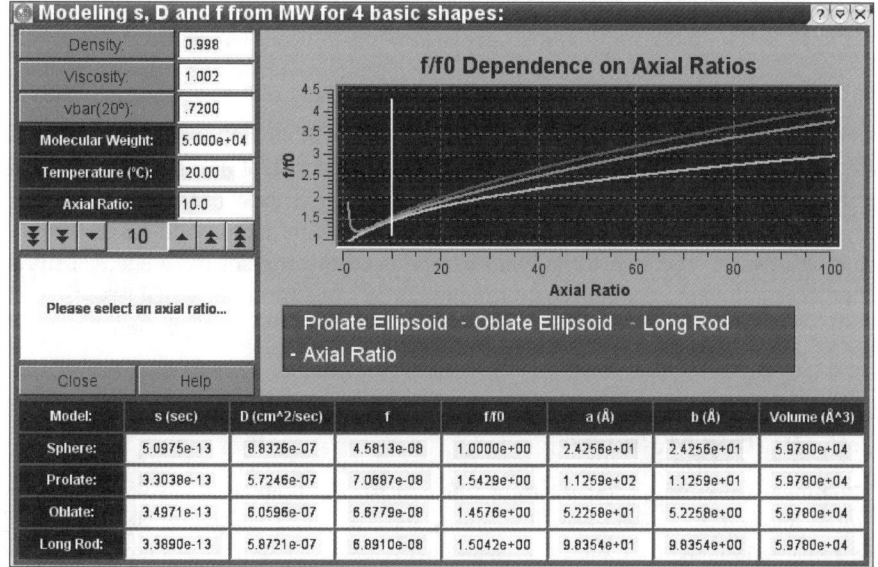

Figure 1 *Module for the modeling of molecular parameters based on molecular weight, axial ratio, hypothetic molecular shape, partial specific volume, and buffer conditions. Hydrodynamic corrections can be imported from predefined buffer files and the partial specific volume of peptides can be estimated from the peptide sequence*

$$\sigma = \frac{M\omega^2(1-\bar{v}\rho)}{2RT} \tag{2}$$

It is recommended to perform equilibrium experiments with 4–5 speeds ranging in σ values between 1 and 4 for sample loading concentrations of 0.3, 0.5, and 0.7 OD. These conditions provide sufficient curvature and variation in the equilibrium gradient, exploit the entire linear range of the absorbance spectrum and provide sufficient variability in the gradient to improve global-fitting statistics. For reversibly self-associating systems, a selection of multiple wavelengths can effectively enhance the concentration range covered by the experiment, so that multiple oligomers are adequately represented in the signal. Next, finite element simulations are used to model the approach to equilibrium. The length of time required to achieve the equilibrium condition is given by satisfying the following equality:

$$k = \sum_{i=1}^{n}\left[L(s, D, r_i, t, C_1, \omega) - C_a \exp\left(\frac{M\omega^2(1-\bar{v}\rho)(r_i^2-r_a^2)}{2RT}\right)\right]^2 \tag{3}$$

where k is a user-selectable constant determining the residual error (typically less than the error resulting from experimental noise), n the number of radial discretization steps, $L(s, D, r_i, t, C_1, \omega)$ is the Lamm equation solution, r_i the radius at the ith position, C_1 the loading concentration, t the time, and C_a, r_a are the concentration and the radius at a reference point in the cell, respectively. Mass conservation is guaranteed by the relationship

$$C_1(r_b - r_m) = \int_{r_m}^{r_b} C_a \exp\left[\frac{M\omega^2(1 - \bar{v}\rho)(r^2 - r_a^2)}{2RT}\right] dr \qquad (4)$$

where r_m is the meniscus position and r_b the position at the bottom of the cell. Once the equilibrium condition in Equation (3) has been satisfied, the resulting concentration distribution is used to initialize the next higher speed and the process is repeated until all speeds have been simulated. The time required to reach equilibrium for each speed is recorded and can be used to program the analytical ultracentrifuge (allowing for additional time required to perform the actual scanning, as well as a separate scan spaced 4–6 h apart to verify that equilibrium has been reached).

3.2 Editing Modules

The process of editing sedimentation data consists of eliminating noisy data regions and to determine meniscus position and in the case of velocity data, determining a baseline and plateau estimates for each scan, and writing all data to a binary copy of the original data, which remain unchanged during editing. This binary representation can be loaded by any UltraScan analysis module, eliminating the need to re-edit experimental data for each analysis method individually. Another advantage of the binary data format is the significant increase in reading speed.

During editing, the software will extract all salient information from the file header, sort all scans according to scan number, cell, centerpiece channel, and wavelength and create a separate binary representation of all scans belonging to the same cell, centerpiece, and wavelength. A separate binary run information file is created to save the rotor speed, time, temperature, plateau, data range, wavelength, and the $\omega^2 t$ integral of each scan. Centerpiece geometry, rotor type, optical system, meniscus, average temperature, as well as rotor acceleration corrections are calculated and saved in this structure as well. If the database module is used, this file also includes the date of the experiment, the database links for the buffer composition file, as well as the links for the peptide and nucleic acid sequences from each channel, and the database name and address used for the experiment.

A special process is required for editing interference velocity data. Such data are often affected by several systematic imperfections, such as shifting baselines, integral fringe offsets, and time-independent noise. The following algorithm is employed to correct these artifacts: initially, all scans are aligned along the air-to-air region. Next, scans are shifted by integral fringe numbers until the numerical integral of each successive scan produces a monotonically decreasing function. In order to correct the baseline shift ('breathing'), often found in interference data, this integral function is then fitted to a polynomial. Each scan's residual of this fit is a reflection of the baseline offset variation in each scan (Figure 2). Subtracting the residuals from each corresponding scan corrects the variations in the baseline offsets. Finally, a provision is made to subtract baseline scans from each scan to correct for time-independent noise caused by heterogeneity in the refractive index of cell windows, which can be substantial for low concentration samples.

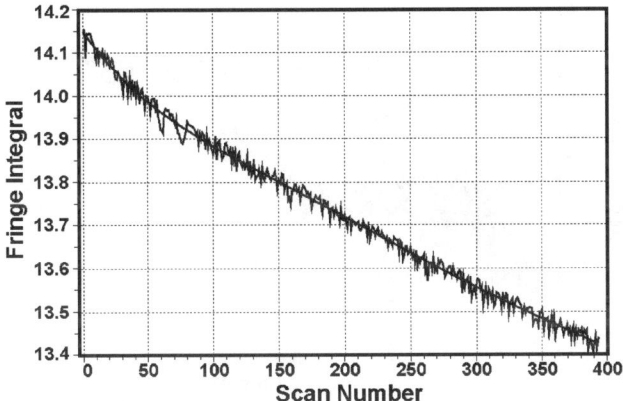

Figure 2 *Fringe integrals of 400 interference scans showing the variation in baseline off-sets fitted to a polynomial function. The frequency of the fluctuation remains relatively constant over time suggesting a systematic cause for this fluctuation. Subtracting the residuals of this fit from the scans corrects the baseline offsets of the scans*

3.3 Velocity Analysis Modules

During sedimentation velocity experiments, the molecules under investigation are subject to two transport processes, sedimentation and diffusion. These processes depend both on molecular weight and the frictional properties of a molecule. The sedimentation coefficient is directly proportional to the molecular weight, and inversely proportional to the frictional coefficient. Diffusion is inversely proportional to the frictional coefficient. For simple systems of one or two components it is therefore possible to obtain both sedimentation and diffusion coefficients, which provide information about molecular weight and shape. However, if multiple components are present, the individual signal strengths from diffusion and sedimentation from each component are reduced, and often not sufficiently resolvable to identify with necessary certainty the shape and molecular weight contributions. In such cases it is still possible to reliably define a sedimentation coefficient distribution, but the corresponding diffusion coefficient distribution may remain hidden.

UltraScan offers methods for the analysis of either situation. A system with just a few noninteracting components can often be well described by whole boundary modeling. UltraScan uses finite element solutions of the Lamm equation[7] to perform this modeling. The finite element solutions are fitted with a nonlinear least-squares fitting algorithm to the concentration distributions from the experimental scans to obtain sedimentation coefficients, diffusion coefficients, partial concentrations, association constants for reversibly self-associating systems, as well as meniscus position and concentration dependency parameters. It is also possible to account for optical artifacts such as baseline drift and sloping plateaus in the model. An example for a fit of a noninteracting two-component system is shown in Figure 3.

Most nonlinear least-squares fitting algorithms rely on steepest descent calculations to find a set of optimal parameters that minimize the χ^2 condition. The calculation of

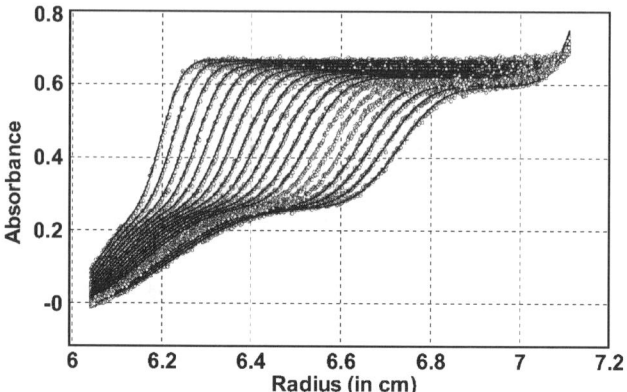

Figure 3 *Finite element fit of a sedimentation velocity experiment containing a two-component, noninteracting system. The continuous black lines show the finite element solution, the grey circles represent the experimental observations*

the steepest descent direction requires the calculation of a Jacobian matrix, which contains the elements of partial derivatives of the solution with respect to the estimated parameters. There exist numerous approaches to obtain the derivatives. An analytical evaluation of the derivatives provides the most accurate result, and facilitates convergence of the least-squares optimization.

However, for finite element solutions of the Lamm equation an analytical evaluation of the partial derivatives of all parameters is not available, and alternative methods have to be used. The finite element fitting modules offer two nonlinear least-squares fitting algorithms for whole boundary fitting of velocity experiments. The first method is based on an approach developed by Ralston and Jennrich[8] which uses first-order tangent approximations to estimate the partial derivatives. While this approach works well for most cases, for complex cases with many floating parameters the inherent error in the tangent approximation approach adversely affects convergence properties, and convergence can stall in a local minimum.

Better convergence properties can be obtained with the second optimization method, which is based on automatic differentiation.[9] In this approach, a tape of chain rule operations is recorded from the evaluation of the finite element solution, which is used to calculate the entries of the Jacobian matrix using the ADOLC C++ library[10] for automatic differentiation. The accuracy of the derivatives evaluated by automatic differentiation is equivalent to analytical solutions. A drawback to this approach is the large storage requirement for the tape, which can slow down computation on smaller computers.

For all systems, including those that are not well described by just a few discrete components, the van Holde–Weischet analysis[11] offers a model-independent graphical transformation of the data that results in diffusion-corrected sedimentation coefficient distributions. This method relies on the realization that while sedimentation is a transport process proportional to the first power of time, diffusion is a transport process proportional to the square-root power of time. In the limit of infinite time the effect on transport by diffusion is negligible compared to transport by sedimentation.

Extrapolation to infinite time can therefore provide diffusion-corrected sedimentation coefficient distributions. In this approach, apparent sedimentation coefficients are calculated from a fixed number of boundary fractions at each scan. Corresponding boundary fractions are then extrapolated to infinite time in a plot of apparent sedimentation coefficients *vs.* the inverse square root of time of the scan (Figure 4).

Special care has to be taken to account for boundary effects at the meniscus and the bottom of the cell, as well as differential radial dilution rates for different components in the system. These effects are taken into consideration in the enhanced van Holde–Weischet algorithm that is implemented in UltraScan. The details of this algorithm are described in ref. 12. The enhanced van Holde–Weischet analysis provides sedimentation coefficient distributions that can be displayed both as integral distribution plots $G(s)$ and as differential distributions $g(s)$ as shown in Figure 5. Given suitable estimates for the partial specific volume and the frictional ratio f/f_0, both distributions can also be transformed into molecular weight distributions. Weight-average sedimentation coefficients are also calculated from the sedimentation coefficient distributions.

An alternative method for obtaining weight-average sedimentation coefficients is provided by the second moment analysis. It provides weight-average sedimentation coefficients for each individual scan. However, only scans with a stable plateau and clear meniscus can provide reliable estimates of second moment weight-average sedimentation coefficients.

3.4 Equilibrium Analysis Module

Sedimentation equilibrium experiments always should include multiple equilibrium scans taken at different rotor speeds, different loading concentrations, and multiple wavelengths. Each individual scan contains a unique distribution of concentration

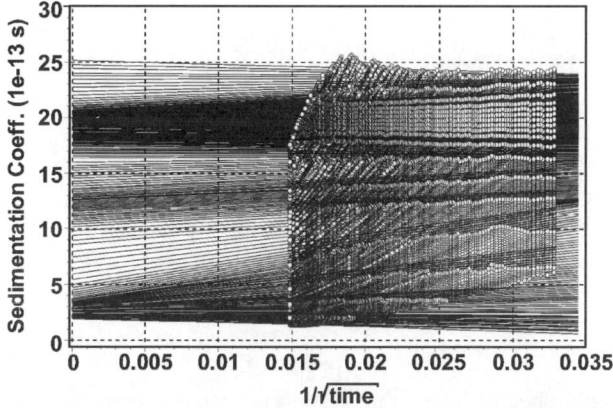

Figure 4 *van Holde–Weischet analysis of a sedimentation velocity data for a heterogeneous system containing multiple degradation products of a protease. Extrapolations to infinite time are made for increasing boundary fractions (bottom to top) and from early scans to late scans (right to left)*

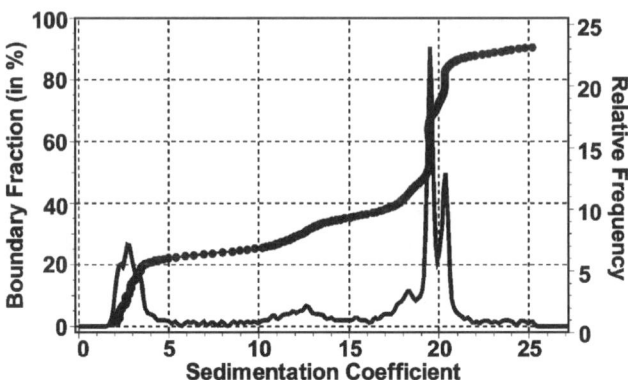

Figure 5 *Integral distribution plot (G(s)) and differential distribution plot (g(s)) for van
Holde–Weischet analysis shown in Figure 4*

observations that reflect the composition of the sample. If the correct model is chosen for a global analysis approach, each observation has to satisfy the same global
parameters such as molecular weight, extinction coefficients, and association constants in the model function. Furthermore, scans taken at different wavelengths
exploit the varying extinction properties of the sample and allow analysis of a sample over a larger range of loading concentrations.

In a reversibly self-associating system a change in the loading concentration will
change the ratio of larger oligomers *vs.* smaller oligomers or monomeric forms of
the sample, thereby enhancing the signal of one species compared to another. A
global analysis approach is essential for a reliable interpretation of data from multicomponent systems. Multiple experiments can be simultaneously analyzed by
UltraScan through nonlinear least-squares fitting to a preset or user-defined global
model as proposed in ref. 13. All models available in UltraScan are based on three
basic forms:

(1) A noninteracting system:

$$C(r) = \sum_{i=1}^{k} \exp\left[\frac{\ln(a_i) + M_i\omega^2(1-\bar{v}_i\rho)(r^2-r_{\text{ref}}^2)}{2RT} \right] + c \qquad (5)$$

(2) A reversibly self-associating system:

$$C(r) = \sum_{i=1}^{k} \exp\left[\frac{i\ln(a_1) + \ln\frac{iK_{1,i}}{(el)^{i-1}} + iM_1\omega^2(1-\bar{v}\rho)(r^2-r_{\text{ref}}^2)}{2RT} \right] + c \qquad (6)$$

(3) A reversibly hetero-interacting system for two components A and B:

$$C(r) = \exp\left[\frac{\ln(a_A) + M_A\omega^2(1-\bar{v}_A\rho)(r^2-r_{\text{ref}}^2)}{2RT} \right]$$

$$+ \exp\left[\frac{\ln(a_B)+M_B\omega^2(1-\bar{v}_B)(r^2-r_{ref}^2)}{2RT}\right]$$

$$+ \exp\left[\frac{\ln\left(a_A a_B K_{A,B}\frac{e_{AB}}{e_A e_B l}\right)+(M_A+M_B)\omega^2\left(1-\left(\frac{\bar{v}_A+\bar{v}_B}{2}\right)\rho\right)(r^2-r_{ref}^2)}{2RT}\right] + c \quad (7)$$

where $C(r)$ is the concentration at radius r, k the maximum number of components or oligomer states, a the concentration of the component at the reference radius r_{ref}, M the molecular weight, $K_{1,i}$ the equilibrium constant for association state i, e the extinction coefficient, l the path length, c a baseline offset, ω the radial velocity, $\bar{v}$ the partial specific volume, and ρ the density of the buffer. UltraScan does not place any limits on the maximum number of scans that can be included in a global analysis. Each scan is allowed a separate entry for ρ and c.

Parameters can be floated or kept fixed. Parameters that can be floated include the concentrations at the reference point for each species, the association constants, the baseline offset, and the molecular weights or the partial specific volumes. Molecular weights, association constants, and partial specific volumes are considered global parameters, all other parameters are local. Global parameters are constrained to be identical for all scans described by the global model, local parameters are allowed to vary for each scan. The concentration a at the reference point r_{ref} is fitted as the natural log of the reference concentration. This effectively constrains the fit to positive values of the reference concentration only, avoiding spurious oscillations of the amplitudes of exponential terms that are common when the model is overdetermined. Molecular weights reported are corrected for density and partial specific volume. Equilibrium constants are displayed as association or dissociation constants, and are reported in molar units.

3.5 Optimization

The models used for fitting of equilibrium experiments shown in Equations (5)–(7) are nonlinear in the parameters and require iterative nonlinear least-squares fitting approaches. For highly nonlinear problems with many parameters the likelihood of convergence is quite dependent on the proper choice of initial parameter estimates. The closer the initial parameter estimates are to the least-squares solution, the greater is the likelihood that the solution will converge at the global minimum.

UltraScan will initialize all parameter estimates with reasonable guesses to improve the convergence properties. This is accomplished by linearizing the model for a single ideal species and fitting the amplitudes for each scan by general linear least squares. The molecular weight estimate is obtained by performing a line search over the nonlinear parameter of the model. For two- or three-component models the scans are divided into two or three equal sections and each section is fitted individually to generate an estimate for one of the components. This approach results in parameter estimates that automatically initialize all fitting parameters to reasonable values that facilitate a more stable convergence of the fit.

Parameter optimization is performed by one of the three optimization methods: for nonlinear least-squares optimization problems, the Levenberg–Marquardt method[14,15,16] and the quasi-Newton method are implemented. The Levenberg–Marquardt method applies a scaling factor to the diagonal of the information matrix to prevent it from becoming singular during optimization. This approach results in a robust method that is not too sensitive to the choice of initial guesses, and therefore useful for obtaining an initial fit.

However, for problems with many nonlinear parameters the Levenberg–Marquardt method has a tendency to converge in a local minimum, and a second approach is needed to find the global minimum. For cases where the solution is close to optimal the quasi-Newton method performs best, and it can help to overcome solutions that are trapped in a local minimum. The quasi-Newton method employs the BFGS formula (named after their developers Broyden,[17] Fletcher,[18] Goldfarb,[19] and Shanno,[20]) and a line search algorithm to update an approximation to the Hessian, which is needed to find the steepest descent direction for locating the global minimum for the least-squares solution. Due to the complexity of the error surface, multi-variate optimization often fails in the task of finding the global minimum solution, and the solution can get trapped in a local minimum. To alleviate this problem, UltraScan employs an automatic convergence algorithm that alternates between the Levenberg–Marquardt and quasi-Newton methods until a global minimum has been found, and neither method can improve the χ^2 value of the fit any further. An example for a global equilibrium fit for multiple speeds and loading concentrations is shown in Figure 6.

For linear least-squares problems, a nonnegatively constrained approach is used (the NNLS algorithm by Lawson and Hanson[21]). A linear problem is given when the molecular weights of each exponential term in Equation (5) are predetermined, and only the coefficients a_i of the linear combination of exponential terms have to be found by the fitting routine:

$$C(r) = \sum_{i=1}^{k} a_i \exp\left[\frac{M_i \omega^2 (1 - \overline{v}_i \rho)(r^2 - r_{\text{ref}}^2)}{2RT} \right] + c \tag{8}$$

where the baseline offset c is simply the zeroth-order term of this linear combination. In such a fitting problem, many terms can be used to account for all M_i present in the sample, and the range and spacing of all M_i species is provided by the user. The nonnegatively constrained Lawson–Hanson algorithm prevents coefficients a_i from turning negative, and coefficients of terms accounting for molecular weight species present in the system are fitted with a nonzero value, while those not present in the sample are assigned a value of zero. Contributions of each molecular weight species from all scans are summed to generate a molecular weight composition histogram which represents the relative concentration of each species in the sample. An example for such a molecular weight distribution is shown in Figure 7.

When studying biological systems the question often arises if the presence of molecular weight heterogeneity in a sample results from the presence of multiple, noninteracting components or from a reversibly self-associating oligomerization. UltraScan provides two diagnostic plots for global equilibrium fits to address this question: (1) a plot of the average molecular weight of all fitted data points *vs.* concentration, and (2) a plot of the average molecular weight of all fitted data points *vs.* the square of the

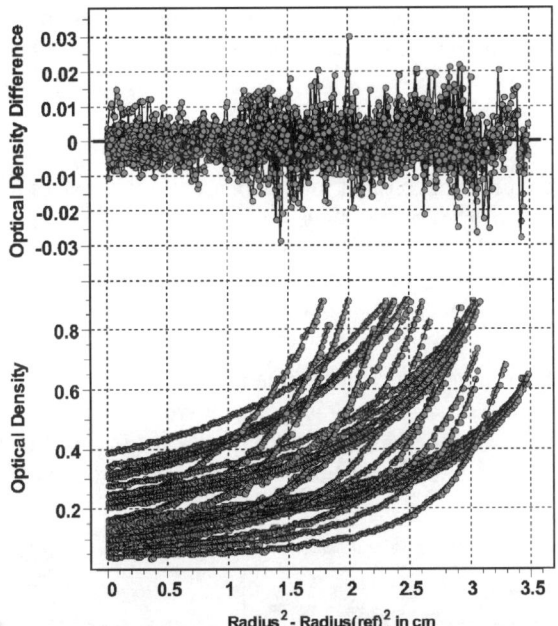

Figure 6 *Global equilibrium fit of 24 scans (six loading concentrations and four speeds). Circles represent experimental observations, solid lines represent the fitted model. Residuals are shown in the top panel. overlays, and the associated fitted model are shown in the bottom panel*

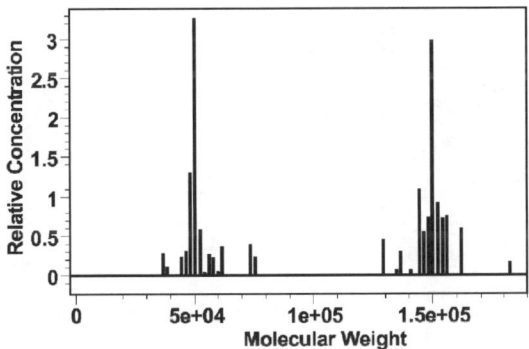

Figure 7 *Molecular weight distribution for a simulated equilibrium experiment containing approximately equal amounts of a 50 and 150 kDa species. The distribution was generated by globally fitting 24 scans for multiple speeds and loading concentrations with the nonnegatively constrained general least-squares fitting method by Lawson and Hanson*

radius. These plots are useful when the data is fitted with models derived from Equation (5) or (8). Since Equation (6) is a subset of Equation (5) or (8), reversibly self-associating systems may also be fitted with Equation (5) or (8), yielding equally good fits.

Due to mass action, for a reversible self-associating system, the ratio of oligomer concentration over monomer concentration stays constant according to

$$K_{1,i} = \frac{[iM]}{[M]^i} \tag{9}$$

Here, $K_{1,i}$ is the equilibrium constant for the ith association state of the monomer, M. Therefore, all scans will follow the same trace in plot (1), regardless of loading concentration or rotor speed. This is shown in Figure 8(a). For a noninteracting system the same plot will show a distribution that is not constrained by the equilibrium constant, and each speed or loading concentration will assume a different distribution. This is shown in Figure 8(c).

However, if the average molecular weights from the self-associating system are plotted against the square of the radius, concentrations at different radial positions will be sufficiently different at different rotor speeds and loading concentrations, and each scans will produce a different trace. This is illustrated in Figure 8(b). For the

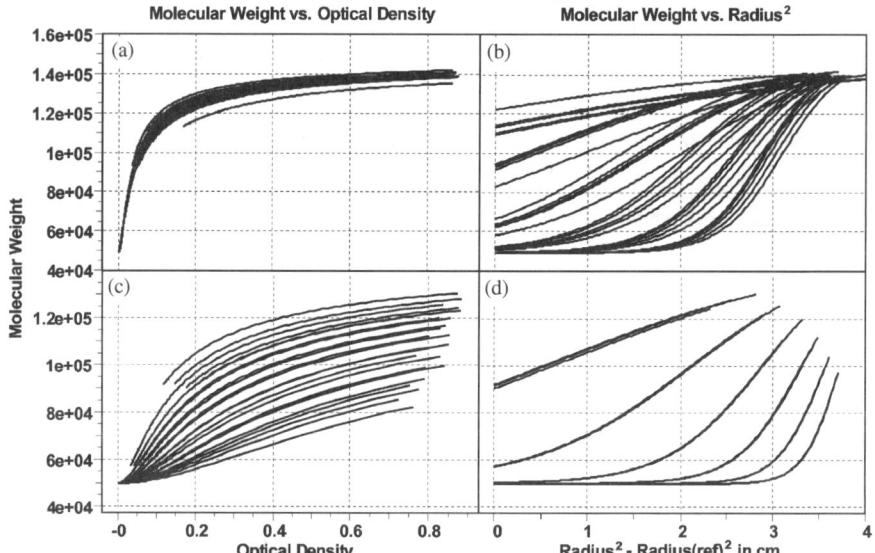

Figure 8 *Equilibrium diagnostic plots in UltraScan for two different simulated multi-component systems. The first system is a reversibly self-associating monomer–trimer system with a monomer molecular weight of 50 kDa (a), (b), and the second system is a two ideal noninteracting species model where component one is 50 kDa and component two is 150 kDa (c), (d). Six speeds and five loading concentrations were simulated for both systems. Both systems were fitted with the model shown in Equation (8) shown in (a), (c) is a plot of average molecular weight vs. concentration for each scan. In graphs (b), (d) a plot of average molecular weight vs. the square of the radius is shown. Note that all traces for a self-associating system overlay when molecular weight is plotted against concentration, but traces overlay for different loading concentrations when molecular weight is plotted against the square of the radius only for a noninteracting system. See text for a more detailed explanation of the diagnostics provided by these plots*

noninteracting system, molecular weight distributions are not the result of mass action, and different loading concentrations have no effect on the ratio of one species over the other. As a result, components will equilibrate solely according to the centrifugal force applied, and all scans at the same rotor speed will follow the same trace. An example for such a case is shown in Figure 8(d). Hence, visual inspection of plots 8(a, c) and 8(b, d) for systems fitted to a noninteracting model allows the investigator readily to distinguish between noninteracting and self-associating systems.

3.6   Extinction Fitting Module

For self-associating systems and hetero-associating systems the measurement of equilibrium experiments at different wavelengths can add important information to the analysis. For self-associating systems, different extinction coefficients at different wavelengths can be exploited to obtain measurements at markedly different concentrations. By bracketing a large concentration range in the experiment, the signal from both monomer and oligomeric species can be enhanced. For hetero-associating systems, the presence of chromophores at different wavelengths for each component can provide important constraints on the fit when multiple wavelengths are fitted in a global fit. When these constraints are included in the fit, it is important that extinction coefficients are correctly identified in the model for each wavelength and for each component to assure internal consistency of the model.

UltraScan offers a global extinction fitting module that generates intrinsic extinction profiles from wavelength scans at different concentration for each component in the fit. These profiles can be normalized with known extinction coefficients at one wavelength and imported into the fitting model such that the appropriate extinction coefficients are applied for each component in the fit. This also helps in cases where hypochromicity is an issue as long as the pure form of the associated species can be measured. The algorithm is applied as follows: for each component, wavelength scans are taken at 3–5 loading concentrations that are chosen so that each chromophore is represented at least once with an optical density between 0.5 and 1.0 absorbance units. All scans are then globally fitted to the following linear combination of Gaussian terms:

$$E_i(\lambda) = c_i \sum_{j=1}^{n} \exp\left[\frac{(-\lambda - a_j)^2}{2\sigma_j^2} \right] + b_i \tag{10}$$

where E_i is the extinction profile of wavelength scan i, c_i the relative concentration of scan i, λ the wavelength, a_j is the position of chromophore j, σ_j the width of the peak produced by chromophore j, n the number of chromophores, and b_i a baseline offset for scan i. a_j and σ_j are global parameters required to be identical for all wavelength scans. Once a global solution has been found, the factor c_i is normalized with a known molar extinction coefficient at a desired wavelength and all other wavelengths are scaled accordingly. An example for a global extinction fit is shown in Figure 9. UltraScan provides the extinction coefficient at 280 nm as an estimate from peptide sequence according to the method of Gill and von Hippel,[22] which can be conveniently used to scale intrinsic extinction coefficient profiles to molar extinction coefficients.

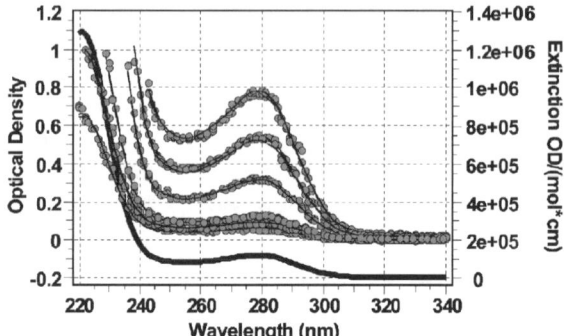

Figure 9 *Global extinction profile for a protein. Six separate concentrations in three repli-*
cate measurements were globally fitted (grey circles fitted with thin lines) result-
ing in an intrinsic extinction profile for the protein (heavy line). The extinction
profile is calibrated at 280 nm with a known extinction coefficient derived from
peptide sequence. The global extinction profile is plotted as the right Y-axis, the
absorbance values from the wavelength scans are plotted on the left Y-axis

3.7   Monte Carlo Module

All experimental data fitted to a linear or nonlinear model contain experimental
noise. The assumption in all fits performed by UltraScan is that the experimental
noise is random and that all systematic deviations are accounted for in the model. If
these assumptions are satisfied, the statistical confidence of the parameter estimates
depends on the magnitude and distribution of the random experimental noise. One
possibility for determining the confidence intervals of the parameter estimates would
be to perform repeat experiments and fit them to the same model in order to generate a distribution of values for each floated parameter. This distribution could be
used to obtain a statistical description of each parameter. Naturally, performing
repeat experiments is impractical because of the time and expense considerations.

Instead, a Monte Carlo approach can be used to simulate a repeat experiment,
where experimental noise is synthetically generated by a random number generator,
and added to the best-fit model values. This will generate a new, synthetic dataset
which can be refit with the same model. Repeating this process a sufficient number
of times will provide the desired parameter distributions from which a statistical
description of the parameters can be obtained.

Two different approaches for generating synthetic noise are possible: (1) generating Gaussian noise comparable to the residuals from the best fit to the experimental
data, and adding it to the best fit; or (2) a bootstrap approach that randomly moves
residuals from the best fit to the experimental data to a different position. UltraScan
supports both approaches. Because noise levels differ nonlinearly with absorbance
(see Figure 6) a true representation of the random noise is often not achieved with the
bootstrap approach, and the Gaussian noise generation is preferable.

In the Gaussian approach, the residuals in a frame of 5–10 points in the neighbor-
hood of each experimental observation are averaged. The average then provides the
standard deviation input for the Box–Muller function.[23] This function will return

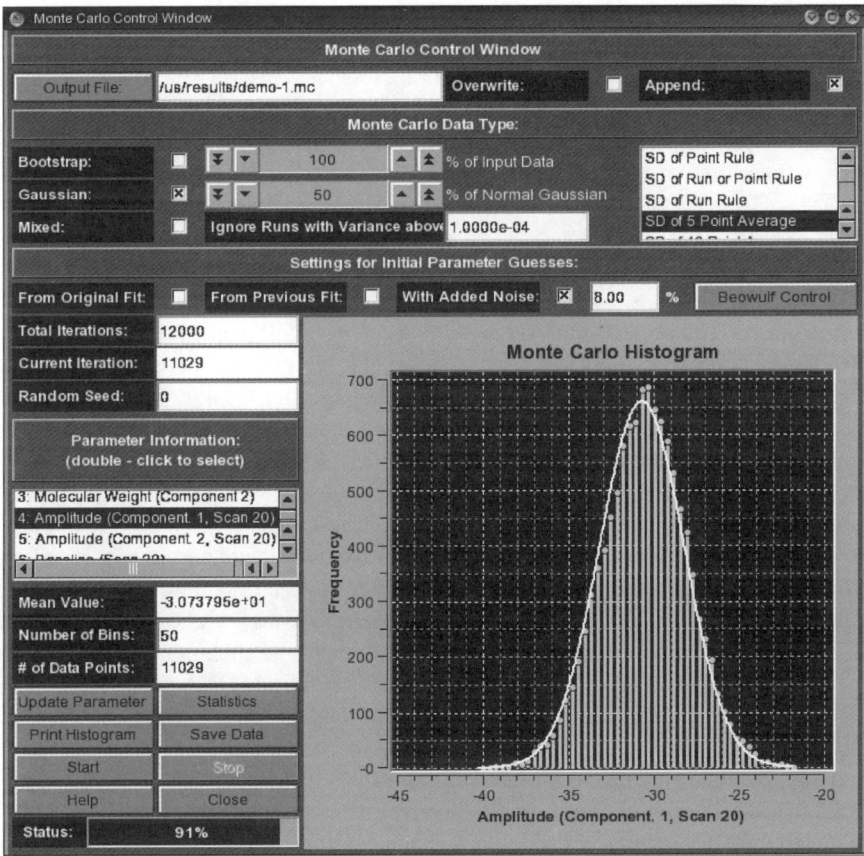

Figure 10 *UltraScan Monte Carlo control window. A parameter distribution for each fitted parameter can be displayed and the analysis can be performed both locally or remotely on a parallel cluster (Unix only)*

Gaussian distributed random variables with a standard deviation that is commensurate with the original residuals. These values are added to the best-fit values predicted by the model function, which are then refit. This method assures that local changes in the size of the residuals are accurately reflected in the synthetically generated dataset. Five thousand Monte Carlo iterations are generally sufficient to obtain a reliable probability distribution of each fitted parameter (Figure 10). The Monte Carlo analysis returns a distribution of parameter values for each fitted parameter.

Once a sufficiently populated distribution has been obtained, the distribution is fitted to a Gaussian function, and the following statistics are reported: minimum and maximum, mean, median, skew, kurtosis, mode, standard error, standard deviation, variance, correlation coefficient, and 95 and 99% confidence intervals. The confidence intervals may be nonsymmetric. One problem with the application of the Monte Carlo analysis is the prohibitive computational expense, especially for large models with many parameters. To address this issue, UltraScan offers a Beowulf module that allows the Monte Carlo analysis to be performed on a cluster of appropriately configured Unix computers. The speed improvement observed by performing the Monte Carlo analysis in parallel scales linearly with each added computational node. To assure that each node starts at a different point in the pseudo-random sequence generated by the random number generator, each node is initialized with a different random seed.

3.8   Utility Modules

UltraScan incorporates a number of utility functions to accomplish frequently performed tasks. The protein analysis module accepts peptide sequence information in Genbank format and calculates from it molecular weight, partial specific volume, and extinction coefficients at 280 nm. The partial specific volume ($\bar{v}$) is calculated by summing the weight fraction of partial specific volume contributions from each amino acid as reported by Durchschlag,[24] and the calculation of the extinction coefficient is based on the partial extinction contribution at 280 nm from the denatured amino acids tyrosine, tryptophane, and cysteine as reported by Gill and von Hippel.[22] A second utility provides for the calculation of hydrodynamic corrections resulting from density and viscosity contributions from a collection of commonly used buffer components according to methods outlined in the reference.[25] Values for density and viscosity are interpolated from polynomial fits to concentration data obtained from the Sednterp database files.[26] Molecular weights of RNA and DNA molecules can be calculated from the sequence, including the contributions from various counterions. Several file utilities assist the user in archiving experimental data, analysis results and experimental reports, to merge data files from different directories and to re-order the files into a single run, and to rename cell descriptions that have been incorrectly entered during data acquisition. A diagnostics module is also available that is useful for reviewing single files or identifying file errors, such as truncated or corrupted data acquisition files.

3.9   Database Modules

To facilitate the management of AUC projects and the large volume of data files collected and generated during analysis, UltraScan includes a LIMS that is designed to address the needs of a multi-user AUC facility. Many analytical ultracentrifuges are employed in multi-user environments, and managing multiple experiments, experimental data, and associated information quickly turns into a complex task. The UltraScan LIMS is based on an Internet-capable relational database with both an UltraScan interface and a web-based interface.

The LIMS addresses multiple objectives: first, it serves as a data repository for all data relevant to an AUC experiment. Second, it serves as a data retrieval and web-based presentation tool, and finally it assists multiple investigators to manage separate experiments. The first goal of the database is to provide logically linked storage for experimental project descriptions, experimental designs, the experimental data and analysis results, for peptide and nucleic acid sequence files, as well as for buffer composition files and images from gels or absorbance spectra. All experimental and result data are stored in a compressed format and are associated with investigator identifications, which facilitates data analysis and retrieval in multi-user environments.

Experimental data are committed to the database by linking an entire run, which may consist of multiple cells and channels in each cell, with an experimental project description and run profile, which are described by the investigator. Each channel is linked with up to three peptide or nucleic acid sequence files, one buffer file, the date of the experiment, the investigator information, as well as the name of the database, the run type (*i.e.*, equilibrium, velocity, diffusion, or wavelength experiment), and the optical measurement method (absorbance, interference, fluorescence, or intensity). Supplemental data needed for data analysis are stored in separate tables and logically linked to the experimental data.

During analysis, any required supplemental data are automatically retrieved from the database, processed on the fly and integrated into the analysis. For example, the appropriate buffer composition is associated with each centerpiece channel and retrieved during the analysis. Using the composition information, viscosity and density corrections are calculated and automatically applied in the analysis. Similarly, peptide sequence data are retrieved and used to calculate an estimate for the partial specific volume of the peptide, which is then applied to the analysis. All automatic values can be manually overridden by the user.

Centerpiece geometry, rotor type, and channel number are also stored in the database and associated with each dataset. This information is used to calculate a precise position for the channel bottom. Each rotor in use at a laboratory can be individually calibrated to determine speed-dependent rotor stretching. Precise information on rotor stretching and channel geometry are needed to determine the position of the channel bottom. A precise value for this position is needed to calculate accurate mass integrals for equilibrium experiments and to set boundary conditions for finite element solutions. Centerpiece geometry, rotor calibration, peptide, nucleic acid, and buffer information only have to be entered once and can be linked from any experiment instead of laboriously recalculating these values by hand for each new experiment.

After data analysis is completed, the LIMS searches the hard drive for all analysis results from each method applied during the analysis and commits the results to the database. An overview of the database structure and the relationships is shown in Figure 11. The relational database engine used in UltraScan is the open source MySQL database.[6] To facilitate data exchange and collaborations, the user can switch among multiple databases and access data from remote sites. Database access is restricted through username and password authentication to prevent unauthorized access to any private data.

3.10 Web-Based Modules

Several web-based PHP applets allow a remote user to interact with the database contents. After authentication, the investigator can enter project requests, peptide sequences, compose buffers, and upload related images. Next, the investigator can enter sample information and specify the type of experiment to be performed. Each item entered by the investigator receives a description which allows the investigator to later search and retrieve or review each item. In order to accommodate a multi-user

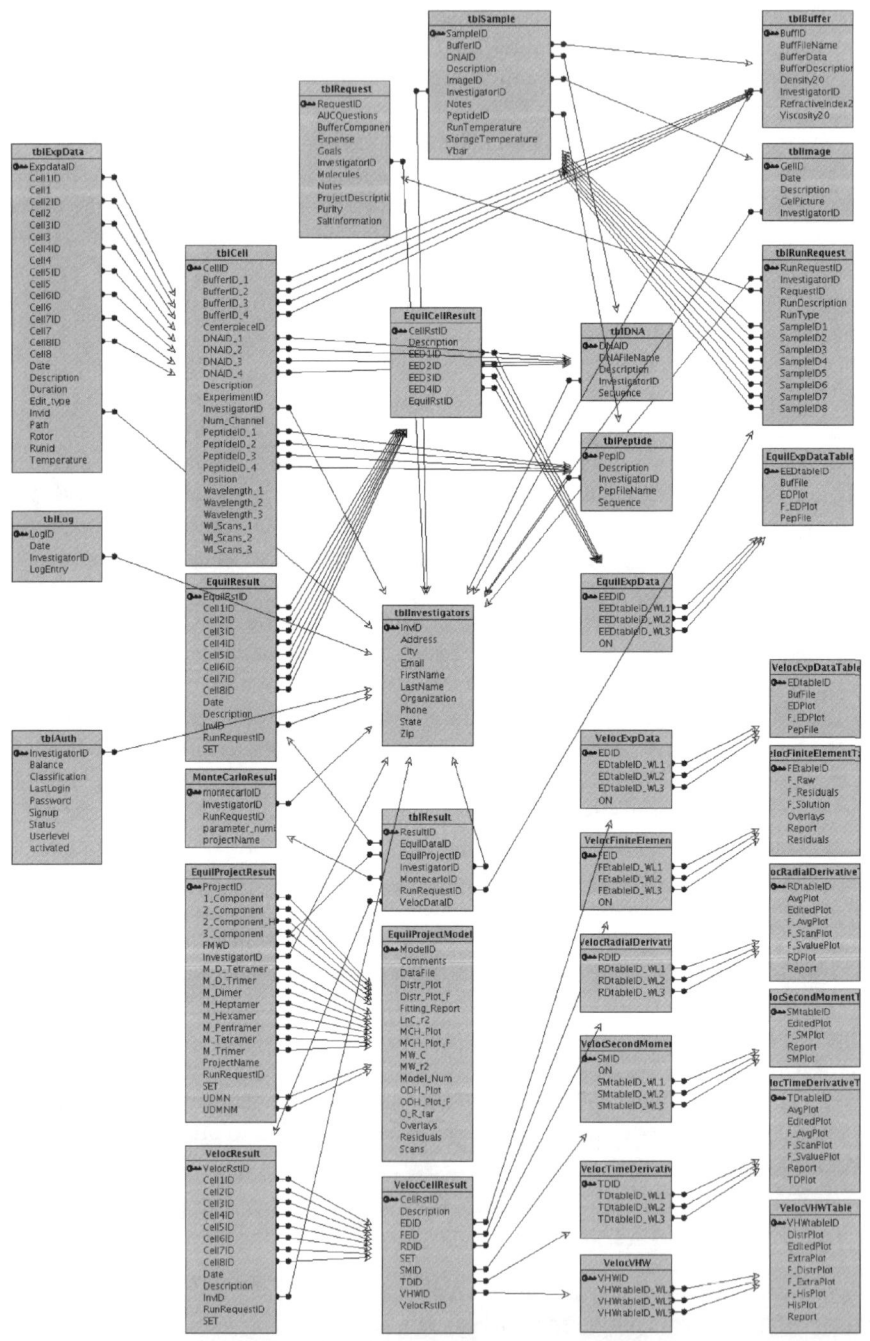

Figure 11 *Overview of the UltraScan LIMS database structure version 1.0. Relationships are shown by arrows*

environment, the LIMS offers different user levels: investigator, technician, supervisor, and data analyst. Each user level has a different authentication status allowing read-only, write, or no access to different data segments.

Once experimental results have been uploaded, the investigator can retrieve the results through the web interface and review or download them to his or her personal computer. A tracking system allows the LIMS users to track the status of each experiment (*i.e.*, designed, scheduled, in progress, uploaded). In addition to the database-oriented data management, analysis results can also be presented as a web page and saved directly to a folder served by a webserver. UltraScan will generate an html encoded report file and provide convenient access to the experimental data and analysis results by linking graphs, data files, and analysis reports to the report file.

4 Summary

UltraScan represents a comprehensive software package addressing a wide range of experimental situations and providing multiple analysis methods for sedimentation velocity and equilibrium experiments. Design and modeling functions assist with the correct design of an experiment, and a Beowulf module allows computationally intensive tasks to be performed on a parallel computer. An Internet-based relational database is available for managing AUC experiments in a multi-user facility environment. The software is written in C++ using the portable QT toolkit. Software packages for Unix, Linux, Microsoft Windows, and Macintosh OS-X can be downloaded for free from http://www.ultrascan.uthscsa.edu.

Acknowledgements

I would like to thank Jeffrey Hansen for critically reviewing this Chapter. The UltraScan software development was supported by the National Science Foundation through grants NSF DBI 9724273 and NSF DBI 9974819, as well as through the Howard Hughes Medical Institute grant 76200-550802.

References

1. B. Demeler, UltraScan – A Software Package for Analytical Ultracentrifugation Experiments. The University of Texas Health Science Center at San Antonio, Department of Biochemistry, San Antonio TX, 78229 USA, 2004, http://www.ultrascan.uthscsa.edu
2. A listing of contributors and references can be found at: http://www.ultrascan.uthscsa.edu/references.html
3. The *gzip* and *tar* utilities are available from ftp://ftp.gnu.org/gnu
4. The *apache* webserver is available from http://www.apache.org
5. The *PHP* scripting language is available from http://www.php.net
6. The MySQL database engine is available from http://www.mysql.com
7. B. Demeler and H. Saber, *Biophys. J.*, 1998, **74**, 444–454.
8. M. L. Ralston and R. I. Jennrich, Technometrics. 1978, **20**, 7–14.
9. A. D. Griewank, Juedes, H. Mitev, J. Utke, O. Vogel and A. Walther, *ACM TOMS* 1990, **22**, 131–167.

10. A. Griewank and A. Walther, ADOL-C. Technische Universität Dresden, Institut für Wissenschaftliches Rechnen, D-01062 Dresden, Germany, 2004, http://www.math.tu-dresden.de/wir/project/adolc/

11. K. E. van Holde and W. O. Weischet, *Biopolymers*, 1978, **17**, 1387–1403.

12. B. Demeler and K. E. van Holde, *Anal. Biochem.*, 2004, **335**(2), 279–288.

13. M. L. Johnson, J. J. Correia, D. A. Yphantis and H. R. Halvorson, *Biophys. J.*, 1981, **36**, 575–588.

14. P. R. Gill, W. Murray and M. H. Wright, in *Practical Optimization*, Academic Press, London, 1981, 136–137.

15. K. Levenberg, *Quart. Appl. Math.*, 1994, **2**, 164–168.

16. D. Marquardt, *SIAM J. Appl. Math.*, 1963, **11**, 431–441.

17. C. G. Broyden, *J. Inst. Maths. Applics.*, 1970, **6**, 76–90.

18. R. Fletcher, *Comput. J.*, 1970, **13**, 317–322.

19. D. Goldfarb, *Math. Comput.*, 1970, **24**, 23–26.

20. D. F. Shanno, *Math. Comput.*, 1970, **24**, 647–656.

21. C. L. Lawson and R.J. Hanson, *Solving Least Squares Problems*, Prentice-Hall, Englewood Cliffs, NJ, 1974.

22. S. C. Gill and P. H. von Hippel, *Anal. Biochem.* 1989, **182**, 319–326.

23. G. E. P. Box and M. E. Muller, *Ann. Math. Stat.*, 1958, **29**, 610–611.

24. H. Durchschlag, in *Thermodynamic Data for Biochemistry and Biotechnology,* H.-J. Hinz, (Ed), Springer, New York, 1986, 45–128.

25. T. M. Laue, B. D. Shah, T. M. Ridgeway and S. L. Pelletier, in *Analytical Ultracentrifugation in Biochemistry and Polymer Science*, S. E. Harding, A. J. Rowe and J. C. Horton (eds), Royal Society of Chemistry, Cambridge, UK, 1992, 90–125.

26. J. Philo, Sednterp database files, personal communication.

CHAPTER 12

Analysis of Polysaccharide Size, Shape and Interactions

STEPHEN E. HARDING

1 Introduction

In the earlier Royal Society of Chemistry book *Analytical Ultracentrifugation in Biochemistry and Polymer Science*, we reviewed the progress and potential of the ultracentrifuge for providing fundamental information about polysaccharides in what for many is their natural state – in solution.[1] Our chapter was reinforced by contributions from Lavrenko and co-workers,[2] who looked in detail at the concentration dependence of the sedimentation coefficient of polysaccharides, Comper and Zamparo,[3] who reviewed the sedimentation analysis of proteoglycans, and Preston and Wik,[4] who examined the non-ideality behaviour of hyaluronan. Since the publication of that book, there has been the launch and establishment of the XL-I analytical ultracentrifuge with full on-line data automatic data capture analysis, concurrent with a general phasing out of the older MOM 2081, MSE Centriscan and Beckman Model E analytical ultracentrifuges, although the latter facilitated off-line automatic data capture and analysis.[5,6]

These developments have facilitated some major advances in software for analysis. Although the focus of these advances has been for the study of protein systems, they also, with some adjustment where appropriate, present possibilities for the study of polysaccharides and related glycopolymers. And alongside these developments in instrumentation and analysis software, there have been some important developments with related techniques such as size-exclusion chromatography and atomic force microscopy. This chapter reflects on these advances and considers how the new generation of analytical ultracentrifugation is contributing, where appropriate in combination with other techniques, to our understanding of the size, shape and interactions of polysaccharides in a solution environment. It will also reflect on some of the special difficulties they still present compared to the study of protein systems, most notably deriving from their polydispersity and non-ideality.

2 Change in Instrumentation

Although a handful of Beckman Model E and MOM ultracentrifuges remain in use, since its launch in 1996 the principal analytical ultracentrifuge used for the study of polysaccharide solutions has become the Optima XL-I from Beckman Instruments (Palo Alto, USA).[7] Six years earlier, the Beckman Optima XL-A ultracentrifuge had been launched with online UV/visible absorption optics providing a direct record of solute concentration (in absorption units) $c(r)$ *vs.* radial displacement r from the axis of rotation.[8] This optical system however had limited relevance for polysaccharides because of the lack of chromophore these substances possess in the near-UV (250–300 nm) and visible region. Nonetheless, some studies on labelled polysaccharides were possible and Cölfen and co-workers in 1996[9] successfully used an XL-A to characterise the sedimentation coefficient and molecular weight of two chitosans labelled with the fluorophore 9-anthraldehyde.

The laser (wavelength 670 nm) on the XL-I instrument provides high-intensity, highly collimated light and the resulting interference patterns (between light passing through the solution sector and reference solvent sector of an ultracentrifuge cell) are captured by a CCD camera. A Fourier transformation converts the interference fringes into a record of concentration $c(r)-c(a)$ relative to the meniscus $(r=a)$ as a function of r. The measurement is in terms of Rayleigh fringe units relative to the meniscus, $j(r)$, with $J(r)=j(r)+J(a)$, $J(r)$, being the absolute fringe displacement and $J(a)$ the absolute fringe displacement at the meniscus. For a standard optical path length cell $(l=1.2$ cm$)$ with laser wavelength $\lambda=6.70\times10^{-5}$ cm, a simple conversion exists from $J(r)$ in fringe displacement units to $c(r)$ in g mL^{-1}:

$$c(r) = \frac{J(r)\lambda}{(\mathrm{d}n/\mathrm{d}c)l} \tag{1}$$

$$= \left\{ \frac{5.58 \times 10^{-5}}{(\mathrm{d}n/\mathrm{d}c)} \right\} \cdot J(r)$$

with similar conversions for $J(a)$ to $c(a)$ and $j(r)$ to $c(r)-c(a)$. dn/dc is the (specific) refractive index increment, which depends on the polysaccharide, solvent and wavelength. A comprehensive list of values for a range of macromolecules has recently been published.[10] In aqueous systems, most values lie between 0.14 and 0.16 mL g^{-1}, although for non-aqueous systems the values can range enormously from 0.044 to 0.218 mL g^{-1}. The data, for example, for κ-carrageenan suggest little temperature dependence although that for dextrans suggests a significant dependence on wavelength. A study on the polycationic chitosan[11] suggested that the degree of substitution of some polysaccharides can strongly affect dn/dc, particularly if ionic groups are involved. Preston and Wik[4] have explored in detail the effect of ionic strength and wavelength on dn/dc for the polyanion hyaluronate. These results show that if a user needs, for whatever reason, an accurate value for dn/dc for a polysaccharide, he should measure it directly in the particular buffer used for the ultracentrifuge experiments. Converting fringe concentrations $\{j(r)$ or $J(r)\}$ into weight concentrations is normally not necessary for most applications. In addition, for sedimentation velocity work it is possible to work with $j(r)$ or $c(r)-c(a)$, *i.e.* concentrations relative to the meniscus without having to worry about measuring the offset or meniscus concentration $J(a)$ or $c(a)$ to convert into absolute $J(r)$ or $c(r)$.

3 Polysaccharide Polydispersity and Simple Shape Analysis by Sedimentation Velocity

Traditional analysis methods on optical records from sedimentation velocity experiments have been based around recording the movement of the radial position of the boundary r_b with time t, from which a sedimentation coefficient, s (s or Svedbergs, S, where $1\,S = 10^{-13}$ s) can be obtained:[12]

$$s = \frac{(dr_b/dt)}{\omega^2 r_b} \qquad (2)$$

Where ω is the angular velocity (rad s^{-1}) and followed by the usual correction to standard conditions – namely the density and viscosity of water at 20.0 °C – to yield $s_{20,w}$[12] and the algorithm SEDNTERP, which also arose out of the 1992 volume[13,14] has been useful facilitating this correction. Non-ideality effects are much more severe for polysaccharides compared with proteins, and the traditional way of correcting for such non-ideality is to measure either s or $s_{20,w}$ for a range of different cell loading concentrations c, and perform an extrapolation to zero concentration. For polysaccharides, this has been conventionally achieved from a plot of $1/s$ (or $1/s_{20,w}$) *vs.* c:[12]

$$\{1/s\} = \{1/s^\circ\} \cdot \{1 + k_s c\} \qquad (3)$$

a relation valid over a limited range of concentration with k_s the Gralén coefficient named after his doctoral dissertation on the analysis of cellulose and its derivatives.[15]

For a wider span of concentrations, a more comprehensive description of concentration dependence has been proposed by Rowe:[16,17]

$$s = s^\circ \left\{ 1 - \frac{\left[k_s c - \left(\frac{(cv_s)^2 (2\phi_p - 1)}{\phi_p^2} \right) \right]}{[k_s c - 2cv_s + 1]} \right\} \qquad (4)$$

Where v_s (mL g^{-1}) is the 'swollen' specific volume of the solute (volume (mL) of a polysaccharide (swollen through solvent association) per gram of the anhydrous molecule) and ϕ_p is the maximum packing fraction of the solute (~ 0.4 for biological solutes.[17]) A least-squares proFit (Quantum Soft, Zurich, Switzerland) algorithm has been developed for fitting s *vs.* c data to Equation (4), and as we will see below, this relation is proving to be of interest for the analysis of polysaccharides.

3.1 Sedimentation Coefficient Distributions: SEDFIT

The ability to acquire multiple data on-line with the XL-A and XL-I has formed the catalysis for advances in software for recording and analysing not only the change in boundary position with time but the change in the whole radial concentration profile, $c(r, t)$ with time t. These advances have in particular facilitated the measurement of *distributions* of sedimentation coefficient.[18–22] The (differential) distribution of sedimentation coefficients can be defined as the population (weight fraction) of species with a sedimentation coefficient between s and $s+ds$. Different symbols exist for this parameter, either $g(s)$ or $c(s)$: despite the choice of symbol for the latter, unlike c and $c(r,t)$, $c(s)$ has units of weight concentration (g mL^{-1}) *per second or Svedberg unit*. A plot of $g(s)$ or $c(s)$ *vs.* s then defines the distribution. Integration of a peak or resolved peaks from these types of plot can then be used to calculate the weight average s of the sedimenting species and their partial loading concentrations.

The simplest way of computationally obtaining a sedimentation coefficient distribution is from time derivative analysis of the evolving concentration distribution profile across the cell.[18,19] More recently, attention has turned to direct modelling of the evolution of the concentration distribution with time for obtaining the sedimentation coefficient distribution.[20,21] The distribution has been related to the experimentally measured evolution of the concentration profiles throughout the cell by a Fredholm integral equation

$$a(r, t) = \int_{s_{min}}^{s_{max}} c(s) \cdot \chi(s, D, r, t)\, ds + a_{TI}(r) + a_{RI}(t) + \varepsilon \qquad (5)$$

In this relation, $a(r, t)$ is the experimentally observed signal, ε represents random noise, $a_{TI}(r)$ the time-invariant systematic noise and $a_{RI}(t)$ the radial-invariant systematic noise. Schuck[20] and Dam, and Schuck[21] describe how this systematic noise is eliminated. χ is the normalised concentration at r and t for a given sedimenting species of sedimentation coefficient s and translational diffusion coefficient D: it is normalised to the initial loading concentration so it is dimensionless. The evolution with time of the concentration profile $\chi(s, D, r, t)$ for a given sedimenting species of sedimentation coefficient s and translational diffusion coefficient D in a sector-shaped ultracentrifuge cell is given by the Lamm[23] equation: although only approximate analytical solutions to this partial differential equation have been available for $\chi(s, D, r, t)$, accurate numerical solutions are now possible using finite element methods first introduced by Claverie and co-workers[24] and recently generalized to permit greater efficiency and stability.[20,21] The algorithm SEDFIT[25] employs this procedure for obtaining the sedimentation coefficient distribution. To solve Equation (5) to obtain $c(s)$ as a function of s requires the limits s_{min} and s_{max} to be carefully chosen and adjusted accordingly: inappropriate choice can be diagnosed by an increase of $c(s)$ towards the limits of s_{min} or s_{max}. However, the contribution from diffusion broadening also has to be dealt with. SEDFIT offers two ways. The first is using a dependence of D on s, via the translational frictional ratio f/f_0:

$$D(s) = \left\{ \frac{\sqrt{2}}{18\pi} \right\} k_B t s^{-1/2} (\eta_o (f/f_0)_w)^{-3/2} \left(\frac{1 - v\rho_o}{v} \right)^{-1/2} \qquad (6)$$

where f is the frictional coefficient of a species, f_0 the corresponding value for a spherical particle of the same mass and (anhydrous) volume (see *e.g.* ref. 26) and k_B is the Boltzmann constant. Although, of course, a distribution of s implies also a distribution in D and f/f_0, for protein work, advantage is taken of the fact that the frictional ratio is a relatively insensitive function of concentration: a single or weight average f/f_0 is taken to be representative of the distribution. Using this assumption, of Equation (5) can be numerically inverted to give the sedimentation coefficient distribution, with the position and shape of the $c(s)$ peak(s) more representative of a true distribution of sedimentation coefficient. $(f/f_0)_w$, where the subscript w denotes a weight average, is determined itera-tively by non-linear regression, optimizing the quality of the fit of the $c(s)$ as a function of $(f/f_0)_w$. It has been shown by extensive simulation that non-optimal values of $(f/f_0)_w$ have little effect on the position of the $c(s)$ peaks, although affects the width and reso-lution, *i.e.* the correct s value is reported. Regularization[25] can be used which provides a measure of the quality of fit from the data analysis. The assumption of a single f/f_0 rep-resenting the whole macromolecular distribution may be reasonable for proteins but it is open to question whether this is so for polysaccharides: this assumption will affect the reliability of distribution widths, but not peaks. Nonetheless, better approximations are currently being sought.

The present version of SEDFIT also offers the option of evaluating the distribution corresponding to non-diffusing particles, *viz* $D \sim 0$, *i.e.* the diffusive contribution to Equation (5) is small compared to the sedimentation contribution. In this case, Equation (5) can be inverted without any assumptions concerning f/f_0. If diffusive effects are sig-nificant, it will lead to an apparent sedimentation coefficient distribution, given as $g^*(s)$ *vs.* s although the correct s value for a peak is still reported. Figure 1 gives a compari-son of the least-squares $g^*(s)$ *vs.* s and $c(s)$ *vs.* s distribution for guar gum.[27]

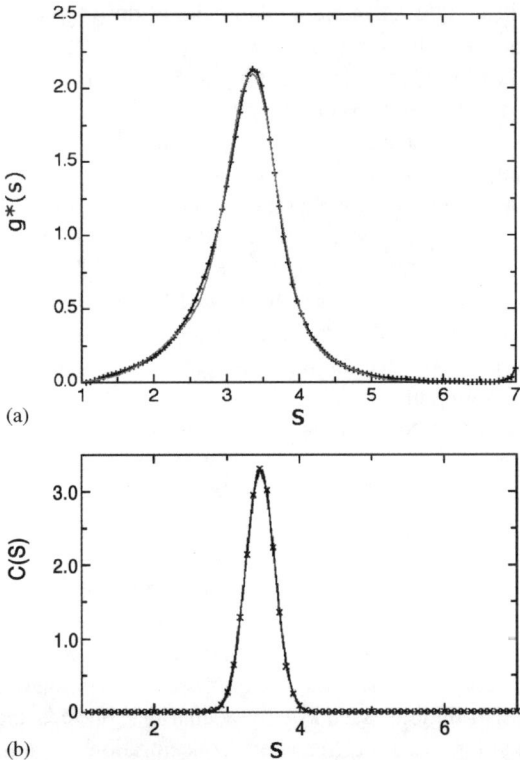

(a)

(b)

Figure 1 *Sedimentation concentration distribution plots for guar gum using SEDFIT. (a) $g^*(s)$ vs. s; (b) c(s) vs. s. A Gaussian fit to the data (lighter line) is also shown in (a). Rotor speed was 40 000 rpm at 20.0 °C, concentration was ~0.75 mg mL^{-1} in 0.02% NaN$_3$. The guar had been heated at 160 °C for 10 min at a pressure of 3 bar (from ref. 27)*

There is a clear shoulder on the low s (lower M) side of the $g^*(s)$ peak – consistent with some lower molecular weight material observed using the technique of SEC–MALLs (size-exclusion chromatrography coupled to multi-angle laser light scattering), whereas the $c(s)$ profile shows only a symmetric peak. The current $c(s)$ procedure in this instance seems to have 'oversmoothed' the data. We would suggest the evaluation of apparent distributions via $g^*(s)$ is preferred at present, particularly for slow-diffusing polysaccharides.

It is possible to get molecular weight from the sedimentation coefficient if we assume a conformation or if we combine with other measurements, namely the translational diffusion coefficient via the Svedberg equation[28]

$$M = RT \; \frac{\{s^o/D^o\}}{(1 - \bar{v}\rho_o)} \tag{7}$$

where ρ_o is the solvent density (if s and D are their normalized values $s^o_{20,w}$, $D^o_{20,w}$, ρ_o will be the density of water at 20.0°C, 0.9981 g mL^{-1}). Equation (7) has been popularly used, for example, to investigate the molecular weights of carboxymethyl-chitins,[29–31] glycodendrimers,[32,33] β-glucans[34] and alginates.[35] The translational diffusion coefficient in Equation (7) can in principle be measured from boundary spreading as manifested, for example, in the width of the $g^*(s)$ profiles: although for monodisperse proteins, this works well, for polysaccharides, interpretation is seriously complicated by broadening through polydispersity. Instead, special cells can be used which allow for the formation of an artificial boundary whose diffusion can be recorded with time at low speed ($\sim$3000 rev per min). This procedure has been successfully employed, for example, in a recent study on heparin fractions.[36] Dynamic light scattering has been used as a popular alternative, and a good demonstration of how this can be performed to give reliable D data has been given by Burchard.[37]

Whereas the s^o is a weight average, the value returned from dynamic light scattering for D^o is a z average. As shown by Pusey,[38] combination of the two via the Svedberg equation (7) yields the weight-average molecular weight M_w, although it is not clear what type of average for M is returned if an estimate for D^o is made from ultracentrifuge measurements.

Another useful combination that has been suggested is $s^o_{20,w}$ with k_s:[16,17]

$$M_w = N_A \left[\frac{6\pi\eta s^o_{20,w}}{(1 - \bar{v}\rho_0)} \right]^{3/2} \left[\frac{3\bar{v}}{4\pi} \left(2\frac{k_s}{\bar{v}} \right) - \left(\frac{v_s}{\bar{v}} \right) \right]^{1/2} \tag{8}$$

s, k_s and v_s can be obtained from fitting s vs. c data to Equation (4). The method was orginally developed for single solutes and where charge effects can be neglected (either because the macromolecular solute is uncharged, or because the double layer or polyelectrolyte behaviour has been 'compressed' by addition of neutral salt). For quasi-continuous distributions, such as polysaccharides, one can apply Equations (5) and (8) to the data, provided that for every concentration one has a 'boundary' to which a weight-averaged s value can be assigned. If the plot of $1/s$ vs. c is essentially linear over the data range, then specific interaction can be excluded, the solute system treated as a simple mixture and Equations (5) and (8) can be applied. Figure 2 shows an example for wheat starch amylopectin, where concentrations for total

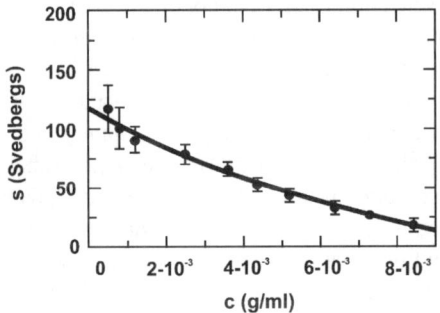

Figure 2 *Concentration dependence of the sedimentation coefficient for wheat amylopectin. The data have been fitted to Equation (4) (see text) yielding $s° = (120 \pm 10)$ S, $k_s = (170 \pm 60)$ mL g^{-1} and $v_s = (40 \pm 4)$ mL g^{-1} (from ref. 39).*

starch have been normalised to amylopectin from the relative areas under the $g^*(s)$ peaks. From this data a value for M_w of $\sim 30 \times 10^6$ g mol^{-1} has been estimated.[39]

Equation (8) is only approximate – any contributions from molecular charge to the concentration dependence parameter k_s are assumed to be negligible or suppressed – but is nonetheless useful when other methods, especially for very large polysaccharides like amylopectin, are inapplicable. The method also provides an estimate for the swollen specific volume v_s: for example, Majzoobi has obtained a value of 40 ± 4 mL g^{-1} for wheat starch amylopectin.[39] For polydisperse materials such as polysaccharides, the question is what sort of average M value is yielded by doing so? In the absence of any obvious analytical solution, computer simulation has been used to determine the form of the average. In their work to be published, Rowe and co-workers have shown that even for 'unfavourable' simulated mixtures (*e.g.* multi-modal, no central tendency), the average M value yielded is very close to an M_w (*i.e.* weight-averaged M). To put this in quantitative terms, the departure from M_w is generally $<1\%$ of the way towards M_z. This is trivial, in terms of the errors present in the raw data. Thus, there is an exact procedure which can be defined for the evaluation of M(average) in a polydisperse solute system under the defined conditions, and simulation demonstrates that for all practical purposes the outcome is an M_w.

A sedimentation coefficient *distribution*, either $c(s)$ *vs.* s or $g^*(s)$ *vs.* s for a polysaccharide can also be converted into an apparent molecular weight distribution if the conformation of the polysaccharide is known or can be assumed, via a power law or 'scaling' relation (see below). An early example of this transformation, assuming a random coil conformation, has been given for a heavily glycosylated mucin glycoprotein with polysaccharide-like properties[42] based on a $g^*(s)$ *vs.* s distribution given by Pain.[43] The assumption was made that the contribution from diffusion broadening of these large molecules was negligible in comparison to sedimentation.

4 Polysaccharide Molecular Weight Analysis by Sedimentation Equilibrium: MSTAR

The final steady-state pattern from a sedimentation equilibrium experiment[44] is a function only of molecular weight and related parameters (non-ideal virial coefficients and association constants were appropriate) and not on molecular shape since at equilibrium there is no net transport or frictional effects: sedimentation equilibrium in the analytical ultracentrifuge provides an absolute way of estimating molecular weight. Since polysaccharides are by their very nature polydisperse, the value obtained will be an average of some sort. With Rayleigh interference and, where appropriate, UV-absorption optics, the principal average obtained is the weight average, M_w.[12] Although relations are available for also obtaining number-average M_n and z-average M_z data, these latter averages are difficult to obtain with any reliable precision. Direct recording of the concentration gradient dc/dr *vs.* radial displacement r using refractive index gradient or 'Schlieren' optics however facilitates the measurement of M_z (see ref. 45). Although present on the older generation Model E and MOM centrifuges, Schlieren optics are not on the present generation XL-A or XL-I ultracentrifuges, except for in-house adapted preparative XL ultracentrifuges.[46]

An important consideration with polysaccharides is that at sedimentation equilibrium there will be a redistribution not only of total concentration of polysaccharide throughout the cell (low concentration at the meniscus building up to a higher concentration at the cell base) but also a redistribution of species of different molecular weight, with a greater proportion of the higher molecular-weight part of the distribution appearing near the cell base. In obtaining a true weight (or number, z averages), it is therefore important to consider the *complete* concentration distribution profile throughout the ultracentrifuge cell. As with our description of sedimentation velocity, for clarity, we will confine our consideration only to the extraction of the two most directly related parameters: the weight average molecular weight and the molecular weight distribution. The extraction of other parameters, such as point average data, are avoided here but can be found in other articles (see refs. 47–49).

4.1   Obtaining the Weight Average Molecular Weight

As stated above, UV-absorption optics, when they can be applied, have the advantage that the recorded absorbances $A(r)$ as a function of radial position are (within the Lambert–Beer law limit of $A(r) \sim 1.4$) directly proportional to the weight concentration $c(r)$ in g mL^{-1}. Although the multiple fringes in interference optics give a much more precise record of concentration, we stress again, these are *concentrations relative to the meniscus*, *i.e.* we obtain directly from the optical records a profile of $c(r) - c(a)$ *vs.* radial displacement r, with the meniscus at $r = a$. In fringe displacement units, this is $J(r) - J(a)$, which we write as $j(r)$ for short. To obtain molecular weight information, we need $J(r)$ and hence, some way of obtaining $J(a)$ is required: this is not such a requirement for sedimentation velocity where relative concentrations are sufficient. Any attempt to deplete the meniscus (rich in the lower molecular-weight part of the distribution) of a polysaccharide solution so that $J(a) \sim 0$ – a method popular for protein work since 1964[50] – is almost guaranteed to result in loss of optical registration of the interference fringes near the bottom of the cell, leading to underestimates for M_w. This means that a procedure for evaluating $J(a)$ is required. It was recently shown by Hall and co-workers[51] that simply floating it as another variable in the procedure for extracting M is not valid, particularly for polydisperse or interacting systems. A convenient procedure for extracting $J(a)$ and then

M_w was given by Creeth and Harding in 1982:[52] the fundamental equation of sedimentation equilibrium can be manipulated to define a new function with dimensions of molar mass (g mol^{-1}) called $M^*(r)$. $M^*(r)$ at a radial position r is defined by

$$M^*(r) = j(r)/\{kJ(a)(r^2-a^2) + 2k \int_a^r rj(r) \, dr\} \tag{9}$$

where $k = (1-\bar{v}\rho_o)\omega^2/2RT$, with ρ_o the solvent density. Equation (9) has the limiting form

$$\lim_{r \to a} \left\{ \frac{j(r)}{(r^2-a^2)} \right\} = kM^*(a)J(a) \tag{10}$$

A plot of $j(r)/(r^2-a^2)$ *vs.* $\{1/(r^2-a^2)\} \int_a^r rj(r) \, dr$ therefore has a limiting slope of $2kM^*(a)$ and an intercept $kM^*(a)J(a)$. Hence $J(a)$ is determinable from 2× (intercept/limiting slope). Other methods of obtaining $J(a)$ have been considered in detail by Teller and co-workers[53] and Creeth and Pain.[47] More recently, Minton[54] has given an almost identical procedure, although unfortunately he appears to have missed the original Creeth and Harding article published 12 years earlier.[52] Once $J(a)$ has been found, M^* as a function of radial position r can be defined. A particularly useful property of the M^* function is that at the cell base ($r=b$),

$$M^*(b) = M_{w,app} \tag{11}$$

the apparent weight average molecular weight of the polysaccharide.[52] It will be an 'apparent' value because it will be affected by thermodynamic non-ideality (molecular co-exclusion and, for charged polysaccharides, polyelectrolyte behaviour), which needs to be corrected for (see below). Optical distortion effects at the cell base means that a short extrapolation of $M^*(r)$ to $M^*(=b)$ is required, but this normally poses no difficulty. Practical details behind the MSTAR algorithm upon which this procedure is based can be found in refs. 48, 49 and 55. It is worth pointing out here that another popular algorithm for analysing molecular weight from sedimentation equilibrium is NONLIN.[56] Whereas this is useful for the analysis of protein systems (monodisperse or associating), for polydisperse system like polysaccharides, it is unsuitable: the estimate for $M_{w,app}$ obtained refers only to a selected region of the ultracentrifuge cell, and provides no rigorous procedure for dealing with the meniscus concentration problem.

4.2 Correcting for Thermodynamic Non-Ideality: Obtaining M_w from $M_{w,app}$

For polysaccharides, non-ideality arising from co-exclusion and polyelectrolyte effects can be a serious problem and, if not corrected for, can lead to significant underestimates for M_w. It was possible with the older generation Model E ultracentrifuges, which could accommodate long (30 mm) optical path length cells to work at very low solute loading concentrations (0.2 mg mL^{-1}). At these concentrations, for some polysaccharides, the non-ideality effect could be neglected: the estimate for $M_{w,app}$ was within a few percent of the true or 'ideal' M_w. However, the new-generation XL-I can only accommodate a maximum 12 mm optical path length cell with a minimum concentration requirement of 0.5 mg mL^{-1}: lower concentrations produce insufficient fringe displacement for meaningful analysis. This makes a large difference to the severity of the non-ideality problem: a concentration extrapolation is now mandatory for sedimentation equilibrium molecular weight determinations on polysaccharides using an equation of the form[47]

$$\{1/M_{w,app}\} = \{1/M_w\} + 2Bc \qquad (12)$$

$$= \{1/M_w\}(1 + 2BM_w c)$$

correct to first order in concentration. In connection with this correction, it is worth mentioning (1) The availability of four- and eight-hole rotors in the XL-A and XL-I means that several concentrations can be run simultaneously. Further multiplexing is possible with the use of style six-channnel ultracentrifuge cells,[50] which permit the simultaneous measurement of three solution/reference solvent pairs, although these tend to return $M_{w,app}$ values of lower accuracy. (2) For polyelectrolytes, the second virial coefficient is very sensitive to ionic strength. Preston and Wik[4] have shown a 10-fold increase in B, from ~50 to ~500 mL mol g^{-2}, upon decreasing the ionic strength from 0.2 down to 0.01 mol L^{-1}. (3) The second virial coefficient B in Equation (12) refers to the static case. In the ultracentrifuge, the measured value can show a speed dependence,[57] an effect which can be minimized by using low speeds and short solution columns. If present, it will not affect the value of M_w after extrapolation to zero concentration. (4) In some extreme cases, third or even higher virial coefficient(s) may be necessary to adequately represent the data, for example, κ-carrageenan[58] and alginate.[59] In a further study on alginates, Straatman and Borchard[60] demonstrated excellent agreement between M_w and B values obtained from sedimentation equilibrium and light-scattering methods.

4.3 Distributions of Molecular Weight: SEC–MALLs and the New Role for Sedimentation Equilibrium

Direct inversions of the concentration distribution profiles to obtain molecular weight distribution information are generally intractable because of complications involving non-ideality. Successful attempts have been given but only for simple discrete forms of polydispersity (two to three macromolecular species.[61] The simplest procedure for avoiding these complications[62] is to use sedimentation equilibrium in conjunction with gel-permeation chromatography (GPC). Fractions of relatively narrow (elution volume) bandwidth are isolated from the eluate and their M_w values evaluated by low-speed sedimentation equilibrium in the usual way: the GPC columns can thereby be 'self-calibrated' and elution volume values converted into corresponding molecular weights – a distribution can therefore be defined in a way which avoids the problem of using inappropriate standards for GPC: the value of multiplexing is clearly indicated. This procedure has been successfully applied, for example, to dextrans, alginates and pectins: for pectins, excellent agreement with analogous procedures involving classical light scattering coupled to GPC has been obtained.[63]

There is now a much easier method available for obtaining molecular weight distribution. The measurement of the angular dependence of the total intensity of light scattered by solutions of polysaccharides provides, like sedimentation equilibrium, a direct and absolute way of measuring the weight average molecular weight, again if allowance for thermodynamic non-ideality is made. (Some researchers tend to prefer 'A_2' as notation for the second virial coefficient rather than B.) Although opinions varied, prior to 1990 (see *e.g.* ref. 64), there was good case for suggesting sedimentation equilibrium as the preferred method of choice for the measurement of molecular weights, simply because of the less stringent requirements on sample clarity: with light scattering, it is essential that solutions are free of supra-molecular aggregates. The inclusion of a flow cell into a light-scattering photometer facilitated the coupling on-line to a gel-permeation chromatography column and SEC–MALLs has now revolutionised the measurement of molecular weight and molecular weight distribution.[65,66] The combined effect of the SEC columns and a pre- or 'guard column' can provide clear fractionated samples to the light-scattering cell, facilitating not only measurement of M_w for the whole distribution, but also the distribution itself. Prior ultracentrifugation of the polysaccharide solution ($\sim$40 000 rpm for 30 min) is still advisable. The first polysaccharides studies were published in 1991[67–69] and it is now regarded by many as the method of choice for polysaccharide molecular weight determination. Furthermore, the angular dependence of the scattered light facilitates measurement of R_g as a function of elution volume and hence molecular weight: the method provides conformation information about the polysaccharide.[70] Nonetheless uncertainties can sometimes remain, particularly if materials have been incompletely clarified or there are problems with the columns (the form of the angular dependence data can usually tell us if things are not well). Sedimentation equilibrium offers a powerful and valuable independent check on the results generated from SEC–MALLs: although it takes a longer time to generate a result, and molecular weight distributions are considerably more difficult to obtain, agreement of M_w from sedimentation equilibrium with M_w from SEC–MALLs gives the researcher increased confidence in some of the other information (molecular weight distribution and R_g–M dependence) coming from the latter.

5 Polysaccharide Conformation Analysis by Sedimentation Velocity

The sedimentation coefficient s^o provides a useful indicator of polysaccharide conformation and flexibility in solution, particularly if the dependence of s^o on M_w is known.[41] There are two levels of approach: (i) a 'general' level in which we are delineating between overall conformation types (coil, rod, sphere), (ii) a more detailed representation where we are trying to specify particle dimensions in the case of rigid structures or persistence lengths for linear, flexible structures.

5.1 The Wales–van Holde Ratio

The simplest indicator of conformation comes not from s^o but the sedimentation concentration dependence coefficient, k_s. Wales and van Holde[71] were the first to show that the ratio of k_s to the intrinsic viscosity $[\eta]$ was a measure of particle conformation. It was shown empirically by Creeth and Knight[72] that this ratio has a value of ~ 1.6 for compact spheres and non-draining coils, and adopted lower values for more extended structures. Rowe[16,17] subsequently provided a derivation for rigid particles, a derivation later supported by Lavrenko and co-workers.[2] The Rowe theory assumed there were no free-draining effects and also that the solvent had sufficient ionic strength to suppress any polyelectrolyte effects. A value of 1.6 was evaluated for spheres, reducing to ~ 0.2 for long rod-shaped molecules.

Lavrenko and co-workers[2] also examined in detail the effects of free draining of solvent during macromolecular motion, demonstrating that this also had the effect of lowering $k_s/[\eta]$. A hydrodynamic intra-chain interaction or 'draining' parameter has been defined[73] with limits $X=\infty$ for the non-free draining case and $X=0$ for the free-draining case. A relation was given between $k_s/[\eta]$ and X:[2,73]

$$\{k_s/[\eta]\} = \frac{8X}{(3 + 8X)} \tag{13}$$

This relation evidently leads to theoretical limits for $k_s/[\eta]=0$ for free draining and 1 for non-free draining. The consequences of this are that unless the draining characteristics of the chain are properly known, one has to be cautious in making conclusions about particle asymmetry, since it has been claimed that draining affects can mimic increase in asymmetry in lowering the $k_s/[\eta]$. Notwithstanding, many non-spherical molecules have empirical values for $k_s/[\eta] > 1.0$: pullulans, for example, considered as a random coil have been shown to have $k_s/[\eta] \sim 1.4$ (see ref. 74). Berth and co-workers[75] have argued that the very low $k_s/[\eta]$ values for chitosans are due to draining effects rather than a high degree of extension. Lavrenko and co-workers[2] have compiled an extensive list of $k_s/[\eta]$ values for a large number of other polysaccharides, complementing a list given by Creeth and Knight:[72] values are seen to range from 0.1 (potato amylose in 0.33 M KCl) to 1.8 (a cellulose phenyl-carbamate in 1,4 dioxane), with some polysaccharides showing a clear dependence on molecular weight.

5.2 Power Law or 'Scaling' Relations

The relation linking the sedimentation coefficient with the molecular weight for a homologous polymer series given above is (see refs. 40 and 76):

$$s = K''M^b \tag{14}$$

(Some researchers (see *e.g.* ref. 2) call the exponent $1-b$.) This relation is similar to the well-known Mark–Houwink–Kuhn–Sakurada relation linking the intrinsic viscosity with molecular weight:

$$[\eta] = K'M^a \tag{15}$$

and also a relation linking the radius of gyration R_g with molecular weight:

$$R_g = K''M^c \tag{16}$$

Table 1 *Power law exponents (from ref. 40)*

	a	b	c
Sphere	0	0.67	0.33
Coil	0.5–0.8	0.4–0.5	0.5–0.6
Rod	1.8	0.15	1.0

The power law or 'MHKS' exponents a,b,c have been related to conformation[40,41] (Table 1).

The coefficients in Table 1 correspond to the non-draining case. If draining effects are present then these will change the values for a and b (see *e.g.* ref. 77). For example, it has been shown that a varies from 0.5 (non-draining case) to 1 (draining), again mimicking the effects of chain elongation.

Another scaling relation exists between the sedimentation coefficient and k_s (see ref. 2):

$$k_s = K''''(s^\circ)^\kappa \qquad (17)$$

and values of κ and K''' have been given for a range of polysaccharides.[2]

Various relations have been proposed linking the various power-law exponents for a homologous series under specified conditions[41] such as

$$\kappa = \frac{(2 - 3b)}{b} \qquad (18)$$

5.3 General Conformation: Haug Triangle and Conformation Zoning

Delineation of the three general conformation extremes (random coil, compact sphere, rigid rod) as indicated by the simple power or scaling laws and Wales/van Holde ratio, have been conveniently represented in the well-known Haug triangle (see ref. 40). An extension of this idea was given by Pavlov and co-workers,[78,79] who suggested five general conformation types or 'zones', all of which could be distinguished using sedimentation measurements. The zones were: A (extra rigid rod), B (almost rigid rod), C (semi-flexible coil), D (random coil) and E (globular/branched). A and B are distinguished by B having a very limited amount of flexibility. The zones were constructed empirically using a large amount of data (s, k_s) accumulated for polysaccharides of 'known' conformation type, and plotted a scaling relation normalised with mass per unit length (M_L) measurements (Figure 3). The latter parameter can be obtained from knowledge of molecular weight from sedimentation equilibrium or light scattering and the chain length L from small-angle X-ray scattering, X-ray fibre diffraction or NMR. Pavlov and co-workers give a comprehensive comparison of methods for heparin.[36] If the molecular weight is known, M_L can also be estimated from electron microscopy.[80] Measurement of a data set (s, k_s, M_L) of any target polysaccharide would then establish its conformation type. The limiting slopes of ~ 4 (extra rigid rod) and ~ 0 (globular/sphere) were shown to be theoretically reasonable. Other normalised scaling relations have been suggested based on viscometry methods.[79]

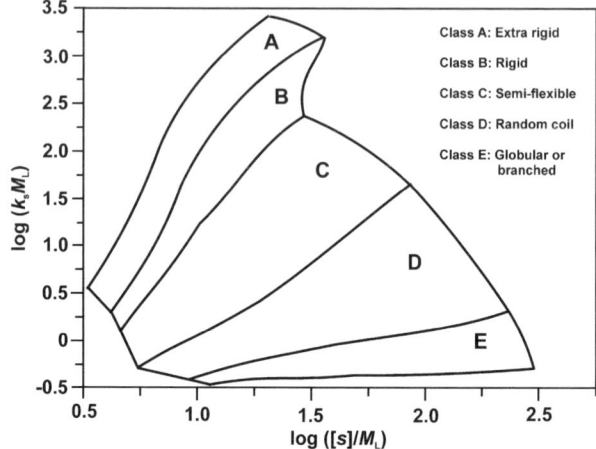

Figure 3 *Conformation zoning of polysaccharides. Empirical plots for various polysaccha-*
rides of known conformation type. This helps to define zones: (A) extra rigid rod, (B):
rigid rod, (C) semi-flexible coil, (D) random coil, (E) globular/heavily branched
structure. Measurement of s, k_s and M_L for a target polysaccharide then define its
conformation zone or type (redrawn and based on ref. 78)

5.4   Rigid Cylindrical Structures

Once a general conformation type or 'preliminary classification' has been established, it is possible to use sedimentation data to obtain more detailed information
about polysaccharide conformation. For example, the low value of $k_s/[\eta] \sim 0.25$
found for the bacterial polysaccharide xylinan has been considered to be due to
asymmetry.[81] If we then assume a rigid structure, the approximate theory of
Rowe[16,17] can be applied in terms of a prolate ellipsoid of revolution to estimate the
aspect ratio p ($\sim L/d$ for a rod, where L is the rod length and d its diameter) ~ 80.

For a cylindrical rod, an expression also exists for the sedimentation coefficient:[82]

$$s^o = \left\{ \frac{M(1-\bar{v}\rho_o)}{(3\pi\eta_o N_A L)} \right\} \{\ln(L/d) + \gamma\} \tag{19}$$

where γ is a function of p and has a limiting value of ~ 0.386 for very long rods ($p \to \infty$). Replacing L by the (molar) mass per unit length $M_L = M/L$ (g mol^{-1} cm^{-1}),
this becomes

$$s^o = \left\{ \frac{M_L(1-\bar{v}\rho_o)}{(3\pi\eta_o N_A)} \right\} \{\ln M - \ln M_L - \ln d + \gamma\} \tag{20}$$

For the cases of finite p (in the range 2–20), the currently accepted expression for
$\gamma(p)$ is that of Tirado and Garcia de la Torre:[83]

$$\gamma(p) = 0.312 + (0.561/p) + (0.100/p^2) \tag{21}$$

Above $p>10$, the limiting value ($\gamma=0.386$) can be used.

From Equations (20) and (21), we can obtain an estimate for the rod length L if we know M or M_L (see the above discussion) and have an estimate for the diameter d. As pointed out by Garcia de la Torre,[84] the choice for d is not so critical since it comes into the equations as the logarithm. It applies only to polysaccharides which are known to be rods.

5.5 Semi-Flexible Chains: Worm-Like Coils

Most linear polysaccharides are not rigid rods at all but are semi-flexible structures. The conformation and hydrodynamics of semi-flexible chains are most usefully represented by worm-like chains (see refs. 85–88), in which the bending flexibility is represented by the persistence length L_p. This is an intrinsic property of a linear macromolecule: the greater the L_p the greater the rigidity and *vice versa*. More precisely, the conformation and flexibility of a macromolecular chain depends directly on L/L_p, the ratio of the contour length to the persistence length. For $L/L_p \ll 1$ the conformation is rod-like and Equations (19)–(21) can be applied. For $L/L_p \gg 0$ the conformation approaches that of a random coil.[85–88] This can be best seen from the dependence of the radius of gyration on chain length, as clearly described by Freire and Garcia de la Torre:[88]

$$R_g^2 = \{L.L_p/3\}\{1 - (3L/L_p) + (6L_p^2/L^2) + 6(L_p^3/L^3)(1 - e^{-L/L_p})\} \quad (22)$$

In the limit $L_p/L \sim 0$, R_g is proportional to $L^{1/2}$ (this is misprinted in ref. 88) – the classical dependence for a random coil – whereas when $L_p/L_o \gg 1$, the classical relation for a rod is obtained: $R_g = L/\sqrt{12}$.

The sedimentation coefficient for worm-like chains was first worked out by Hearst and Stockmayer,[89] later improved by Yamakawa and Fujii[90] to give this expression for s^o

$$s^o = \{M(1 - \bar{v}\rho_o)/(3\pi\eta_o N_A L)\}\{1.843 \ln\{L/2L_p\}^{1/2} + \alpha_2 + \alpha_3(L/2L_p)^{-1/2} + \cdots\} \quad (23)$$

If the persistence length L_p is much larger than the mean chain diameter d, Yamakawa and Fujii gave limiting values for $\alpha_2 = - \ln(d/2L_p)$ and $\alpha_3 = 0.1382$. Freire and Garcia de la Torre[88] have considered further these coefficients. The factor $2L_p$ appears rather than L_p simply because $2L_p$ is equivalent to the statistical Kuhn segment length λ^{-1}.

A fundamental problem with the sedimentation coefficient is that it is the least sensitive parameter to conformation when compared with the intrinsic viscosity $[\eta]$ and the radius of gyration R_g. This lower sensitivity is offset by the ease of measurement and the ability to obtain s^o to a higher accuracy (to better than 1%) compared with the other parameters. Nonetheless it is advisable not to use s in isolation but in conjunction with R_g and $[\eta]$ *vs.* M. Two recent examples are a comparative study using ultracentrifugation, viscometry and light scattering on the relative conformations and flexibilities of galactomannans (guar, tara gum and locust bean gum), after pressure-assisted solubilisation procedures[27] and a study using ultracentrifugation, viscometry and small-angle X-ray scattering to investigate the conformation and flexibility of heparin.[36]

6   Analysis of Polysaccharide Interactions

There are many instances where (associative) interactions involving polysaccharides, whether they be self-association, complex formation or with small ligands are important (see ref. 91). Examples of self-association are dimerisation or trimerisation of helical types of polysaccharides, such as schizophyllan, scleroglucan are good examples (we could mention also xanthan and κ-carrageenan although that has been the subject of some disagreement). Examples of complex formation include the use of cellulose derivatives as dental adhesives, and an example of small ligand interactions is the intercalation of iodine by amylose or amylopectin. There has been considerable attention focussed on the use of polysaccharide systems as encapsulation agents for flavours and drugs, and this invokes both macromolecular and small ligand interactions involving polysaccharides. The analytical ultracentrifuge would appear to offer considerable potential for the analysis of these and other types of interaction. Indeed one of the main reasons behind the renaissance of analytical ultracentrifugation in the 1990s[92,93] was the simmering need of molecular biologists and protein chemists for non-invasive solution-based methods for studying biomolecular interactions, particularly the weaker ones involved in molecular recognition phenomena (see *e.g.* refs. 94 and 95). The analytical ultracentrifuge – its clean, medium-free (no columns or membranes) and absolute nature has indeed proven a highly attractive tool for characterising the stoichiometry, reversibility and strength (as represented by the molar dissociation constant K_d) of an interaction between well-defined systems: protein–protein, protein–DNA, protein–small ligand. With polysaccharides, we are generally dealing with a different situation. Firstly, a polysaccharide does not have a single, clearly defined molecular weight: it is polydisperse with a distribution of molecular weights. Secondly, weak interactions ($K_d > 50$ μM), at least as far as we know, do not play a crucial functional role with polysaccharides as they do with proteins. Interactions, particularly involving polyelectrolytes of opposite charge (chitosan–alginate for encapsulation systems, chitosan–DNA for gene therapy) tend to be very strong or irreversible: the complexes tend to be much larger than for the simple associative protein–protein interactions. This means the main ultracentrifuge tool used for investigating protein–protein interactions, namely sedimentation equilibrium, has only limited applicability: sedimentation equilibrium has an upper limit of molecular weight of $\sim$50 million g mol^{-1}. Examples of the use of the analytical ultracentrifuge to assay interactions involving polysaccharides are a study on mixtures of alginate with bovine serum albumin,[96,97] a study of galactomannan incubated with gliadin (as part of an ongoing investigation into the possible use of galactomannans to help intestinal problems),[98] chitosan with lysozyme[99] and synergistic interactions involving xanthan.[100]

For large irreversible complexes involving polysaccharides, a more valid assay procedure is to use sedimentation velocity (which can cope with complexes as large as 10^9 g mol^{-1}), with change in sedimentation coefficient s (normalised to standard conditions or not) or as our marker for complex formation. If we so wish, we can then convert this to a change in molecular weight if we assume a conformation and use the power-law relation (Equation (13)). Alternatively, we can simply use s directly as our size criterion (this is not unusual; it is used, for example, in ribosome size representations, 30S, 50S, ..., or in seed globulin, the 7S, 11S soya bean glob-ulins, *etc.*[101]

A good example of where sedimentation velocity has played a valuable role in assaying large polysaccharide complexes is in the assessment of polysaccharides as mucoadhesives (see *e.g.* ref. 102 and references cited therein): a drug administered orally or nasally tends to be washed away from the site of maximum absorption by the body's natural clearance mechanisms before being absorbed. Incorporating the drug into a polysaccharide material which interacts with epithelial mucus in a controllable way has been proposed as a method of increasing the residence time and enhancing the absorption rate. The key macromolecule in mucus is mucin glycoprotein – a lin-ear polypeptide backbone with linked saccharide chains to the extent >80% of the molecule is carbohydrate (see *e.g.* ref. 42). The carbohydrate has potential sites for ionic interaction (clusters of sialic acid or sulphate residues) and also hydrophobic interaction (clusters of methyl groups offered by fucose residues). The sedimentation ratio ($s_{complex}/s_{mucin}$) – the ratio of the sedimentation coefficient of the complex to that of the pure mucin itself – is used as the measure for effectiveness of a candidate mucoadhesive, supported by imaging (electron microscopy, scanning tunnelling microscopy and atomic force microscopy) and macroscopic studies. The UV absorp-tion optics on the XL-A or XL-I ultracentrifuge have been used as the main optical detection system. Although the polysaccharide is generally invisible in the near UV ($\sim$280 nm), at the concentrations normally employed, the mucin, in uncomplexed and complexed form, is detectable. Experiments on a series of neutral and polyanionic polysaccharides revealed no significant change in the sedimentation coefficient (sedi-mentation ratio $s_{complex}/s_{mucin} \approx 1$) reinforcing macroscopic observations on whole mucus using tensiometry.[103] A contrasting picture is seen for polycationic deacetylated aqueous (pH<6) soluble derivatives of chitin known as chitosans with sedimentation coefficient ratios $s_{complex}/s_{mucin}$ approaching $\sim$40. Interestingly, altering the degree of acetylation did not seem to affect the interaction greatly, suggesting that hydrophobic as well as electrostatic interactions play a mucoadhesive role. The demonstration of large-size interaction products by the analytical ultracentrifuge used in this manner is reinforced by images from the powerful imaging techniques of electron microscopy and atomic force microscopy. Conventional transmission electron microscopy clearly demonstrates large complexes of the order of $\sim$1 μm in size,[104] and if we label the chitosan with gold we can see that the chitosan is distributed throughout the complex with 'hot spots' in the interior.[105] Images from atomic force microscopy, visualized in topographic and phase modes, again shows complexes of this size. Control experi-ments revealed a loose coiled structure for pig gastric mucin and a shorter, stiffer con-formation for the chitosan, consistent with solution measurements.[106] The analytical ultracentrifuge procedure has also facilitated investigation of the effect of the solvent environment (pH, bile salts, *etc.*) on the extent of interaction.

6.1 Sedimentation Fingerprinting

A further modification of the procedure has been developed for the investigation of the interactions of human mucin from specific regions of the alimentary tract, generally extractable in only very small quantities, with chitosan. In this method, introduced in

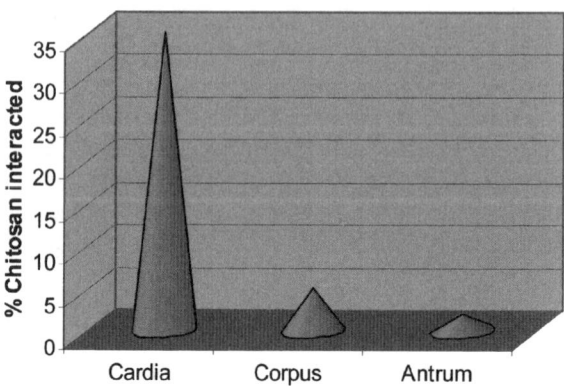

Figure 4   *Sedimentation fingerprinting analysis of the comparative mucoadhesiveness of a chitosan to mucins from different parts of the stomach (adapted from ref. 107)*

1999,[107] the Schlieren optical system is used to record the concentration (refractive index) gradient dn/dr as a function of radial position r in the ultracentrifuge cell. The area under a 'Schlieren peak' provides a measure of the sedimenting concentration. Alternatively, if interference optics on the XL-I ultracentrifuge are used, the area under a $g^*(s)$ *vs.* s or $c(s)$ *vs.* s plot would provide similar concentration information. Although the mucins from human stomach are at too low a concentration to be detected, we can assay for interaction from the loss of area under the chitosan peak caused by interaction. In this way, Deacon and co-workers[107] have shown it is possible to demonstrate significant differences in mucoadhesive interactions for different regions of the stomach (Figure 4).

This type of information obtained with the ultracentrifuge reinforced with other data is helping us design effective mucoadhesive systems. An example is the use of tripolyphosphate to cross-link chitosan into a sphere, and these have been shown to give good mucoadhesion.[108] If this is done in the presence of a drug, the drug can be encapsulated. Work is now in progress based around the principle of co-sedimentation[109] to investigate the encapsulation of drugs by these and similar delivery systems.

References

1. S. E. Harding, in *Analytical Ultracentrifugation in Biochemistry and Polymer Science*, S. E. Harding, A. J. Rowe and J. C. Horton (eds), Royal Society of Chemistry, Cambridge, UK, 1992, 495.
2. P. N. Lavrenko, K. J. Linow and E. Görnitz, in *Analytical Ultracentrifugation in Biochemistry and Polymer Science*, S. E. Harding, A. J. Rowe and J. C. Horton (eds), Royal Society of Chemistry, Cambridge, UK, 1992, 517.
3. W. D. Comper and O. Zamparo, in *Analytical Ultracentrifugation in Biochemistry and Polymer Science*, S. E. Harding, A. J. Rowe and J. C. Horton (eds), Royal Society of Chemistry, Cambridge, UK, 1992, 533.
4. B. N. Preston and K. O. Wik, in *Analytical Ultracentrifugation in Biochemistry and Polymer Science*, S. E. Harding, A. J. Rowe and J. C. Horton (eds), Royal Society of Chemistry, Cambridge, UK, 1992, 549.

5. S. E. Harding and A. J. Rowe, Opt. Laser. Eng., 1988, **8**, 83–96.
6. A. J. Rowe, S. Wynne-Jones, D. G. Thomas and S. E. Harding, in *Analytical Ultracentrifugation in Biochemistry and Polymer Science*, S. E. Harding, A. J. Rowe and J. C. Horton (eds), Royal Society of Chemistry, Cambridge, UK, 1992, 49–62.
7. A. J. Furst, *Eur. Biophys. J.*, 1997, **25**, 307.
8. R. Giebeler, in *Analytical Ultracentrifugation in Biochemistry and Polymer Science*, S. E. Harding, A. J. Rowe and J. C. Horton (eds), Royal Society of Chemistry, Cambridge, UK, 1992, 16.
9. H. Cölfen, S. E. Harding and K. M. Vårum, *Carbohyd. Polym.*, 1996, **30**, 5.
10. C. Theisen, C. Johann, M. P. Deacon and S. E. Harding, *Refractive Increment Data Book for Polymer and Biomolecular Scientists*, Nottingham University Press, Nottingham, UK, 2000.
11. M. W. Anthonsen, K. M. Vårum and O. Smidsrød, *Carbohyd. Polym.*, 1993, **22**, 193.
12. H. Schachman, *Ultracentrifugation in Biochemistry*, Academic Press, New York, 1959.
13. T. M. Laue, B. D. Shah, T. M. Ridgeway and S. L. Pelletier, in *Analytical Ultracentrifugation in Biochemistry and Polymer Science*, S. E. Harding, A. J. Rowe and J. C. Horton (eds), Royal Society of Chemistry, Cambridge, UK, 1992, 90.
14. http://www.jphilo.mailway.com/download.htm and http://www.rasmb.bbri.org/rasmb/windows/sednterp-philo/
15. N. Gralén, Sedmentation and diffusion measurements on cellulose and cellulose derivatives, Ph.D. Dissertation , University of Uppsala, Uppsala, Sweden, 1994.
16. A. J. Rowe, *Biopolymers*, 1977, **16**, 295.
17. A. J. Rowe, in *Analytical Ultracentrifugation in Biochemistry and Polymer Science*, S. E. Harding, A. J. Rowe and J. C. Horton (eds), Royal Society of Chemistry, Cambridge, UK, 1992, 394.
18. W. Stafford, in *Analytical Ultracentrifugation in Biochemistry and Polymer Science*, S. E. Harding, A. J. Rowe and J. C. Horton (eds), Royal Society of Chemistry, Cambridge, UK, 1992, 359.
19. J. S. Philo, *Anal. Biochem.*, 2000, **279**, 151.
20. P. Schuck, *Biophys. J.*, 1998, **75**, 1503.
21. J. Dam and P. Schuck, *Method. Enzymol.* 2003, **384**, 121.
22. http://www.jphilo.mailway.com/download.htm
23. O. Lamm, *Ark. Mat. Astr. Fys.*, 1923, **21B**(2), 1.
24. J. M. Claverie, H. Dreux and R. Cohen, *Biopolymers*, 1975, **14**, 1685.
25. http://www.analyticalultracentrifugation.com/download.htm
26. S. E. Harding, *Biophys. Chem.*, 1995, **55**, 69.
27. T. Patel, S. E. Harding, S. B. Ross-Murphy, D. R. Picout and G. Pavlov, 2005, in preparation.
28. T. Svedberg and K. O. Pedersen, *The Ultracentrifuge*, Oxford University Press, Oxford, UK, 1940.
29. E. V. Korneeva, G. A. Vichoreva, S. E. Harding and G. M. Pavlov, *Abstr. Am. Chem. Soc.*, 1996, **212**, 75.
30. G. M. Pavlov, E. V. Korneeva, S. E. Harding and G. A. Vichoreva, *Polymer*, 1998, **39**, 6951.
31. G. M. Pavlov, E. V. Korneeva, G. A. Vikhoreva and S. E. Harding, *Polym. Sci. Ser. A*, 1998, **40**, 1275 and *Vysokomolekulyarnye Soed. Ser. A.*, 1998, **40**, 2048.
32. G. M. Pavlov, E. V. Korneeva, S. A. Nepogod'ev, K. Jumel and S. E. Harding, *Polym. Sci. Ser. A*, 1998, **40**, 1282 and *Vysokomolekulyarnye Soed. Ser. A*, 1998, **40**, 2056.
33. G. M. Pavlov, E. V. Korneeva, K. Jumel, S. E. Harding, E. W. Meyer, H. W. I. Peerlings, J. F. Stoddart and S. A. Nepogodiev, *Carbohyd. Polym.* 1999, **38**, 195.
34. W. R. Sharman, E. L. Richards and G. N. Malcolm, *Biopolymers*, 1978, **17**, 2817.

35. D. J. Wedlock, B. A. Fasihuddin and G. O. Phillips, *Food Hydrocolloid.* 1987, **1**, 207.
36. G. Pavlov, S. Finet, K. Tatarenko, E. Korneeva and C. Ebel, *Eur. Biophys. J.*, 2003, **32**, 437.
37. W. Burchard, in *Laser Light Scattering in Biochemistry*, S. E. Harding, D. B. Sattelle and V. A. Bloomfield (eds), Royal Society of Chemistry, Cambridge, UK, 1992, 3.
38. P. N. Pusey, in *Photon Correlation and Light Beating Spectroscopy*, H. Z. Cummings and E. R. Pike (eds), Plenum Press, New York, 1974, 387.
39. M. Majzoobi, Ph.D. Dissertation, University of Nottingham, Nottingham, UK, 2004.
40. O. Smidsrød and I. L. Andresen, Biopolymerkjemi, Tapir, Trondheim, Norway, 1979.
41. V. N. Tsvetkov, V. Eskin and S. Frenkel, *Structure of Macromolecules in Solution*, Butterworths, London, 1970.
42. S. E. Harding, *Adv. Carbohyd. Chem.*, 1989, **47**, 345.
43. R. H. Pain, *Symp. Soc. Exp. Biol.*, 1980, **34**, 359.
44. T. Svedberg and R. Fåhraeus, *J. Am. Chem. Soc.*, 1926, **48**, 430.
45. A. C. Clewlow, N. Errington and A. J. Rowe, *Eur. Biophys. J.*, 1997, **25**, 305.
46. W. Mächtle, *Prog. Coll. Polym. Sci.*, 1999, **113**, 1.
47. J. M. Creeth and R. H. Pain, *Prog. Biophys. Mol. Biol.*, 1967, **17**, 217.
48. S. E. Harding, J. C. Horton and P. J. Morgan, in *Analytical Ultracentrifugation in Biochemistry and Polymer Science*, S. E. Harding, A. J. Rowe and J. C. Horton (eds), Royal Society of Chemistry, Cambridge, UK, 1992, 275.
49. H. Cölfen and S. E. Harding, *Eur. Biophys. J.*, 1997, **24**, 333.
50. D. A. Yphantis, *Biochemistry*, 1964, **3**, 297.
51. D. R. Hall, S. E. Harding and D. J. Winzor, *Prog. Coll. Polym. Sci.*, 1999, **113**, 62.
52. J. M. Creeth and S. E. Harding, *J. Biochem. Biophys. Meth.*, 1982, **7**, 25.
53. D. C. Teller, J. A. Horbett, E. G. Richards and H. K. Schachman, *Ann. New York Acad. Sci.*, 1969, **164**, 66.
54. A. P. Minton, in *Modern Analytical Ultracentrifugation*, T. M. Schuster and T. M. Laue (eds), Birkhäuser, Boston, 1994, 81.
55. http://www.nottingham.ac.uk/ncmh/unit/method.html#Software
56. http://www.biotech.uconn.edu/uaf
57. H. Fujita, *Foundations of Ultracentrifugal Analysis*, Wiley, New York, 1975.
58. S. E. Harding, K. Day, R. Dhami and P. M. Lowe, *Carbohyd. Polym.*, 1997, **32**, 81.
59. J. C. Horton, S. E. Harding, J. R. Mitchell and D. F. Morton-Holmes, *Food Hydrocolloid.*, 1991, **5**, 125.
60. A. Straatman and W. Borchard, *Prog. Coll. Polym. Sci.*, 2002, **119**, 64–69.
61. S. E. Harding, *Biophys. J.*, 1985, **47**, 247.
62. A. Ball, S. E. Harding and J. R. Mitchell, *Int. J. Biol. Macromol.*, 1988, **10**, 259.
63. S. E. Harding, G. Berth, A. Ball, J. R. Mitchell and J. Garcia de la Torre, *Carbohyd. Polym.*, 1991; **16**, 1.
64. S. E. Harding, *Gums Stablisers Food Ind.*, 1988, **4**, 15.
65. P. J. Wyatt, in *Laser Light Scattering in Biochemistry*, S. E. Harding, D. B. Sattelle and V. A. Bloomfield (eds), Royal Society of Chemistry, Cambridge, UK, 1992, 35.
66. http://www.wyatt.com/
67. J. C. Horton, S. E. Harding and J. R. Mitchell, *Biochem. Soc. Trans.*, 1991, **19**, 510.
68. J. E. Rollings, *Biochem. Soc. Trans.*, 1991, **19**, 493.
69. J. E. Rollings, in *Laser Light Scattering in Biochemistry*, S. E. Harding, D. B. Sattelle and V. A. Bloomfield (eds), Royal Society of Chemistry, Cambridge, UK, 1992, 275.
70. D. Wolff, S. Czapla, A. G. Heyer, S. Radosta, P. Mischnick and J. Springer, *Polymer*, 2000, **41**, 8009.
71. M. Wales and K. E. van Holde, *J. Polym. Sci.*, 1954, **14**, 81.

72. J. M. Creeth and C. G. Knight, *Biochim. Biophys. Acta*, 1965, **102**, 549.
73. K. F. Freed, *J. Chem. Phys.*, 1976, **65**, 4103.
74. K. Kawahara, K. Ohta, H. Miyamoto and S. Nakamura, *Carbohyd. Polym.*, 1984, **4**, 335.
75. G. Berth, H. Cölfen and H. Dautzenberg, *Progr. Colloid Polym. Sci.*, 2002, **119**, 50.
76. S. E. Harding, K. M. Vårum, B. T. Stokke and O. Smidsrød, *Adv. Carbohyd. Anal.*, 1991, **1**, 63.
77. G. M. Pavlov, *Progr. Colloid Polym. Sci.*, 2002, **119**, 84.
78. G. M. Pavlov, A. J. Rowe and S. E. Harding, *Trends Anal. Chem.*, 1997, **16**, 401.
79. G. M. Pavlov, S. E. Harding and A. J. Rowe, *Prog. Coll. Int. Sci.*, 1999, **113**, 76.
80. B. T. Stokke and A. Elgsaeter, *Adv. Carbohyd. Anal.*, 1991, **1**, 195.
81. S. E. Harding, G. Berth, J. Hartmann, K. Jumel, H. Cölfen, B. E. Christensen, *Biopolymers*, 1996, **39**, 729.
82. S. Broesma, *J. Chem. Phys.*, 1960, **32**, 1626.
83. M. M. Tirado and J. Garcia de la Torre, *J. Chem. Phys.*, 1979, **71**, 2581.
84. J. Garcia de la Torre, in *Analytical Ultracentrifugation in Biochemistry and Polymer Science*, S. E. Harding, A. J. Rowe and J. C. Horton (eds), Royal Society of Chemistry, Cambridge, UK, 1992, 333.
85. H. Yamakawa, *Modern Theory of Polymer Solutions*, Harper and Row, New York, 1971.
86. V. A. Bloomfield, D. M. Crothers and L. Tinoco, *Physical Chemistry of Nucleic Acids*, Harper and Row, New York, 1974.
87. C. R. Cantor and P. R. Schimmel, *Biophysical Chemistry*, Freeman, New York, 1979.
88. J. J. Freire and J. Garcia de la Torre, in *Analytical Ultracentrifugation in Biochemistry and Polymer Science*, S. E. Harding, A. J. Rowe and J.C Horton (eds), Royal Society of Chemistry, Cambridge, UK, 1992, 346.
89. J. E. Hearst and W. H. Stockmayer, *J. Chem. Phys.*, 1962, **37**, 1425.
90. H. Yamakawa and M. Fujii, *Macromolecules*, 1973, **6**, 407.
91. M. P. Tombs and S. E. Harding, *An Introduction to Polysaccharide Biotechnology*, Taylor and Francis, London, 1997.
92. H. K. Schachman, *Nature*, 1989, **941**, 259.
93. H. K. Schachman, in *Analytical Ultracentrifugation in Biochemistry and Polymer Science*, S. E. Harding, A. J. Rowe and J. C. Horton (eds), Royal Society of Chemistry, Cambridge, UK, 1992, 3.
94. J. D. Watson, *The Molecular Biology of the Gene*, 2nd edn, Benjamin, New York, 1970.
95. H. Silkowski, S. J. Davis, A. N. Barclay, A. J. Rowe, S. E. Harding and O. Byron, *Eur. Biophys. J.*, 1997, **25**, 455.
96. S. E. Harding, K. Jumel, R. Kelly, E. Gudo, J. C. Horton and J. R. Mitchell, in *Food Proteins: Structure and Functionality*, K. D. Schwenke and R. Mothes (eds), VCH Verlagsgesellschaft, Weinheim, Germany, 1993, 216.
97. R. Kelly, E. S. Gudo, J. R. Mitchell and S. E. Harding, *Carbohyd. Polym.*, 1994, **23**, 115.
98. A. Seifert, L. Heinevetter, H. Cölfen and S. E. Harding, *Carbohyd. Polym.*, 1995, **28**, 239.
99. H. Cölfen, S. E. Harding, K. M. Vårum and D. J. Winzor, *Carbohyd. Polym.*, 1996, **30**, 45.
100. R. O. Mannion, C. D. Melia, B. Launay, G. Cuvelier, S. E. Hill, S.E. Harding and J. R. Mitchell, *Carbohyd. Polym.* 1992, **19**, 91.
101. K. D. Schwenke, *Eiweißquellen der Zukunft*, Urania-Verlag, Leipzig, German Democratic Republic, 1985.
102. S.E. Harding, *Biochem. Soc. Trans.*, 2003, **31**, 1036.
103. C. M. Lehr, J. A. Bouwstra, E. H. Schacht and H. E. Junginger, *Int. J. Pharmaceut.*, 1992, **78**, 43.
104. I. Fiebrig, S. E. Harding, A. J. Rowe, S. C. Hyman and S. S. Davis, *Carbohyd. Polym.*, 1995, **28**, 239.

105. I. Fiebrig, K. M. Vårum, S. E. Harding, S. S. Davis and B. T. Stokke, *Carbohyd. Polym.*, 1997, **33**, 91.
106. M. P. Deacon, S. McGurk, C. J. Roberts, P. M. Williams, S. J. B. Tendler, M. C. Davies, S. S. Davis and S. E. Harding, *Biochem. J.*, 2000, **348**, 557.
107. M. P. Deacon, S. S. Davis, R. J. White, H. Nordman, I. Carlstedt, N. Errington, A. J. Rowe and S. E. Harding, *Carbohyd. Polym.*, 1999, **38**, 235.
108. P. He, S. S. Davis and L. Illum, *Int. J. Pharm.*, 1998, **166**, 75.
109. S. E. Harding and D. J. Winzor, in *Protein–Ligand Interactions: Hydrodynamics and Calorimetry*, S. E. Harding and B. Z. Chowdhry (eds), Oxford University Press, Oxford, 2001, 75–103.

CHAPTER 13

Analytical Ultracentrifugation as a Key Complementary Technique in Structural Biology

ROBERT J. C. GILBERT

1 Overview

The interactions and oligomeric states of macromolecules are often the source of their biological function. These can be transient or permanent self-interactions, or interactions with other molecules – protein–protein, protein–nucleic acid and protein–lipid. Sometimes the existence of these interactions becomes apparent when an X-ray crystal structure is solved, and then analytical ultracentrifugation (AUC) can prove a vital accessory technique to show that this observed oligomerisation occurs in solution, *i.e.*, in a state unconstrained by crystal contacts. On this point we should however note that (i) protein crystals are frequently 50–70% solvent, so the argument that the crystalline state presents the molecule in an unnatural environment where its conformation and interactions can be expected to be artefactual may be specious and (ii) protein–protein interactions in a biologically significant oligomer can be expected to be stronger than the crystal contacts and therefore to be preserved accurately in the crystalline state. Nevertheless, by showing, for example, that a dimer is present in the crystal structure and in solution, a major insight into the mechanism of the protein as a complex is obtained, and AUC is much the best method of doing this. AUC wins over other hydrodynamic techniques, such as dynamic light scattering and analytical gel filtration, because analysis of the data obtained is not hobbled, as in these techniques it can be, by the assumption that the macromolecule of interest is globular, or by interaction with a solid matrix. Oligomerisation can be measured either using sedimentation equilibrium to get a molecular weight, or sedimentation velocity to get sedimentation coefficients. Indeed, for more complex mixtures, the velocity mode is likely to be the best one, especially with the widespread availability of $g(s^*)$, $c(S)$ and similar methods of data analysis.[1] Techniques such as $g(s^*)$ and $c(S)$ allow a complex sedimentation boundary to be deconvoluted into peaks representing the differentially sedimenting species in the mixtures: for proteins that assemble into different sizes of oligomer or for mixtures of more than two interacting partners this is perhaps the most sensible approach.[1-7]

Velocity approaches are also able to provide information on the conformation of macromolecules, which is another key application of AUC in structural biology. Because the conformation of a macromolecule affects its frictional properties, the difference between the empirically determined behaviour of a species and its projected behaviour (according to its mass and partial specific volume) assuming a spherical conformation can be directly related to its degree of elongation via its frictional coefficient. For example, the arrangement of subunits in a protein dimer[8] or the behaviour of regions of a polypeptide lacking itself secondary or tertiary structure[9] can be inferred using this approach. A more detailed method of analysing velocity data to obtain conformational information is to represent the molecule(s) of interest as models for which hydrodynamic parameters can be computed. These "bead models" have been used to predict behaviour from very low-resolution models, where perhaps a single large bead represents a protein or domain thereof,[10-14] but increasingly such models will be based on atomic models relating either to the protein in question or to some homologue.[15-17] An especially well-developed set of algorithms for generating bead models has been developed by Garcia de la Torre and colleagues, who have now produced programs which are capable of computing the hydrodynamic properties of electron microscopy reconstructions as well as the more traditional atomic coordinates.[18] Neutron and X-ray scattering are used to generate similar bead models, and there is naturally significant cross-validation and synergy to be found in comparing solution scattering and AUC approaches in understanding the shapes and interactions of macromolecules in solution.

The more sophisticated one's approach to AUC data analysis, the more one needs in general to take care not to over-interpret the available data. While it is safe to go as far as suggesting a certain level of elongation for a molecule from a frictional coefficient, to specify on the basis of a model calculation the physical extension of it in emphatic, rather than suggestive, terms is a mistake. A bead model can be shown to be consistent with a particular shape for a sedimenting species but it is, of course, one among many possible models: our interpretation needs to be constrained by other known facts about a macromolecule – its crystal structure, or appearance in an electron microscope, or its solution scattering profile. One key problem in modelling hydrodynamic parameters of macromolecules when a model is based on crystallographic or NMR data is the need to account explicitly for the hydration water of a structure, which will significantly affect its empirically determined biological properties. In general, we account for this by applying a range of different hydrations to experimental values to correct them to a "dry" value as if they were moving through an inert medium; or alternatively to correct bead models – *in silico* entities which are therefore necessarily dry – to different levels of potential hydration. Equation (1) describes how to "dry out" a frictional ratio to get a Perrin function;[19] Equation (2) describes how to dry out a sedimentation coefficient. This can be done for a series of values (say $\delta = 0.1$, 0.3 and 0.5 $g_{H_2O}/g_{protein}$) which are then reported when the work is written up along with an estimate provided by a program such as AtoB or HYDROPRO[15,17] and then the comment that a value of 0.3–0.5 is quite likely for most soluble proteins. It should be noted that the water layer at macromolecular surfaces is reckoned to be ~10% more dense than that in bulk solution:[20]

$$P_{exp} = \frac{f}{f_0}\left[1+\left(\frac{\delta}{\bar{v}\rho}\right)\right]^{-1/3} \qquad (1)$$

$$s_\delta = s\left[1+\left(\frac{\delta}{\bar{v}\rho}\right)\right]^{1/3} \qquad (2)$$

where f/f_0 is the frictional ratio (between the friction experienced by the sedimenting object and that which would be experienced by a sphere of the same mass), s the sedimentation coefficient, P_{exp} the Perrin function, $\bar{v}$ the partial specific volume, δ the hydration and ρ the solvent density.

Another way to deal with hydration is to add a thin layer of hydrodynamic beads over the surface of the model, or performing a uniform expansion across the surface of the structure of the order of 3 Å or so. What explicitly modelling the hydration layer can achieve, which correction by a formula cannot, is a reflection of the uneven distribution of water over the surface. Accurately accounting for hydration can be a special concern when a structure is not globular and surely when it contains complex branched polar regions, such as the glycans found on extracellular and cell-surface proteins. The example of glycans is particularly appropriate because, if material is prepared *ex vivo* or in a mammalian cell line, for example, then the target protein may not only be glycosylated but also heterogeneous in the composition of each one of its glycan sidechains at each one of its glycosylation sites.[9] This raises another problem, which is to ensure one accounts accurately for the composition of a sedimenting species in terms of its partial specific volume ($\bar{v}$). However, a simple weighted sum of the $\bar{v}$ terms for the components of an assembly should suffice – if the complex is 35% RNA and 65% protein, an estimate of $0.35 \times 0.55 + 0.65 \times 0.73 = 0.67$. Finally, if one species under study has a known sedimentation coefficient, then the effects of hydration can be simulated for related species by normalising their observed sedimentation coefficients to it. For example, AUC of ribosomes and the complexes which they form up into: correcting the observed sedimentation coefficient of the eukaryotic ribosomal small subunit to its known value of 40 S gives a correction factor to apply to observed parameters for eukaryotic initiation complexes. On the other hand, if one calculates bead models from cryo-EM (electron microscopy) reconstructions determined for such complexes then the reconstructions can be validated by AUC where they show new conformations for the ribosomal subunits, with normalisation such that the small subunit bead model has a calculated sedimentation coefficient of 40 S and the initiation complexes values determined by the same correction factor.

In this chapter, we shall take a series of systems and analyse the way in which analytical ultracentrifugation has been used to show the biological basis of their activity. We start with examples of the use of AUC to backup insights gained from crystallography, to provide structural data, to assist in studies of membrane proteins, and to measure the effects of crowding phenomena, and then go on to consider systems in which it has been applied systematically.

2 Examples: General Strategies

2.1 Demonstrating Oligomerisation and Quality Control

When crystal structures are found to consist of dimers or higher order oligomers it is necessary to show that the oligomers also occur in solution. Sometimes this is rather straightforward[21-24] but it can require a significant amount of work, since the concentrations involved in crystallisation trials ($\sim$5 up to 100 mg mL^{-1}) can allow very low-affinity interactions to be manifested as oligomers in crystals. Even a low-affinity interaction can nevertheless be biologically important, especially if one factors in the effects of concentration in a plane if the protein is a cell surface one, or the crowded environment of the cell interior (see below). In both cases the degrees of freedom and therefore the diffusion coefficient of a protein are severely restricted and consequently a lower dissociation constant will serve to keep binding partners together. AUC can often be augmented with light scattering or (analytical) gel filtration to get a fuller picture of the oligomerisation state of a molecule. However AUC is more sophisticated than these techniques: less concentration-limited, and less affected by assumptions concerning the globularity of the sample (sedimentation equilibrium analysis giving a molecular weight is unaffected by shape; velocity analysis is affected by shape but this can be empirically dealt with – or used to advantage).[25] In the case of gel filtration there is the further problem that interaction with a matrix may induce anomalous behaviour. A good example is provided by the myelin sheath-associated protein P_0, the structure of which was solved at 1.9 Å by Hendrickson and colleagues.[26] Three different kinds of oligomerisation states were observed in crystals of P_0: a cyclic tetrameric association and two kinds of dimers, with three different contact interfaces (*i.e.*, the dimer is not half the tetramer and the dimers are unalike). One dimer interface is constituted by six hydrogen bonds, sidechain stacking and van der Waals contacts; the other by van der Waals contacts between a tryptophan sidechain and the main chains of the opposing protomer – the first interface (burying 2200 Å^2) is therefore much more likely to be biologically relevant than the second (which buries 1431 Å^2). Further characterisation showed that both the dimer and the tetramer assemblies occur naturally. Under gel filtration P_0 behaved anomalously, eluting "long after size standards of comparable molecular weight, as if its molecular weight were near zero".[26] With dynamic light scattering, at concentrations of 2 mg mL^{-1} a molecular weight intermediate between that of a monomer and a dimer was obtained and higher concentrations were impractical due to sample quantity limitations. AUC succeeded where the other approaches failed: P_0 at two concentrations (1.2 and 100 mg mL^{-1}) was subjected to sedimentation equilibrium analysis at 4, 20 and 37 °C. Performing an AUC experiment with protein at 100 mg mL^{-1} is a daunting task: it was made possible by the use of a 0.1 mm centrepiece thickness and by measuring the absorbance of the sample at 297 nm in an XL-A AUC;[26] with the availability of XL-I machines nowadays interference optics could be used instead. The data obtained could be analysed for sample molecular weight at all three temperatures and both concentrations except that the 100 mg mL^{-1} sample displayed inexplicable nonideality at 37 °C. At lower concentration, P_0 was in a monomer–dimer equilibrium, with a K_d (K_{12}, $c_{50}(2)$) of $\sim$1 mM. At higher concentration, the protein behaved as a monomer–dimer/monomer–tetramer equilibrium with the same monomer–dimer affinity and a tetramer $c_{50}(4)$ of 1 mM at 4 °C and 7 mM at 25 °C (extrapolated to a $c_{50}(4)$ of 15 mM at 37 °C). A van't Hoff analysis showed that the monomer–dimer interaction was temperature-independent ($\Delta H° = -3.2$ kcal mol^{-1} and $\Delta S° = 3$ cal deg mol^{-1}, $\Delta G° = -4.0$ kcal mol^{-1}), while the monomer-tetramer oligomerisation was strongly temperature-dependent, with $\Delta H° = -43.2$ kcal mol^{-1} and $\Delta S° = -113$ cal deg mol^{-1}, $\Delta G° = -9.5$ kcal mol^{-1}.[26] In the case of P_0, showing that the tetramer and dimer are real entities biologically is very important since they then provide a convincing explanation for the structuring of myelin sheaths by P_0; for example, the height of a dimer-of-tetramers is 46 Å, which is the same as the spacing between concentric lipid bilayers of the myelin sheath, and freeze-fracture studies reveal structures peppering the membrane with a similar size to the tetramer.[26]

In another case, the structure of the DNA Holliday junction-processing protein, RuvA, has been shown to be tetrameric in some crystals and octameric (a dimer-of-tetramers) in others, while neutron scattering also indicated an octameric assembly. AUC (sedimentation equilibrium) was used to show that the tetramer pertains at lower concentrations and the octamer occurs at higher concentrations.[26] Agreement between different methods of determining the oligomerisation state of proteins is found in the case of Ca^{2+}/calmodulin-dependent kinase II, where the protein was tetradecameric in its crystal form, as determined by sedimentation equilibrium, gel filtration and light scattering.[27] Nevertheless, it is worth noting that AUC gave a more accurate estimate of the molecular weight of the assembly, equivalent to 14.01 subunits per sedimenting species, than did a combination of gel filtration and light scattering (14.19 subunits).

Differences in the oligomerisation state of proteins under different conditions are also shown by the example of an extremophile dehydrogenase which forms an octameric assembly (a tetramer of tight dimers).[28] In solution the protein was shown to be a dimeric only, using sedimentation velocity AUC, gel filtration and neutron scattering. Even mimicking as closely as possible the conditions under which the dehydrogenase was crystallised (up to 9% PEG 8000) failed to produce evidence of octamerisation in solution.[28] Another example where AUC seems to undermine the significance of a crystallographic oligomer is the chemokine IP10; in crystals a dimeric interface and three different kinds of tetrameric assembly were observed, but sedimentation equilibrium AUC showed only a monomer–dimer equilibrium in solution.[29] Particularly interesting experiments have been done in situations where the addition of some small molecule ligand or ionic ligand brings about a change in the behaviour of a protein that can then be monitored by AUC. For example, the SH3 signalling module Mona/Gads was observed in its crystal form as a dimer where the monomer–monomer interface was formed by mutual chelation of a single ion. The size of the electron density observed for the dimer-forming ion meant it could be either Ca^{2+} or Zn^{2+}; subsequently it was shown using sedimentation equilibrium that Mona/Gads is a monomer in the presence of Ca^{2+} but a dimer in the presence of Zn^{2+}, with a K_d of 0.73 mM and hence the ion in the electron density map was identified as a zinc.[30] In the realm of biotechnology, bivalent small molecule ligands were shown to convert a protein domain from a monomer into a dimer by sedimentation equilibrium AUC.[31] Finally, a comment should be made about the usefulness of AUC at the beginning of a crystallographic study rather than its end: if a protein is to crystallise it is most useful if it is monodisperse, and this can be efficiently judged using AUC (in either velocity or equilibrium modes), although light scattering is more often applied to the same purpose since, for the requirement of estimating monodispersity, it is often sufficient. An example of the use of AUC in this manner (and where standard dynamic light scattering would have too low a resolution to be useful) is provided by the cores of hepatitis B virus (HBV), which exist both as 180- and 240-copy assemblies of the HBV core antigen.[32]

2.2   Modelling Systems of Unknown Conformation and Arrangement

Techniques in which the structure of a macromolecule is studied in solution are even more sensitive to whether the molecule is monodisperse or not than crystallisation. As a result, performing AUC on a sample is often a key preparation for techniques such as small-angle X-ray scattering and neutron scattering (SAXS and SANS). For example, the co-chaperone GrpE was shown to be a monodisperse dimer in solution prior to scattering experiments that demonstrated the dimer to be symmetrical (by determining a scattering envelope for the sample *ab initio*).[33] By the same token, an immunomodulatory protein (ES-62) from a filarial nematode was shown to be a tightly bound tetramer, but in this case AUC was further harnessed to the validation of a directly determined (from SAXS data) low-resolution structure by showing that its sedimentation coefficient was consistent with the experimentally determined one.[34] Furthermore, due to the fact that sedimentation equilibrium had previously been used to determine the tetramerisation of ES-62, the structural model could be constrained with four-fold symmetry. Thus, in this case, AUC both facilitated and validated an *ab initio* small-angle scattering model determination. The same is true of another study in which the conformation of a DNA Holliday junction was determined in the absence and presence of magnesium from SAXS data; it appeared that its conformation became significantly different on addition of Mg^{2+}, and this could be backed-up with AUC analysis and the calculation of the sedimentation coefficient expected for the structure.[35] A more direct way of determining a low-resolution model of a structure is the use of electron microscopy – the membrane complex I of the electron transport chain was so studied and a hydrodynamic model consisting of a small number of beads used to show that its conformation was consistent with the solution structure of the complex.[14] Being a membrane complex, the AUC was performed in a detergent solution – in this case dodecyl maltoside. The use of detergents raises special issues in AUC, and these are discussed below. It is clear that the application of AUC to the modelling of macromolecular structures is often a hybrid approach, making use of several techniques, in particular circular dichroism (CD), SAXS, SANS, differential scanning calorimetry (DSC) and electron microscopy. It is also possible to use AUC as the sole experimental probe of conformation,[36] although clearly this will be a weaker approach to the problem of what the structure of a molecule is than the use of several techniques in synergy.

2.3   Studying Membrane Proteins

The standard approaches to the structural study of membrane proteins make use of detergent to solubilise hydrophobic regions that would normally be in contact with the membrane bilayer interior. This raises a serious problem for AUC analyses since the partial specific volume of the detergent will be different from the protein and in general non-unity. Furthermore, an accurate measure of the amount of detergent bound per protein molecule will be difficult to obtain. Another problem can be that detergents such as Triton X-100 have high absorbances at the wavelengths where proteins are usually measured in absorbance AUC – around 280 nm. A solution to these problems is to use the neutrally buoyant detergent pentaoxyethylene octyl ether (C8E5).[37] The alternative is to render the detergent neutrally buoyant by matching the density of the buffer to its density, with the addition of heavy water ($^2H_2O/D_2O$)[38] or sucrose.[39] Examples of the application of AUC to membrane proteins are: measurement of membrane helix–helix interactions within detergent micelles;[38] demonstrating the oligomerisation state of a protein for which a projection structure has been determined by two-dimensional electron crystallography;[39,40] and the demonstration that the influenza haemagglutinin can be refolded *in vitro* in detergent to form a stable, monodisperse trimer as in its native state.[41]

2.4 Macromolecular Crowding and Aggregation

The high concentration of solutes within the cytoplasm is considered to have a significant effect on the effective affinities of molecules for each other, and to pose special problems for the prevention of non-specific aggregation.[42–44] One of the effects of aggregation of proteins is that they sometimes form amyloid, a generic insoluble fibrillar protein assembly.[45] Both the basis of crowding phenomena and the formation of amyloid have been measured using AUC. Crowding can be mimicked by the inclusion of high concentrations of other solutes in the sample buffer – such as dextran T10,[46] trimethylamine *N*-oxide and betaine.[47] The formation of amyloid from insulin induced by low pH has been studied using sedimentation equilibrium approaches[48] and the way in which serum amyloid P component and apolipoprotein E affect amyloid fibre aggregation[49] and the formation of the fribrils influenced by macromolecular crowding[46] by sedimentation velocity techniques

3 Examples: Case Histories

3.1 Pore-forming Toxins

Pore-forming protein toxins are produced by organisms in all realms of life. The most avidly studied are bacterial pore formers, for reasons of ease of production and genetic manipulation, but examples from higher organisms include enterolobin from *Enterolobium contortisiliquum* (a South American plant),[50] metridiolysin from *Metridium senile* the sea anemone, and whole families of equinatoxins from sea anemones,[51] and the human complement terminal pore-forming complex. Indeed, enterolobin is homologous to the toxin aerolysin from *Aeromonas hydrophila*, and metridiolysin seems similar to the cholesterol-binding toxins of Gram-positive bacteria, so the evolutionary history of pore formers is a long one.

Pore-forming proteins represent an intriguing class because they are capable of moving from a soluble state in which they freely move in aqueous solution, to a membrane-inserted state when they form a pore. This must involve two separate processes;

firstly recognition of and binding to a membrane receptor, and secondly pore formation. In practice, pore formation itself breaks down into a further set of steps; often proteins will oligomerise into ring-shaped structures consisting of 4–50 subunits (depending on the toxin) that define the pore, and this oligomerisation is coupled to the refolding and membrane-insertion necessary for pore formation to occur. A key question in working out the mechanism of action of these pore forming proteins is: how do they start out in solution, *i.e.*, from what state do they bind to the membrane they will permeabilise? This is related to another question: if pore forming toxins have the innate ability to refold into an alternate, oligomeric structure, what prevents them doing so in solution? AUC has proved central to working this out.

The X-ray crystal structure of aerolysin and a model of its pore-forming heptameric oligomer were published in 1994.[52] Aerolysin (or rather its pro-toxin, requiring proteolytic activation at the membrane surface by furin, whereby a C-terminal peptide is removed) was revealed to be a dimeric structure [Figure 1(a)]. Intriguingly, the dimer is in a head-to-tail (antiparallel) topology; since the toxin forms a parallel heptameric pore in its target membrane, doing so must involve separation of the dimer subunits, reorientation of the subunits and then oligomerisation, before membrane insertion. Alternatively, the dimer could dissociate before binding to the membrane, which might be a simpler scheme. Nevertheless, as shown by AUC, proaerolysin remains a dimer even at 5 μg mL^{-1} concentration – and in that dimeric state is competent to bind its receptor.[53] Enterolobin, too, is a dimer in an antiparallel arrangement.[50] So, in the case of aerolysin the more complicated of two conceivable

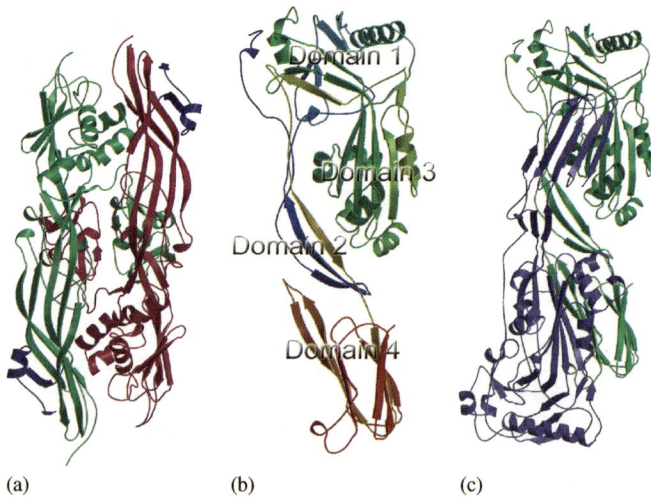

(a)   (b)   (c)

Figure 1   *(a) Crystallographic structure of the proaerolysin dimer.[52] The two antiparallel protomers are coloured aquamarine and cerise, respectively. The propeptide is coloured blue in both cases. (b) Crystallographic structure of perfringolysin.[58] The structure is coloured from blue at its N-terminus to red at its C-terminus. (c) The dimer of perfringolysin found in protein crystals.[58] The antiparallel protomers are colour aquamarine and blue; the aquamarine molecule is in the same orientation as in (b)*

pathways pertains and the crystal dimer is shown not only to be a native state of the protein but to be its preferred state prior to activation at the membrane surface.

Another pore-forming toxin to have been crystallised is perfringolysin from *Clostridium perfringens* [Figure 1(b)]. Perfringolysin is one of the cholesterol-binding toxins that includes pneumolysin from *Streptococcus pneumoniae*, listeriolysin from *Listeria monocytogenes* and streptolysin from *Streptococcus pyogenes*.[54,55] These toxins bind to membranes via cholesterol or some other receptor[56] and then oligomerise into a prepore state before inserting themselves into the membrane, forming a large lesion and leading to cell death. The oligomers of these toxins can contain up to 50 subunits, forming in the end a gigantic pore-defining barrel in the membrane. AUC has played a particularly important role in our understanding of how these toxins work.

Hydrodynamic studies of pneumolysin began in the early 1990s, really as an effort towards crystallising the protein. The first studies indicated that pneumolysin was mostly monomeric in solution, since its measured molecular weight from sedimentation equilibrium equalled its polypeptide chain mass;[11,12] its sedimentation coefficient was determined using velocity analysis and, on the basis of metal-shadow electron microscopy of toxin monomers, a hydrodynamic model for pneumolysin as an L-shaped structure was proposed, and so a structure for its oligomeric pore-forming ring.[11,12] Subsequent electron microscopy studies of the pore confirmed the general scheme for the shape of the protein and its arrangement in pores; the basic plan for the action of pneumolysin indicated by this exercise has been shown to be more-or-less correct and these were really the first efforts at describing the structure of a cholesterol-binding toxin.

Further exploration of the solution behaviour of pneumolysin indicated that the toxin was capable of self-associating in solution to form ring-shaped structures that appeared similar to its pore-forming state. AUC was used to measure the underlying self-interaction; because it is rather weak, in dilute solutions the $g(s^*)$ method for obtaining a sedimentation coefficient distribution was used to demonstrate the presence of self-interaction.[57] Electron microscopy showed that, in addition to oligomers of the ring-shaped pore-like variety, there were helical structures present and the existence of these two oligomeric forms in solution together show that, for pneumolysin, self-interaction is an innate process.[3,57] Pneumolysin is primed to form ring-shaped assemblies, with the potential to form pores if assembly occurs on a membrane, in the presence of enough protein. Pneumolysin has a single cysteine residue, towards its C-terminus; it was serendipitously found that reaction with dithionitrobenzoate produced adduction of the cysteine with a benzyl moiety, which can be observed directly as a result of its absorbance of 337 nm light. Subsequent AUC showed that simply adducting the toxin at this single cysteine rendered it incapable of self-interaction and therefore inert.[3] In part, this demonstration benefited from the use of different wavelengths of light to measure absorbance from different species (adducted and non-adducted protein) in a mixture. Subsequent AUC experiments have been used to follow the process of oligomerisation in solution by pneumolysin, showing over time the progressive appearance of differently sized assemblies including octamers, 16mers, 26mers and complete rings.[6]

The crystal structure of perfringolysin showed it to be an antiparallel dimer; three crystal forms have been solved and in each case the dimer is present (either as a crystallographic or non-crystallographic axis of symmetry)[58] (PDB Refs 1PFO, 1M3I, 1M3J) [Figure 1(c)]. This marks it out as being very different from pneumolysin in the way it starts out in solution, despite the fact that these are very similar proteins with highly homologous structures.[58,59] Nevertheless, models for the action of perfringolysin, including models for how it is activated, are predicated on the fact that it is monomeric.[60,61] Recent AUC experiments, using the $c(s)$ mode of analysis, have confirmed what is apparent from the crystal structures, that this assumption is wrong and perfringolysin is a dimeric protein in solution.[6] The antiparallel topology of the dimer (confirmed in solution by use of scattering methods and bead modelling[6]) means that its C-terminal domain, through which the toxin binds the membrane, is not available for interaction in the predominant solution state. This is the origin of perfringolysin's inactivity in solution; it protects itself from the danger of prematurely oligomerising by existing in a different, incompatible, assembly state. It is likely, however, that the small amount of monomer present in solution is the form competent to bind membranes, drawing the monomer–dimer equilibrium gradually towards monomers bound to the membrane that then interact, from the predominant inert dimer in solution.

3.2   Assigning Function to a SARS Virus Protein

SARS coronavirus (CoV) emerged as a human pathogen in 2002. CoVs are large, enveloped positive-strand RNA viruses which produce a replicase complex constituted by a number of non-structural proteins (NSPs) from a single replicase gene initially translated as a polyprotein.[62] The recent renewal of interest in coronaviruses arising from the appearance of SARS means that rather little is known in detail about their protein biochemistry, for example, the function of many of the proteins and how they interact with each other. AUC was used in this case as a way of probing the functions of some SARS proteins and in validating observations made from a crystal structure; as will probably be apparent from the ensuing discussion, here AUC is being used to interrogate a poorly understood system at a fairly low level of molecular descriptiveness.

Following synthesis, the SARS polyprotein is extensively processed by endogenous viral proteases to generate a series of isolated polypeptide products, the nsps.[63] These proteins assemble or interact to form the replicase complex. The structure of nsp9 was solved independently by two groups[7,64] and revealed a dimeric structure. Analytical ultracentrifugation was used to show that the protein indeed dimerises in solution, with a K_d of ~0.1 mM.[7] This demonstration made use of the classic approach to determining the behaviour of a protein in solution, measuring sedimentation equilibrium data at three speeds with the squares of the first two being related by a factor of 1.4 or more and the squares of the first and third speeds being related by a factor of 3 or more.[65] In this scenario, while both polydispersity and a specific self-association are apparent in an increasing apparent molecular weight with increasing protein concentration, polydispersity gives rise to decreasing observed weights at increased speeds. It was found that the protein as expressed with a His$_6$ tag (for ease of purification) displayed both polydispersity and a specific self-interaction, whereas proteolytic removal of the tag produced a form of nsp9 that only dimerised and did not otherwise non-specifically aggregate [Figure 2(a)]. This is apparent in the linearly increasing observed mass at lower speeds where aggregates have yet to be removed by the increased gravitational field at high speed, and the plateau-forming rectangular hyperbola that describes nsp9 dimerisation at high speed and when denuded of the His_6 tag.[7]

Having determined the structure and oligomerisation state of nsp9, it was necessary to try and assign a function to it on the basis of its interactions with other binding partners – nucleic acid, for example, or other proteins that derive from the replicase polyprotein. The binding partners investigated were nsp8 and nsp5 (which is the 3C-like protease of SARS CoV), in binary mixtures and when all three were combined. Nsp8 behaved in a non-ideal manner in sedimentation equilibrium experiments but did appear to be dimeric, with a solution molecular weight of ~50 kDa. Nsp5 had a molecular weight consonant with it being monomeric, and behaved in an ideal manner. Here, ideality means simply that one can observe one or more apparently independently sedimenting species; non-ideality means that some interparticle interactions are present – perhaps a charge:charge interaction, aggregation due to misfolding or a crowding effect. It was found that nsp8 behaved ideally in combination with nsp9, which suggests that nsp9 alters the behaviour of nsp8 and thus that they can interact. Use of $g(s^*)$ analysis, to probe this insight further, seemed to confirm it – nsp9 showed two peaks, equal to the monomeric and dimeric species; nsp8 showed a broad $g(s^*)$ spectrum, consistent with its non-ideality; a mixture of the two gave three discrete peaks, indicating a conversion of the behaviour of nsp8 in solution from a disordered melée to interacting with nsp9 in some ordered way to produce ~3 differentially sedimenting species [Figure 2(b)]. AUC here provides a

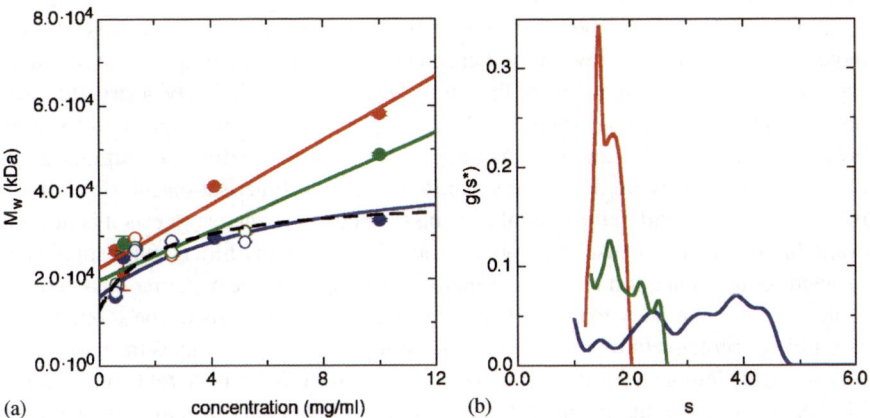

(a)   concentration (mg/ml)   (b)   s

Figure 2   *(a) Plots of apparent molecular weight against concentration for tagged SARS-CoV nsp9 (closed symbols, solid lines) at 12 000 rpm (red symbols), 15 000 rpm (green) and 22 000 rpm (blue) derived from AUC experiments. For untagged nsp9 (open symbols, colours as for tagged), only the plot at 22 000 rpm is shown (blue broken line) for clarity as the measured values were similar at all three speeds. (b) g(s*) profiles of nsp9 (red), SARS-CoV nsp8 (green), and an equimolar mixture of the two (blue) showing a change in the behaviour of nsp8 on addition of nsp9*

glimpse of the behaviour of a protein from a large multi-component complex in altering the behaviour of its interaction partners, and so makes a contribution to the effort to map out structures and interactions for the coronavirus replicase complex.

3.3 Co-receptors in the Immune System

The immunological synapse is a complex structure formed at the interface of a T cell and an antigen presenting cell during the immune response. Thus, recognition by a T-cell receptor of an antigen presented within the groove of a major histocompatibility complex (MHC) molecule results in the assembly of the synapse in the apposed membranes through the recruitment to the apposed regions of each membrane of complementary co-receptors. The immunological synapse functions as "a dynamic multitasking system that integrates multiple cellular processes required for T-cell activation and the execution of T-cell effector activities".[66] Important among the co-receptors are B7-1 and B7-2 on the antigen presenting cell (APC) and CD28 and CTLA-4 on the T cell. CD28 binding to B7-1 or B7-2 co-stimulates the T-cell response and consolidates immunological synapse formation; on the other hand, binding by B7-1 or B7-2 of CTLA-4, a similar molecule to CD28, inhibits T-cell activity. B7-1 and CTLA-4 are bivalent, whereas B7-2 and CD28 are monovalent; it is thought that B7-2 is more important for stimulation *in vivo* and that conversely B7-1 is more important for inhibition of T-cell response.[67]

AUC has played an important role in the model just outlined. The crystal structure of B7-1 revealed a dimeric molecule; sedimentation equilibrium analysis showed it to be a dimer in solution, with a K_d of 20–50 μM.[23] This immediately provided a model for the bivalency of the B7-1 interaction with, for example, CTLA-4: as a dimer, B7-1 presents two identical CTLA-4 binding sites. Since CTLA-4 is itself a dimer, this allows the proposal of a zippering mechanism for the interaction of B7-1 and CTLA-4 whereby they form larger structures of intercalated dimers of each molecule, giving rise to "avidity enhancement" of their signalling.[23] A subsequent crystal structure showed this to be the case[68] [Figure 3(a)]. Similarly, a structure was obtained for the complex formed of B7-2 and CTLA-4,[69] which appeared to show bivalency for both B7-2 and CTLA-4 and the existence therefore of a similar avidity enhancement [Figure 3(b)]. However, the apparent dimerisation interface of the B7-2 molecules found in the crystal was unsymmetrical, which suggests it is not relevant biologically because the spontaneous formation of dimers philosophically demands equivalence (on the same principles that equivalence or rather quasi-equivalence in combination with conformational switching is the basis of the spontaneous assembly of proteins into viral capsids, for example).[70] To show this is the case, AUC was used to demonstrate the monomeric nature of B7-2 and a related molecule LICOS in solution at up to 7 mg mL^{-1}.[71] A more recent crystal structure of the receptor-binding domain of B7-2 also showed a monomer in both the crystal and in solution.[72] Some still maintain that B7-2 is dimeric within signalling complexes[72] but the available data do not support this contention. Thus the monovalency of B7-2 derives from its monomeric nature. The reason why CD28 binds monovalently though it is itself a dimer like CTLA-4 is that occupation of one CD28 dimer-binding face positions its ligand protein at such an angle that it obscures the other binding face.[73]

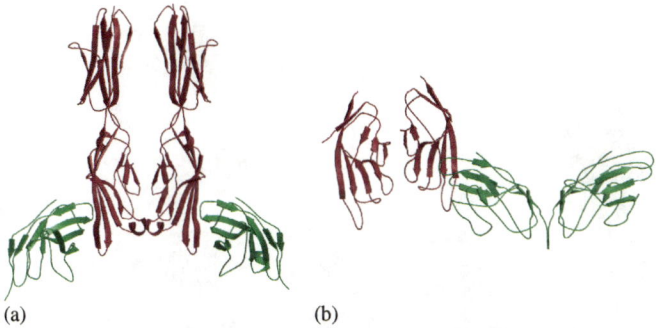

(a)   (b)

Figure 3   *(a) The crystallographic complex formed between B7-1 and CTLA-4 complex. B7-1 is coloured cerise and monomers of CTLA-4 aquamarine. CTLA-4 is normally a constitutive dimer via a disulfide bridge between monomers, which was mutated away in the construct used for crystallography.[68] (b) The crystallographic complex formed between B7-2 (receptor-binding domain only) and CTLA-4, which in this case was a dimer. B7-2 is cerise, CTLA-4 aquamarine: note the asymmetry in the B7-2 dimerisation interface[69]*

Another molecule involved in enhancing the stimulation of B- and T-cell activity is known as SLAM, for signalling lymphocytic activation molecule. SLAM is part of the CD2 subset of the immunoglobulin superfamily and had been thought to be homophilic – that is, to form dimers through its receptor-binding surfaces, head-to-head. To show this it was necessary to use sedimentation velocity with data analysis by $g(s^*)$ because the interaction was too weak to be detected at available concentrations in equilibrium mode.[8] The K_d of the self-interaction of SLAM was estimated, using the relative areas of the Gaussian distributions of monomer and dimer into which the total apparent sedimentation coefficient distribution in the sample ($g(s^*)$) could be deconvoluted, as 0.1–1 mM. Furthermore, it could be shown using the frictional ratio of the monomeric and dimeric species (expressed as Perrin functions) in combination with their sedimentation coefficients and bead modelling that the interaction produced an elongated structure – that is, that SLAM is indeed a self-ligand interacting head to head.[8] The frictional coefficient can be calculated directly[8] or, for example, within software such as SEDFIT,[9,74,75] using the equation[76]

$$\frac{f}{f_0} = \frac{M(1-\bar{v}\rho_0)}{N_A 6\pi\eta_0 s_{20,w}^0 \left(\dfrac{3\bar{v}M}{4\pi N_A}\right)^{1/3}} \tag{3}$$

where M is molecular weight, N_A the Avogadro's number and the η viscosity. A Perrin function can then be calculated as described in Equation (1): this is a measure of particle shape (particularly elongation) independent of weight. Then for bead models the same value can be computed theoretically using SOLPRO, for example,[77] and comparison will indicate to what extent the bead model is consistent with the behaviour of the protein in hand [Figure 4(a)].

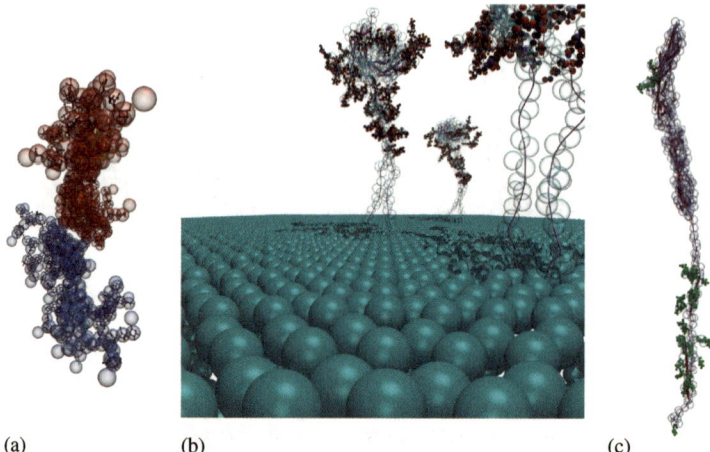

(a)   (b)   (c)

Figure 4   *(a) An atomic model for a SLAM dimer with modelled glycan sidechains within a*
semi-transparent representation of a bead model used to compute the expected
behaviour of this low-affinity homodimer in solution.[8] (b) A depiction of CD8 atop
its extended linker regions with modelled glycans; AUC was used to show that the
level of extension shown by these linkers is maintained even in the presence of
merely core sugar sidechains. Lipid headgroups defining the membrane surface:
cyan spheres; CD8 is encased in its hydrodynamic bead model as in (a).[9] (c) A
depiction of DAF, which consists of four domains in an extended arrangement sup-
ported by an extended linker, like CD8. AUC was used to show that the folded
domains adopt this arrangement in solution and to probe the role of the linker
regions in supporting them. Sugars are shown in green, the polypeptide backbone
in red[78]

A similar approach was used to address the role of glycosylation in the structure
of the linker regions by which CD8, in assisting T-cell receptors to recognise MHC
molecules, reaches up to interact with the MHC close to the cell membrane of the
APC. CD8 has extensive N-linked glycosylation on its head regions and O-linked
glycosylation on its extended linkers by which it stands above the T-cell membrane.
Since these linkers have no secondary structure it was thought that the branched O-
glycans would have a scaffolding role. To investigate whether that is indeed the case,
forms of CD8 were generated in which only the core sugars were added during synthesis at the O-gycan sequons; surprisingly, using the computation of Perrin functions from sedimentation velocity data compared to bead models it could be shown
that in fact even with relatively small sugar sidegroups the linker regions remain
extended.[9] Furthermore, by generating a series of entirely theoretical conformations
for the stalk regions it could be shown that in this case the Perrin function and sedimentation coefficient indicated the extension of the stalks with a fair degree of discrimination compared to other possibilities[9] [Figure 4(b)].

Sedimentation velocity has also been used to show that the molecule decay accel-
erating factor (DAF; also known as CD55), a four-domain protein, possesses a conformation in solution similar to that seen in its crystal. DAF is an elongated molecule
[Figure 4(c)] and a natural concern is that the domain arrangement derives from

crystal-packing effects; yet it was shown that, with a sedimentation coefficient of 2, DAF behaves the same in solution as would be predicted from its crystal structure and therefore innately adopts the elongated conformation observed there. It is clearly important with signalling molecules to know their conformations if any kind of useful models for the way they mediate interaction between cells are to be developed.[78]

Acknowledgement

R.J.C.G. is a Royal Society University Research Fellow.

References

1. A. Balbo, K. H. Minor, C. A. Velikovsky, R. A. Mariuzza, C. B. Peterson and P. Schuck, *Proc. Natl. Acad. Sci. USA*, 2005, **102**, 81.
2. R. J. Gilbert, J. Rossjohn, M. W. Parker, R. K. Tweten, P. J. Morgan, T. J. Mitchell, N. Errington, A. J. Rowe, P. W. Andrew and O. Byron, *J. M. Biol.*, 1998, **284**, 1223.
3. R. J. Gilbert, R. K. Heenan, P. A. Timmins, N. A. Gingles, T. J. Mitchell, A. J. Rowe, J. Rossjohn, M. W. Parker, P. W. Andrew and O. Byron, *J. M. Biol.*, 1999, **293**, 1145.
4. W. F. Stafford and P. J. Sherwood, *Biophys. Chem.*, 2004, **108**, 231.
5. W. F. Stafford and E. H. Braswell, *Biophys. Chem.*, 2004, **108**, 273.
6. A. S. Solovyova, M. Nollmann, T. J. Mitchell and O. Byron, *Biophys. J.*, 2004, **87**, 540.
7. G. Sutton, E. Fry, L. Carter, S. Sainsbury, T. Walter, J. Nettleship, N. Berrow, R. Owens, R. Gilbert, A. Davidson, S. Siddell, L. L. Poon, J. Diprose, D. Alderton, M. Walsh, J. M. Grimes and D. I. Stuart, *Structure*, 2004, **12**, 341.
8. N. Mavaddat, D. W. Mason, P. D. Atkinson, E. J. Evans, R. J. Gilbert, D. I. Stuart, J. A. Fennelly, A. N. Barclay, S. J. Davis and M. H. Brown, *J. Biol. Chem.*, 2000, **275**, 28100.
9. A. H. Merry, R. J. Gilbert, D. A. Shore, L. Royle, O. Miroshnychenko, M. Vuong, M. R. Wormald, D. J. Harvey, R. A. Dwek, B. J. Classon, P. M. Rudd and S. J. Davis, *J. Biol. Chem.*, 2003, **278**, 27119.
10. J. Garcia de la Torre and B. Carrasco, *Biopolymers*, 2002, **63**, 163.
11. P. J. Morgan, P. G. Varley, A. J. Rowe, P. W. Andrew and T. J. Mitchell, *Biochem. J.*, 1993, **296**, 671.
12. P. J. Morgan, S. C. Hyman, O. Byron, P. W. Andrew, T. J. Mitchell and A. J. Rowe, *J. Biol. Chem.*, 1994, **269**, 25315.
13. F. G. Diaz, A. Iniesta and J. Garcia de la Torre, *Biopolymers*, 1990, **30**, 547.
14. B. Bottcher, D. Scheide, M. Hesterberg, L. Nagel-Steger and T. Friedrich, *J. Biol. Chem.*, 2002, **277**, 17970.
15. J. Garcia De La Torre, M. L. Huertas and B. Carrasco, *Biophys. J.*, 2000, **78**, 719.
16. J. Garcia de la Torre, B. Carrasco and S. E. Harding, *Eur. Biophys. J.*, 1997, **25**, 361.
17. O. Byron, *Biophys. J.*, 1997, **72**, 408.
18. J. Garcia de la Torre, O. Llorca, J. L. Carrascosa and J. M. Valpuesta, *Eur. Biophys. J.*, 2001, **30**, 457.
19. P. G. Squire and M. E. Himmel, *Arch. Biochem. Biophys.*, 1979, **196**, 165.
20. D. I. Svergun, S. Richard, M. H. Koch, Z. Sayers, S. Kuprin and G. Zaccai, *Proc. Natl. Acad. Sci. USA*, 1998, **95**, 2267.
21. J. H. Morais Cabral, C. Petosa, M. J. Sutcliffe, S. Raza, O. Byron, F. Poy, S. M. Marfatia, A.H. Chishti and R.C. Liddington, *Nature*, 1996, **382**, 649.
22. B. A. Appleton, A. Loregian, D. J. Filman, D. M. Coen and J. M. Hogle, *Mol. Cell*, 2004, **15**, 233.

23. S. Ikemizu, R. J. Gilbert, J. A. Fennelly, A. V. Collins, K. Harlos, E. Y. Jones, D. I. Stuart and S. J. Davis, *Immunity*, 2000, **12**, 51.

24. M. J. van Raaij, E. Chouin, H. van der Zandt, J. M. Bergelson, and S. Cusack, *Struct. Fold Des.*, 2000, **8**, 1147.

25. S. N. Savvides, H. J. Yeo, M.R. Beck, F. Blaesing, R. Lurz, E. Lanka, R. Buhrdorf, W. Fischer, R. Haas and G. Waksman, *EMBO J.*, 2003, **22**, 1969.

26. L. Shapiro, J. P. Doyle, P. Hensley, D. R. Colman and W. A. Hendrickson, *Neuron*, 1996, **17**, 435.

27. A. Hoelz, A. C. Nairn and J. Kuriyan, *Mol. Cell*, 2003, **11**, 1241.

28. A. Irimia, D. Madern, G. Zaccai and F. M. Vellieux, *EMBO J.*, 2004, **23**, 1234.

29. G. J. Swaminathan, D. E. Holloway, R. A. Colvin, G. K. Campanella, A. C. Papageorgiou, A. D. Luster and K. R. Acharya, *Structure (Camb)*, 2003, **11**, 521.

30. M. Harkiolaki, M. Lewitzky, R. J. Gilbert, E. Y. Jones, R. P. Bourette, G. Mouchiroud, H. Sondermann, I. Moarefi and S. M. Feller, *EMBO J.*, 2003, **22**, 2571.

31. C. T. Rollins, V. M. Rivera, D. N. Woolfson, T. Keenan, M. Hatada, S. E. Adams, L. J. Andrade, D. Yaeger, M. R. van Schravendijk, D. A. Holt, M. Gilman and T. Clackson, *Proc. Natl. Acad. Sci. USA*, 2000, **97**, 7096.

32. A. Zlotnick, N. Cheng, J. F. Conway, F. P. Booy, A. C. Steven, S. J. Stahl and P. T. Wingfield, *Biochemistry*, 1996, **35**, 7412.

33. J. C. Borges, H. Fischer, A. F. Craievich, L. D. Hansen and C. H. Ramos, *J. Biol. Chem.*, 2003, **278**, 35337.

34. C. J. Ackerman, M. M. Harnett, W. Harnett, S. M. Kelly, D. I. Svergun and O. Byron, *Biophys. J.*, 2003, **84**, 489.

35. M. Nollmann, W. M. Stark and O. Byron, *Biophys. J.*, 2004, **86**, 3060.

36. M. Smolle, R. T. Hay and O. Byron, *Biophys. Chem.*, 2004, **108**, 259.

37. K. G. Fleming, A. L. Ackerman and D. M. Engelman, *J. Mol. Biol.*, 1997, **272**, 266.

38. J. D. Lear, H. Gratkowski, L. Adamian, J. Liang and W. F. DeGrado, *Biochemistry*, 2003, **42**, 6400.

39. C. Ziegler, S. Morbach, D. Schiller, R. Kramer, C. Tziatzios, D. Schubert and W. Kuhlbrandt, *J. Mol. Biol.*, 2004, **337**, 1137.

40. I. Ubarretxena-Belandia and C.G. Tate, *FEBS Lett.*, 2004, **564**, 234.

41. S. E. Swalley, B. M. Baker, L. J. Calder, S. C. Harrison, J. J. Skehel and D. C. Wiley, *Biochemistry*, 2004, **43**, 5902.

42. B. van den Berg, R. J. Ellis and C. M. Dobson, *EMBO J.*, 1999, **18**, 6927.

43. C. Ebel and G. Zaccai, *J. Mol. Recog.*, 2004, **17**, 382.

44. R. J. Ellis, *Curr. Opin. Struct. Biol.*, 2001, **11**, 114.

45. C. M. Dobson, *Nature*, 2003, **426**, 884.

46. D. M. Hatters, A. P. Minton and G. J. Howlett, *J. Biol. Chem.*, 2002, **277**, 7824.

47. N. A. Chebotareva, I. E. Andreeva, V. F. Makeeva, N. B. Livanova and B. I. Kurganov, *J. Mol. Recog.*, 2004, **17**, 426.

48. J. L. Whittingham, D. J. Scott, K. Chance, A. Wilson, J. Finch, J. Brange and G. Guy Dodson, *J. Mol. Biol.*, 2002, **318**, 479.

49. C. A. MacRaild, C. R. Stewart, Y. F. Mok, M. J. Gunzburg, M. A. Perugini, L. J. Lawrence, V. Tirtaatmadja, J. J. Cooper-White and G. J. Howlett, *J. Biol. Chem.*, 2004, **279**, 21038.

50. S. E. Bittencourt, L. P. Silva, R. B. Azevedo, R. B. Cunha, C. M. Lima, C. A. Ricart and M. V. Sousa, *FEBS Lett* 2003, **549**, 47.

51. A. W. Bernheimer and B. Rudy, *Biochim. Biophys. Acta*, 1986, **864**, 123.

52. M. W. Parker, J. T. Buckley, J. P. Postma, A. D. Tucker, K. Leonard, F. Pattus and D. Tsernoglou, *Nature*, 1994, **367**, 292.

53. R. Barry, S. Moore, A. Alonso, J. Ausio and J. T. Buckley, *J. Biol. Chem.*, 2001, **276**, 551.
54. R. J. Gilbert, *Cell Mol. Life Sci.*, 2002, **59**, 832.
55. R. K. Tweten, M. W. Parker and A. E. Johnson, *Curr. Top. Microbiol. Immunol.*, 2001, **257**, 15.
56. K. S. Giddings, J. Zhao, P. J. Sims and R. K. Tweten, *Nat. Struct. Mol. Biol.*, 2004, **11**, 1173.
57. R. J. Gilbert, J. Rossjohn, M. W. Parker, R. K. Tweten, P. J. Morgan, T. J. Mitchell, N. Errington, A. J. Rowe, P. W. Andrew and O. Byron, *J. Mol. Biol.*, 1998, **284**, 1223.
58. J. Rossjohn, S. C. Feil, W. J. McKinstry, R. K. Tweten and M. W. Parker, *Cell*, 1997, **89**, 685.
59. J. Rossjohn, R. J. Gilbert, D. Crane, P. J. Morgan, T. J. Mitchell, A. J. Rowe, P. W. Andrew, J. C. Paton, R. K. Tweten and M. W. Parker, *J. Mol. Biol.*, 1998, **284**, 449.
60. D. M. Czajkowsky, E. M. Hotze, Z. Shao and R. K. Tweten, *EMBO J.*, 2004, **23**, 3206.
61. R. Ramachandran, R. K. Tweten and A. E. Johnson, *Nat. Struct. Mol. Biol.*, 2004, **11**, 697.
62. J. Ziebuhr and S. Siddell, *Nidovirales*, in *The Encyclopaedia of Life Sciences*, Stockton Press, London, 2002.
63. E. J. Snijder, P. J. Bredenbeek, J. C. Dobbe, V. Thiel, J. Ziebuhr, L. L. Poon, Y. Guan, M. Rozanov, W. J. Spaan and A. E. Gorbalenya, *J. Mol. Biol.*, 2003, **331**, 991.
64. M. P. Egloff, F. Ferron, V. Campanacci, S. Longhi, C. Rancurel, H. Dutartre, E. J. Snijder, A. E. Gorbalenya, C. Cambillau and B. Canard, *Proc. Natl. Acad. Sci. USA*, 2004, **101**, 3792.
65. T. M. Laue, in *Short Column Sedimentation Equilibrium Analysis for Rapid Characterization of Macromolecules in Solution*, Beckmann, Palo Alto, CA, 1992.
66. S. J. Davis and P. A. van der Merwe, *Curr. Biol.*, 2001, **11**, R289.
67. S. J. Davis, S. Ikemizu, E. J. Evans, L. Fugger, T. R. Bakker and P. A. van der Merwe, *Nat. Immunol.*, 2003, **4**, 217.
68. C. C. Stamper, Y. Zhang, J. F. Tobin, D. V. Erbe, S. Ikemizu, S. J. Davis, M. L. Stahl, J. Seehra, W. S. Somers and L. Mosyak, *Nature*, 2001, **410**, 608.
69. J. C. Schwartz, X. Zhang, A. A. Fedorov, S. G. Nathenson and S. C. Almo, *Nature*, 2001, **410**, 604.
70. A. Klug, *Point Groups and the Design of Aggregates*, in *Nobel Symposium II: Symmetry and Function of Biological Systems of the Macromolecular Level*. A. Engstroem and B. Strandberg, eds., 1969, Nobel Foundation, Sweden, pp 425–436.
71. A. V. Collins, D. W. Brodie, R. J. Gilbert, A. Iaboni, R. Manso-Sancho, B. Walse, D. I. Stuart, P. A. van der Merwe and S. J. Davis, *Immunity*, 2002, **17**, 201.
72. X. Zhang, J. C. Schwartz, S. C. Almo and S. G. Nathenson, *Proc. Natl. Acad. Sci. USA*, 2003, **100**, 2586.
73. E. J. Evans, R. M. Esnouf, R. Manso-Sancho, R. J. C. Gilbert, J. James, C. Yu, C. Vowles, T. Hanke, B. Walse, T. Hunig, P. Sorenson, D.I. Stuart and S.J. Davis, *Nat. Immunol.* 2005, **6**, 271.
74. P. Schuck, C. E. MacPhee and G. J. Howlett, *Biophys. J.*, 1998, **74**, 466.
75. P. Schuck, *Anal. Biochem.*, 1999, **272**, 199.
76. B. Carrasco, J. G. de la Torre, O. Byron, D. King, C. Walters, S. Jones and S. E. Harding, *Biophys. J.*, 1999, **77**, 2902.
77. J. Garcia de la Torre, S. E. Harding and B. Carrasco, *Eur. Biophys. J.*, 1999, **28**, 119.
78. P. Lukacik, P. Roversi, J. White, D. Esser, G. P. Smith, J. Billington, P. A. Williams, P. M. Rudd, M. R. Wormald, D. J. Harvey, M. D. Crispin, C. M. Radcliffe, R. A. Dwek, D. J. Evans, B. P. Morgan, R. A. Smith and S. M. Lea, *Proc. Natl. Acad. Sci. USA*, 2004, **101**, 1279.

CHAPTER 14

Introduction to Differential Sedimentation

IAN LAIDLAW AND MARC STEINMETZ

1 Introduction

Differential centrifugal sedimentation (DCS), sometimes also called "two-layer" sedimentation, is a widely used analysis method that produces extremely high resolution size distributions of microscopic to sub-microscopic particles. The normal measurement range for the method is from ~0.01 μm (10 nm) to ~50 μm, although it is possible with some types of materials to extend the range to below 0.003 μm (3 nm) or up to 120 μm or more. This chapter provides some background information on particle size analysis by sedimentation, explains how the DCS method works and describes the advantages and limitations of the method. Several sample analyses are presented to help illustrate the capabilities of DCS.

2 Basic Theory of Particle Size Analysis by Sedimentation

Sedimentation of particles in a fluid has long been used to characterise particle size distribution. Stokes' law[1] is used to determine an unknown distribution of spherical particle sizes by measuring the time required for the particles to settle a known distance in a fluid of known viscosity and density. Sedimentation can be either gravitational (1 *g* force), or centrifugal (many *g* force).

Gravitational sedimentation is normally limited to particles of relatively large size, because the rate of sedimentation for small particles is too low to give a practical analysis time, and because Brownian motion of small particles becomes too large to allow effective settling. A very narrow distribution of small particles will be reported as a broad distribution when the rate of particle diffusion is comparable to the sedimentation rate. Very small particles (<0.1 μm) never settle by gravity unless they are extremely dense, so most types of very small particles cannot be measured by gravitational sedimentation. Sedimentation in a centrifuge extends the range of sedimentation analysis to much smaller particles. High *g* force makes sedimentation of small particles much faster than Brownian diffusion, even for very small particles.

When a centrifuge is used, Stokes' law must be modified to account for the variation in **g** force with distance from the centre of rotation:

$$D = \left\{ \frac{(18\eta \ln(R_f/R_0))}{((\rho_p - \rho_f)\omega^2 \, t)} \right\}^{0.5}$$

(1)

where D is the particle diameter (cm), η the fluid viscosity (poise), R_f the final radius of rotation (cm), R_0 the initial radius of rotation (cm), ρ_p particle density (g ml^{-1}), ρ_f the fluid density (g ml^{-1}), ω the rotational velocity (rad/s) and t the time required to sediment from R_0 to R_f (s).

For a centrifuge running at constant speed and temperature, all of the parameters except time are constant during an analysis. The values for these are either well known or can be accurately measured. Within a broad range of analysis conditions, the modified form of Stokes' law accurately measures the diameter of spherical particles based on arrival time at the detector.

3 Methods of Sedimentation Analysis

There are two common sedimentation methods: integral and differential. The following discussion explains the differences between these methods.

3.1 Integral Sedimentation

The integral method (Figure 1) is the oldest of the sedimentation methods. A detector beam (a light beam or X-ray beam) passes through the fluid at a known distance from the fluid surface and measures particle concentration. The initial intensity of light or X-rays reaching the detector is a minimum, corresponding to the maximum concentration of particles. As particles settle through the fluid, the concentration of particles remaining in the dispersion falls, and the intensity of light or X-rays that reach the detector increases. Stokes' law is used to calculate the size of particles that sediment out of the fluid as a function of time, and a particle size distribution is generated by plotting the measured concentration of particles against the calculated particle diameter. The result of the analysis is an integral representation of the particle size distribution. The method is called integral sedimentation because the sum

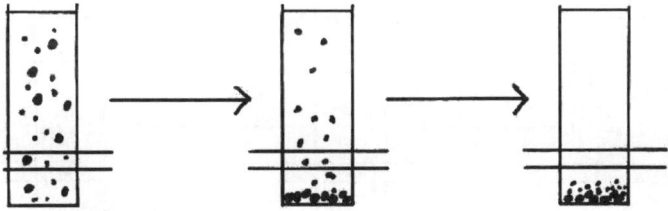

Figure 1 *Integral sedimentation method*

(the "integral") of all particles smaller than a particular size is being continuously measured during the analysis. A differential particle size distribution can be generated from the integral results by applying mathematical differentiation with respect to diameter.

Integral sedimentation can also be applied to particles lower in density than the fluid in which they are suspended. In this case, the particles have a net buoyancy, so they sediment towards the surface of the fluid rather towards the bottom.

There are three significant operational problems with integral sedimentation in a centrifuge. First, the initial conditions of the analysis are difficult to characterise. If the sample is added to a centrifuge that is already spinning, then there will be turbulent mixing of the sample dispersion as it is added to the centrifuge, which makes accurate measurement of sedimentation time difficult. If a sample is added to a centrifuge that is not spinning, and is later accelerated to high speed, then it is necessary to accurately measure and account for the changing speed during the acceleration period. It is also necessary to use a centrifuge of a design that ensures that there is no mixing of the sample during acceleration. Second, convection currents can develop during an analysis unless the temperature of the sample is held constant; any convection currents in the fluid can reduce both resolution and the accuracy of results. High-speed centrifuges generate frictional heat, which makes it more difficult to maintain constant temperature in the sample fluid. Third, the sedimentation chamber must be emptied and cleaned following each sample, which increases operator labour.

3.2   Differential Sedimentation

Differential sedimentation (see Figure 2) was first reported in 1930.[2] A sample of particles to be analysed is placed on top of a column of clear liquid at the start of the analysis, and particles settle according to Stokes' law, just as in integral sedimentation. The detector initially reads maximum intensity, but the signal is reduced when particles reach the detector beam. The reduction in intensity indicates the concentration of particles in the detector beam. When an X-ray beam is used, the reduction in intensity is proportional to particle concentration. When a monochromatic light source is used the Mie theory of light scattering can be applied to the intensity data to calculate particle concentration.

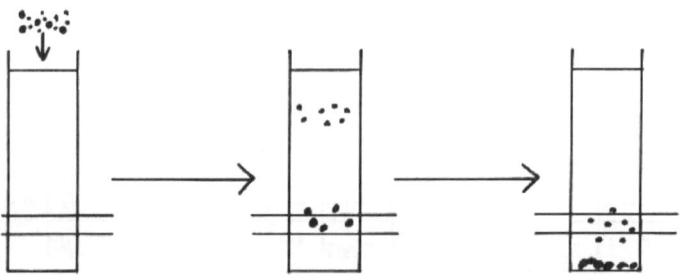

Figure 2   *Differential sedimentation method*

When all particles have passed the detector, the signal returns to the original level. A plot of the particle concentration against the calculated particle diameter produces a differential distribution. At any time during the analysis, only particles of one particular size range are being measured by the detector beam; all larger particles have already passed the beam, and none of the smaller particles have arrived yet. The method is called differential sedimentation because only a tiny part of the distribution (a "differential") is being measured by the detector beam at any time. An integral distribution can be generated from a differential distribution by applying mathematical integration with respect to particle diameter. A differential size distribution and its corresponding integral distribution are shown in Figure 3.

Actually running a differential sedimentation is a little more complicated than suggested by the above description. When a sample of dispersed particles that are more dense than the fluid in the column is placed on top of the column, the particles do not settle individually according to Stokes' law. Instead, the entire sample suspension rapidly settles as a bulk fluid through the liquid column, in exactly the same way as a homogeneous liquid of higher density (like 10% sodium chloride in water) would settle through a column of another liquid of lower density (like water). The bulk settling of a sample in differential sedimentation is commonly called "streaming" or "sedimentation instability".[3] All information about the particle size distribution can be lost when streaming takes place. Several methods[4–6] have been developed to eliminate streaming. Each of these methods is effective because a slight density gradient is formed within the fluid column, prior to starting analyses. A wide range of fluids can be used to form a density gradient. In aqueous systems, gradually changing concentrations of methanol, ethanol, glycerine, sucrose and many other materials have been used. In nonaqueous systems, many mixtures of fluids of different density can be used.

A density gradient eliminates streaming because at all times during the analysis the net density of the fluid, which is the average density of fluid plus any suspended

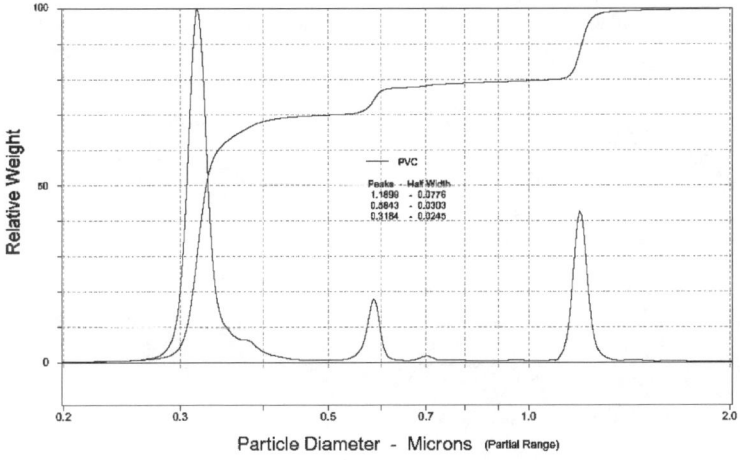

Figure 3 *Differential and integral distributions*

particles, increases continuously from top to bottom in the fluid column. The condition that guarantees stable sedimentation is given by

$$\frac{\delta \rho_{net}}{\delta R} \geq 0 \qquad (2)$$

where ρ_{net} is the net fluid density (including liquid plus any suspended particles) and R the distance from the centre of rotation.

When a small volume of a particle suspension is placed on the surface of the fluid column, the net density of the suspension is very slightly higher than the pure fluid; but the fluid just under the surface is also slightly higher in density than the pure fluid owing to the density gradient. There is no driving force for bulk settling of the particle suspension, so there is no instability, and the particles sediment through the fluid according to Stokes' law. The required steepness of the density gradient depends upon the net density of the sample to be measured. A sample with higher net density (higher particle concentration or higher particle density) requires a steeper density gradient than a sample with lower net density. Most samples are diluted to low concentration, so only a very slight density gradient is required to ensure stability. Density gradients of less than 0.01 g mL^{-1} cm^{-1} of fluid height are normally sufficient to ensure complete stability.

A density gradient also eliminates thermal convection, so sedimentation is not disrupted by slight changes in fluid temperature during an analysis. Relatively large temperature changes (>0.5 °C) can cause some loss of accuracy unless they are accounted for, because fluid viscosity changes with temperature.

4 Differential Centrifugal Sedimentation

4.1 DCS Instrument Design

The most common design for DCS instruments is a hollow, optically clear disc that is driven by a variable speed motor. A typical disc cross-section is shown in Figure 4. The disc can be of virtually any size, but manufacturers have settled on a diameter of

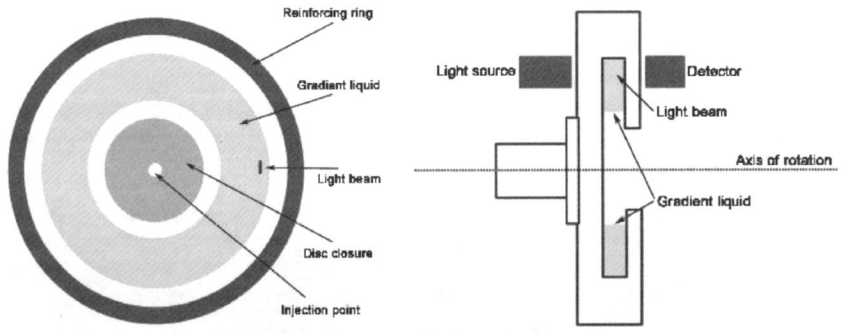

Figure 4 *Hollow disc centrifuge design*

approximately 125 to 150 mm. The detector beam is usually monochromatic light of relatively short wavelength (400–500 nm); some instruments use a longer wavelength (~650 nm), or X-rays. Shorter wavelength light gives better detector sensitivity when particles smaller than 100 nm are measured.

To prepare the instrument for analysis, the disc is set in motion at constant speed, and then the disc chamber is filled with a fluid that contains a slight density gradient. Samples are prepared for analysis by dilution in a fluid of slightly lower density than the least dense fluid in the disc. The lower density fluid used for the sample reduces initial mixing of the fluid inside the disc with the sample. When a sample is injected (normally using a small syringe), it strikes the back inside face of the disc, and forms a thin film, which spreads as it accelerates radially towards the surface of the fluid. When the sample dispersion reaches the fluid surface, it quickly spreads over the surface, because it is of lower density (it "floats" on the higher density fluid). Once a sample is on the fluid surface, sedimentation of individual particles begins. The injection of a sample is rapid (typically <50 ms), so the starting time for an analysis is well defined, and the precision of sedimentation time is quite good.

When an analysis is complete, the instrument is ready for the next sample. There is no need to empty and clean the centrifuge, so many samples can be run in sequence without stopping the centrifuge. The only limitation on continuous run time is that the density gradient slowly degrades owing to molecular diffusion. When the density gradient is no longer steep enough to maintain stable sedimentation, the instrument must be stopped, emptied and a new gradient formed. Typical gradient lifetime is 2 to 72 h, depending on the molecular weight and viscosity of the materials that form the gradient.

4.2 Advantages and Limitations of the DCS Method

All methods of particle size analysis can be characterised by three parameters:

- the *accuracy* of the reported size distribution;
- the *repeatability* of the reported size distribution; and
- the *resolution* of the distribution.

4.2.1 Accuracy and Repeatability

Accuracy and repeatability of the DCS method are very good in nearly all cases. Any significant inaccuracy in the results is caused either by inaccurate values for the physical parameters of the system (densities, viscosity, rotational speed, *etc.*), instability in the sedimentation, or by deviation of the sedimentation from Stokes' law.

4.2.1.1 Physical parameters. The overall accuracy of the analysis depends upon the combined accuracy of each of the values in Equation (1). For example, if the viscosity of the fluid is actually 2% higher than entered in Equation (1), then the reported particle size will be about 1% smaller than the correct one. It is possible to achieve nearly any desired level of accuracy by improving the accuracy of the parameters in Equation (1). An alternative method to improve accuracy is to use a narrow calibration standard of precisely known size to determine the effective combined value, K, for all the parameters in Equation (1). Equation (1) then reduces to

$$D = K\left(\frac{1}{t}\right)^{0.5} \tag{3}$$

where K is a combination of constants and t is the time to reach the detector.

A calibration standard can be used externally, where it is analysed just before or after an unknown sample to determine K, or internally, where a small amount of the calibration standard is added to the unknown sample. Instrument software finds the calibration peak within the distribution of the unknown sample, and adjusts the value for K so that the calibration standard peak is exactly the correct diameter. The adjusted value for K is applied to the entire distribution, so the accuracy of the analysis improves. Internal calibration gives extremely high accuracy and repeatability: the peak sizes in replicate analyses of an unknown sample are usually within ±0.25% when an internal standard is used.

4.2.1.2 Sedimentation stability. Any instability (streaming) during an analysis reduces both accuracy and resolution. Streaming causes the reported size distribution to be larger than correct, because during streaming, particles move towards the detector faster than they would in normal sedimentation. Streaming usually takes place near the beginning of an analysis, when the entire sample is contained in a thin, fluid layer near the surface. A small amount of streaming will cause the sample to form a broad initial band, followed by normal sedimentation; the result is both lower resolution and larger than correct reported sizes.

Commercial DCS instruments are normally set up to operate under conditions that always yield stable sedimentation. However, to verify that the sedimentation is stable, a direct means of confirming stability is needed. Some DCS instruments are equipped with a strobe light that is synchronised with the rotation of the centrifuge. This allows direct visual observation of the stability of sedimentation. With experience, an operator can judge whether there is any instability on the basis of the appearance of the sedimentation. Other instruments rely on a narrow calibration standard to verify stability. When a calibration standard is used (either internal or external), evaluation of sedimentation stability can be made automatic; the instrument software can compare the measured width and shape of the calibration standard peak with the known width and shape for that calibration standard. Any significant change in distribution width or shape indicates instability in the sedimentation.

4.2.1.3 Deviation from Stokes' law. Stokes' law does not accurately describe the sedimentation process if the Reynolds number for the system becomes too high. The Reynolds number increases with larger particles, faster sedimentation rate and lower fluid viscosity. Most sedimentation analyses are run at low Reynolds numbers (<0.02), at which deviation from Stokes' law is less than 0.5%. For example, at a centrifuge speed of 10 000 rpm, analysis of acrylic latex particles of 3 μm (density 1.13 g mL^{-1}) in water will produce a Reynolds number of ~0.007, and a deviation from Stokes' law of ~0.25%. In cases where the Reynolds number is higher, deviation from Stokes' law can be taken into account by the instrument software so that the reported particle size distribution is accurate, regardless of the Reynolds number.

4.2.2 Resolution

Compared with most other particle size analysis methods, DCS gives distributions that have excellent resolution. Calibration standards with very narrow distributions can be routinely resolved when the ratio of diameters is $\sim$1.05, and partially separated when the ratio is as low as $\sim$1.02.

4.3 What is Resolution?

The resolution of a size measurement method is the ability of the method to visualise a size distribution clearly. All size measurement methods report a distribution that is more or less "fuzzy" compared with the true distribution, much as an out-of-focus lens produces a fuzzy image. A lens can be very slightly out of focus (higher resolution) or far out of focus (lower resolution). Different particle sizing methods and different instruments have vastly different resolutions, even though nearly all particle sizing instrument manufacturers claim that their instruments have "high resolution". In order to rationally evaluate instrument resolution, we must first have a clear definition of resolution. For this document, resolution of a size measurement method is defined in two ways, both giving the same result. First, resolution is defined as the minimum fractional size difference between two perfectly narrow families of particles that allows the two reported peaks to overlap by less than 5% of their total area. Resolution is stated as a percentage:

$$\text{Resolution} = 200 \frac{(D_1 - D_2)}{(D_1 + D_2)} \qquad (4)$$

where D_1 is the diameter of the larger family and D_2 the diameter of the smaller family.

For example, if we find that two families can be resolved with $D_1 = 1.05\ \mu\text{m}$ and $D_2 = 0.95\ \mu\text{m}$, then the instrument resolution is 10%. Second, we can express the same resolution value in terms of the reported peak width for a single family of particles that are perfectly uniform in size, compared with their reported median diameter (D_{50}):

$$\text{Resolution} = 100 \frac{(D_{95} - D_5)}{D_{50}} \qquad (5)$$

where D_{95} is the diameter larger than 95% of the entire reported distribution and D_5 the diameter larger than 5% of the entire reported distribution.

These two resolution calculations give the same value for resolution.

4.4 Theoretical Resolution of the DCS

Particles sediment in the DCS according to Stokes' law. Particles sediment at rates that are proportional to the square of the particle diameter; 1 μm particles sediment four times faster than 0.5 μm particles. At the start of a typical analysis, all particles are located in a thin band at the fluid surface. When particles arrive at the instrument's detector beam, they have separated from particles of different sizes, so the detector beam measures only a small slice (a "differential") of the whole size distribution. This is why we call the disc centrifuge method "differential sedimentation". Figure 5 shows a close-up of the sedimentation process.

The theoretical resolution depends on three factors:

- the width of the detector beam;
- the thickness of the initial sample band; and
- the sedimentation depth.

As the sedimentation depth increases, the theoretical resolution increases as well, because the physical separation of particles of different sizes becomes larger. The following equation calculates the resolution as a function of the three variables shown in Figure 5:

$$R = 100\left(\left(\left(1+\frac{(2T+W)}{D}\right)^{0.5}\right)-1\right) \qquad (6)$$

The detector beam width is approximately 0.5 mm. The sedimentation depth depends on how much fluid is added to the centrifuge, but with a typical set-up of the instrument the depth is in the range of 10 mm. The initial sample bandwidth depends on the volume of sample that is injected into the disc. With a sample volume of 0.1 mL, the initial sample ring has a thickness of approximately 0.066 mm. Using these typical values, we can calculate a theoretical resolution of ~3.11%.

The theoretical resolution can be improved by reducing the detector beam width, increasing the sedimentation depth and reducing the thickness of the initial sample band. For example, if the sedimentation depth is increased to 20 mm and the sample volume is reduced to 0.05 mL, then the theoretical resolution improves to ~1.4%. This means that two perfectly narrow peaks with only a 1.4% difference in diameter could be completely resolved.

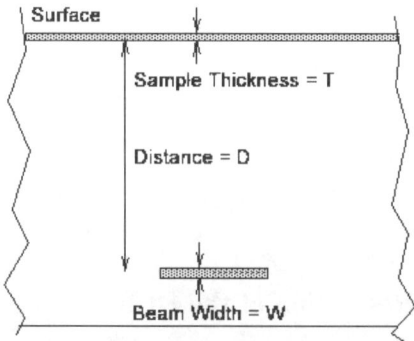

Figure 5   *Resolution of the DCS*

4.5 Factors that Reduce Resolution of the DCS

Actual instrument resolution is always slightly worse than the theoretical resolution described above. There are three factors that can reduce resolution. These factors are: Brownian motion of the particles during sedimentation, sedimentation instability (streaming) and a broader than expected initial sample band that comes from the injection process. Each of these potential broadening factors is discussed below.

4.5.1 Brownian Motion

Random diffusion of particles during sedimentation will cause some particles (larger apparent diameter) to arrive at the detector beam earlier than expected, and some particles (smaller apparent diameter) to arrive later than expected. Brownian motion is a true diffusion process, with a calculable diffusion constant that depends on both particle size and fluid viscosity. In general, the mean absolute diffusion distance during a brief time (say 1 s) is proportional to the inverse square root of the particle diameter. A "random-walk" simulation of Brownian motion shows how the diffusion progresses.

If we were to measure a perfectly narrow family of 0.3 μm particles that required 12 min to reach the detector beam, then the band would reach the detector with an increase in band width equal to ~0.125 mm. In order to estimate the effect of Brownian motion on resolution, we can add the Brownian diffusion to the initial sample thickness. This yields an estimated resolution of

$$100\left(\left(\left(1+\frac{(2T+W)}{D}\right)^{0.5}\right)-1\right)=100\left(\left(\left(1+\frac{(2(0.033+0.125)+0.5)}{20}\right)^{0.5}\right)-1\right)=2.02\%$$

After accounting for the effects of Brownian motion over 12 min, the DCS should still resolve peaks near 0.3 μm diameter that differ by as little as 2%, but by 1.4% in the absence of Brownian motion.

For particles larger than 0.3 μm, or particles with sedimentation times less than 12 min, the effect of Brownian motion on resolution will be considerably less. For example, 0.5 μm particles arriving at the detector after 4 min form a band less than 0.02 mm wider than the initial sample thickness, so the resolution in this case would be ~1.5%, only very slightly different than the resolution would be without Brownian motion.

For particles that are significantly smaller than 0.3 μm, or that reach the detector more slowly, the effect of Brownian motion will be considerably more. For example, 0.05 μm particles that require 45 min to reach the detector will arrive as a band ~0.4 mm wide; at a total sedimentation depth of 20 mm, resolution in this case would be ~3.8%. Particles of 0.05 μm diameter that reach the detector after 90 min form a band that is about 0.75 mm wide; at a total sedimentation depth of 20 mm, the resolution in this case would be ~7.9%.

4.5.2 Wall Effects

Particles that are near a wall of the centrifuge chamber sediment more slowly than particles that are far from either wall. There appear to be two different reasons for this slower sedimentation:

- deviation from Stokes' law due to the wall, or
- attraction/adhesion of particles to the wall, plus microscopic imperfections that can temporarily hold particles.

The combined effects are difficult to quantify accurately, but are still significant.

4.5.3   Initial Sedimentation Instability

All analyses in the DCS must be conducted in the presence of a density gradient, where the fluid at the outside edge of the disc chamber is of slightly higher density than the fluid near the surface. In the absence of a density gradient, differential sedimentation is unstable: an injected sample sediments *"en-masse"* rather than as individual particles. This instability is sometimes called "streaming". The instability is caused by the effect of the (more dense) suspended particles on the net density of the fluid in which they are suspended. If the net density of the sample suspension is higher than the fluid inside the rotating disc, then the sedimentation will become unstable. During the entire analysis, the fluid that is just "below" a band of particles (*i.e.* fluid slightly further from the centre of rotation) must be equal to or higher in density than the net density of the fluid that holds the band of particles. This requirement for stability can be expressed mathematically as in Equation (2) and suggests that it is impossible to have an instantaneous, "step-like" increase in suspended particle concentration without inducing instability. In fact, there will always be some (very brief) instability immediately following sample injection until the above equation is satisfied. The effect of instability is a broader than the expected initial sample band, and so lower than the expected resolution.

We can estimate the effect of instability by comparing the net sample density with the steepness of the density gradient inside the disc centrifuge. For example, suppose we inject a sample of polystyrene particles with a concentration of 0.05% by wt (typical for a polystyrene sample), and that the fluid in the centrifuge ranges from 1.0178 g mL^{-1} (5% sucrose solution) to 0.9981 g mL^{-1} (water) over a sedimentation distance of 20 mm. The steepness of the gradient is

$$(1.0178-0.9981)/20 = 0.000985 \ (\text{g mL}^{-1}) \ \text{mm}^{-1}$$

The density of polystyrene is 1.050 g mL^{-1}, so, a 0.05% dispersion in water at 20 °C has a density of 0.998126 g mL^{-1}, or 0.000026 g mL^{-1} higher than pure water. The distance over which this increase in density can be supported by the density gradient is

$$0.000026/0.000985 = 0.264 \ \text{mm}$$

In other words, the leading edge of the sample band cannot be less than 0.0264 mm wide in order to maintain stable sedimentation if the polystyrene concentration in the sample is 0.05%. The initial sample thickness (based on injected sample volume of 0.05 mL) is ~0.033 mm. Initial instability will add about 0.0264 to the initial band thickness. Higher or lower sample concentrations will lead to a proportionally larger or smaller contribution from initial instability.

Materials with higher density (*e.g.* polyvinyl chloride, density 1.385 g mL^{-1}) provoke additional instability unless a proportionally steeper density gradient is used. In nearly all cases, the effects of instability can be kept quite small by using relatively low sample concentration and an appropriate density gradient.

With 0.05% of 0.3 μm polystyrene particles, 0.1 mL sample volume, a 20 mm sedimentation distance, and the above-described density gradient, the expected resolution of the instrument (including the effect of Brownian motion over 10 min of sedimentation) is ~1.8–1.9%.

4.5.4 Injection Effects

The injection process can impact resolution in two ways. First, the injection is not instantaneous, but actually takes place over a period of about 0.1 s. This means that all particles do not start the sedimentation process at exactly the same time. Second, the physical impact of the sample striking the fluid surface inside the disc can cause some initial mixing of the sample with the gradient fluid, so that the initial sample band is not as narrow as the volume of the injected sample would suggest.

4.5.4.1 Injection timing. The effect of injection timing on the reported width of a perfectly narrow family of particles depends on the total sedimentation time. The percentage increase in reported peak width is given by:

$$100(((1 + (T_i/T_s))^{0.5}) - 1) \qquad (7)$$

where T_i is the time required for injection and T_s the time required for the particles to reach the detector beam.

At a sedimentation time of 60 s and with an injection time of 0.1 s, the increase in width is ~0.083%. This contribution is very small compared with the other factors that impact resolution. At sedimentation times longer than about 1 min, injection timing will always have a negligible impact on resolution. With much shorter sedimentation times, the effect can be significant. For example, if a peak reaches the detector in 10 s, the increase in reported width from injection timing will be ~0.5% of the peak diameter, or about 25% of the total reported width.

The overall impact of injection timing is actually a little less than indicated by the above equation, because if particles begin sedimentation at slightly different times, the effect of initial instability (as described in the above section) will be reduced. Reduced initial instability partially offsets the effect of injection timing.

4.5.4.2 Physical impact. It is difficult to predict the effect of physical impact of the sample on the fluid surface. However, experience has shown that the initial mixing (band broadening) is relatively small in nearly all cases, especially when the total sedimentation distance is ~20 mm. The rotating disc can be viewed using a synchronised strobe light, and the mixing from physical impact of the injection can be seen; it is clearly <1 mm, although an exact value is difficult to measure. When the sample is prepared using a fluid that is significantly lower in density than the fluid at the top of the density gradient, the mixing is drastically reduced ($\ll 1$ mm). For example, if the gradient consists of sucrose in water, the sample can be prepared in a mixture of 8% ethanol in water, with a density of ~ 0.985 g mL^{-1}. With this type of sample preparation fluid, the sample does not penetrate the density gradient surface very far; the sample fluid tends to quickly "float" and spread across the fluid surface. While it is not possible to exactly predict the effect of physical impact, the contribution to reported width of a perfectly narrow beam should be $<1\%$ in all cases, and will likely be well under 0.5% if the sample is prepared in a lower density fluid.

4.6 Actual Resolution of the DCS

When all of the factors that impact resolution are taken into account (Brownian motion, initial instability, injection effects), the expected resolution at a particle diameter of 0.3 μm is in the range of 1.9 to 2.5% when the instrument is set up with a sedimentation depth of 20 mm. This means that two perfectly narrow distributions of particles of the same material that differ in diameter by 1.9–2.5% should overlap by not more than 5% of their peak area.

4.7 Enhancing Resolution in the DCS

The most important factors that impact instrument resolution are known (detector beam width, sample thickness, Brownian motion) and can be mathematically modelled. It is possible to enhance the instrument's resolution by mathematically treating the distribution data that comes from the instrument to remove the effects of these factors. The process of removing a known effect from an unknown distribution is sometimes called "deconvolution".

Of course, for almost all measurements, the basic resolution of the DCS method is more than adequate, and deconvolution is not needed.

5 Sensitivity and Sample Size

The DCS method is very sensitive, especially in the size range of 0.01–10 μm diameter, where the efficiency of light scattering is high. For larger and smaller diameters, sensitivity gradually decreases. Broad distributions require more sample weight, but any sample with a total dry weight of 50–100 μg usually produces a good distribution. Some other analysis methods require much larger sample sizes.

5.1 Speed of Analysis and Dynamic Range

Total analysis time depends on centrifuge speed, particle density, fluid density, fluid viscosity, minimum particle size, maximum particle size and data-collection rate. Different commercial instruments often show large differences in total analysis time for the same sample. A higher data-collection rate (more data readings per second) allows a wider dynamic range to be measured in the same total analysis time, because larger (faster moving) particles can be measured more accurately. A higher maximum centrifuge speed reduces total analysis time for samples with very small particles. Dynamic size range has a very strong effect on total analysis time. Using a constant speed centrifuge and constant detector position, and measuring a dynamic size range of 25 (ratio of largest size to be measured to smallest size in the distribution), the total analysis time will normally range from ~10 to ~40 min, depending on the instrument. If the dynamic range is 50, then analysis time for most samples will be from ~40 to ~160 min, depending on the instrument. If the dynamic range is relatively narrow (<15), most samples can be analysed within ~4 to ~16 min.

Some types of samples contain a very wide range of particle sizes. Historically, these samples have been very difficult to measure using the differential sedimentation sizing method, because differential sedimentation has been limited to a dynamic range of about 70. A modified disc design and ramping of centrifuge speed can now be used to greatly increase dynamic range with the DCS. A dynamic range of over 1000 can be easily attained.

5.2 Dynamic Range Limit for Fixed Speed

Suppose that one wanted to measure particles between 20 and 0.05 μm in a sample. One must select a centrifuge speed so that the 20 μm particles (fastest moving) arrive at the detector beam no faster than ~0.75 s after injection, because it is not possible to accurately time the sample injection and collect distribution data if the sedimentation time is below ~0.75 s. The smallest particles will sediment at a much slower rate than the largest. The ratio of sedimentation speeds is $(D_1/D_2)^2$. If the 20 μm particles reach the detector in 0.75 s, then the 0.05 μm particles reach the detector at $0.75(20/0.05)^2$, or 120 000 s, or 33 h and 20 min. This is clearly not a practical analysis time, and even if one were willing to wait 33 h for results, Brownian motion of the smallest particles over 33 h of sedimentation would cause substantial errors in the reported distribution.

The dynamic range depends in reality upon how long one can wait for results. If the longest run one can tolerate is 30 min, then the practical dynamic range is

$$\left(\frac{(30*60)}{0.75} \right)^{0.5} = 48.99$$

If one can wait only 15 min, the dynamic range falls to 34.64; if one can wait 60 min, the practical dynamic range is 69.28.

Many types of samples (if not most types) are easily measured with a dynamic range of 30 to 70, and so present no problem for the DCS. However, there are some types of samples that really do have a broader dynamic size range, and these have previously been difficult or impossible to completely characterise using the DCS.

5.3 Ramping the Disc Speed

The g force inside the centrifuge is proportional to the square of the rotational speed, and so the sedimentation velocity of a particle is also proportional to the square of the rotational speed. If the rotational speed at the beginning of an analysis were low, and then gradually increased during the analysis, then the problem of limited dynamic range would be resolved: the lower initial speed allows analysis of the large end of the distribution, while the higher final speed allows one to measure the smallest particles in a practical run time.

Consider the 20 to 0.05 μm sample discussed in the above section (dynamic range of 400). If the rotational speed were increased by a factor of 12 over the first 6–10 min of the analysis, then the whole size range could be measured in a run lasting only ~20 to 23 min, which is a reasonable run time for most particle-sizing applications.

To make speed ramping practical, a couple of operational problems have to be resolved.

5.3.1 Disrupting the Gradient

Stable sedimentation (no "streaming") in the DCS depends on having a density gradient inside the disc chamber such that the fluid at the outside edge of the disc is slightly higher in density than the fluid at the surface, and the fluid density changes gradually between these two extremes. As long as the disc is turning at constant speed, the fluid inside the disc is "quiescent"; there is no reason for the gradient to be disrupted. If the disc speed is changed after the gradient is established, the change in speed will cause turbulent mixing of the fluids inside the disc, and hence, the total disruption of the density gradient. In other words, the fluid inside the disc will become uniform in composition, and stable sedimentation will not be possible.

5.3.2 Calculating the Particle Size

If it were possible to avoid disruption of the density gradient, it would still be necessary to account for the changing speed so that an accurate particle size distribution can be calculated.

5.4 Resolving the Speed Ramp Problems

Figure 6 shows how the disc is modified to avoid disruption of the gradient. A "separator wall" is placed inside the disc chamber. During acceleration or deceleration of the disc, this separator keeps the fluid inside the disc from moving at a speed different from the disc speed, and so eliminates virtually all mixing due to acceleration/deceleration.

The problem of calculating an accurate size distribution with changing speed is solved by modifying the operating software to account for changing disc speed.

The equation below is Stokes' law, modified to account for the changing g forces inside a centrifuge:

$$D = \{(18 \, \eta \ln (R_f/R_0))/((\rho_p - \rho_f) \int \omega_t^2 \, dt)\}^{0.5} \tag{8}$$

As long as we know how the rotational speed ω_t varies with time we can continuously integrate with respect to time during an analysis and generate an accurate distribution.

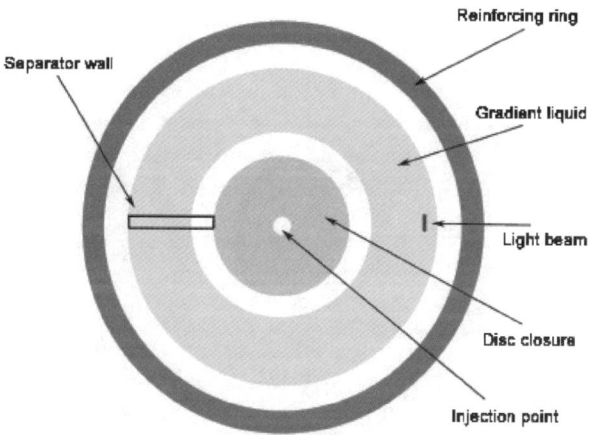

Figure 6 *Disc design for speed ramping*

6 Low-Density and Neutral Buoyancy Particles

The most important historical limitation for differential centrifugal sedimentation has been the requirement that the particles to be measured be significantly higher in density than the fluid in the centrifuge. A minimum density difference of 0.05 g mL^{-1} is desirable for most samples, and a difference of 0.1 g mL^{-1} or more is better. Some aqueous dispersions, such as polymer latexes and oil emulsions, often have particle densities near or below 1 g cm^{-3}. It is possible to use a mixture of water and methanol or ethanol, which has a density lower than water, to measure some types of low-density samples, but many are not compatible with the required alcohol concentration. Many low-density dispersions have been impossible to measure using the DCS method.

6.1 A New Differential Method

A new method[7] has been developed for differential sedimentation of low-density materials. The new method uses a centrifuge design that deposits a low-density sample at the bottom of a spinning centrifuge chamber, rather than at the surface of the fluid in the chamber. This method requires that the particles be lower in density than the fluid in which they are suspended; the particles move from the bottom of the chamber towards the top during the analysis. The implementation of the new method in a centrifuge of the hollow disc design is shown in Figure 7. A "V"-shaped groove is machined into the front face of the hollow disc, and four or more small capillary channels go radially from the base of the "V" groove to connect with the bottom of the centrifuge chamber. The level of the base of the "V" must be at least slightly above the level of fluid in the centrifuge (a lesser distance from the centre of rotation) to keep the groove free of liquid.

A sample is injected into the groove at the start of an analysis. A typical injection volume is in the range of 20 to 50 μL. When a sample is injected into the "V"-shaped

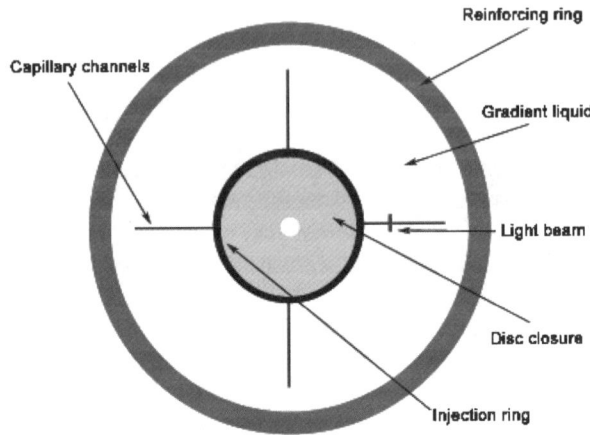

Capillary channels

Reinforcing ring

Gradient liquid

Light beam

Disc closure

Injection ring

Figure 7   *Modified disc design for new method*

groove, it is quickly (<0.1 s) carried by centrifugal force to the bottom of the centrifuge chamber via the small radial channels. The combined volume of the channels can be less than 10 μL , so even a small sample volume is sufficient to displace the liquid in the channels.

Any sample that remains in the channels may be flushed to the bottom of the centrifuge by immediately following the sample with a small volume (10–20 μL) of the same fluid that was used to prepare the sample for injection.

The large central opening to the disc chamber may be covered with a removable insert. With the insert in place, density gradient fluids and samples can be injected directly onto the centre of the rotating disc, rather than into the "V"-shaped groove. Rotation of the disc quickly carries any injected fluid or sample to the base of the "V"-shaped groove, and then to the bottom of the chamber via the capillary channels. When a removable insert is used in the centre of the disc, the density gradient may be formed by injecting a series of fluids with slightly different densities: the lowest density fluid first and the highest density last. The lower density fluids float upon the higher density fluids, and there is only a small amount of mixing as the gradient is formed.

Samples are prepared for analysis by dilution in a fluid that is more dense than the fluid at the bottom of the centrifuge chamber. The net density of the sample dispersion (average of particles and fluid) must be higher than the density of the fluid at the bottom of the centrifuge chamber so that the dispersion of particles quickly spreads to form a thin layer at the bottom of the chamber. Sedimentation of the particles proceeds in the normal fashion, except that the particles move towards the surface of the fluid rather than towards the bottom of the centrifuge chamber. Multiple analyses can be run without stopping the centrifuge, and it is even possible to alternate analyses between samples that are higher in density than the fluid, which are injected onto the surface, and samples that are lower in density than the fluid, which are injected into the "V"-shaped groove.

By using either the conventional differential method or the new differential method reported here, virtually any sample that is an aqueous dispersion can be measured by differential sedimentation. If the particles are significantly higher in density than water, they can be analysed using the conventional differential method. If the particles are significantly lower in density than water, they can be analysed using the new method with water in the centrifuge. If the density of the particles is near the density of water, then the new method can be used with deuterium oxide partially or totally substituted for water in the centrifuge.

The new differential method can be extended to centrifuges of nearly any design, and to many nonaqueous solvent systems as well, so as long as the fluid within the centrifuge has a density gradient and so as long as the samples are prepared in a fluid that is both higher in density than and miscible with the fluid at the bottom of the centrifuge chamber. A sample may be prepared in a fluid that is not miscible with the fluid at the bottom of the centrifuge chamber as long as the interfacial surface tension between the fluid phases does not prevent particles from passing from one phase to the other.

6.2 Eliminating Injection Artefacts

The new differential method yields distributions that may include small injection artefacts. These artefacts have no connection to the actual particle size distribution; they are seen even when a blank (particle-free) sample is analysed. The injection artefacts are of two types:

- relatively large-diameter particles that are actually air bubbles entrained when a sample is injected; and
- a relatively broad baseline deflection that comes from a slight change in optical density of the fluid in the centrifuge when a sample is injected.

The injection artefacts can be minimised or eliminated using one or more of the techniques discussed below.

Entrained air bubbles show up as large "particles" because they rise rapidly through the fluid: they are both relatively large in size and much lower in density than the fluid in the centrifuge. The volume of entrained air bubbles can be minimised by having the level of the fluid within the centrifuge close to the top of the capillary channels that transport samples to the bottom of the centrifuge (refer to Figure 7), and by using capillary channels that are of the smallest practical diameter. If the distance between the top of the fluid and the top of the capillary channel is small, and the diameter of the capillary channel is also small, then the artefact from entrained air bubbles is minimised.

The mechanism for production of the second type of injection artefact (that due to a change in optical density of the fluid) is not obvious. Before a sample is injected, the fluid in the centrifuge chamber is moving at the same rotational speed as the centrifuge. When a small sample is injected at the bottom of the centrifuge chamber, the total volume of the fluid in the chamber increases very slightly. All of the fluid in the chamber is raised slightly when a sample is injected because the sample is higher in density than the fluid in the chamber, and enters at the bottom of the chamber. When the fluid is raised, it rotates at a very slightly smaller radius than it rotated at before the injection. The absolute linear velocity of all of the fluid in the chamber is not immediately changed when a sample is injected, but the radius of rotation for the fluid in the chamber is suddenly (very slightly) reduced when a sample is injected. This means that the rotational velocity of all of the fluid in the chamber increases slightly relative to the rotational speed of the centrifuge at the moment a sample is injected. The physical effect of the injection is similar to a small, instantaneous reduction in centrifuge speed. Inside a hollow disc-type centrifuge, the fluid cannot suddenly change in speed; it must gradually catch up with the speed of the centrifuge disc.

This difference in speed between the centrifuge and fluid causes slight mixing to take place within the fluid as its rotational velocity recovers to match the rotational velocity of the centrifuge. The fluid in the chamber is not uniform in composition: its composition changes owing to the presence of the density gradient. If the refractive index of the fluid also changes as the composition of density gradient changes, then mixing caused by injection of a sample will cause some of the detector light beam to be scattered: the optical transmission of the fluid in the chamber is slightly reduced owing to optical inhomogeneity during mixing. As the homogeneity of the fluid gradually recovers (owing to diffusion), the optical transmission returns to the original level.

The artefact due to changing rotational speed can be minimised in two ways. First, the smallest practical sample volume can be used. The smaller the sample volume (relative to the volume of the centrifuge chamber) the smaller the effect of the injection. Second, a density gradient can be prepared that is constant in refractive index. If all of the fluid in the chamber has the same refractive index (even though the composition does change), then mixing will not cause the optical transmission of the fluid to change. Density gradients with virtually constant refractive index can be formed using mixtures of three components. For example, an aqueous density gradient that goes from 2 to 0% (by wt) sucrose and at the same time from 0 to 5% ethanol has nearly constant refractive index over the entire composition range.

Both types of injection artefacts can be mathematically subtracted from a particle size distribution. A blank sample (free of particles) can be run to record only the injection artefacts, and then subtracted from the distribution of an unknown sample. The distribution that remains after the subtraction is the distribution for the unknown, free of injection artefacts.

6.3 Conclusion with Regard to Low-Density and Neutral Buoyancy Particles

By using the new method reported here, DCS can be applied to measure the size distributions of materials that are lower in density than the fluid in which they are suspended. This eliminates the single most important limitation of the differential method, while maintaining the high resolution, accuracy and operational advantages of the differential method.

7 Non-Spherical Particles

The weight distribution reported by the DCS method is a "Stokes-equivalent" distribution: the weight distribution of spherical particles that would yield the reported distribution. The Stokes-equivalent distribution is equal to the true weight distribution only if the particles in the distribution are spherical. Particles with other geometries are reported as smaller than their actual weight distribution. For particles that closely approximate spheres (for example, icosahedrons), the measured distribution will be very nearly correct, while geometries very different from spheres, such as long, thin rods, will be reported as significantly smaller than their actual weight distribution.

Cylindrical rods with an aspect ratio of ~2 (length/width) produce a reported weight distribution about 5% smaller than correct, while rods with an aspect ratio of ~3 produce a reported weight distribution about 10% lower than correct. Disc-shaped particles ~2 times wider than they are thick are reported as about 6% smaller than correct. For all non-spherical particles, no matter what the geometry, the DCS method produces very consistent and repeatable results, even if those results are not exactly correct in absolute weight sense. The DCS method is commonly used for characterisation and quality control with a wide range of inorganic pigments, fillers and abrasives, even though the particles being measured are not spherical in shape.

8 Introduction to Differential Sedimentation – An Overall Conclusion

DCS is an extremely powerful tool for high-resolution particle characterisation, especially in the size range of 0.003 μm (3 nm)–10 μm. It enables very narrow

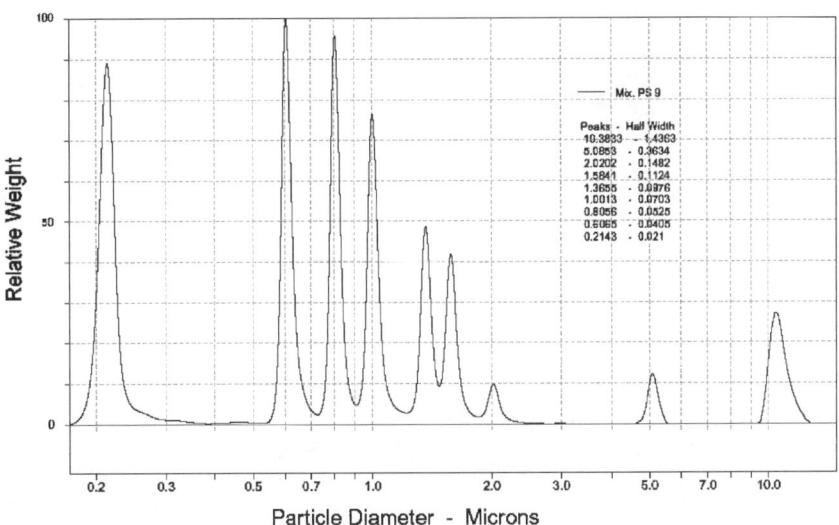

Figure 8 *Resolution of multi-modal distribution of polystyrene particles*

distributions of particles differing in diameter by less than 2% to be resolved and, hence, extremely small differences, changes or shifts in particle size to be accurately and reproducibly detected and measured. The new method described in this chapter for measurement of low-density neutral buoyancy particles also addresses the only previous technical benefit of integral sedimentation over DCS.

The DCS instrumentation used in this chapter is the CPS Disc Centrifuge DC24000. More information regarding this instrument can be found at www.cpsin-struments-eu.com and www.analytik.co.uk.

References

1. G. G. Stokes, *Mathematical and Physical Papers*, Cambridge University Press, 1880, **1**.
2. C. E. Marshall, *Proc. Roy. Soc.*, A, 1930, **126**, 427.
3. T. Allen, *Particle Size Measurement*, Chapman and Hall, London, 1968, 120.
4. M. K. Brakke, *Arch. Biochem. Biophys.*, 1953, **45**, 275–290.
5. M. H. Jones, *U.S. Patent 3,475,968*, November 4, 1969.
6. H. Puhk, *U.S. Patent 4,699,015*, October 13, 1987.
7. S. T. Fitzpatrick, *U.S. Patent 5,786,898*, July 28, 1998.

CHAPTER 15

Relating Small Angle Scattering and Analytical Ultracentrifugation in Multidomain Proteins

STEPHEN J. PERKINS, HANNAH E. GILBERT, YIE CHIA LEE, ZHE SUN AND PATRICIA B. FURTADO

1 Introduction

1.1 Scope of This Chapter

This chapter compares the major features of X-ray and neutron solution scattering[1-4] with analytical ultracentrifugation in application to multidomain proteins.[5,6] It illustrates the utility of joint scattering and ultracentrifugation studies by presenting recent examples from our own work. This chapter expands previous brief reviews on multidomain proteins in the complement system of immune defence.[4,5] The proteins of interest can be categorised into four types. Type 1 proteins possess no covalent linker between domains within the protein. Type 2, Type 3 and Type N proteins comprise two-, three- or multidomain proteins in that order, with one, two or more linkers, respectively, connecting these domains. The word 'domain' usually refers to a single, independently folded protein subunit. In this chapter, this term is also used to describe an essentially rigid multidomain fragment (such as the four domains within the Fab or Fc fragments of an antibody, but not the intact antibody itself) or an oligomer containing as many as four or five protomers.

From a biological standpoint, large multidomain proteins with many linkers between many domains represent an important and common set of proteins. The existence of long linkers between the domains is thought to preclude the crystallisation of many multidomain proteins. If crystals are obtained, the resulting multidomain structure may possess large conformational artefacts which need to be characterised. Alternative structural techniques are needed in order to identify the solution structures in question, and this justifies the use of scattering and ultracentrifugation methods. The strengths and limitations of these methods require appraisal. Analytical ultracentrifugation provides a quick and useful tool for characterising macromolecular structures and compositions within the home laboratory, in particular the state of oligomerisation. The strength of solution scattering is its ability to provide additional structural detail. At best, scattering permits biologically useful medium resolution structures to be determined, and these can be deposited in the Protein Data Bank accessible at http://www.rcsb.org/pdb/.[7] The combination of both methods with crystallography sometimes reveals some unusual results, and these are discussed.

1.2 What is Solution Scattering?

Solution small-angle scattering is a diffraction technique that is used to study the overall structure of biological macromolecules in random orientations.[1-4] A macromolecular sample is irradiated with a collimated, monochromatic beam of X-rays or neutrons produced by high-flux X-ray synchrotrons or at nuclear reactors or pulsed sources, respectively. As a result of this irradiation, a circularly symmetric diffraction pattern is recorded on a two-dimensional area detector placed behind the sample. Intensities are measured in a small angular range defined by the scattering vector Q, where $Q = 4\pi \sin \theta/\lambda$ (2θ is the scattering angle and λ the wavelength). Data acquisition times range from seconds to hours. Radial integration of this two-dimensional pattern yields the scattering curve $I(Q)$ in one dimension [Figures 1(a) and (b)]. While X-rays are diffracted by electrons, and neutrons are diffracted by nuclei, the physical principles are the same in both types of scattering experiments. A typical solution scattering curve views structures at a low structural resolution of about 2–4 nm from data obtained in a Q range between about 0.05 to over 2 nm^{-1} [Figures 1(a) and (b)].[8] Analyses of $I(Q)$ at the lowest Q values by Guinier plots of $\ln I(Q)$ *vs.* Q^2 lead to the overall molecular weight M_r from the $I(0)/c$ value, at which Q is 0, and c is the macromolecular concentration. The radius of gyration R_G monitors the macromolecular elongation and is also determined from the Guinier plot. In certain cases, if the protein is elongated, the radius of gyration of the cross-section (R_{XS-1} and sometimes R_{XS-2}) can be determined from a plot of $\ln (I(Q) \cdot Q)$ *vs.* Q^2 at larger Q values. For lamellar shapes, the radius of gyration of the thickness can be determined (not shown). The Fourier transformation of $I(Q)$ gives the distance distribution function $P(r)$ [Figures 1(c) and (d)]. The maximum dimension of the macromolecule L is determined from the point where $P(r)$ intersects zero at large r.[9] The macromolecular shape can be deduced from the appearance of the $P(r)$ curve and the positions of the maxima M, M1 and M2 seen in this curve relative to the value of L.

A number of approaches have been developed in order to visualise the macromolecular structure responsible for the appearance of the scattering curve. A unique structure cannot be determined only by scattering for reason of the randomised orientations of the macromolecule in solution. Our approach discussed below uses curve-fit methods based on tight constraints that are based on known crystal structures.[1,2,4] Known domain structures are taken either directly from that for the macromolecule of interest, or are re-adapted by appropriate homology (or 'comparative') modelling of related crystal structures. The constraints enable automated curve-fit procedures to be developed. They are sufficiently tight such that only a small fraction of all the stereochemically feasible conformations will give scattering curve fits. In this way, structural information can be extracted to a precision of as high as 0.5–1.0 nm, and this corresponds to a medium resolution structure determination. The degree of bend between adjacent domains can be assessed, but not the rotational orientation between them. The advantage of this constrained approach is that it permits direct comparisons with crystal structures and provides molecular insights into biological function.

1.3 What is Analytical Ultracentrifugation?

Analytical ultracentrifugation is a dynamic method. The movement of biological macromolecules in solution is monitored inside a specially designed sample cell when this is subjected to a high centrifugal force inside an ultracentrifuge rotor.[10,11] Spectroscopic methods based either on the absorption or refractive index difference between the sample cell and a reference buffer cell are used to follow the sedimentation behaviour of the sample within the cell. At the beginning of the experiment,

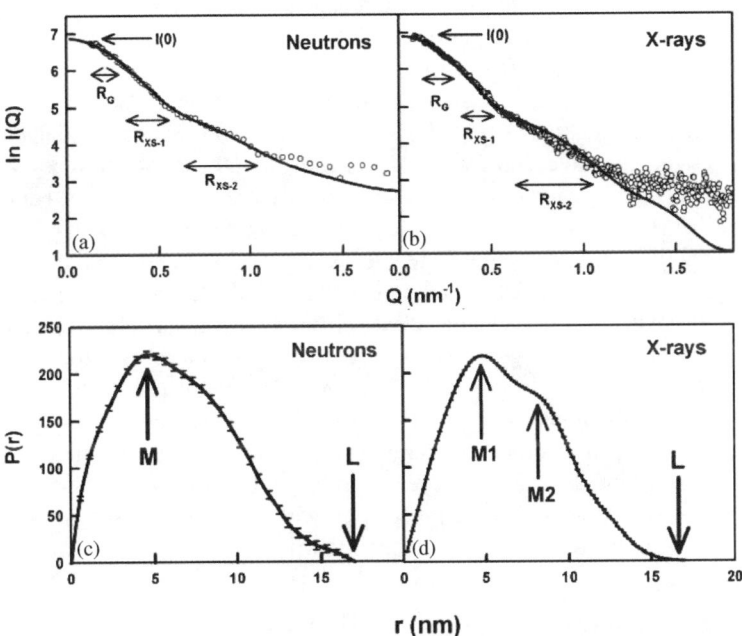

Figure 1 *A typical scattering curve analysis for an antibody to show the main parameters determined by scattering. The data are for human IgA2 (adapted from ref. 20). (a) The experimental neutron data I(Q) for IgA2 in 100% ^2H_2O buffer from Instrument LOQ at ISIS are shown as open circles, and the continuous line shows the constrained modelling fit to the data. The Q ranges used to analyse the R_G, R_{XS-1} and R_{XS-2} values are indicated both here and in (b) by arrowed ranges as labelled. The I(0) parameter at Q=0 is likewise labelled. (b) The experimental X-ray data I(Q) for IgA2 from Instrument ID02 at ESRF are shown as open circles, and the continuous line shows the constrained modelling fit to the data. (c) Indirect Fourier transformation of the neutron I(Q) data in reciprocal space into the distance distribution function P(r) revealed a single major peak at M of 4.5 nm and a maximum length L of 17 nm. (d) The corresponding X-ray P(r) curve revealed two peaks M1 and M2 at 4.5 and 6.8 nm, respectively, and a maximum length L of 17 nm*

the solute is present in a uniform concentration throughout the sample cell. By appropriate selection of a rotor speed, two types of experiments are performed:

(i) In a sedimentation velocity experiment, the use of high rotor speeds causes the solute to sediment to the outermost side of the cell in several hours. This process is generally observed with a column height of about 11 mm [Figure 2(a)]. The rate at which this sedimentation occurs gives the sedimentation coefficient $s_{20,w}^o$ in standard conditions at 20 °C in water. The processing of the difference between pairs of scans shown in Figure 2(a) leads to the $g(s^*)$ plot. A typical $g(s^*)$ analysis is shown in Figure 2(b), from which the $s_{20,w}^o$ value is determined from the centre of the Gaussian peak. The value of $s_{20,w}^o$ depends on the elongation of the macromolecular shape, where higher values correspond to more compact structures. The Svedberg equation relates the $s_{20,w}^o$ value and molecular weight M_r to the frictional coefficient f, which is an alternative measure of structural elongation.

(ii) In a sedimentation equilibrium experiment, the rotor speed is set to be lower in order to create an equilibrium between the sedimentation of the solute to the outermost side of the cell and the diffusion of the solute within the cell. The column height is much shorter at 2 mm in order to enable equilibrium to be reached after about 10 h for a given rotor speed. At equilibrium, an exponential distribution of solute is measured. Standard curve-fit methods using single- or multiple-species exponential functions directly provide the molecular weight M_r determination for the solute. They will also give the relative proportion of the species present if there is an associative equilibrium between them [Figures 2(c)–(e)]. Provided that the buffer density ρ and the macromolecular partial specific volume $\bar{v}$ are known, accurate molecular weights M_r can be determined in solution by sedimentation equilibrium. These values are more reliable than the approximate determinations from gel-filtration chromatography or the denaturing conditions of reducing or nonreducing polyacrylamide gel electrophoresis.

Computational approaches relate the macromolecular structure with the observed $s_{20,w}^o$ value. Again a unique structure cannot be determined, and several approaches have been developed in order to interpret $s_{20,w}^o$ values. The use of the constrained scattering modelling approach[1,2,4] is readily adapted to model $s_{20,w}^o$ values, meaning that automated trial-and-error fit procedures will show which conformations are compatible with the experimental $s_{20,w}^o$ value within a specified error range.

1.4   Rationale for Combining Both Approaches

Scattering and ultracentrifugation each provides similar compositional information. This identifies the oligomeric state of the protein, given that protein molecular weights are nowadays reliably calculated from its translated DNA sequence. Compositional information from scattering is derived from the $I(0)$ value, the intensity of forward scattering at zero Q [Figures 1(a) and (b)]. Normalisation by the sample concentration c in mg mL^{-1} gives the molecular weight M_r.[3] Neutron $I(0)$ values are standardised using either a 1-mm-thick water sample (Institut Laue Langevin) or a standard deuterated polymer (ISIS facility). Absolute molecular weights are readily derived from the resulting neutron $I(0)/c$ values. The calibration graph for Instrument LOQ at ISIS is shown in Figure 3.[12] X-ray $I(0)$ values are not generally calibrated to give absolute values, as standards for this purpose are not

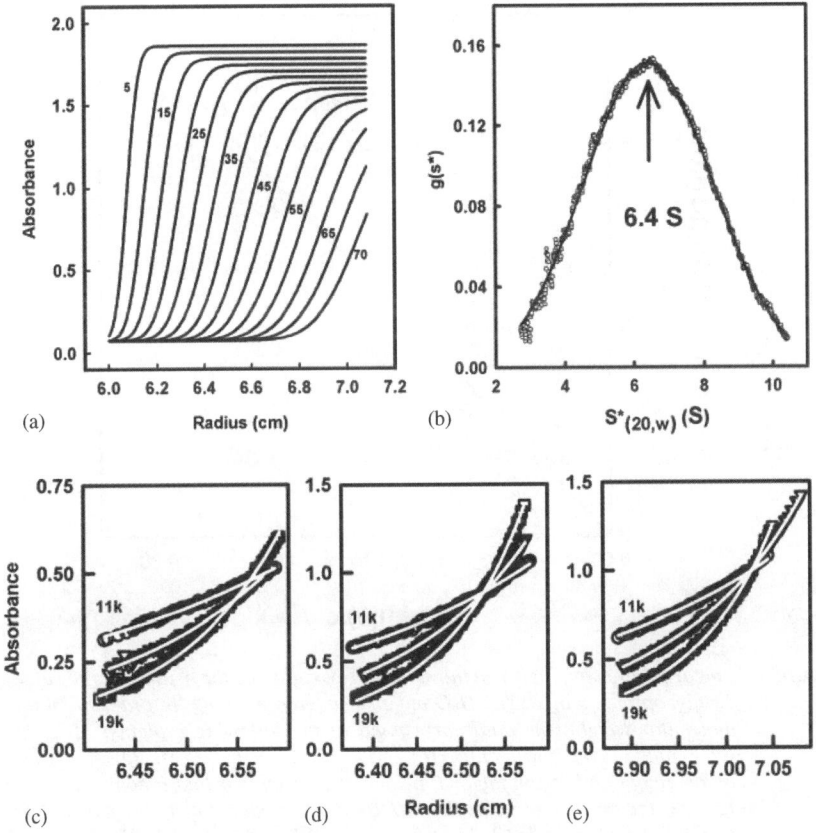

Figure 2 *A typical sedimentation analysis for two proteins. (a) In a sedimentation velocity experiment for human IgA2, every fifth scan of 70 sedimentation scans measured at 25 000 rpm using absorbance optics are shown as numbered lines. Scan fit analysis using the SEDFIT program resulted in a $s^0_{20,w}$ value of 6.48 S (adapted from ref. 20). (b) The difference between pairs of sedimentation scans are analysed using g(s*) distributions to result in a Gaussian function as shown. The DCDT+ program used 10 scans recorded at 20 000 rpm at 8 min intervals to give a similar $s^0_{20,w}$ value of 6.42 S. The width of the Gaussian function gave a molecular weight estimate of 147 000±25 000 Da. (c)–(e) In a sedimentation equilibrium analysis of the single-chain Fv fragment MFE-23, non-linear least-squares regression analysis was used to fit simultaneously the nine observed curves to a monomer–dimer equilibrium. The curves were obtained at three concentrations of 1.1, 2.2 and 3.0 mg mL^{-1} and three rotor speeds of 11 000, 15 000 and 19 000 rpm. The continuous white line is the result of the simultaneous fit (adapted from ref. 15)*

generally available. However, relative molecular weights from the $I(0)$ values measured during a single synchrotron X-ray beam-time session are calculable, provided that the sample concentrations c have been measured. These $I(0)/c$ values enable an association to be followed during a dilution series. In the special case of neutron contrast variation, the observed $\sqrt{I(0)}/c$ values in different heavy-water buffers result in neutron contrast variation matchpoint graphs, from which both the total molecular

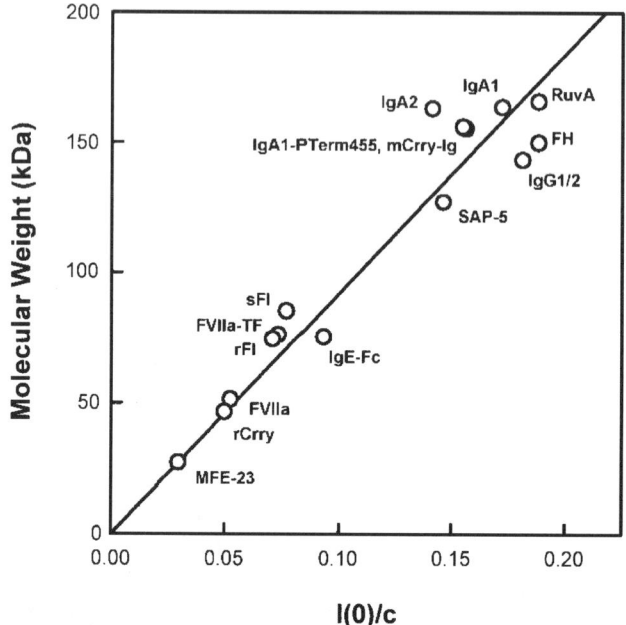

Figure 3   *Linear relationship between the molecular weight and the neutron I(0)/c values for 15 glycoproteins in 100% ^2H_2O buffer measured on LOQ. In order of increasing molecular weight, the data correspond to the following proteins: MFE-23, rat Crry, factor VIIa, the factor VIIa–tissue factor complex, the IgE–Fc fragment, recombinant and serum factor I, pentameric serum amyloid P component, bovine IgG1/2, factor H, the recombinant PTerm455 fragment of IgA1, mouse Crry-Ig, human recombinant IgA2, human serum IgA1 and octameric RuvA (this graph was first published in ref. 12)*

weight and the relative proportion of the components in a multicomponent system (*e.g.* protein–lipid or protein–DNA) can be identified.[13] The proportion of components is obtained from the matchpoint itself, at which the observed value of $I(0)$ is zero. From the ultracentrifuge, equilibrium runs lead to molecular weight determinations. It is more straightforward to monitor a wider range of concentrations by ultracentrifugation than by scattering, as runs can be repeated at short notice using the former. The use of different rotor speeds provides a useful control of the quality of the data, *e.g.* as a check of sample polydispersity that may occur because of variable glycosylation levels.[14] These dilution series will quantitatively characterise any associative processes that may occur. For example, in the case of a monomer–dimer equilibrium, if molecular weight data reveal the presence of both the monomer and dimer forms, the dissociation constant K_D can be determined by a simultaneous fit of all the curves, typically at three different speeds and three concentrations [Figures 2(c)–(e)].[15,16]

Both scattering and ultracentrifugation characterise macromolecular structures in solution, where both methods yield the degree of macromolecular elongation. This corresponds to the radius of gyration R_G (scattering) or the sedimentation coefficient $s^o_{20,w}$ (ultracentrifugation). From either of these, the anisotropy ratio is determined to show the degree to which the shape deviates from that of a perfect sphere. For scattering, this is given by R_G/R_O, where R_O is the R_G of the sphere of volume equal to that of the hydrated protein (X-rays) or the dry protein (neutrons). For ultracentrifugation, this is given by the frictional ratio f/f_0, where f_0 is the frictional coefficient of the sphere of volume equal to that of the hydrated macromolecule. For both the R_G/R_O and f/f_0 values, a value of 1.0 would correspond to a spherical protein structure, the most compact one possible. For globular single-domain proteins, a literature survey of 47 proteins showed that the average R_G/R_O value is 1.28 ± 0.10. Many compact globular proteins show f/f_0 values of about 1.2–1.3.[17] A comparative survey of R_G/R_O and f/f_0 values for multidomain proteins with a 10-fold range in molecular weights from 14.7 to 159 kDa is shown in Figure 4. The structure of these proteins can include as many as 20 domains (see Section 3.4 below). While the values are correlated with each other, it is clear that no linear relationship exists between the R_G/R_O and f/f_0 values. Empirically (dashed line in Figure 4), a rectangular hyperbola function will replicate the observed trend as the R_G/R_O values go from 1 to 4, and this is useful as a check of the self-consistency of the data collection.

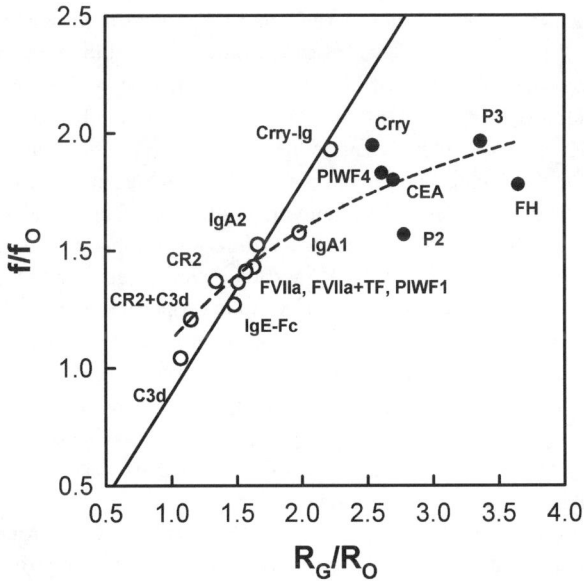

Figure 4 *Relationship between the f/f_0 and R_G/R_O anisotropy ratios for 16 X-ray R_G determinations for proteins ranging in molecular weight between 14.7 and 159 kDa. The open circles correspond to data points up to a R_G/R_O of about 2, up to which the relationship is approximately linear. The filled and open circles were fitted empirically to a rectangular hyperbola function of the form $y = ax/(b+x)$. Symbols were used to identify the data points as follows: CR2, complement receptor type 2 SCR domains 1 and 2; C3d, fragment of complement component C3; PIWF1 and PIWF4, the three- and six-domain forms of anosmin-1; IgA1 and IgA2, two subclasses of immunoglobulin A; FVIIa, activated factor VII; TF, tissue factor; P2 and P3, the dimer and trimer of properdin; Crry, complement receptor-related Y protein; Crry-Ig, chimaeric complex of Crry with the mouse IgG1 Fc fragment; CEA, carcinoembryonic antigen; FH, factor H*

The $I(0)$ and R_G parameters from the Guinier plots are equivalent to the molecular weight and $s^o_{20,w}$ parameters, respectively, of ultracentrifugation. In distinction to ultracentrifugation, scattering provides further structural and compositional information from data obtained beyond the Guinier R_G region at low Q [Figures 1(a) and (b)]. For example, using the cross-sectional Guinier R_{XS} measurements, the length of the macromolecule is calculated by combining the R_{XS} and R_G values.[8] The distance distribution function $P(r)$ yields macromolecular shape information in real space. This includes the macromolecular length and any notable spatial distributions within the molecule. For example, the double peak seen in the X-ray $P(r)$ curve for the antibody IgA2 [Figure 1(d)] was interpreted as evidence to show that the IgA2 solution structure is a flexible structure with a relatively rigid extended averaged arrangement of two Fab arms. If the $P(r)$ curve is symmetric about its maximum M, and if L is double the value of M, this shows that the macromolecule is spherical.

The molecular modelling of the full scattering curve $I(Q)$ is more informative than the corresponding procedure based on the R_G or the $s^o_{20,w}$ values (Figure 5). In a typical randomised modelling search, 10000 trial-and-error models are generated. After removal of the models that show steric clashes or overlaps between the domains, about 5–15% of the models will be compatible with both the R_G and $s^o_{20,w}$ values. If the modelling is satisfactory, the joint use of these R_G and $s^o_{20,w}$ values is complementary and will not significantly reduce the number of structurally compatible structures. This is the expectation from the correlation of Figure 4. However, when either the R_{XS} values are included as additional filters, or the 5–15% of compatible models are ranked in order of their R factors (the measure of the goodness-of-fit with the full scattering curve), the number of models showing good agreement with the scattering data is much reduced to less than 1% of the starting trial-and-error models. These final models usually correspond to a small family of related structures compatible with the scattering data.

There is the possibility that, as the R_G and $s^o_{20,w}$ parameters correspond to the independent monitoring of different macromolecular properties, the measurements will not agree with each other. This justifies the comparative analysis of neutron and X-ray scattering data alongside ultracentrifugation data. Two examples of where this may occur may be caused by high hydration levels or pronounced interdomain flexibility.

(i) A standard hydration level of 0.3 g H_2O (g protein)$^{-1}$ corresponds approximately to a monolayer of water molecules on the surface of the protein.[18,19] Generally, this gives a good account of X-ray scattering and ultracentrifugation results. The hydration level of a protein is largely invisible to neutron scattering. An increase in the amount of bound water is possible if the protein structure facilitates this. For example, the presence of voids within the structure may facilitate the formation of unusually larger well-ordered hydration shells. This can cause the R_G and $s^o_{20,w}$ values to deviate from each other, the reason being that the $s^o_{20,w}$ values are more sensitive to increases in hydration levels than the R_G values.[20,21] The hydration level modifies $\bar{v}$, and in turn the $s^o_{20,w}$ value.[18,19]

(ii) Multidomain proteins may exhibit flexibility at the linkers between the domains. If this flexibility becomes excessive, a range of solution structures can simultaneously co-exist. Consequently, while the averaged R_G and the averaged $s^o_{20,w}$ values may agree with each other (Figure 4), these models may deviate from the curve fits analyses for the full $I(Q)$ scattering curve.[22–24]

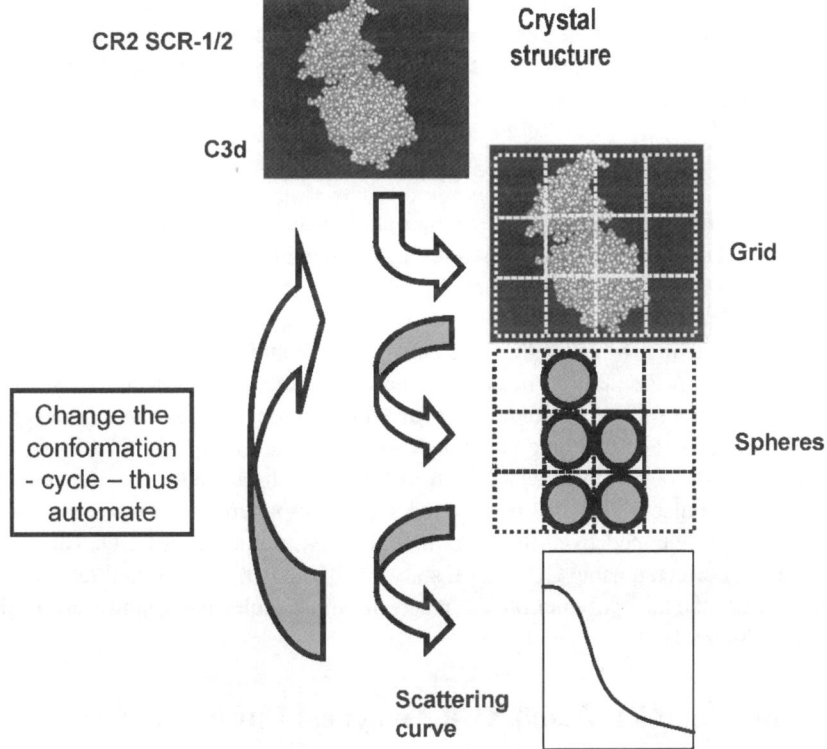

Figure 5 *Cartoon illustrating the algorithm used for constrained sphere modelling of multidomain proteins. A calibrated procedure based on tests with known crystal structures is used, in which the atomic coordinates in the crystal structure is converted into spheres by placing these coordinates within a 3D grid of cubes. A Debye sphere is assigned to each cube containing enough atoms, and this typically results in 200 to 1000 spheres. Only five spheres are shown in this cartoon for clarity. The neutron scattering curve is calculated using an unhydrated model. The X-ray scattering curve and the sedimentation coefficient are calculated from the hydrated sphere model*

2 Methods

This section summarises the source of the experimental data and their analyses. Fuller accounts of these procedures are given elsewhere.[3,10,11]

2.1 X-ray ESRF Beamline at ID02

The current state of the art is exemplified by the camera at Station ID02 of the ESRF in Grenoble, France.[25] A monochromated collimated beam of X-rays is directed at the sample of volume 20 μL, and the diffraction pattern is recorded on an area detector (1024$\times$1024 pixels) in acquisition times of 1–10 s. The detector is calibrated for the Q range using a silver behenate standard and its efficiency using a Lupolen standard.

The scattering pattern for the sample is converted to a one-dimensional representation by radial averaging about the main beam centre. Subtraction of the buffer background run from the sample run, also processed in the same way, gives the scattering curve $I(Q)$ [Figure 1(b)].

2.2 Neutron ISIS Beamline at LOQ

Neutron scattering can be performed either at high-flux reactor sources, such as on Instruments D11 and D22 at the Institut Laue Langevin, Grenoble,[26,27] or on pulsed neutron sources, such as on Instrument LOQ at the ISIS facility at the Rutherford Appleton Laboratory.[28] At ISIS, neutrons are generated in pulses, each pulse containing neutrons in a wavelength range between 0.22 and 1.0 nm. On LOQ, monochromatisation of the neutron beam is achieved by time-of-flight techniques. After diffraction by the sample of volume 300 μL, the diffraction pattern in a Q range of 0.06–2.4 nm^{-1} is recorded on an area detector with an active area of 64 cm$\times$64 cm and a resolution of 5 mm in acquisition times of 1–2 h depending on the sample. Each neutron pulse yields 102 time-sliced scattering patterns for each sample and buffer that are obtained from the time-of-flight measurements on LOQ. These are merged and corrected using COLETTE software.[28] The $I(0)$ value is calibrated using a deuterated standard polymer, as the result of which molecular weights are forthcoming (Figure 4).

2.3 Beckman XL-A and XL-I Analytical Ultracentrifuges

Analytical ultracentrifugation is most commonly performed on Beckman Optima XL-A or XL-I instruments,[29,30] more recently upgraded to the Proteome XL-A or XL-I instruments with digitised optics. Sedimentation velocity runs are performed in two-sector cells with 400 μL sample in one sector and 420 μL buffer in the other, observing the sedimentation across a column height of 11 mm [Figure 2(a)]. Sedimentation equilibrium runs are performed in six-sector cells with three 100 μL samples, observing the equilibrium distribution across a column height of 2 mm [Figures 2(c)–(e)]. Detection on the XL-A is achieved by monitoring the sample absorption. That on the XL-I is achieved by simultaneously measuring both the absorbance and refractive index differences between the sample and the buffer reference. For proteins, the latter is effective in covering a wide range of concentrations from low (absorption optics) to high (interference optics).

2.4 Experimental Analyses and Modelling Calculations

Scattering analyses proceed in two stages, first the analysis of the experimental data, then the modelling of the data in terms of macromolecular structures. The importance of the modelling is to confirm that the experimental data has been correctly interpreted. Scattering data processing of the $I(Q)$ curves at the lowest Q values yields the R_G and $I(0)$ parameters, together with cross-sectional analyses at medium Q values for the R_{XS} parameter if these are possible.[3] The transformation of the $I(Q)$ curves to distance distribution functions $P(r)$ utilise software such as GNOM.[9]

GNOM yields the scattering curve in real space $P(r)$, together with another calculation of the R_G and $I(0)$ parameters. Scattering curve modelling can proceed in several ways.[31] If this is done without any structural constraints, the scattering curve $I(Q)$ is converted into a shape function using a spherical harmonics representation, and this can be visualised using molecular graphics. Alternatively, again without using any constraints, one of several published *ab initio* methods based on genetic algorithms, simulated annealing or Monte Carlo algorithms can be used to generate a best-fitting shape. This is represented by small spheres that corresponds to the observed scattering curve for visualisation purposes. Because scattering does not observe unique structures, the outcome of these methods can sometimes be arbitrary. Both these approaches work best with relatively compact structures.

If atomic structures are used to constrain the scattering analyses (Figure 5), this enables relatively elongated structures to be modelled. The structures are first converted into about 200–1000 small spheres using a grid transformation in order to shorten the computations. The resulting dry model is inputted directly into neutron scattering modelling. For X-ray and sedimentation coefficient modelling, the model requires the addition of a hydration shell, most usually corresponding to 0.3 g H_2O (g protein)$^{-1}$.[19] Hydration is implemented by adding extra spheres to the surface of the model. The scattering curve is then calculated from these spheres using the Debye equation. The spheres need to be small enough not to affect the calculation of the scattering curves $I(Q)$ in the desired Q range. Typically a diameter of 0.5 nm is satisfactory. For the study of multidomain proteins, between 5000 and 12000 trial models are generated in random domain arrangements by molecular dynamics, then these are tested against the experimental $I(Q)$ curve to identify a small subset of the best-fit models. This latter procedure can be largely automated (Figure 5).

Ultracentrifugation analyses also proceed in two stages: firstly the processing of the experiment, then its modelling. After correction for the temperature, the partial specific volume of the solute $\bar{v}$ calculated from its amino acid and carbohydrate composition and the buffer density ρ calculated from its composition,[18,32] the $s^o_{20,w}$ value is obtained from the scans shown in Figure 2(a). Analysis of multiple sedimentation scans by the use of powerful PC programs such as DCDT+ or SEDFIT enables these scans to be averaged and corrected to standard conditions.[33,34] An example of a typical $g(s^*)$ plot is shown in Figure 2(b). The effect of diffusion processes on the $s^o_{20,w}$ determination is removed by these programs. The structural modelling of the $s^o_{20,w}$ values is often performed using small sphere models that correspond to the appearance of the protein. If the diameter of the hydrodynamic spheres matches that of the scattering spheres, both the scattering and ultracentrifugation modelling can be achieved using identical sphere models. This strategy unifies both techniques. It is achieved by the use of software from the Garcia de la Torre group, most notably GENDIA and HYDRO.[35] The use of small spheres should not mask the fact that they offer more detail than the structural resolution of the $s^o_{20,w}$ values. The use of simple ellipsoidal shapes for modelling purposes using software such as ELLIPS[36] is less effective as many protein structures such as antibodies are not easily represented in this way, although the simplicity of these ellipsoidal models helps to ensure that the $s^o_{20,w}$ values are not over-interpreted.

The exponential distributions obtained from sedimentation equilibrium data are analysed using ORIGIN (Microcal Inc.) and add-ons provided by Beckman or by the use of other software such as the classic WINNONLIN program or the more recent SEDPHAT program.[37,38] Multiple species are analysed by altering the assumptions used for the fits. For example, a monomer–dimer equilibrium would be analysed by fixing the M_r for the monomer, specifying the presence of a dimer in a two-species fit algorithm, and allowing a dissociation constant K_D for the equilibrium to be optimised by the curve-fit analysis for data obtained at different speeds and concentrations. If this is successful, the r.m.s. deviation of the curve fits will be significantly lower than that the corresponding fit by assuming a single species. Sometimes baseline corrections are required to obtain a satisfactory outcome.

3   Results

Four distinct procedures for multidomain protein analyses can be distinguished. These are defined by the domain arrangement and the nature of the linker peptides (if any) between the domains.

3.1   Type 1: Proteins with No Covalent Linkers

Type 1 proteins are essentially defined by the absence of a flexible protein linker joining the subunits or domains. As crystal structures are invariably known for many of these proteins, scattering and ultracentrifugation methods are less useful for determining their solution structures. This has meant that these proteins have been useful for calibration studies for the structural modelling of larger proteins. Scattering and ultracentrifugation techniques are however important to establish the state of oligomerisation in solution. Proteins in their crystallographic unit cells generally exist in a high salt or precipitant concentration used to grow the crystals. The oligomeric state of the protein in physiological buffer condtions is often unknown. At other times, oligomer formation can be unexpected, and may sometimes play a role in the biological functioning of the protein. Gel filtration methods to detect oligomers do not give unambiguous results, in particular when there are unusual features that may perturb the standard protein calibrations of the gel column. There is no control over the protein concentration within the column. Both scattering and ultracentrifugation give the required oligomerisation information, provided that sufficient instrument time is available to perform sufficiently detailed $I(0)$ scattering or equilibrium ultracentrifugation measurements. Data over at least a 10-fold concentration range is required for both the dissociated and associated forms. As seen from the following examples, either technique alone will work.

Monomer–dimer equilibria constitute the most common example of association. Figure 6 summarises those found in the complement fragment C3d, the two-domain single-chain antibody Fv fragment MFE-23, dimethylarginine dimethylaminohydrolase (DDAH) and the RuvA tetramer. While both scattering and ultracentrifugation detected these equilibria for C3d and MFE-23 in routine dilution series, in practice, all four equilibria were most readily studied in the home laboratory by a combination of gel filtration and analytical ultracentrifugation. The observed K_D values were 40 μM for C3d, 450 mM for MFE-23, 0.45 μM for DDAH, and 4 μM for RuvA, indicating the range of K_D values that can be studied by these methods.

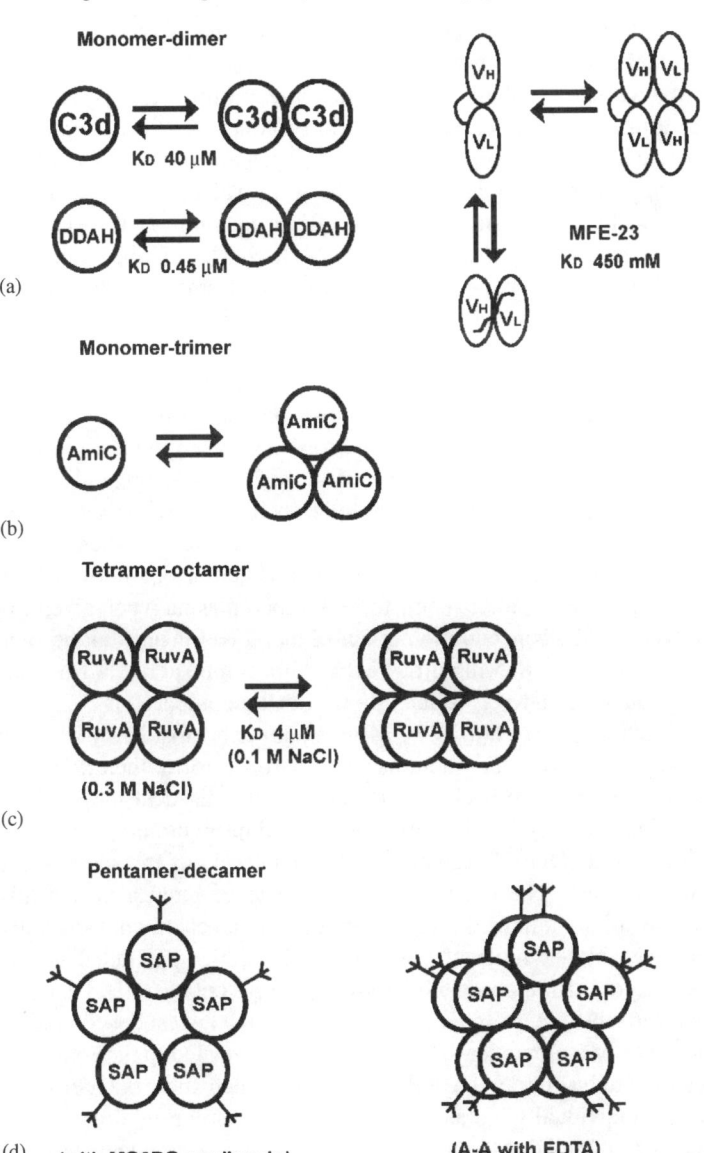

Figure 6 *Protein oligomers studied by scattering, ultracentrifugation and other methods. (a) Three examples of monomer–dimer equilibria, with the measured dissociation constants K_D indicated as shown. (b) One example of a monomer–trimer equilibrium. (c) An example of a tetramer–octamer equilibrium, in which the tetramer prevails in 0.3 M NaCl and an equilibrium with the octamer is established in 0.1 M NaCl. (d) SAP exists as a pentamer in the presence of MOβDG or human serum albumin. SAP becomes a decamer with the A faces (A for α-helices) of the pentamer in contact with each other in the presence of EDTA, and the B faces (B for Ca^{2+} binding sites) in contact with each other in the presence of dAMP. The single N-linked oligosaccharide chain per protomer is denoted by a 'ψ' symbol*

C3d is a 35 kDa cleavage fragment of C3, the major complement protein of immune defence. This is a ligand for the complement receptor type 2 (CR2) and its crystal structures showed that C3d was a monomer.[22] Surprisingly C3d was observed to exist as a monomer–dimer equilibrium in solution from scattering and ultracentrifugation data. The K_D was estimated to be 40 μM from the midpoint of the concentration dependence of the R_G and the $s^o_{20,w}$ values.[22] A full sedimentation equilibrium analysis gave an improved K_D determination of 20 μM from a multi-concentration multispeed experiment (V. A. A. Birch, H. E. Gilbert and S. J. Perkins, unpublished results). Since the C3 concentration in plasma is 1.3 mg mL^{-1}, and that of C3d is expected to be much less except in localised areas of increased concentration, C3d is expected to be monomeric in plasma.

The interest in the MFE-23 equilibrium rests in the therapeutic importance of this antibody fragment. MFE-23 has an nanomolar binding affinity for its ligand carcinoembryonic antigen (CEA), and is used to detect or eliminate colon cancer, for which CEA is a marker. ScFv fragments of antibodies consisting only of the V_H and V_L antibody domains in the heavy and light chains, respectively (see Section 3.3 below) show a prospensity to dimerise and this may affect its therapeutic application. The K_D of 450 mM determined by sedimentation equilibrium showed that MFE-23 will exist in the monomeric form in blood plasma when given to patients. The effective MFE-23 concentration in 5 L of blood within an adult becomes 0.2 ng mL^{-1} when a dose of 0.815 mg mL^{-1} MFE-23 is administered.[15] The combination of neutron scattering and $s^o_{20,w}$ values showed that a compact MFE-23 monomer is formed in solution [Figure 6(a)]. The MFE-23 dimer is most likely to be formed by the unfolding of this compact monomer into an open form, then its association to form a symmetric back-to-back dimer that was visualised in the MFE-23 crystal structure.[39] This pathway is schematically outlined in Figure 6(a).[15]

Enzymes such as DDAH regulate the cellular levels of metabolites that inhibit nitric oxide synthase. The DDAH equilibrium was of interest in that this protein crystallised as a homodimer through inter-α-helix sidechain contacts between two monomers. The higher molecular weight of the dimer precluded NMR studies. Analytical size-exclusion chromatography in a range between 0.05 and 5 mg mL^{-1} DDAH showed a difference in its elution profile at the lowest concentrations, showing the presence of a monomer–dimer equilibrium. This led to the equilibrium ultracentrifugation study of four DDAH mutants in which these inter-α-helix contact residues were individually mutated. The K_D values monitoring dimer formation was weakened with single DDAH mutants. The double DDAH mutant R40E and R98H was identified to be monomeric over a wide concentration range, and therefore was suitable for high-resolution multinuclear NMR studies.[40]

More complex equilibria that have been observed for Type 1 proteins include the monomer–trimer equilibrium in AmiC.[41] AmiC is the negative regulator of the amidase operon that binds to amides. Acetamide is the substrate and butyramide is an anti-inducer. Structurally AmiC resembles the two-domain periplasmic binding proteins that possess a large cleft between the two domains and undergo a large conformational change to close the cleft when binding to ligands. In order to explore these conformational changes on the binding of acetamide and butyramide ligands to AmiC, X-ray scattering experiments were performed. Unexpectedly, AmiC exhibited a monomer–trimer equilibrium [Figure 6(b)]. Sufficient protein was available for two X-ray beam sessions in which sufficient $I(0)/c$ data were acquired up to 17 mg mL^{-1}. Analyses of the $I(0)/c$ values for the acetamide- and butyramide-bound forms showed that the monomer–trimer association is stronger in the presence of acetamide. Scattering curve fits gave an association constant K_{a3} of 2×10^{10} M^{-2}. This illustrates how an equilibrium can be analysed using scattering data alone.

The tetramer–octamer equilibrium in RuvA illustrates the complementary nature of scattering and ultracentrifugation [Figure 6(c)]. The RuvA tetramer participates in DNA recombination and repair in bacteria by forming a complex with the four-way crossover intermediate of DNA called a Holliday junction. Crystal structures for RuvA alone and in complexes with four-way junctions revealed both tetrameric and octameric complexes of RuvA. Scattering studies were required to clarify which of these corresponded to their solution structures.[13] A protein–DNA complex cannot be readily interpreted by X-ray scattering as a result of the significantly different electron densities of protein and nucleic acid compared to that of the buffer. Neutron contrast variation studies were accordingly performed on RuvA, the DNA four-way junction and their complex in 65% ^{2}H$_2$O. In this buffer, the DNA component is invisible. In the standard concentration range and buffers of neutron scattering, unbound and complexed RuvA was octameric [Figure 6(c)]. The DNA in the complex was shown to be sandwiched between the two tetramers, and this was confirmed by constrained molecular modelling. This agreed with one of the crystal structures, but then begged the question of how the complex is assembled from its components.

Analytical ultracentrifugation was convenient to explore other conditions in detail to identify how tetrameric RuvA could be formed.[16] This work was not made easy by the lack of any detectable chromophores in RuvA, apart from the peptide signal at 230 nm. Nonetheless, RuvA was found to be tetrameric at the most dilute 0.2–0.5 mg mL^{-1} concentrations in 0.1 M NaCl buffer. The octamer that was seen at higher RuvA concentrations dissociated into tetramers in 0.3 M NaCl. This result explained how the other crystal structures showing tetrameric RuvA had been formed (see Lee *et al.*[16] for details). Furthermore, estimates of the amounts of RuvA present in bacterial cells suggested that RuvA exists in an equilibrium between the two forms *in vivo*, hence providing an explanation of a possible molecular mechanism for the assembly of this complex.

The three forms of serum amyloid P component (SAP) constitute a group in its own right for reason of the complexity of its associative behaviour [Figure 6(d)]. Scattering and other methods provided key insights. SAP is a plasma glycoprotein that is a calcium-dependent ligand-binding protein. SAP binds to all forms of amyloid fibrils *in vitro* and is present in amyloid deposits including those implicated in Alzheimer's disease. It also binds to DNA and chromatin. SAP is normally purified as a soluble decamer, but only in the presence of EDTA. A crystal structure for this form has not been forthcoming. Even though 2.5 mM calcium is found in plasma, purified SAP autoaggregates and precipitates in the presence of calcium. Neutron scattering showed that in the presence of a small ligand, methyl 4,6-O-(1-carboxyethylidene)-β-D-galactopyranoside (MOβDG), SAP is pentameric and soluble.[42] The pentamer could be crystallised and its structure determined.[43] One face of the pentamer contains an α-helix and an oligosaccharide chain and is termed the A-face. The other face contains the calcium binding site and is termed the B-face. Subsequently, SAP in the presence of dAMP was found to form decamers which could be crystallised. The crystal structure of this third form showed that the pentamers associated at this B-face.[44]

$I(0)/c$ scattering measurements on SAP demonstrated that the decamer in EDTA associated through its A-face.[45] These $I(0)/c$ values were unexpectedly lower when compared with the pentamer. It turned out that each A-face contained about 20 Trp residues. The interaction of 40 Trp residues at the interface in the A–A decamer altered the absorption coefficient of SAP used to determine c for the $I(0)/c$ values. The anomalous precipitation of SAP with calcium could not be resolved by scattering or ultracentrifugation. Instead the use of density gradients formed in undiluted human serum showed that SAP sedimented as pentamers not associated with any macromolecular ligands, irrespective of the presence or absence of calcium. The calcium-dependent aggregation of SAP was completely inhibited by physiological concentrations of human serum albumin (>2000-fold molar excess over SAP), in which SAP remained pentameric.[46] To date, no equilibrium constants for any pentamer–decamer conversions have been determined, indicating that the K_D is undetectably low.

3.2 Type 2: Two Rigid Protein Domains Joined by a Flexible Linker

The proteins considered here possess two rigid domains connected by a long flexible linker. Here, the importance of scattering and ultracentrifugation lies more in determining linker structures rather than the state of oligomerisation.

The single-chain Fv fragment MFE-23 is a candidate for a Type 2 classification, as this is formed from V_L and V_H domains linked by a 15-residue $(Gly_4Ser)_3$ linker [Figure 6(a)]. However, this predominantly behaves as a monomeric compact structure, as the two domains form close β-sheet-to-β-sheet contacts with each other, and this resulted in the Type 1 classification of this protein. Its weak tendency to dimerise was discussed above. This compact structure in solution was supported by both neutron scattering and $s^0_{20,w}$ modelling analyses.[15]

If the two domains of a two-domain protein do not form contacts with each other, then scattering and ultracentrifugation become important to define their solution structure. This is illustrated by the two proteins below. The first of these corresponds to the largest protein superfamily found in the complement proteins of the immune system, that of the short consensus/complement repeat (SCR). Each SCR domain possesses about 61 amino acid residues and two disulfide bridges. The SCR linker is defined by the number of residues between the final Cys residue of the preceding SCR domain and the first Cys residue of the following SCR domain. Most inter-SCR linkers are three or four residues in length. The CR2 SCR-1 and SCR-2 domains are of functional interest for reason of their binding with the C3d fragment. Interestingly, the linker between SCR-1 and SCR-2 in CR2 is eight residues, one of the longest in the complement proteins. In the intact CR2 protein with 15 SCR domains, four linkers are eight residues in length, and two are seven residues in length.[5]

Crystal structures for free CR2 SCR-1/-2 and its complex with C3d showed that the long linker permits SCR-1 and SCR-2 to fold back upon each other. In these, side-by-side contacts between SCR-1 and SCR-2 are observed [Figures 7(b) and 8(b)].[47,48] Many other linkers in crystal or NMR structures of SCR proteins are approximately linear or bent by varying degrees [Figure 7(b)]. In contrast to this, it was readily apparent from $s^o_{20,w}$ modelling that free CR2 SCR-1/-2 showed an extended arrangement.[49] This was investigated in more detail by scattering and ultracentrifugation modelling. Gilbert and co-workers analysed 9950 trial models for SCR-1 and SCR-2 connected in all orientations. This showed that both the R_G value and the full scattering curve were best explained by open V-shaped structures that are schematically represented by the cartoon of Figure 8(a).[22] The same result was also determined for the complex with C3d. CR2 SCR-1/-2 also showed an open V-shaped structure and both SCR domains were likely to be in contact with the surface of C3d [Figure 8(a)]. This agreed well with the outcome of mutagenesis studies that showed that both SCR-1 and SCR-2 interacted with C3d.[50] However, the best-fit scattering model showed a deviation when this was used to calculate the $s^o_{20,w}$ values. Discrepancies of 0.31 and 0.46 S between the calculated and experimental values for free CR2 SCR-1/-2 and its complex with C3d, respectively, were seen. This was best explained by postulating that flexibility in the SCR-1/-2 linker permitted many structures to exist [Figure 8(a)]. The more extended arrangements for the double-domain structure give a better account of modelling the $s^o_{20,w}$ values than the scattering curves. Models with a bend of 63° to 121° between SCR-1 and SCR-2 give a better account of the scattering curve [Figure 8(a)].

A more extreme example of flexibility between two linked domains is that shown by the fungal cellulases. Cellulases are enzymes that digest cellulose, and are comprised of a large catalytic domain attached by a long *O*-glycosylated linker to a small cellulose binding domain. Little is known about the linker, apart from noting that any shortening or deletion drastically reduced its enzymatic activity. X-ray scattering was used to show that these enzymes possess flexible and extended linkers. However, the small sizes of the linker and the cellulose-binding domain made it difficult to identify their conformations.[51] To rectify this, a recombinant cellulase chimera Cel6BA was synthesised, in which the N-terminal Cel6B catalytic domain and linker was fused with a C-terminal Cel6A linker and catalytic domain arranged in that order [Figure 8(c)]. The O-glycosylated linker is now 88 residues long. The $P(r)$ curve for such a two-domain molecule is expected to show two peaks, one at low r corresponding to all the inter-atomic distances within each catalytic domain, and one at large r corresponding to all the inter-domain inter-atomic distances. That this was not seen experimentally showed that Cel6BA demonstrated great variability in its linker conformation. This is schematically illustrated in Figure 8(c). The modelling of Cel6BA structures was achieved with randomized linkers generated by molecular dynamics. The weighting of Cel6BA models with interdomain separations of 0.6–12.9 nm in 0.1 nm steps and peaking at 1.5 nm successfully replicated most of the features seen experimentally.[23] The most compact linkers are the most stable, but this can unwind into longer conformations at a relatively low energy cost. This property would enable the cellulose-binding domain to remain fixed to the surface of its substrate, while permitting the catalytic domain to diffuse at will to hydrolyse fresh glycosidic bonds. Hence scattering is able to analyse conformational disorder.

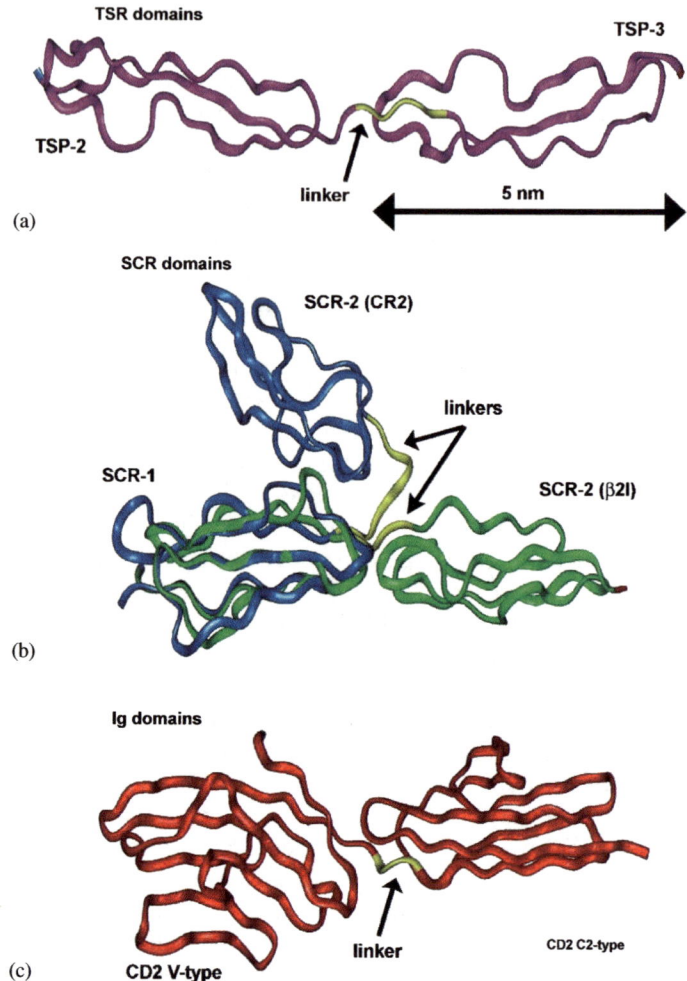

Figure 7 *Domain pairs in proteins studied by crystallography. Ribbon traces are shown, all drawn to the same scale. The three structures are arranged in order of the number of strands running from tip to tip. TSR domains are three-stranded, SCR domains are five-stranded, and Ig domains are seven or nine-stranded. In all cases, the linker peptide is highlighted in yellow. (a) Two TSR domains observed in human throm-bospondin, and used to construct homology models of the properdin dimer and trimer. A single TSR domain has two long antiparallel β-strands running the length of the TSR domain and a third disordered peptide chain, their arrangement being stabilised by three disulfide bridges. (b) Two SCR domains observed in the crystal structures of human complement receptor type 2 SCR-1/-2 and human β2-glycopro-tein I SCR-1/-2. These correspond to the most folded back and the most extended inter-SCR arrangement seen so far for SCR proteins. A single SCR domain has three pairs of short antiparallel β-strands, whose arrangement is stabilised by two disul-fide bridges at the ends of the SCR structure. (c) Two immunoglobulin folds seen in the crystal structure of CD2. The N-terminal V-type domain contains nine β-strands denoted A to G and arranged as two β-sheets DEBA and GFCC'C". The C-terminal C2-type domain contains seven β-strands arranged as two β-sheets DEBA and GFC*

Solution structures   **Crystal structures**

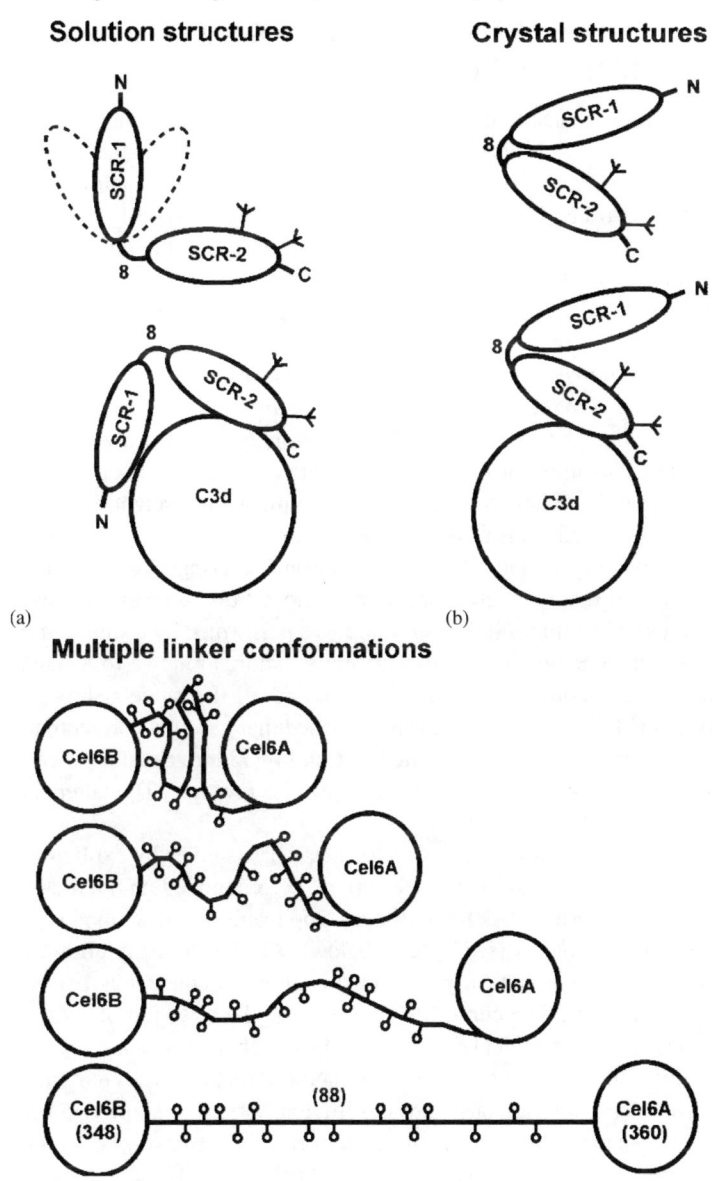

(a)   (b)

(c)

Figure 8 *Two domain proteins studied by scattering and ultracentrifugation. (a) A schematic outline of the solution structures of complement receptor type 2 SCR-1/-2 when free and when complexed with the complement fragment C3d. The oligosaccharide chains are denoted by 'ψ' symbols. The dashed outline of SCR-1 in free CR2 shows the range of fits obtained by constrained scattering modelling (adapted from ref. 22). (b) The corresponding outlines of crystal structures for the same two proteins. (c) A schematic outline of the solution structures of a recombinant two-domain protein created from two cellulases, namely the N-terminus of Cel6B and the C-terminus of Cel6A. The 88-residue linker peptide is O-glycosylated (denoted by O) at about 16 positions (adapted from ref. 22)*

3.3   Type 3: Three Rigid Protein Domains Joined by Two Flexible Identical Linkers

Antibody molecules constitute Type 3 proteins with great immunological and biotech-nological significance. An antibody is comprised of two light chains (with V_L and C_L domains) and two heavy chains (with V_H, C_H1, C_H2 and C_H3 domains). Overall, anti-bodies possess two Fab fragments that are usually joined to one Fc fragment by two hinge peptides (Figure 9). The IgE class is different in that the hinge is replaced by another domain pair. Generally for antibodies, their overall structure can be repre-sented by three relatively rigid structures connected by two flexible identical linkers. The averaged arrangement in solution will exhibit two-fold symmetry about the Fc fragment. This symmetry is not observed in crystal structures of intact antibodies, as these reveal only a single snapshot of one possible conformation. This symmetry means that only one linker conformation requires analysis, not two. As for Type 2 pro-teins, scattering and ultracentrifugation studies are most usefully focussed on structure determinations of the linker structure, rather than molecular weight determinations.

The first automated constrained scattering analysis of an intact antibody was reported for bovine IgG1 and IgG2.[52] The neutron scattering analyses at ISIS yielded R_G and R_{XS} values that were comparable with those from other antibodies. The con-strained modelling of this solution structure was performed by a simple translational search in two dimensions of two Fab crystal structures relative to a fixed Fc frag-ment. The hinge peptide was not considered, and all the models showed two-fold symmetry. An Fab–Fc arrangement was identified that gave good agreement with the observed neutron scattering curve. The IgG1 and IgG2 hinge peptides were then fit-ted to this arrangement using a molecular dynamics protocol. The latter showed that this solution structure was feasible.

The analysis of IgA antibodies extended these first studies by explicitly consider-ing the molecular structure of the hinge peptide before modelling the intact antibody. This became a powerful modelling strategy applicable to all antibodies as well as other multidomain proteins (see Type N below). All the sterically allowed antibody conformations were automatically generated using tight constraints before proceed-ing with the curve fits. The curve fits then become the selection of those structures that best fitted the experimental data. For the IgA1 subclass with a 23-residue O-gly-cosylated hinge, the experimental $P(r)$ curve showed that this has a maximum dimen-sion of 21 nm. This solution structure is sufficiently rigid to show two peaks in the $P(r)$ curve, unlike the case of the cellulases above. To model IgA1, 12 000 randomised hinge structures were created using molecular dynamics. These connected the Fab and Fc structures to make 12 000 trial two-fold symmetric IgA1 models in all orien-tations. Filtering based on (i) the neutron and X-ray R_G values and (ii) the absence of steric overlap between the Fab and Fc fragments initially reduced the 12 000 models to 867 models. This total was reduced further to 104 models by use of the R-factor goodness-of-fit parameter. A flexible T-shaped structure was determined, however, the hinge is sufficiently rigid to maintain this as an extended structure [Figures 9(a) and 11(a)].[12] For the IgA2 subclass with a shorter 10-residue hinge, a similar strategy was followed to analyse the IgA2m(1) isotype using 10 000 models. The use of filters for steric overlap, R_G and R_{XS} resulted in 104 best-fit models. IgA2 differs from IgA1

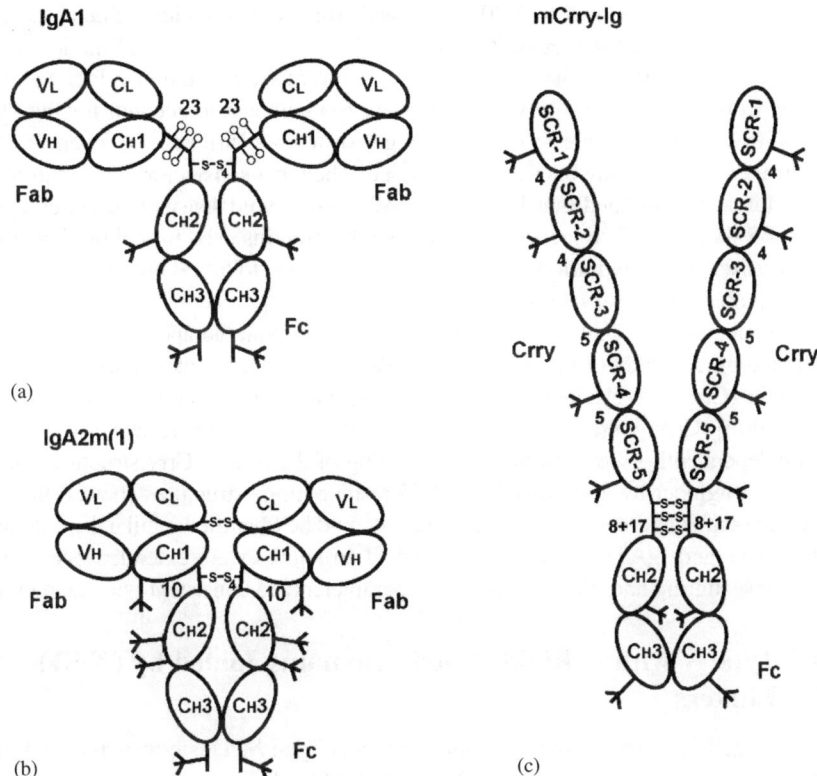

Figure 9 *Three fragment proteins studied by scattering and ultracentrifugation. All three are human or recombinant antibody molecules. An antibody is comprised of two light chains (with V_L and C_L domains) and two heavy chains (with V_H, C_H1, C_H2 and C_H3 domains). In solution, a vertical two-fold axis of symmetry about the Fc fragment can often be assumed. N-linked oligosaccharide chains are denoted by a 'ψ' symbol, while the O-glycosylation at the IgA1 hinge is denoted by 'O' at about 10 positions. (a) Human IgA1 subclass of immunoglobulin A, comprised of two Fab and one Fc fragments, at the centre of which four disulfide bridges ($S–S_4$) link the two halves of the structure together. (b) Human IgA2 subclass of immunoglobulin A, likewise comprised of two Fab and one Fc fragments. (c) A mouse chimaera antibody composed of a heavy chain with a five-domain N-terminal mouse complement-related receptor gene/protein Y (Crry) and a C-terminal mouse IgG1 Fc fragment linked by three disulfide bridges (adapted from refs. 12, 20 and 53)*

in that a disulfide bridge connects the two constant light-chain C_L domains in the Fab fragments [Figure 9(b)]. To handle this, the disulfide bridge was incorporated within all 104 best-fit models, after which its conformation was energy-refined to optimise its stereochemistry. About 75% of these best-fit models were lost because of excessive movement of the Fab fragments during the disulfide bridge refinement. The final best-fit models showed a predominantly T-shaped structure, but included some Y-shaped structures [Figure 9(b)].[20]

The best-fit scattering IgA1 and IgA2 structures were compared with the $s^o_{20,w}$ values.[20] The calculated value of 6.81 S for the IgA1 model was noticeably larger than the observed value of 6.15 to 6.20 S obtained from sedimentation velocity experiments. The calculated values of 7.32 to 7.82 S for the 104 best IgA2 models were larger than the observed value of 6.42 S obtained from sedimentation velocity experiments. Both the IgA1 and IgA2 models were too compact to account for the $s^o_{20,w}$ values. It was not possible to extend the IgA structures further to compensate for this, as they were already maximally extended. The most satisfactory explanation of this difference postulated that IgA1 and IgA2 possess void spaces into which extra hydration shells could be incorporated, unlike the proteins used for calibration studies. Increases in the hydration of IgA of as much as 300 to 500% brought the calculated and observed $s^o_{20,w}$ values into agreement.

The application of scattering and ultracentrifugation to a therapeutic antibody was illustrated by the study of mouse Crry-Ig.[53] Two mouse complement receptor-related Y protein (Crry) were joined to a mouse IgG1 Fc fragment by 25-residue hinges [Figure 9(c)]. Crry possesses a flexible five SCR-domain structure which was modelled independently (see Section 3.4). Modelling of the best-fit Crry structure joined to the Fc fragment by 2000 randomised 25-residue hinge structures showed that no preferred conformation for the Crry fragments could be identified, unlike the outcome of the IgA1 and IgA2 modelling. The best-fit structures were extended ones. This time, the scattering and ultracentrifugation parameters are consistent with each other.

3.4 Type N with N Rigid Protein Domains Joined by $(N-1)$ Linkers

The most general case of multidomain proteins is Type N. This covers the analyses of three or more domains (a total of N) joined by $(N-1)$ linkers in a linear arrangement. The aim of solution structural modelling for the Type N proteins is to define an averaged interdomain linker conformation as best as possible by constrained methods. Unlike Type 2 and Type 3 proteins, it is no longer possible to define a single linker conformation.

Nine diverse structures are illustrated in Figure 10. In contrast to the two- and three-fragment proteins considered above, these structures exhibit great variety. These proteins have three domains (PIWF1 fragment of anosmin-1), four domains (factor VIIa; FVIIa), five domains (Crry), six domains (PIWF4 fragment of anosmin-1), seven domains (properdin and CEA) and more, such as complement factor H (FH) with 20 SCR domains. Many multidomain proteins are constructed from small domains containing from about 40 to 100 residues. However, FVIIa is notable for having a large serine protease domain (SP) compared to the small EGF and Gla domains [Figure 10(a)]. Domains can be packed side by side as well as being in an extended linear arrangement, as exemplified by antibodies (Figure 9) and IgE–Fc [Figure 10(b)]. CEA is heavily glycosylated, containing 28 N-linked oligosaccharide chains and is 50% carbohydrate by weight [Figure 10(c)]. The anosmin-1 domains possess long interdomain linkers between 9 and 31 residues [Figure 10(d)], compared to other shorter linkers between 2 and 8 residues in Figure 10. The majority of SCR proteins possess linkers of 4 or 5 residues [Crry in Figure 10(e)]. However, FH contains longer linkers between SCR-10 and SCR-14 [Figure 10(f)]. Unusual oligomers can be formed, such as that of properdin which can form cyclic dimers and trimers [Figures 10(g) and (h)].

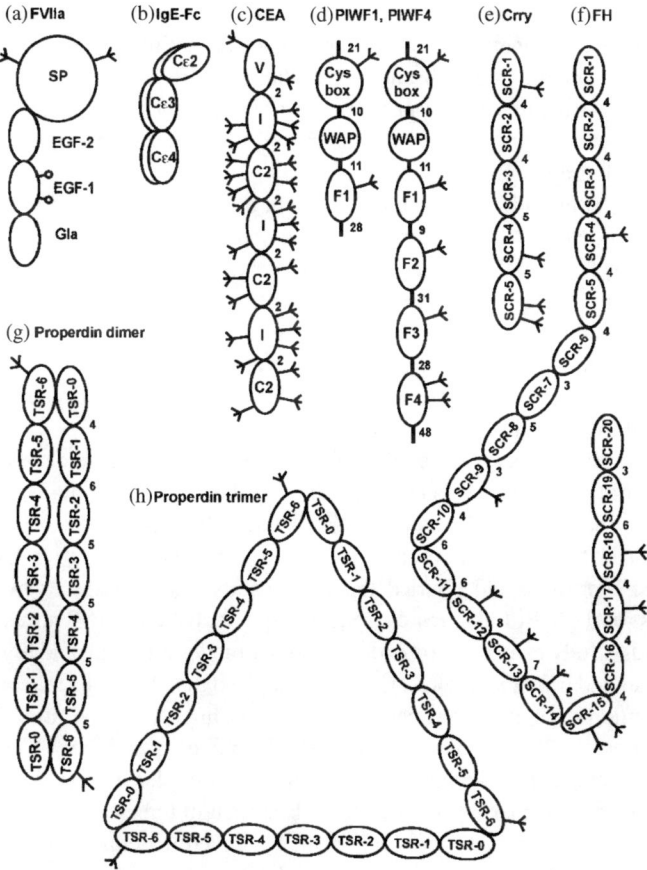

Figure 10 *Domain structures of multidomain proteins studied by constrained scattering and ultracentrifugation modelling. Where relevant, the residue length of the linkers joining two consecutive domains are indicated by numbers. N-linked oligosaccharide chains are indicated by a 'ψ' symbol. (a) Factor VIIa of the blood coagulation cascade is composed of a Gla and two epidermal growth factor (EGF) domains, followed by a catalytic serine protease (SP) domain. (b) The Fc fragment of the immunoglobulin E (IgE) antibody is schematically shown in the bent conformation first identified by fluorescence measurements, then determined quantitatively by scattering and ultracentrifugation fits, and subsequently confirmed by its crystal structure. (c) Carcinoembryonic antigen (CEA) is comprised of an N-terminal V-type immunoglobulin (Ig) domain, followed by an alternation of three I-type and C2-type Ig domains. There are 28 N-linked oligosaccharide chains in CEA. (d) The recombinant PIWF1 and PIWF4 forms of anosmin-1 are shown. Full-length anosmin-1 contains a whey acidic protein (WAP) domain, followed by a Cys-rich domain (Cys-box) and four fibronectin type III domains (F1 to F4). (e) The SCR domains of rat complement-related receptor gene/protein Y (Crry) contains five SCR domains that are similar to several SCR domains in complement receptor type 1. (f) Human factor H contains 20 SCR domains joined by linkers of lengths between 3 and 8 residues, schematically shown in a folded-back conformation as observed by scattering analyses. (g), (h) A human properdin dimer and trimer are shown, each comprised of a monomer with up to seven TSR domains*

Solution models for all these proteins have been developed by a combination of scattering and ultracentrifugation data with constrained modelling. Often these models are the only ones available in the absence of crystal structures. In the cases of FVIIa and IgE–Fc, crystal structures have appeared.[54,55] For both of these, the agreement between the solution scattering models[56–58] and the crystal structures are good, thereby validating the crystallographic models.

For the largest multidomain proteins, the joint use of scattering and ultracentrifugation facilitated determinations of their solution conformations. One practical problem is that their large size makes difficult the observation of the appropriate Guinier region at low Q values. The R_G values require careful determinations, and confirmatory agreement between both the X-ray and neutron values is desirable. Ultracentrifugation $s^0_{20,w}$ values are not limited in this way, and provide an independent test of the R_G values through the use of Figure 4. The addition of further R_G/R_O and f/f_0 pairs of values to Figure 4 will improve this check of consistency.

Short linkers are expected to be relatively inflexible, and to lead to a reasonably well-defined solution structure. This is exemplified by the TSR and Ig domains of Figures 7(a) and (c). The TSR domain in properdin is three-stranded with two β-strands. The TSR linker can be as long as four or six residues, being defined by a Cys disulfide bridge at its N-terminus and a buried Trp residue at its C-terminus as its two anchoring points. Most of the TSR linker residues are incorporated into the TSR structure at its C-terminal end. A fully extended properdin monomer model was built, and this was linked to form dimer and trimer structures. Inspection showed that the conformational searches for TSR solution structures in properdin dimers and trimers could be simplified into six rotational steps of 72° each about one ψ main chain angle per linker.[21] The final models showed extended TSR arrangements that indicated moderate inter-domain flexibility. The view of the trimer in Figure 11(e) is best seen in three-dimensions as the TSR domains do not form contacts with each other. The seven Ig domains in CEA belong to one of the V-type, I-type or C2-type groups of this superfamily [Figure 10(c)]. As illustrated in Figure 7(c), Ig domains are either nine-stranded (V-type) or seven-stranded (C2-type). CEA possesses short two-residue linkers that are defined as the non-β-strand residues found between the two adjacent Ig domains. A similar but less sophisticated inter-domain rotational strategy was successfully used to create trial CEA models for testing against scattering and ultracentrifugation data.[59,60] A simplified rotational and translational strategy was likewise used for modelling the IgE–Fc structure.[58]

Longer linkers are expected to be relatively unrestrained in their interdomain arrangements. The resulting solution structure can be a combination of many different structures that form a dynamic equilibrium with each other. The SCR proteins provide examples of this. Figure 7(b) shows how SCR domains are formed as five-stranded structures that incorporate three pairs of antiparallel β-strands in each. Figure 7(b) also demonstrates the extremes of inter-SCR conformational variability observed in crystal structures. Accordingly the constrained modelling of multidomain SCR proteins requires the use of molecular dynamics to define many randomised linker conformations. These conformations were then used to assemble randomised SCR arrangements for comparison with the scattering and ultracentrifugation data. In the two-domain case of CR2 SCR-1/-2, good curve fits could be obtained [Figure 11(c)]. In the five-domain case of rat Crry, a largely extended SCR arrangement was determined from these scattering fits [Figure 11(d)], that

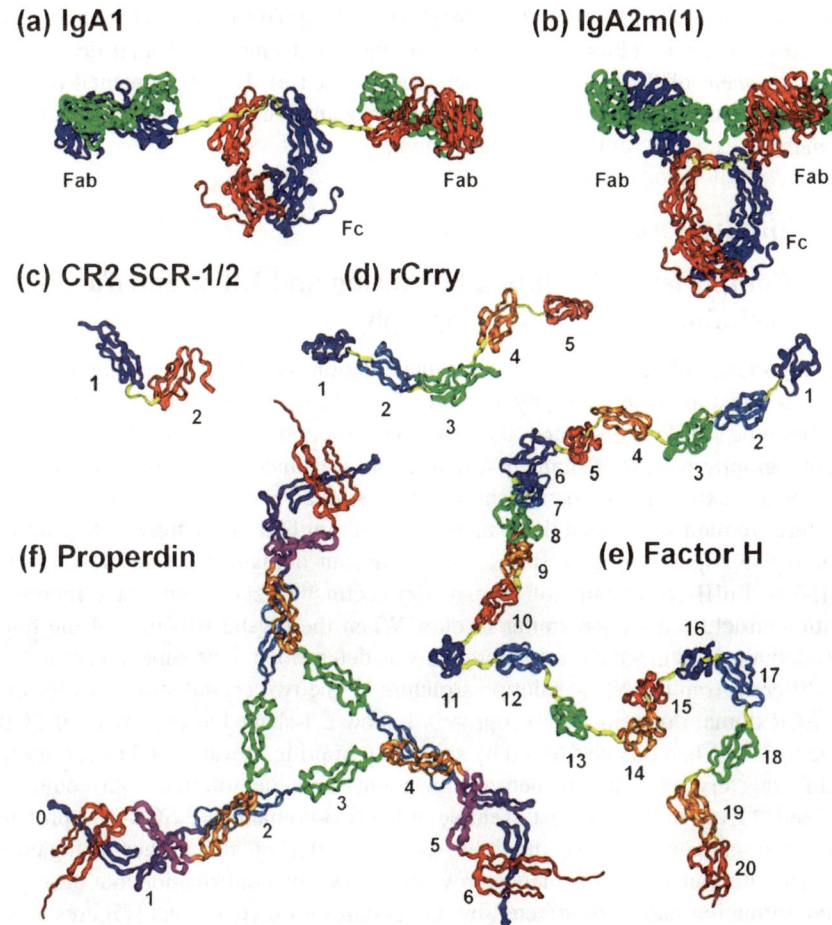

(a) IgA1

Fab Fab Fc

(b) IgA2m(1)

Fab Fab Fc

(c) CR2 SCR-1/2

1 2

(d) rCrry

1 2 3 4 5

(f) Properdin

0 1 2 3 4 5 6

(e) Factor H

1 2 3 4 5 6 7 8 9 10 11 12 13 14 15 16 17 18 19 20

Figure 11   *Types 2, 3 and N protein solution structures resulting from the constrained modelling of the scattering and ultracentrifugation data summarised in this chapter. All the structures are drawn to the same scale, unlike the cartoons of Figures 6, 8, 9 and 10. The two-fold symmetric T-shaped human IgA1 and IgA2 structures are shown at the top, with their Fab and Fc fragments labelled. Three SCR proteins are shown, namely human CR2 SCR-1/-2, rat Crry and human factor H. The SCR domains are numbered to correspond to the schematic cartoons of Figures 8 and 10. The properdin trimer is shown at the bottom, with the seven TSR domains numbered to correspond to that of Figure 10*

agreed well with the NMR structure of the corresponding SCR domains in human complement receptor type 1.[53] The complexity of the modelling for the 19 linkers in FH was reflected by the limited number of models that provided a reasonable curve fit (about 31 out of 2010 trial models, of which 4 were published).[61] In addition, good curve fits only extended up to a smaller Q value of about 0.4 nm^{-1} compared to fits obtained for other structures [such as that for IgA2 up to a Q value of about 1.3 nm^{-1} in Figures 1(a) and (b)]. Nonetheless it was possible to show that the FH structure is sufficiently flexible to form predominantly folded-back structures [Figure 11(e)]. The maximum length of these structures is 43 nm in solution, compared to the maximum length of 70 nm if FH is a linear arrangement of SCR domains. Other proteins such as the developmental protein anosmin-1 have even longer linkers. Here, scattering analyses suggested that anosmin-1 has multiple extended conformations in solution.[24]

4 Conclusions

4.1 Comparison of Solution Scattering and Ultracentrifugation Techniques with Crystallography

The importance of scattering and ultracentrifugation methods for structural analyses, in contrast to crystallography, is shown from the comparison between the crystal structures of Figure 7 and the solution structures of Figure 11. A recent crystallography review of domain structure and organisation in multidomain proteins of the extracellular matrix showed that single- and double-domain crystal structures remain very much the standard (as exemplified in Figure 7). Only occasionally was it possible to crystallise and solve four-domain structures such as the FnIII-7 to FnIII-10 domains of human fibronectin.[62] Other groups have reported solution structure findings similar to ours. When the crystal structure of the four SCR domain protein vaccinia coat protein was determined, it became apparent that this differed from the NMR solution structure.[63] The two crystal structures for the five SCR domain protein β2-glycoprotein I show a J-shaped arrangement of SCR domains.[64,65] When this was tested by solution scattering, it was found necessary to modify this crystal structure before agreement with the solution data could be obtained.[66] Even in the simplest example of the two-domain CR2 SCR-1/-2 protein (discussed in detail above), the joint use of scattering and ultracentrifugation together with mutagenesis evidence provided a powerful confirmation that the crystal conformation had been affected by its crystallisation conditions [Figures 7(b), 8(a), (b) and 11(c)]. Scattering and ultracentrifugation do not only provide the first views of the structures of intact multidomain proteins (Figure 11), but also often provide the opportunity to refine and improve the understanding of crystal structures for these multidomain proteins.

4.2 What Do We Learn from Combining Both Approaches?

The combination of scattering and ultracentrifugation data improves the reliability of the information content provided by the use of either technique on its own. This relates to the common parameters, namely the comparison of the R_G and the $s^o_{20,w}$ values, or the $I(0)/c$ and equilibrium M_r determinations. Given the ease of performing ultracentrifugation measurements in the home laboratory, compared to the need to book and travel for neutron and X-ray beamtime at multiuser facilities, the ultracentrifuge has obvious advantages. Thus the study of associations in the Type 1 proteins is easier using the ultracentrifuge alone. In contrast, the study of the Types 2, 3 and N multidomain proteins benefit from a joint scattering and ultracentrifugation strategy. The additional structural parameters provided outside the Guinier R_G range make scattering the more powerful structural tool. Here, the utility of the ultracentrifuge is to provide confirmatory insight into the solution structures determined by scattering.

The combination of both scattering and ultracentrifugation data provides unexpected insight into the protein when the modelling of the independent data sets using tried and tested algorithms result in discrepancies. So far, two have arisen, each of which will require further analysis. One is based on the postulate of a higher than expected hydration level of 0.3 g H_2O (g protein)$^{-1}$. This was proposed for data on IgA1 and IgA2 and for the properdin dimer. It may be correlated with voids within the structure or with an abundance of charged groups that are heavily hydrated.[20,21] The other is when the assumption of a single well-defined solution structure breaks down, whereupon multiple structures may freely co-exist with each other in solution. This was inferred to be the case for discrepancies seen with CR2 SCR-1/-2 and for the PIFW1 and PIWF4 fragments of anosmin-1. There the interdomain linkers are longer than usual and presumably permit greater interdomain conformational flexibility.[22,24]

Acknowledgements

We thank with gratitude our biochemical collaborators with whom we have published our analyses, whose names are listed in the references. We thank the Biotechnology and Biological Sciences Research Council (H. E. G.), the Clement Wheeler Bennett Trust (Y. C. L.) and the Wellcome Trust (Z. S. and P. B. F.) for grant support, Dr S. Finet (ESRF, Grenoble), Dr J. G. Grossman (SRS, Daresbury), Dr R. K. Heenan (ISIS) and Dr S. M. King (ISIS) for excellent scattering instrumental support, Dr J. T. Eaton for excellent computational support, and Mr J. Gor for excellent assistance with the analytical ultracentrifuge.

References

1. S. J. Perkins, A. W. Ashton, M. K. Boehm and D. Chamberlain, *Int. J. Biol. Macromol.*, 1998, **22**, 1.
2. S. J. Perkins, C. G. Ullman, N. C. Brissett, D. Chamberlain and M. K. Boehm, *Immunol. Rev.*, 1998, **163**, 237.
3. S. J. Perkins, in *Protein-Ligand Interactions: A Practical Approach*, B. Chowdhry and S. E. Harding (eds), Oxford University Press, Oxford, 2000, 223.
4. S. J. Perkins, *Fibre Diffr. Rev.*, 2001, **9**, 51.
5. S. J. Perkins, H. E. Gilbert, M. Aslam, J. P. Hannan, V. M. Holers and T. H. J. Goodship, *Biochem. Soc. Transact.*, 2002, **30**, 996.
6. S. J. Perkins and P. B. Furtado, in *Structural Biology of the Complement System*, D. Morikis and J.D. Lambris (eds), Marcel Dekker, New York, 2005, in press.
7. H. M. Berman, J. Westbrook, Z. Feng, G. Gilliland, T. N. Bhat, H. Weissig, I. N. Shindyalov and P. E. Bourne, *Nucl. Acids Res.*, 2000, **28**, 235.
8. O. Glatter and O. Kratky, *Small-Angle X-ray Scattering*, Academic Press, New York, 1982.
9. A. V. Semenyuk and D. I. Svergun, *J. Appl. Crystallogr.*, 1991, **24**, 537.
10. S. E. Harding and D. J. Windzor, in *Protein-Ligand Interactions: A Practical Approach*, B. Chowdhry and S. E. Harding (eds), Oxford University Press, Oxford, 2000, 75.
11. D. J. Windzor and S. E. Harding, in *Protein-Ligand Interactions: A Practical Approach*, B. Chowdhry and S. E. Harding (eds), Oxford University Press, Oxford, 2000, 105.

12. M. K. Boehm, J. M. Woof, M. A. Kerr and S. J. Perkins, *J. Mol. Biol.*, 1999, **286**, 1421.
13. D. Chamberlain, A. Keeley, M. Aslam, J. Arenas-Licea, T. Brown, I. R. Tsaneva and S. J. Perkins, *J. Mol. Biol.*, 1998, **284**, 385.
14. D. K. McRorie and P. J. Voelker, *Self-Associating Systems in the Analytical Ultracentrifuge*, Beckman Instruments, Palo Alto, 1993.
15. Y. -C. Lee, M. K. Boehm, K. A. Chester, R. H. J. Begent and S. J. Perkins, *J. Mol. Biol.*, 2002, **320**, 107.
16. Y.-C. Lee, R. Flora, J. McCafferty, J. Gor, I. R. Tsaneva and S. J. Perkins, *J. Mol. Biol.*, 2003, **333**, 677.
17. H. Durchschlag and P. Zipper, *J. Appl. Crystallogr.*, 1997, **30**, 1112.
18. S. J. Perkins, *Eur. J. Biochem.*, 1986, **157**, 169.
19. S. J. Perkins, *Biophys. Chem.*, 2001, **93**, 129.
20. P. B. Furtado, P. W. Whitty, A. Robertson, J. T. Eaton, A. Almogren, M. A. Kerr, J. M. Woof and S. J. Perkins, *J. Mol. Biol.*, 2004, **338**, 921.
21. Z. Sun, K. B.M. Reid and S. J. Perkins, *J. Mol. Biol.*, 2004, **343**, 1327.
22. H. E. Gilbert, J. T. Eaton, J. P. Hannan, V. M. Holers and S. J. Perkins, *J. Mol. Biol.*, 2005, **346**, 859.
23. I. von Ossowski, J. T. Eaton, M. Czjzek, S. J. Perkins, T. P. Frandsen, M. Schülein, P. Panine, B. Henrissat and V. Receveur-Bréchot, *Biophys. J.*, 2005, **88**, 2823.
24. Y. Hu, Z. Sun, J. T. Eaton, P. M. G. Bouloux and S. J. Perkins, *J. Mol. Biol.*, 2005, **350**, 553.
25. T. Narayanan, O. Diat and P. Bosecke, *Nucl. Instrum. Methods Phys. Res. A*, 2001, **467–468**, 1005.
26. P. Lindner, R. P. May and P. A. Timmins, *Physica B*, 1992, **180**, 967.
27. http://www.ill.fr/YellowBook/D22/
28. R. K. Heenan, J. Penfold and S. M. King, *J. Appl. Crystallogr.*, 1997, **30**, 1140.
29. R. Giebeler, in *Analytical Ultracentrifugation in Biochemistry and Polymer Science*, S. E. Harding, A. J. Rowe and J.C. Horton (eds), Royal Society of Chemistry, Cambridge, UK, 1992, 16.
30. A. Furst, *Eur. Biophys. J.*, 1997, **35**, 307.
31. P. Zipper and H. Durchschlag, *J. Appl. Crystallogr.* 2003, **36**, 509.
32. T. M. Laue, B. D. Shah, T. M. Ridgeway and S. L. Pelletier, in *Analytical Ultracentrifugation in Biochemistry and Polymer Science*, S.E. Harding, A. J. Rowe and J. C. Horton (eds), Royal Society of Chemistry, Cambridge, UK, 1992, 90.
33. J. Philo, *Anal. Biochem.*, 2000, **279**, 151.
34. P. Schuck, *Biophys. J.*, 1998, **75**, 1503.
35. B. Carrasco and J. Garcia de la Torre, *Biophys. J.*, 1999, **76**, 3044.
36. S. E. Harding, J. C. Horton and H. Cölfen, *Eur. Biophys. J.*, 1997, **25**, 347.
37. M. L. Johnson, J. A. Correia, D. A. Yphantis and H. R. Halvorton, *Biophys. J.*, 1981, **36**, 575.
38. P. Schuck, *Anal. Biochem.*, 2003, **320**, 104.
39. M. K. Boehm, A. L. Corper, T. Wan, M. K. Sohi, B.J. Sutton, J. D. Thornton, P. A. Keep, K. A. Chester, R. H. J. Begent and S. J. Perkins, *Biochem. J.*, 2000, **346**, 519.
40. M. J. Plevin, B. S. Magalhaes, A. Sankar, S. J. Perkins and P. C. Driscoll, *J. Mol. Biol.*, 2004, **341**, 171.
41. D. Chamberlain, B. P. O'Hara, S. A. Wilson, L. H. Pearl, and S. J. Perkins, *Biochemistry*, 1997, **36**, 8020.
42. S. P. Wood, G. Oliva, B. P. O'Hara, H. E. White, T. L. Blundell, S. J. Perkins, I. Sardharwalla and M.B. Pepys, *J. Mol. Biol.*, 1988, **202**, 169.
43. J. Emsley, H. E. White, B. P. O'Hara, G. Oliva, N. Srinivasan, I. J. Tickle, T. L. Blundell, M. B. Pepys and S. P. Wood, *Nature*, 1994, **367**, 338.
44. E. Hohenester, W. L. Hutchinson, M. B. Pepys and S. P. Wood, *J. Mol. Biol.*, 1997, **269**, 570.

45. A. W. Ashton, M. K. Boehm, J. R. Gallimore, M. B. Pepys and S. J. Perkins, *J. Mol. Biol.*, 1997, **272**, 408.
46. W. L. Hutchinson, E. Hohenester and M. B. Pepys, *Mol. Med.*, 2000, **6**, 482.
47. G. Szakonyi, J. M. Guthridge, D. Li, K. Young, V. M. Holers and X. S. Chen, *Science*, 2001, **292**, 1725.
48. A. E. Prota, D. R. Sage, T. Stehle and J. D. Fingeroth, *Proc. Natl. Acad. Sci. USA*, 2002, **99**, 10641.
49. J. M. Guthridge, J. K. Rakstang, K. A. Young, J. Hinshelwood, M. Aslam, A. Robertson, M. G. Gipson, M.-R. Sarrias, W.T. Moore, M. Meagher, D. Karp, J. D. Lambris, S. J. Perkins and V. M. Holers, *Biochemistry*, 2001, **40**, 5931.
50. J. P. Hannan, K. A. Young, J. M. Guthridge, R. Asokan, G. Szakonyi, X. S. Chen and V.M. Holers, *J. Mol. Biol.*, 2005, **346**, 845.
51. V. Receveur, M. Czjzek, M. Schulein, P. Panine and B. Henrissat, *J. Biol. Chem.*, 2002, **277**, 40887.
52. M. O. Mayans, W. J. Coadwell, D. Beale, D. B. A. Symons and S. J. Perkins, *Biochem. J.*, 1995, **311**, 283.
53. M. Aslam, J. M. Guthridge, B. K. Hack, R. J. Quigg, V. M. Holers and S. J. Perkins, *J. Mol. Biol.*, 2003, **329**, 525.
54. D. W. Banner, A. D'Arcy, C. Chene, F. D. Winkler, A. Guha, W. H. Konigsberg, Y. Nemerson and D. Kirchofer, *Nature*, 1996, **380**, 41.
55. T. Wan, R. L. Beavil, S. M. Fabiane, A. J. Beavil, M. K. Sohi, M. Keown, R. J. Young, A. J. Henry, R. J. Owens, H. J. Gould and B. J. Sutton, *Nature Immunol.*, 2002, **3**, 681.
56. A. W. Ashton, G. Kemball-Cook, D. J. D. Johnson, D. M. A. Martin, D. P. O'Brien, E. D. G. Tuddenham and S. J. Perkins, *FEBS Lett.*, 1995, **374**, 141.
57. A.W. Ashton, M. K. Boehm, D. J. D. Johnson, G. Kemball-Cook and S. J. Perkins, *Biochemistry*, 1998, **37**, 8208.
58. A. J. Beavil, R. J. Young, B. J. Sutton and S. J. Perkins, *Biochemistry*, 1995, **34**, 14449.
59. M. K. Boehm, M. O. Mayans, J. D. Thornton, R. H. J. Begent, P. A. Keep and S. J. Perkins, *J. Mol. Biol.*, 1996, **259**, 718.
60. M. K. Boehm and S. J. Perkins, *FEBS Lett.*, 2000, **475**, 11.
61. M. Aslam and S. J. Perkins, *J. Mol. Biol.*, 2001, **309**, 1117.
62. E. Hohenester and J. Engel, *Matrix Biol.*, 2002, **21**, 115.
63. K. H. M. Murthy, S. A. Smith, V. K. Ganesh, K. W. Judge, N. Mullin, P. N. Barlow, C. M. Ogata and G. J. Kotwal, *Cell*, 2001, **104**, 301.
64. R. Schwarzenbacher, K. Zeth, K. Diederichs, A. Gries, G. M. Kostner, P. Laggner and R. Prassl, *EMBO J.*, 1999, **18**, 6228.
65. B. Bouma, P. G. de Groot, J. M. H. van den Elsen, R. B. G. Ravelli, A. Schouten, M. J. A. Simmelink, R. H. W. M. Derksen, J. Kroon and P. Gros, *EMBO J.*, 1999, **18**, 5166.
66. M. Hammel, M. Kriechbaum, A. Gries, G.M. Kostner, P. Laggner and R. Prassl, *J. Mol. Biol.*, 2002, **321**, 85.

CHAPTER 16

Modelling of Biopolymers

PETER ZIPPER, HELMUT DURCHSCHLAG AND
ANGELIKA KREBS

1 Introduction

Over the years, the investigation of biopolymers in aqueous solution has provided valuable information on the size, shape and function of many classes of molecules, and the interest in this topic is still increasing. The fabrication of tailored pharmaceuticals is only one concrete example in a long wish list of feasible biotechnical and medicinal applications. Therefore, detailed knowledge of the biochemical and physiological processes in cells is mandatory. This, of course, also requires that the structural details and mechanisms of all species involved in the interplay are known.

The discovery and inspection of structural refinements have also stimulated the desire for comparing the results obtained by different techniques, thereby revealing molecular properties and data in solution or the solid state (in the latter case, preferably in the form of crystals). Among the techniques supplying low- and high-resolution information, methods such as small-angle scattering (SAS), light scattering, classical electron microscopy (EM) and the hydrodynamic techniques (analytical ultracentrifugation (AUC) and viscometry), on the one hand, and crystallography, nuclear magnetic resonance (NMR) spectroscopy, and advanced EM techniques, on the other, are the most important, as is the desire for possibilities to critically assess the results obtained by these techniques. This demand also determines the claim for highly efficient modelling approaches.[1,2]

Owing to many technical improvements and an arsenal of new instrumentation, the results obtained by modern solution techniques are much more accurate and comprehensive than in former times. For example, solution scattering employing X-rays may be performed nowadays by using synchrotron radiation, and AUC is currently enjoying a renaissance. Above all, modern computer facilities offer a gigantic potential of possibilities, including the performance of many modelling approaches in the area of solution biophysics. Currently, we can dare to tackle *ab initio* problems even by using personal computers.

Proteins of different size and nature (monomeric cellobiose dehydrogenase, dimeric phosphorylase and a multimeric annelid haemoglobin) and one representative of the class of nucleic acids (RNA from the bacteriophage MS2) have been selected for demonstrating the efficacy of advanced modelling techniques for compiling small-angle X-ray scattering (SAXS) and hydrodynamic data.[2] The previously mentioned biopolymers were subjected to some *ab initio* modelling programs (DAMMIN,[3] DALAI_GA[4,5] and SAXS3D[6]), which are based on SAXS data, and to program HYDRO,[7,8] an outstanding representative of a hydrodynamic bead modelling approach for the prediction of hydrodynamic molecular properties. The present study shows how efficiently these programs work, in the reconstruction of 3D structures and the retrieval of both structural and hydrodynamic properties.

2 Modelling Approaches

In the last decades, a variety of modelling techniques have been described, both for the X-ray field (diffraction and solution scattering) and the area of hydrodynamics (AUC, viscometry etc.). The approaches applied include simple whole-body (WB) approaches, multibody (MB) procedures, and, more recently, advanced *ab initio* modelling techniques (for summaries see refs. 2, 9, 10). In the following we confine ourselves primarily to the description of approaches that are based on SAS data and are followed by predictions of hydrodynamic data. Above all, the applied procedures aim at the reconstruction of particle shapes in solution and the comparison of data from SAS and hydrodynamic analyses for establishing the solution structure of biopolymers, preferably proteins.

2.1 Conventional Whole-Body and Multibody Modelling Approaches

As a first approximation, biopolymers can be modelled as spheres, prolate or oblate ellipsoids of revolution (PE, OE) or tri-axial bodies with unequal axes. Approaches along these lines are called WB modelling.[9,11–18] They consider hydration contributions by using experimental molecular properties such as the hydrated volume, V, obtained by SAXS or by applying realistic assumptions for the amount of hydration, δ_1;[10,19] alternatively, hydration-independent size-functions may be adopted.[13] In this context, it has to be emphasised that the surface of WB models is entirely smooth.

Since proteins and other types of biopolymers are composed of a number of various constituents, some kind of MB modelling seems to meet this demand more appropriately, *e.g.* see refs. 1, 2, 7–9, 20–49. Modelling the structures of these biological objects naturalistically means to use an assembly of spheres (beads) instead of a single geometrical body. The target experimental profile is thus simulated by a finite element method ("bead modelling"). The approach may be applied to both small numbers of objects (*e.g.* protein subunits) and large numbers of spheres at definite coordinates (*e.g.* of atoms or groups of atoms). Scattering profiles, $I(h)$, are usually calculated by means of Debye's formula.[50]

Since currently the atomic coordinates of thousands of biopolymers are available in a number of databases, modern modelling approaches can utilise this kind of information advantageously for modelling purposes, *e.g.* see refs. 1, 8, 29, 32, 35–38, 40–42, 44, 51, 52. Calculation problems, however, may arise from some circumstances: (i) a preceding data reduction step is required, if a huge number of beads, N_b, has to be processed for the hydrodynamic prediction of data ($N_b>10\ 000$); (ii) handling of differently sized spheres may require application of a special type of interaction tensor for hydrodynamic modelling if the spheres overlap; and (iii) hydration contributions have to be considered for both scattering and hydrodynamic modelling, particularly if small biopolymers are under analysis. Data reduction steps may be managed by merging several spheres (*e.g.* the atoms belonging to one amino acid residue or a definite number of units along a protein backbone) into one bigger sphere, advantageously by a cubic or hexagonal grid approach.[1,42] The usage of overlapping spheres of different size may become inevitable if both scattering and hydrodynamic data have to be predicted by filling-model strategies;[9,10] the numerical problem occurring may be overcome by application of our *ad hoc* expression for the interaction tensor.[24,42] Hydration can be taken into account by rescaling procedures, by modelling an additional water shell or by modelling individual water molecules.[10,19,53–55]

In the absence of given coordinates, however, the modelling problem stringently leads to a trial-and-error approach, to some kind of educated guess at best. For many years SAXS made a virtue of necessity by applying to the modelling procedure as much information as possible and including also the information content of other techniques (*e.g.* from EM or hydrodynamics).[56] A fine tuning of bead models can be achieved by changes of the input variables, preferably of the arrangement (coordinates), number N_b and radius r_b of beads.

Density distributions derived from 3D reconstructions from cryo-EM may also be exploited favourably for MB bead modelling intentions.[2,45–48,57] A direct application of the 3D-EM reconstructions implies an elaborate analysis of the density distribution to differentiate between anhydrous and hydrated models by use of definite density thresholds. In contrast, the use of 3D-EM information for creating structural templates used in advanced modelling approaches is an indirect application of EM density data.

2.2 Structure Reconstruction from SAXS Profiles by *Ab Initio* Modelling Approaches

A fascinating aspect of modern evaluation procedures for SAXS data is the possibility of establishing *ab initio* reconstructions of low-resolution biopolymer shapes (*cf.* refs. 2, 58–63) without resorting to any kind of spatial information, in contrast to the procedures used in the past. Among these highly advanced methods, in particular, the approaches based on simulated annealing (SA),[3,64,65] a genetic algorithm (GA),[4,5] Monte Carlo approaches (MC)[6,66] or a molecular dynamics algorithm (MD)[67] have to be addressed.

(i)   The program DAMMIN[3] generates models from densely packed dummy atoms (beads). In the case of SAXS (and absence of contrast-variation data) each dummy atom is ascribed either to the particle or the solvent. Starting from a random initial configuration in a chosen search space (in general, a sphere of diameter slightly exceeding the particle diameter), SA is used as a global minimization algorithm to find a configuration matching the SAXS data. To calculate scattering curves, spherical harmonics are used. The system is continuously cooled down until no further improvement is observed.

The comfortable advantage of the DAMMIN program is well founded in the possibility of including additional information (expert mode) such as particle symmetry (point group), anisometry (prolate/oblate/unknown) and use of structural templates (replacement of the default spherical search space), obtained, *e.g.* from EM or atomic data. DAMMIN prefers compact dummy configurations; loose and disconnected configurations are banned. The program permanently uses the whole scattering curve for fitting and equally sized spheres for modelling.

The program GASBOR[64,65] employs SA similar to DAMMIN, but uses a chain-like ensemble of dummy residues to mimic the backbone of the protein structure and dummy water molecules to simulate hydration. The input of experimental data for the programs DAMMIN and GASBOR is entered by means of a GNOM file.[68]

(ii)   An iterative fitting of scattering curves may also be accomplished by a GA, using the program DALAI_GA.[4,5] A population of genes codifying a given mass distribution on a hexagonal lattice in a confined volume (*e.g.* an ellipsoidal search space of selected dimensions) is randomly generated. The obtained model structure from each genotype is compared with the observed SAXS data (by calculating the scattering profile of the model body by the Debye approach), and a fitness criterion is used to generate the next population by genetic operators (crossover and mutation), until, as a consequence of the selection pressure, the system converges, *i.e.,* the best-fitting model is found. For fitting, DALAI_GA starts with a low number of large spheres and uses incipiently only the innermost portion of the scattering curve; during the run, the size of the spheres is gradually scaled down and the limit of resolution is increased by addition of further portions of the scattering curve. Consequently, the volume of the model may change with the progress of the GA procedure.

(iii)   Two programs, SAXS3D[6] and SASMODEL,[66] are based on MC-type approaches without limitations of the search space.

The program SAXS3D[6] represents an MC-type reconstruction algorithm (an MC simulation at zero temperature) and uses a "give-and-take" algorithm to add and remove beads on a hexagonal lattice (lattice parameter l) until an optimum fit to the experimental SAXS profile is reached and the value of the score is at its minimum. Scattering curves are calculated according to Debye's formula, however, by employing a pair-distance histogram quantised to multiples of $1/20$ of l and neglecting the form factor of the spheres. For calculations, the entire scattering curve is used throughout.

Another MC-based approach, which tests a large number of models against the experimental scattering data, is the program SASMODEL.[66] It does not use a fixed grid, but generates a chain of ellipsoids whose total volume is constrained to be approximately that of the protein to be modelled. The conformation of the chain is randomly selected by rotating each ellipsoid about its origin, defined as the tip of the previous ellipsoid. The model is constrained to be contiguous, but is flexible enough to allow the creation of various random shapes. A later development, program GA_STRUCT,[69] combines GA and MC techniques. The program produces a family of several *ab initio* models composed of spheres of arbitrary radii that are not constrained to lie on a grid.

(iv) The restrained MD algorithm[67] utilises the sum of squared differences between calculated and experimental SAXS intensities as a constrained energy function, and starts from a set of given coordinates. Currently, the method may be used for further refinement of a pre-determined structure, *e.g.* of models obtained by other *ab initio* methods. However, it should also be adaptable for independent *ab initio* modelling after incorporating elements such as a residue-based force field.

All the programs under consideration take hydration into account eventually, by fitting the experimental SAXS profile that already comprehends hydration contributions.

Developing a structural model from SAXS data requires the execution of multiple runs for each condition chosen (*e.g.* bead radius, *i.e.*, resolution) to avoid misinterpretations due to unfavourable or ill-posed calculation conditions. A comparison of the results requires visualisation, alignment and superimposition of the obtained models, followed by some kind of averaging and filtering. In any case, the resultant models can by no means guarantee absolute uniqueness.[70] In particular, in the case of very complex structures, or if considerable particle inhomogeneities exist, a note of caution is advised.

2.3   Model Selection and Generation of Averaged Models

For automated best-matching alignment and superimposition of the bead models obtained and subsequent averaging, the program package DAMAVER[70] was applied. This package contains the programs DAMSEL, DAMSUP, DAMAVER and DAMFILT, and makes use of SUPCOMB.[71] All model pairs are compared; a reference model is selectively given the smallest average dissimilarity measure and mavericks are discarded. All remaining models are then superimposed onto the reference model, and the assemblage of chosen models is averaged and mapped onto a densely packed grid of equally sized spheres. The occupancy factors of the grid points are used for filtering the averaged model at a given cut-off volume by neglecting grid points of extremely low occupancy.

However, averaging was also performed by alternative procedures. Our programs DAMMIX (for direct averaging) and DAMHEXMIX (for averaging by mapping onto a grid of hexagonal cells) were used to generate averaged models where the different probabilities of spheres are expressed either by different densities of equally sized spheres or by different sphere volumes (implying different r_b). The volume, V, and the radius of gyration, R_G, of the latter models are identical with the corresponding mean values of the models used for averaging. Optionally, beads of low probability (below an adjustable threshold) may be sorted out; in particular, in the context of models composed of equally sized spheres, this filtering is used to decrease the initially large volume. Our averaging approach was used without a preceding alignment step if the models were already in a fixed orientation, *e.g.* owing to the use of templates in the DAMMIN analysis; if this was not the case, the alignment was performed by DAMSEL and DAMSUP.

The philosophy of the applied averaging procedures has to be scrutinised to avoid loss of information, on the one hand, and over-interpretation of alleged structural details, on the other. In doing so, model averaging may be a legitimate procedure to accumulate overall information, to emphasise the most persistent particle features and to cope with the problem of uniqueness of shape reconstructions.[48]

2.4 Visualisation and Surface Representation of Models

The various models obtained can be visualised by any 3D molecular graphics program, preferably by application of the program RASMOL.[72] In addition, surface representation of the models turns out to be a helpful tool, particularly with respect to the hydrodynamic features of the molecules under analysis. Surface renderings were generated by means of programs from the CCP4 suite,[73] primarily designed for structure determination by protein crystallography. Electron density maps were created from the coordinates of the respective spheres and a threshold was chosen in order to simulate a model volume that is identical to the volume of the bead model or the experimentally derived SAXS volume. Finally, the electron densities were displayed by the program VOLVIS.[74]

2.5 Prediction of Scattering and Hydrodynamic Parameters by Bead Modelling Strategies

The structural parameters R_G and V and many other quantities can be derived directly from the SAXS profiles. The scattering intensity, $I(h)$, can be converted to a pair-distance distribution function (PDDF), $p(r)$, by Fourier transformation[50] (where $h = (4\pi \sin \theta)/\lambda$, θ = half the scattering angle, λ = wavelength of the incident X-rays, r = the distance between pairs of scatterers and $p(r) \equiv 0$ for $r > d_{max}$ (maximum particle diameter)). Scattering curves of MB models can be calculated by means of Debye's formula;[50] an alternative calculation procedure by means of spherical harmonics[75] is employed, for instance, by the programs DAMMIN[3] and GASBOR.[64,65] PDDFs may also be computed directly from the coordinates of the bead models.[76]

Owing to the different computation procedures used in the various *ab initio* modelling approaches, a critical comparison of the SAXS profiles generated by the modelling programs with the experimental profiles, and with model profiles obtained independently by alternative calculations starting from the coordinates of the bead models, can be important for assessing the computational consistency and plausibility of models and SAXS profiles.

For hydrodynamic modelling, the hydrated models had to be reduced to a manageable number of beads if the number of beads exceeded 10 000. Sedimentation coefficients, s, translational diffusion coefficients, D, and, in the case of MS2 RNA, intrinsic viscosities, $[\eta]$, were calculated by means of a modified version of the program HYDRO,[7,8] making allowance for the overlapping of nonequal spheres.[24,42] To derive both scattering and hydrodynamic quantities, we used "filling models" strategies.[8–10,23] Shell models,[8,22,23,25] obtained by removal of the internal beads, would have provided a more accurate prediction of some hydrodynamic properties (viscosities and rotational quantities), but were not suitable for our purposes.

2.6 Computational Requirements

For our model calculations, personal computers with Intel Pentium IV (3.2 GHz and 2 GB RAM) or AMD 3000+ (2.2 GHz and 1 GB RAM) processors were used; for surface rendering, a SGI workstation was applied. The CPU time for calculations with the *ab initio* programs depended on the program type and the input variables (size of molecule and number of beads, calculation modes). Runs of SAXS3D amounted to up to 2*d*; typical CPU times for application of DAMMIN and DALAI_GA varied between 2 h and 1 d. HYDRO runs for models with 8000 beads took about 4 h; the time required for calculations with smaller sphere numbers was much lower.

3 Biopolymers

In order to test different modelling approaches, proteins exhibiting different molecular features (size, shape, compactness, number of domains and subunits) have been chosen. The same procedures used to study compact proteins have been applied to one representative of the group of nucleic acids (MS2 RNA), which is known to have a rather loose solution structure as compared with typical globular proteins. Only examples for which both SAXS results and hydrodynamic properties have been described have been chosen. In particular, scattering intensity, $I(h)$; PDDF, $p(r)$; and the molecular parameters, molar mass, M; partial specific volume, $\bar{v}$; radius of gyration, R_G; hydrated volume, V; maximum particle diameter, d_{max}; sedimentation coefficient, s; and translational diffusion coefficient, D, have been presented for the biomolecules used. If available, the atomic coordinates from crystallographic work (CDH domains) or the results from 3D reconstructions from cryo-EM (*L.t.* Hb) were additionally exploited. In the case of MS2 RNA, the intrinsic viscosity, $[\eta]$, has been considered as well.

3.1 Phosphorylase

Phosphorylases are key enzymes in carbohydrate metabolism. Glycogen phosphorylases catalyse the degradation of glycogen and structurally related storage polysaccharides.[77–79] They are homodimers (M about 180 kg mol^{-1}), and the cofactor pyridoxal 5′-phosphate is indispensable for enzymatic activity. The phosphorylase from the soil bacterium *Corynebacterium callunae* differs in several aspects from structurally similar phosphorylases, *e.g.* by its preference for polysaccharides (starch) over small maltodextrin.

Starch phosphorylase from *C. callunae* (StP) has been investigated by SAXS,[80] and its solution structure was characterised by a flat, slightly elongated, ellipsoidal model exhibiting two-fold symmetry, generally similar to the maltodextrin phosphorylase 3D model for the *E. coli* enzyme,[81,82] but slightly different from the SAXS model for phosphorylase from rabbit muscle.[83] While crystallographic studies[81,82,84,85] on several glycogen phosphorylases from various sources, including rabbit muscle, yeast and *E. coli*, revealed their 3D structure, attempts to characterise the enzyme *C. callunae* have not yet been successful. There exist no hydrodynamic results for the enzyme from *C. callunae*, whereas various sedimentation velocity data for enzymes from other sources have been reported, *e.g.* refs. 83, 86, 87.

For phosphorylases, no WB or MB modelling attempts applying both structural and hydrodynamic data have been undertaken till now.

3.2 Cellobiose Dehydrogenase

Cellobiose dehydrogenase (CDH) is suggested to play a multifunctional role in the breakdown of lignocellulose by degrading both lignin and cellulose, presumably by a hydroxyl-radical-generating mechanism.[88] The extracellular redox enzyme from the white-rot fungus *Phanerochaete chrysosporium* is a monomeric glycoprotein (*M* about 100 kg mol^{-1}) consisting of two well-defined domains: an *N*-terminal haeme-containing domain and a *C*-terminal dehydrogenase domain with flavin adenine dinucleotide (FAD) as a cofactor. The two domains are connected via a peptide hinge of considerable extension, which acts as a linker ("spacer") and provides the flexibility required for the enzymatic function of this flavocytochrome.

In solution, linear as well as kinked conformations have been found to be compatible with SAXS data.[89] The SAXS studies were performed on both the holoprotein and the isolated haeme and FAD domains. Combining the results for these three investigated entities resulted in a CDH model of elongated structure. The SAXS studies also revealed the existence of the linker region between the two collinearly arranged domains. AUC experiments[90] allowed the establishment of an *s* value for the holoprotein. While the crystal structures of the isolated domains have been solved,[91,92] atomic coordinates for the holoprotein and the linker are not available to date. Uncertainties regarding the extent of protein glycosylation and the localisation of glycan chains and linker obviously complicate the straightforward elucidation of clear-cut 3D structures.[91]

Previous modelling studies included both WB and MB approaches to simulate linear and bent CDH solution structures. Apart from application of the WB trials,[1,9,16,18] SAXS trial-and-error[1,89] and advanced *ab initio* modelling approaches (DAMMIN, DALAI_GA, SAXS3D)[2,45] were applied to predict the scattering behaviour of CDH and to generate low-resolution 3D reconstructions. A recent modelling attempt also included application of DAMMIN to the FAD and haeme domains.[49] To anticipate the hydrodynamic behaviour of the MB models, program HYDRO was used.

3.3 Multisubunit Haemoglobin

Haemoglobin (Hb) is encountered in all the kingdoms of living organisms.[93] The Hbs of invertebrates represent a special branch in the group of oxygen-carrying haeme proteins. In contrast to human Hb (*M*=64 kg mol^{-1}, 4 subunits: $\alpha_2\beta_2$), invertebrate Hbs range from small, single-chain globins to large, multisubunit, multidomain Hbs. Among the latter group, the extracellular annelid Hbs are giant molecules (*M* about 3500 kg mol^{-1}) composed of both globin and nonglobin chains, and feature a hexagonal bilayer (HBL) appearance.[94] The Hb from the common earthworm *Lumbricus terrestris* (*L.t.* Hb) is the most extensively studied representative of this class. The crystal structure of *L.t.* Hb at a resolution of 5.5 Å confirmed the D$_6$ symmetry[95] and revealed a hierarchical organisation of 144 oxygen-binding Hb chains assembled into 12 dodecamers arranged at the periphery of the complex around a central scaffold formed by 36 nonHb subunits.[96] A recently completed 2.6Å resolution crystal structure of the dodecameric subassembly[97] represents the first step in understanding the strong allosteric properties of *L.t.* Hb.

SAXS investigations of *L.t.* Hb[57] aimed at establishing a detailed model for the HBL complex. The obtained consensus model (consisting of spheres of unequal weights, thereby representing an individual probability of the spheres and simulating an inhomogeneous density distribution) was compared with 3D reconstructions from cryo-EM.[98–101] The SAXS-based and EM-based models showed far-reaching conformity and led to similar particle dimensions. Furthermore, SAXS studies of the oxy and deoxy forms of *L.t.* Hb and of its dodecameric subunit revealed a lack of significant structural alterations between the two oxygenation states.[102,103] The consensus model for the dodecamer[103] established a design similar to the 3D structure obtained by X-ray crystallography.[97] Recent SAXS studies on other types of annelid Hbs (*Eudistylia vancouverii* chlorocruorin, *Macrobdella decora* Hb) revealed slight, but significant differences between these giant molecules with respect to the structure/hollowness in the central part of the complexes and their subunit architecture;[47,48] surface rendering turned out to be helpful for visually comparing the different architectures.[48] A plethora of AUC analyses have been performed on *L.t.* Hb, both on the intact complex and its subunits. In particular, numerous values for sedimentation and diffusion coefficients have been published and date back to the early work of the Svedberg group; for a compilation of the most relevant values see ref. 48. AUC experiments also included sedimentation velocity studies of the dodecamer under varying conditions.[103]

Because of its special architecture and its eminent importance, the annelid Hb complexes have already been subject of many WB and MB modelling attempts. In the case of *L.t.* Hb, WB approaches were applied to both the complex[1,9] and the dodecameric subunit.[16,18] Apart from SAXS trial-and-error modelling attempts[1,46,48,57] and several 3D-EM reconstructions,[98–101] the validity of advanced modelling procedures has been scrutinised in the case of the three previously mentioned annelid Hb complexes.[1,2,45–48] Again, HYDRO was used to predict the hydrodynamic properties of the Hb species under analysis.

3.4 Viral Ribonucleic Acid

The phage MS2 is a small, roughly spherical RNA virus that infects a variety of bacteria, including *Escherichia coli*. The genome of the RNA phages encodes the amino acid sequence of the phage proteins and regulates translation and replication, among other tasks. The RNA of MS2 is fully single stranded, 3569 nucleotides long (M about 1200 kg mol^{-1}) and contains a high degree of secondary structure (about 73% base pairing).[104] The secondary structure of the whole viral RNA resembles a "bouquet of flowers", obviously caused by foldings in several parts of the molecule.[104]

SAXS investigations[105] on free MS2 RNA established an elongate, flat particle of 62 nm diameter and of remarkable loose substructure, suggesting some kind of flexibility. Two cross-sectional regions and one thickness range were observed in the scattering curves; the cross-sectional region at low angles and the thickness range were ascribed to the overall shape of the molecule, and the cross-sectional region at large angles to the substructure presumably consisting of single-stranded and double-helical regions. In the course of further scattering and AUC experiments, the presence of ions turned out to influence the compactness of the RNA molecule; in particular, Mg^{2+} induced a considerable tightening of the structure.[106,107] Recent SANS studies[108] on the MS2 virion aimed at elucidating the spatial relationship between the protein shell and genomic RNA: the RNA was found to be tightly compacted within the virion. The 3D structure of the virion has been solved;[109,110] many interactions between the RNA and the inside of the protein shell can occur.

Hydrodynamic experiments of both the virion and its components comprised both sedimentation velocity runs and viscosity measurements; in the case of the MS2 RNA molecule, they, of course, refer to the free molecule.[111]

WB approaches have been applied for numerous modeling attempts on the free MS2 RNA.[9,16,18,105,111] To be compatible with experimental hydrodynamic data, however, they necessitated the treatment as a special case. The volume of a model corresponding to the overall shape instead of the correlation volume obtained by SAXS had to be taken into account in order to consider the hydrodynamically effective hydration influenced by cavities and particle inhomogeneities properly. A sophisticated MB approach yielded more realistic coil models,[105] mimicking both the tertiary structure and the substructure of the RNA. The coordinates of the beads were generated on the basis of random number configurations and a given set of geometrical parameters; the scattering curves calculated for a large number of models were averaged to match the experimental scattering profile.

4   Results and Discussion

The *ab initio* MB modelling approaches DAMMIN, DALAI_GA and SAXS3D were applied to the previously specified proteins (StP, CDH, *L.t.* Hb) and one nucleic acid (MS2 RNA). The applied reconstruction techniques have already proved to be able to yield more or less reliable models as well as convergence criteria.[2,63] They were tested in their ability to fit given target SAXS profiles and to generate realistic bead models of molecules of different natures. Scattering curves were computed on the basis of the multibody models and were compared with the curves provided directly by the reconstruction programs as well as with the experimental curves. The fine structure of the models achieved was also compared with previous trial-and-error reconstructions. To illustrate the progress of the modelling procedures in an even more persuasive manner, snapshots of the evolving models have been depicted in some cases. To quantify the results, both scattering and hydrodynamic parameters were predicted and compared with observed data and the results of simple WB approaches.

4.1   Phosphorylase from *C. callunae*: Probing the Applicability of *Ab Initio* Modelling Approaches

In the course of SAXS studies and conventional trial-and-error modelling attempts, the enzyme StP turned out to have a relatively simple shape in solution[80] (Figure 1). Therefore, this protein was tested in a case study to illustrate and compare the effectiveness of different *ab initio* modelling approaches.

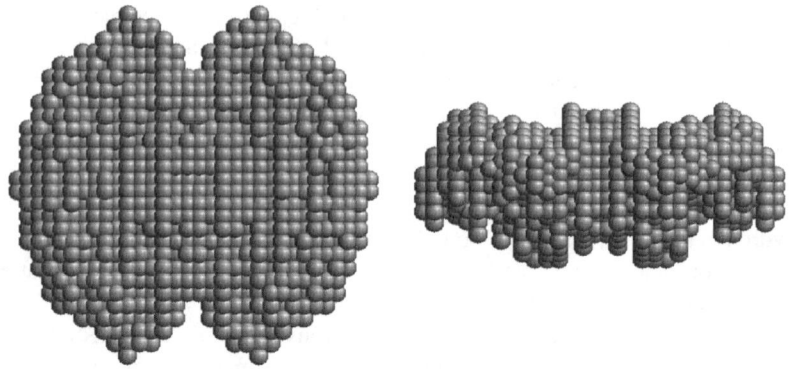

Figure 1 *Top and side views of a SAXS-based model for the StP structure ($N_b=4612$, $r_b=0.248$ nm), generated by previous trial-and-error modelling approaches*[80]

The evolution of DAMMIN models, recorded at selected consecutive cycles, impressively shows the transition from the initial spherical configuration to a loosened structure, and finally yields an oblate model, without significant shape improvements during the last steps of the SA process (Figure 2). This gradual change in the structure is paralleled by characteristic changes in the number of beads, N_b, constituting the respective model: a pronounced diminution of N_b during the initial cycles is followed by a moderate increase in the late phase of evolution. As may be expected, the corresponding SAXS profiles, the scattering intensities, $I(h)$, and PDDFs, $p(r)$, exhibit drastic changes during the first 50 cycles, but converge afterwards perfectly to the experimental functions (Figure 3). Consequently, the quantification of the results in terms of molecular properties (V, R_G, d_{max}, s, D) discloses major changes in most parameters under consideration during the initial cycles, similar to the alterations of N_b (Figure 4).

A tour through selected DALAI_GA models created in turn unveils the advancements in the development of these models, consisting of increased numbers of beads, N_b, and a concomitant decrease in bead radii r_b (Figure 5). Of course, the progress in modelling by DALAI_GA is accompanied by the creation of more detailed structures as compared with the start model. A visual inspection of the resulting SAXS profiles (Figure 6) yields a situation similar to that observed for the SA approach: marked alterations of the $I(h)$ and $p(r)$ profiles at the beginning of the GA process (# <9000), but highly converging profiles afterwards. Again, the alterations of the molecular properties (V, R_G, d_{max}, s, D) during the evolution process reveal the most important changes in the parameters during the initial steps (Figure 7). In contrast to the DAMMIN procedure, however, the progress of modelling is accompanied by drastic changes in both N_b and r_b.

Modelling StP by the MC approach SAXS3D and analysing the intermediary results bespeaks some aspects of the program proceeding. Increasing step numbers are attended by elevated numbers of beads and diminished scores (Figure 8). In this context, however, it has to be noted that the step numbers count the number of successful iterations but not the time required. Indeed, reaching a more or less final stage shapes up as a very time-consuming and tedious task.

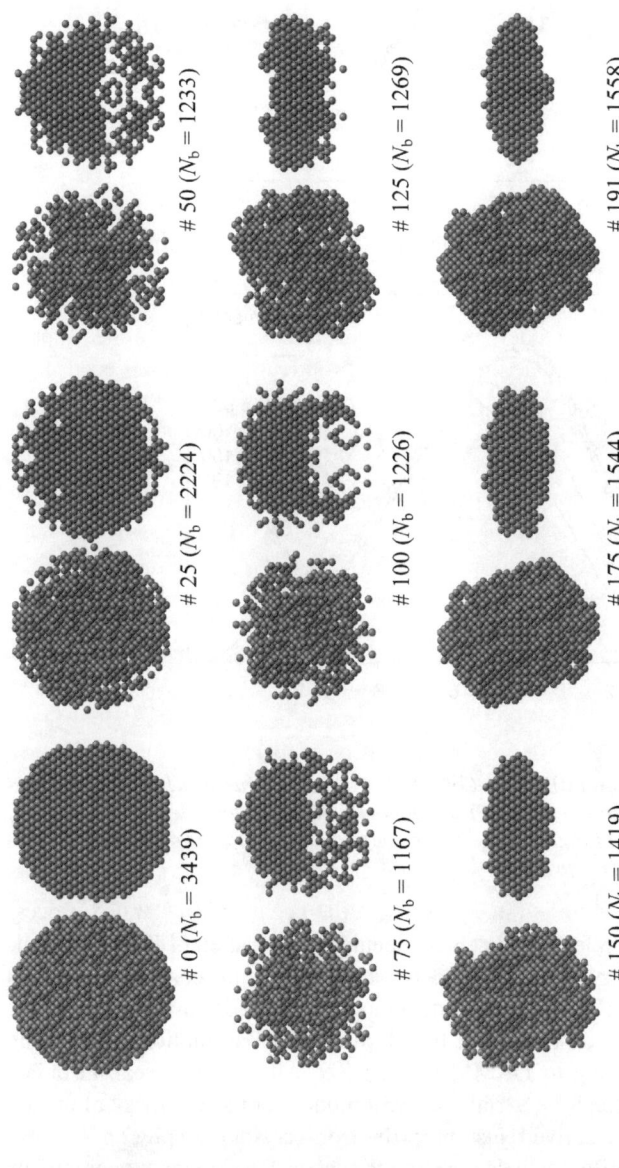

Figure 2 *Evolution of a DAMMIN model of StP. The figure shows snapshots of the bead models, taken at selected steps of the SA process, in two different views (along z and y). The gradual change of the model structure from the initial spherical configuration (# 0), composed of a high number of beads (N_b) with radius $r_b=0.3$ nm, to the final optimised configuration (# 191) is convincingly demonstrated. The considered DAMMIN analysis of the experimental scattering curve of the protein was performed by using C_2 symmetry (around z) and preference to oblate shape as constraints (code P2OL)*

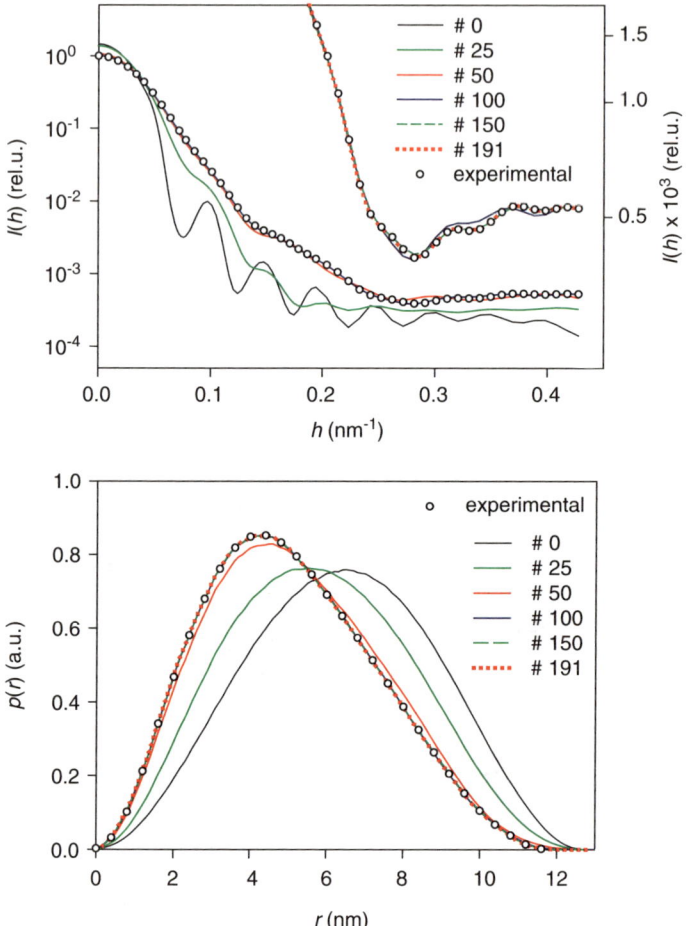

Figure 3   *Evolution of a DAMMIN model of StP. The scattering curves I(h) (top) and pair-distance distribution functions p(r) (bottom) calculated for selected snapshots of the bead models of Figure 2 at different steps of the SA process are compared with the corresponding experimental functions*

Summarising the most relevant models of a detailed application of DAMMIN calculations furnishes a gallery of similar models, irrespective of the constraints or averaging procedures applied (Figure 9: individual models; Figure 10: averaged models), provided that the applied constraints are based on realistic assumptions concerning symmetry (C_2 corresponding to DAMMIN code P2). All models reveal an oblate particle shape, similar to the SAXS trial-and-error model[80] or the 3D image of related glycogen phosphorylases derived just recently from crystallography.[81,82,84,85] By contrast, the application of unrealistic constraints such as P3–P6 may reveal models that are irrelevant from the biological point of view. Surface representations of several models under analysis add weight to the previous considerations concerning the shape of the protein. The overall impression is that of an oblate spheroid without

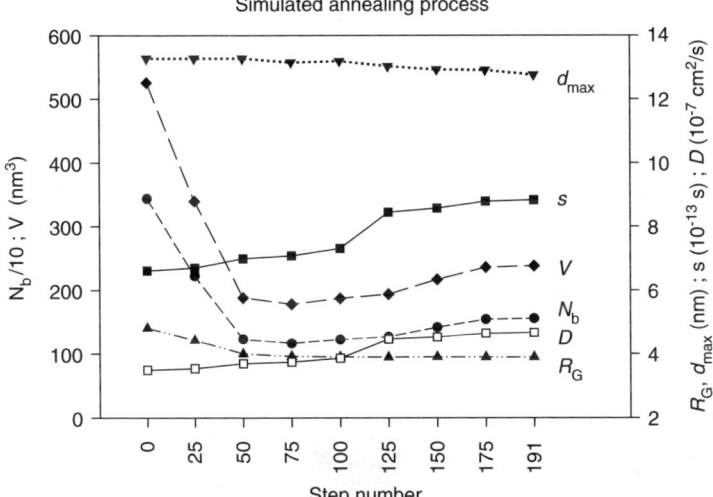

Figure 4 *Evolution of the number of beads (N_b) during the SA process illustrated in Figure 2 for DAMMIN modelling of StP, together with the alterations of the resulting molecular parameters V, R_G, d_{max}, s and D*

significant structural peculiarities (Figure 11). A thorough analysis of the SAXS profiles shows convincing agreement between experimental profiles, $I(h)$ and $p(r)$, and the SAXS patterns derived from the various models outlined (except the models representing averaged structures) (Figure 12). Similar results were obtained in the case of testing the GA procedure. When applying the SAXS3D algorithm, however, we were confronted with a series of acute problems, probably owing to the fact that the algorithm used in the available program is suboptimal for certain types or sizes of molecules. After applying some modifications of the initial algorithm with respect to the internal calculation of scattering curves (by using a smaller bin width for the distance histogram and inclusion of the form factor of the beads) and performing numerous runs, the best models obtained for StP, again, supply the expected oblate structure (Figure 13). The experimental scattering curve can only be approximated reasonably by models if modifications of the algorithm are applied (Figure 14).

A quantification of the results obtained by DAMMIN, DALAI_GA and SAXS3D corroborate the statements concerning tracking putative shapes and predicting scattering and hydrodynamic molecular parameters (Table 1). All *ab initio* approaches applied yield proper predictions of the molecule properties under consideration, and the same holds true for the adoption of a WB approach.

4.2 Cellobiose Dehydrogenase from *P. chrysosporium*: Modelling the Holoprotein and its Domains and Docking Manoeuvres

Based on the results of SAXS studies[89] of the holoprotein and the isolated domains, previous trial-and-error models[1,89] of the holoprotein were generated by uniting the

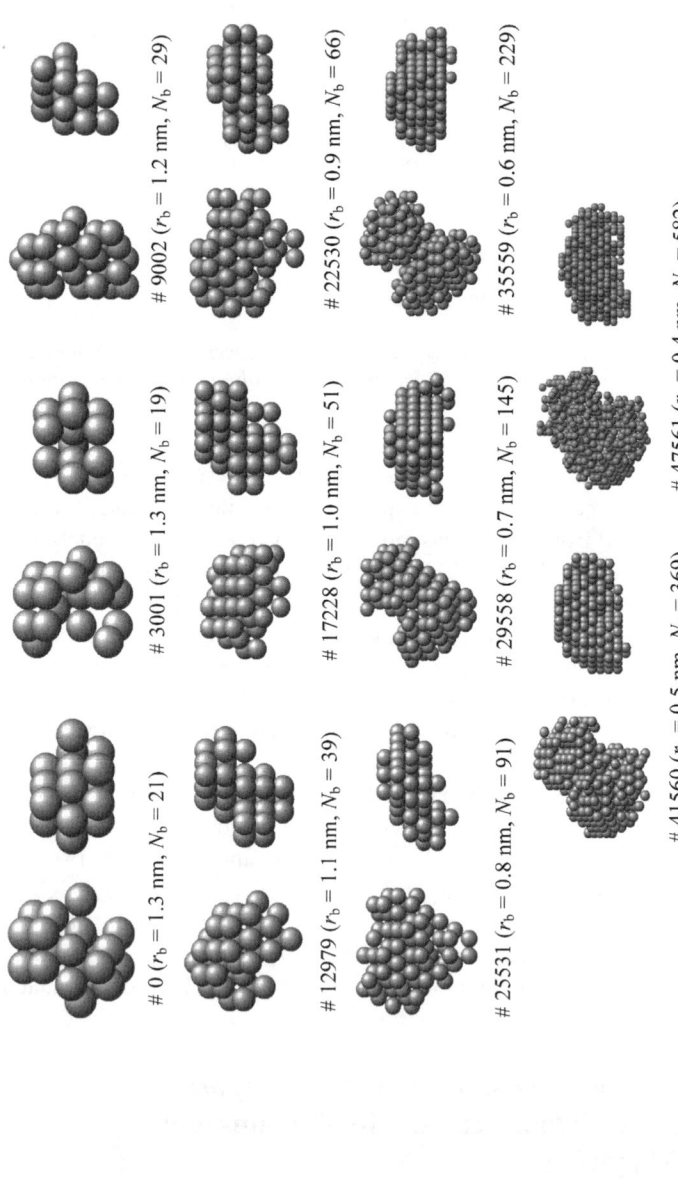

0 (r_b = 1.3 nm, N_b = 21)

3001 (r_b = 1.3 nm, N_b = 19)

9002 (r_b = 1.2 nm, N_b = 29)

12979 (r_b = 1.1 nm, N_b = 39)

17228 (r_b = 1.0 nm, N_b = 51)

22530 (r_b = 0.9 nm, N_b = 66)

25531 (r_b = 0.8 nm, N_b = 91)

29558 (r_b = 0.7 nm, N_b = 145)

35559 (r_b = 0.6 nm, N_b = 229)

41560 (r_b = 0.5 nm, N_b = 369)

47561 (r_b = 0.4 nm, N_b = 582)

Figure 5 *Evolution of DALAI_GA models of StP. The figure shows snapshots of bead models, taken at crucial steps of the GA process, in two different views (along z and y). The images labelled # 0 show the initial model configuration. The images labelled # 3001 represent the optimised model generated by the GA at the initial level of resolution (defined by the initial bead radius, r_b). The optimised model at a given resolution provides the basis for a new initial model configuration at the next higher resolution level, using beads of increasingly smaller size, and the GA process is repeated to generate a new optimised model if the new resolution is within the limit imposed by the largest abscissa value of the experimental scattering curve*

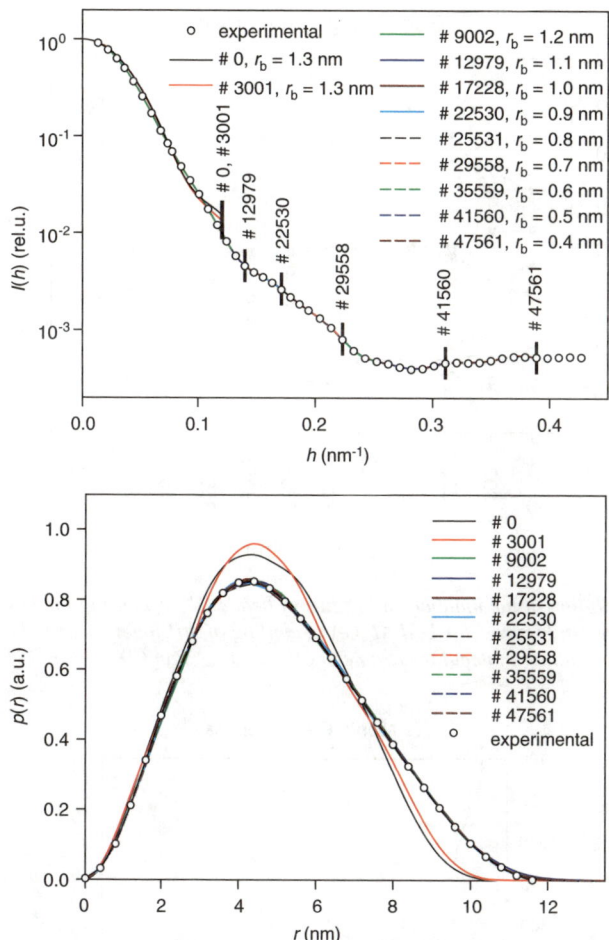

Figure 6   *Evolution of DALAI_GA models of StP. The scattering curves I(h) (top) and pair-distance distribution functions p(r) (bottom) calculated for selected snapshots of bead models of Figure 5 at different steps of the GA process are compared with the corresponding experimental functions. The vertical bars (top) indicate the respective upper limit of the portion of the experimental I(h) curve that is analysed in the GA process at the given level of resolution*

models of the isolated haeme and FAD domains with a hypothetical linker to provide an elongate model for the entire protein. Since, however, *ab initio* DAMMIN calculations[2,49] yielded models for all major components under consideration [holoprotein, haeme and FAD components; Figure (15a)], it was tempting to combine the two isolated domains and thus obtain a model for the complete entity. This model was compared with the crystallographic data of the isolated domains and used to derive properties of hydrodynamic interest.

As opposed to the modus operandi used in the trial-and-error approaches, we abandoned modelling the linker, but made allowance for appropriate alignment and

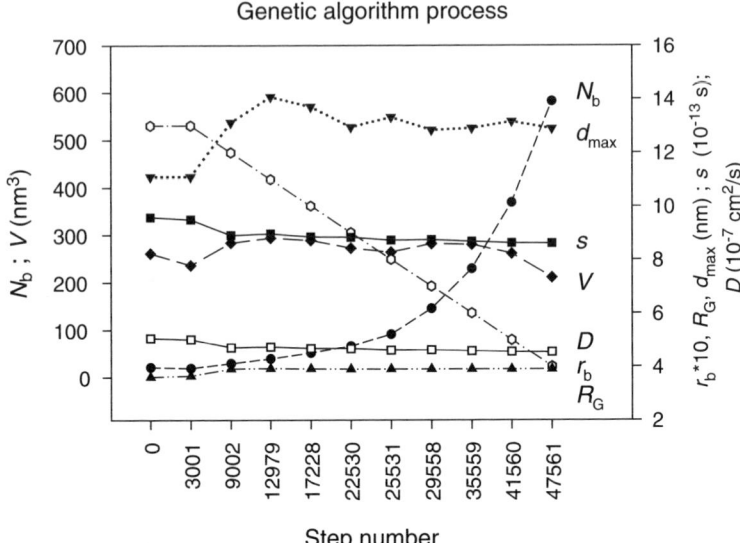

Figure 7   *Evolution of the number and radius of beads (N_b, r_b) during the GA process illustrated in Figure 5 for DALAI_GA modelling of StP, together with the alterations of
the resulting molecular parameters V, R_G, d_{max}, s and D*

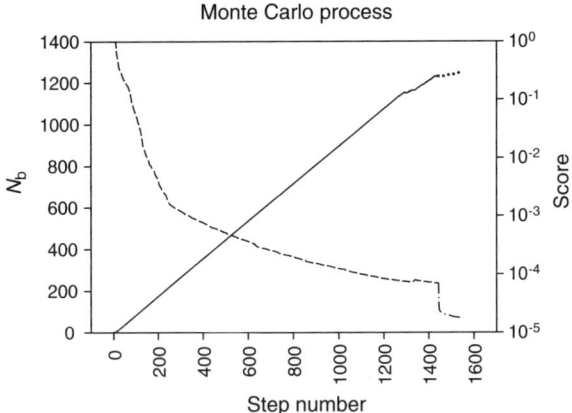

Figure 8   *Evolution of the number of beads (N_b) during the modelling process of StP by
SAXS3D (solid line) or a modified version of it (dotted line), together with the resulting score value (dashed and dot-dashed lines) achieved at various step numbers*

rotation of the domains under consideration and the correct distance between their
centres of gravity. Figures 15(b) and (c) depict two examples out of a great variety
of possible docking manoeuvres, all of them giving elongated shapes for the holoprotein. Figure 15(b) shows the result of a docking procedure neglecting any additional adjustments, and Figure 15(c) gives the best-fitting docked model as proved
by the distance criterion of the domains.

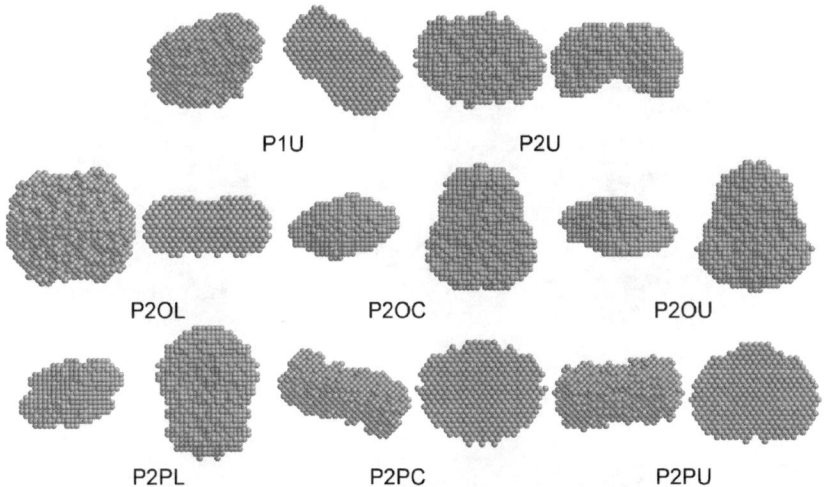

P1U   P2U

P2OL   P2OC   P2OU

P2PL   P2PC   P2PU

Figure 9 *Side and top views of selected reconstructions of the StP structure, generated by DAMMIN modelling approaches using the absence of symmetry constraints (P1: unknown shape) or plausible assumptions (P2: unknown, oblate or prolate shape, and further classification concerning the direction of the C_2 axis: along, across or unknown)*

An inspection of the two domains as derived from the crystal data [Figure 16(a)] reveals differently sized shapes for the haeme and FAD domains, similar to the findings of the DAMMIN approaches. Considering the correct spacing between the two domains, a comparison of the shapes for the two domains clearly shows coincidence between the *ab initio* solution models and the crystal structures of the domains [Figure 16(b)].

The comparison of experimental SAXS $p(r)$ functions of the holoprotein (Figure 17) with those obtained from various DAMMIN models reveals good agreement. By contrast, the functions derived from the crystal structures of the domains bear exceptional features owing to the lack of mass in the linker region.

The influence of the docking manoeuvres on predictions of scattering and hydrodynamic properties is listed in Table 2. After applying appropriate rescaling procedures, the prediction of SAXS parameters turns out to be satisfactory. Similarly, predictions of s come up to our expectations. Owing to the peculiar structure of CDH and its domains, WB approaches using spheroids can only provide crude anticipations of hydrodynamic parameters.

4.3   *Lumbricus terrestris* Haemoglobin: Modelling the Dodecameric Subunit and Reassembling the Complex from Subunits

While the entire *L.t.* Hb molecule (the native HBL complex) has already been modelled successfully by *ab initio* approaches,[46–48] no attempts have been made to date to model the individual dodecameric subunit or the whole complex from its building

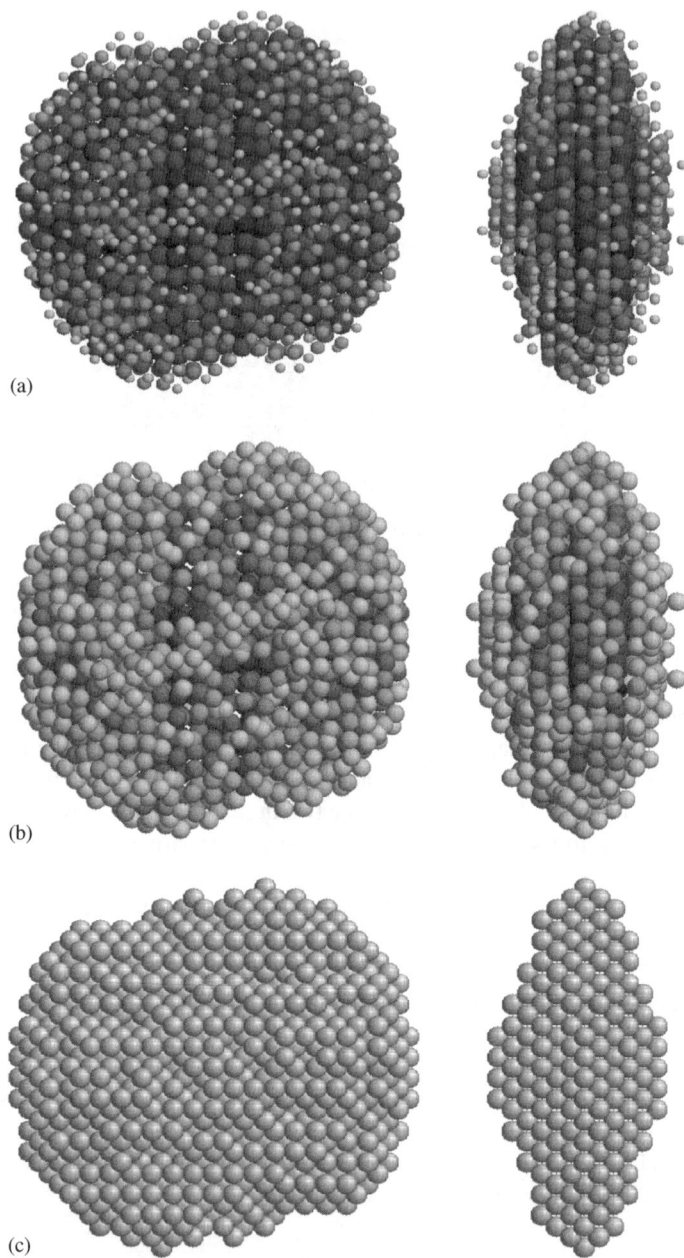

(a)

(b)

(c)

Figure 10 *Top and side views of averaged DAMMIN models of StP (constraints: P2 and*
oblate shape and along), utilising different averaging approaches: DAMHEXMIX
((a) variable r_b, and (b) r_b=0.3 nm; in both presentations the different probabili-
ties of beads are reflected by different shades of grey, in (a) additionally by r_b), or
the DAMAVER suite ((c) r_b=0.3 nm). The different averaging procedures provide
the basis for subsequent calculations of hydrodynamic parameters by HYDRO
[(a), (c)] or surface renderings [(b), (c)]

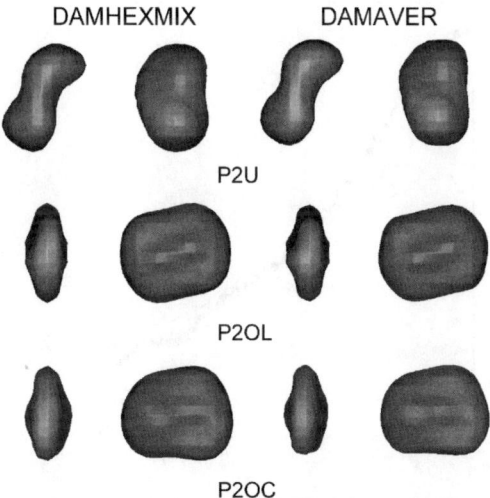

Figure 11 *Surface representations of averaged DAMMIN models of StP in two orthogonal views. The models were created with different combinations of symmetry and shape constraints: P2 and unknown shape (P2U), P2 and oblate shape and along (P2OL) or P2 and oblate shape and across (P2OC). The surface representations are based on averages obtained by DAMHEXMIX (left) or the DAMAVER suite (right)*

blocks (subunits). In this respect, it was highly challenging to attempt such a project, because SAXS data as well as SAXS-based trial-and-error consensus models of both the dodecameric subunit[103] and the HBL complex[57] are on hand. Since, in previous work on *L.t.* Hb,[47,48] the *ab initio* modelling program DAMMIN turned out to be most effective for modelling, this approach was used exclusively for the envisaged endeavour.

Application of DAMMIN to the dodecameric subunit of *L.t.* Hb revealed several suitable models for further modelling attempts. The averaged DAMMIN models obtained bear an obvious resemblance to the trial-and-error consensus model[103] ascertained earlier, mainly if the models were generated without any constraints, and also if C_3 symmetry and oblate shape are implied (Figure 18). Surface renderings of the averaged models emphasise the outcomes of the DAMMIN approaches applied (Figure 19). All relevant *ab initio* models disclose a similar domed, oblate shape and, in addition, are similar to averaged DAMMIN models that were indirectly biased by 3D-EM reconstructions of the HBL complex. The averaged scattering curves, $I(h)$, and the PDDFs, $p(r)$, of the original DAMMIN models agree quite well with the experimental profiles (Figure 20), whereas the SAXS functions calculated on the basis of the coordinates of the averaged models may differ considerably from the experiment. Pronounced discrepancies between the profiles are observed if the automatic subtraction of a constant scattering background by DAMMIN is suppressed. Evident improvements, however, can be achieved if the background subtraction is allowed. Table 3 compares the molecular parameters observed for the dodecameric subunit of *L.t.* Hb with the parameters calculated from SAXS models or DAMMIN

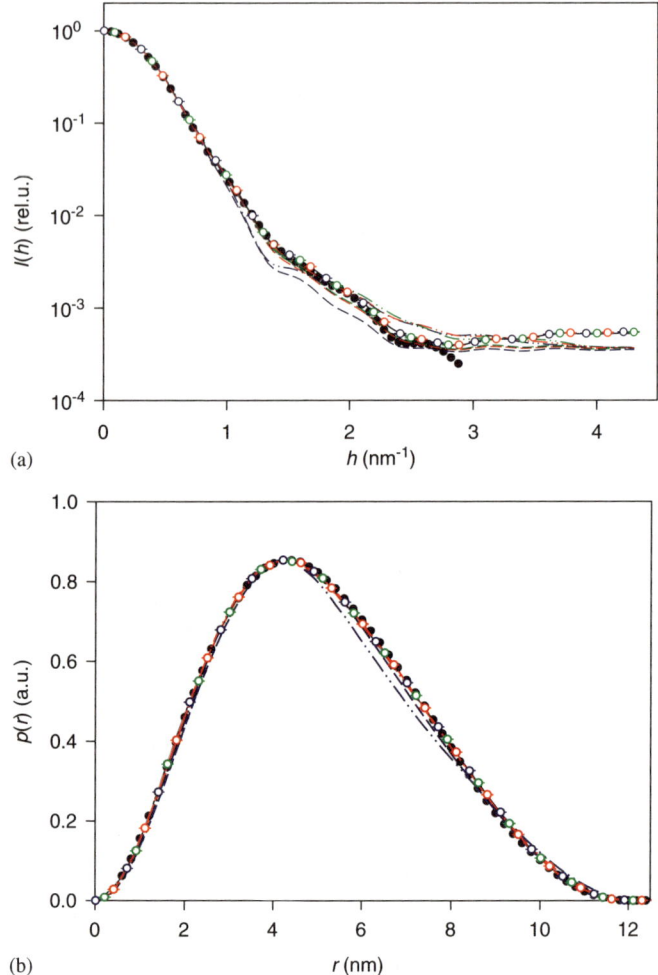

Figure 12 *Comparison of experimental and model I(h) (a) and p(r) (b) curves of StP. The experimental functions (solid black lines) are nearly completely overlaid by functions calculated for various models: the black circles represent the functions obtained for the previously established trial-and-error model;[80] the open circles show the averaged functions for DAMMIN models, which were generated on the assumption of C_2 symmetry without any shape constraint (blue circles) or with preference to oblate shape with the symmetry axis along the shortest dimension (green circles) or across to it (red circles). In general, the averages were calculated in each case from six independent models; the corresponding standard deviations are indicated by error bars which are, however, very small and poorly visible. The SAXS functions obtained for averaged structures are shown as dashed and dot-dashed lines (the colour codes are the same as for the circles): the dashed lines refer to averaged models composed of unequal beads, the bead size reflecting the probability of occupation; averaged structures obtained by means of the DAMAVER suite consist of equal-sized beads, their scattering functions are presented as dot-dashed lines. (b) The green lines have been omitted because of their close coincidence with the red lines*

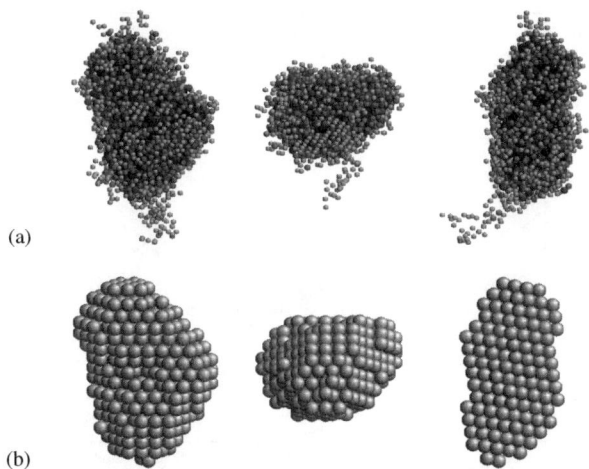

(a)

(b)

Figure 13 *Three orthogonal views of selected reconstructions of the StP structure, generated by SAXS3D modelling approaches (including the refinement of initial models by means of a modified version of the original program) and averaging by
DAMHEXMIX ((a), variable r_b) or the DAMAVER suite ((b), r_b=0.5 nm)*

models based on different symmetry constraints, shape biases and templates. The
check-up demonstrates that all approaches applied yield satisfactory agreement of
the parameters calculated, provided that a reasonable scattering background and
some upscaling of the bead radii are taken into account; in some cases, a rescaling
of the coordinates had to be employed as well. For the calculation of the hydrodynamic parameters of the dodecameric subunit, the WB approach is also acceptable;
however, the prolate ellipsoid is clearly ruled out because of its high d_{max} value.

Averaged DAMMIN models for the native HBL complex, generated on the bias of
EM templates, reveal the well-known *L.t.* Hb structure composed of two hexagonal
bilayers, with a reduced amount of mass but no distinct hole in the central part of the
complex [*cf.* the slab in Figure 21(a)]. Unlike this attribute, the HBL models constructed from an assembly of dodecameric subunits [Figures 21(b)–(e)] bespeak a
major hollowness in the particle centre. Accordingly, in the SAXS profiles, the lack
of mass in the centre of the models is reflected by perceptible deviations from the
experimental functions for *L.t.* Hb, as opposed to the proper behaviour of the EM-
biased DAMMIN model for the native HBL complex (Figure 22). While the averaged
functions of the HBL models built up from subunit models and the functions of models averaged by the DAMAVER suite exhibit drastic deviations from the experimental curves, the functions of averaged models obtained by means of DAMMIX are less
deviant. Interpretation of the results in terms of molecule parameters points in the
same direction (Table 4). While both the SAXS models and the EM-biased DAMMIN
models of the native HBL complex yield correct values for V, d_{max}, s and D, the sub-
unit-based DAMMIN constructions provide appropriate numbers for d_{max}, but underestimations of V and s and overestimations of D. The agreement of the d_{max} values can
be traced back to correct configurations and extensions of the constituent subunit

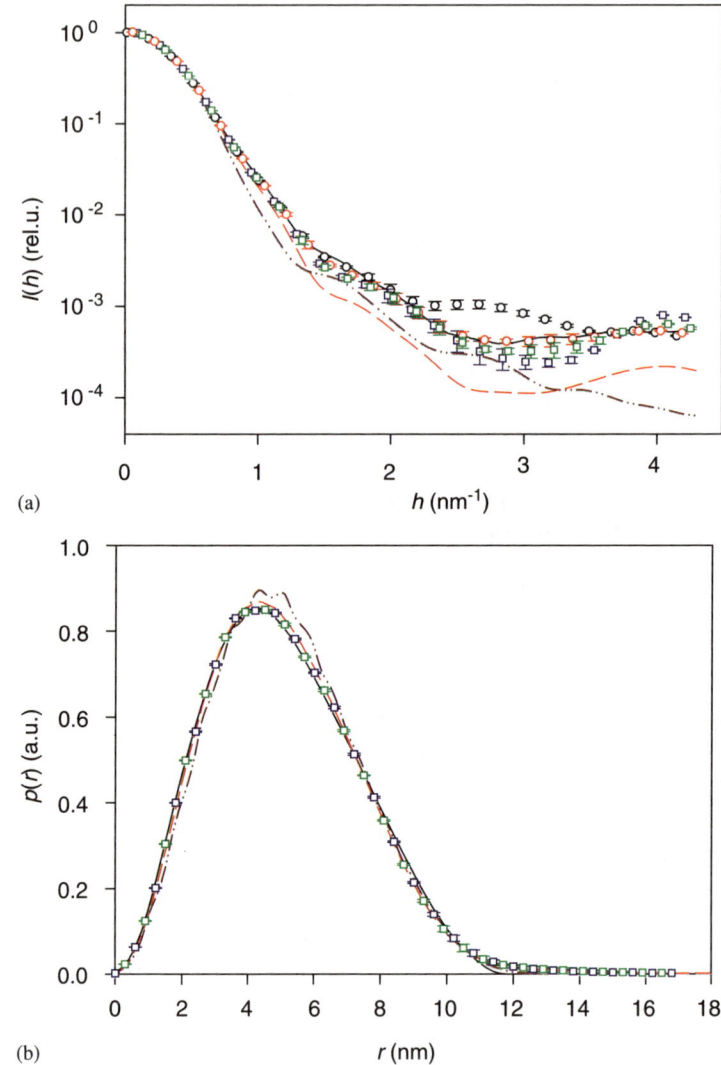

(a)

(b)

Figure 14 *Comparison of the experimental I(h) and p(r) curves (solid black lines in (a) and (b)) of StP with the averaged functions of models established by different versions of SAXS3D. In (a), the black and blue symbols represent averaged intensities of four models created by means of the original version of SAXS3D; a refinement of these initial models by using a modified program version finally resulted in the averaged intensities shown by the red and green symbols. The error bars symbolise the standard deviation of the averaged intensities. The circles are based on the intensities provided by SAXS3D itself, whereas the squares represent intensities that were calculated afterwards from the bead coordinates by exact usage of Debye's formula. (b) The symbols represent the averaged p(r) functions of the initial (blue squares) and refined (green squares) models. The scattering intensities (a) and p(r) functions (b) computed for averaged models are shown as dashed (averaging by DAMHEXMIX) and dot-dashed (averaging by DAMAVER) coloured lines*

Table 1 *Comparison of experimental and model parameters of phosphorylase from C. callunae*

	N_m	N_b / $r_b (nm)^a$	R_G (nm)	V (nm³)	d_{max}^b (nm)	D (10^{-7} cm² s⁻¹)	s (10^{-13} s)
Experimental[c]			3.89±0.02	290±14	11.8±0		8.3–8.7
Trial-and-error model		4612 / 0.248	3.87	293	11.9[d] / 12.4	4.67	8.83
DAMMIN models (BC, symmetry constraint P2)[e]							
No shape bias	6	1823±6 / 0.3	3.897±0.001	278±1	12.7±0.1	4.61±0.03	8.73±0.05
No shape bias, averaged (DAMAVER)		1149 / 0.3	3.897[f]	278	12.8[f]	4.68[f]	8.85[f]
No shape bias, averaged (DAMHEXMIX)	4	2655–8081 / 0.6–0.3[g]	3.890±0.002	278±0	13.0±0.0	4.43±0.12 / 4.58[h]	8.37±0.22 / 8.66[h]
Shape bias oblate (across)	6	1845±11 / 0.3	3.897±0.001	282±2	12.6±0.1	4.61±0.01	8.73±0.05
Shape bias oblate (across), averaged (DAMAVER)		1846 / 0.3	3.897[f]	282	12.6[f]	4.67[f]	8.84[f]
Shape bias oblate (across), averaged (DAMHEXMIX)	4	2360–8068 / 0.6–0.3[g]	3.892±0.002	282±0	13.0±0	4.52±0.08 / 4.62[h]	8.56±0.14 / 8.74[h]
Shape bias oblate (along)	6	1838±9 / 0.3	3.897±0.001	281±1	12.8±0.1	4.59±0.01	8.69±0.01
Shape bias oblate (along), averaged (DAMAVER)		1840 / 0.3	3.897[f]	281	12.8[f]	4.66[f]	8.81[f]
Shape bias oblate (along), averaged (DAMHEXMIX)	4	2342–6461 / 0.6–0.3[g]	3.892±0.002	281±0	12.5±0.0	4.52±0.07 / 4.60[h]	8.56±0.13 / 8.70[h]
DALAI_GA models (unconstrained)[e]							
NBC	16	29–369 / 1.2–0.5	3.91±0.01	278±10	13.3±0.4	4.63±0.08	8.76±0.14

(continued)

Table 1 (continued)

	N_m	N_b $r_b\,(nm)^a$	R_G (nm)	V (nm³)	d_{max}^b (nm)	D (10^{-7} cm² s⁻¹)	s (10^{-13} s)
BC	7	30–227 1.2–0.6	3.93±0.03	286±7	13.2±0.4	4.61±0.08	8.73±0.16
SAXS3D models (NBC, unconstrained)[e]							
Initial	4	1245±69 0.3	3.94±0.02	190±11	16.0±1.4	4.39±0.12	8.31±0.23
Initial, averaged (DAMAVER)		610 0.5[i]	3.94[f]	190	13.2[f]	4.52[f]	8.54[f]
Initial, averaged (DAMHEXMIX)		2305 0.6[g]	3.945	190	17.8	4.01	7.59
Refined	4	1306±71 0.3	3.93±0.02	199±11	16.2±1.1	4.41±0.13	8.33±0.24
Refined, averaged (DAMAVER)		601 0.5[i]	3.93[f]	199	13.2[f]	4.54[f]	8.58[f]
Refined, averaged (DAMHEXMIX)		2293 0.6[g]	3.93	199	18.3	3.97	7.52
WB model (OE: R_G, V)[j]					12.0	4.69	8.91

[a]For the calculation of parameters. in general the nominal bead radii r_b were upscaled by a factor 1.105 to account for the packing density of the beads.
[b]If not stated otherwise, for bead models the value corresponding to the maximum vertex-to-vertex distance of the beads is given.
[c]Experimental values were taken from the literature: SAXS data from ref. 80 and hydrodynamic data for the enzyme from other sources from refs. 83, 86 and 87; the molar mass of phosphorylase as estimated from SAXS is M=170 kg mol⁻¹.
[d]Value corresponding to the maximum centre-to-centre distance of the beads.
[e]BC signifies that a background subtraction was performed and NBC denotes no background subtraction.
[f]Parameter calculated after rescaling the coordinates of the beads to match the mean R_G of the models used for averaging.
[g]The edge length of the hexagonal lattice cells (voxel size) is given instead of r_b.
[h]Value referring to the model with voxel size 0.3 nm.
[i]For the calculation of parameters the nominal bead radius was rescaled to match the mean V of the models used for averaging.
[j]OE symbolises an oblate ellipsoid of revolution; the WB approach is based on the experimental values for R_G and V.

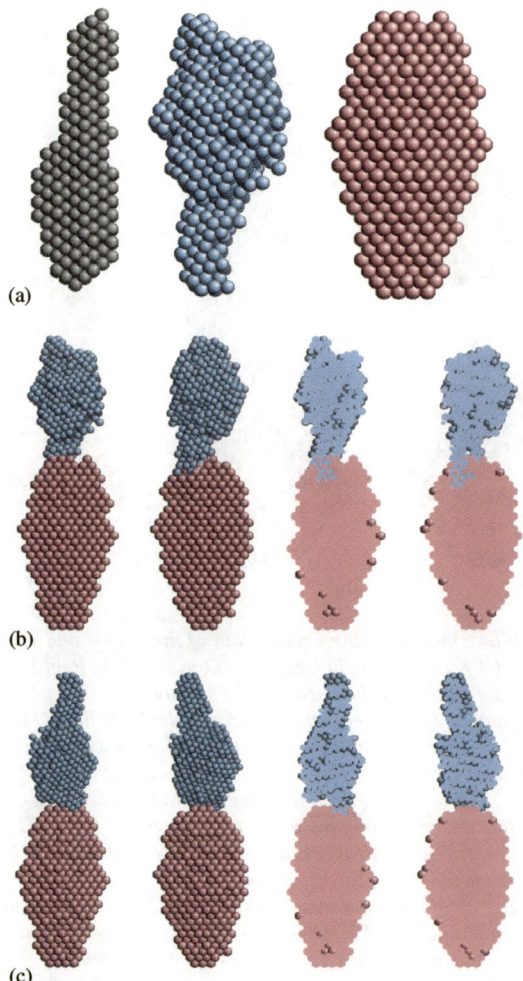

Figure 15 *Reconstructions of the structure of the CDH holoprotein and its isolated domains. (a) Aligned averaged models for the entire holoprotein (grey; N_b=431, r_b=0.4 nm), the haeme domain (blue; N_b=909, r_b=0.2 nm), and the FAD domain (pink; N_b=1324, r_b=0.25 nm); the images of the models are not to scale. The initial models are based on experimental scattering profiles and ab initio DAMMIN calculations in the 'expert mode'. Averaging and filtering of the DAMMIN models was performed by means of the DAMAVER program suite. The models for the domains were aligned along the y direction and rotated around z or x, to make the domains ready for the docking manoeuvres to be executed. (b) and (c) Two different views (90° rotation around the y axis) of models for the docked haeme and FAD domains, together with the corresponding central slabs. The required distance (about 7.4 nm) between the centres of gravity of the two averaged ab initio models of the domains has been computed by considering their values for R_G and V, in order to achieve the R_G value of the averaged model for the holoprotein. Model (b) shows the results of the first trial of a docking procedure (docking based on the initial orientation of the domains). The best-fitting model (c) is characterised by a minimum of overlapping of the domains and a maximum value for d_{max}*

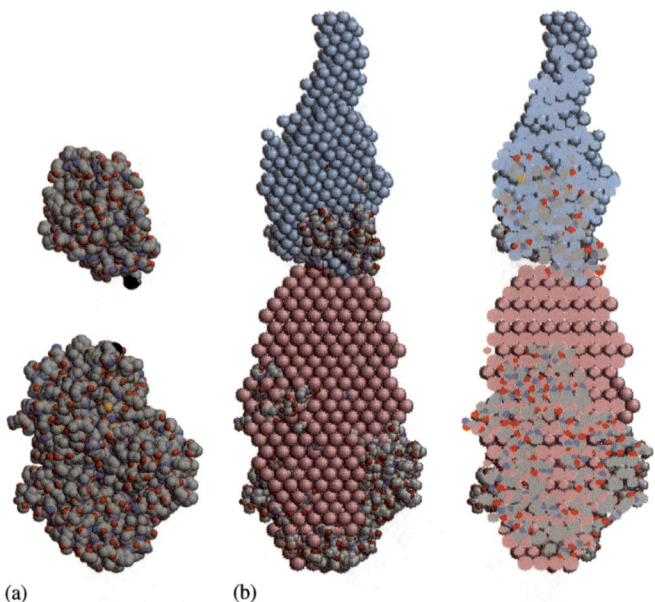

(a)   (b)

Figure 16   *Space-filling models of the crystal data of the haeme and FAD domains of CDH in usual CPK colours (C in light grey, O in red, N in light blue, and S in yellow). (a) Aligned models of the haeme and FAD domains. The coordinates of the basic protein atoms were taken from the available PDB files (1D7C, 1KDG); hetero atoms of the prosthetic groups, glycan chains and water molecules were neglected. The domains were arranged similar to the docked DAMAVER models described in Figure 15, i.e. at the required distance and facing each other. The FAD domain of the crystal structure contains a few residues of the linker. The terminal atoms of the crystal structures are marked in black (enlarged spheres); the gap between the two domains is caused by the distance specifications between the respective centres of gravity of the two domains. (b) Comparison of the aligned models obtained from crystal data (in CPK colours) with DAMMIN reconstructions (blue and pink spheres), together with the corresponding central slabs*

models, and the disagreement of the other parameters to the lack of mass in the centre and other regions of the molecule. It is plausible to attribute the missing mass to the linker chains forming a kind of scaffold in the native *L.t.* Hb architecture. As may be expected, the WB approach of the HBL complex supplies acceptable values for all parameters mentioned, since it does not use the subunit data; by contrast, use of the molar mass corresponding to the sum of the subunit masses yields an *s* value as low as with the HBL models constructed from DAMMIN models of the subunits.

4.4   RNA from the Bacteriophage MS2: Modelling the Solution Structure of a Free Nucleic Acid

Previous SAXS data[105] of MS2 RNA were exploited for tentatively modelling the structure of a free RNA in solution by *ab initio* approaches, similar to the procedures adopted for proteins.

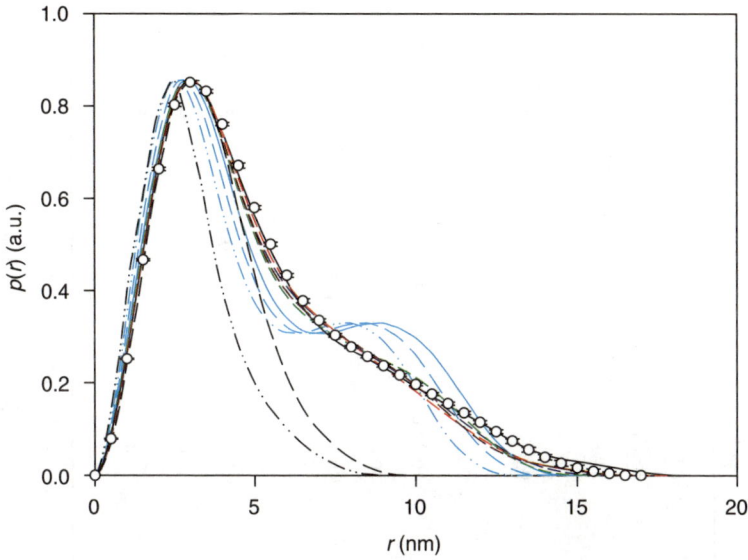

Figure 17 *Comparison of the experimental pair-distance distribution, p(r), (solid black line) of the CDH holoprotein with the p(r) functions of various models. The black circles with error bars show the mean p(r) function and the correspon-ding standard deviations obtained by averaging the functions of 30 different ab initio DAMMIN models of the holoprotein. The dashed blue, red, green and brown lines represent the p(r) functions calculated for models composed of ab initio DAMMIN models of the two isolated domains (haeme domain, FAD domain). The domains were aligned along the y direction, with a constant dis-tance of about 7.4 nm between their centres of gravity; the mutual orientation of the domains was systematically changed by rotations around y and z, respec-tively. The original arrangement of the domains yielded the p(r) function drawn in blue, rotations of the haeme domain around y by ±30° did not lead to any noticeable change in the p(r) function. Red symbolises the p(r) function obtained after rotating the haeme domain by 180° around the z direction; the green line was obtained after rotating the FAD domain by 180° around z; rota-tions of both domains by 180° around z yielded the brown line. The lines coloured in cyan represent p(r) functions calculated for models derived from the crystal structures of both domains; for these models only the coordinates of protein atoms were evaluated, and atoms belonging to hetero groups (FAD, haeme, carbohydrates) were neglected. Again at a distance of about 7.4 nm between the centres of gravity, the domains were arranged in such a way that the C-terminus of the haeme domain and the N-terminus of the FAD domain were facing each other. The dot-dashed cyan line is the p(r) function obtained without rescaling the coordinates of the crystal structure. Since the values for both R_G and d_{max} of this model are quite low if compared to the experimental data for CDH or the parameters corresponding to the DAMMIN models, the coordinates of the crystal structure were upscaled to match the mean R_G value of the DAMMIN models (dashed cyan line) or the experimental R_G of CDH (solid cyan line). For convenience, the mean p(r) function of the DAMMIN mod-els for the isolated domains (dot-dashed black line: haeme domain; dashed black line: FAD domain) are additionally shown in the figure. To facilitate the comparison, all p(r) functions presented have been normalised to the same height of the maximum*

Table 2 *Comparison of experimental and model parameters of CDH from Phanerochaete chrysosporium and its isolated domains*

	N_m	N_b / r_b (nm)[a]	R_G (nm)	V (nm³)	d_{max}[b] (nm)	D (10^{-7} cm² s⁻¹)	s (10^{-13} s)
Haeme domain							
Experimental[c]			2.30±0.03	35.2±3.5	8.8		
DAMMIN models							
Ab initio	24	75±1 / 0.46	2.29±0.03	41.1±0.4	9.2±0.4	8.23±0.09	2.50±0.03
		909 / 0.2					
Averaged (DAMAVER)			2.11	41.1	8.8	8.44	2.57
WB model (OE: R_G, V)[d]					7.2	8.40	2.56
(PE: R_G, V)					9.6	9.12	2.79
FAD domain							
Experimental[c]			2.75±0.07	110±9	9.7±0.3		
DAMMIN models							
Ab initio	24	212±1 / 0.46	2.73±0.01	117±1	10.2±0.1	6.47±0.03	4.75±0.02
		1324 / 0.25					
Averaged (DAMAVER)			2.70	117	10.0	6.58	4.83
WB model (PE: R_G, V)[d]					10.6	6.75	4.97
Holoprotein							
Experimental[c]			4.37±0.11	165±5	17±1		5.6±0.1
DAMMIN models							
Ab initio	30	127–216 / 0.6–0.5	4.32±0.04	156±4	17.5±0.6	5.04±0.07	5.74±0.08
		431 / 0.4					
Averaged (DAMAVER)			4.15	156	17.0	5.31	6.04
Docked domains		2233 / 0.2, 0.25	4.15	158	17.8	5.10	5.80
WB model (PE: R_G, V)[d]					18.7	5.17	5.90

[a]For the calculation of parameters the nominal bead radii r_b were upscaled by a factor 1.105 to account for the packing density of the beads.
[b]For bead models the value corresponding to the maximum vertex-to-vertex distance of the beads is given.
[c]Experimental values were taken from the literature: SAXS data for CDH and its domains from ref. 89 and the s value for the holoprotein from ref. 90. As obtained from SAXS, the molar masses of haeme domain, FAD domain, and holoprotein are $M=28.6\pm6$, 69 ± 19, and 107 ± 22 kg mol⁻¹, respectively.
[d]OE and PE symbolise oblate and prolate ellipsoids of revolution, respectively; the applied WB approaches are based on the experimental values for R_G and V.

Trial-and-error consensus model:

Averaged DAMMIN ab initio models:
P1U

P3U: prolate shape

P3U: oblate shape

P3O

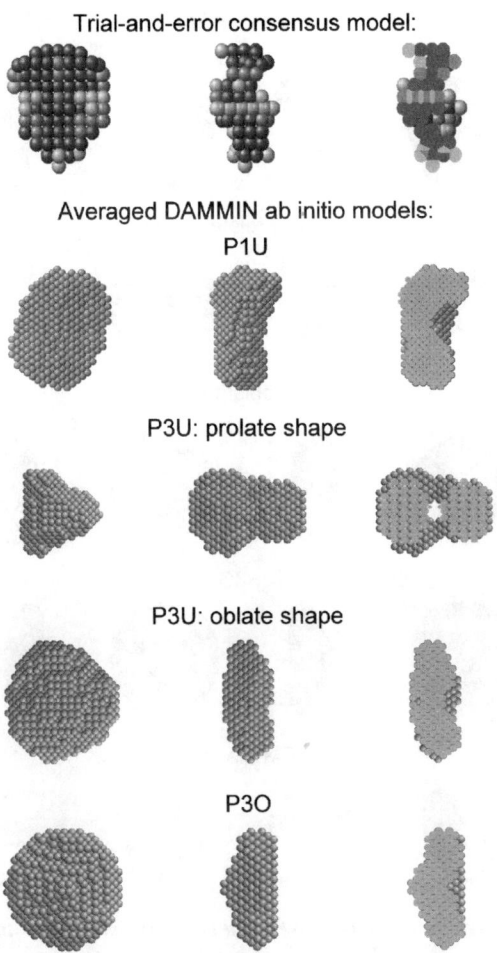

Figure 18 *Averaged models for the dodecameric subunit of L.t. Hb, presented in two orthog-*
onal views and as central slabs. The previously established consensus model[103]
is an average of 33 different models all of them created by trial and error; the
different shades of grey reflect the different probabilities of the beads. The aver-
aged DAMMIN models are based on ab initio models generated without any con-
straint (P1U) or with symmetry constraint P3 (C_3 axis) and no shape bias (P3U)
or preference to oblate shape (P3O). The oblate and prolate models resulting in
the case of constraint P3U were averaged separately. All averages of DAMMIN
models shown in the figure were obtained by means of the DAMAVER suite; the
number of models used for averaging ranged from 12 to 36

The DAMMIN procedure turned out to be highly effective for demonstrating the
progress of modelling trials. Starting from a spherical search space, after several
cycles the solution structure commutes, more or less abruptly, the spherical model
assemblage into a rather loose structure characteristic of free nucleic acids, concomitantly reducing the number of beads of the constitutive model from about 3500
to values around 200 for the steps shown in Figure 23. In agreement with the changes

Ab initio models:

P1U

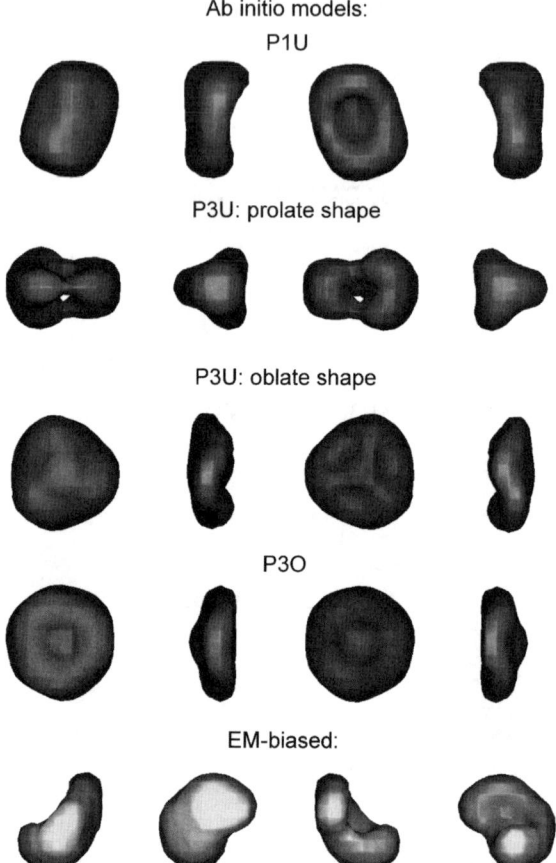

P3U: prolate shape

P3U: oblate shape

P3O

EM-biased:

Figure 19 *Surface representations of averaged DAMMIN models for the dodecameric sub-unit of L.t. Hb. Each model is shown in four views corresponding to successive 90° rotations around the vertical axis. The ab initio models were created without any constraints (P1U) or with different combinations of symmetry and shape constraints: P3 and unknown shape (P3U, oblate or prolate models resulting), P3 and oblate shape (P3O). For comparison, also the representation of the averaged structure of DAMMIN models, which were indirectly biased by 3D-EM reconstructions of the entire HBL complex[57] (EM-biased) is added; the underlying models were created by means of templates derived from recently established EM-biased DAMMIN models of L.t. Hb.[47] All surface representations are based on averages obtained by the DAMAVER suite*

documented in this figure, the $p(r)$ functions of the respective models alter their appearance in the same step, until ultimately, under the given conditions, no further improvements are made (Figure 24).

All the models created by DAMMIN, DALAI_GA and SAXS3D share some common characteristics, irrespective of the averaging procedures applied (Figure 25). They cannot be displayed by simple geometrical bodies, as they exhibit some kind of bent shape and feature several protuberances. The bent structure has already been identified

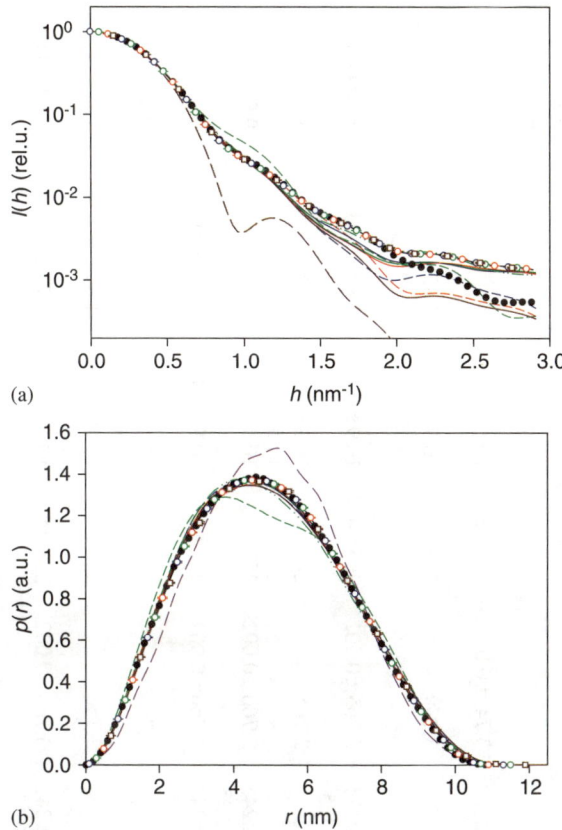

Figure 20 *SAXS curves I(h) (a) and pair-distance distribution functions p(r) (b) of the dode-cameric subunit of L.t. Hb. The experimental functions (solid black lines) are nearly completely overlaid by functions calculated for various models: the black circles represent the functions obtained for the previously established trial-and-error model;[103] the open circles show the averaged functions for DAMMIN models which were generated without any a priori information (blue circles) or with the assumption of C_3 symmetry, without shape constraint (green circles) or with preference to oblate shape (red circles). The brown squares represent the averaged functions obtained from indirectly EM-biased DAMMIN models. The averages were calculated from 7 to 48 independent models; the corresponding standard deviations are indicated by error bars which are, however, very small and poorly visible. The SAXS functions obtained for averaged structures are shown as coloured lines (the colour codes are the same as for the circles). The blue, green and red lines refer to DAMAVERaged models (obtained by means of the DAMAVER suite) consisting of equal-sized beads; the dashed lines are based on models which were generated by DAMMIN without the automatic subtraction of a constant background from the experimental scattering curve; the solid and dot-dashed lines refer to models where this background subtraction was applied (for the presentation, the background was readded to the I(h) curves); the green solid lines refer to models of oblate shape and the dot-dashed lines to models of prolate shape. The solid brown lines represent the functions obtained for DAMAVERaged biased models, the dashed brown lines refer to averaged structures (obtained by means of DAMMIX) composed of unequal beads, where the bead size reflects the probability of occupation*

Table 3 *Comparison of experimental and model parameters of the dodecameric subunit from Lumbricus terrestris haemoglobin*

	N_m, shape	N_b, r_b (nm)	R_G (nm)	V (nm³)	d_{max} (nm)	D (10^{-7} cm² s⁻¹)	s (10^{-13} s)
Experimental[a]			3.74±0.01	255±10	10.6±0.0		9.38[b] 9.18[c] 9.44[d]
SAXS models							
Trial-and-error[e]	33 Oblate Oblate	206±2 0.661 265 0.661	3.74 ±0.01	249±2	10.6±0.0		
Consensus[f,g]			3.739	249[h]	10.7[i] 11.1[j]	4.82	10.04
DAMMIN models[k]							
No symmetry constraint, no shape bias[l]	12 Oblate	1548±171 0.3–0.325	3.765±0.001	265±4	10.9±0.1[i]	4.80±0.03	10.00±0.06
No symmetry constraint, no shape bias, DAMAVERaged[m]	Oblate	1369 0.325	3.765[n]	266	10.8[i,n] 11.5[i,n]	4.82[n]	10.05[n]
Symmetry constraint P3, no shape bias[l]	13 Oblate 23 Prolate	1351±245 0.30–0.35 1326±249 0.30–0.35	3.766±0.002 3.766±0.002	257±4 252±5	10.9±0.1[i] 10.8±0.1[i]	4.76±0.02 4.75±0.05	9.92±0.05 9.90±0.10
Symmetry constraint P3, no shape bias, DAMAVERaged[m]	Oblate Prolate	1325 0.325 1299 0.325	3.766[n] 3.766[n]	257 252	11.1[i,n] 11.8[i,n] 11.0[i,n] 11.8[i,n]	4.77[n] 4.84 [n]	9.93[n] 10.09[n]
Symmetry constraint P3, shape bias oblate[l]	36 Oblate	1357±243 0.30–0.35	3.766±0.002	258±5	10.9±0.1[i]	4.77±0.03	9.94±0.06
Symmetry constraint P3, shape bias oblate, DAMAVERaged[m]	Oblate	1066 0.35	3.766[n]	259	11.0[i,n] 11.8[i,n]	4.78[n]	9.95[n]
No symmetry constraint, usage of templates[o,p,q]	48	179±1 0.60	3.798±0.008	267±2	11.0±0.3[i]	4.77±0.02	9.94±0.05
No symmetry constraint, usage of templates, averaged[f,g,q]		483 0.60	3.798[n]	267	11.2[i,n] 12.8[i,n]	4.09[n] 4.52±0.03[s]	8.52[n] 9.43±0.06[s]

No symmetry constraint, usage of templates, DAMAVERaged[m,q]		900		267	3.798[n]	11.3[j,n] 12.1[j,n]	4.77[n] 9.93[n]
WB model (OE: R_G, V)[t] (PE: R_G, V)	Oblate	0.35				11.5 14.6	4.89 10.22 5.05 10.56

[a] Experimental SAXS and hydrodynamic data for the oxygenated dodecamer are taken from ref. 103. SAXS data refer to the dodecamer (from urea dissociation) at infinite dilution ($c=0$) and represent mean values and standard deviations of the results derived from the $I(h)$ and $p(r)$ functions obtained from two independent measurements; the molar mass of the dodecamer as obtained from SAXS is $M=190\pm19$ kg mol^{-1}.

[b] Dodecamer from urea dissociation, $s_{w,20}$ from $c=12.4$ mg mL^{-1} at 1 °C.

[c] Dodecamer from SiW dissociation, $s_{w,20}$ from $c=4.0$ mg mL^{-1} at 1 °C.

[d] Dodecamer from SiW dissociation, $s_{w,20}$ from $c=0.04$ mg mL^{-1} at 1 °C.

[e] The data represent mean values and standard deviations of the SAXS parameters calculated from a set of 33 models, [113] which were selected from a total of about 300 trial-and-error bead models because of their good fit to the experimental SAXS data.

[f] The data represent the parameters calculated from the averaged structure of the 33 selected trial-and-error models.

[g] The probabilities of positions were taken into account by different weightings of equal-sized beads or preferably by different bead volumes (for the calculation of V and d_{max} and the prediction of D and s); in the latter case the bead radius r_b given in column three only applies to positions of probability=1.

[h] Neglect of the probabilities of positions would result in a volume $V=320.6$ nm^3.

[i] Value corresponding to the maximal centre-to-centre distance of the beads.

[j] Value corresponding to the maximal vertex-to-vertex distance of the beads.

[k] If not stated otherwise, the experimental $I(h)$ function was modified by DAMMIN by subtraction of a constant background, and for the calculation of parameters the nominal bead radii r_b given in column 3 were upscaled by a factor of 1.105 to account for the packing density of the beads.

[l] Mean values and standard deviations of the parameters of N_m models.

[m] Parameters calculated from the averaged structure of the preceding N_m models; averaging was performed by means of DAMAVER.

[n] Parameter calculated after rescaling the coordinates of the beads to match the mean R_G of the models used for averaging.

[o] The experimental $I(h)$ function was not modified by DAMMIN.

[p] Mean values and standard deviations of the parameters of N_m models obtained by using different templates derived from eight EM-biased DAMMIN models of the HBL complex of *L.t.* Hb.[47] The templates contain 1/12 of each HBL complex, starting at two different angular positions (22.5° and 30°).

[q] The nominal bead radius r_b given in column 3 was upscaled by a factor of 1.183 to account for the packing density of the beads in these models.

[r] Parameters calculated from the averaged structure of the preceding N_m models; averaging was performed by means of DAMMIX.

[s] These parameters were derived from averaged structures composed of equal-sized beads. Only beads occurring in at least f_{min} models were taken into account; the weightings of these beads were neglected. The mean values and standard deviations were obtained by varying f_{min} from 12 to 24, thereby applying separate optimised scaling factors for the coordinates and radii of the beads to match the mean V and R_G of the N_m models used for averaging.

[t] OE and PE symbolise oblate and prolate ellipsoids of revolution, respectively; the WB approaches are based on the experimental values for R_G and V.

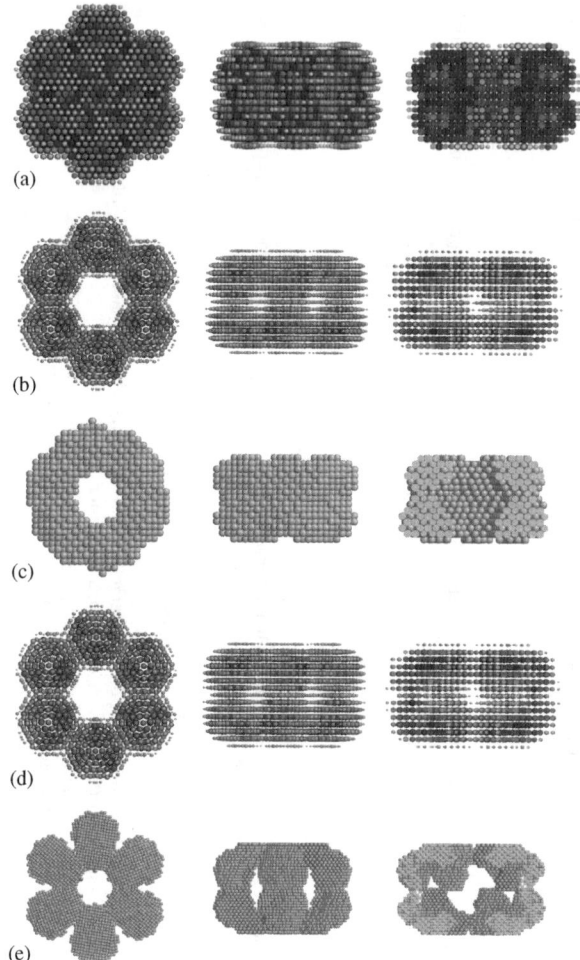

(a)

(b)

(c)

(d)

(e)

Figure 21 *Comparison of the average of eight EM-biased DAMMIN models for the HBL complex of L.t. Hb (a) with HBL models built from DAMMIN models for the dodecameric subunit [(b)–(e)]. The 48 subunit models under consideration were created by using 16 templates that were derived from the eight EM-biased models for the native HBL complex. Models (b) and (c) represent averages of the 48 different HBL models constructed from the different subunit models; averaging was performed by our program DAMMIX (b) or by means of the DAMAVER suite (c). Models (d) and (e) were built from averages of the subunit models, obtained again by DAMMIX (d) and DAMAVER (e), respectively. In models (a), (b) and (d) the different probabilities of the beads are expressed by different bead volumes and, for better visualisation, by different shades of grey; beads of low probability were incorporated. On the contrary, for models (c) and (e) the beads of low probability have been discarded by the DAMFILT routine in the DAMAVER suite. Since by application of DAMAVER the initially fixed orientation of the models included is altered, the obvious loss of structural details in model (c) is not surprising; for creating model (e), the averaged subunit structure had to be reoriented appropriately to minimise the overlapping of the subunits in the HBL model. All models are presented in top and side views and as central slabs right through the subunits*

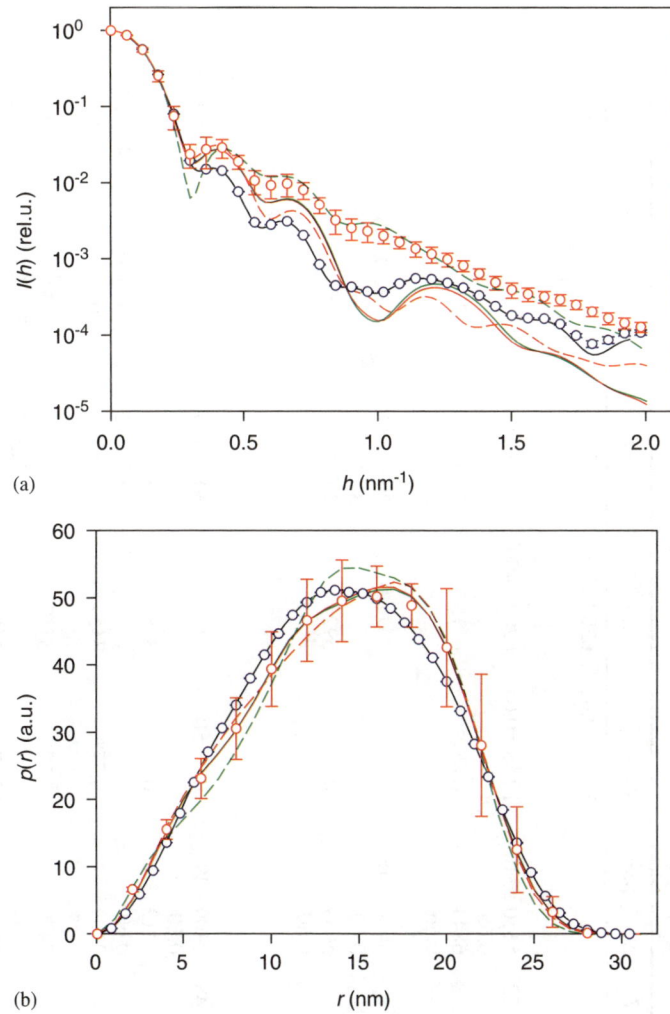

Figure 22 *Comparison of I(h) (a) and p(r) (b) functions of various HBL models based on DAMMIN models for the dodecameric subunit of L.t. Hb with the experimental functions (black lines) and the functions for EM-biased DAMMIN models for the native HBL complex (blue circles). The red circles represent the averaged functions of the HBL models built from the 48 different subunit models; the functions of the corresponding averaged models are shown as red lines: the solid lines refer to averaging by DAMMIX, the dashed lines to averaging by the DAMAVER suite. The green lines represent the functions of HBL models constructed from averaged models for the subunit; again the averaging procedure applied is indicated by solid (DAMMIX) and dashed lines (DAMAVER), respectively*

in other types of RNA (L-form of tRNA or rRNA[112]), and the surface peculiarities are in accord with secondary structure details in terms of "flower" or "bouquet" models.[104] Owing to the modus operandi of the averaging procedures applied, the DAMHEXMIX procedure accentuates the peculiar features of nucleic acid solution structure, whereas

Table 4 Comparison of experimental and model parameters of the HBL complex from Lumbricus terrestris haemoglobin

	N_m	N_b, r_b (nm)	R_G (nm)	V (nm³)	d_{max} (nm)	D (10^{-7} cm² s⁻¹)	s (10^{-13} s)
Experimental[a]			10.71±0.02	6200±200	29.37±0.21	1.3–1.81 / 1.66±0.05	58.9–61.1
SAXS models							
Trial-and-error[b]	22	5150±310[c] / 0.66	10.68±0.04	6200±400	29.61±0.05		
Consensus[d]		6844 / 0.66	10.68	6215	31.6[e] / 32.4[f]	1.627	62.4
DAMMIN models							
No symmetry constraint, usage of EM templates[g,h]	8	3959±80 / 0.60	10.70±0.01	5924±120	29.8±0.2[e] / 31.2±0.2[f]	1.639±0.008	62.9±0.3
No symmetry constraint, usage of EM templates, averaged (by DAMMIX)[d,h]		5939 / 0.60	10.69	5924	30.4[e] / 31.2[f]	1.636	62.8
HBL models constructed from DAMMIN models of the dodecameric subunit[i]							
HBL models from individual subunit models[h,j]	48	2109±48 / 0.60	10.77±0.56	3157±72	30.0±0.9[f]	1.74±0.04	43.5±1.1
HBL models from individual subunit models, averaged (by DAMMIX)[d,h]		5412 / 0.60	10.77	3157	30.6[f] / 29.7±0.2[f,k]	1.674 / 1.73±0.01[k]	41.8 / 43.1±0.3[k]
HBL models from individual subunit models, DAMAVERaged[l]		2299 / 0.75	10.77[m]	3157[m]	30.1[f,m]	1.704[m]	42.6[m]
HBL model from the DAMMIX-averaged model of the subunit[n]		5400 / 0.60	10.77[o]	3157[o]	30.6[f,o] / 29.7±0.2[f,o,p]	1.675[o] / 1.73±0.02[o,p]	41.9[o] / 43.2±0.4[o,p]
HBL model from the DAMAVERaged model of the subunit[q]		10680 / 0.35 ; 2598[r] / ≤0.75 nm	10.77[o,s]	3157[o,s]	29.3[f,o] / 29.2[f,o,s]	1.77[o,s]	44.3[o,s]

| WB model (OE: R_G, V)[t] | 33.0 | 1.70 | 65.5 |
| | | | 42.7 |

[a]Experimental SAXS and hydrodynamic data of *L.t.* Hb are taken from ref. 48; the molar mass of the HBL complex as obtained from SAXS is $M = 3500 \pm 200$ kg mol^{-1}.

[b]The data represent mean values and standard deviations of the SAXS parameters calculated from a set of 22 models. The probabilities of positions were taken into account by different and-error bead models because of their good fit to the experimental SAXS data.

[c]Calculated from the data given in ref. 113.

[d]The data represent the parameters calculated from the averaged structure of the preceding N_m models. The probabilities of positions were taken into account by different weightings of equal-sized beads or preferably by different bead volumes (for the calculation of V and d_{max} and the prediction of D and s); in the latter case the bead radius r_b given in column 3 only applies to positions of probability = 1.

[e]Value corresponding to the maximal centre-to-centre distance of the beads.

[f]Value corresponding to the maximal vertex-to-vertex distance of the beads.

[g]Mean values and standard deviations of the parameters of eight DAMMIN models obtained by using two templates of slightly different size that were derived from 3D-EM reconstructions of *L.t.* Hb.[47] The templates consisted of beads with r_b=0.6 nm, placed at hexagonal lattice points.

[h]If not stated otherwise, the nominal bead radii r_b given in column 3 were upscaled by a factor of 1.183 to account for the packing density of the beads in the hexagonal models.

[i]For the calculation of s values, the molar mass of the HBL models assembled from 12 dodecameric subunits was assumed to be 2280 kg mol^{-1} (corresponding to the 12-fold value of the dodecameric subunit of 190 kg mol^{-1}).

[j]Mean values and standard deviations of the parameters of the 48 HBL models built from the individual EM-biased DAMMIN models of the dodecameric subunit presented in Table 3.

[k]These parameters were derived from averaged structures composed of equal-sized beads. Only beads occurring in at least f_{min} models were taken into account; the weightings of these beads were neglected. The mean values and standard deviations were obtained by varying f_{min} from 12 to 24, thereby applying separate optimised scaling factors for the coordinates and radii of the beads to match the mean V and R_G of the N_m models used for averaging.

[l]Parameters calculated from the averaged structure of the preceding N_m models; averaging was performed by means of DAMAVER.

[m]Parameter obtained by applying separate optimised scaling factors for the coordinates and radii of the beads to match the mean V and R_G of the N_m models used for averaging.

[n]Parameters of the HBL models built from the averaged structures (obtained by DAMMIX) of the 48 EM-biased DAMMIN models of the dodecameric subunit (Table 3).

[o]Parameter obtained by applying separate optimised scaling factors for the coordinates and radii of the beads to match the mean V and R_G of the preceding N_m HBL models built from individual subunit models.

[p]These parameters are based on averaged subunit structures composed of equal-sized beads. Only beads occurring in at least f_{min} subunit models were taken into account; the weightings of these beads were neglected. The mean values and standard deviations were obtained by varying f_{min} from 12 to 24.

[q]Parameters of the HBL models built from the averaged structure (obtained by DAMAVER) of the 48 EM-biased DAMMIN models of the dodecameric subunit (Table 3).

[r]Reduced model consisting of beads of unequal size; the smaller N_b was required for the prediction of hydrodynamic parameters.

[s]Parameter based on the reduced model.

[t]OE symbolises an oblate ellipsoid of revolution; the WB approach is based on the experimental values for R_G and V; for M, values of 3500 kg mol^{-1} (upper line) or 2280 kg mol^{-1} (lower line) were applied.

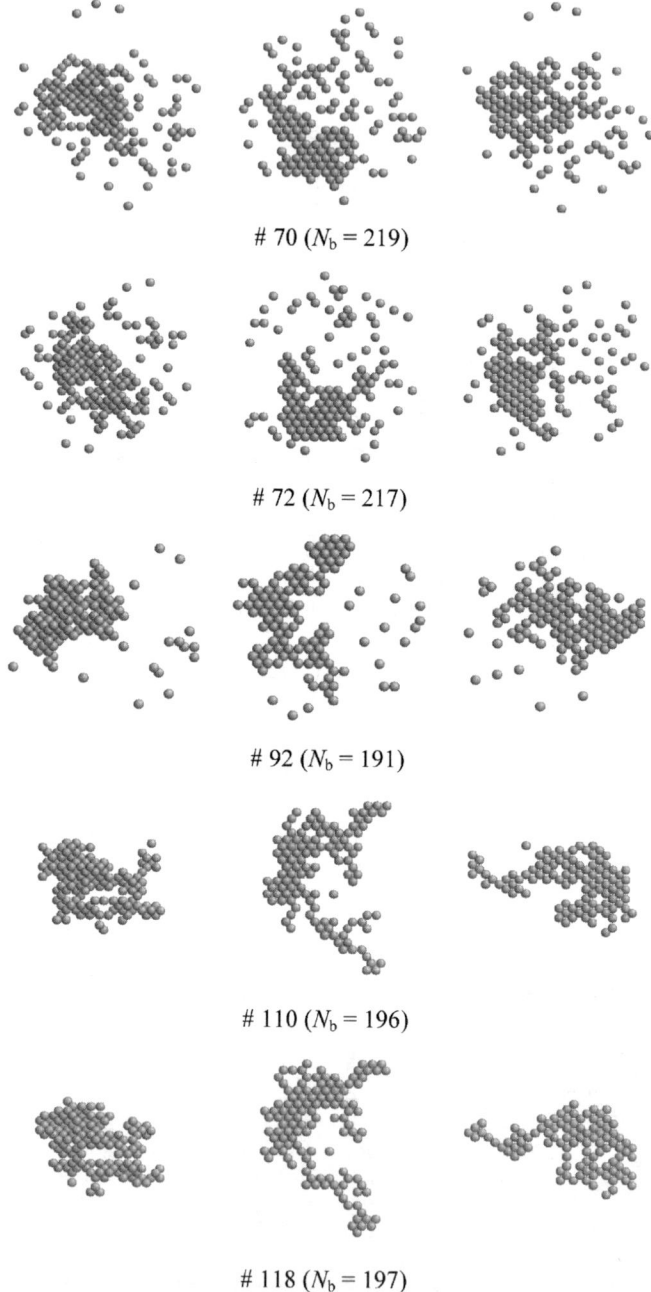

70 ($N_b = 219$)

72 ($N_b = 217$)

92 ($N_b = 191$)

110 ($N_b = 196$)

118 ($N_b = 197$)

Figure 23 *Evolution of dummy atoms configurations for the reconstruction of MS2 RNA, as created by subsequent runs with the program DAMMIN and shown for selected cycles. The models tend to be shaped at cycle about 70, get rid of isolated beads at cycle about 100, until finally, after about 120 cycles, no further significant improvements can be achieved*

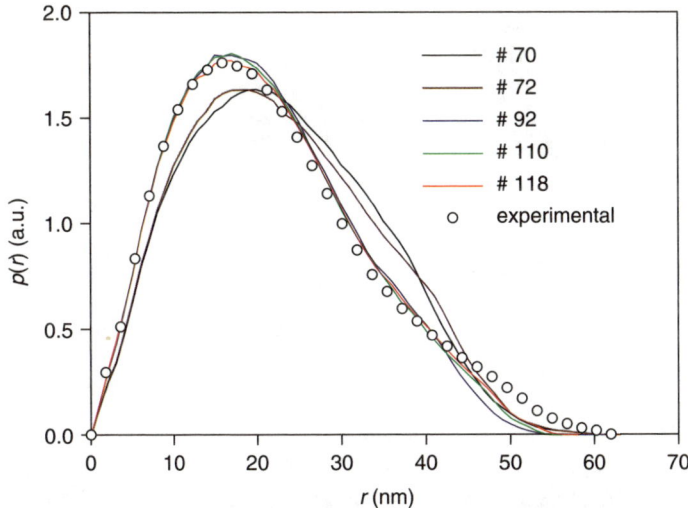

Figure 24 *Evolution of the PDDFs of the DAMMIN models for MS2 RNA shown in Figure 23*

application of the DAMAVER suite rather polishes the particle surface, thus eliminating characteristic details of the secondary structure of RNA in solution.

Surface representation techniques, such as those inherent in the CCP4 package, mirror the most essential features of the surface characteristics of RNA in solution. The surface renderings (Figure 26) corroborate the impression of the structural peculiarities (uneven surface, bent structure) already gained from the inspection of the models depicted in Figure 25. The less-detailed molecular features of these representations, however, characterise and govern the hydrodynamic behaviour more purposively.

Figures 27–29 illustrate the SAXS profiles of experimental and model data created by the applied *ab initio* approaches and the subsequent averaging procedures. The DAMMIN approach (Figure 27) convincingly demonstrates excellent agreement between observed $I(h)$ profiles and DAMMIN calculations up to high angles, and fair agreement between experimental data and model calculations based on bead coordinates by means of Debye's formula. The models obtained by averaging the restored structures exhibit marked deviations from the experimental data at rather small angles ($h < 0.2$ nm^{-1}). Averaging of the models by DAMHEXMIX was superior to that of the DAMAVER suite; in the latter case, however, discrepancies can be reduced slightly by rescaling procedures. Although less pronounced, similar findings are encountered when looking at the corresponding PDDFs. A comparison of the scattering curves and PDDFs of DALAI_GA models with the experimental data (Figure 28) reveals a behaviour similar to that found for DAMMIN: excellent agreement between experimental data and DALAI_GA calculations, and marked deviations, even at small angles, between the experiment and the calculations applying the averaged models. The agreement between experimental PDDFs and those obtained by DALAI_GA is inferior to that produced by DAMMIN. Application of the SAXS3D

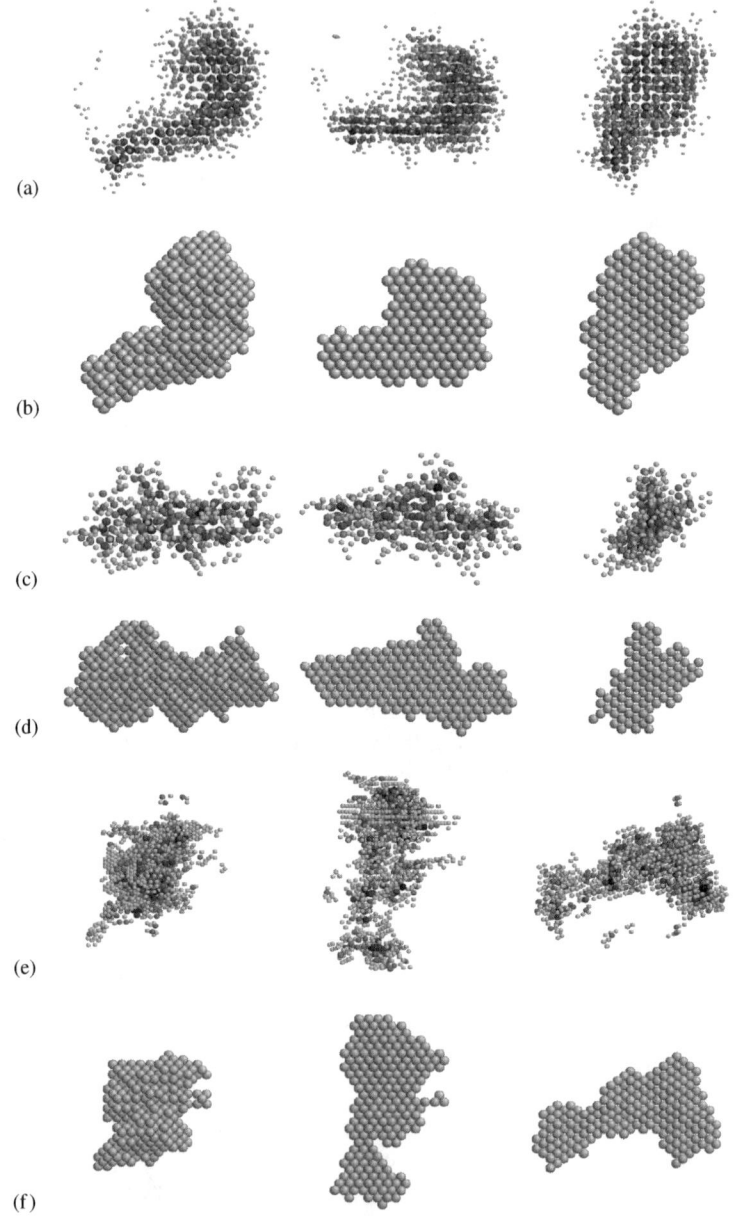

(a)

(b)

(c)

(d)

(e)

(f)

Figure 25   *Averaged models of MS2 RNA as obtained by different analytical approaches and averaging procedures. Each model is presented in three orthogonal views. Models (a) and (b) are based on 20 DAMMIN models, (c) and (d) on five DALAI_GA models, and (e) and (f) on three SAXS3D models. Models (a), (c), and (e) were averaged from individual ab initio models by means of our program DAMHEXMIX, models (b), (d), and (f) were averaged and filtered by means of Svergun's DAMAVER suite*

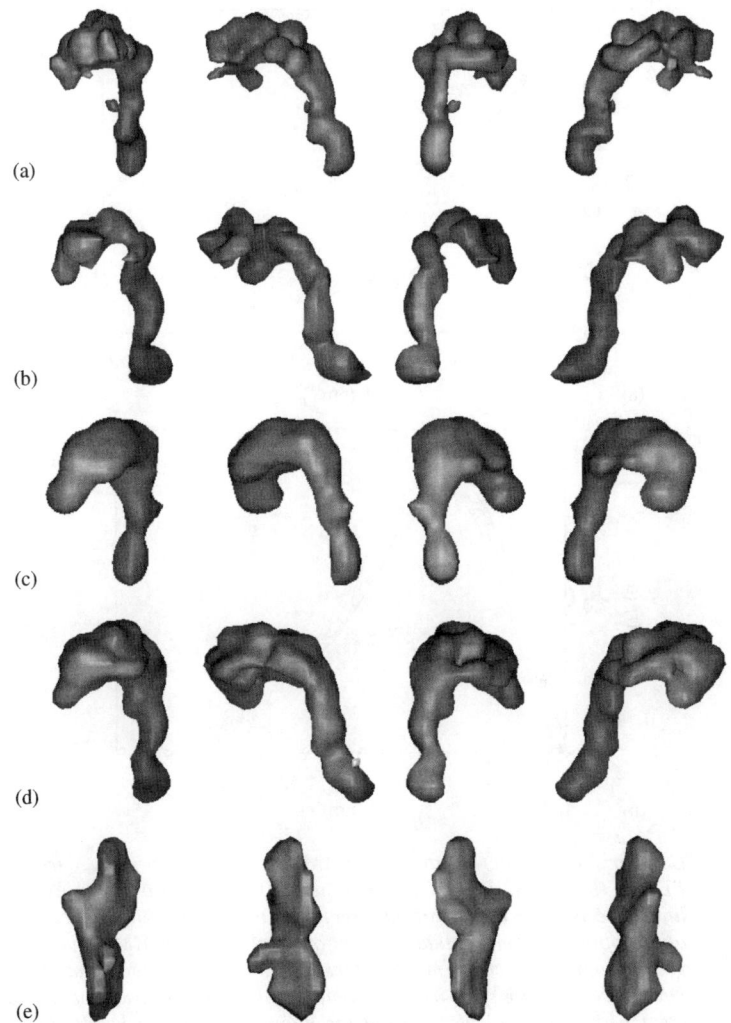

(a)

(b)

(c)

(d)

(e)

Figure 26 *Surface representations of averaged models of MS2 RNA. Each model is shown in four views corresponding to successive 90° rotations around the vertical axis. Models [(a)–(d)] are averages of ab initio DAMMIN models: (a) is based on four models created without any shape preference; for the eight models included in (b) or (c), preference to prolate or oblate shape was used as constraint; (d) was averaged from all 20 ab initio models irrespective of shape constraints. Model (e) was derived from five DALAI_GA ab initio models. All averages were computed by means of DAMHEXMIX*

procedure (Figure 29) reveals a lack of agreement between experimental data and models. For the scattering curves, the agreement may be improved by using the modified program version instead of the original SAXS3D algorithm. The occurrence of a marked shoulder in the outer part of the PDDFs, however, is obviously not dependent on the program version applied.

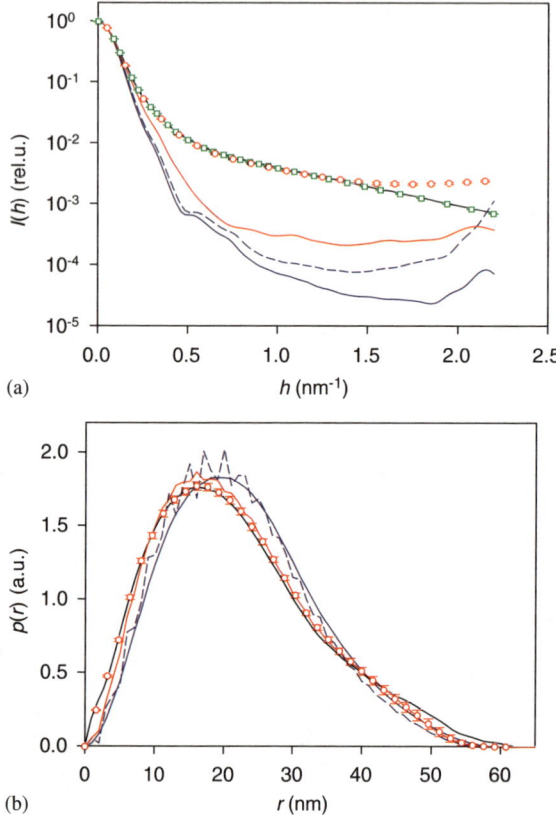

Figure 27 *Comparison of experimental and DAMMIN model scattering curves (a) and PDDFs (b) of MS2 RNA. The experimental functions are shown as solid black lines. The symbols with error bars represent the mean values and standard deviations obtained from evaluating a set of 19 ab initio DAMMIN models. (a) The green squares reflect the mean values of the I(h) curves calculated directly by DAMMIN (by using spherical harmonics) and the red circles symbolise the averaged I(h) curves calculated from the coordinates of the beads (by means of Debye's formula); to achieve maximum coincidence of the I(h) curves, the nominal bead radius r_b had to be reduced by a factor 0.5 in the Debye calculations. The coloured lines represent the I(h) (a) and p(r) (b) curves of averaged models; these models were obtained by averaging the structures of the 19 ab initio DAMMIN models by using our program DAMHEXMIX (red lines) or Svergun's DAMAVER suite (blue lines). Since the volume of the initial averaged model generated by DAMAVER was too big, this model was rescaled by applying two empirical factors to the bead radii and coordinates, respectively; the SAXS functions of the rescaled model are shown as dashed lines*

Quantification of the results of the *ab initio* modelling approaches in terms of structural and hydrodynamic parameters (Table 5) shows fair agreement between observables and predicted parameters for all procedures applied. Irrespective of the constraints selected, the DAMMIN calculations yield reasonable anticipations of the

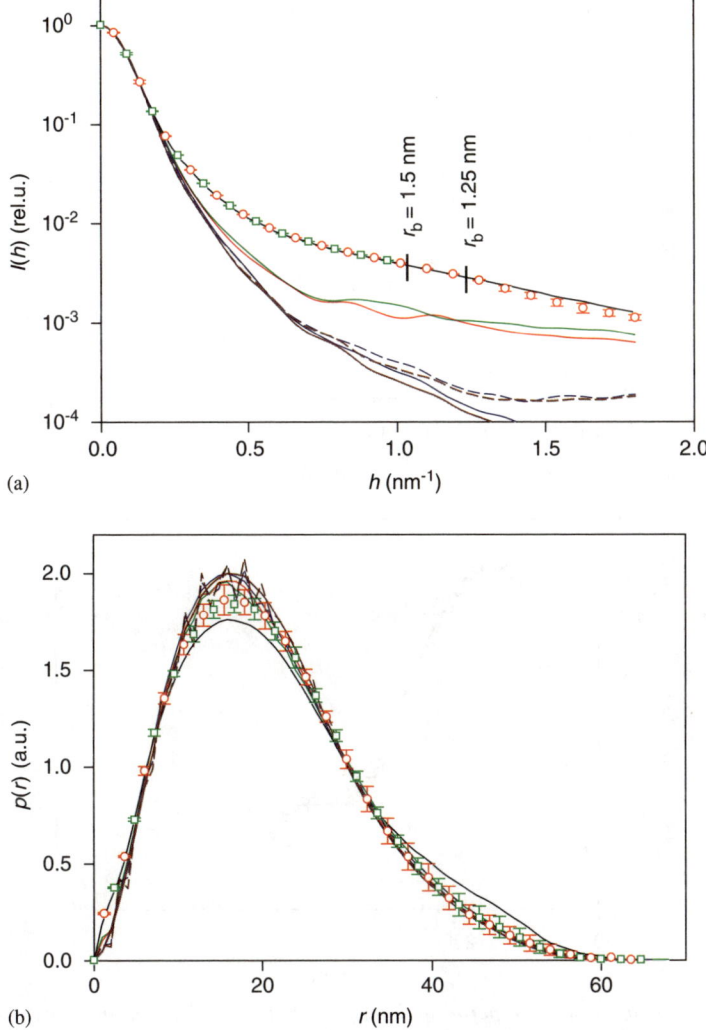

Figure 28 *Comparison of experimental and DALAI_GA model scattering curves (a) and PDDFs (b) of MS2 RNA. The experimental functions are shown as solid black lines. The symbols with error bars represent the mean values and standard deviations obtained from evaluating two related sets of ab initio DALAI_GA models. The coloured lines reflect the functions calculated for averaged models: these models were obtained by averaging the structures of five ab initio models by means of our program DAMHEXMIX (green and red lines) or Svergun's DAMAVER suite (blue and brown lines). The functions drawn in green or blue refer to models based on a nominal bead radius of 1.5 nm, and those drawn in red or brown to models based on a nominal r_b of 1.25 nm. The vertical bars (a) indicate the corresponding upper limits of the portions of experimental I(h) used in the DALAI_GA analysis. Since the volumes of the initial averaged models generated by DAMAVER were too big, these models were rescaled by applying empirical factors to the bead radii and coordinates, respectively; the SAXS functions of the rescaled models are shown as dashed lines*

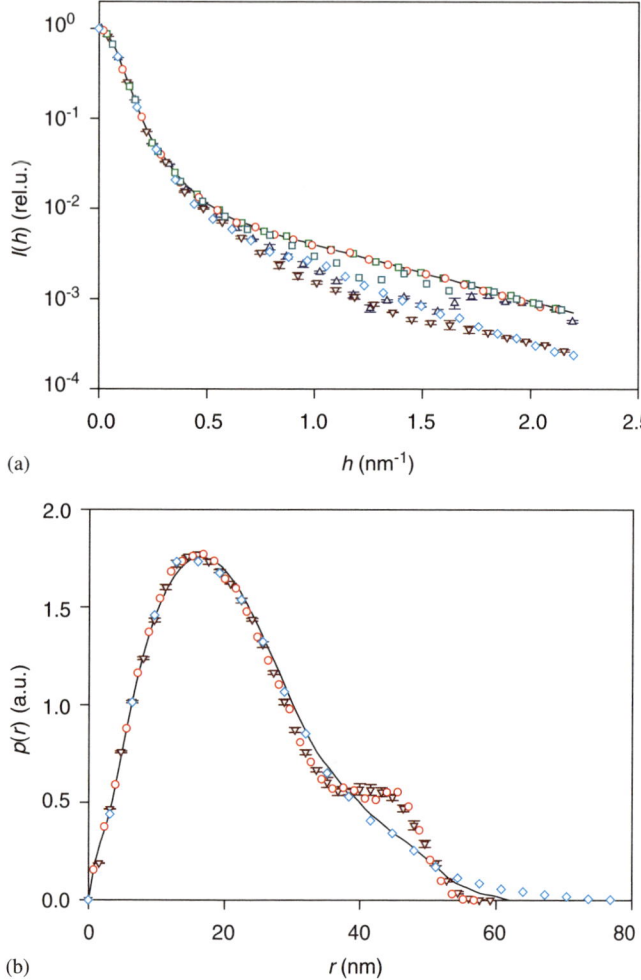

(a)

(b)

Figure 29 *Comparison of experimental and SAXS3D model scattering curves (a) and PDDFs (b) of MS2 RNA. The experimental functions are shown as solid black lines. The symbols with error bars represent the mean values and standard deviations obtained from evaluating three similar ab initio models generated by using the original version of SAXS3D; symbols without error bars refer to single models. (a) The blue triangles reflect the mean values of the I(h) curves calculated directly by the original version of SAXS3D (using Debye's formula in combination with an approximate distance histogram) and the brown triangles symbolise the average of the I(h) curves that were calculated from the coordinates of the beads by an exact application of Debye's formula; similarly, the dark cyan squares and the cyan diamonds symbolise the SAXS3D and the Debye scattering curve, respectively, of an alternative model, also created by means of the original program version. On the contrary, the green squares represent the SAXS3D scattering curve of another model that was established by using a modified program version, and the red circles correspond to the exact Debye scattering curve for that model. (b) The brown triangles, red circles and cyan diamonds refer to the same models as in (a)*

Table 5 Comparison of experimental and model parameters of MS2 RNA from E. coli

	N_m	N_b, r_b (nm)[a]	R_G (nm)	V (nm^3)	d_{max}[b] (nm)	D (10^{-7} cm^2 s^{-1})	s (10^{-13} s)	$[\eta]$[c] (cm^3 g^{-1})
Experimental[d]			18.1	2097[e] 21200[f]	62.0		26.6	44.0
DAMMIN models, unconstrained								
Ab initio	4	189±2 1.5	17.4±0.1	3600±40	61.4±1.2	1.27±0.01	30.8±0.2	33.8±0.6
Averaged (DAMAVER)[g]		612 1.5	17.4	3600	63.7	1.16	28.1	42.8
Averaged (DAMHEXMIX)		510 Variable	17.4	3600	62.9	1.15	27.8	38.5
DAMMIN models, prolate								
Ab initio	8	190±2 1.5	17.2±0.2	3620±40	59.8±1.4	1.29±0.01	31.2±0.3	32.6±0.7
Averaged (DAMAVER)[g]		789 1.5	17.2	3620	61.1	1.17	28.3	42.1
Averaged (DAMHEXMIX)		745 Variable	17.2	3620	63.2	1.12	27.2	46.0
DAMMIN models, oblate								
Ab initio	8	184±4 1.5	16.9±0.2	3520±70	58.6±1.6	1.24±0.03	30.0±0.6	36.4±2.2
Averaged (DAMAVER)[g]		824 1.5	16.9	3520	57.8	1.14	27.7	43.9
Averaged (DAMHEXMIX)		766 Variable	16.9	3520	60.5	1.05	25.5	54.4
DAMMIN models, all								
Ab initio	20	187±4 1.5	17.2±0.3	3580±70	59.6±1.7	1.26±0.03	30.6±0.7	34.3±2.2
Averaged (DAMAVER)[g]		804 1.75	17.2	3580	61.0	1.12	27.0	46.6
Averaged (DAMHEXMIX)		1144 Variable	17.2	3580	65.8	1.00	24.2	62.5

(continued)

Table 5 (Continued)

	N_m	N_b, r_b (nm)[a]	R_G (nm)	V (nm³)	d_{max}[b] (nm)	D (10^{-7} cm² s⁻¹)	s (10^{-13} s)	$[\eta]$[c] (cm³ g⁻¹)
DALAI_GA models, penultimate stage								
Ab initio	5	144±21.5 / 1.5	16.6±0.4	2750±30	62.8±2.0	1.31±0.02	31.8±0.6	30.0±1.5
Averaged (DAMAVER)[g]		614	16.6	2750	65.7	1.19	28.8	38.6
Averaged (DAMHEXMIX)		485 / Variable	16.7	2730	67.9	1.13	27.4	43.8
DALAI_GA models, last stage								
Ab initio	5	195±6 / 1.25	16.5±0.5	2150±60	62.6±1.5	1.32±0.02	32.1±0.4	28.6±1.1
Averaged (DAMAVER)[g]		666 / 1.5	16.5	2150	63.7	1.19	28.9	37.4
Averaged (DAMHEXMIX)		695 / Variable	16.6	2140	65.4	1.13	27.3	43.2
SAXS3D models								
Ab initio	3	675±8 / 1.0	17.8±0.0	3820±40	60.0±0.1	1.19±0.01	28.8±0.2	40.2±0.6
Averaged (DAMAVER)[g]		797 / 1.5	17.8	3820	63.4	1.12	27.1	47.1
Averaged (DAMHEXMIX)		1488 / Variable	17.8	3820	62.8	1.06	25.6	53.8
WB model (OE/SC: R_G, V)[h] p=0.279					63.2	1.09	26.6	44.5

[a]For the calculation of parameters, in general the nominal bead radii r_b were upscaled by a factor 1.105 to account for the packing density of the beads.

[b]Values for bead models correspond to the maximum vertex-to-vertex distance of the beads.

[c]For the calculation of [η] a volume correction was applied.[21,24]

[d]Experimental values were taken from the literature: SAXS data from ref. 105 and hydrodynamic data from ref. 111. The molar mass as obtained from SAXS is M = 1088 kg mol⁻¹.

[e]Correlation volume as cited in ref. 16.

[f]Volume of a model calculated from the dimensions.[16]

[g]For this model bead radii and coordinates were rescaled by applying two empirical scaling factors to match the mean R_G and V of the underlying ab initio models.

[h]OE/SC symbolises an oblate ellipsoid of revolution and WB treatment as a special case; the axial ratio, $p=a/b$, is defined as the ratio of the semiaxis of revolution, a, to the equatorial semiaxis, b, of the ellipsoid; the applied approach uses the experimental value for R_G and the calculated volume and axial ratio of a dimensions-based model.

maximum particle diameter and the hydrodynamic parameters. DALAI_GA and SAXS3D seem to be equally well suited for making predictions. For the averaging process, the DAMAVER suite seems to be of advantage in the case of the DAMMIN calculations, while DAMHEXMIX might be used more favourably when applying DALAI_GA or SAXS3D; the systematically lower values for D and s obtained upon application of DAMHEXMIX averaging may be a consequence of the enhanced rugosity of the models consisting of unequal beads. For the simple prediction of hydrodynamic parameters, the WB approach is also effective, provided that special assumptions concerning the hydrodynamically effective volume are made.

5 Conclusions

Since both scattering and hydrodynamic techniques are able to provide valuable information on the properties of biopolymers in solution, these complementary methods can be exploited to the best advantage for establishing appropriate models of the molecules under analysis. This endeavour also involves the prediction of reliable scattering and hydrodynamic molecule properties on the basis of SAXS data, irrespective of the ultimate perfection of the models derived.

If solution X-ray scattering data of biopolymers are known, *ab initio* modelling of biopolymers may be applied advantageously, in addition to the rather conventional trial-and-error multibody or WB approaches. In the case of simple WB approaches, knowledge of R_G and V values is sufficient, apart from information on M and $\bar{v}$. A necessary requirement for the application of the more sophisticated calculation procedures, such as the *ab initio* approaches, is the availability of precise SAXS profiles in a wide angular range. Usually, model targeting requires repeated application of the modelling procedures chosen. Further structural information, for instance particle symmetry, can be utilised accessorily, to speed up the finding of realistic particle shapes and to avoid structures unsuitable from the biological point of view.

In the absence of symmetry constraints, all *ab initio* approaches tested (DAMMIN, DALAI_GA, SAXS3D) turned out to yield similar models for the biopolymers under investigation (proteins, nucleic acids). The accuracy of the individual model trials can be checked by superimposing the models achieved. For further evaluation of the molecule reconstructions, averaging procedures have to be adopted. This can be achieved by means of the DAMAVER suite or our programs DAMHEXMIX or DAMMIX; the quality of the results, however, will depend on the problem to be considered.

The problem of shape uniqueness is pertinent to all models applied.[1,70] Major differences in the results of the *ab initio* approaches used occur if inclusion of constraints is desired. In this connection, the program DAMMIN is clearly superior to the other approaches because it is capable of processing other types of information, such as symmetry specifications. While both DAMMIN and DALAI_GA provide excellent fits of experimental $I(h)$ profiles, even in the high angular range, the application of SAXS3D seems to be less purposive, particularly in its original version. Similar disagreements between the results of the different approaches are found when looking at the $p(r)$ functions. In this case as well, the use of SAXS3D yields the most obvious distinctions. Discrepancies between the scattering profiles provided by the *ab initio* approaches and Debye scattering profiles of the resulting beaded model structures reveal a similar nonconformity. On the other hand, the combined use of reconstruction approaches based on different modelling philosophies enhances the maximum likelihood of the models generated, thereby minimising the problem of particle uniqueness.

All the *ab initio* approaches tested allow the reasonable prediction of structural and hydrodynamic parameters (in particular, R_G, V, d_{max}, D and s), even in those cases where modelling attempts resulted in less realistic models. For the prediction of hydrodynamic parameters, HYDRO is usually the program of choice. If only hydrodynamic quantities are the target of modelling trials, the use of simple WB approaches may be sufficient, provided that the biopolymers under investigation exhibit average properties. Rather complex molecules (*e.g.* molecules containing considerable cavities or branching, or molecules with shapes deviating markedly from simple tri-axial bodies), however, require the more advanced modelling approaches. Among them, the *ab initio* approaches turn out to be the most promising ones. However, while application of pure *ab initio* conditions may lead, in certain cases, to unsatisfactory shape reconstructions (which nevertheless provide good fits to the experimental scattering profiles), the concomitant use of plausible constraints regarding symmetry and shape or, alternatively, the use of appropriate templates, is a valuable tool for optimising the outcome of the modelling process.

Acknowledgements

The authors are much obliged to several scientists for the use of their computer programs: to D. I. Svergun for DAMMIN, the DAMAVER suite, GNOM and SUPCOMB; to J. F. Díaz for DALAI_GA; to D. Walther for SAXS3D; to J. García de la Torre for HYDRO; to R. A. Sayle for RASMOL; to the SERC Daresbury Laboratory for the CCP4 suite; and to the Research Foundation of the State University of New York for VOLVIS. This work was supported by the Austrian Academy of Sciences (APART fellowship granted to A. K.).

References

1. P. Zipper and H. Durchschlag, *J. Appl. Crystallogr.*, 2000, **33**, 788.
2. P. Zipper and H. Durchschlag, *J. Appl. Crystallogr.*, 2003, **36**, 509.
3. D. I. Svergun, *Biophys. J.*, 1999, **76**, 2879.
4. P. Chacón, F. Morán, J. F. Díaz, E. Pantos and J. M. Andreu, *Biophys. J.*, 1998, **74**, 2760.
5. P. Chacón, J. F. Díaz, F. Morán and J. M. Andreu, *J. Mol. Biol.*, 2000, **299**, 1289.
6. D. Walther, F. E. Cohen and S. Doniach, *J. Appl. Crystallogr.*, 2000, **33**, 350.
7. J. García de la Torre, S. Navarro, M. C. López Martínez, F. G. Díaz and J. J. López Cascales, *Biophys. J.*, 1994, **67**, 530.
8. J. García de la Torre, M. L. Huertas and B. Carrasco, *Biophys. J.*, 2000, **78**, 719.
9. H. Durchschlag and P. Zipper, *Prog. Colloid Polym. Sci.*, 2002, **119**, 121.
10. H. Durchschlag and P. Zipper, *Eur. Biophys. J.*, 2003, **32**, 487.
11. T. F. Kumosinski and H. Pessen, *Arch. Biochem. Biophys.*, 1982, **219**, 89.
12. S. E. Harding, in *Dynamic Properties of Biomolecular Assemblies*, S. E. Harding and A. J. Rowe (eds), Royal Society of Chemistry, Cambridge, UK, 1989, 32.
13. S. E. Harding, *Biophys. Chem.*, 1995, **55**, 69.

14. S. E. Harding, *Prog. Biophys. Mol. Biol.*, 1997, **68**, 207.
15. S. E. Harding, J. C. Horton, H. Cölfen, *Eur. Biophys. J.*, 1997, **25**, 347.
16. H. Durchschlag and P. Zipper, *J. Appl. Crystallogr.*, 1997, **30**, 1112.
17. H. Durchschlag and P. Zipper, *Prog. Colloid Polym. Sci.*, 1997, **107**, 43.
18. H. Durchschlag and P. Zipper, *Prog. Colloid Polym. Sci.*, 1999, **113**, 87.
19. H. Durchschlag and P. Zipper, *Biophys. Chem.*, 2001, **93**, 141.
20. J. García de la Torre and V. A. Bloomfield, *Q. Rev. Biophys.*, 1981, **14**, 81.
21. J. García de la Torre, in *Dynamic Properties of Biomolecular Assemblies*, S. E. Harding and A. J. Rowe (eds), Royal Society of Chemistry, Cambridge, UK, 1989, 3.
22. J. García de la Torre, B. Carrasco and S. E. Harding, *Eur. Biophys. J.*, 1997, **25**, 361.
23. B. Carrasco and J. García de la Torre, *Biophys. J.*, 1999, **75**, 3044.
24. B. Carrasco, J. García de la Torre and P. Zipper, *Eur. Biophys. J.*, 1999, **28**, 510.
25. J. García de la Torre, *Biophys. Chem.*, 2001, **93**, 159.
26. J. García de la Torre and B. Carrasco, *Biopolymers*, 2002, **63**, 163.
27. J. García de la Torre, H. E. Pérez Sánchez, A. Ortega, J. G. Hernández, M. X. Fernandes, F. G. Díaz and M. C. López Martínez, *Eur. Biophys. J.*, 2003, **32**, 477.
28. E. Longman, K. Kreusel, S. B. Tendler, I. Fiebrig, K. King, J. Adair, P. O'Shea, A. Ortega, J. García de la Torre and S. E. Harding, *Eur. Biophys. J.*, 2003, **32**, 503.
29. O. Byron, *Biophys. J.*, 1997, **72**, 408.
30. O. Byron, *Method. Enzymol.*, 2000, **321**, 278.
31. B. Spotorno, L. Piccinini, G. Tassara, C. Ruggiero, M. Nardini, F. Molina and M. Rocco, *Eur. Biophys. J.*, 1997, **25**, 373, and erratum **26**, 417.
32. M. Yu. Pavlov and B. A. Fedorov, *Biopolymers*, 1983, **22**, 1507.
33. J. J. Müller, *J. Appl. Crystallogr.*, 1983, **16**, 74.
34. J. J. Müller, *Biopolymers*, 1991, **31**, 149.
35. J. J. Müller and H. Schrauber, *J. Appl. Crystallogr.*, 1992, **25**, 181.
36. D. Svergun, C. Barberato and M. H. J Koch, *J. Appl. Crystallogr.*, 1995, **28**, 768.
37. A. J. Beavil, R. J. Young, B. J. Sutton and S. J. Perkins, *Biochemistry*, 1995, **34**, 14449.
38. S. J. Perkins, A. W. Ashton, M. K. Boehm and D. Chamberlain, *Int. J. Biol. Macromol.*, 1998, **22**, 1.
39. S. J. Perkins, *Biophys. Chem.*, 2001, **93**, 129.
40. H. Durchschlag, P. Zipper, R. Wilfing and G. Purr, *J. Appl. Crystallogr.*, 1991, **24**, 822.
41. H. Durchschlag, P. Zipper, G. Purr and R. Jaenicke, *Colloid Polym. Sci.*, 1996, **274**, 117.
42. P. Zipper and H. Durchschlag, *Prog. Colloid Polym. Sci.*, 1997, **107**, 58.
43. P. Zipper and H. Durchschlag, *Biochem. Soc. Trans.*, 1998, **26**, 726.
44. P. Zipper and H. Durchschlag, *Physica A*, 2002, **304**, 283.
45. P. Zipper and H. Durchschlag, *Physica A*, 2002, **314**, 613.
46. P. Zipper, A. Krebs and H. Durchschlag, *Prog. Colloid Polym. Sci.*, 2002, **119**, 141.
47. P. Zipper, A. Krebs and H. Durchschlag, *Prog. Colloid Polym. Sci.*, 2004, **127**, 126.
48. A. Krebs, H. Durchschlag and P. Zipper, *Biophys. J.*, 2004, **87**, 1173.
49. P. Zipper and H. Durchschlag, *Physica Scripta*, 2005, **T118**, 228.
50. O. Glatter and O. Kratky (eds), *Small Angle X-ray Scattering*, Academic Press, London, 1982.
51. D. I. Svergun, *J. Appl. Crystallogr.*, 1997, **30**, 792.
52. D. I. Svergun, S. Richard, M. H. J. Koch, Z. Sayers, S. Kuprin and G. Zaccai, *Proc. Natl. Acad. Sci. USA*, 1998, **95**, 2267.
53. H. Durchschlag and P. Zipper, *J. Phys.: Condens. Matter*, 2002, **14**, 2439.
54. H. Durchschlag and P. Zipper, *Prog. Colloid Polym. Sci.*, 2002, **119**, 131.
55. H. Durchschlag and P. Zipper, *Prog. Colloid Polym. Sci.*, 2004, **127**, 98.
56. O. Kratky and I. Pilz, *Q. Rev. Biophys.*, 1978, **11**, 39.

57. A. Krebs, J. Lamy, S. N. Vinogradov and P. Zipper, *Biopolymers*, 1998, **45**, 289.
58. D. I. Svergun, *J. Appl. Crystallogr.*, 2000, **33**, 530.
59. D. I. Svergun and M. H. J. Koch, *Curr. Opin. Struct. Biol.*, 2002, **12**, 654.
60. D. I. Svergun and M. H. J. Koch, *Rep. Prog. Phys.*, 2003, **66**, 1735.
61. M. H. J. Koch, P. Vachette and D. I. Svergun, *Q. Rev. Biophys.*, 2003, **36**, 147.
62. M. Hammel, M. Kriechbaum, A. Gries, G. M. Kostner, P. Laggner and R. Prassl, *J. Mol. Biol.*, 2002, **321**, 85.
63. Y. Takahashi, Y. Nishikawa and T. Fujisawa, *J. Appl. Crystallogr.*, 2003, **36**, 549.
64. D. I. Svergun, M. V. Petoukhov and M. H. J. Koch, *Biophys. J.*, 2001, **80**, 2946.
65. M. V. Petoukhov and D. I. Svergun, *J. Appl. Crystallogr.*, 2003, **36**, 540.
66. D. Vigil, S. C. Gallagher, J. Trewhella and A. E. García, *Biophys. J.*, 2001, **80**, 2082.
67. M. Kojima, A. A. Timchenko, J. Higo, K. Ito, H. Kihara, and K. Takahashi, *J. Appl. Crystallogr.*, 2004, **37**, 103.
68. D. I. Svergun, *J. Appl. Crystallogr.*, 1992, **25**, 495.
69. W. T. Heller, E. Abusamhadneh, N. Finley, P. R. Rosevear and J. Trewhella, *Biochemistry*, 2002, **41**, 15654.
70. V. V. Volkov and D. I. Svergun, *J. Appl. Crystallogr.*, 2003, **36**, 860.
71. M. B. Kozin and D. I. Svergun, *J. Appl. Crystallogr.*, 2001, **34**, 33.
72. R. A. Sayle and E. J. Milner-White, *Trends Biochem. Sci.*, 1995, **20**, 374.
73. Collaborative Computational Project, Number 4, *Acta Crystallogr.*, 1994, **D50**, 760.
74. R. Avila, T. He, L. Hong, A. Kaufman, H. Pfister, C. Silva, L. Sobierajski, S. Wang, in *Proceedings IEEE Visualization '94*, R. Bergeron and A. Kaufman (eds), IEEE Computer Society, Washington DC, USA, 1994, 31.
75. D. I. Svergun and H. B. Stuhrmann, *Acta Crystallogr.*, 1991, **A47**, 736.
76. O. Glatter, *Acta Phys. Austriaca*, 1980, **52**, 243.
77. L. N. Johnson, *FASEB J.*, 1992, **6**, 2274.
78. V. L. Rath and R. J. Fletterick, *Nature Struct. Biol.*, 1994, **1**, 681.
79. R. Schinzel and B. Nidetzky, *FEMS Microbiol. Lett.*, 1999, **171**, 73.
80. A. Weinhäusel, R. Griessler, A. Krebs, P. Zipper, D. Haltrich, K. D. Kulbe and B. Nidetzky, *Biochem. J.*, 1997, **326**, 773.
81. K. A. Watson, R. Schinzel, D. Palm and L. N. Johnson, *EMBO J.*, 1997, **16**, 1.
82. K. A. Watson, C. McCleverty, S. Geremia, S. Cottaz, H. Driguez and L. N. Johnson, *EMBO J.*, 1999, **18**, 4619.
83. G. Puchwein, O. Kratky, C. F. Gölker and E. Helmreich, *Biochemistry*, 1970, **9**, 4691.
84. L. N. Johnson and D. Barford, *J. Biol. Chem.*, 1990, **265**, 2409.
85. K. Lin, V. L. Rath, S. C. Dai, R. J. Fletterick and P. K. Hwang, *Science*, 1996, **273**, 1539.
86. C. C. Childress and B. Sacktor, *J. Biol. Chem.*, 1970, **245**, 2927.
87. N. A. Chebotareva, S. E. Harding and D. J. Winzor, *Eur. J. Biochem.*, 2001, **268**, 506.
88. G. Henriksson, G. Johansson and G. Pettersson, *J. Biotechnol.*, 2000, **78**, 93.
89. D. Lehner, P. Zipper, G. Henriksson and G. Pettersson, *Biochim. Biophys. Acta*, 1996, **1293**, 161.
90. A. R. Ayers, S. B. Ayers and K.-E. Eriksson, *Eur. J. Biochem.*, 1978, **90**, 171.
91. B. M. Hallberg, T. Bergfors, K. Bäckbro, G. Pettersson, G. Henriksson and C. Divne, *Structure*, 2000, **8**, 79.
92. B. M. Hallberg, G. Henriksson, G. Pettersson and C. Divne, *J. Mol. Biol.*, 2002, **315**, 421.
93. R. E. Weber and S. N. Vinogradov, *Physiol. Rev.*, 2001, **81**, 569.
94. J. N. Lamy, B. N. Green, A. Toulmond, J. S. Wall, R. E. Weber and S. N. Vinogradov, *Chem. Rev.*, 1996, **96**, 3113.
95. W. E. Royer Jr. and W. A. Hendrickson, *J. Biol. Chem.*, 1988, **263**, 13762

96. W. E. Royer Jr., K. Strand, M. van Heel and W. A. Hendrickson, *Proc. Natl. Acad. Sci. USA*, 2000, **97**, 7107.

97. K. Strand, J. E. Knapp, B. Bhyravbhatla and W. E. Royer Jr., *J. Mol. Biol.*, 2004, **344**, 119.

98. M. Schatz, E. V. Orlova, P. Dube, J. Jäger and M. van Heel, *J. Struct. Biol.*, 1995, **114**, 28.

99. F. de Haas, A. Kuchumov, J.-C. Taveau, N. Boisset, S. N. Vinogradov and J. N. Lamy, *Biochemistry*, 1997, **36**, 7330.

100. J.-C. Taveau, N. Boisset, S. N. Vinogradov and J. N. Lamy. *J. Mol. Biol.*, 1999, **289**, 1343.

101. F. Mouche, N. Boisset and P. A. Penczek, *J. Struct. Biol.*, 2001, **133**, 176.

102. A. Krebs, P. Zipper and S. N. Vinogradov, *Biochim. Biophys. Acta*, 1996, **1297**, 115.

103. A. Krebs, A. R. Kuchumov, P. K. Sharma, E. H. Braswell, P. Zipper, R. E. Weber, G. Chottard and S. N. Vinogradov, *J. Biol. Chem.*, 1996, **271**, 18695.

104. W. Fiers, R. Contreras, F. Duerinck, G. Haegeman, D. Iserentant, J. Merregaert, W. Min Jou, F. Molemans, A. Raeymaekers, A. Van den Berghe, G. Volckaert and M. Ysebaert, *Nature*, 1976, **260**, 500.

105. P. Zipper, W. Folkhard and J. Clauwaert, *FEBS Lett.*, 1975, **56**, 283.

106. P. Zipper, in *Small Angle X-ray Scattering*, O. Glatter and O. Kratky (eds), Academic Press, London, 1982, 295.

107. G. Ribitsch, R. De Clercq, W. Folkhard, P. Zipper, J. Schurz and J. Clauwaert, *Z. Naturforsch.*, 1985, **40c**, 234.

108. D. A. Kuzmanovic, I. Elashvili, C. Wick, C. O'Connell and S. Krueger, *Structure*, 2003, **11**, 1339.

109. K. Valegård, L. Liljas, K. Fridborg and T. Unge, *Nature*, 1990, **345**, 36.

110. R. Golmohammadi, K. Valegård, K. Fridborg and L. Liljas, *J. Mol. Biol.*, 1993, **234**, 620.

111. H. Slegers, J. Clauwaert and W. Fiers, *Biopolymers*, 1973, **12**, 2033.

112. S. S. Funari, G. Rapp, M. Perbandt, K. Dierks, M. Vallazza, C. Betzel, V. A. Erdmann and D. I. Svergun, *J. Biol. Chem.*, 2000, **275**, 31283.

113. A. Krebs, Thesis, University of Graz, Austria, 1996.

Fitting of Thermodynamic Data from Equilibrium Ultracentrifugation by Robust Least-Squares with Reduced Parameter Cross-Correlation Coefficients

MARC S. LEWIS, MICHAEL M. REILY AND
LESLIE A. HOLLADAY

1 Introduction

Equilibrium analytical ultracentrifugation is based entirely upon the first principles of reversible thermodynamics.[1] As such, it permits the direct measurement of the masses and the relative concentrations of the reactants and the products of reversible chemical reactions as a function of radial position at both chemical and ultracentrifugal equilibrium. In addition to permitting obtaining the stoichiometries of the reactions, this permits the calculation of the equilibrium constants of the reactions, and from these, the values of the changes of the standard Gibbs free energies, $\Delta G°_T$ for these reactions as a function of temperature. The temperature dependence of the changes of free energy permits calculation of the corresponding changes in enthalpy, $\Delta H°_T$, entropy, $\Delta S°_T$, and specific heat content, $\Delta C°_{P,T}$, for the reactions being studied, values which can be of considerable significance in understanding the nature of the reactions.

Examination of the literature pertaining to the thermodynamics of biochemical reactions clearly reveals that such studies are in the minority. Microcalorimetry, and in particular, isothermal titration microcalorimetry (ITC) is the predominantly used technique. While each technique has its advantages and disadvantages, it is not the intention of this chapter to compare their relative merits, since they are actually complementary approaches to the same problem. Analytical ultracentrifugation has the advantage of being very parsimonious of material and, in our experience, is capable of measuring equilibrium constants over a very wide range of values, 10^2–10^8 for 1:1 associations, which may be either homogeneous (self-associations) or heterogeneous, and with corresponding values for other stoichiometries. It has proved capable of analyzing reactions which defied calorimetric analysis, such as the thermodynamics of antiparallel hairpin-double helix equilibria in oligonucleotides.[2] It is capable of dealing with complex reactions and with appropriate labeling, has the potential for analyzing reactions involving more than two reactants. However, the time required to attain equilibrium and the time required to perform multiple experiments when thermal stability has been a problem[3] can be daunting and frequently renders such analysis impractical. In spite of these, ultracentrifugal thermodynamic analysis can be an optimal technique for certain problems, such as when the values of ΔH_T° are small in the temperature range studied, and thus does have a significant role to play in biothermodynamics.

It is the intent of this chapter to deal primarily with one particular problem in the thermodynamic analysis of equilibrium ultracentrifugal data, that of obtaining optimal values of the parameters and their standard errors by the minimization of parameter cross-correlation or dependency values and the application of appropriate weights when performing linear least-squares fitting of the data. Because of the significant role that error in the data plays in such analyses, there will also be a limited discussion of the sources of such error.

2 Mathematical Models for Thermodynamic Analysis

The Gibbs standard free energy change, ΔG_T°, as determined by the value of $\ln K_{a,T}$ that is obtained from the analysis of the equilibrium ultracentrifugal data of the association under investigation at temperature T, and its relationship to the values of the standard changes in enthalpy, ΔH_T°, and entropy, ΔS_T°, are given by

$$\Delta G_T^\circ = - RT \ln K_{a,T} = \Delta H_T^\circ - T\Delta S_T^\circ \tag{1}$$

These values of ΔG_T° can be fit using a function derived from the standard definitions of the thermodynamic parameters. Beginning with the change in the specific heat content, which is critical because it is the primary factor for determining the temperature dependence of the reaction:

$$\Delta C_{P,T}^\circ = \mathrm{d}\Delta H_T^\circ/\mathrm{d}T = \Delta C_{P,T_\mathrm{R}}^\circ + \int_{T_\mathrm{R}}^{T} (\mathrm{d}\Delta C_{P,T_\mathrm{R}}^\circ/\mathrm{d}T)\,\mathrm{d}T = \Delta C_{P,T_\mathrm{R}}^\circ + \delta'C_P^\circ(T - T_\mathrm{R}) \tag{2}$$

where T_R is a reference temperature and $\delta'C_P^\circ$ denotes the first derivative of the specific heat content, $\Delta C_{P,T}^\circ$ with respect to temperature at T_R and will be assumed to be temperature-invariant. Then,

$$\Delta H_T^\circ = \Delta H_{T_\mathrm{R}}^\circ + \int_{T_\mathrm{R}}^{T} \Delta C_{P,T}^\circ\,\mathrm{d}T = \Delta H_{T_\mathrm{R}}^\circ + \int_{T_\mathrm{R}}^{T} \Delta C_{P,T_\mathrm{R}}^\circ\,\mathrm{d}T + \int_{T_\mathrm{R}}^{T} \delta'C_P^\circ(T - T_\mathrm{R})\,\mathrm{d}T$$

$$= \Delta H_{T_\mathrm{R}}^\circ + \Delta C_{P,T_\mathrm{R}}^\circ(T - T_\mathrm{R}) + \delta'C_P^\circ(T^2/2 + T_\mathrm{R}^2/2 - TT_\mathrm{R}) \tag{3}$$

and

$$\Delta S_T^\circ = \Delta S_{T_R}^\circ + \int_{T_R}^T (\Delta C_{P,T}^\circ / T) dT$$

$$= \Delta S_{T_R}^\circ + \int_{T_R}^T (\Delta C_{P,T_R}^\circ / T) dT + \int_{T_R}^T \delta' C_P^\circ dT - \int_{T_R}^T T_R (\delta' C_P^\circ / T) dT \qquad (4)$$

$$= \Delta S_{T_R}^\circ + \Delta C_{P,T_R}^\circ \ln(T/T_R) + \delta' C_P^\circ (T - T_R - T_R \ln(T/T_R))$$

Substituting the integrated forms of Equations (3) and (4) in Equation (1) and rearranging then gives

$$\Delta G^\circ{}_T = \Delta H_{T_R}^\circ - T \Delta S_{T_R}^\circ + T \Delta C_{P,T_R}^\circ (1 - T_R/T - \ln(T/T_R))$$

$$+ T T_R \delta' C_P^\circ (T_R/T - T/T_R + 2\ln(T/T_R)) \qquad (5)$$

Extensive experience with both real and simulated data has demonstrated that it is not possible to successfully fit for four parameters since such a model is extremely ill-conditioned and gives quite meaningless parameter values with equally meaningless and excessively large associated errors and very large parameter dependencies. This makes it necessary to truncate at the $\delta' C_P^\circ$ term and to assume that ΔC_P° is temperature-invariant. Thus, the operative fitting function is

$$\Delta G_T^\circ = \Delta H_{T_R}^\circ - T \Delta S_{T_R}^\circ + T \Delta C_P^\circ (1 - T_R/T - \ln(T/T_R)) \qquad (6)$$

It can be demonstrated that as long as T_R is not a fitting parameter, Equation (6) is linear in the parameters. Since one obtains that $\partial \Delta G_T^\circ / \partial \Delta H_{T_R}^\circ = 1$, $\partial \Delta G_T^\circ / \partial \Delta S_{T_R}^\circ = -T$, and $\partial \Delta G_T^\circ / \partial \Delta C_P^\circ = (T - T_R) - T \ln(T/T_R)$, their second derivatives are all equal to zero, which is requisite for an equation being linear in its parameters. However, if T_R is a parameter, then $\partial \Delta G_T^\circ / \partial T_R = -1 + (T/T_R) \Delta C_P^\circ$, and $\partial^2 \Delta G_T^\circ / \partial T_R^2 = -(T/T_R^2) \Delta C_P^\circ$, which is obviously not equal to zero, demonstrating that the inclusion of T_R as a fitting parameter makes Equation (6) non-linear. By inference, Equation (5) is also either linear or non-linear by the same criteria.

There are significant advantages to fitting linear equations.[4] In the Levenberg–Marquardt algorithm, which was used here, and which is a modification of the Gauss–Newton least-squares procedure, each of the i data points is approximated as a first-order series expansion of the fitting function about an estimation of the parameter values p'.

$$Y_i/\sigma_i = F(X_i, p)/\sigma_i = F(X_i, p')/\sigma_i + (1/\sigma_i) \sum_j [(\partial F(X_i, p')/\partial p_j')(p_j - p_j')] + \cdots \qquad (7)$$

where Y_i is the dependent variable, X_i the independent variable, σ_i the standard error of the mean of Y_i, F the fitting function, p the vector of parameter values that are to be determined by the weighted least-squares fitting procedure, and p' an estimated vector of fitting parameters with j denoting a specific parameter. The higher order derivatives in Equation (7) are all zero for a linear fitting function. Thus, the first-order series expansion is exact and Equation (7) will give exact values for the parameters. If the fitting function is non-linear, then the series expansion in this equation is approximate and the corresponding parameter values will also be approximate. These returned values are used as the new parameter estimates and the procedure is repeated until the changes in parameter estimates returns a sum of squares which differs from the sum of squares of the preceding iteration by less than a determined convergence limit.

3 Optimized Fitting of Noiseless Data

We have chosen to begin the exploration of optimization with noiseless data since this eliminates the need for weighting the data, permitting the assumption that σ_i equals one, and also eliminates any effects attributable to noise in the optimization process and lets us defer that consideration to a later section. For purposes of exposition, we have chosen to simulate data with a 1:1 stoichiometry which can be either a reversible $A+B=AB$ reaction or a reversible monomer–dimer reaction with values of ΔG_T° in the range of -5.8 to -9.8 kcal mol^{-1}. Equation (6) has been used to generate this data with values of $\Delta H_{T_R}^\circ = 20\,000$ cal mol^{-1}, $\Delta S_{T_R}^\circ = 100$ cal mol^{-1} K^{-1}, $\Delta C_P^\circ = -3000$ cal mol^{-1} K^{-1}, and $T_R = 293.15$ K. These values are fairly typical of those we have observed in this laboratory; the relatively large negative value of ΔC_P° was used because we wished to have data that exhibited marked temperature dependence. Such a value is usually attributed to the association resulting in the extensive shielding of non-polar residues from the solvent.[5,6] The resulting data are presented in Table 1 and illustrated in Figure 1.

Table 1 *Values of ΔG_T° over the temperature range of 275.15–311.15 K (2–38 °C) with 3 °C increments generated using Equation (6) with the parameter values $\Delta H_{T_R}^\circ = 20000$ cal mol^{-1}, $\Delta S_{T_R}^\circ = 100$ cal mol^{-1} K^{-1}, $\Delta C_P^\circ = -3000$ cal mol^{-1} K^{-1}, and $T_R = 293.15$ K. No noise has been added to the data, which has been rounded to the nearest 0.1 cal mol^{-1}*

Temperature (K)	ΔG_T° (cal mol^{-1})
275.15	-5822.1
278.15	-6643.6
281.15	-7367.9
284.15	-7996.2
287.15	-8529.4
290.15	-8968.8
293.15	-9315.0
296.15	-9569.1
299.15	-9712.0
302.15	-9804.7
305.15	-9788.0
308.15	-9682.9
311.15	-9490.1

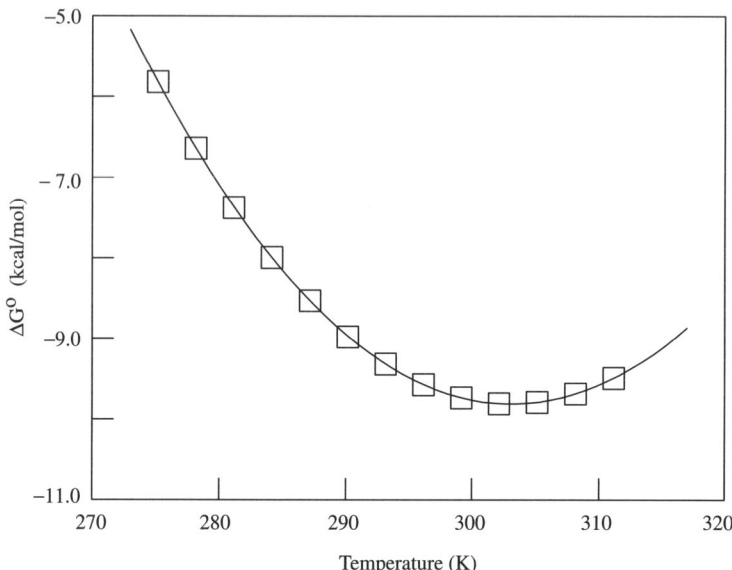

Figure 1 *Values of $\Delta G°$ as a function of temperature using the error-free data presented in Table 1 with the fit obtained using Equation (6) without weights and using the parameter values given in Table 2*

The results obtained by linear least-squares fitting of these data using Equation (6) and a value of T_R of 293.15 K, the mean value of the temperatures, are presented in Table 2 and are illustrated by the fitting line in Figure 1.

The fact that the returned parameter values in Table 2 are quite close to the generating parameter values and have small errors attributable to the rounding off of the generated values is to be expected. What is of critical importance here is the magnitude of the dependency values. In MLAB, the mathematical modeling system used for these analyses, the dependency value of a parameter is defined as that fraction of the variance of the parameter that can be attributed to uncertainties in the other parameters.[7] Denoting the parameter of interest as P, then the dependency of P is 1- var$(P)_{\text{other parameters fixed}}$/var$(P)$. If var$(P)_{\text{other parameters fixed}}$ = var(P), there is no dependency and the dependency value will be zero. As the dependency becomes greater the value will approach one. The dependency values do not depend on the noise in the data, as can be seen from our example, but do depend on the mathematical model, the parameter values and the range of the data. The fact that the dependency values for $\Delta H°_{T_R}$ and $\Delta S°_{T_R}$ are both 0.9985 is clearly indicative that a change in one value can be virtually perfectly compensated by a change in the other value and must diminish our confidence in these values. The fact that they are essentially identical to the generating values is rather fortuitous, but should not give confidence that this will always be the case. The value of 0.5600 for the dependency of $\Delta C°_P$ is acceptable and, while not optimal, does give reasonable confidence in that parameter value. The balance of this section will be devoted to two possible solutions to this problem.

Table 2 *Final parameter values obtained fitting Equation (6) to the data in Table 1 with the value of T_R fixed at 293.15 K*

Parameter	Value	Standard error	Dependency
$\Delta H^\circ_{T_R}$	19992.568	21.024	0.9985
$\Delta S^\circ_{T_R}$	99.980	0.072	0.9985
ΔC°_P	-3004.205	4.210	0.5600

Note: Presentation of the large number of significant figures in this table and in those which follow is for the sole purpose of permitting the comparison of small parameter value differences and small standard error values and is not intended to imply that such a level of precision is attainable with experimental data. The standard errors given here and in Tables 3–5 are those returned by the Levenberg–Marquardt algorithm used for least-squares curve-fitting in MLAB. Where the fitting function is linear in the parameters, these values, which are influenced by the dependency values, are correct. When the function is non-linear, the values of the standard errors are only an approximation.

The first approach to this problem is based upon a publication by Johnson on parameter cross-correlations while curve fitting.[4] The cross-correlation coefficient of any two parameters, j and k, is defined by:

$$c - cc_{jk} = (J^T J)^{-1}_{jk} / [(J^T J)^{-1}_{jj} (J^T J)^{-1}_{kk}]^{1/2} \qquad (8)$$

The subscripts refer to a particular element of the $J^T J$ matrix. If all of the off-diagonal elements of the inverse of this matrix are zero, then the parameters are orthogonal to each other and the individual parameters may be evaluated without simultaneously evaluating the other parameters or having any other knowledge of them. As will be seen, this permits a unique approach to determining the individual parameter values. The elements of the various $J^T J$ matrices for Equation (6) are defined by

$$(J^T J)_{jk} = \sum_i [(1/\sigma_i^2)(\partial \Delta G^\circ_T / \partial p_j)(\partial \Delta G^\circ_T / \partial p_k)] \qquad (9)$$

The various partial derivatives of ΔG°_T as described by Equation (6) are given by

$$\partial \Delta G^\circ_T / \partial \Delta H^\circ_{T_R} = 1, \qquad \partial \Delta G^\circ_T / \partial \Delta S^\circ_{T_R} = -T \qquad (10a,b)$$

$$\partial \Delta G^\circ_T / \partial \Delta C^\circ_P = T - T_R - T \ln(T/T_R) \qquad (10c)$$

The appropriate $J^T J$ matrices are then given by

$$(J^T J)_{\Delta H \Delta S} = \sum_i - T_i \qquad (11a)$$

$$(J^T J)_{\Delta H \Delta C_P} = \sum_i (T - T_R - T \ln(T/T_R)) \qquad (11b)$$

$$(J^T J)_{\Delta S \Delta C_P} = \sum_i (-T)(T - T_R - T \ln(T/T_R)) \qquad (11c)$$

Since orthogonality of the parameters requires that the inverse of the $J^T J$ matrices be equal to zero, the $J^T J$ matrices must also be equal to zero. It is obvious that none of the summations in Equations (11) can meet this condition.

We can overcome this problem by rewriting Equation (6) in a form that will permit attainment of the desired results:

$$G_T = H - S(T - \alpha) + C((T - \beta) - T_R - (T - \beta)\ln((T - \gamma)/T_R)) \quad (12)$$

and we can now obtain values of α, β, and γ that will permit the conditions of orthogonality. The partial derivatives of G_T as described by Equation (12) are given by

$$\partial G_T/\partial H = 1, \quad \partial G_T/\partial S = \alpha - T,$$

$$\partial G_T/\partial C = (T - \beta) - T_R - (T - \beta)\ln((T - \gamma)/T_R) \quad (13)$$

We can now write the following summations which can be used to obtain the values of α, β, and γ which will make the summations zero as indicated and make Equation (12) orthogonal in the parameters, keeping in mind that, because the data are error-free, all the values of σ_i are 1:

$$\sum_i (1/\sigma_i^2)(\partial G_T/\partial H)(\partial G_T/\partial S) = \sum_i 1(\alpha - T_i) = 0 \quad (14)$$

$$\sum_i (1/\sigma_i^2)(\partial G_T/\partial H)(\partial G_T/\partial C) = \sum_i 1((T_i - \beta) - T_R - (T_i - \beta)\ln((T_i - \gamma)/T_R)) = 0 \quad (15)$$

$$\sum_i (1/\sigma_i^2)(\partial G_T/\partial S)(\partial G_T/\partial C) = \sum_i (\alpha - T_i)((T_i - \beta) - T_R - (T_i - \beta)\ln((T_i - \gamma)/T_R)) = 0 \quad (16)$$

The solution of Equation (14) is quite simple. Since $\sum_i \alpha = \sum_i T_i$, $13\alpha = \sum T_i$ and $\alpha = T_R$. However, the simultaneous solution of Equations (15) and (16) in order to obtain the values of β and γ is quite cumbersome. A much easier and entirely adequate approximation is to use Equation (12), setting the value of α to T_R and the values of β and γ to zero and by iteratively fitting the data with small incremental changes in β, finding the value that minimizes the dependency values of the parameters. With the value of β now set, the iterative process is now used to find the value of γ that further minimizes the dependency values of the parameters, bringing them virtually to zero. It was found that the optimal values were $\beta = 0$ and $\gamma = 0.215$. The fitting equation thus becomes:

$$G_T = H - S(T - 293.15) + C(T - 293.15 - T \ln((T - 0.215)/293.15)) \quad (17)$$

The results obtained when using this model are shown in Table 3. While the dependency values for H, S, and C are not exactly zero, their values are so small that we can consider the parameters to be orthogonal, since zero value dependencies are the equivalent of zero value cross-correlation coefficients. As noted in the table, since the values are orthogonal and hence independent of each other, it is appropriate to

Table 3

Parameter	Value	Standard error	Dependency
(A) *Final parameter values obtained fitting Equation (17) to the data in Table 1*			
H	-8670.527	0.803	1.677×10^{-7}
$S\ (\Delta S^\circ_{T_R})$	99.901	0.072	6.602×10^{-4}
$C\ (\Delta C^\circ_P)$	-3004.207	4.210	6.604×10^{-4}
(B) *Final parameter value for* $\Delta H^\circ_{T_R}$ *obtained when fitting the data in Table 1 with Equation (6) with the values of* $\Delta S^\circ_{T_R}$ *and* ΔC°_P *fixed at the values of S and C, respectively. This is appropriate since H, S, and C are orthogonal and hence their values are independent of each other*			
$\Delta H^\circ_{T_R}$	19992.805	0.733	0.000

substitute the values of S and C for the values of $\Delta S^\circ_{T_R}$ and ΔC°_P in Equation (6) and fit to obtain the value of $\Delta H^\circ_{T_R}$ with zero dependency.

While the method described above is quite robust and would appear to be applicable to a wide range of data, it is a bit laborious in application, particularly if weights are used. The method which follows is almost as robust, easier to use with weights, and is a great deal more rapid in application. It is based upon the observation of Kelly and Holladay in a study on the thermodynamics of the unfolding of alligator heart metmyoglobin that the analysis was facilitated by finding a reference temperature where the value of $\Delta H^\circ_{T_R}$ was equal to zero, thus permitting fitting for one fewer parameter value.[8] We have extended this concept to the use of two reference temperatures, one where $\Delta H^\circ_{T_R}$ is equal to zero and the other where $\Delta S^\circ_{T_R}$ is equal to zero, resulting in global fitting of the data with two fitting functions whose only common parameter is ΔC°_P. Such a system is strongly constrained and will have markedly reduced dependency values. The two fitting functions are given by

$$\Delta G^\circ_{T,\Delta H=0} = - T\Delta S^\circ_{T_R,\Delta H=0} + T\Delta C^\circ_P(1 - T_{R,\Delta H=0}/\ T - \ln(T/T_{R,\Delta H=0})) \quad (18)$$

$$\Delta G^\circ_{T,\Delta S=0} = \Delta H^\circ_{T_R,\Delta S=0} + T\Delta C^\circ_P(1 - T_{R,\Delta S=0}/T - \ln(T/T_{R,\Delta S=0})) \quad (19)$$

We initially fit globally with the three thermodynamic parameters and the two reference temperatures as fitting parameters, making this model non-linear in the parameters. The values of the reference temperatures, which have very small estimated errors and also have moderate dependencies, are set as fixed values. The data is now globally fit with a model that is linear in the parameters. In order to compare the values of $\Delta H^\circ_{T_R}$ and $\Delta S^\circ_{T_R}$ at the two different reference temperatures with the values they would have at the reference temperature of 293.15 K, we must use the equations:

$$\Delta H^\circ_{293.15} = \Delta H_{T_R,\Delta S=0} + \Delta C^\circ_P(293.15 - T_{R,\Delta S=0}) \quad (20)$$

$$\Delta S^\circ_{293.15} = \Delta S^\circ_{T_R,\Delta H=0} + \Delta C^\circ_P\ln(293.15/T_{R,\Delta H=0}) \quad (21)$$

Table 4

Parameter	Value	Standard error	Dependency

(A) *Final parameter values obtained by global fitting of Equations (18) and (19) to the data in Table 1 with the values of the thermodynamic parameters and the reference temperatures varied*

Parameter	Value	Standard error	Dependency
$\Delta H_{T_R, \Delta S=0}$	−9809.832	1.053	0.4471
$\Delta S_{T_R, \Delta H=0}$	32.544	0.003	0.3514
ΔC_P°	−3004.205	2.905	0.8573
$T_{R, \Delta S=0}$	303.070	0.012	0.8110
$T_{R, \Delta H=0}$	299.805	0.010	0.6110

(B) *Final parameter values obtained by global fitting of Equations (18) and (19) to the data in Table 1 with the values of the thermodynamic parameters varied and the values of the reference temperatures fixed*

Parameter	Value	Standard error	Dependency
$\Delta H_{T_R, \Delta S=0}$	−9809.832	0.917	0.3333
$\Delta S_{T_R, \Delta H=0}$	32.544	0.003	0.2134
ΔC_P°	−3004.205	1.395	0.4354

(C) *Final parameter values from Table 4A adjusted to a reference temperature of 293.15 K using Equations (20) and (21). The corresponding values from the orthogonal model (OM) are included for comparison*

Parameter	Value	Standard error	OM value	Standard error
$\Delta H^\circ_{T_R}$	19992.568	0.917	19994.805	0.732
$\Delta S^\circ_{T_R}$	99.980	0.003	99.901	0.072
ΔC_P°	−3004.205	1.395	−3004.207	4.210

The results of this fitting are shown in Table 4.

Some interesting observations can be made regarding the data presented in Table 4. It is in no way surprising that fixing the reference temperatures in Equations (18) and (19) significantly reduces the error estimates of two of the three thermodynamic parameters and the dependency values of all of them. This can be attributed to the facts that there are now only three parameters totally, that each fitting function has only two parameters, that the value of ΔC_P° serves as a constraint on each function, and that both functions are linear in their parameters. It is also of great importance that both methods give almost identical values for the parameter values and for the values of the estimated standard errors. Tables 3 and 4 clearly demonstrate that reducing the cross-correlation coefficients or, equivalently, the dependencies, very significantly reduces the parameter standard errors. There is no real significant difference between the two methods in terms of results.

4   Optimized Fitting of Data with Added Noise

Having begun with the exploration of optimization with noiseless data, we will now extend these analytical methods to data with added noise. We can no longer make the assumption that $\sigma_i = 1$ and must now give consideration to appropriate weights. In least-squares fitting the appropriate weight is given by $w_i = 1/\mathrm{var}_i = 1/\sigma_i^2$. For purposes of comparison to unweighted fitting, it has been our practice to normalize all weights to unity by the relationship $w_{i,\mathrm{norm}} = w_i/\overline{w_i} = w_i n/\sum_{i=1}^{n} w_i$, and for simplicity we will refer to normalized weights as w_i regardless of how they were obtained. It has been our observation that the values of σ_i for $\ln K_{a,i}$ obtained during fitting generally lie in the range of 0.05–0.20, giving values for σ_i of $\Delta G_{T_i}^\circ$ of approximately 30–120 cal mol^{-1}. However, we have observed that data points which are obviously outliers may have very good values of σ, indicating that this is not necessarily a good method for generating appropriate weights even if it is the theoretically appropriate method for least-squares analysis. Using these weights or using no weights at all accentuates the problem that least-squares analysis has with outliers. Because we are minimizing the sum of the squares of the residuals, outliers have a disproportionate effect on parameter values and their estimated standard errors. A practical way of dealing with this is to use either the normalized reciprocals of the squares of the residuals from an unweighted fitting or the normalized reciprocals of the absolute values of these residuals as weights. Extensive simulation studies have clearly demonstrated that the use of the absolute values of the residuals is optimal; the use of the squares of the residuals overcompensates and has the effect of essentially making even moderate outliers ineffective in determining parameter values and their estimated standard errors. The examples of the effects of weighting cited here use the method of residual absolute values, which may be considered a form of robust regression.

The data that we will be analyzing here is based on that in Table 1, except that random error with a root-mean-square (RMS) value of 260 cal mol^{-1} has been added. This is 3% of the mean value of $\Delta G_T^\circ = -8671$ cal mol^{-1} and is typical of experimental results that we have obtained. The results obtained by unweighted and weighted linear least-squares fitting of this data using Equation (6) and a value of T_R of 293.15 K are presented in Table 5A. It should be noted that weighting has little effect on the parameter values, significantly reduces the standard error and makes the dependency values slightly worse. Thus, while the values appear to be reasonable, we cannot have much confidence in them.

Table 5B presents the data using the orthogonal polynomial approach. When we performed the unweighted fits, we were able to use the same values of α, β, and γ as were used for the non-noisy data. However, for the weighted fits, new values of α, β, and γ must be calculated. The value of α is again obtained by using Equation (14), giving $\sum_i \alpha = \sum w_i T_i$, which gives $\alpha = 296.594$. The values of β and γ were obtained iteratively, as before, and had the values of 0.008 and 0.215, respectively. It should be noted that while the values of the parameters are similar for weighted and unweighted fits, the standard errors are better with weighting. When compared to the data in Table 5A, the standard errors are significantly lower and the dependencies are very low, giving a significant level of confidence in the values obtained.

Table 5C presents the results of unweighted and weighted global fitting of the data. As before, Equations (18) and (19) were used to determine the values of the three thermodynamic parameters and the two reference temperatures. The latter were then fixed and the data was refit for the thermodynamic parameters. Equations

Table 5

Parameter	Unweighted value	Error	Dependency	Weighted value	Error	Dependency

(A) *Results from unweighted and weighted fits of data with normally distributed error with a root-means-square (RMS) error of 3% of the mean value of ΔG°_T added using Equation (6) as the fitting function*

Parameter	Unweighted value	Error	Dependency	Weighted value	Error	Dependency
$\Delta H^\circ_{T_R}$	21830.35	1934.16	0.9985	21646.82	1367.24	0.9990
$\Delta S^\circ_{T_R}$	106.60	6.58	0.9985	106.10	4.70	0.9990
ΔC°_P	−3123.80	387.32	0.5600	−3250.80	297.77	0.6753

(B) *Results from unweighted and weighted fits of data with normally distributed error with a root-means-square (RMS) error of 3% of the mean value of ΔG°_T added using Equation (12) as the fitting function. For the unweighted fit, values of $\alpha= 293.15$, $\beta=0.000$, $\gamma=0.215$, and $T_R=293.15$ were used. For the weighted fit, values of $\alpha=296.594$, $\beta=0.008$, $\gamma=0.215$, and $T_R=293.15$ were used*

Parameter	Unweighted value	Error	Dependency	Weighted value	Error	Dependency
H	−8747.67	73.83	1.677×10^{-7}	−9147.25	43.39	9.350×10^{-8}
$\Delta H^\circ_{T_R}$	21830.59	67.40	0.000	21620.76	39.61	0.0000
$S\,(\Delta S^\circ_{T_R})$	106.60	6.58	6.600×10^{-4}	106.01	4.69	0.2276
$C\,(\Delta C^\circ_P)$	−3127.80	387.32	6.604×10^{-4}	−3250.71	297.76	0.2276

(C) *Results from unweighted and weighted fits of data with normally distributed error with a root-means-square (RMS) error of 3% of the mean value of ΔG°_T added using equations (18) and (19) as the functions for global fitting of the data and equations (20) and (21) to convert the values into a reference temperature of 293.15 K*

Parameter	Unweighted value	Error	Dependency	Weighted value	Error	Dependency
$\Delta H^\circ_{T_R,\Delta S=0}$	−9959.06	97.68	0.4559	−9968.11	52.56	0.3553
$\Delta S^\circ_{T_R,\Delta H=0}$	33.01	0.31	0.3529	33.08	0.18	0.3694
ΔC°_P	−3127.80	267.27	0.8637	−3250.68	193.16	0.8027
$T_{R,\Delta S=0}$	303.31	1.09	0.8208	302.88	0.58	0.6899
$T_{R,\Delta H=0}$	300.13	0.87	0.6371	299.81	0.45	0.3852
$\Delta H^\circ_{T_R,\Delta S=0}$	−9959.06	84.25	0.3323	−9968.11	48.58	0.3109
$\Delta S^\circ_{T_R,\Delta H=0}$	33.01	0.26	0.2138	33.08	0.15	0.2197
ΔC°_P	−3127.80	125.42	0.4349	−3250.68	107.92	0.4229
$\Delta H^\circ_{T_R=293.15}$	21830.34	84.25		21651.11	48.58	
$\Delta S^\circ_{T_R=293.15}$	106.60	0.26		106.11	0.15	

(20) and (21) were then used to calculate those values for a reference temperature of 293.15 K. These values were then used in Equation (6) to obtain the values of the residuals used to calculate the weights. These weights were then used with Equations (18) and (19), as before, and Equations (20) and (21) were used to calculate the values for the reference temperature of 293.15 K. A comparison of the results in Tables 5B and C shows a slight advantage for the latter method, but it is really of little practical consequence and the two methods should be considered essentially equivalent in effectiveness. Figure 2 illustrates the data fit with Equation (6) using the final parameter values from Table 5C.

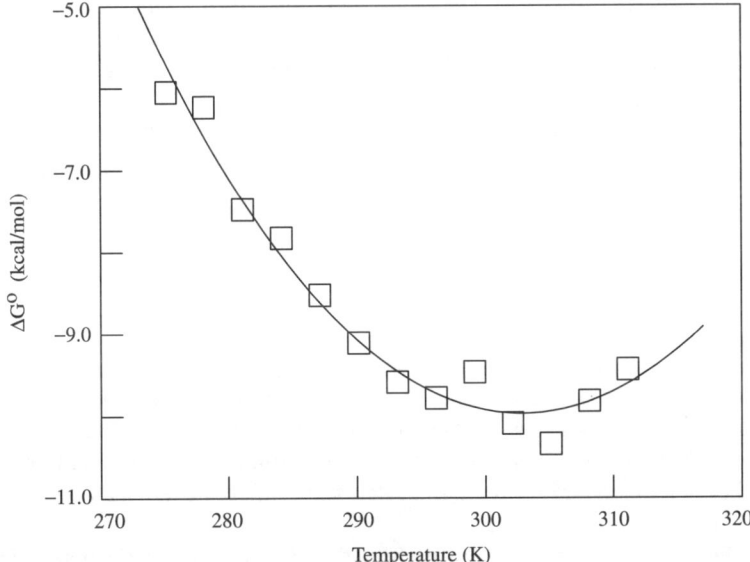

Figure 2   *Values of $\Delta G°$ as a function of temperature using the data presented in Table 1 with a RMS error of 260 cal mol^{-1} (3% of -8671 cal mol^{-1}, the mean value of $\Delta G°$), with weighted fits using Equations (18) and (19) for global fitting and Equations (20) and (21) for parameter value transformation and generating the fitting line with Equation (6) using the parameter values as given in Table 5C*

5 Sources of Error in Ultracentrifugal Thermodynamic Data

Having examined the application of the two methods of analysis, first without error and then with added error, it is now appropriate to discuss some of the factors that might lead to error in the values of $\Delta G°_T$ used in the analyses. Such errors fall into two categories, systematic and random. In the first category, errors in temperature measurement, errors in molar extinction coefficients, and errors that are model-dependent would appear to be the most important. In the random category, errors arising from methodology of data analysis, failure to attain equilibrium, and failure of reversibility appear not only to be the most important, but also appear to be significantly more important than most systematic errors.

In their work on the optical measurement of temperature, Liu and Stafford note that the actual temperature in the ultracentrifuge may be offset by as much as 1° from the indicated value, but that this offset appears to be essentially constant over the operative temperature range of the instrument and that fluctuations of temperature during a run appear to be about an order of magnitude smaller.[9] In Table 6 we present the results of the actual temperature being 1° higher and 1° lower than the indicated temperature on the values of the thermodynamic parameters determined using the global analysis method with noise-free data.

Table 6   *Effect of errors in actual temperatures on returned parameter values when the assumed temperature is correct, when the actual temperature is 1 °C higher and when it is 1 °C lower. The global analysis method with noise-free data was used for the calculations*

Temperature parameter	Correct value	+1 °C value	−1 °C value
$\Delta H^{\circ}_{T_R}$	19992.57	20093.86	19891.28
$\Delta S^{\circ}_{T_R}$	99.98	99.98	99.98
ΔC°_P	−3004.21	−3014.47	−2993.94

From these data it can be readily seen that systematic errors in temperature will have little effect on the values obtained for the thermodynamic parameters. These effects will be less marked when there is a lower temperature dependence of ΔG°_T as a consequence of smaller values of ΔC°_P. It also follows that small random differences in the temperature from the indicated value when the temperature is changed should have a negligible effect.

Another systematic effect might arise from the use of incorrect extinction coefficients when calculating the natural logarithm of the molar equilibrium constant when using the absorbtion optical system. For a monomer–dimer-equilibrium, the fitting function is

$$c_r = c_{b,1}\exp(AM_1(r^2-r_b^2)) + c_{b,1}^2 \exp(\ln K_{a,T,12} - \ln E_{1,2} + 2AM_1(r^2-r_b^2)) + \varepsilon \quad (22)$$

where the terms have their usual meaning with $A=(1-\bar{v}\rho)\omega^2/2RT$ and $\ln E_{1,2} = \ln(E_1/2)$ with E_1 being the molar extinction coefficient of the monomer at the recording wavelength. This term is needed to convert the molar $\ln K$ into an absorbance $\ln k$ for consistency with the measuring units. For an A+B=AB reaction, then the appropriate fitting function is

$$c_r = c_{b,A}\exp(A_A M_A(r^2-r_b^2)) + c_{b,B}\exp(A_B M_B(r^2-r_b^2))$$

$$+ c_{b,A}c_{b,B}\exp(\ln K_{a,T,AB} - \ln E_{AB} + (A_A M_A + A_B M_B)(r^2-r_b^2)) + \varepsilon \quad (23)$$

where $\ln E_{AB} = \ln(E_A E_B/(E_A + E_B))$. It is assumed for both cases that the molar extinction coefficient of the complex is the sum of the molar extinction coefficients of the individual components and it is also assumed that the values of the reduced masses, $M(1-\bar{v}\rho)$, of the components are additive.

We can now illustrate the effect of an error in the molar extinction coefficient on the results of a monomer–dimer reaction. Let us assume that the correct value of E_1 is 40 000 mol^{-1} cm^{-1}, but the value measured in a spectrophotometer is 30 000 mol^{-1} cm^{-1}. Then, the values of $\ln E_{12}$ will be 9.90349 and 9.61581, respectively. The difference between these values represents an error of 2.992% in the value of the $\ln E_{12}$ term, which will be a constant amount in each value of ΔG°. Using a mean value of $\Delta G^{\circ} = -8671$ cal mol^{-1}, the error will be 259 cal mol^{-1}, a fixed amount and not a percentage of the total value of ΔG°. The effect of this is illustrated in Table 7, which

Table 7   *Effect of a 25% error in the molar extinction coefficient in a monomer–dimer reaction on the values of the thermodynamic parameters. The global analysis method with noise-free data was used for the calculations*

Extinction coefficient parameter	Correct value	25% Low value
$\Delta H^{\circ}_{T_R}$	19992.57	20252.57
$\Delta S^{\circ}_{T_R}$	99.98	99.98
ΔC°_{P}	-3004.21	-3004.21

clearly demonstrates that an error of 25% in the molar extinction coefficient results in trivial differences in the thermodynamic parameters when analyzing noise-free data. This holds equally well for comparable errors in an $A+B=AB$ reaction. This error can be avoided by obtaining the spectra of the solutes at identical concentrations in buffer and in 6 M guanidine HCl where any shielding effects will be absent and using the ratios of these spectra to correct the apparent extinction coefficients.[10] The comparable error when the interference optical system is used will be even smaller since the specific refractive increment of proteins exhibits very little variability.

There are a number of sources of error arising from experimental conditions that are, to a significant degree, model-dependent. If one is working with a monomer–n-mer association, then it is usually necessary to calculate the value of the partial specific volume of the monomer from the amino acid composition. We prefer the consensus values of Perkins for this purpose.[11] The values of the partial specific volume at temperatures other than the table value can be calculated using the equation $\Delta\bar{v}/\Delta T=0.000425$. It is our opinion that for best results, solvent densities should be measured experimentally rather than calculated from tables. The Anton Paar density meters are admirably suited for this purpose, giving excellent precision over the temperature range of interest. Lacking such an instrument, constructing a pycnometer from a 10 mL volumetric flask and a portion of a 0.1 mL pipette, which will provide calibration marks, should be well within the abilities of an amateur glassblower. Such a pycnometer is easily calibrated with pure water and an analytical balance and will give results that are quite adequate for experimental purposes and definitely superior to calculated densities from tables. Because the values of the partial specific volume and density are used in the form of $1-\bar{v}\rho$, and since protein partial specific volumes are usually $0.72+/-0.03$ cm^3 g^{-1}, an error in the partial specific volume is multiplied approximately three-fold in the value of $\ln k_a$. The error is diminished proportionally according to what fraction of the value of the molar term, $\ln K_a$, this represents. The magnitude of this error is difficult to estimate.

This problem is readily avoided for heterogeneous associations such as $mA+nB= A_mB_n$. All that is required is that, depending on the rotor used, at least one cell is used for each reactant alone. This permits direct measurement of the value of the $M(1-\bar{v}\rho)$ term for each reactant at the same temperature and possibly the same rotor speed as is used for the reacting mixture. Molar masses and/or stoichiometries may dictate that this be done in a separate experiment at a different rotor speed. The errors here should be minimal if the experiments are carried out properly.

An intermediate situation exists for mixed associations such as the situation where $2A = A_2$ and $A + B = AB$. Under these conditions, the value of $M_B(1 - \bar{v}\rho)_B$ can be determined experimentally but the value of $M_A(1 - \bar{v}\rho)_A$ must be calculated. Additionally, one or more cells should be loaded with component A alone so that the homogeneous association can be studied under the same conditions as the heterogeneous association. If the molar masses of A and B are similar, then a conservation of mass constraint may be necessary.[12] The magnitude of the errors here are also difficult to estimate. All of the above assume that there is no pressure effect on the partial specific volume of the complex. If this effect is present, it is usually small; it can be measured by using multiple rotor speeds or several different solution column heights with identical total amounts of protein in each so that different hydrostatic pressures are obtained.

Considered by itself, allowing insufficient time to attain equilibrium leads to systematic rather than random error. If one is in the temperature range where the absolute magnitude of ΔG°_T is increasing, then it may be expected that not being at equilibrium will lead to too small an absolute value for ΔG°_T. If the temperature is such that the absolute magnitude of ΔG°_T is decreasing, then the reverse will be true. As is well known, the time required to attain equilibrium is proportional to the square of the solution column height and to the reciprocal of the diffusion coefficient, which is inversely proportional to the solution viscosity which is a function of temperature. Thus, a compromise must be made between the greater resolution of longer columns and the time required to attain equilibrium. This becomes really critical if one is planning to make measurements at 10–14 different temperatures. We believe that 3–4 mm columns appear to be the best compromise. As a criterion of equilibrium we require that scans be invariant within the limits of experimental precision over a period of at least 24 h initially at the lowest temperatures. As the temperature is increased we gradually reduce this time because of the lower solution viscosity, but always require invariance over at least 12 h. In addition to invariance, we also require minimal differences in noise and parameter error. Increased noise appears to be a sign of either thermally caused convective perturbations and/or reactant degradation. Convection can lead to truly random error, and if it is suspected, then more time should be allowed for reequilibration. If this does not occur, then the temperature should be lowered to the initial value to see if the system is still capable of thermodynamic reversibility. If this is not attained then other experiments should be performed to determine the highest temperature that permits reversibility. This should be done as the termination of any experiment even if the data look entirely satisfactory throughout. We have observed more problems with what appears to be convection when using the six-channel centerpieces and have abandoned their use. In the past, our best results have been observed using the Kel-F-coated aluminum double-sector centerpieces. We attribute this to the fact that they have better thermal conductivity than the carbon-filled epon centerpieces. It is regrettable that they are no longer available.

The final potential source of random error that we will discuss is that error which can arise from the method used to analyze the data. If the absorbtion optical system is used, it has been clearly demonstrated that the use of L-1 robust regression, which minimizes the sum of the absolute values of the residuals, gives optimal results.[13] As discussed in that reference, this is because the noise distribution of the absorbencies is demonstrably non-Gaussian and least-squares analysis is statistically inappropriate. Parameter errors are best estimated in this case by the use of a balanced bootstrap method. The advantages of this method become greater as the complexity of the model increases and/or the data becomes noisier. In the examples cited in ref. 13, with generating values of $\ln k_{12} = \ln k_{AB} = 12.000$ and with 5% RMS Gaussian error in the incident and transmitted intensities, for the monomer–dimer system, a value of $\ln k_{12} = 12.057 \pm 0.206$ was obtained using L-1 regression as compared to a value of $\ln k_{12} = 11.940 \pm 0.235$ when the Levenberg–Marquardt algorithm was used. For the $A + B = AB$ reaction, the respective results were $\ln k_{AB} = 11.989 \pm 0.063$ and $\ln k_{AB} = 11.740 \pm 0.201$.

If the interference optical system is used, then least-squares analysis is optimal since the noise distribution of that optical system is Gaussian. The balanced bootstrap is a useful and appropriate means of estimating parameter error for such analyses as well, although there are other options here that are not suitable for L-1 regression. Thus, it would appear, at least for the absorbtion optical system, random error can be reduced by the use of L-1 robust regression for data analysis.

6 Discussion

In Sections 3 and 4 we have described two different approaches to reducing the cross-correlation coefficients or the dependencies of the parameter values in thermodynamic analyses. Both methods offer enhanced confidence in the values obtained. The question now is which one should be used. Because of its greater simplicity, we would suggest that an unweighted global analysis should be attempted first. If the parameter values, the standard errors, and the dependencies all appear reasonable, then this approach can be pursued further. In particular, the reference temperatures should probably lie within the range of 260–330 K and should have standard errors of less than 3 K. If this is not the case, probably because of the value of ΔC_p°, then the orthogonal polynomial analysis would definitely be more appropriate. If in doubt, do both and compare them.

Summarizing the section on error, it would appear that there are only two major factors contributing to error leading to unsatisfactory thermodynamic analyses. The first of these is the failure to attain true equilibrium at any specific temperature either by virtue of failure to allow adequate time to attain equilibrium or because thermal gradients perturb equilibrium distributions. Stringent criteria for the attainment of equilibrium can minimize this problem but cannot necessarily eliminate it. The second major factor is the thermal stability of the reactants. Thermodynamic analyses require exposing the reactants to temperatures where they may have limited stability for the periods of time required to attain equilibrium. For some reactants this may severely limit the maximum temperature that can be attained. If this is the case, then the experiment must be performed with smaller temperature increments so that the number of usable data points obtained will permit a meaningful analysis. Additionally, it can be fruitful to conduct a series of experiments in which one attains equilibrium at a low temperature, jumps to a higher temperature where questionable data were previously obtained and then returns to the low temperature to assess reversibility.[3] A variation on this theme is to make measurements at more than one elevated temperature if the data appear to justify this option. As a study proceeds, the requirements for attaining the best results usually become obvious. These are costly procedures in terms of time, effort, and material, but can be the best options with solutes of limited thermal stability. In our experience, oligonucleotides, protein domains of limited size, synthesized polypeptides, and proteins expressed in thermophilic bacteria have demonstrated the best thermal stability and are the best candidates for study from 0 to 40 °C, the temperature range of the XL-A/XL-I instruments. Thermodynamic analysis using the analytical ultracentrifuge is a very time-consuming, laborious, and a demanding procedure. It should not be undertaken lightly, but only if the anticipated results justify the investment required to attain them.

References

1. T. Svedberg and K. O. Pedersen, *The Ultracentrifuge*, Clarenden Press, Oxford, 1940. Johnson Reprint Corporation, New York.
2. P. D. Ross, F. B. Howard and M. S. Lewis, *Biochemistry*, 1991, **30**, 6269.
3. M. S. Lewis and R. J. Youle, *J. Biol. Chem.*, 1986, **261**, 11571.
4. M. L. Johnson, *Method. Enzymol.*, 2000, **321**, 424
5. J. R. Livingstone, R. S. Spolar and T. M. Record, Jr., *Biochemistry,* 1991, **30**, 4237.
6. R. S. Spolar, J. R. Livingstone and T. M. Record, Jr., *Biochemistry,* 1992, **31**, 3947.
7. G. D. Knott and D. Kerner, MLAB Applications Manual, 1998, 10.
8. L. Kelly and L. A. Holladay, *Biophys. Chem.* 1987, **27**, 77.
9. S. Liu and W. F. Stafford, III, *Anal. Biochem.*, 1995, **224**, 199.
10. A. J. Rowe, Personal communication.
11. S. J. Perkins, *Eur. J. Biochem.*, 1986, **157**, 169.
12. M. S. Lewis, Appendix to S. P. Becera, A. Kumar, M. S. Lewis, S. G. Widen, J. Abbotts, E. M. Karawaya, S. H. Hughes, J. Shiloach and S. H. Wilson, *Biochemistry*, 1991, **30**, 11707.
13. M. S. Lewis and M. M. Reily, in *Modern Analytical Ultracentrifugation: Techniques and Methods*, Scott *et al.* (eds), Royal Society of Chemistry, Cambridge, UK, 2005, 152.

CHAPTER 18

Calculation of Volume, Surface, and Hydration Properties of Biopolymers

HELMUT DURCHSCHLAG AND PETER ZIPPER

1 Introduction

The application and interpretation of hydrodynamic and other low-resolution methods and results frequently require some information on volume, surface, and hydration properties of biomolecules. In principle, the experimental determination of these properties is possible by complementary techniques, but is frequently not feasible for various reasons (scarce amounts of material, instability of the samples, unavailability of special techniques, *etc.*). Moreover, the required experiments are commonly tedious and often provide only rough estimates of the properties under consideration. In all these cases, it is desirable to replace experiments with calculations or reliable approximations. Calculations of parameters should be performed preferably for application in two-component solutions (*e.g.* a simple protein in water or dilute buffer). In some cases, calculative procedures even allow parameter predictions of multicomponent systems: study of complex molecules, on the one hand, and of molecules in multicomponent solutions, on the other. Among the group of complex biopolymers, conjugated biopolymers (*e.g.* nucleoproteins, glycoproteins, lipoproteins), biopolymer–ligand complexes (*e.g.* between enzyme and substrate, cofactor or product), and associations of different types of biopolymers (*e.g.* protein–nucleic acid complexes) should be mentioned. The analysis of the behaviour of biopolymers under differing environmental and denaturing conditions necessitates knowledge of the parameters in multicomponent solutions.

A variety of physicochemical techniques such as analytical ultracentrifugation (AUC), small-angle X-ray and neutron scattering (SAXS, SANS), and light scattering (LS) require partial specific volumes and other volumetric properties of both biopolymers and low-molecular compounds for the evaluation of several parameters.[1–5] In this context, the determination of molar masses of biopolymers deserves special concern, both in two- or multicomponent solutions. Similarly, for the correction of sedimentation coefficients to standard conditions (20 °C, water) accurate volume values are needed, particularly when working in multicomponent solutions where drastic volume changes may occur (*e.g.* in the presence of elevated concentrations of electrolytes or strong denaturants).

The thorough interpretation of the behaviour of biopolymers in aqueous solution calls for the consideration of preferentially bound water molecules to biopolymers. Although the contribution of hydration to hydrodynamic and scattering behaviour is rather small, an exact interpretation of the results of the respective techniques requires strict inclusion of all hydration contributions.[6-8] This matter may be settled by the assumption of overall hydration values for the polymer under consideration, use of appropriate hydration shells or by modelling eligible water molecules on the biopolymer surface. For localizing hypothetical water molecules on the biopolymer envelope, the last-mentioned approach turned out to be highly promising. However, the realization of sophisticated hydration algorithms is based on the preceding accurate calculations of biopolymer surfaces. The analytical calculation of biopolymer surface areas complies with this requirement.

2   Volumetric Properties

2.1   Theoretical Background and Definitions

Detailed definitions concerning volumetric and densimetric properties have been given elsewhere.[1-4,9-13] Considerations related to AUC evaluations have also been described in several articles and textbooks,[1,2,5,13-22] and those relevant for the interpretation of SAXS and SANS data are outlined in refs. 2, 14–17, 23–27. In the following, therefore, only some relations pertinent to this study will be given.

2.1.1   Volume and Density

The total volume, V, of a solution at constant temperature, T, and pressure, P, may be defined in terms of partial specific volumes, $\bar{v}$, and the masses in grams, g, of each of the N components:

$$V = \sum_{i=1}^{N} \bar{v}_i g_i \tag{1}$$

The partial specific volume, $\bar{v}_i$, of the ith component of a solution is defined as the change in total volume, ∂V, per unit mass upon addition of an infinitesimal amount, ∂g_i, of component i at constant T and P and masses in grams, g_j, of all other components j:

$$\bar{v}_i = \left(\frac{\partial V}{\partial g_i} \right)_{T,P,g_j} \quad (j \neq i) \tag{2}$$

Substitution of g by the number of moles, n, leads to an analogous expression for the partial molar volume, $\bar{V}$:

$$\bar{V}_i = \left(\frac{\partial V}{\partial n_i} \right)_{T,P,n_j} \quad (j \neq i) \tag{3}$$

Partial specific and molar volumes, $\bar{v}$ and $\bar{V}$, are commonly given in cm³ g⁻¹ and cm³ mol⁻¹, respectively. These two volumetric quantities are related by

$$\bar{v}_i = \frac{\bar{V}_i}{M_i} \tag{4}$$

where M_i is the molar mass of the ith component, expressed in g mol⁻¹.

The density, ρ, of a solution at given T and P is related to the total volume, V, and the concentration, c:

$$\rho = \frac{\sum\limits_{i=1}^{N} g_i}{V} = \frac{\sum\limits_{i=1}^{N} g_i}{\sum\limits_{i=1}^{N} \bar{V}_i g_i} = \sum_{i=1}^{N} c_i \tag{5}$$

where ρ and c are given in g cm⁻³.

2.1.2 Density Increments

As may be derived from the above definitions, specific volumes of the macromolecular component 2 are obtained by measurement of the densities of a series of solutions (preferably by digital densimetry[3,4,28]), in which only the mass of the macromolecule is varied. Following considerations deduced from the thermodynamics of multicomponent systems,[1,2,9] mass–density increments, $(\partial\rho/\partial c_2)_m$ and $(\partial\rho/\partial c_2)_\mu$, may be obtained from density measurements at constant masses or constant chemical potential of added solvent components, respectively. Evaluation of the limiting slopes yields simple expressions, from which, by rearrangement, isomolal or isopotential specific volumes may be derived.

The partial specific volume at $c_2 \to 0$, $\bar{v}_2^0$, is defined by the relation

$$\bar{v}_2^0 = \frac{1}{\rho_s}\left[1 - \left(\frac{\partial\rho}{\partial c_2}\right)_m^0\right] \tag{6}$$

where the subscript m refers to constant molality of the nonmacromolecular components in a multicomponent solution. Instead of the measured densities of a series of solutions, however, in practice, apparent quantities are often obtained with measurement of only the density at one fixed value for c_2, ρ, and the density of the solvent, ρ_s. Hence, the apparent specific volume, ϕ_2, is given by

$$\phi_2 = \frac{1}{\rho_s}\left[1 - \left(\frac{\rho - \rho_s}{c_2}\right)_m\right] \tag{7}$$

In a similar way, the isopotential specific volume at $c_2 \to 0$, $v_2'^0$, or the apparent isopotential specific volume, ϕ_2', may be obtained from density measurements of solutions that have been dialysed to equilibrium:

$$v_2'^0 = \frac{1}{\rho_s}\left[1 - \left(\frac{\partial\rho}{\partial c_2}\right)_\mu^0\right] \tag{8}$$

and

$$\phi_2' = \frac{1}{\rho_s}\left[1 - \left(\frac{\rho - \rho_s}{c_2}\right)_\mu\right] \qquad (9)$$

where the subscript μ denotes that each of the diffusible components has the same chemical potential in every phase, and the prime designates isopotential conditions.

As follows from the above definitions, v_2' and ϕ_2' are only operational quantities that have no precise physical meaning *per se*. However, they include solvent interactions with the macromolecules, and may be used efficaciously for evaluation of sedimentation or scattering experiments.[2]

2.1.3 Relations between Different Volume Quantities

Differences between apparent volume quantities (ϕ_2, ϕ_2') and true quantities ($\bar{v}_2$, $\bar{v}_2'$) are unimportant, at least at low c_2, where $\phi_2 \approx \bar{v}_2$ and $\phi_2' \approx \bar{v}_2'$.

In two-component solutions as well as in multicomponent solutions containing negligible amounts of third components, the differences between isopotential and isomolal volume quantities are negligible if Donnan effects are absent. By contrast, in typical multicomponent solutions (aqueous solutions of charged macromolecules in the presence of substantial amounts of third components such as salts or buffer substances), marked differences between $\bar{v}_2$ and v_2' may occur, since in this case Donnan effects cause an asymmetric distribution of small ions across the dialysis membrane.[1,29]

At $c_2 \to 0$, the difference between isopotential and isomolal mass–density increments is given by

$$\left(\frac{\partial\rho}{\partial c_2}\right)_\mu^0 - \left(\frac{\partial\rho}{\partial c_2}\right)_m^0 = (1 - v_2'^0\rho_s) - (1 - \bar{v}_2^0\rho_s) = \xi_1^0(1 - \bar{v}_1\rho_s) = \xi_3^0(1 - \bar{v}_3\rho_s) \quad (10)$$

where ξ_1 and ξ_3 represent preferential interaction parameters designating the number of grams of component 1 (or 3) that must be added per gram of component 2 to maintain components 1 and 3 at constant chemical potential; using the weight molalities w_i (in g of component i per g of component 1) the interaction may be expressed as $\xi_1 = (\partial w_1/\partial w_2)_\mu$ and $\xi_3 = (\partial w_3/\partial w_2)_\mu$; $\bar{v}_1$ and $\bar{v}_3$ are partial specific volumes of components 1 and 3, respectively. At vanishing c_2, the parameters ξ_1 and ξ_3 are connected: $\xi_1 = -\xi_3/w_3$.

2.1.4 Different Density Increments

As pointed out by Eisenberg[1,2,16] in detail, the application of different physical techniques yields (or needs) different increments, either at constant solvent composition (m) or at constant chemical potential (μ): the mass–density increments $(\partial\rho/\partial c_2)_m$ and $(\partial\rho/\partial c_2)_\mu$ may be obtained by densitometry and the latter quantity also by AUC, whereas for SAXS, SANS and LS experiments, the electron-density increment, $(\partial\rho_{el}/\partial c_2)_\mu$, the scattering-length-density increment, $(\partial\rho_N/\partial c_2)_\mu$, and the refractive index increment, $(\partial n/\partial c_2)_\mu$, are usually required.

2.1.5 Relations between Different Density Increments

The electron-density increment, $(\partial \rho_{el}/\partial c_2)_\mu$, and the scattering-length-density increment, $(\partial \rho_N/\partial c_2)_\mu$, for multicomponent systems may be calculated from the experimentally accessible mass–density increment at constant μ, $(\partial \rho/\partial c_2)_\mu$, and from the preferential interaction parameter, ξ_i $(i \neq 2)$. Rearrangement of Equation (10) shows that interaction parameters may be obtained from the mass–density increments, $(\partial \rho/\partial c_2)_m = (1 - \bar{v}_2 \rho_s)$ and $(\partial \rho/\partial c_2)_\mu = (1 - v'_2 \rho_s)$. For the calculation of $(\partial \rho_{el}/\partial c_2)_\mu$ or $(\partial \rho_N/\partial c_2)_\mu$, an approximate value of ξ_i is sufficient. However, the scattering-length-density increment, $(\partial \rho_N/\partial c_2)_\mu$, can also be evaluated directly by neutron interferometry. For further details, the reader is referred to papers by Eisenberg, Zaccai and co-workers (*e.g.* refs. 2, 14, 16).

2.1.6 Specific Volumes of Nonmacromolecular Components

Partial and apparent specific volumes of nonmacromolecular (third) components, $\bar{v}_3$ and ϕ_3, may be obtained by the same procedure as outlined previously for macromolecules at isomolal conditions. Measuring the densities, ρ_s, of a series of aqueous solutions of component 3 at varying c_3, in the absence of component 2 (macromolecule), yields

$$\bar{v}_3^0 = \frac{1}{\rho_1}\left[1 - \left(\frac{\partial \rho_s}{\partial c_3}\right)^0\right] \qquad (11)$$

and

$$\phi_3 = \frac{1}{\rho_1}\left[1 - \left(\frac{\rho_s - \rho_1}{c_3}\right)\right] \qquad (12)$$

Since, in contrast to macromolecules, small molecules, especially electrolytes, show a pronounced concentration dependence on the volumetric properties, $\bar{v}_3^0$ values should preferably be used.

2.1.7 Volume Changes

As shown above and proven by many experiments,[3,4,10] pronounced differences between isomolal and isopotential volumes may occur, especially with polyelectrolytes such as nucleic acids or proteins. In addition, both types of volumes exhibit significant changes: especially severe alterations of isopotential volumes have been observed. Such changes occur as a consequence of several environmental changes (*e.g.* in the presence of various additives such as buffer components, salts, sugars, ligands, denaturants, reductants), but may also arise in the absence of additives (*e.g.* due to changes of temperature and pressure or upon onset of denaturation). Therefore, conditions that favour the absence of pronounced volume changes are usually preferred for both experiments and calculations.

2.2 Specific Volumes Used in Analytical Ultracentrifugation

For two-component solutions (*e.g.* a macromolecule in water or dilute buffer), the formulae for the determination of molecular parameters (s, D, M) are well established.[18–22] Partial specific volumes need to be known for the correction of sedimentation and diffusion coefficients to standard conditions (standardization of s and D values to pure water at 20 °C as solvent: s_w, D_w) and for the determination of molar masses from sedimentation and diffusion data ($M_{s,D}$), or from the sedimentation equilibrium (M_{SE}).

In multicomponent solutions the interactions of the macromolecules with the solvent have to be taken into consideration. Hence, the buoyancy term has to be modified to enable the estimation of true molecular parameters. While the formalism for M_{SE} is straightforward, the correction of s values and the determination of $M_{s,D}$ are much more sophisticated, particularly in the case of halophilic macromolecules.

The following formulae are given for 20 °C; Equations (13)–(24) have been taken from ref. 13:

(a) Two-component solutions:

$$s_w = s_s \left(\frac{\eta_s}{\eta_w} \right) \frac{(1 - \bar{v}_w \rho_w)}{(1 - \bar{v}_s \rho_s)} \tag{13}$$

$$D_w = D_s \left(\frac{\eta_s}{\eta_w} \right) \tag{14}$$

$$M_{s,D} = \frac{s_w}{D_w} \frac{RT}{(1 - \bar{v}_w \rho_w)} = \frac{s_s}{D_s} \frac{RT}{(1 - \bar{v}_s \rho_s)} \tag{15}$$

$$M_{SE} = \frac{2RT}{(1 - \bar{v}_w \rho_w)\omega^2} \frac{d \ln c}{dr^2} \tag{16}$$

In a two-component solution the subscripts w and s denote water and any other simple solvent (without additives), respectively; η symbolizes the relative viscosity, and ρ the density. For viscosity and density, the solvent quantities may be used in good approximation instead of solution quantities.

(b) Multicomponent solutions:

$$s_w = s_s \left(\frac{\eta_s}{\eta_w} \right) \frac{(1 - \bar{v}_w \rho_w)}{(1 - v'_s \rho_s)} \tag{17}$$

$$D_w = D_s \left(\frac{\eta_s}{\eta_w} \right) \tag{18}$$

$$M_{s,D} = \frac{s_w}{D_w} \frac{RT}{(1 - \bar{v}_w \rho_w)} = \frac{s_s}{D_s} \frac{RT}{(1 - v'_s \rho_s)} \tag{19}$$

$$M_{SE} = \frac{2RT}{(1 - v'_s \rho_s)\omega^2} \frac{d \ln c}{dr^2} \tag{20}$$

Special case of halophilic biopolymers:

$$s_w = s_s \left(\frac{\eta_s}{\eta_w} \right) \frac{(1 - v_s' \rho_w)}{(1 - v_s' \rho_s)} \tag{21}$$

$$D_w = D_s \left(\frac{\eta_s}{\eta_w} \right) \tag{22}$$

$$M_{s,D} = \frac{s_w}{D_w} \frac{RT}{(1 - v_s' \rho_w)} = \frac{s_s}{D_s} \frac{RT}{(1 - v_s' \rho_s)} \tag{23}$$

$$M_{SE} = \frac{2RT}{(1 - v_s' \rho_s)\omega^2} \frac{d\ln c}{dr^2} \tag{24}$$

In a multicomponent solution subscripts w and s refer to water and any other solvent (including additives). For simplicity, subscripts 1, 2, 3, *etc.* used in multicomponent thermodynamics have been omitted in the above formulae.

The estimation of molecular parameters by AUC analysis is sensitive to the value of the partial specific volume applied. Error propagation calculations show that the error in M is greater than the original error in $\bar{v}$. A factor of 3 is typical of simple proteins in two-component solutions; for glyco- and nucleoproteins the error is slightly reduced, but for lipoproteins it may become substantially larger. The size of the error depends on the percentage and the nature of the nonprotein components and on the density of the solvent.[13] Of course, at high values of solvent density, mass determinations are extremely prone to error.

2.3 Calculation of Partial Volumes

According to the above-mentioned considerations, many physicochemical studies urgently require specific volume quantities, both of macromolecular and low-molecular compounds ($v_2', \bar{v}_2, \bar{v}_1, \bar{v}_3$). While the assumption of values for $\bar{v}_1$ offers no serious problems (≈ 1 for water as the usual solvent), knowledge of the other volume quantities is beset by many experimental or calculative problems.

2.3.1 *Survey of Calculation Procedures*

Usually, some kind of additivity principle is applied for calculating partial specific or partial molar volumes of low-molecular and macromolecular compounds. In 1899, Traube[30] developed a simple but suitable procedure for estimating the partial molar volumes of simple organic compounds mainly built up from C, H, N, O, and S. Subsequently, many scientists tried to adapt his procedure for individual classes of both small molecules and biopolymers in aqueous solution. The most prominent example is the well-known method of Cohn and Edsall[31] for calculating the partial specific volumes of native, nonconjugated proteins by summing up the contributions of the constituent amino acid (AA) residues (whose volumes, however, were designed only partly by means of the Traube concept). Similarly, the increments for particular groups of homologous series of compounds (such as the methylene group) were used by many authors. Approaches making use of van der Waals volumes[32] gave systematic underestimates of volumes, but may be improved to some extent by special corrections. Also, procedures using crystallographic volumes[25] did not exactly match the situation in solution; they tended to yield overestimates. Earlier procedures striving for some universality[33-36] obviously suffer from the complexity of the calculation scheme to be applied, lack certain group increments, and, in general, do not allow one to compute ionic compounds.

A few years ago, we succeeded in developing a simple, rapid and universal approach for the *ab initio* calculation of partial molar and partial specific volumes of diverse organic and biochemical compounds in aqueous solution.[37] Our approach is based on Traube's initial concept,[30] but had to be improved considerably. Owing to deficiencies in precise density data at the time of Traube, several figures had to be adapted and corrected; in particular, the increments for molecule groups containing N turned out to be afflicted with considerable errors. In our approach, for the calculation of nonionic organic compounds, only a minimum of volume and special increments is required, and these increments can be coupled to ionization contributions and tabulated values for inorganic ions if the volumes of ionic organic compounds are in demand. The validity of the approach has been tested by comparing a broad variety of predicted and observed volumes of molecules of quite different chemical structure and composition.[4,37-39] Because of its usefulness and applicability for manifold volume problems, the universal approach will be described in the following section in more detail.

2.3.2   The Universal Approach

Partial molar and partial specific volumes, $\bar{V}$ and $\bar{v}$, of both low-molecular and macromolecular organic and biochemical compounds can be calculated by means of a simple calculation scheme.[37] The applied approach considers compounds of nonionic or ionic nature, mono- and polyfunctional molecules, aliphatic and aromatic compounds, and homo- and heterocycles of different ring size, micellization effects, *etc.* The calculated volumes are valid for aqueous solutions at 25 °C and for infinite dilution of the solutes; for volumes at other temperatures, a temperature coefficient, $\Delta\bar{v}/\Delta T$, of 5×10^{-4} cm^3 g^{-1} K^{-1} may be applied.

Partial molar volumes of organic compounds can be obtained by exploiting the respective volume increments for atoms and atomic groups and considering some special corrections:

$$\bar{V}_c = \Sigma V_i + V_{CV} - \Sigma V_{RF} - \Sigma V_{ES} \qquad (25)$$

where $\bar{V}_c$ denotes the calculated partial molar volume, V_i symbolizes the volume increment for any atom or atomic group, and V_{CV}, V_{RF}, and V_{ES} correspond to volume changes caused by covolume, ring formation and electrostriction (ionization), respectively. If the monomeric unit of polymers is the object of the volume calculation, the contribution of V_{CV} and end groups has to be disregarded.

Table 1 summarizes various volume increments, V_i, for some fundamental atoms in organic and biochemical compounds, some increments, V_{ion}, for common inorganic

Table 1 *Partial molar volume increments for selected atoms and inorganic ions, valid for aqueous solutions at 25 °C[a]*

Increments for atoms	V_i^b $(cm^3\ mol^{-1})$	Comment
C	9.9[c]	
H	3.1	
O	0.4	2nd or further neighbouring OH; OH or O$^-$ in a carboxylic group, phosphate, sulfate, or sulfonate
	0.9	2nd =O at a ring, provided a tautomerism is possible
	2.3	1st OH; O$^-$ in a phenolate
	5.5	$-$O$-$ in an ether, ester or anhydride; O in a ring; =O, except 2nd =O at a ring; O in an amine oxide or nitro group
N	0.5	primary N in diamides
	2.0[d]	primary N in monoamides; diamines, polyamines; N in amine oxides; unsubstituted urea; α- and ω- amino groups in aminoacids and peptides; amines at hydrated rings
	4.0	primary, secondary, and tertiary monoamines; aminoalcohols; secondary and tertiary N in mono- and diamides; N in substituted ureas; 2nd ammonium N; NH in hydrated rings
	7.0	N without H in hydrated rings; N or NH in aromatic rings; nitro groups and nitriles
	8.0[e]	guanidinium
	12.2	1st ammonium N; N in betaines
S	15.5	
P	17.0	P(III)
	28.5	P(V)
Cl	15.0	
Br	19.7	

Atomic volumes	V_A $(cm^3\ mol^{-1})$
Mg	14.0
Ca	26.0
Fe	7.1

Increments for inorganic ions	V_{ion}^f $(cm^3\ mol^{-1})$
H$^+$	-3.8
Na$^+$	-5.0
K$^+$	5.2
Mg^{2+}	-28.8
Ca^{2+}	-25.4
Fe^{2+}	-32.3
Fe^{3+}	-55.1
OH$^-$	-0.2
Cl$^-$	21.6
Br$^-$	28.5
SO$_4^{2-}$	21.6

(continued)

Table 1   (*continued*)

Special increments and decrements	V_j (cm³ mol⁻¹)	Comment
Covolume (V_{CV})	12.4	
Ring formation (V_{RF})	6.1[g]	5-membered ring
	8.1[g]	6-membered ring
	10.1[g]	7-membered ring
Ionization (V_{ES})	6.8	Contribution for one positive charge in organic moieties (V_{ES+})
	6.7	Contribution for one negative charge in organic moieties (V_{ES-})
Micellization of ionic and steroid surfactants (V_{mic})	0.7[h]	Alkyltrimethylammonium halides
	0.9[h]	Salts of fatty acids
	1.5[h]	Alkyl sulfates and sulfonates
	−23.0	Steroid surfactants

[a]The values were essentially adopted from previous papers.[37-39] Details and further increments may be found in comprehensive compilations.[4,37]
[b]The V_i values mentioned do not contain the contribution of any ionization.
[c]The same value holds for single-, double-, or triple-bonded C.
[d]The value does not apply to the ring-N in Pro.
[e]The value refers to each N in a guanidinium group; in Arg only to the two terminal N.
[f]The V_{ion} values include already the contribution for ionization.
[g]The value may also be applied to heterocycles containing N, O and/or S.
[h]Contribution per carbon in the variable part of the surfactant.

ions, and the special increments and decrements, V_j, for covolume, ring formation, electrostriction, and micellization (V_{CV}, V_{RF}, V_{ES}, and V_{mic}). The atomic volumes, V_A, of certain elements have also been listed to allow the volume prediction of metal-containing biomolecules. Comprehensive tables considering a lot of further atoms and groups and addressing a variety of volume aspects have been given elsewhere.[4,37-39]

Partial molar volumes, $\overline{V}_{c,el}$, of inorganic electrolytes may be calculated by simply summing up the values, V_{ion}, for cation and anion:

$$\overline{V}_{c,el} = \Sigma V_{ion} \tag{26}$$

The volumes of organic electrolytes, composed of both organic and inorganic moieties, can be derived from the corresponding volume contributions:

$$\overline{V}_{c,el} = \Sigma \overline{V}_c + \Sigma V^*_{ion} \tag{27}$$

where $\overline{V}_c$ corresponds to the organic part of the electrolyte and V^*_{ion} to the inorganic one. $\overline{V}_c$ can be obtained via Equation (25), and V^*_{ion} from the following relations, which differ for inorganic cations and anions:

$$V^*_{ion} = V_{ion} \quad \text{(inorganic cations)} \tag{28}$$

$$V^*_{ion} = V_{ion} - ZV_{CV} \quad \text{(inorganic anions)} \tag{29}$$

where Z is the number of electric charges.

The calculation of partial molar volumes of compounds forming micelles[38] necessitates consideration of special volume increments, V_{mic}. This has to be taken into account in the case of certain ionic and steroid surfactants, whereas most non-ionic surfactants do not require treatment as a particular case. Thus, the partial molar volume, $\overline{V}_{c,mic}$, of micellar ionic and steroid surfactants can be calculated according to

$$\overline{V}_{c,mic} = \overline{V}_c + xV_{mic} \quad \text{(ionic surfactants)} \tag{30}$$

$$\overline{V}_{c,mic} = \overline{V}_c + V_{mic} \quad \text{(steroid surfactants)} \tag{31}$$

where x is the number of carbons in the variable part of certain ionic surfactants (C_xTMABr, C_xTMACl, NaC_xSO_4, NaC_xSO_3, C_xCOONa). For monomeric surfactants the additional corrections are not required.

The performance of volume calculations from the increments according to Table 1 demands knowledge of both the empirical and the structural formulae of the compound under analysis. This is imperative for distinguishing between different bonding states and different functional and neighbouring groups and for considering the contributions of ring formation and ionization. In the case of polymers, only the structure of the monomeric unit(s) is required. The 3D structure of the molecules does not have to be known.

As outlined previously, the partial volume of a solute consists of contributions caused by the intrinsic volume (size of atoms) and volume effects of solute–solvent interactions (hydration and hydrophobic effects). Hence, all kinds of interactions are taken into account by the increments mentioned, and the applied values for V_i, V_{CV}, V_{RF}, V_{ES}, and V_{mic} represent only operational quantities. Nevertheless, they allow reliable predictions of true volumes (V and $\overline{V}$).

For volume calculations, the volume increments for atoms (Table 1) are normally used. However, the increments for atomic groups or already calculated volumes of simple compounds can be used instead.[4,37] In particular, the latter approach might be used advantageously if a series of similar compounds has to be calculated; it requires, however, the consideration of several pitfalls (*e.g.* problems caused by the proximity of certain groups in a molecule).

2.3.3 Special Approaches for Proteins

In the case of proteins, special approaches have been developed for calculating volume quantities of simple and conjugated proteins in their native or denatured states. If one has the choice, calculations should be performed for the native state and two-component solutions (protein in water or dilute buffer). Consideration of preferential interactions, however, also allows the prediction of volumes.[3,4]

The partial specific volume of a native, nonconjugated (simple) protein or large peptide can be calculated according to Cohn and Edsall[31] from partial specific volume increments for the individual AA residues:

$$\bar{v}_c = \frac{\sum_{i=1}^{N} W_i \bar{v}_i}{\sum_{i=1}^{N} W_i} = \frac{\sum_{i=1}^{N} N_i M_i \bar{v}_i}{\sum_{i=1}^{N} N_i M_i} \tag{32}$$

where $\bar{v}_c$ is the calculated partial specific volume of the protein (or peptide) in aqueous solution at 25 °C, W_i, M_i and $\bar{v}_i$ the weight percentage, molecular weight and partial specific volume of the ith AA residue, respectively, and N_i the number of residues. A comparison of different approaches for proteins prompted Perkins[25] to derive a consensus volume set for AA residues that was very close to the volume values given by Cohn and Edsall.[31] Starting from experimental values for AAs, Zamyatnin[40,41] proposed slightly different sets of increments, which resulted in slightly lower $\bar{v}_c$ values for proteins; these values represented the unfolded state of proteins in the absence of a denaturant. The volume increments emerging from different approaches (Cohn and Edsall, Perkins, Zamyatnin) are summarized in Table 2.

Conjugated proteins (such as nucleoproteins, glycoproteins, and lipoproteins) are composed of a protein part consisting of AA residues and a nonprotein moiety (nucleic acids, carbohydrates, lipids, *etc.*). Again, the volume can be calculated by summing up the volumes of the constituents; frequently, use of the major components is sufficient:

$$\bar{v}_c = \sum_{i=1}^{N} f_i \bar{v}_i = f_p \bar{v}_p + \sum f_{np} \bar{v}_{np} \tag{33}$$

where f_p and f_{np} are the weight fractions of protein or nonprotein components, respectively, and $\bar{v}_p$ and $\bar{v}_{np}$ their partial specific volumes. For the principal components of conjugated proteins, average values for the partial specific volumes have been derived:[13,42] $\bar{v}_a = 0.54$, 0.61, 0.735 and 1.02 $cm^3\,g^{-1}$ for the nucleic acid, carbohydrate, protein, and lipid moieties, respectively. For conjugated proteins with usual AA composition, these default values already yield reasonable volume estimates, $\bar{v}_c$, for the proteins in their native state.[13] If the precise composition of the conjugated protein is known, use of the individual volume increments may improve the results.

Similarly, the partial specific volumes of protein complexes (*i.e.* proteins associated with detergents, lipids, pigments, or dyes; and enzymes liganded with substrates, coenzymes, products, and analogues) can be estimated assuming the validity of an additivity approach for the corresponding volume increments. The partial specific volume of the complex, $\bar{v}_c$ (complex), may be calculated via Equation (33) or, if the amounts of bound ligands are given in grams per gram of protein, by the following relation:

$$\bar{v}_c(\text{complex}) = \frac{\bar{v}_p + \delta_d \bar{v}_d + \sum \delta_l \bar{v}_l}{1 + \delta_d + \sum \delta_l} \tag{34}$$

where $\bar{v}_p$, $\bar{v}_d$, $\bar{v}_l$ are the partial specific volumes of protein, detergent, and lipid (or any other bound or associated ligand), respectively; δ_d and δ_l are the corresponding amounts in g per g of protein. Preferential interactions are usually neglected in such calculations. In this terminology, the molar mass of the complex, M (complex), is given by

$$M(\text{complex}) = M_p(1 + \delta_d + \sum \delta_l) \qquad (35)$$

where M_p refers to the molar mass of the protein in the absence of the ligands.

Special approaches have been devised for the prediction of the isopotential specific volume of proteins at vanishing polymer concentration, v'_c, in diverse multicomponent solutions, such as proteins in salt, sugar, and AA solutions,[43,44] as well as proteins denatured in guanidinium chloride (GdmCl) or urea.[45–49]

In principle, the isopotential specific volume, v'_2, of proteins in multicomponent solutions can then be calculated from the partial specific volume, $\bar{v}_2$, provided that the interaction parameter, ξ_j ($j=1$ or 3), the partial specific volume of a small molecule, $\bar{v}_j$ ($j=1$ or 3), and the density of the complex solvent, ρ_s, are known:

$$v'_c = \bar{v}_2^0 + w_3\xi_1\left(\frac{1}{\rho_s} - \bar{v}_3\right) = \bar{v}_2^0 + (w_3\delta_1 - \delta_3)\left(\frac{1}{\rho_s} - \bar{v}_3\right) \qquad (36)$$

The values for $\bar{v}_2^0$ (protein) and $\bar{v}_3$ (additive, *e.g.* salt) can be determined experimentally or by calculation, ρ_s is the solvent density, and δ_1 and δ_3 characterize the corresponding number of g of water or component 3 bound per g of protein. The molality of the additive in g g^{-1}, w_3, may be calculated from the concentration in g cm^{-3}, c_3, of the additive: $w_3 \approx c_3\bar{v}_1/(1-c_3\bar{v}_3)$. In these calculations, the preferential hydration of proteins is generally assumed to be relatively independent of the additive concentration and is proportional to the protein surface area.

The interaction parameter, ξ_1, is not very sensitive to the molar masses of globular proteins, since the surface area of a protein molecule is proportional to $M_2^{2/3}$. Therefore, within a close approximation, approximate values of molar masses[43,44] suffice. The value of ξ_1 for any protein B can be calculated from that determined for a specific protein A by

$$\xi_1^B = \xi_1^A\left(\frac{M_2^A}{M_2^B}\right)^{1/3} \qquad (37)$$

Interaction parameters for many proteins and conditions are given elsewhere,[11,12,43,44,50–56] and those for nucleic acids have also been summarized recently.[10] A series of representative examples of interaction parameters are outlined in Table 3, valid for different proteins and DNA and a variety of environmental conditions. In this context it has to be stressed that the interaction parameters for different additives vary over a wide range, and may even have different signs, as may be taken from a comparison of the ξ values for salts, sugars, polyols, AAs, on the one hand, and strong denaturants such as GdmCl, on the other. It may be totally incorrect to replace the interaction parameter of a certain additive by another one, whereas the transfer of interaction parameters from one protein to another is unproblematic (Equation 37).

Timasheff and co-workers[45–49] developed special approaches to characterize the volume of simple (nonconjugated) proteins denatured in 6 M GdmCl or 8 M urea. The calculations of the isopotential specific volume, ϕ'_c, of proteins in these two

Table 2 Partial specific volumes and hydration of AA residues[a]

3-letter code	1-letter code	M (g mol⁻¹)	$\bar{v}$[b] (cm³ g⁻¹) Cohn–Edsall[31d] 'CE'	Perkins[25e] 'P'	Zamyatnin[40] 'Z1'	Zamyatnin[41] 'Z2'	h[c] (mol mol⁻¹) Kuntz[57]	pH 4	pH 6–8	pH 12
Ala	A	71.07	0.74	0.744	0.748	0.732			1.5	
Arg	R	156.18	0.70	0.726	0.666	0.756			3	
Asn	N	114.10	0.62	0.634	0.619	0.610			2	
Asp	D	115.08	0.60	0.604	0.579	0.573		2	6	
Cys	C	103.14	0.63[f]	0.615	0.631	0.630	2		1	
2 Cys	2 C	204.26	0.63	0.615[f]	0.676	0.676[f]			1[f]	
Gln	Q	128.13	0.67	0.682	0.674	0.667			2	
Glu	E	129.11	0.66	0.657	0.643	0.605	2		7.5	
Gly	G	57.05	0.64	0.632	0.632	0.610			1	
His	H	137.14	0.67	0.686	0.670	0.659			4	
Ile	I	113.16	0.90	0.884	0.884	0.876			1	
Leu	L	113.16	0.90	0.894	0.884	0.876			1	
Lys	K	128.17	0.82	0.811	0.789	0.775			4.5	
Met	M	131.19	0.75	0.758	0.745	0.739			1	
Phe	F	147.17	0.77	0.776	0.774	0.766			0	
Pro	P	97.11	0.76	0.765	0.758	0.748			3	
Ser	S	87.07	0.63	0.634	0.613	0.596			2	
Thr	T	101.10	0.70	0.705	0.689	0.676			2	

		M	$\bar{v}$					h	
Trp	W	186.21	0.74	0.737	0.734	0.728		2	
Tyr	Y	163.17	0.71	0.706	0.712	0.703		3	7.5
Val	V	99.13	0.86	0.843	0.847	0.831		1	
Asx[f,g]	B	114.59	0.61	0.619	0.599	0.592	2	4	
Glx[f,g]	Z	128.62	0.665	0.670	0.659	0.636	2	4.75	
Xaa[f,h]	X	118.9	0.72	0.725	0.715	0.708	1.95	2.4	2.65

Adapted from previous compilations.[3,4,78]

[a] M, $\bar{v}$, and h refer to molar mass, partial specific volume, and hydration number of the AA residues, respectively.

[b] Values are given for 25°C.

[c] Values are given for $-35\ °C$ ("non-freezable water"); at acidic and alkaline pH only the data which differ from the value at neutrality are mentioned. The h values at pH 6–8 and 4 are required for ϕ'_c calculations of proteins in 6 M GdmCl and 8 M urea, respectively.

[d] The value given originally by Cohn and Edsall[31] for one cystine (code: 2 Cys or 2 C) of 0.61 $cm^3\ g^{-1}$ was corrected later on to 0.63 $cm^3\ g^{-1}$.[94,95]

[e] Original values (consensus volumes) were given as partial molecular volumes in nm^3.

[f] A few values are completed by conclusions from analogy.

[g] For Asx and Glx the mean values of acid and amide forms are taken.

[h] For unspecified AA residues (Xaa), averages of all 20 residues are given.

Table 3 *Preferential interaction parameters ξ_1 and ξ_3 of proteins and DNA in various solvent systems at 20–25 °C, together with the volume properties of the corresponding macromolecules (component 2) and additives (component 3); water is used as principal solvent (component 1)[a]*

Co-solvent			Macromolecule		Specific volume		Interaction parameters		Reference
Additive and conditions	w_3 (g g^{-1})	$\bar{v}_3$ (cm^3 g^{-1})	Protein or DNA	M_2 (kg mol^{-1})	ϕ_2^0 or $\bar{v}_2^0$ (cm^3 g^{-1})	$\phi_2'^0$ or $\bar{v}_2'^0$ (cm^3 g^{-1})	ξ_1 (g g^{-1})	ξ_3 (g g^{-1})	
1 M NaCl, pH 4.5, 20 °C	0.0597	0.331	Lysozyme	14.3	0.707	0.723	0.424	−0.0253	52, 53
1 M NaCl, pH 4.5, 20 °C	0.0596	0.331	BSA	66.4	0.734	0.744	0.265	−0.0158	52
1 M NaCl, pH 5.6, 20 °C	0.0597	0.331	BSA	66.4	0.735	0.744	0.243	−0.0145	52, 53, 56
1 M NaOAc, pH 4.7, 20 °C	0.0906	0.516	Lysozyme	14.3	0.712	0.731	0.478	−0.0433	52
1 M NaOAc, pH 5.6, 20 °C	0.0906	0.516	BSA	66.4	0.735	0.747	0.312	−0.0270	53, 56
0.5 M Na$_2$SO$_4$, pH 4.5, 20 °C	0.0718	0.165	BSA	66.4	0.734	0.750	0.287	−0.0206	52
1 M Na$_2$SO$_4$, pH 4.5, 20 °C	0.145	0.208	BSA	66.4	0.735	0.781	0.459	−0.0669	52
1 M Na$_2$SO$_4$, pH 5.6, 20 °C	0.141	0.208	BSA	66.4	0.735	0.788	0.524	−0.0739	52, 53, 56
1 M MgCl$_2$, pH 4.5, 20 °C	0.0975	0.176	Lysozyme	14.3	0.712	0.721	0.122	−0.0119	53
0.5 M MgCl$_2$, pH 4.5, 20 °C	0.0479	0.150	BSA	66.4	0.733	0.737	0.102	−0.0049	52
1 M MgCl$_2$, pH 4.5, 20 °C	0.0975	0.176	BSA	66.4	0.736	0.739	0.041	−0.0040	53, 56
1 M MgCl$_2$, pH 5.6, 20 °C	0.0975	0.176	BSA	66.4	0.737	0.740	0.043	−0.0042	52
1 M MgSO$_4$, pH 4.5, 20 °C	0.121	0.136	Lysozyme	14.3	0.710	0.748	0.412	−0.0498	53
1 M MgSO$_4$, pH 4.5, 20 °C	0.121	0.136	BSA	66.4	0.734	0.769	0.388	−0.0469	52, 53, 56
1 M Glucose, pH 3.0, 20 °C	0.203	0.628	BSA	66.4	0.722	0.745	0.371	−0.0755	12, 51
1 M Glucose, pH 6.0, 20 °C	0.203	0.628	BSA	66.4	0.726	0.741	0.250	−0.0509	12, 51
1 M Saccharose, 20 °C	0.435	0.620	RNase A	13.7	0.692	0.743	0.437	−0.190	12, 56, 96
40% Ethylene glycol, 25 °C	0.731	0.883	BSA	66.4	0.733	0.740	0.130	−0.097	12, 97, 98
40% Glycerol, 25 °C	0.833	0.772 (20 °C)	BSA	66.4	0.727	0.746	0.185	−0.154	4, 12, 97
10% PEG-1000, pH 2.0, 20 °C	0.108	0.836	RNase A	13.7	0.695	0.723	1.780	−0.194	99
10% PEG-1000, pH 7.0, 20 °C	0.108	0.836	Lysozyme	14.3	0.715	0.729	0.827	−0.095	99

Condition	w_3	$\bar{v}_3$	Macromolecule	M_2	$\bar{v}_2^0$	ϕ_2^0	ξ_1	ξ_3	References
10% PEG-1000, pH 7.0, 20 °C	0.108	0.836	BSA	66.4	0.734	0.752	1.185	−0.128	12, 99
2 M Gly, pH 6.2, 20 °C	0.165	0.614	Lysozyme	14.3	0.700	0.726	0.474	−0.0784	100
2 M Gly, pH 6.2, 20 °C	0.165	0.614	BSA	66.4	0.735	0.758	0.416	−0.0687	56, 100
1 M GdmCl, pH 4.5, 20 °C	0.103	0.732	BSA	66.4	0.735	0.729	−0.239	0.0246	56, 101
3 M GdmCl, pH 4.5, 20 °C	0.365	0.747	BSA	66.4	0.728	0.718	−0.147	0.0536	101
0 M NaCl, 25 °C		0.284	NaDNA		0.499ᵇ				4, 10, 102
0.01 M NaCl, 25 °C	0.000586		NaDNA			0.515	36.3	−0.0213	10
0.01 M NaCl, 25 °C	0.005874	0.298	NaDNA			0.527	5.79	−0.0340	10
0.2 M NaCl, 25 °C	0.01176		NaDNA		0.503	0.535	3.44	−0.0405	10, 102
0.5 M NaCl, 25 °C	0.02958	0.318	NaDNA			0.544	1.437	−0.0425	4, 10
1.0 M NaCl, 25 °C	0.05976	0.331	NaDNA		0.528	0.568	1.118	−0.0668	4, 10, 102
2.0 M NaCl, 25 °C	0.1219	0.357	NaDNA			0.584	0.710	−0.0865	4, 10
5.0 M NaCl, 25 °C	0.3266		NaDNA		>0.543	0.596	0.361	−0.118	10, 102
0 M CsCl, 25 °C		0.233	CsDNA		0.440ᵇ				4, 10, 102
0.2 M CsCl, 25 °C	0.0340		CsDNA		0.446	0.503	1.471	−0.050	2, 10, 102
0.5 M CsCl, 25 °C	0.0859		CsDNA			0.517	0.873	−0.075	2, 10
1.0 M CsCl, 25 °C	0.1743		CsDNA		0.460	0.530	0.654	−0.114	2, 10, 102
2.3 M CsCl, 25 °C	0.4288		CsDNA		0.467	0.563	0.443	−0.190	2, 10, 102
5.0 M CsCl, 25 °C	1.0927		CsDNA		>0.471	0.584	0.282	−0.308	2, 10, 102

ᵃThe interaction parameters $\xi_1 = (\partial w_1/\partial w_2)_\mu$ and $\xi_3 = (\partial w_3/\partial w_2)_\mu$ designate "preferential hydration" of the respective macromolecule and "preferential binding" of additives to it, respectively; $\bar{v}_2^0$ and ϕ_2^0 are the isomolal specific volume, and $v_2^{\prime 0}$ and $\phi_2^{\prime 0}$ the isopotential specific volume of the macromolecule of molar mass M_2, as defined in the text, and $\bar{v}_3$ marks the partial specific volume of the additive of weight molality w_3.

ᵇFor details regarding the difference of $\bar{v}_2^0$ values for NaDNA and CsDNA at $c_3 = 0$ see Eisenberg and co-workers.[2,10,102]

denaturants are based on the AA composition and simplifying assumptions concerning preferential interactions between protein and solvent components (water and denaturant as cosolvent). The hydration values derived for single AAs[57] can be used for estimating the extent of protein hydration. Furthermore, it is implied that each aromatic AA side chain (including His) and each pair of peptide bonds interact with one molecule of the denaturant. In the case of conjugated proteins, of course, predictions may yield misleading results.

Calculated values of the isopotential volume, ϕ'_c, of the protein in solvents containing 6 M GdmCl or 8 M urea can be estimated from:

$$\phi'_c = \bar{v}_2 - \xi_3\left(\frac{1}{\rho_s} - \bar{v}_3\right) = \bar{v}_2 - (\delta_3 - w_3\delta_1)\left(\frac{1}{\rho_s} - \bar{v}_3\right) \tag{38}$$

where $\bar{v}_2$ and $\bar{v}_3$ are the partial specific volumes of protein and denaturant (GdmCl or urea), respectively, and ρ_s the solvent density; δ_1 and δ_3 characterize the corresponding number of g of water or denaturant bound per g of protein, and w_3 is a measure of the number of g of denaturant per g of water. The value of $\bar{v}_2$ is usually obtained using the Cohn–Edsall method.[31] For 6 M GdmCl at 20 °C the values for ρ_s, $\bar{v}_3$ and w_3 are 1.1418 g cm^{-3}, 0.763 cm^3 g^{-1} and 1.007 g g^{-1}, respectively,[46] and the corresponding values for 8 M urea are 1.1152 g cm^{-3}, 0.763 cm^3 g^{-1} and 0.752 g g^{-1}.[49]

The amount of denaturant binding to the protein, δ_3, in g of denaturant per g of protein, is assumed to be:

$$\delta_3 = \frac{M_d}{M_p}\left(\frac{N_t-1}{2} + N_{ar}\right) \tag{39}$$

where M_d and M_p denote the molar masses of denaturant and protein subunit, respectively; for GdmCl and urea the values of M_d are 95.53 and 60.06 g mol^{-1}. N_t and N_{ar} are the total number of AA residues and the number of aromatic residues, respectively.

The degree of hydration, δ_1, in g of water per g of protein, can be calculated on the basis of the AA content and the hydration of the constituent AAs according to Kuntz:[57]

$$\delta_1 = \frac{18}{M_p}\sum h_i N_i \tag{40}$$

where h_i is the hydration in mol of water per mol of the ith AA, and N_i is the number of residues i. The values for h_i (Table 2) are dependent of the pH range used (about 4 or 6–8 or 12). The previously mentioned approach rests on the assumption that all AAs are exposed to the solvent.

Calculation procedures for proteins denatured in 6 M GdmCl or 8 M urea revealed that the values predicted for these denaturing conditions agree satisfactorily with experimental values.[45–49] Combining the results found for different simple proteins, the calculated isopotential specific volumes of simple proteins are only slightly smaller in these denaturants than those observed in the native state (about 2% for 6

M GdmCl).[3,4] This approximation does not apply for low denaturant concentrations where different volume effects are superimposed.[58,59]

In the case of glycoproteins denatured in 6 M GdmCl, slight modifications of the previous approach also deliver a practicable approximation.[5] For glycoproteins, it is suggested that each carbohydrate moiety binds one denaturant molecule, and that δ_1 is not severely affected by the presence of the carbohydrate:

$$\delta_3 = \frac{M_d}{M_p}\left(\frac{N_t - 1}{2} + N_{ar} + N_{ch}\right) \tag{41}$$

where N_{ch} stands for the number of carbohydrate residues.

Because of a lack of experimental data, similar approaches for glycoproteins in 8 M urea or other types of conjugated proteins in denaturing agents do not exist to date.

2.4 Examples: Small Molecules and Biopolymers

2.4.1 Use of the Universal Approach

Figure 1 illustrates the modus operandi for calculating the partial molar volume, $\overline{V}_c$, and the partial specific volume, $\overline{v}_c$, of some typical model compounds from the volume increments of Table 1 or more comprehensive tables given in previous papers.[4,37] As examples, two DNA-intercalating agents, actinomine and the antibiotic actinomycin C_3, and the complex zwitterionic surfactant, bisCHAPSO, have been chosen. In all cases, the calculations posed no problems, and the results are in full accord with the experimental data.

Numerous values for calculated and experimental values of partial molar and partial specific volumes of organic and biochemical compounds in aqueous solution have been given earlier.[3,4,37,39] The published volumes cover a broad spectrum of small molecules and polymers with quite different features: nonionic and ionic organic compounds, inorganic electrolytes and denaturants used as additives to the principal solvent; monomeric and micellar detergents (ionic, zwitterionic and nonionic) used for the solubilization of biopolymers; low-molecular organic compounds that may act as ligands of enzymes or other types of proteins (substrates, coenzymes, products and/or analogues); constituents of biopolymers such as AAs, peptides, nucleobases, nucleosides and nucleotides, carbohydrates, and lipids; biopolymers like polyaminoacids, nonconjugated and conjugated proteins, polynucleotides, nucleic acids, polysaccharides, complexes between different types of molecules, and, for comparison, some synthetic polymers in aqueous and nonaqueous solution.

Figure 2 comprises a variety of examples of calculated and experimental values for biomolecules of special interest, to demonstrate the ability of our approach. Indeed, the vast majority of the predicted values for small molecules and polymers is within a range of $\pm 2\%$, if compared to the observed data.[37,39] Larger deviations could only be found with small compounds consisting of certain, in particular, substituted heterocycles (*e.g.* certain nucleobases), whereas in the case of more complex molecules (*e.g.* nucleotides, nucleic acids) the deviations usually cancel out each other.[39] Pronounced deviations between predicted and observed data have only been found for very small compounds (*e.g.* the smallest representatives of a homologous series).

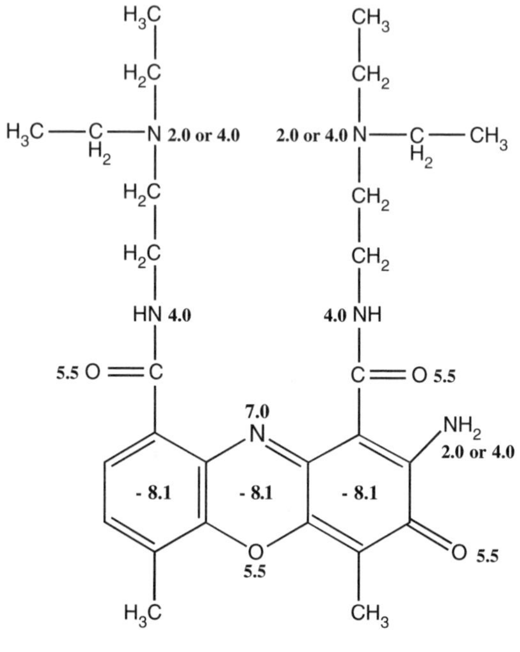

H₃C—C H₂—N 2.0 or 4.0   2.0 or 4.0 N—C H₂—CH₃

(a)						**ACTINOMINE**

Figure 1 *Selected examples of calculations applying the universal volume approach.[37] The calculation of the partial molar and partial specific volumes, $\bar{V}_c$ and $\bar{v}_c$, makes use of Equations (25)–(31) and the volume increments given in Table 1, together with the structural formula (and the empirical formula and molar mass, M, which may be derived from it). Knowledge of the stereochemical constitution is not needed. If the choice between different volume increments is not clear from the structural formula (e.g. distinction between polyamines and simple amines as in (a), the lower and upper increment limits should be used, to give the minimum and maximum volumes. The calculated values refer to aqueous solution and 25 °C (a) Actinomine; empirical formula: $C_{28}H_{40}N_6O_4$; M=524.7 g mol⁻¹. $\bar{V}_c$=28×9.9 (C)+40×3.1 (H)+ 3×2.0 or 4.0 (N)+2×4.0 (N)+1×7.0 (N)+4×5.5 (O)+12.4 (V_{CV})−3×8.1 (V_{RF})= 432.3 or 438.3 cm³ mol⁻¹; $\bar{v}_c$=$\bar{V}_c$ /M=0.824 or 0.835 cm³ g⁻¹. The calculated $\bar{v}_c$ values are in agreement with the experimental value of 0.829 cm³ g⁻¹ as obtained after applying a temperature correction of 5×10⁻⁴ cm³ g⁻¹ K⁻¹ to the value of 0.824 cm³ g⁻¹ derived at 15 °C by SAXS.[89] (b) Actinomycin C_3; empirical formula: $C_{64}H_{90}N_{12}O_{16}$; M=1283.5 g mol⁻¹. $\bar{V}_c$=64×9.9 (C)+90×3.1 (H)+8×2.0 (N)+ 1×4.0 (N)+3×7.0 (N)+16×5.5 (O)+12.4 (V_{CV})−3×8.1 (V_{RF})−2×6.1 (V_{RF})= 1017.5 cm³ mol⁻¹; $\bar{v}_c$=$\bar{V}_c$ /M=0.793 cm³ g⁻¹. The calculated value $\bar{v}_c$ complies with the experimental value of 0.796 cm³ g⁻¹ as obtained after applying a temperature correction to the value of 0.786 cm³ g⁻¹ obtained at 5 °C by pycnometry[90] or values of 0.799 and 0.790 cm³ g⁻¹ after correcting the values of 0.794 and 0.785 cm³ g⁻¹ that had been determined at 15 °C by SAXS.[89] (c) BisCHAPSO; empirical formula: $C_{40}H_{75}N_3O_{12}S_2$; M=854.2 g mol⁻¹. $\bar{V}_c$=40×9.9 (C)+75×3.1 (H)+2×4.0 (N)+1×12.2 (N)+2×0.4 (O)+5×2.3 (O)+5×5.5 (O)+2×15.5 (S)+12.4 (V_{CV})− 3×8.1 (V_{RF})−1×6.1 (V_{RF})−2×6.7 ($V_{ES−}$)−2×6.8 (V_{ES+})=674.5 cm³ mol⁻¹; $\bar{V}_{c,mic}$= $\bar{V}_c$−23 (V_{mic})=651.5 cm³ mol⁻¹; $\bar{v}_c$=$\bar{V}_c$ /M=0.790 cm³ g⁻¹; $\bar{v}_{c,mic}$=$\bar{V}_{c,mic}$ /M= 0.763 cm³ g⁻¹. The calculated value for the micellar state, $\bar{v}_{c,mic}$, of the surfactant compares well with the observed value of 0.75 cm³ g⁻¹ determined at ambient temperature by pycnometry[91]*

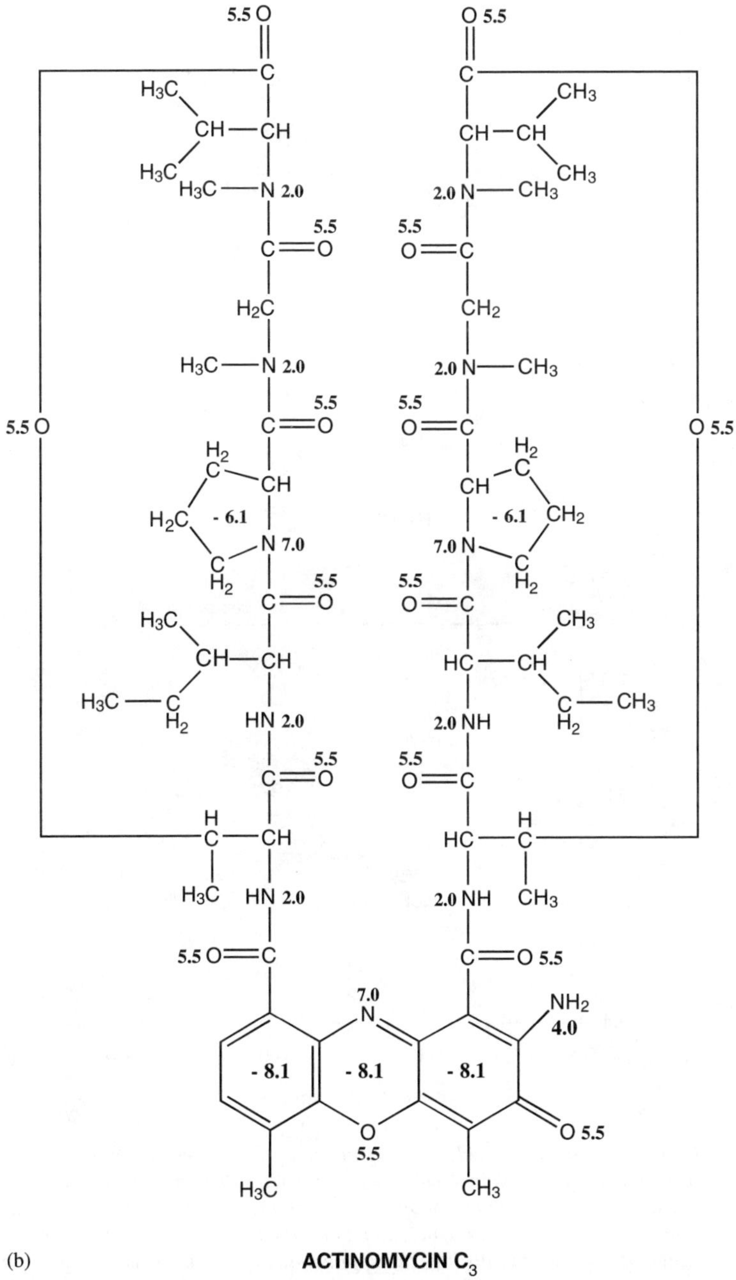

(b)   **ACTINOMYCIN C₃**

Figure 1   *(continued)*

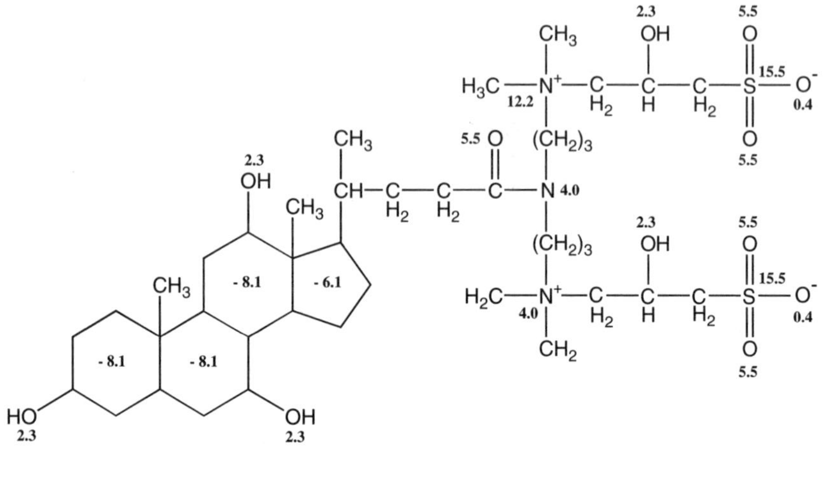

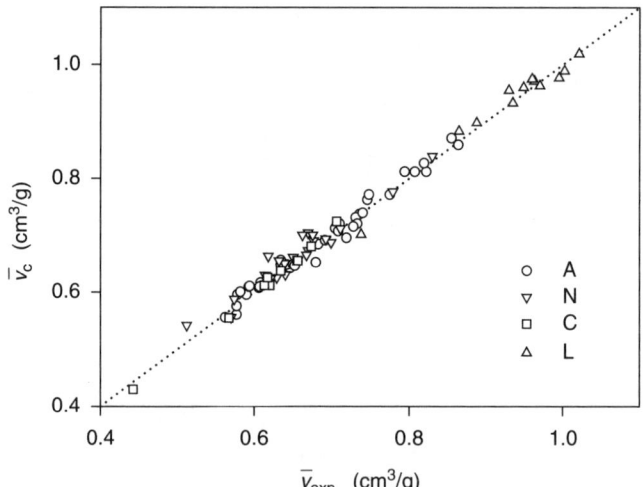

(c)   **BisCHAPSO**

Figure 1 *(continued)*

Figure 2 *Comparison of calculated and experimental partial specific volumes, $\overline{V}_c$ and, $\overline{V}_{exp}$, of different types of biomolecules (A, amino acids, betaines and peptides; N, nucleobases, nucleosides and nucleotides; C, carbohydrates; L, lipids and constituents). The agreement between predicted and observed data is remarkable over the whole range of volumes and justifies the applicability of the universal volume approach[37] for the calculation of different types of biomolecules, including biopolymers*

Of course, the prediction of polymer volumes in organic solvents can only provide rough estimates, since our approach has been adapted for use in aqueous solution. Experimental data of volumes in nonaqueous media may deviate substantially from the volumes observed in water.[3,4,37,60,61]

The broad range of possible applications of the universal volume approach has also been demonstrated for the surfactant area, where both monomeric and micellar surfactants may be anticipated correctly.[38] Some typical examples are depicted in Figure 3, emphasizing the broad range of volume data for surfactants and the substantial influence of the chain length on the values for $\bar{V}$.

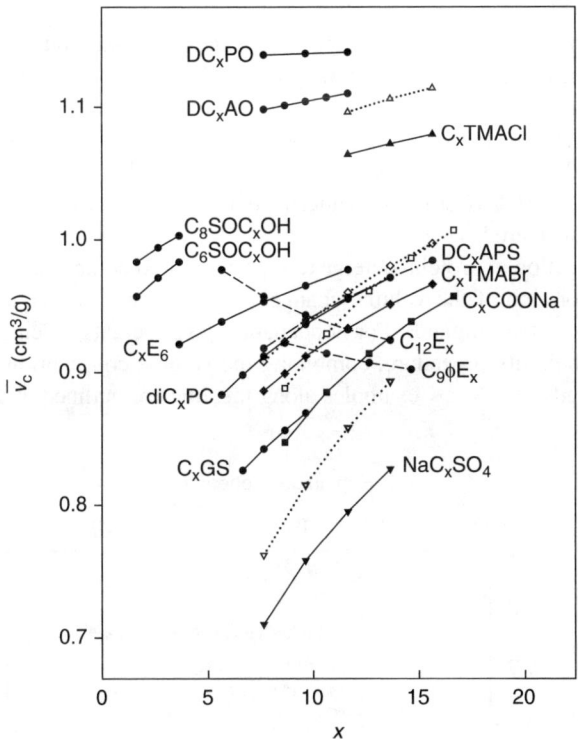

Figure 3   *Calculated partial specific volumes, $\bar{V}_c$, of selected surfactants, depending on the change in the chain length x, where x is the number of C or E in the variable part of the surfactant. Calculations of surfactants were performed according to the universal volume approach,[37] exploiting, if required, corrections for micellization.[38] In the case of ionic surfactants, the volumes for micelles (open symbols) are significantly different from those of the monomeric units (full symbols). The applied surfactants belong to different classes of surfactants: cationic detergents (alkyltrimethylammonium bromides and chlorides, C_xTMABr and C_xTMACl, respectively), anionic detergents (sodium salts of alkyl sulfates, NaC_xSO_4), zwitterionic detergents (dimethylalkylammoniopropane sulfonates, DC_xAPS), nonionic detergents (polyoxyethylene alkylphenols, $C_9\phi E_x$, polyoxyethylene alcohols, C_xE_6 and $C_{12}E_x$, dimethylalkyl amine oxides, DC_xAO, dimethylalkyl phosphine oxides, DC_xPO, alkylsulfinyl alkanols, C_6SOC_xOH and C_8SOC_xOH, and alkylglucosides, C_xGS), and lipids/constituents (diacyl phospholipids, diC_xPC, and sodium salts of carboxylic acids, C_xCOONa). The values for experimental data[3,4,37-39,92] (not shown) are in respectable agreement with the predicted values (±2 %), similar to those for the biomolecules shown in Figure 2*

2.4.2   Use of Special Approaches

The simple estimation of volumes from the increments of the major components already provides volume data for conjugated proteins of high reliance (Figure 4). The accuracy obtained for the predicted volumes is rather high (mostly better than ±2%) and comparable to the experimentally accessible accuracy. Obviously, in the majority of cases, extraordinary interactions fail to appear and possible deviations are compensated.

The influence of specific and unspecific ligands (*e.g.* substrates or cofactors) on protein volumes is generally rather marginal,[3,4,39] although the number and mass of the ligands under consideration can differ considerably. In a few cases, owing to complex interactions between protein and ligand(s), the observed volume data suggest relevant differences. Only when we aim at extremely precise determinations of molar masses by AUC or solution scattering experiments, must the occurring volume changes be considered.

The investigation of protein–detergent or protein–lipid complexes, however, necessitates incorporation of the volume changes occurring as a consequence of binding large amounts of these ligands. For nondenaturing surfactants, calculations make use of a simple additivity procedure, combining the volume contributions of native proteins and ligands.[3,4,39] A few examples along this line are outlined in Table 4.

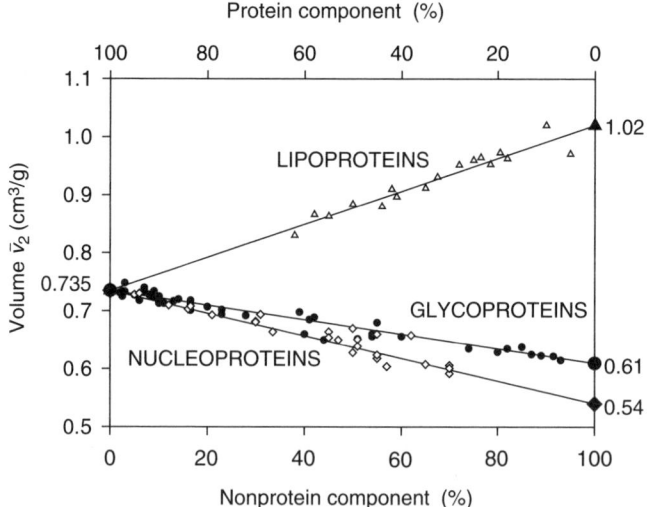

Figure 4   *Partial specific volumes, $\bar{V}_2$, of some conjugated proteins (nucleo-, glyco- and lipoproteins) consisting of two principal components (one protein and one nonprotein component) at varying weight percentages of these components. Experimental values for conjugated proteins were taken from previous data compilations[3,4,13,42] and are given as symbols. Calculated values are represented by the full lines; for calculations the default values for the main components (proteins, nucleic acids, carbohydrates, lipids) have been used which are given on the ordinate intercepts on the left and right axes. The figure can also be used for a rough estimation of $\bar{V}_2$ of an unknown sample, provided its approximate composition is known*

Table 4   *Calculated partial specific volumes of selected proteins, ligands and protein–ligand complexes. Calculations were performed on the basis of an additivity approach, using the calculated or experimental values for the partial specific volumes of the protein and the ligands. The values are valid for aqueous solutions at about 20–25 °C[a]*

Protein	Source	Ligand(s)	Partial specific volumes[b]			
			$\bar{v}$ (protein) ($cm^3\ g^{-1}$)	$\bar{v}$ (ligand) ($cm^3\ g^{-1}$)	$\bar{v}$ (complex) ($cm^3\ g^{-1}$)	Reference
Complexes between proteins and specific ligands:						
Citrate synthase	Pig heart	2 Oxaloacetate per dimeric protein	0.740	0.477	0.739	103
		2 CoASAc per dimeric protein		0.640[c]	0.738	103
		2 (CoASH.citrate) per dimeric protein		0.634,[c] 0.461	0.737	103
Malate synthase	Baker's yeast	3 Glyoxylate per trimeric protein	0.745	0.548	0.745	103
		3 CoASAc per trimeric protein		0.640[c]	0.744	103
		3 (CoASAc.pyruvate) per trimeric protein		0.640,[c] 0.644	0.743[c]	103
Glyceraldehyde-3-phosphate dehydrogenase	Baker's yeast	4 NAD$^+$ per tetrameric protein	0.737	0.62	0.735	3,104
Complexes between proteins and detergents or lipids:						
Band 3 protein	Human erythrocytes	0.4 g DC$_{12}$AO g^{-1} protein	0.740	1.122	0.849	3, 38
		0.64 g C$_{12}$E$_9$ g^{-1} protein		0.950	0.822	3
ApoB	Human serum	0.64 g DOC g^{-1} protein	0.725	0.778	0.746	3, 38
ApoC-III	Human serum	3.8 g di-C$_{14}$PC g^{-1} protein	0.723	0.963	0.913	3, 38
		12.5 g di-C$_{14}$PC g^{-1} protein		0.963	0.945	3, 38
Rhodopsin	Bovine retina	0.92 g DC$_{12}$AO g^{-1} protein	0.71	1.122	0.907	3, 38
		1.46 g Triton X-100 g^{-1} protein		0.915	0.832	3, 38

Adapted from ref. 39.

[a]Values were calculated on the basis of the weight fractions or amounts in g and partial specific volume data cited in the given literature.

[b]For details of experimental conditions the reader is referred to the original literature.

[c]Slightly refined value.

Table 5 *Comparison of calculated and experimental partial and isopotential specific volumes of selected nonconjugated proteins in two- and multicomponent solutions. The values are valid for aqueous solutions at about 20–25 °C[a]*

Protein	Calculated volumes						Experimental volumes			
	Solvent conditions	$\bar{v}_c$ or v'_c (cm³ g⁻¹)					Solvent conditions	$\bar{v}_{exp}$ or v'_{exp} (cm³ g⁻¹)	Reference	$\Delta v^{c,d}$ (%)
		CE	P	Z1	Z2					
Serum albumin (bovine)										
	Water	0.735	0.735	0.725	0.717	($\bar{v}_c$)	Water or dilute buffer	0.735 ($\bar{v}, v'$)	3	0 (CE, P)
	1 M NaCl	0.744	0.744	0.734	0.726	(v'_c)	1 M NaCl	0.744 (v')	52	0 (CE, P)
	6 M GdmCl, pH 4	0.713	0.712	0.703	0.695	(v'_c)	6 M GdmCl, pH 7.1	0.717 (v')	11	+0.1 (Z1)
	6 M GdmCl, pH 6–8	0.728	0.728	0.718	0.710	(v'_c)				
	8 M urea, pH 4	0.723	0.723	0.714	0.705	(v'_c)				
	8 M urea, pH 6–8	0.737	0.736	0.727	0.719	(v'_c)				
α-Lactalbumin (bovine)										
	Water	0.730	0.727	0.718	0.706	($\bar{v}_c$)	Dilute buffer	0.735 ($\bar{v}$)	105	−0.7 (CE)
	1 M NaCl	0.745	0.743	0.734	0.722	(v'_c)				
	6 M GdmCl, pH 4	0.705	0.703	0.694	0.682	(v'_c)	6 M GdmCl, pH 7.1	0.698 (v')	11	+1.0 (Z2)
	6 M GdmCl, pH 6–8	0.728	0.725	0.716	0.705	(v'_c)				

| 8 M urea, pH 4 | 0.716 | 0.713 | 0.705 | 0.693 (v'_c) | 0.699 (v') | 47 | ±0.9 (Z1, Z2) |
| 8 M urea, pH 6-8 | 0.736 | 0.733 | 0.725 | 0.713 (v'_c) | | | |

Adapted from ref. 39.

[a] For details of the calculation procedures the reader is referred to the references cited in the text and summarizing reviews.[3,4,12] In the mentioned reviews also comprehensive compilations of partial specific volumes, hydration and molar mass increments of AA residues, and interaction parameters required for performing such calculations, together with the theoretical formalism, are given. Calculation of parameters for proteins are based on the AA compositions (for the two mentioned proteins: *cf.* refs. 106, 107).

[b] Calculations are based on the following auxiliary parameters: $\xi_1 = 0.25$ g g⁻¹ (serum albumin) or 0.42 g g⁻¹ (α-lactalbumin), $\bar{V}_1 = 1.0$ cm³ g⁻¹, $\bar{V}_3 = 0.331$ cm³ g⁻¹, $\rho_s = 1.0413$ g cm⁻³, $M_3 = 58.44$ g mol⁻¹, $w_3 = 0.0597$ g g⁻¹ for 1 M NaCl; $\bar{V}_3 = 0.763$ cm³ g⁻¹, $\rho_s = 1.1418$ g cm⁻³, $M_3 = 95.53$ g mol⁻¹, $w_3 = 1.007$ g g⁻¹ for 6 M GdmCl; $\bar{V}_3 = 0.763$ cm³ g⁻¹, $\rho_s = 1.1152$ g cm⁻³, $M_3 = 60.06$ g mol⁻¹, $w_3 = 0.752$ g g⁻¹ for 8 M urea, and $M_2 = 66267$ g mol⁻¹ in the case of serum albumin; and $\delta_1 = 0.296$ g g⁻¹ (pH 4) or 0.494 g g⁻¹ (pH 6-8), $\delta_1 = 0.314$ g g⁻¹ (pH 4) or 0.447 g g⁻¹ (pH 6-8), $\delta_3 = 0.512$ g g⁻¹ (GdmCl) or 0.322 g g⁻¹ (urea), and $M_2 = 14176$ g mol⁻¹ in the case of α-lactalbumin. In the listed quantities ξ, $\bar{V}$, ρ, M, w, δ designate preferential interaction parameter, partial specific volume, mass density, molar mass, concentration and amount of bound additive, respectively, and subscripts 1, 2, 3, s refer to component 1 (water), component 2 (protein), and component 3 (salt and denaturant), and complex solvent (buffer+nonmacromolecular additives).

[c] Abbreviations of calculation methods for proteins in two-component solutions: CE, Cohn and Edsall;[31] P, Perkins;[25] Z1, Zamyatnin;[41] Z2, Zamyatnin.[42] In the case of multicomponent solutions, the mentioned calculation schemes for two-component schemes were combined with the special corrections for the corresponding multicomponent solution according to Timasheff and co-workers[45-49] (*cf.* Section 2.3.3).

[d] $\Delta v = 100(v_c - v_{exp})/v_{exp}$. Only Δ for the best suitable calculation method is mentioned.

Table 5 resumes the results of calculation approaches for proteins in their native and denatured states. As may be expected, the prediction of volumes of native proteins offers no problems and yields sound results when using the increments given by Cohn and Edsall or Perkins (Table 1). Application of the increments given by Zamyatnin rather refers to the denatured state of proteins. Both types of increments cope with the situation in multicomponent solutions, if combined with known interaction parameters.

3   The Surface of Biopolymers

From the macroscopic point of view, area definitions seem to be obvious, whereas on the level of individual molecules definitions become rather vague.[62-66] The same is true for experimental measurements of surface areas. The observed or calculated molecule surface depends on the method used for its determination.

The basis for methods developing smooth surface envelopes of proteins is connected with the work of Richards and co-workers[62-64] on solvent-accessible areas (Figure 5). Initially, all atoms of a protein are formed as spheres of corresponding van der Waals radii. The resulting surface, the 'van der Waals surface', is of no practicable meaning. A more realistic concept of surface combines solvent molecules in contact with protein, thereby forming a 'solvent-accessible surface area'. Solvent-accessible surfaces are defined by tracing out the area with the centre of a probe sphere representing a solvent molecule as it rolls over the surface of a protein. For water molecules, a value of 0.14–0.15 nm is appropriate for the radius of the spherical probe. Atoms that are not contacted by the probe (owing to lacking accessibility) are considered to be interior atoms, not part of the molecule surface. The part

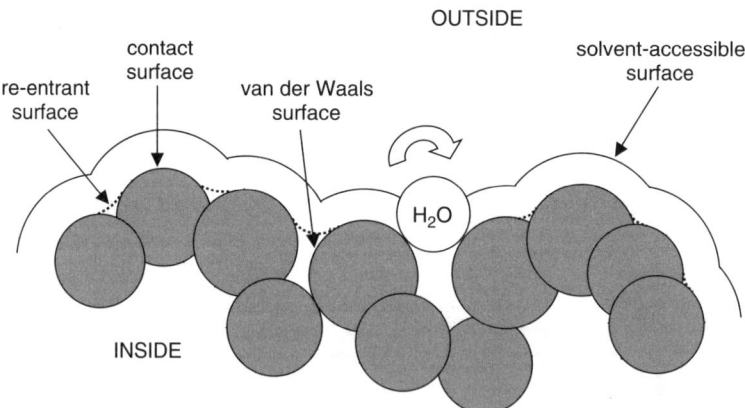

Figure 5 *Schematic representation of basic types of surfaces using the "rolling-ball mechanism" for analysing the surface area of biopolymers, in particular of proteins. A spherical solvent probe (usually water) of radius r_{probe} is allowed to roll on the outside of the molecule, while maintaining contact with the van der Waals surface. The contact and re-entrant surfaces together represent a continuous sheet called the molecular surface. The continuous sheet produced by the locus of the centre of the probe defines the accessible surface*

of the surface that makes contact with the probe is named the 'contact surface', comprising a series of disconnected patches. If the probe sphere is simultaneously in contact with more than one atom (as in some crevices), the resulting interior surface is designated 're-entrant surface'. The continuous sheet formed by contact surface and re-entrant surface together is called 'molecular surface'. For obvious reasons, the majority of calculations strive to calculate the accessible area. When comparing the data obtained or cited by different authors, one has to bear in mind that sometimes, slightly different surface and volume definitions are in use.

The total accessible surface area of proteins is approximately proportional to the two-thirds power of M, while the volume of proteins is directly proportional to M. The comparison of these parameter combinations and various other relationships, obtained by solution techniques (SAXS or hydrodynamics) on the one hand, and crystallography, on the other, may be used for tracking surface roughness (rugosity) effects.[67–69]

3.1 Calculation of Accessible Surface Areas

The accurate calculation of even the static surface areas is a mathematically awkward problem. Many approximate solutions can be obtained by using different algorithms for calculating solvent-exposed areas (*cf.* ref. 64). However, several trials, achieved success in deriving analytical expressions for describing mathematically correct accessible areas.[70]

The program MSROLL by Connolly[71–73] computes a piecewise quartic molecular-surface envelope surrounding the molecule. The program yields different types of areas (contact, re-entrant, molecular and accessible) and solvent-excluded volumes of the entire molecule and of each cavity of the interior, in addition to a 'dot surface' and a polyhedral surface. In a similar way, the program SIMS by Vorobjev and Hermans[74] is able to create a smooth molecular dot surface by rolling two probe spheres. The generated dot surface is invariant with molecular rotation and stable under changes in the molecular conformation. The major difference between the results of the two programs is the different homogeneity of the dot distributions obtained.

3.2 Examples: Protein Surfaces

In the past, we have tested several surface calculation programs,[6–8,75–79] including the approximate program ACCESS by Lee and Richards[62] as well as the analytically accurate programs MSROLL[71–73] and SIMS,[74] with respect to their ability to calculate accessible areas. Beyond doubt, the analytical programs, which directly yield a dot surface, are superior to the simpler programs, particularly in the context of modelling solvation effects.[6–8,75–79]

In the following, we will focus on two representative examples, namely a small, monomeric protein, ribonuclease A (RNase A), and a larger, tetrameric protein, lactate dehydrogenase (LDH). Starting points for calculations are the atomic coordinates of these enzymes as stored in the protein data bank (PDB).[80] Current information on AA sequences and M may be obtained from the SWISS-PROT data bank.[81] For the calculation of dot surface points, the program SIMS was applied. The results reveal that a huge number of dot surface points were generated on the protein

Table 6   *Comparison of calculated parameters for anhydrous RNase as obtained by the surface calculation program SIMS and different input parameters*

Input		Output		
$r_{probe}(nm)$	$d_{dot}\ (nm^{-2})$	N_{dot}	$S\ (nm^2)$	$V\ (nm^3)$
0.150	100	6989	55.44	15.93
0.150	300	18158	55.40	15.60
0.150	500	28370	55.47	15.59
0.1525	100	6946	55.44	15.93
0.1525	300	17975	55.29	15.61
0.1525	500	28161	55.36	15.60

Table 7   *Comparison of calculated parameters for anhydrous LDH as obtained by the surface calculation program SIMS and different input parameters*

Input		Output		
$r_{probe}\ (nm)$	$d_{dot}\ (nm^{-2})$	N_{dot}	$S\ (nm^2)$	$V\ (nm^3)$
0.140	100	68585	495.80	179.72
0.140	200	116728	495.72	177.72
0.140	300	166007	495.47	176.04
0.140	400	215110	495.17	175.40
0.140	500	260584	495.61	175.76
0.145	100	65939	479.08	181.70
0.145	200	111599	479.77	179.39
0.145	300	157681	480.02	177.50
0.145	400	215678	480.01	177.16
0.145	500	259724	480.35	177.17

envelope. The number of points generated can be governed chiefly by the dot density, d_{dot}, used as input parameter (Tables 6 and 7).

4. Biopolymer Hydration

Water binding is an important aspect of biopolymer structure and function, but extremely difficult to determine. In principle, a variety of physicochemical techniques, including hydrodynamic and scattering techniques, as well as modelling approaches, can be applied.[6,82] Since hydration is a dynamic process, in general, only time-averaged properties may be obtained. Accordingly, the hydration values found for different molecules by different methods vary over a wide range.

In the case of simple proteins the situation is a bit more promising, since in the last decades many scientists concentrated on investigating various aspects of the hydration problem. A unified picture of protein hydration has begun to emerge. Evidently, preferred hydration sites around protein surfaces seem to exist, at least for a short time. The preferentially bound water molecules are positioned in the first layer of water on the protein surface and have characteristics different from those of the bulk water: they are more ordered, less mobile and exhibit a higher (average) density (detailed citations of articles dealing with hydration problems are given in refs. 6, 8, 78). This hydration layer contains approximately 0.35 g of water per g of protein,[6,7] and corresponds to about two water molecules per AA residue. Because of the particular distribution of AAs on the protein surface, the water layer around the protein is not uniform. Not only can charged and polar AA residues be found on the protein surface, some hydrophobic patches may be found to exist as well. In this respect, the assumption of a uniform water layer for modelling purposes represents only a crude approximation. Biophysically, more realistic models of protein–water complexes require consideration of this fact. Fortunately, hydration values for individual AA residues have been presented by several authors, the values by Kuntz[57] (Table 2) and Hopfinger[83] being the most relevant ones for modelling studies of protein hydration.[7,78] In conjunction with coordinate and sequence information data and modern surface calculation programs, these values may be used for deriving special hydration algorithms.[6–8,78] For rough approximations of the overall hydration of proteins, use of the AA composition and the hydration numbers given for the individual AA residues (Table 2) is sufficient;[6] for these calculations, all AA residues are used, irrespective of their localization in the protein interior or on the protein surface exposed to the solvent.

4.1 Calculation of Hydration Contributions

Atomic data or coordinates of AA residues of proteins were converted to dry (anhydrous) models as described earlier,[84–86] *i.e.* all atoms of the crystal structure or all AA residues were represented by spheres ("beads") of appropriate volume. By means of surface calculation programs such as MSROLL or SIMS a huge number of surface points (N_{dot}) and the corresponding normal vectors were created. For surface calculations by means of the program SIMS we applied a version adapted to handle up to 25000 atoms and used dot densities, d_{dot}, varying between 10 and 500 nm^{-2}, a probe radius, r_{probe}, of 0.14 nm, and a smoothing probe sphere, r_{sm}, of 0.04 nm.

The huge number of dot surface points and normal vectors of the surface topography calculated in this way were utilized as an ample supply of starting points for potential positions of ordered water molecules on the protein surface. Special hydration algorithms,[6–8,78] such as our program HYDCRYST for atomic coordinates, or the program HYDMODEL for AA coordinates, can now be applied for the selection of preferential positions for bound water molecules. The selected water molecules were represented by spheres of average volumes V_w between 0.0220 and 0.0299 nm^3. The number of water molecules, N_w, assigned to each accessible AA residue, can be obtained from the values given by Kuntz[57] or Hopfinger,[83] for neutral or slightly acidic or alkaline pH values.[78] In previous calculations, a radius of water molecules r_w of 0.14 nm and a minimum distance between adjacent water molecules d_w of 0.28 nm were applied. Subsequently, the water molecules selected that way were added to the dry models.

To achieve different levels of hydration (minimum, intermediate and maximum), scaling factors, f_K and f_H, acting on the hydration values given by Kuntz[57] and Hopfinger,[83] respectively, and, for further fine tuning, many other tools (process of the selection of water molecules, sorting mechanisms, *etc.*) can be used.[7,8,77,78]

Finally, for scattering and hydrodynamic modelling, the usual approaches can be applied. For example, scattering curves, $I(h)$, can be calculated by Debye's formula, and can be converted to distance distribution functions, $p(r)$, by Fourier transformation; from both profiles several structural parameters such as radius of gyration, R_G, hydrated volume, V, and maximum particle diameter, d_{max}, can be obtained.[27] As usual, sedimentation coefficients, s, translational diffusion coefficients, D, and further hydrodynamic properties such as intrinsic viscosities, $[\eta]$, can be computed by program HYDRO.[87,88] For further details and references the reader is referred to Chapter 16.

4.2   Examples: Protein Hydration

Previous studies have shown that calculation of overall hydration values is possible by merely exploiting the AA composition and the hydration numbers given for individual AA residues.[6] Application of the surface calculation programs MSROLL or SIMS yielded detailed surface topographies that could be used for the adoption of our special hydration algorithms, HYDMODEL or HYDCRYST, to create realistic protein–water models.

In this chapter, modelling of proteins of different size and subunit composition is demonstrated for the previously mentioned enzymes, RNase and LDH. Initially, the images for both proteins exhibit a large number of dot surface points [Figures 6(a) and 7(a)] engulfing the dry protein model with a cloud of points. These surface points and the normal vectors in these points may be used for the localization of potential water points [Figures 6(b) and 7(b)]. Relating to the size of the proteins, the difference between dot surface points and the potential water points can be recognized more clearly in the case of the small protein RNase.

Selecting water molecules on the protein surfaces by HYDCRYST yields the distribution of water molecules around the proteins under investigation [Figures 6(c) and 7(c)]. Clear differences between the behaviour of the small, monomeric and the large, tetrameric protein can be explored. In the case of RNase, the water molecules are only bound to the envelope of the protein. LDH, which is built up from four subunits, exhibits water molecules bound to each subunit, *i.e.* water molecules are found in the protein interior (in crevices between the subunits) and the exterior of the protein; this can be observed convincingly from the central slabs shown in the figure. In all cases, binding of water molecules to protein surfaces did not lead to uniform water shells, but provided characteristic protein–water patterns.

To enable hydrodynamic calculations by means of the bead-modelling program HYDRO,[87,88] the number of beads N_b must not exceed a practicable size (currently about 10000). In the case of LDH, the sum of beads representing protein atoms and water molecules exceeds such a manageable number. Therefore, the original crystal data had to be converted to reduced structures [Figure 7(d)]. The model of the highest reduction state (#4) represented the basis for the application of HYDRO and led to the protein–water models shown in Figure 7(e).

Quantifying the results for RNase and LDH in terms of structural and hydrodynamic parameters yielded the results listed in Tables 8 and 9.

The experimental value of R_G of RNase is best predicted by a hydrated model (V_w=0.0270 nm^3 and f_K=0.75–0.80 or V_w=0.0284 nm^3 and f_K=1.3–1.5); a total of about 200–300 water molecules is preferentially bound to the protein consisting of 124 AA residues. The comparison of predicted and observed *s*, *D* and [*η*] values, again, proves that the anhydrous model is not suitable for correct parameter predictions, whereas a hydrated model (V_w=0.0270 or 0.0284 nm³ and f_K=1.3–1.5) is capable of an appropriate prediction.

The results obtained for LDH also demonstrate the necessity of applying hydrated models. The anhydrous models were not able to simulate scattering or hydrodynamic

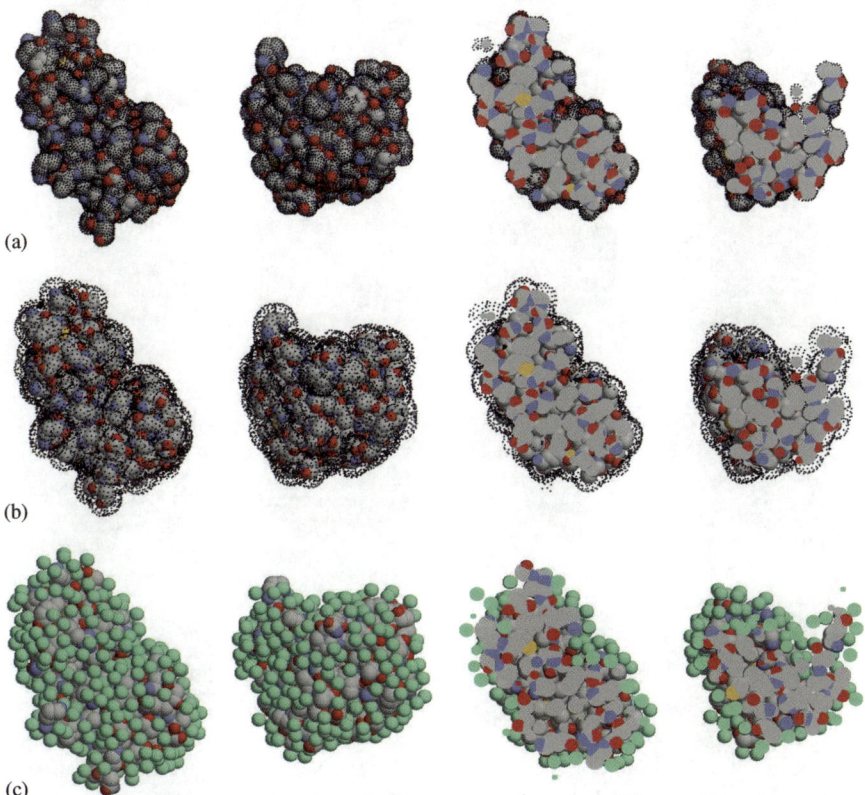

Figure 6 *Space-filling models for anhydrous and hydrated RNase in two different views (90° rotation around the x axis), together with the corresponding central slabs. (a) Molecular dot surface points created by the program SIMS for the anhydrous contour (d_{dot}=300 nm⁻², and applying r_{probe}=0.1525 nm). All dot surface points were calculated from the atomic coordinates of the enzyme. The basic protein atoms derived from the Protein Data Bank (file 1RBX.pdb) are shown in space-filling format and in the usual CPK colours (C in light grey, O in red, N in light blue, and S in yellow) and the dot surface points are coloured black (in a few cases they are located slightly underneath the surface of the spheres). (b) Molecular dot surface points created by HYDCRYST for the contour of potential water points; these water points are located at a distance r_{probe} from the initially obtained surface points. (c) Model for the hydrated enzyme as obtained by HYDCRYST (applying V_w=0.0284 nm³, Kuntz hydration numbers, f_K=1.5); bound water molecules are displayed in green. Graphics were made with the program RASMOL[93]*

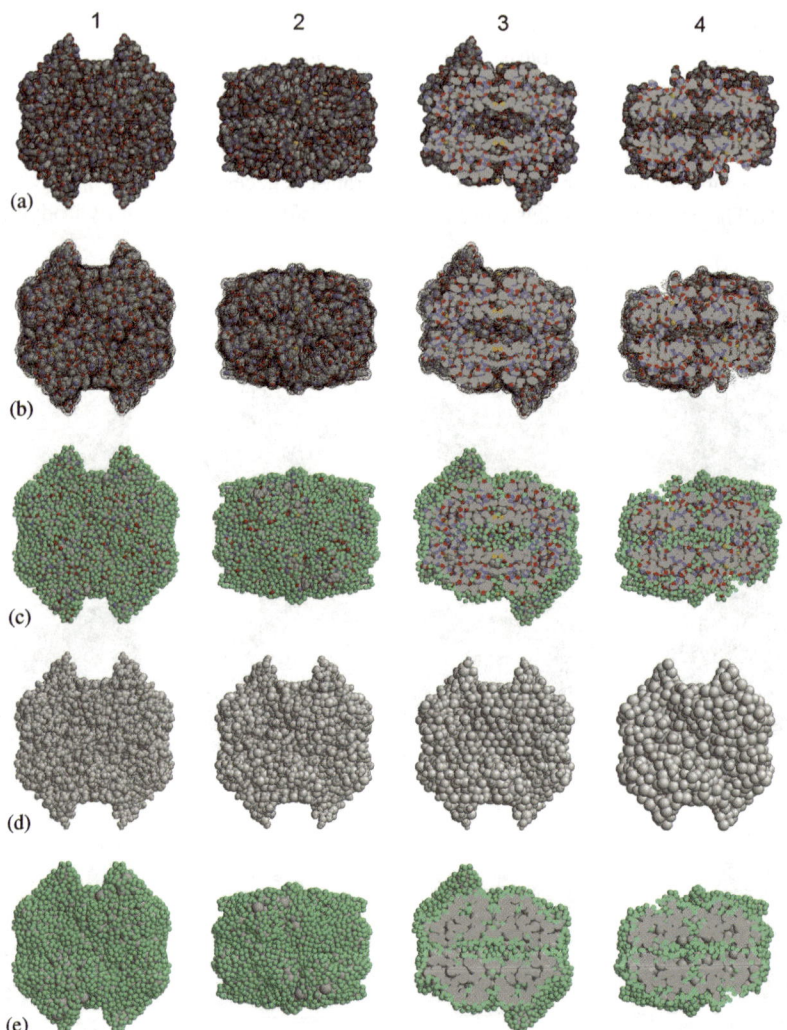

Figure 7 *Space-filling models for anhydrous and hydrated LDH in two different views (90° rotation around the x axis), together with the corresponding central slabs. (a) Molecular dot surface points created by the program SIMS for the anhydrous contour (d_{dot}=300 nm^{-2}, and applying r_{probe}=0.140 nm). All dot surface points were calculated from the atomic coordinates of the enzyme. The basic protein atoms derived from the Protein Data Bank (file 6LDH.pdb1) are shown in space-filling format and in the usual CPK colours and the dot surface points are coloured black. (b) Molecular dot surface points created by HYDCRYST for the contour of potential water points; these water points are located at a distance r_{probe} from the initially obtained surface points. (c) Model for the hydrated enzyme as obtained by HYDCRYST (applying V_w=0.0220 nm^3, Kuntz hydration numbers, f_K=4.0); bound water molecules are displayed in green. (d) Space-filling models for anhydrous LDH at different stages of model reduction (for details see Table 9): N_b=8732 (1), 4537 (2), 2161 (3), and 1316 (4). (e) Model for the hydrated enzyme, based on the reduced structure shown in (d) (4)*

Table 8 Comparison of calculated parameters for hydrated RNase models as obtained by the hydration program HYDCRYST and different input parameters, together with the experimental parameters

Experiment/model	Input		Output							
	V_w (nm³)	f_K	N_w	N_b	V (nm³)	$R_G{}^a$ (nm)	$s \times 10^{13}$ (s)	$D \times 10^7$ (cm² s⁻¹)	$[\eta]^b$ (cm³ g⁻¹)
Experiment[c]					22.0	1.48	1.78	10.68	3.30
Anhydrous model[d]	0.0270	0	0	951	16.11	1.451 / 1.428	2.049	12.23	2.46
Hydrated models[e]									
	0.0270	0.75	206	1157	21.67	1.481 / 1.516	1.850	11.04	3.36
	0.0270	0.80	215	1166	21.92	1.484 / 1.523	1.841	10.99	3.40
	0.0270	0.85	226	1177	22.21	1.487 / 1.529	1.836	10.96	3.44
	0.0270	0.90	231	1182	22.35	1.487 / 1.529	1.835	10.96	3.44
	0.0270	0.95	232	1183	22.37	1.487 / 1.530	1.834	10.95	3.45
	0.0270	1.00	245	1196	22.73	1.490 / 1.535	1.824	10.89	3.51
	0.0270	1.20	257	1208	23.05	1.492 / 1.540	1.819	10.86	3.54
	0.0270	1.30	291	1242	23.97	1.500 / 1.556	1.800	10.74	3.66
	0.0270	1.50	310	1261	24.48	1.504 / 1.565	1.792	10.70	3.71
Mean±S.D. ($V_w = 0.0270$ nm³, $f_K = 1.30$ and 1.50) HYDCRYST			301±13	1252±13	24.2±0.4	1.502±0.003			

(continued)

Table 8 (*continued*)

Experiment/model	Input				Output				
	V_w (nm³)	f_K	N_w	N_b	V (nm³)	$R_G{}^a$ (nm)	$s \times 10^{13}$ (s)	$D \times 10^7$ (cm² s⁻¹)	$[\eta]^b$ (cm³ g⁻¹)
HYDRO	0.0284	0.75	207	1158	21.98	1.561±0.006 / 1.468	1.80±0.01	10.72±0.03	3.69±0.04
	0.0284	0.80	213	1164	22.15	1.520 / 1.469	1.839	10.98	3.42
	0.0284	0.85	223	1174	22.44	1.524 / 1.470	1.834	10.95	3.45
	0.0284	0.90	224	1175	22.46	1.530 / 1.471	1.830	10.93	3.47
	0.0284	0.95	225	1176	22.49	1.530 / 1.472	1.830	10.93	3.47
	0.0284	1.00	238	1189	22.86	1.537 / 1.473	1.830	10.92	3.47
	0.0284	1.20	250	1201	23.20	1.540 / 1.478	1.824	10.89	3.51
	0.0284	1.30	286	1237	24.22	1.560 / 1.481	1.821	10.87	3.53
	0.0284	1.50	305	1256	24.76	1.568	1.796	10.72	3.69
							1.786	10.66	3.75
Mean±S.D. ($V_w = 0.0284$ nm³, $f_K = 1.30$ and 1.50)									
HYDCRYST			296±13	1247±13	24.5±0.4	1.480±0.002	1.79±0.01	10.69±0.04	3.72±0.04
HYDRO						1.564±0.006			

[a] The value in the first line of each model refers to R_G obtained by HYDCRYST, and the value in the second line to R_G obtained by HYDRO. The first value is calculated from bead radii and coordinates and the number of excess electrons, and the second value from bead radii and coordinates only.

[b] No volume correction for $[\eta]$ was applied when using HYDRO.

[c] Experimental values for bovine RNase were taken from ref. 108; for $\bar{v}_2$ a value of 0.702 cm³ g⁻¹ was applied.

[d] The atomic coordinates were obtained from the Protein Data Bank (accession code 1RBX)[80] and the molar mass ($M = 13.690$ kg mol⁻¹) was retrieved from the AA sequence data stored in the SWISS-PROT data bank (entry P61823).[81]

[e] For comparison, the values for maximum hydration based on the values by Kuntz[57] and for the hydration of accessible AA residues are given: $N_w = 301.5$ ($\delta_1 = 0.397$ g g⁻¹) and $N_{w,acc} = 287.5$ ($\delta_1 = 0.378$ g g⁻¹).

Table 9 *Comparison of calculated parameters for hydrated LDH models as obtained by the hydration program HYDCRYST and different input parameters, together with the experimental parameters*

Experiment/model	Input			N_b	V (nm^3)	Output			
	V_w (nm^3)	f_K	N_w			$R_G{}^a$ (nm)	$s \times 10^{13}$ (s)	$D \times 10^7$ $(cm^2\ s^{-1})$	$[\eta]^b$ $(cm^3\ g^{-1})$
Experiment^c					253.3	3.47	7.54	5.05	3.8
Anhydrous model^d			0	10168	178.45	3.150 3.137	8.726	5.65	2.87
Reduced anhydrous models^e			0	8732	178.45	3.137	8.542	5.53	2.88
			0	6884	178.45	3.137	8.428	5.46	2.90
			0	5601	178.45	3.137	8.359	5.41	2.90
			0	4537	178.45	3.137	8.294	5.37	2.90
			0	3613	178.45	3.137	8.248	5.34	2.90
			0	2717	178.45	3.136	8.198	5.31	2.90
			0	2161	178.45	3.135	8.183	5.30	2.90
			0	1316	178.45	3.136	8.211	5.32	2.82
Hydrated models^f,g	0.0220	1.00	2229	12397 3545	227.38 227.38	3.263 3.237	7.892	5.11	3.39
	0.0220	2.00	2976	13144 4292	243.77 243.77	3.325 3.294	7.835	5.08	3.58
	0.0220	3.00	3205	13373 4521	248.80 248.80	3.346 3.314	7.820	5.07	3.63
	0.0220	4.00	3302	13470 4618	250.93 250.93	3.354 3.320	7.827	5.07	3.64

(continued)

Table 9 (*continued*)

Experiment/model	Input			Output					
	V_w (nm^3)	f_K	N_w	N_b	V (nm^3)	R_G^a (nm)	$s \times 10^{13}$ (s)	$D \times 10^7$ (cm^2 s^{-1})	$[\eta]^b$ (cm^3 g^{-1})
	0.0220	5.00	3350	13518	251.98	3.357			
				4666	251.98	3.324	7.829	5.07	3.65
	0.0220	6.00	3366	13534	252.33	3.359			
				4682	252.33	3.325	7.825	5.07	3.66
	0.0220	7.00	3376	13544	252.55	3.360			
				4692	252.55	3.326	7.819	5.07	3.67
	0.0220	8.00	3380	13548	252.64	3.360			
				4696	252.64	3.327	7.819	5.07	3.67
Mean±S.D. (V_w=0.0220 nm^3, f_k=3–8)			3330±67	13498±67	251.54±1.48	3.356±0.005			
HYDCRYST				4646±67	251.54±1.48	3.322±0.005	7.82±0.00	5.07±0.00	3.65±0.01
HYDRO	0.0244	1.00	2163	12331	231.23	3.241			
				3479	231.23	3.253	7.862	5.09	3.43
	0.0244	2.00	2824	12992	247.36	3.288			
				4140	247.36	3.311	7.779	5.04	3.64
	0.0244	3.00	2999	13167	251.63	3.303			
				4315	251.63	3.328	7.771	5.03	3.67
	0.0244	4.00	3076	13244	253.51	3.307			
				4392	253.51	3.333	7.775	5.04	3.68

0.0244	5.00	3120	13288	254.58	3.310	7.775	5.04	3.69
			4436	254.58	3.337			
0.0244	6.00	3134	13302	254.92	3.311	7.776	5.04	3.69
			4450	254.92	3.338			
0.0244	7.00	3146	13314	255.21	3.313	7.768	5.03	3.70
			4462	255.22	3.340			
0.0244	8.00	3146	13314	255.21	3.313	7.768	5.03	3.70
			4462	255.22	3.340			

Mean±S.D. ($V_w = 0.0244$ nm^3, $f_K = 3$–8)

HYDCRYST	3104±57	13272±57	254.18±1.40	3.310±0.004	7.77±0.00	5.04±0.00	3.69±0.01
HYDRO		4420±57	254.18±1.40	3.336±0.005	7.77±0.00	5.04±0.00	3.69±0.01

[a] If two values are given for R_G, the value in the first line of each model refers to R_G obtained by HYDCRYST, and the value in the second line to R_G obtained by HYDRO. The first value is calculated from bead radii and coordinates and the number of excess electrons, and the second value from bead radii and coordinates only. If only one value for R_G is mentioned, this value refers to the calculation from bead radii and coordinates.
[b] No volume correction for $[\eta]$ was applied when using HYDRO.
[c] Experimental values for dogfish LDH were taken from ref. 108; for D only the value for the pig-heart enzyme was available; for $\bar{v}_2$ a value of 0.741 cm^3 g^{-1} was applied.
[d] The atomic coordinates were obtained from the Protein Data Bank (accession code 6LDH)[80] and the molar mass ($M = 145.17$ kg mol^{-1}) was retrieved from the AA sequence data stored in the SWISS-PROT data bank (entry P00341).[81]
[e] The reduced models mirror different extents of reduction; the highest reduction state ($N_b = 1316$) corresponds to AA residues.
[f] The hydrated models were constructed by applying HYDCRYST to the unreduced crystal structure. For the purpose of hydrodynamic predictions, however, AA coordinates had to be used.
[g] For comparison, the values for maximum hydration based on the values by Kuntz[57] and for the hydration of accessible AA residues are given: $N_w = 3112$ ($\delta_l = 0.386$ g g^{-1}) and $N_{w,acc} = 2936$ ($\delta_l = 0.364$ g g^{-1}).

behaviour correctly. However, slight underestimations for R_G were still predicted, even if lower values for V_w (0.022 or 0.024 nm^3) and high values for f_K (3–8) are assumed. Best predictions for s, D and [η] are also found for the previous values for V_w and f_K. Again, the total number of water molecules of about 2000–3000 for this 1316 AA-containing protein is in fair agreement with the previously mentioned ratio of two water molecules per AA residue.

5   Conclusions

As follows from a comparison of predicted and observed volume data for quite different compounds, the anticipation of partial specific volumes is possible in good approximation. Calculations generally provide reliable results for all groups of chemical compounds under analysis comprising small molecules, (bio)polymers, and complexes between different types of molecules. Apart from the obvious success of *ab initio* calculations, special calculation approaches are at our disposal these days to estimate the volume behaviour of certain biomacromolecules, even in multi-component solutions. Approaches dealing with simple and conjugated proteins and protein–ligand complexes in two- and multicomponent solutions allow the prediction of both isomolal and isopotential volumes under a variety of native and denaturing conditions, including the presence of high amounts of additives or denaturants. The results of the volume predictions show a high degree of reliability, if compared with experimental data.

Application of hydration contributions allows a target-oriented prediction of hydrodynamic parameters. This requires realistic assumptions concerning the extent of water binding. Modern surface calculation programs allow the determination of the exact surface topography of biomacromolecules, preferably of proteins. In conjunction with advanced hydration algorithms, as implemented in HYDCRYST, this yields the biophysically relevant simulation of protein–water models. Starting from the atomic coordinates, these hydrated models may be used for predicting hydrodynamic parameters more accurately. Owing to the differences in proportion to size, the influence of water molecules is of less importance if large molecules are to be analysed.

Acknowledgements

The authors are much obliged to Y. N. Vorobjev, J. García de la Torre and R. A. Sayle for use of the programs SIMS, HYDRO and RASMOL, respectively.

References

1. H. Eisenberg, *Biological Macromolecules and Polyelectrolytes in Solution*, Clarendon Press, Oxford, 1976.
2. H. Eisenberg, *Q. Rev. Biophys.*, 1981, **14**, 141.
3. H. Durchschlag, in *Thermodynamic Data for Biochemistry and Biotechnology*, H.-J. Hinz (ed.), Springer-Verlag, Berlin, 1986, 45.
4. H. Durchschlag, in *Landolt–Börnstein New Series Biophysics – Proteins*, Vol. VII/2A, H.-J. Hinz (ed.), Springer-Verlag, Berlin, 2003, 4/1.

5. T. M. Laue, B. D. Shah, T. M. Ridgeway and S. L. Pelletier, in *Analytical Ultracentrifugation in Biochemistry and Polymer Science*, S. E. Harding, A. J. Rowe and J. C. Horton (eds), Royal Society of Chemistry, Cambridge, UK, 1992, 90.

6. H. Durchschlag and P. Zipper, *Biophys. Chem.*, 2001, **93**, 141.

7. H. Durchschlag and P. Zipper, *J. Phys.: Condens. Matter*, 2002, **14**, 2439.

8. H. Durchschlag and P. Zipper, *Eur. Biophys. J.*, 2003, **32**, 487.

9. E. F. Casassa and H. Eisenberg, *Adv. Protein Chem.*, 1964, **19**, 287.

10. H. Eisenberg, in *Landolt–Börnstein New Series Biophysics – Nucleic Acids*, Vol. VII/1C, W. Saenger (ed.), Springer-Verlag, Berlin, 1990, 257.

11. J. C. Lee, K. Gekko and S. N. Timasheff, *Method. Enzymol.*, 1979, **61**, 26.

12. S. N. Timasheff, in *Landolt-Börnstein New Series Biophysics – Proteins*, Vol. VII/2A, H.-J. Hinz (ed.), Springer-Verlag, Berlin, 2003, 5/1.

13. H. Durchschlag, *Colloid Polym. Sci.*, 1989, **267**, 1139.

14. G. Zaccai, E. Wachtel and H. Eisenberg, *J. Mol. Biol.*, 1986, **190**, 97.

15. H. Eisenberg and E. J. Wachtel, *Annu. Rev. Biophys. Biophys. Chem.*, 1987, **16**, 69.

16. H. Eisenberg, *Biophys. Chem.*, 1994, **53**, 57.

17. C. Ebel, *Prog. Colloid Polym. Sci.*, 1995, **99**, 17.

18. C. H. Chervenka, *A Manual of Methods for the Analytical Ultracentrifuge*, Spinco Division of Beckman Instruments, Palo Alto, 1972.

19. S. E. Harding, A. J. Rowe and J. C. Horton (eds), *Analytical Ultracentrifugation in Biochemistry and Polymer Science*, Royal Society of Chemistry, Cambridge, UK, 1992.

20. T. M. Schuster and T. M. Laue (eds), *Modern Analytical Ultracentrifugation*, Birkhäuser, Boston, 1994.

21. C. R. Cantor and P. R. Schimmel, *Biophysical Chemistry, Part II: Techniques for the Study of Biological Structure and Function*, W. H. Freeman and Co., San Francisco, 1980.

22. K. E. van Holde, W. C. Johnson and P. S. Ho, *Principles of Physical Biochemistry*, Prentice Hall, Upper Saddle River, 1998.

23. B. Jacrot and G. Zaccai, *Biopolymers*, 1981, **20**, 2413.

24. G. Zaccai and B. Jacrot, *Annu. Rev. Biophys. Bioeng.*, 1983, **12**, 139.

25. S. J. Perkins, *Eur. J. Biochem.*, 1986, **157**, 169.

26. S. J. Perkins, in *Modern Physical Methods in Biochemistry, Part B*, A. Neuberger and L. L. M. van Deenen (eds), Elsevier, Amsterdam, 1988.

27. O. Glatter and O. Kratky (eds), *Small Angle X-ray Scattering*, Academic Press, London, 1982.

28. O. Kratky, H. Leopold and H. Stabinger, *Method. Enzymol.*, 1973, **27**, 98.

29. C. R. Cantor and P. R. Schimmel, *Biophysical Chemistry, Part III: The Behavior of Biological Macromolecules*, W. H. Freeman and Co., San Francisco, 1980.

30. J. Traube, *Samml. Chem. Chem.-Tech. Vortr.*, 1899, **4**, 255.

31. E. J. Cohn and J. T. Edsall (eds), *Proteins, Amino Acids and Peptides as Ions and Dipolar Ions*, Reinhold, New York, 1943, and reprint by Hafner, New York, 1965.

32. J. T. Edwards, *J. Chem. Educ.*, 1970, **47**, 261.

33. S. Cabani, P. Gianni, V. Mollica and L. Lepori, *J. Solution Chem.*, 1981, **10**, 563.

34. P. Gianni, V. Mollica and L. Lepori, *Z. Phys. Chem. N. F.*, 1982, **131**, 1.

35. P. Gianni and L. Lepori, *J. Solution Chem.*, 1996, **25**, 1.

36. D. P. Kharakoz, *J. Solution Chem.*, 1992, **21**, 569.

37. H. Durchschlag and P. Zipper, *Prog. Colloid Polym. Sci.*, 1994, **94**, 20.

38. H. Durchschlag and P. Zipper, *Jorn. Com. Esp. Deterg.*, 1995, **26**, 275.

39. H. Durchschlag and P. Zipper, *J. Appl. Crystallogr.*, 1997, **30**, 803.

40. A. A. Zamyatnin, *Prog. Biophys. Mol. Biol.*, 1972, **24**, 107.

41. A. A. Zamyatnin, *Annu. Rev. Biophys. Bioeng.*, 1984, **13**, 145.

42. H. Durchschlag, *Biochem. (Life Sci. Adv.)*, 1988, **7**, 181.
43. T. Arakawa and S. N. Timasheff, *Method. Enzymol.*, 1985, **117**, 60.
44. T. Arakawa, *J. Biochem. (Tokyo)*, 1986, **100**, 1471.
45. J. C. Lee and S. N. Timasheff, *Arch. Biochem. Biophys.*, 1974, **165**, 268.
46. J. C. Lee and S. N. Timasheff, *Method. Enzymol.*, 1979, **61**, 49.
47. V. Prakash, C. Loucheux, S. Scheufele, M. J. Gorbunoff and S. N. Timasheff, *Arch. Biochem. Biophys.*, 1981, **210**, 455.
48. V. Prakash and S. N. Timasheff, *Anal. Biochem.*, 1981, **117**, 330.
49. V. Prakash and S. N. Timasheff, *Method. Enzymol.*, 1985, **117**, 53.
50. E. Reisler, Y. Haik and H. Eisenberg, *Biochemistry*, 1977, **16**, 197.
51. T. Arakawa and S. N. Timasheff, *Biochemistry*, 1982, **21**, 6536.
52. T. Arakawa and S. N. Timasheff, *Biochemistry*, 1982, **21**, 6545.
53. T. Arakawa and S. N. Timasheff, *Biochemistry*, 1984, **23**, 5912.
54. T. Arakawa and S. N. Timasheff, *Method. Enzymol.*, 1985, **114**, 49.
55. T. Arakawa, R. Bhat and S. N. Timasheff, *Biochemistry*, 1990, **29**, 1924.
56. S. N. Timasheff and T. Arakawa, in *Protein Structure, A Practical Approach*, 2nd edn., T. E. Creighton (ed.), IRL Press at Oxford University Press, Oxford, 1997, 349.
57. I. D. Kuntz, *J. Am. Chem. Soc.*, 1971, **93**, 514.
58. H. Durchschlag and R. Jaenicke, *Biochem. Biophys. Res. Commun.*, 1982, **108**, 1074.
59. H. Durchschlag and R. Jaenicke, *Int. J. Biol. Macromol.*, 1983, **5**, 143.
60. C. Tziatzios, H. Durchschlag, C. H. Weidl, C. Eschbaumer, W. Maechtle, P. Schuck, U. S. Schubert and D. Schubert, in *ACS Symposium Series No. 812: Synthetic Macromolecules with Higher Structural Order*, I. M. Khan (ed.), American Chemical Society, Oxford University Press USA, Washington DC, 2002, 185.
61. C. Tziatzios, A. A. Precup, C. H. Weidl, U. S. Schubert, P. Schuck, H. Durchschlag, W. Mächtle, J. A. van den Broek and D. Schubert, *Prog. Colloid Polym. Sci.*, 2002, **119**, 24.
62. B. Lee and F. M. Richards, *J. Mol. Biol.*, 1971, **55**, 379.
63. F. M. Richards, *Annu. Rev. Biophys. Bioeng.*, 1977, **6**, 151.
64. F. M. Richards, *Method. Enzymol.*, 1985, **115**, 440.
65. http://www.netsci.org/Science/Compchem/feature14.html
66. M. Gerstein and R. M. Lynden-Bell, *J. Mol. Biol.*, 1993, **230**, 641.
67. D. C. Teller, *Nature*, 1976, **260**, 729.
68. C. Chothia, *J. Mol. Biol.*, 1976, **105**, 1.
69. H. Durchschlag and P. Zipper, *Prog. Colloid Polym. Sci.*, 2002, **119**, 121.
70. M. L. Connolly, *J. Appl. Crystallogr.*, 1983, **16**, 548.
71. M. L. Connolly, *Science*, 1983, **221**, 709.
72. M. L. Connolly, *J. Am. Chem. Soc.*, 1985, **107**, 1118.
73. M. L. Connolly, *J. Mol. Graph.*, 1993, **11**, 139.
74. Y. N. Vorobjev and J. Hermans, *Biophys. J.*, 1997, **73**, 722.
75. P. Zipper and H. Durchschlag, *Physica A*, 2002, **304**, 283.
76. P. Zipper and H. Durchschlag, *Physica A*, 2002, **314**, 613.
77. H. Durchschlag and P. Zipper, *Prog. Colloid Polym. Sci.*, 2002, **119**, 131.
78. H. Durchschlag and P. Zipper, *Prog. Colloid Polym. Sci.*, 2004, **127**, 98.
79. H. Durchschlag, T. Hefferle and P. Zipper, *Radiat. Phys. Chem.*, 2003, **67**, 479.
80. H. M. Berman, J. Westbrook, Z. Feng, G. Gilliland, T. N. Bhat, H. Weissig, I. N. Shindyalov and P. E. Bourne, *Nucleic Acids Res.*, 2000, **28**, 235.
81. B. Boeckmann, A. Bairoch, R. Apweiler, M.-C. Blatter, A. Estreicher, E. Gasteiger, M. J. Martin, K. Michoud, C. O'Donovan, I. Phan, S. Pilbout and M. Schneider, *Nucleic Acids Res.*, 2003, **31**, 365.
82. S. E. Harding (ed.), *Biophys. Chem.*, 2001, **93**, 87.

83. A. J. Hopfinger, *Intermolecular Interactions and Biomolecular Organization*, Wiley, New York, 1977.

84. P. Zipper and H. Durchschlag, *Prog. Colloid Polym. Sci.*, 1997, **107**, 58.

85. P. Zipper and H. Durchschlag, *Biochem. Soc. Trans.*, 1998, **26**, 726.

86. P. Zipper and H. Durchschlag, *J. Appl. Crystallogr.*, 2000, **33**, 788.

87. J. García de la Torre, S. Navarro, M. C. López Martínez, F. G. Díaz and J. J. López Cascales, *Biophys. J.*, 1994, **67**, 530.

88. J. García de la Torre, M. L. Huertas and B. Carrasco, *Biophys. J.*, 2000, **78**, 719.

89. P. Zipper and H. Bünemann, *Eur. J. Biochem.*, 1975, **51**, 3.

90. W. Müller and I. Emme, *Z. Naturforsch.*, 1965, **20b**, 835.

91. L. M. Hjelmeland, D. W. Nebert and J. C. Osborne Jr., *Anal. Biochem.*, 1983, **130**, 72.

92. K.-J. Tiefenbach, H. Durchschlag and R. Jaenicke, *Prog. Colloid Polym. Sci.*, 1999, **113**, 135.

93. R. A. Sayle and E. J. Milner-White, *Trends Biochem. Sci.*, 1995, **20**, 374.

94. T. L. McMeekin, M. L. Groves and N. J. Hipp, *J. Am. Chem. Soc.*, 1949, **71**, 3298.

95. T. L. McMeekin and K. Marshall, *Science*, 1952, **116**, 142.

96. J. C. Lee and S. N. Timasheff, *J. Biol. Chem.*, 1981, **256**, 7193.

97. K. Gekko and T. Morikawa, *J. Biochem.*, 1981, **90**, 39.

98. S. N. Timasheff and H. Inoue, *Biochemistry*, 1968, **7**, 2501.

99. R. Bhat and S. N. Timasheff, *Protein Sci.*, 1992, **1**, 1133.

100. T. Arakawa and S. N. Timasheff, *Arch. Biochem. Biophys.*, 1983, **224**, 169.

101. T. Arakawa and S. N. Timasheff, *Biochemistry*, 1984, **23**, 5924.

102. G. Cohen and H. Eisenberg, *Biopolymers*, 1968, **6**, 1077.

103. H. Durchschlag, P. Zipper, G. Purr and R. Jaenicke, *Colloid Polym. Sci.*, 1996, **274**, 117.

104. H. Durchschlag and P. Zipper, *J. Appl. Crystallogr.*, 1997, **30**, 1112.

105. W. G. Gordon and W. F. Semmett, *J. Am. Chem. Soc.*, 1953, **75**, 328.

106. T. Peters Jr., *Adv. Protein Chem.*, 1985, **37**, 161.

107. W. G. Gordon, in *Milk Proteins, Chemistry and Molecular Biology*, Vol. II, H. A. McKenzie (ed.), Academic Press, New York 1971, 331.

108. H. Durchschlag and P. Zipper, *Prog. Colloid Polym Sci.*, 1997, **107**, 43.

CHAPTER 19

Analysis of Membrane Proteins Using Analytical Ultracentrifugation

KAREN G. FLEMING

1 Introduction

In recent years, analytical ultracentrifugation (AUC) has played an important role in the analysis of membrane proteins. Sedimentation velocity continues to be an extremely useful tool for assessing the homogeneity of membrane protein preparations in detergent micelle solutions; however, because of the multiple components present in a membrane protein:detergent micelle particle, the hydrodynamic interpretation of the sedimentation coefficient is not as straightforward as it is with soluble proteins. As in the past, sedimentation equilibrium continues to be used to determine the molecular weights of many membrane proteins when dissolved in detergent micelles. More recently, the method has played a major role in studies on the thermodynamics of subunit interactions between membrane proteins. The development of these protocols has spurred theories describing the dependence of a membrane protein oligomeric population on the amount of detergent as well as novel studies aimed at designing membrane proteins with desired oligomerization properties. The bulk of the contents of this chapter will therefore mainly focus on the issues related to measuring and interpreting the buoyant molecular weight of a membrane protein in complex with detergent.

1.1 General Considerations for Sedimentation Equilibrium Experiments to be Carried Out on Membrane Proteins

The general considerations for an experiment to be carried out on a membrane protein are similar to those applicable to a soluble protein. Since sedimentation equilibrium experiments can be very sensitive to small amounts of contaminating material, the best results will be obtained if the membrane protein preparation is >95% pure as judged by polyacrylamide gel electrophoresis. Due to their hydrophobic nature, purified integral membrane proteins in solution must be studied in buffers containing a detergent micelle or lipid vesicle cosolvent, and the vast majority of AUC experiments are executed on membrane proteins dissolved in detergent micelles. Membrane proteins are typically analyzed using the absorbance optics because the interference optics would also contain a signal from the detergent, which would introduce uncertainty into monitoring the sedimentation of the particle of interest. Therefore, care must be taken to avoid buffer components that have significant absorbance at the detection wavelength of interest. Unless the membrane protein binds nucleic acids, this is not typically a concern for detection at 280 nm since most commonly used buffer components are transparent at 280 nm. However, some buffering components and reducing agents do have significant absorbance at 230 nm. Their use should be carefully monitored to avoid saturation of the absorbance optics system by the background buffer. Membrane proteins have been analyzed in the standard 2- and 6-sector cells using either quartz or sapphire windows. Sapphire windows are not always transparent at 230 nm, and the user should empirically determine the transparency for each set of windows.

The absorbance properties of the detergent micelles must also be considered. Most detergents are transparent at 280 nm; however Triton X-100 is a notable exception, and this detergent is not usually the optimal choice for this reason, although there have been studies carried out in low concentrations of Triton X-100.[1] Reduced Triton X-100 has been used more frequently because it has greater transparency at 280 nm. Neither of these detergents are a problem for membrane proteins containing chromophores, like heme, when the absorbance of the sedimenting particle can be monitored at a longer wavelength. Depending on the chemical structure, some detergents also have a modest absorbance at 230 nm. If the detergent concentration can be maintained such that this contribution is small (typically <0.4OD), then their use does not usually pose a significant experimental problem.

If a density-matching strategy (described below) is to be used, the detergent micelle environment should also be as homogeneous as possible. A mixture of detergent micelle types with different partial specific volumes could compromise the integrity of the density matching method, since a membrane protein may preferentially bind one type of detergent over the other. To obtain a homogeneous detergent environment, the membrane protein may be purified using that detergent. Alternatively, ion exchange chromatography can be used to exchange one detergent for another.[2–5] In some cases, if the critical micelle concentration is high enough, the detergent can be brought to equilibrium with the solution by dialysis.

1.2 Choice of Detergent Micelle Environment

The chemical diversity of detergent micelles types is large, and not surprisingly the effects of any particular detergent or lipidic environment upon any particular protein can vary widely.[6] Both the type and concentration can have profound effects on the integrity of a membrane protein and on the stability of any oligomers that the protein may form.[5–7] In addition to considerations of the effects of the detergent on the membrane protein functional properties, there may be experimental constraints on the choice of detergent micelles. These will depend on the type of analysis to be carried out. For a simple molecular weight determination of a membrane protein population that is expected to be homogeneous, the available detergent choices are large since many of the strategies described below can be optimized for this purpose. However, in recent years, much progress has been made using sedimentation equilibrium to determine the thermodynamics of interactions between membrane proteins in detergent micelle environments. For these thermodynamic experiments, the detergent choices are limited to those that can be used with the density-matching approach. This is because the monomeric and oligomeric forms of the membrane protein may bind different amounts of detergent, in which case the density increment would be expected to be concentration-dependent.

2 Expressions for the Buoyant Molecular Weight

2.1 The Buoyant Molecular Weight Expressed in Terms of the Protein Moiety Alone

In the case of a membrane protein, the sedimenting particle will contain contributions from the sedimentation of the protein as well as from any bound detergent or lipid molecules. As is the case for soluble proteins, the sedimenting particle also contains a contribution from bound water, although it is negligible under most experimental conditions. By consideration of the thermodynamics of multicomponent systems it has been shown that the buoyant molecular weight of a membrane protein in complex with its hydrophobic solvent is measured in a sedimentation equilibrium experiment and equals[8]

$$M_P(1 - \phi'\rho) = \frac{2RT}{\omega^2}\left(\frac{d\ln c}{dr^2}\right) \tag{1}$$

where the buoyant molecular weight equals $M_P(1-\phi'\rho)$. In the above equation M_P is the molecular weight of only the protein portion of the sedimenting particle, which excludes the molecular weight of bound detergent, lipid and water, ϕ' the effective partial specific volume of the protein moiety in the sedimenting particle and takes into account the contributions of bound detergent, lipid and water; ρ the solvent density, R the gas constant, T the absolute temperature, ω the angular velocity (rad s^{-1}), c the concentration of the sedimenting particle, and r the radial distance from the center of rotation.

In a thermodynamically ideal solution, it has been shown that the buoyant molecular weight can be expressed in terms of its components[9,10]

$$M_P(1 - \phi'\rho) = M_P(1 - \bar{v}_p\rho) + \Sigma n_i M_i(1 - \bar{v}_i\rho) \tag{2}$$

where Equation (2) is the general equation, and M_i and $\bar{v}_i$ are the molecular weight and partial specific volumes (mL g^{-1}) of the ith component, and n_i is the number of molecules of any ith component bound to the protein. Since most analysis of membrane proteins is carried out on purified proteins dispersed in a detergent micelle solution, we can explicitly write the buoyant molecular weight as

$$M_P(1 - \phi'\rho) = M_P(1 - \bar{v}_p\rho) + n_{Det}M_{Det}(1 - \bar{v}_{Det}\rho) + n_{H_2O}M_{H_2O}(1 - \bar{v}_{H_2O}\rho) \tag{3}$$

where the subscripts Det and H_2O indicate the contributions from bound detergent and water molecules, respectively. This equation is also written in the form

$$M_P(1 - \phi'\rho) = M_P[(1 - \bar{v}_P\rho) + \delta_{Det}(1 - \bar{v}_{Det}\rho) + \delta_{H_2O}(1 - \bar{v}_{H_2O}\rho)] \quad (4)$$

where δ_i represents the amount of the ith component bound in grams per gram of protein. In Equations (3) and (4) it is assumed that the contribution from lipid is negligible, however in some cases, tightly bound lipid molecules will co-purify with a membrane protein. If the number of lipids is large, then an additional term accounting for the bound lipid must also be included in Equations (3) and (4); however small amounts of lipid can usually be ignored because the partial specific volumes for lipids are generally near unity.[10] For clarity in this initial discussion, we will assume that negligible amounts of lipid are present, and examples where this is not the case will be discussed later. In addition, it is important to note that the $\bar{v}_{Det}$ term in Equations (3) and (4) represents the partial specific volume of the detergent when it is bound to the protein, which is one of the fundamental assumptions in this analysis. Tanford and Reynolds[10] has argued that the partial specific volume of detergents above the critical micelle concentration should be used because this represents the $\bar{v}_{Det}$ when detergent is in a self-associated state, as it would be when associated with a membrane protein.

From Equations (3) and (4), it can be seen that the experimentally measured buoyant molecular weight contains information about both the membrane protein molecular weight and the molecular weight of any bound detergent. Depending on the scientific question, the experimental conditions can be optimized for determining the value of one or both of these parameters.

2.2 The Buoyant Molecular Weight Expressed in Terms of the Protein–Detergent Complex

An alternative theoretical expression for the buoyant molecular weight of the sedimenting particle was first used by Hersh and Schachman[9] in 1958 and assumes that the partial specific volume of the complex is approximated by the sum of the volumes of the protein and detergent. This theoretical formalism expresses the quantity measured in a sedimentation experiment more directly in terms of the protein–detergent particle:

$$M_C(1 - \bar{v}_C\rho) = \frac{2RT}{\omega^2}\left(\frac{d\ln c}{dr^2}\right) \quad (5)$$

where here the buoyant molecular weight is $M_C(1 - \bar{v}_C\rho)$, with M_C being the molecular weight of the entire protein detergent particle which equals $M_P(1 + \delta_{Det})$ and $\bar{v}_C$ equals the partial specific volume of the entire complex and can be written as

$$\bar{v}_C = \frac{\bar{v}_P + \delta_{Det}\bar{v}_{Det}}{1 + \delta_{Det}} \quad (6)$$

Additional terms can be added if a contribution from lipid needs to be taken into account. It is important to note that the buoyant molecular weights in Equations (1) and (5) must be equal, which emphasizes the point that $\phi' \neq \bar{v}_C$ because $M_P \neq M_C$. Expression (5) is included here because there is sometimes confusion in the literature about the definition of the buoyant molecular weight for a membrane protein in detergent solutions. In practice, the theoretical formalism in Equation (5) is not used as widely and will not be discussed extensively in this chapter. Tanford and Reynolds[10] has pointed out that, using the expression in Equation (1), the quantity M_P may be determined with greater accuracy than the quantity M_C from Equation (5) because $\bar{v}_C$ is not easily measured. In contrast, as will be discussed below, the quantity $(1 - \phi'\rho)$ can be directly determined by density measurements.

3 Sedimentation Equilibrium Methods to Determine Protein Molecular Weights in Protein–Detergent Complexes

3.1 Determination of Membrane Protein Molecular Weights by Classical Density Matching

The molecular weight of only the protein portion of the buoyant molecular complex can be determined by adjusting the experimental conditions such that the net contribution of the detergent becomes negligible. This method is cartooned in Figure 1 and has been referred to as "density matching" or the condition of rendering the detergent gravitational transparent.[11] Density matching can be accomplished by modifying the solvent density such that it becomes equal to the effective density of the bound detergent, *e.g.* $\rho = 1/\bar{v}_{\text{Det}}$. When this is the case, the term $(1 - \bar{v}_{\text{Det}}\rho)$ approaches zero, and the contribution from the detergent term in Equation (3), $n_{\text{Det}}M_{\text{Det}}(1 - \bar{v}_{\text{Det}}\rho)$ becomes negligible regardless of the number of bound detergent molecules.

Thermodynamically, the most favorable experimental condition can be obtained when the density of the solvent can be modulated using a mixture of light and heavy water. This experimental condition produces the most minimal perturbation to the contribution of water in the buoyant molecular weight of the sedimenting species. When the density matching of detergent is carried out using mixtures of water, the average effective density of any bound water is assumed to be equal to the density of the solvent. In this case, the water contribution term from Equation (3), $\bar{v}_{\text{H}_2\text{O}}M_{\text{H}_2\text{O}}(1 - \bar{v}_{\text{H}_2\text{O}}\rho)$, will also be essentially equal to zero. The result of such a density matching strategy is that Equation (3) can be reduced to

$$M_P(1 - \phi'\rho) = M_P(1 - \bar{v}_P\rho) \tag{7}$$

Assuming that the partial specific volume of the protein can be calculated from the amino acid composition using standard procedures,[12] the buoyant molecular weight of the sedimenting complex can be directly interpreted in terms of the molecular weight of the protein portion alone.

To match the effective density of the detergent micelles, the most commonly used water mixture is $\text{H}_2\text{O}/\text{D}_2\text{O}$. Experiments at higher solvent densities can be carried out using $\text{D}_2{}^{18}\text{O}$, although this has been used less extensively due to the expense of

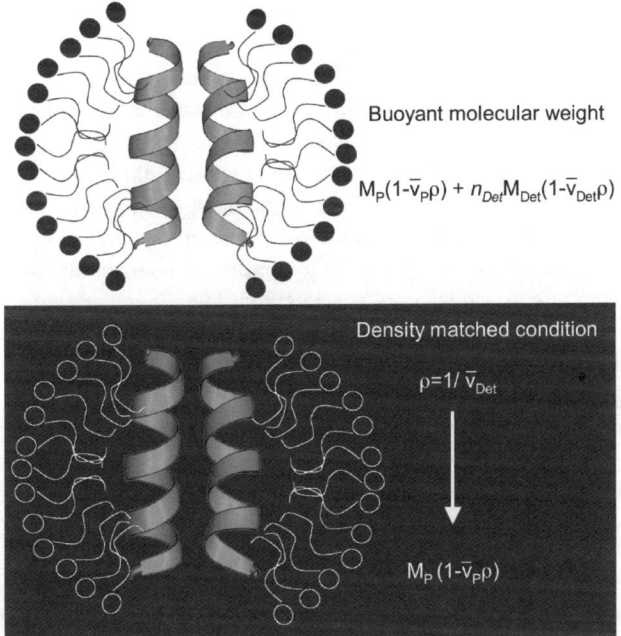

Figure 1 *Cartoon of density matching strategy. Away from the density match point, both the protein and the detergent contribute to the sedimentation of the particle at equilibrium. If the solvent density is adjusted such that it equals the effective density of the bound detergent as in lower panel of the figure, the bound detergent becomes essentially invisible to the gravitational field, and the sedimentation of the particle is determined by the protein moiety alone. This cartoon represents the classic density matching strategy where the density is matched with heavy water, and so no distinction is made for bound water, which is density-matched in both cases in this figure*

the $D_2^{18}O$ compound.[13] Using deuterium water mixtures, there is a small mass correction that must be incorporated into the analysis due to the exchange of deuterium for hydrogen on exchangeable protons.[13,14] The density of heavy water is limited to 1.10392 and 1.21465 g mL^{-1} for 100% D_2O and 100% $D_2^{18}O$, respectively (at 20 °C).[12] For this reason, the use of the density-matching strategy places limitations on the detergents that can be employed in sedimentation equilibrium experiments with this method. To match the effective density, the partial specific volumes of the detergent must be greater than 0.9058 mL g^{-1} to match with D_2O and greater than 0.8232 g mL^{-1} to match with $D_2^{18}O$. It is also important to keep in mind that detergents and lipids with partial specific volumes greater than unity will have effective densities that are less than the density of normal water. This will cause a negative contribution to the buoyant molecular weight and be manifested as a tendency to float in the ultracentrifugation experiment.[15] Due to these density considerations, the detergents most commonly used with the classic density-matching strategy are C_8E_4, C_8E_5, $C_{12}E_8$, C_{14} sulfobetaine, and dodecylphosphocholine (DPC).

The first step in a density-matching experiment is to establish the conditions under which the detergent micelles are gravitationally transparent. A good starting point is the consideration of the partial specific volume for the detergent above the critical micelle concentration, and the partial specific volumes for many detergents are known.[10,16–18] When the partial specific volume is not known, it can be determined by at least two methods. Using a high-precision densimeter, direct density measurements can be used to measure the partial specific volume of a detergent micelle solution in a buffer of interest.[17] Alternatively, the analytical ultracentrifuge can be used to measure the buoyant molecular weight of the detergent micelles alone in a series of solvents whose densities differ. Since many partial specific volumes have been tabulated, this approach is typically used to confirm the percentage of D_2O required for matching the effective density of the detergent micelle in the context of the other buffering components. At the density-match point, the detergent micelles will be uniformly dispersed within the cell. At slightly lower or higher solution densities, the distributions are usually linear with positive and negative slopes, respectively. For a rigorous analysis, the value for the detergent micelle buoyant molecular weight can be determined using Equation (1), and the detergent micelles are density matched when this buoyant molecular weight equals zero. It is optimal to use interference optics to detect the detergent micelle radial distribution when the detergent micelles are transparent at accessible wavelengths. Although the micelles can be visualized using a hydrophobic dye that partitions into them, we have found that even a small amount of dye can alter the micelle density.

3.2   Examples of Density Matching

Several groups have used the density-matching strategy to determine molecular weights of membrane proteins. Classic work by Reynolds and Tanford first demonstrated the usefulness of the density-matching strategy.[10,18,19] Since then, many studies have been carried out with increasing frequency in the past few years. Several groups have taken advantage of the properties of the C_8E_4 or C_8E_5 detergents, which are neutrally buoyant and require no D_2O for density matching. The molecular weights of membrane proteins with diverse folds including monotopic and polytopic α-helical membrane proteins as well as transmembrane β-barrel proteins have been analyzed using these neutrally buoyant detergent micelles.[20–26] $C_{12}E_8$ micelles have also been used in the analysis of membrane proteins by sedimentation equilibrium. In one study, its effective density was matched with 18% D_2O.[27] In a different study the relative mass contribution due to bound $C_{12}E_8$ was assumed to be small and no D_2O was added.[28] More recently, two groups have used C_{14} sulfobetaine micelles matched with 13% D_2O or DPC micelles matched with 50% D_2O to determine the protein molecular weights of both natural and designed membrane protein sequences.[5,29–36] The density-matching strategy has especially found utility in the thermodynamic analysis of membrane protein interactions.

3.3   When Heavy Water is not Heavy Enough

The effective densities of some of the most widely used types of detergent micelles exceed that of D_2O and therefore cannot be matched using heavy water. This class includes dodecylmaltoside (DDM) and β-D-octylglucoside (βOG), which are both commonly used to dissolve membrane proteins for functional and structural studies.[6] When it is optimal for the membrane protein of interest to be solubilized in these types of detergents or other detergents with partial specific volumes ≤ 0.9 mL g^{-1}, alternative strategies to density matching must be used to disentangle the relative contributions of the protein and detergent contributions to the buoyant molecular weight. Two strategies for the analysis of membrane proteins with density agents will be discussed.

3.3.1   Extrapolation to the Match Point

The first method involves determining the buoyant molecular weight in several different mixtures of heavy water followed by extrapolation to the density at which the detergent micelles are gravitationally transparent.[19] If the protein–detergent complex is not altered by changing the H$_2$O/D$_2$O (or D$_2$^{18}O) ratio, the buoyant molecular weight of the sedimenting complex will be a linear function of solvent density. A line can be fitted and at the match point for the effective density of the detergent, the value of the buoyant molecular weight can be interpreted in terms of the protein portion alone using Equation (3) since the two right terms in Equation (3) are negligible at this match point. A good example of this method is shown in Figure 2. Suarez *et al.*[13] used this extrapolation approach to measure the partial specific volume of DDM micelles as well as the molecular weights of both heart and rat liver cytochrome oxidase dissolved in the same detergent. As can be seen in Figure 2, the extrapolated line for the buoyant molecular weight of DDM micelles crosses zero at a solvent density of 1.22 g mL^{-1}. At this same point, which is indicated by the dotted vertical line, the buoyant molecular weights of the cytochrome oxidase proteins are still positive, reflecting the sedimentation of the protein portion. At this extrapolated match point the molecular weight of the protein portion alone can be calculated using Equation (4) by knowing the protein partial specific volume, which can be calculated from the amino acid composition.[12, 37] This extrapolation method has the disadvantage of requiring an extrapolation to a match point that is not experimentally accessible, however it can be accurate if sufficient data are collected to describe the line and if the extrapolation is not too long. In fact, Reynolds and McCaslin[18] have even used this method to estimate the molecular weight of bovine rhodopsin in sodium cholate micelles, whose partial specific volume is nearly that of proteins.[18]

3.3.2   Determination of Membrane Protein Molecular Weights by Density Matching the Hydrated Detergent Micelle with Density Agents

A second experimental approach recently employed is the use of molecules to increase the solution density, which will be referred to as densifier molecules. This experimental approach is most successfully implemented when moderate amounts of the density agents can successfully match the sedimentation of the micelles. Nycodenz, sucrose, and glycerol have all been employed in several different studies.[11,38,39] Since the addition of these densifier compounds can lead to preferential binding or exclusion of water, the effective density of the detergent usually contains a contribution from bound water. For this reason, this method is sometimes referred to as the density-matching of

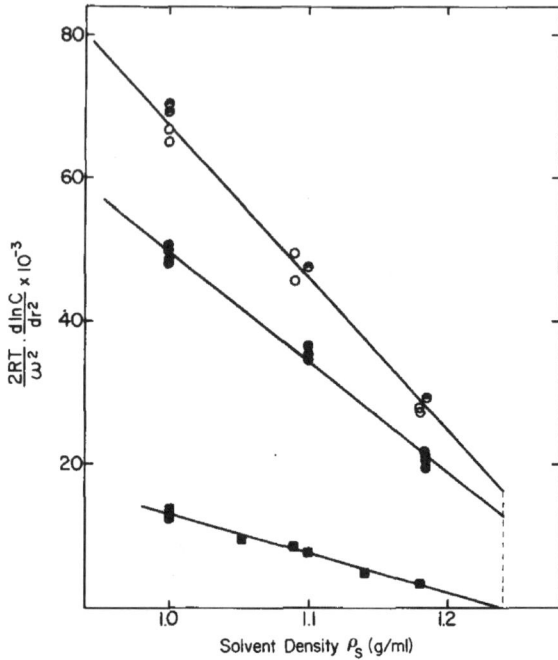

Figure 2   *An example of the method of extrapolation to the density match point using heavy water. The buoyant molecular weights for beef heart cytochrome oxidase (open circles), rat liver cytochrome oxidase (closed circles), and DDM micelles (squares) were determined as a function of solvent density. The dotted vertical line indicates the X-intercept for the linear fit to the sedimentation of DDM micelles; at which point the DDM micelles are density-matched. The increased mass for the two protein samples at this solvent density reflects the sedimentation due to the protein portion of the sedimenting complex for those oxidase proteins. Using the protein molecular weight and partial specific volume for DDM determined at the match point for DDM, the amount of detergent bound to each protein can be calculated from the buoyant molecular weights at the other solvent densities using Equation (3). Reproduced with permission from Suarez et al.[13]*

the hydrated detergent micelles. Like the classic density-matching approach carried out with H_2O/D_2O mixtures, the first step with this experimental approach is to establish the density matching condition. As shown in Figure 3, this is accomplished by carrying out a sedimentation equilibrium experiment on detergent micelles alone as a function of density, where the density is modulated by the concentration of the density agent. Like the original method, the effective density of the hydrated detergent micelle is matched when the buoyant molecular weight equals zero. It has been found that the effective density of the hydrated detergent micelle varies significantly as a function of density agent, and it is generally lower than the effective density in H_2O/D_2O solutions, demonstrating the effect of the bound water.

In principle, the protein moiety may also experience preferential hydration or exclusion of water, and this will be reflected in the effective partial specific volume term for the protein. In practice, this contribution has been observed to be small as long as

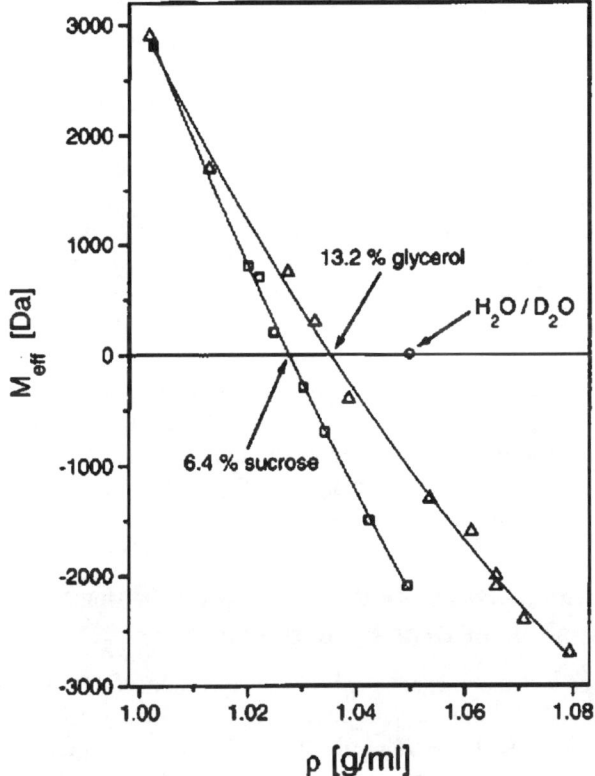

Figure 3 *An example of establishing the density-matching conditions for $C_{12}E_9$ micelles using sucrose or glycerol. The equilibrium sedimentation of 0.3% (w/v) $C_{12}E_9$ micelles was measured using interference optics in the presence of varying amounts of sucrose (open squares), glycerol (open triangles) or DwO (open circle). At equilibrium, the effective mass of the hydrated $C_{12}E_9$ micelles was determined as*

$$M_{\text{eff}} = M_{\text{Det}}(1 - \overline{V}_{\text{Det}}\rho) = \frac{2RT}{\omega^2}\left(\frac{\text{d ln } c}{\text{d}r^2}\right)$$

Reproduced with permission from Mayer et al.[38]

moderate levels of the density agent are used. Lustig *et al.*[11] used an empirical approach to estimate the contribution of water to the partial specific volumes of six different membrane proteins and obtained protein molecular weights that agreed within ±15% of the known values in both DDM and βOG micelles. Lebowitz and co-workers used reduced Triton X-100 micelles density matched with sucrose to demonstrate the oligomeric state for simian immunodeficiency viral envelop protein particles.[39] These authors assumed a negligible contribution of bound water to the protein moiety and still were able to demonstrate a trimeric stoichiometry for the viral envelop protein particles. Mayer *et al.*[38] showed in $C_{12}E_9$ and reduced Triton X-100 that the partial specific volumes for the cytochrome *c* oxidase protein depended only slightly on the nature of the density agent and were similar to the value obtained in H_2O/D_2O mixtures.

However, this same group found that experiments in DDM micelles were difficult to interpret when matched with either sucrose or glycerol alone in water.[38] Even in a subsequent study[40] the combination of sucrose with D_2O, which reduced the sucrose requirement, still required some assumptions about the contribution of hydration to the protein moiety in the presence of DDM micelles since the use of the partial specific volume calculated from the composition using the values of Cohn and Edsall[37] returned a protein molecular weight differing from the known value by a significant amount. In sum, if the hydrated detergent micelles can be matched by moderate amounts of density agent, this density-matching approach appears a promising method for molecular weight determinations in a wide variety of detergent micelle types.

Mayer and co-workers have pointed out that many of these densifier agents are known to stabilize proteins, and this may offer an additional advantage due to the relatively long periods of time required for sedimentation equilibrium experiments.[38] On the other hand, significant amounts of densifier agent can increase the solvent viscosity, which can increase the time to equilibrium. Thus, the advantages of this approach may depend on the molecular properties of the membrane protein in a particular detergent micelle environment.

3.4 Determination of Membrane Protein Molecular Weights by the Analysis of Density Increment

An alternative to density matching is a consideration of the buoyant molecular weight of the sedimenting particle directly in terms of the density increment. As defined by Casassa and Eisenberg,[8] in a thermodynamic consideration of multicomponent systems, the density increment for the macromolecule at constant chemical potential of water and all diffusible solutes (including detergent) equals

$$\left(\frac{\partial \rho}{\partial c_2} \right)_\mu = (1 - \phi' \rho_0) \qquad (8)$$

where c_2 is the weight concentration of the protein alone, and the change in the solution density as a function of changing the protein concentration gives the density increment. This term $(1 - \phi' \rho_0)$ is identical to the buoyancy factor given in Equation (1) where ϕ' is defined as before. When multiplied by the molecular weight of the protein moiety alone as in Equation (1), this quantity returns the buoyant molecular weight of the entire sedimenting complex. The advantage of analysis using the density increment method is that the quantity $(\partial \rho / \partial c_2)_\mu$ can be directly and independently measured on a protein solution at dialysis equilibrium with all other components using a high-precision densimeter. As little as two density measurements on solutions in the presence and absence of protein has been used to calculate the subsequent slope, which is applied to the sedimentation results to calculate the accurate of the molecular weights membrane protein. As long as accurate and precise density measurements can be made, this method greatly expands the repertoire of detergent micelles that can be used in a sedimentation equilibrium experiment since no density-matching considerations are necessary. In addition, this approach works well for membrane proteins dispersed in mixtures of different types of detergent micelles as well as detergent/lipid mixtures, since there is no uncertainty due to preferential binding of one of the hydrophobic components.

3.5   Examples of Use of the Density Increment

Butler and co-workers have used this approach to determine the average molecular weights of membrane proteins dispersed in a wide variety of detergent micelles. The versatility of the method has been demonstrated for membrane proteins samples having effective partial specific volumes ranging from –0.14 to 0.734 mL g^{-1}.[41,42] The density increment method overcomes many of the difficulties described above associated with using the class of micelles containing maltoside head groups, as several membrane proteins have been measured in either DDM or decylmaltoside (DM).[43] The stoichiometry of the bovine major intrinsic protein was shown to be tetrameric in DM, one of the few detergents that stabilized the complex,[44] and the versatility of the density increment approach is demonstrated by recent studies aimed at determining the oligomeric state of the EmrE multidrug transporter solubilized in DDM detergent.[41] Two forms of the protein were prepared. A "high molecular weight" form that was purified using minimal detergent concentrations as well as a "lower molecular weight" form purified using higher concentrations of DDM. Sedimentation equilibrium demonstrated that both forms of the EmrE were in fact dimeric. In complementary work that determined the amount of DDM bound to EmrE, it was found that the "high molecular weight" form of EmrE was not a different oligomeric state, but in fact had ~10-fold more lipid bound. This additional lipid increased the apparent size as measured by gel filtration and sedimentation velocity and obscured true molecular weight analysis using those methods.

3.6   Thermodynamics of Membrane Protein Association

The origins of the specificity and free energy of interaction between membrane proteins are not well understood. Probably the largest impact that sedimentation equilibrium has had on membrane proteins in recent years has been due to its use in experimental measurements in this area of research. While it had been widely appreciated that the properties of micelles influenced the functional and oligomeric states of membrane proteins, this knowledge was derived from empirical observations of the effects of particular detergents on specific proteins.[6] Using sedimentation equilibrium, Fleming and co-workers have shown that the free energy of association between membrane proteins can be measured in detergent micelle solutions as long as the kinetics of rearrangement are fast relative to the time for sedimentation equilibrium.[2,4,27]

The experimental advances lead to several theories describing the concentration dependence of interactions in micelles.[4,5,28,45] Under the assumption that a self-association reaction for an integral membrane protein will take place within the micellar environment, it has been shown in two different detergents that the apparent equilibrium constant for a monomer–dimer reaction varies in a systematic way as a function of detergent concentration.[4] This apparent equilibrium constant is derived from the sedimentation equilibrium data using the protein concentration in bulk molar units. As can be seen in Figure 4, the free energy of association will become less favorable if the

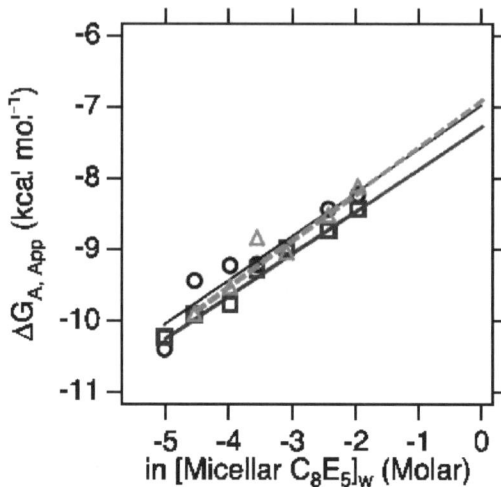

Figure 4 *The apparent free energy of association for a membrane protein monomer–dimer reaction decreases as the size of the micelle phase increases. The apparent free energy of association is plotted as a function of the logarithm of the micellar detergent concentration expressed on the aqueous scale. Under the ideal dilute assumption, the data should vary linearly with a slope of RT. The linear regressions of three independent experiments an average slope of 0.62 (RT = 0.59 at 25 °C) Experiments in C_8E_5 micelles were carried out on the glycophorin A transmembrane helix expressed as a fusion protein with Staphylococcal nuclease. Reproduced with permission from Fleming, 2002[4]*

size of the micellar phase increases because the membrane protein population experiences an increase in volume of the reaction phase. The large number of data points available in a sedimentation equilibrium experiment coupled with modern computational power allowing global fits of different initial starting conditions are two advantages of using the analytical ultracentrifuge for these studies over alternative methods.

The availability of protocols to quantify the free energy of association between integral membrane proteins in detergent solutions has catalyzed a new area of research in the membrane protein field. Several mutagenesis studies on model systems as well as natural membrane proteins have begun to address the physical principles underlying oligomeric stability of membrane proteins.[2,25,26,36] These principles are being tested by membrane protein design efforts whose success are evaluated in large part by sedimentation equilibrium analysis of the designed sequences.[29–33,46]

3.7 Determination of the Amount of Detergent Bound in a Membrane Protein Complex

In the characterization of a membrane protein complex, the amount of detergent bound to a membrane protein may be of interest for a particular scientific question. Implicit in many of the discussions above is the fact that sedimentation equilibrium experiments carried out at densities away from the detergent match point contain a measurable contribution from the bound detergent. Once the molecular weight of the protein moiety has been determined, Equation (3) can be used to calculate the quantity n_{Det} as long as the contribution from hydration is small.[13] More recently, modern computers have been used to globally fit the primary sedimentation equilibrium data collected at different D_2O concentrations directly for this parameter.[47]

4 Sedimentation Velocity Analysis

While most of the recent advances in experimental methods for analysis of membrane proteins have come in the area of sedimentation equilibrium, sedimentation velocity still remains an extremely useful tool. Since finding optimal conditions for solubilization of a membrane protein can be an arduous task, sedimentation velocity continues to be a fairly quick experimental approach for evaluating the homogeneity of a sample.[11,48,49] A good example of this approach is shown in Figure 5, where Musatov *et al.* used the van Holde–Weischet analysis method to evaluate the sedimentation coefficient distribution for cytochrome *bo3* as a function of the concentration of Triton X-100. The sedimentation coefficient distribution showed that the protein aggregated when the detergent concentration was low (open squares). In contrast, a homogeneous sedimentation coefficient distribution was found at detergent/protein ratios >2:1 (w/w). Subsequent sedimentation equilibrium analysis demonstrated that the experimental conditions producing the homogeneous population represented a monomeric form of the cytochrome *bo3* membrane protein.

A similar question of homogeneity was of interest in a study by MacPhee *et al.*, who used sedimentation velocity to characterize the interactions of lipoprotein lipase

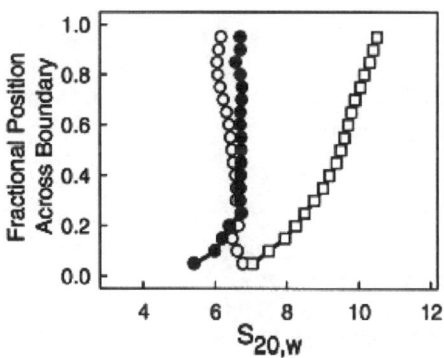

Figure 5 *An example of sedimentation velocity analysis to determine homogeneity in a membrane protein preparation. Sedimentation velocity data collected at 27 000 rpm were analyzed using the van Holde–Weischet method. Each line of symbols represents the distribution of sedimentation coefficients across the boundary. Cytochrome bo3 (0.5 mg protein mL^{-1}) was solubilized in 50 mM potassium phosphate (pH 8.1) containing (a) 1.0 mg mL^{-1} Triton X-100 (open squares), (b) 5 mg mL^{-1} Triton X-100 (filled circles) or (c) 10 mg mL^{-1} Triton X-100 (open circles). The near identity of sedimentation coefficients across the boundary as in samples b and c indicate sample homogeneity. Subsequent sedimentation equilibrium analysis demonstrated a monomeric population for cytochrome bo3 under conditions where the sedimentation coefficient indicated homogeneity. Reproduced with permission from Musatof et al.[49]*

with homogeneous lipid emulsions.[15] An interesting aspect of this work is that the density of the lipid emulsions is less than that of water. Therefore, the sedimentation velocity experiment was used to evaluate the flotation gradients formed upon application of the gravitational potential. This resulted in the quantification of complexes in a specialized version of the time derivative software[50] by defining a flotation co-efficient, which corresponds to the flotation equivalent of a sedimentation coefficient. Flotation equilibrium studies were also carried out in this study.

5 The Future for Membrane Proteins and the Analytical Ultracentrifuge

As the development of expression systems to generate purified membrane proteins continues, an increasing number of membrane protein samples will be available for molecular weight and thermodynamic analysis using sedimentation equilibrium. The increase in availability of modern ultracentrifuges will make the method more accessible, and sedimentation velocity will continue to find utility as a method for screening membrane protein preparations for homogeneity. The exponential increase in high-resolution structural studies on membrane proteins,[51] most of which are carried out in detergent micelles, will stimulate solution studies on this class of proteins in order to test functional questions arising from the structures. The ultracentrifuge will continue to be a biophysical tool of choice for answering many of them.

Acknowledgments

The author gratefully acknowledges the General Medical Sciences Institute of the US National Institutes of Health, the US National Science Foundation, and the US Department of Defense Breast Cancer Research Fund for support of past and current research on membrane proteins.

References

1. H. Hackenberg and M. Klingenberg, *Biochemistry*,1980, **19**, 548–555.
2. K. G. Fleming and D. M. Engelman, *Proc. Natl. Acad. Sci., USA*, 2001, **98**, 14340–14344.
3. K. G. Fleming, A. L. Ackerman and D. M. Engelman, *J. Mol. Biol.*, 1997, **272**, 266–275.
4. K. G. Fleming, *J. Mol. Biol.*, 2002, **323**, 563–571.
5. K. G. Fleming, C. C. Ren, A. K. Doura, F. J. Kobus, M. E. Eisley and A. M. Stanley, *Biophys. Chem.*, 2004, **108**, 43–49.
6. M. le Maire, P. Champeil and J. V. Møller, *Biochim. Biophys. Acta*, 2000, **1508**, 86–111.
7. M. le Maire, K. Kwee, J. P. Anderson and J. V. Møller, *Eur. J. Biochem.*, 1983, **129**, 525–532.
8. E. F. Casassa, and H. Eisenberg, *Adv. Prot. Chem.*, 1964, **19**, 287–395.
9. R. T. Hersh and H. K. Schachman, *Virology*, 1958, **6**, 234–243.
10. C. Tanford and J. A. Reynolds, *Biochim. Biophys. Acta*, 1976, **457**, 133–170.
11. A. Lustig, A. Engel, G. Tsiotis, E. M. Landau and W. Baschong, *Biochim. Biophys. Acta*, 2000, **1464**, 199–206.
12. T. M. Laue, B. Shah, T. M. Ridgeway and S. L. Pelletier, in *Analytical Ultracentrifugation in Biochemistry and Polymer Science*, S. E. Harding, A. J. Rowe and J. C. Horton, (eds), Royal Society of Chemistry, Cambridge, UK, 1992, 90–125.

13. M. D. Suarez, A. Revzin, R. Narlock, E. S. Kempner, D. A. Thompson and S. Ferguson-Miller, *J. Biol., Chem.*, 1984, **259**, 13791–13799.
14. S. J. Edelstein, and H. K. Schachman, *Method. Enzymol.*, 1973, **27**, 82–99.
15. C. E. MacPhee, R. Y. Chan, W. H. Sawyer, W. F. Stafford and G. J. Howlett, *J. Lipid Res.*, 1997, **38**, 1649–1659.
16. H. Durshlag, in *Thermodynamic Data for Biochemistry and Biotechnology*, H.-J. Hinz, (ed), Springer, Berlin, 1986, 45–128.
17. C. Tanford, Y. Nozaki, J. A. Reynolds and S. Makino, *Biochemistry*, 1974, **13**, 2369–2376.
18. J. A. Reynolds and D. R. McCaslin, *Method. Enzymol.*, 1985, **117**, 41–53.
19. J. A. Reynolds and C. Tanford, *Proc. Natl. Acad. Sci., USA*, 1976, **73**, 4467–4470.
20. B. Ludwig, M. Grabo, I. Gregor, A. Lustig, M. Regenass and J. P. Rosenbusch, *J. Biol. Chem.*, 1982, **257**, 5576–5578.
21. K. P. Locher and J. P. Rosenbusch, *Eur. J. Biochem.*, 1997, **247**, 770–775.
22. S. Hellstern, S. Pegoraro, C. B. Karim, A. Lustig, D. D. Thomas, L. Moroder and J. Engel, *J. Biol., Chem.*, 2001, **276**, 30845–30852.
23. A. M. Stanley and K. G. Fleming, *J. Mol. Biol.*, 2005, in press.
24. F. J. Kobus and K. G. Fleming, *Biochemistry*, 2005, in press.
25. A. K., Doura, F. J. Kobus, L. Dubrovsky, E. Hibbard and K.G. Fleming, *J. Mol. Biol.*, 2004, **341**, 991–998.
26. A. K. Doura and K. G. Fleming, *J. Mol. Biol.*, 2004, **343**, 1498–1497.
27. K. G. Fleming, *Method. Enzymol.*, 2000, **323**, 63–77.
28. D. Josse, C. Ebel, D. Stroebel, A. Fontaine, F. Borges, A. Echalier, D. Baud, F. Renault, M. Le Maire, E. Chabrieres and P. Masson, *J. Biol. Chem.*, 2002, **277**, 33386–33397.
29. C. Choma, H. Gratkowski, J. D. Lear and W. F. DeGrado, *Nat. Struct. Biol.* 2000, **7**, 161–166.
30. J. D. Lear, H. Gratkowski, L. Adamian, J. Liang and W. F. DeGrado, *Biochemistry*, 2003, **42**, 6400–6407.
31. J. D. Lear, A. L. Stouffer, H. Gratkowski, V. Nanda and W. F. Degrado, *Biophys. J.*, 2004, **87**, 3421–3429.
32. H. Gratkowski, J. D. Lear, and W. F. DeGrado *Proc. Natl. Acad. Sci., USA*, 2001, **98**, 880–885.
33. H. Gratkowski, Q. H. Dai, A. J. Wand, W. F. DeGrado and J. D. Lear, *Biophys. J.* 2002, **83**, 1613–1619.
34. D. Salom, B. R. Hill, J. D. Lear and W. F. DeGrado, *Biochemistry*, 2000, **39**, 14160–14170.
35. R. Li, C. R. Babu, J. D. Lear, A. J. Wand, J. S. Bennett and W. F. DeGrado, *Proc. Natl. Acad. Sci., USA*, 2001, **98**, 12462–12467.
36. K. P. Howard, J. D. Lear and W. F. DeGrado, *Proc. Natl. Acad. Sci., USA*, 2002, **99**, 8568–8572.
37. E. J. Cohn and J. T. Edsall, in *Proteins, Amino Acids and Peptides*, E. J. Cohn and J. T. Edsall, (eds), Reinhold Publishing Corporation, New York, 1943, 370–381.
38. G. Mayer, B. Ludwig, H.-W. Muller, J. A. van den Broek, R. H. Friesen and D. Schubert, *Prog. Coll. Polym. Sci.*, 1999, **113**, 176–181.
39. R. J. Center, P. Schuck, R. D. Leapman, L. O. Arthur, P. L. Earl, B. Moss and J. Lebowitz, *Proc. Natl. Acad. Sci., USA*, 2001, **98**, 14877–14882.
40. G. Mayer, O. Anderka, B. Ludwig and D. Schubert, *Prog. Coll. Polym. Sci.*, 2002, **119**, 77–83.
41. P. J. Butler, I. Ubarretxena-Belandia, T. Warne and C. G. Tate, *J. Mol. Biol.*, 2004, **340**, 797–808.
42. P. J. Butler and W. Kuhlbrandt, *Proc. Natl. Acad. Sci., USA*, 1988, **85**, 3797–3801.

43. E. H. Heuberger, L. M. Veenhoff, R. H. Duurkens, R. H. Friesen and B. Poolman, *J. Mol. Biol.,* 2002, **317**, 591–600.
44. N. Konig, G. A. Zampighi and P.J. Butler, *J. Mol. Biol.,* 1997, **265**, 590–602.
45. L. E., Fisher, D. M. Engelman and J. N. Sturgis, *Biophys. J.,* 2003, **85**, 3097–3105.
46. J. D., Lear, H. Gratkowski and W. F. DeGrado, *Biochem. Soc. Trans.,* 2001, **29**, 559–564.
47. D. Noy, J. R. Calhoun and J. D. Lear, *Anal. Biochem.,* 2003, **320**, 185–192.
48. A. Musatov and N. C. Robinson, *Biochemistry,* 1994, **33**, 13005–13012.
49. A. Musatov, J. Ortega-Lopez, B. Demeler, J. P. Osborne, R. B. Gennis and N. C. Robinson, *FEBS Lett.,* 1999, **457**, 153–156.
50. W. F. Stafford III, *Anal. Biochem.,* 1992, **203**, 295–391.
51. S. H. White, *Protein Sci.,* 2004, **13**, 1948–1949.

CHAPTER 20

Solution Properties of Flexible Macromolecules: Theoretical and Computational Approaches

J. GARCIA DE LA TORRE, A. ORTEGA AND
H. E. PEREZ SANCHEZ

1 Introduction

1.1 Background

Macromolecules are, in principle, flexible entities. They are essentially composed of linear chains of chemical, covalent bonds joining atoms such as carbon, oxygen, nitrogen, *etc.*, to which side groups are usually anchored. The length of the bonds between successive atoms, b, and the angle between consecutive bonds are more or less fixed. All, or at least some of the bonds in the chain skeleton are single (not double or triple), *i.e.*, they are sigma bonds, in the terminology of physical chemistry. Such bonds (unlike double or triple ones) have the remarkable feature of allowing the internal rotation of the neighboring bonds around them. Assuming four consecutive atoms, i, $i+1$, $i+2$, and $i+3$, if we fix atom i, then atom $i+1$ is fixed by the bond length, and $i+2$ is fixed by the bond angle. Owing to the internal rotation, or torsion, around the bond joining $i+1$ and $i+2$, atom $i+3$ can rotate on a cone, as illustrated in Figure 1. There is a variety of positions, characterized by the rotational angle, ϕ. Then, for a given position of atom $i+3$ there is a variety of positions for atom $i+4$, and so on. As a consequence, such a molecule is continuously changing from one possible conformation to another, and it therefore behaves like a flexible entity. It is usually regarded as a *random coil*. The solution properties of flexible macromolecular chains reflect both their interaction with the surrounding solvent (the frictional drag) and the conformational variability of the solute itself, and thus make both hydrodynamic and statistical contributions.

However, many biological macromolecules are considered as rigid particles; globular proteins are typical examples. The reason is that in addition to the obvious constraints of bond lengths and angles, there are many other specific, strong, and long-ranged interactions that make one special conformation much more stable than

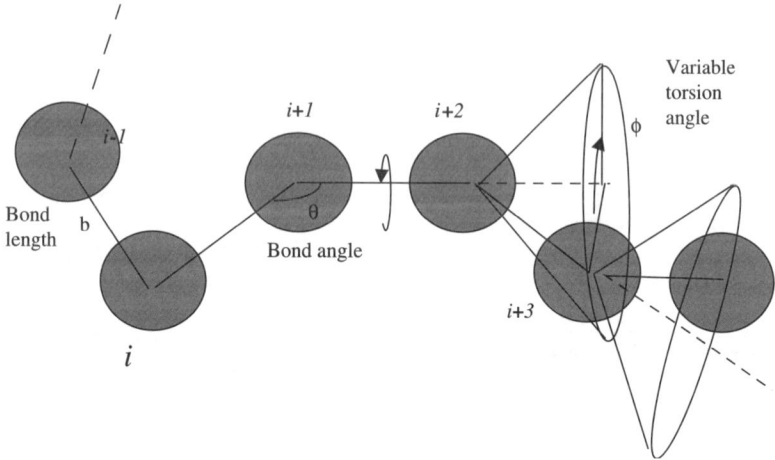

Figure 1 *Scheme of a flexible molecular chain*

any other, because it minimizes the internal energy associated with those interactions. In the absence of denaturing agents, such a conformation is maintained and the macromolecule behaves like a rigid entity.

The rigid-body paradigm is thus deeply implemented in molecular biology: molecules are regarded as particles with a well-defined form, and the intricacies of their shape are considered to determine their association with other macromolecules or ligands. Indeed, there are many instances in which this point of view is useful and correct. But there are other important situations in which it fails, because the macromolecule has some type and degree of flexibility.

1.2 Types of Macromolecular Flexibility

The flexible, random-coil chain represents an extreme case of flexibility. The many units (bonds, monomers, residues…) composing the chain are joined in a fully flexible manner. Models with restricted, partial flexibility can be broadly classified into two categories (see Figure 2):

- Semiflexible molecules with uniform flexibility. This is the case of the worm-like macromolecules, representable by the Kratky–Porod[1] model, which is a chain or filament with some degree of flexibility (manifested by some local curvature) that is the same along the whole contour length of the molecule.
- Segmentally flexible macromolecules, in which a few rigid subunits are joined by small regions that act as partially flexible joints (or even fully flexible swivels). The term "segmentally flexible" was coined by Yguerabide et al.[2] for antibodies that where suspected to have some flexibility in the hinge that joins the two Fab and the Fc subunits. There are cases in which the connection between the rigid parts is made by moderately long and flexible molecular chains; multidomain proteins like calmodulin,[3] with a semiflexible central helix

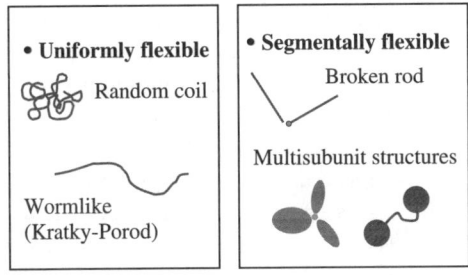

Figure 2 *Types of semiflexible macromolecules*

joining two globules, or rubredoxin,[4] with two domains joined by a short but flexible chain, are good examples.

1.3 Global and Internal Dynamics

The simplest type of macromolecular dynamics in solution is the Brownian displacement of the particle as a whole. The mean square displacement of the center of mass (or, more properly, the hydrodynamic center), according to the famous Einstein law, gives the translational diffusion coefficient, D, which in turn determines the sedimentation coefficient, s, through the Svedberg equation. In the time scale of observation of diffusion or sedimentation, the flexible particle adopts practically every possible conformation, and thus these quantities correspond to the conformational average. Also, the intrinsic viscosity is measured in a time that is molecularly very long, and under moderately low shear intensities, which do not essentially change the conformational statistics, and maintain the Newtonian behavior of the solution viscosity. Thus, $[\eta]$ can also be regarded as a conformational average.

However, there are other dynamic techniques like fluorescence anisotropy decay or NMR relaxation that monitor the reorientational dynamics of some specific vectors (in NMR relaxation of proteins, for instance, the amide N–H bonds). If the macromolecular particle is rigid, then any vector, or in general any part of the particle moves solidarily with it. Obviously, it makes sense to talk about rotation of rigid bodies, and any part, vector, or whatever, within the particle moves solidarily with it. The rotational hydrodynamics[5] and Brownian dynamics (BD)[6] of rigid bodies determine the outcome of such techniques.

Now, consider the dynamics of a fully flexible, random coil polymer. Apart from the translational BD, the polymer chain experiences an incessant conformational change. The chain does not have a fixed shape, and therefore the idea of an overall rotation is incorrect. Still, specific vectors (for instance, the skeletal bonds) have a reorientational dynamics, and the above mentioned techniques can be applied, but we remark that the concept of "rotation" is misleading; it would be more appropriate to talk about "internal dynamics." The situation is similar for segmentally flexible macromolecules. Consider, for instance, a broken rod formed by two rigid, straight arms connected by a fully flexible joint. Rotation of the two subunits is well defined, but there is no such thing for the whole molecule.

The internal BD, or the reorientational dynamics of its characteristic vectors, can be characterized by a set of relaxation times. The pioneering theories of Rouse[7] and Zimm,[8] predict the existence of a series of relaxation times for fully flexible macromolecules. Later Hagerman and Zimm[9] showed that this is also the case for simple segmentally flexible molecules, like the semiflexible broken rod. Many experimental observations are mainly related to the first, longest relaxation time, τ_1. These theories made some simplifying assumptions about hydrodynamic interactions, but it seems that a set of characteristic relaxation times really exists for every flexible molecule; indeed, it has been shown that τ_1 is the same regardless of the property or technique used in its measurement.[10,11]

All these considerations illustrate the complexity of describing the full dynamics (simultaneously translational, internal, *etc.*) of flexible macromolecules. Since the pioneering works of Rouse,[7] Kirkwood and Riseman,[12] and Zimm,[8] approximate theories were proposed for the most simple flexible model, the randomly coiled chain. Such theoretical results are also available for semiflexible wormlike chains. Other, more particular cases (or any case, when hydrodynamic interaction is rigorously considered) require some kind of computer simulation. The most powerful, general and complicated – approach is BD simulation. Another useful approach is the Monte Carlo rigid-body treatment (MCRB), which is simpler to implement because it is partially based on rigid-body hydrodynamics. The latter procedure is rather adequate for translational properties, and therefore is applicable to predict the diffusion and sedimentation coefficients determined by analytical ultracentrifugation. Therefore, this will be the procedure on which we shall concentrate in this chapter.

1.4   Models for Flexible Macromolecules

For rigid macromolecules of arbitrary shape we usually employ bead models.[13,14] The spherical beads represent the size and shape of the macromolecule, and act as friction elements in the hydrodynamic description. Our flexible models will also be composed of beads, but we must include other features to specify the connections, interactions, and freedoms within the structure. These features can be of the following types:

(a) Bonds which are links between neighboring beads. Although the usual concept considers a fixed bond length, in practical instances a stiff "spring" whose length fluctuates slightly about the equilibrium length may be used.

(b) Short-ranged interactions, of which the simplest and most important ones are bond angles. Other short-ranged interactions may include torsions or internal rotations, that are proper of atomic models of real molecules.

(c) Long-ranged interactions, perhaps between every pair not involved in bond or angular interactions. The most important class is that of excluded volume (EV). Other long-ranged interactions can be included in the model; for instance, Coulombic or Debye–Hückel interactions needed to represent charged macromolecules, polyelectrolytes, *etc.*

A more detailed description of the physical features of these interactions will be given below, in the section devoted to simulation methodologies.

2 Basic Models

The main aim of this chapter is the description of the computational methodologies available for flexible structures with arbitrary conformations. However, for the sake of completeness, we also include a summary of the most basic models, which are indeed very useful for a variety of macromolecules that are essentially linear chains, fully or partially flexible.

2.1 The Fully Flexible Chain

The basic model for a fully flexible chain molecule is the freely jointed chain, composed of N segments with length b. The classical random-walk treatment predicts that the mean square end-to-end distance is $\langle r^2 \rangle_0 = Nb^2$. For large N, the mean square radius of gyration is $R_g^2 = \langle s^2 \rangle_0 = \langle r^2 \rangle / 6$. If short-ranged interactions between neighboring segments (bond angles, internal torsion, *etc.*) are included, then the mean square dimensions are just increased by a numerical factor:[15] $\langle r^2 \rangle_0 = C_\infty Nb^2$.

Regardless of how the "segment" is defined, it is obvious that for a series of chains of varying molecular weight, M and N are proportional and we can write $M = NM_1$, where M_1 is the molecular weight of the segment. Then, we have

$$R_g = \left(\frac{C_\infty b^2}{6M_1} \right)^{1/2} M^{1/2} \tag{1}$$

Thus, the radius of gyration (root-mean-square radius) is proportional to molecular weight.

This model, with short-ranged interactions only, is not valid in general cases. Owing to the flexibility of the chain, parts of it that are widely separated along the contour may occasionally get close to each other, producing a long-ranged interaction, which is a balance between steric (excluded volume) repulsion and some molecular attraction. The former effect dominates, so these long-ranged interactions are the cause of a chain expansion. The radius of gyration is given by a power law

$$R_g = K_g M^{a_g} \tag{2}$$

where $a_g = (1+v)/2$ and v is a value smaller than but usually close to 0.2, so that $a_g \cong 0.6$. Only at a fixed temperature, called Θ temperature, which is usually low, just above the point of limit of polymer solubility, the two effects cancel each other and we recover the result in Equation (1), with $a_g = 1/2$. The Θ point can be reached for synthetic polymers in certain solvents. However, biological macromolecules are usually near the other, good-solvent limit, with $v \cong 0.2$.

The hydrodynamics of flexible chains has been the subject of extensive study over many years. Theories predict power laws for the dependence of properties on molecular weight:

$$D = K_D M^{a_D}, \quad s = K_s M^{a_s}, \quad [\eta] = K_\eta M^{a_\eta} \tag{3}$$

The power-law exponents are related to each other; thus, $a_D=-a_G$, $a_s=1-a_D=1+a_G$, and $a_\eta=3a_g/2-1$. The corresponding numerical values in the two limits are given in Table 1.

From the relationships among the a's, it turns out that some dimensionless combinations can be formulated, in such a way that they are independent not only of M, but also of the chemical structure, thus being "universal" constants. These are the Flory–Fox,[16] Φ, constant, which relates dimensions and viscosity:

$$\Phi = \frac{[\eta]M}{6^{3/2}R_g^3} \tag{4}$$

and the Flory, P, constant,[17] which relates dimensions to translational friction properties:

$$P = \frac{f}{\eta_0\langle r^2\rangle^{1/2}} = \frac{k_B T}{\eta_0 6^{1/2}DR_g} = \frac{k_B T(1-\bar{v}\rho)M}{RT\eta_0 6^{1/2}sR_g} \tag{5}$$

While the "universality" of the P and Φ constants was known long ago, their numerical values had to be determined more recently (precisely, using MonteCarlo simulation methods[18,19] such as those described below). Values for these two constants, in the two limits of the random-coil model for flexible macromolecule chains, are indicated in Table 1.

Another dimensionless combination is the Scheraga–Mandelkern constant

$$\beta = \frac{M^{1/3}[\eta]^{1/3}\eta_0}{100^{1/3}f} = \frac{M^{1/3}[\eta]^{1/3}\eta_0 D}{100^{1/3}k_B T} = \frac{[\eta]^{1/3}\eta_0 sN_A}{100^{1/3}M^{2/3}(1-\bar{v}\rho)} \tag{6}$$

which has the peculiarity of having practically the same value $\beta=2.1$–2.3×10^{23} for flexible and rigid macromolecules, being practically independent of macromolecular conformation.

In Equations (4)–(6) and elsewhere in the chapter, k_B is Boltzmann constant, N_A the Avogadro's constant, T the absolute temperature, $\bar{v}$ the solute specific volume, and ρ the solution density.

Table 1 *Characteristic exponents and universal parameters for random-coil polymers*

	Θ point ($v=0$)	Good solvent ($v=0.2$)
a_g	0.5	0.6
a_D	−0.5	−0.6
a_s	0.5	0.4
a_η	0.5	0.8
P	6.0	5.3
$\Phi\times10^{-6}$	2.5	1.9
$\beta\times10^{-23}$	2.3	2.3

Note that the K constants for the various properties are related to each other by expressions that are the same regardless of solvent (Θ or good) conditions. As examples, we give just two of them:

$$K_D = \frac{k_B T}{P \eta_0 6^{1/2}} \frac{1}{K_g}$$

(7)

$$K_\eta = 6^{3/2} \Phi K_g$$

(8)

2.2 The Wormlike Chain

There are many relevant (mainly biological) macromolecules that can be represented as a chain or filament with contour (fully extended) length L, which is locally rather stiff, but not absolutely rigid. These cases can be described in terms of the famous wormlike chain of Kratky and Porod.[1] The degree of flexibility is determined by the persistence length P, which measures the local stiffness. The overall aspect or conformation is determined by the ratio L/P. For small, short chains, with contour lengths L smaller than the persistence length, the local stiffness makes them appear as nearly straight rods. On the other hand, for very long molecules, when L is much longer than P, the residual flexibility becomes important and, over such a long contour length, the filament looks like a random coil. Thus, the wormlike model has the fully flexible chain as a limiting case when $L/P \to \infty$, and the other limit is the rigid rod, corresponding to $L/P \to 0$. In practical instances, the coil region could be for $L/P > 10$, and the rod-like region for $L/P < 1$, with the intermediate range corresponding to the typical wormlike behavior. In the region of moderate or small L/P, the local molecular structure has some influence on the solution properties. In the wormlike model, this is simply accounted for by a finite diameter of the filament, d, which is usually handled as a d/L ratio.

The conformational statistics of the wormlike chain are well known. Benoît and Doty[20] derived, over 50 years ago, an exact expression for the radius of gyration.

$$\langle s^2 \rangle_0 = R_g^2 = \frac{LP}{3} - P^2 + \frac{2P^3}{L} \left[1 - \frac{P}{L}(1 - e^{-L/P}) \right]$$

(9)

However, the hydrodynamics of wormlike chains presents theoretical difficulties. Classical results for s or D and $[\eta]$ are those of Yamakawa and Fujii,[21] who arrived at explicit expressions for these hydrodynamic properties. Bohdanecky[22] improved the results for $[\eta]$. Some approximations made in the early works were removed in subsequent studies by García Molina *et al.*,[23] based on computer simulation (using the RBMC procedure described below).

For the molecules representable as wormlike chains (nucleic acids, polysaccharides, *etc.*), one usually has samples with varying molecular weight, M. The contour length is related to M as $L = M/M_L$, where M_L is the mass-per-unit length (for instance, about 1950 Da nm^{-1} for B-DNA).[24] Thus, the properties can be regarded as functions of an independent variable, M, and three molecular parameters. For instance, for the sedimentation coefficient,

$$s \equiv s(M; M_L, L, d)$$

(10)

A full description of the various theoretical and computational studies of wormlike chains is obviously outside the scope of this chapter. However, we have employed the results of those studies frequently. Here, we announce for the first time the availability of a new computer program, WORMPROPS, which provides values for the various properties (R_g, D, s, $[\eta]$, and τ_j), for given values of the parameters, and for a range of M.

Tables of properties *vs.* M will not be the ultimate goal; instead, in practical instances, one may have values for various properties of various samples, and the final purpose is the determination of the molecular parameters, M_L, L, and d. Some procedures have been traditionally employed for this purpose, which were perhaps specific for each property or particular to one of the regions (near rod or near coil) of the model. At the time of writing this chapter, we are working on a global analysis procedure, implemented in a user-friendly computer program (preliminary name is HYDFIT) that, for a collection of data for various properties and samples of different M, will be able to find the optimum values of M_L, P, and d.

3   Computer Simulation

In this section, we review briefly the simulation methodology applicable to flexible macromolecules, with an emphasis on the Monte Carlo method. A useful previous reference, which also covers BD, is Garcia de la Torre *et al.*[25]

3.1   Bead-and-Connector Models

As mentioned above, our models for flexible entities are bead models. We consider a flexible array of N beads, with friction coefficients $\zeta_i = 6\pi\eta_0\sigma_i$. Thus, the first part of the model data are the bead hydrodynamic (Stokes) radii, σ_i.

Beads will usually be connected in an arbitrary (not necessarily linear) topology. The connecting bonds may be simply expressed as distance constrains: the distance between the bead centers i and j, r_{ij}, should be equal to some specified value, the bond length, d_{ij}. Alternatively, a stiff "spring" whose length fluctuates slightly about the equilibrium length, with an associated Hookean potential $V_{bond} = (1/2)H_{ij}$ $(r_{ij} - d_{ij})^2$ and a large value for H_{ij} (for instance, $H_{ij} = 100\ k_B T/d_{ij}^2$), can be employed in some cases. Next are short-ranged interactions, like possible bond angles. As in the former case, a bond angle θ can be kept fixed, or allowed to have minor fluctuations about an equilibrium value, θ_0, with a potential such as $V_{ang} = (1/2)Q(\theta - \theta_0)^2$ (this potential depends on three centers, but we omit the triplet of subindices, ijk, for better clarity). A common situation would be that of three consecutive beads representing a nearly rigid part of the molecule; an arbitrarily high value can be used for the constant, for instance, $Q = 100\ k_B T$. If, regardless of the degree of local flexibility, the local conformation is straight, one would set $\theta_0 = 0$.

We may also consider excluded volume (EV) effects acting between every non-bonded pair of beads. Our models with spherical elements allow a simple way of introducing the EV restriction by means of a hard-spheres potential: the distance between any pair of spheres must be larger than the sum of their EV radius: $r_{ij} > a_i + a_j$ (the EV radius may differ from the hydrodynamic radius σ_i). This EV potential is

$V_{EV}(r_{ij})=0$ if that happens, or infinite otherwise. Instead of this hard, discontinuous potential, other potential energy functions can be alternatively employed; the Lennard–Jones potential is a usual choice. As mentioned above, suitable and simple functions are available to describe other long-ranged interactions, such as Coulombic or Debye–Hückel electrostatics. The whole set of bonded, short, and long-ranged interactions constitute what is called a force field. The term comes from the field of molecular (or Brownian) dynamics simulation, where forces are employed, while Monte Carlo simulation is made in terms of potential energy.

3.2 The Monte Carlo Rigid-body (MCRB) Approach

The Monte Carlo procedure is widely used to predict properties of systems with conformational variability. It can be implemented in different ways, of which we will consider two extreme but simple ones. In the so-called "importance sampling" procedure,[26] a series of conformations are sequentially generated in such a way that the Maxwell–Boltzmann statistics of the internal energy is fulfilled. Starting from the previous conformation, the new one is generated by making small, random displacements of the system coordinates. The potential energy of the new conformation, V, is compared with that of the previous one, V_{prev}. The new conformation is accepted if $V < V_{prev}$. Otherwise, the ratio of the Boltzmann exponentials and a random number, u, uniform in $(0,1)$ is generated. If the ratio is smaller than the random number

$$\exp(-V/k_{B}T)/\exp(-V_{prev}/k_{B}T) = \exp[-(V - V_{prev})/k_{B}T] < u \qquad (11)$$

then the new conformation is accepted, and if not, a copy of the previous conformation is taken as the new one. The final values of the properties are the averages over the Monte Carlo conformations (the averages are simple, unweighted means; the importance sampling procedure has already accounted for the statistical weights).

A "brute force" Monte Carlo can be envisioned in which one would generate N_{conf} conformations in an absolutely random manner. Then, the final values of the properties are the weighted averages:

$$p = \Sigma\, p_{k} \exp(-V_{k}/k_{B}T)/\Sigma \exp(-V_{k}/k_{B}T), \qquad k = 1, ..., N_{conf} \qquad (12)$$

The great incovenience of this procedure is that it is much less efficient than importance sampling. However, it is physically correct, and can be adopted for simplicity in some circumstances. For instance, for a simple model including only fixed connectors and hard-spheres interactions, one can easily build a procedure for generating conformations that just satisfy the connectivity. Once the conformation is fully generated, it is checked for EV interactions; if there is any overlap, the conformation is rejected. Conformation is forbidden with $V_{k}=\infty$ and $\exp(-V_{k}/k_{B}T)=0$. Otherwise, conformations are accepted with uniform probability. Then, the final Monte Carlo average is just a simple average over the N_{acc} accepted conformations.

$$p = (1 - N_{acc}) \sum_{k} p_{k}, \qquad k = 1, ..., N_{acc} \qquad (13)$$

We remark that for models with many units this procedure is rather inefficient, but it can be a choice for simple models.

The Monte Carlo method is absolutely accurate for calculating equilibrium properties. For macromolecular solutions, this would include the radius of gyration and other indicators of average size or shape, the scattering intensities, *etc.* Its application to hydrodynamic properties, *i.e.* what we call the MCRB calculation of hydrodynamic coefficients, was suggested by Zimm,[27] and developed by other workers.[18] Fixman[28] has shown that the MCRB approach is not fully exact, but it has been shown that, even in the extreme case of a fully flexible random coil, the effect of the approximation[18] is not greater than typical experimental errors. It can be safely used to calculate translational (frictional, diffusion, and sedimentation) coefficients and intrinsic viscosities.

The main drawback of Monte Carlo methods is that they do not consider the internal dynamics of the continuously changing molecular conformation. Thus, they are not able to predict properties related to reorientational motions and the relaxation times of flexible entities. In some instances, exceptions may be made in the case of scarcely flexible molecules, like weakly bending rods (wormlike chains near the rod limit), and other models with rather limited possibility. In such special cases, the MCRB method can be applied to estimate the longest relaxation time.[9,23]

3.3 MULTIHYDRO: A Computer Tool Interfacing Monte Carlo and HYDRO

It is clear that the RBMC simulation requires:

(a) the generation of a number of possible conformations,
(b) the calculation of hydrodynamic and other solution properties for each conformation, and
(c) the final calculation of averages over the Monte Carlo sample.

HYDRO[29] and other computer programs of the HYDRO suite can be useful for the calculation of properties, *i.e.* step (b). Indeed, the program is prepared to make the calculations for a series of cases in a single run provided that the data on the solution and the structure are given in files with some format. Step (a) includes, apart from the details of the case being considered, some common, repetitive tasks. If it is to be interfaced with HYDRO, one of these tasks would be to build the properly formatted files required by HYDRO. Also, the set of results for each conformation should be organized somehow to prepare step (c), either for MCRB averages or for conformational search. We have written a computer code, named MULTIHYDRO, in which these repetitive tasks are accomplished.

MULTIHYDRO[30] is a Fortran program in which the user has to insert two pieces of code corresponding to the specific case: (a) one with physical data and information on the model; and (b) another in which the coordinates of the beads, for each individual conformation of the model, are generated. Execution of this program produces all the files needed for a subsequent execution of a multi-case HYDRO

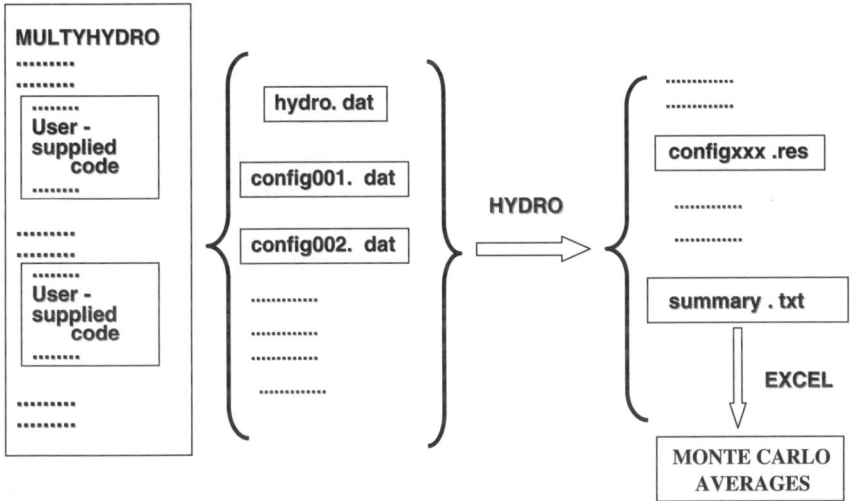

Figure 3 *Scheme of a MULTIHYDRO + HYDRO calculation*

calculation that gives the properties of all the generated (accepted) conformations, compiled in a tabular file that can be exported to Excel or any other spreadsheet for the final calculation of the Monte Carlo averages. A scheme of this process is displayed in Figure 3.

We now give an example of a MULTIHYDRO + HYDRO calculation. Suppose that we have a molecule composed of two, approximately spherical, globular domains with hydrodynamic diameters of 12 and 6 nm, respectively, and a connector with length 16 nm and thickness 2 nm. The connector is represented by beads with a radius of 1 nm. The molecule is supposedly flexible because a portion of the connector is unstructured and acts as a hinge with a complete, swivel-like flexibility. We call this model "the broken dumb-bell".* Suppose that the hinge is not at the middle of the connector, but near the end to which the larger bead is attached. The broken dumbbell model is depicted in Figure 4 (note that the hinge is the third bead in the connector string). The universal swivel permits in principle any angle between the two arms of the particle; however, conformations with bead overlapping are obviously forbidden.

The two pieces of code that the user of MULTIHYDRO has to supply are presented in Figure 5. The first one, with model and simulation data, is trivial. The second one contains the calculation of bead position (Cartesian coordinates) for the 10 bead that compose the model. The shorter straight arm, including the largest sphere, is placed on the negative side of z-axis while the second straight arm, including the second globule, is placed along a direction characterized by polar θ angles and ϕ (Figure 4). The position vector of a bead in the second arm is calculated from the

---

*Dumb-bell: a short bar with a weight at each end used for exercising the muscles, *The Oxford Reference Dictionary*.

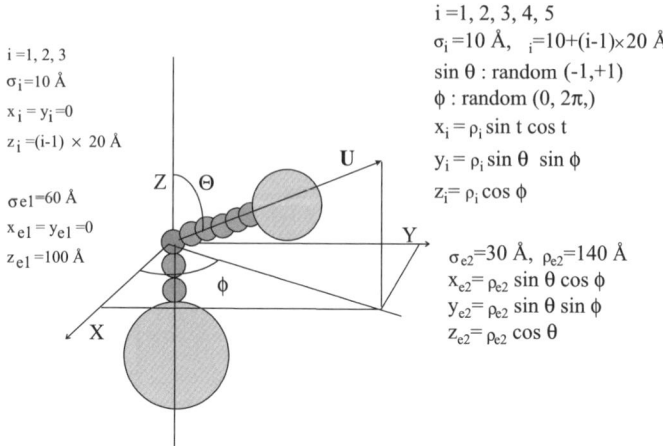

$i = 1, 2, 3$
$\sigma_i = 10$ Å
$x_i = y_i = 0$
$z_i = (i-1) \times 20$ Å

$\sigma_{e1} = 60$ Å
$x_{e1} = y_{e1} = 0$
$z_{e1} = 100$ Å

$i = 1, 2, 3, 4, 5$
$\sigma_i = 10$ Å, $_i = 10 + (i-1) \times 20$ Å
$\sin \theta$: random $(-1, +1)$
ϕ : random $(0, 2\pi,)$
$x_i = \rho_i \sin t \cos t$
$y_i = \rho_i \sin \theta \ \sin \phi$
$z_i = \rho_i \cos \phi$

$\sigma_{e2} = 30$ Å, $\rho_{e2} = 140$ Å
$x_{e2} = \rho_{e2} \sin \theta \cos \phi$
$y_{e2} = \rho_{e2} \sin \theta \sin \phi$
$z_{e2} = \rho_{e2} \cos \theta$

Figure 4   *The broken dumb-bell, with the formulas to calculate the coordinates (dimensions in Å)*

distance *r* to the origin (easily evaluated from the bead sizes), and a unitary vector from the direction of those angles,

$$\mathbf{u} = (\sin\theta \, \cos\phi, \ \sin\theta \, \sin\phi, \ \cos\theta)$$

For a vector with a uniformly distributed random orientation, the two polar angles have different distributions: ϕ is a uniform random number in the interval $(0, 2\pi)$, while $\cos\theta$ (not θ) is obtained as a uniform random number in the interval $(-1, +1)$. Note that our MULTIHYDRO code already contains a subroutine (RANDOM_UNIF_VECTOR) to generate uniformly distributed random vectors. The generated conformations have to be tested for bead overlapping. Our code also includes a subroutine (CHECK_OVERLAP) for this purpose.

Once the whole MULTIHYDRO program is compiled and executed, a certain number (100, in this case) of accepted conformations are generated. Running HYDRO next, the solution properties are evaluated from each of them. From the individual values, the final results are evaluated as simple averages, according to Equation (13). An estimate of the statistical error can be made if the sample is subdivided into a few (say, 5) subsamples: the standard deviation of the averages for the subsamples can be used as an estimate of the statistical uncertainty. The final results are presented in Table 2.

3.4   MONTEHYDRO: Importance-Sampling MCRB Simulation

The procedure outline in the previous section is absolutely general, since situations of arbitrary complexity can be implemented in the user-supplied lines of code in MULTIHYDRO. However, the method has some drawbacks. One obvious inconvenience is the need of writing a Fortran code for the user's model. Another is that the simple "brute force" Monte Carlo may be inefficient in some situations; we have

```
! Physical data
temp=293.      ! Temperature, Kelvin
eta0=0.010     ! Solvent viscosity
rm=830000.     ! Molecular weigth
vbar=0.760     ! Specific volume of macromolecule
solden=1.0     ! Solution density
title='The broken dumb-bell'
filename='config'

! Data for simulation and model (in Angs.)
iseed=654321 !seed for random numbers
nconf=100    !number of accepted conformations
n1=3 ; n2=5 ;
ntotal=n1+n2+2; sigma=10.
sigma1=60. ; sigma2=30.
_____

! Generate conformation
10 CONTINUE

i=0
! First arm, aligned with axis Z
DO i1=1,n1
  i=i+1 ; e(i)=sigma ;
  x(i,3)=-(i1-1)*2.*sigma ; x(i1,1)=0. ; x(i1,2)=0. ;
ENDDO

! One of the big spheres
i=i+1; e(i)=sigma1
x(i,3)=x(i-1,3)-sigma-sigma1 ; x(i,1)=0. ; x(i,2)=0. ;

! Second arm, in random direction
rad=i1*2.*sigma
CALL RANDOM_UNIF_VECTOR(iseed,u)

DO i2=1,n2
  i=i+1 ; e(i)=sigma ; rad=i2*2.*sigma
  x(i,1)=rad*u(1) ;x(i,2)=rad*u(2) ;x(i,3)=rad*u(3) ;
ENDDO
! Second big sphere
i=i+1 ;e(i)=sigma2 ;rad=rad+sigma+sigma2
x(i,1)=rad*u(1) ;x(i,2)=rad*u(2) ;x(i,3)=rad*u(3) ;

!Check overlap of one of the big bead with other beads
i=n1+1 ;    CALL CHECK_OVERLAP(x,e,ntotal,i,iflag)
IF(iflag.EQ.1) GOTO 10

!Check overlap of the other big bead with other beads
i=ntotal ; CALL CHECK_OVERLAP(x,e,ntotal,i,iflag)
IF(iflag.EQ.1) GOTO 10
```

Figure 5 *Two user-supplied pieces of MULTIHYDRO Fortran code needed for the broken dumb-bell*

already commented that more efficient Monte Carlo simulations can be performed using the importance-sampling algorithm. In order to circumvent these difficulties, we have developed an alternative MCRB procedure, implemented in the executable program MONTEHYDRO.[30]

Table 2   *Results for the solution properties of the "broken dumb-bell" molecule*

Property	MULTIHYDRO	MONTEHYDRO
$D_t \times 10^7 (\text{cm}^2\text{s}^{-1})$	2.79 ± 0.03	2.84 ± 0.01
R_g (nm)	7.4 ± 0.2	7.26 ± 0.02
$[\eta]$ (cm^3 g^{-1})	4.5 ± 0.2	4.23 ± 0.02
s, (S)	24.7 ± 0.3	23.2 ± 0.1
Covolume, $u \times 10^{-4}$ (nm^3)	1.58 ± 0.04	—

This program works with a "generalized" bead-and-connector model, which is not absolutely general, but may be applicable to most situations. The model includes:

- Bond, or connector, interactions between some pairs (i,j) of beads, represented as described above by stiff springs with a given value of the equilibrium length, d_{ij}.
- Angular, bending interactions between three beads (i,j,k) such that a bond exists between i and j, and another between j and k (j is the central bead). As described previously, a quadratic potential is associated with the angle θ, and a bending constant Q gauges the intensity of the interaction, determining a more or less wide fluctuation of the angle about its equilibrium value θ_0.
- Non-bonded interactions, needed to represent excluded volume effects, acting on some pairs that are not previously involved in bond or angular interactions. This may be a "hard-spheres" interaction, which forbids bead overlapping, or a smoothly distance-varying potential.

Note that many cases of practical interest are covered by this rather general model. For instance, a discrete representation of a wormlike chain can be made in terms of a string of beads with adequate bending potential.[9,31] In our model, we would have for this case a linear chain of N nearly touching beads with radius σ. Bond interactions are introduced between successive beads, i and $j=i+1$, with equilibrium length $d_{ij}=2\sigma$ and a high value for the spring constant so that, as indicated above, the fluctuation of the bond length is rather small. In this way, one of the two parameters of the wormlike chain, the contour length, is just $L=2N\sigma$. This would adequately represent a freely jointed chain. The wormlike behavior is introduced by means of angular bending interactions between every three consecutive beads, i, $j=i+1$, and $k=i+2$, with equilibrium angle $\theta_0=0$ (the equilibrium conformation is the straight one) and a bending constant, Q, which is related to the other wormlike-chain parameter, the persistence length P, as $Q=k_BTP/2\sigma$.[32] This representation has been indeed employed to simulate properties of the wormlike chain (mainly internal dynamics), which are beyond the reach of the Yamakawa–Fujii and related studies.

Segmentally flexible entities, composed of rigid domains and flexible linkers, with domains of arbitrary shape and linkers of arbitrary flexibility are also covered by our generalized model. A rigid domain can be represented, as for rigid particles, as an array of beads. The rigid shape can be maintained in the simulation by placing hard springs between every pair of beads in the domain. On the other hand, the flexible linker can be modeled as chains of beads with the adequate connectivity and bending potentials.

The MONTEHYDRO executable program includes the three main aspects in the MCRB calculation with importance sampling:

- the generalized model, with the potentials associated to the various kinds of interactions,
- the importance-sampling algorithm, for the generation of the Monte Carlo sample, and
- the calculation of the solution properties, by means of an embedded HYDRO subroutine.

The program is driven simply by the contents of a user data file.

In order to show a simple example, we consider again the broken dumb-bell in Figure 4. A description of the data file for this case, presented in Figure 6, illustrates the general form of this file and also indicates how the program works. After some lines that give the basic physical data of solute and solvent, and other data regarding the simulation procedure, there are four blocks with information about beads (radii and initial position), bonds (involved beads, equilibrium length, and spring constant), angles (involved beads, equilibrium angle, and bending constant), and non-bonded pairs (pair of beads, type, and parameters of the potential). A more detailed description of the program has been presented elsewhere.[30]

Some features of this example are noteworthy. Thus, the equilibrium distance or the bond between successive spheres is equal to the sum of the radii. Also, all the angles are straight ($\theta_0=0$ and a large Q), except for that corresponding to the hinge, between beads 4, 5, and 6, for which $Q=0$, as it corresponds to a freely hinged particle. Note that cases with intermediate flexibility could be studied varying the bending constant Q at the hinge. Then, hard-sphere EV is imposed on every pair of spheres not involved in bonds or angles, with an overlap distance equal, again, to the sum of the radii.

With such a simple data file, MONTEHYDRO performs the RBMC calculation of the solution properties, yielding in this example the values reported in Table 2, where we can observe how the results of this procedure and of the alternative method based on MULTIHYDRO are practically equivalent.

3.5 Brownian Dynamics

The most comprehensive simulation technique for the study of flexibility is BD. As indicated in the Introduction, for the purpose of calculating hydrodynamic coefficients related to the overall or mean size, the simpler Monte Carlo approaches are sufficient. However, BD is the method of choice when one looks at the internal motion and related properties. In this chapter, we will not be covering this topic in detail, but for the sake of completeness, we will describe some of its basic aspects.

The basis of this technique is the Langevin equation for Brownian motion, which is a generalized form of Newton's law of classical mechanics, with additional terms due to friction and Brownian forces. The Langevin equation is a stochastic differential equation that, as Newton's law, has to be integrated to describe the time evolution of the Brownian particle. A simple integration scheme is that of Ermak and McCammon,[33,34] which results in an algorithmic, step-by-step solution that generates

```
293.      !Kelvin temperature
0.01      !Solvent viscosity
830000.   !Molecular weight
0.71      !Specific volume
1000000   !Number of steps
100000    !Number of conformations
0.001     !Delta, step amplitude
6543231   !Seed for random numbers
10    !N, number of beads
2.0   0.0   0.0 -4.5 !radius and coords
0.5   0.0   0.0 -2.0
0.5   0.0   0.0 -1.0
0.5   0.0   0.0  0.0
0.5   0.0   0.0  1.0
0.5   0.0   0.0  2.0
0.5   0.0   0.0  3.0
0.5   0.0   0.0  4.0
0.5   0.0   0.0  5.0
1.5   0.0   0.0  7.0
9    !Nbonds
 1    2   2.5   100. !i,j, d, H
 2    3   1.0   100.
 3    4   1.0   100.
   . . . . . . . . . . . : . . . .
 8    9   1.0   100.
 9   10   2.0   100.
 8   !Nangles
 1 2   3   0. 100.    !i,j,k, alfa0, Q
 2 3   4   0. 100.
 3 4   5   0. 100.
 4 5   6   0.  0.
 5 6   7   0. 100.
   . . . . . . . . . . . . . . . .
 8 9 10   0. 100.
28   !Nnonbon
 1    4   1   2.5   !i,j,itype,dij
 1    5   1   2.5
   . . . . . . . . . . . .
 1 10   1   4.0
 2   5   1   1.0
   . . . . . . . . . . . .
 2 10   1   2.0
   . . . . . . . . . . . .
 6   9   1   1.0
 6 10   1   2.0
 7 10   1   2.0
```

Figure 6 *Data file for the MONTEHYDRO calculation of the properties of the broken dumb-bell*

the trajectory of the particle. At each step, the new coordinates are evaluated from the preceding ones adding two kinds of displacements:

- A deterministic displacement due to forces acting on the model's beads, both the internal forces associated to the bonds, angles, excluded volume, *etc.*, and eventually including external forces. This term is determined by mobility of the beads in the viscous solvent.

- A random displacement caused by the Brownian motion of the beads, determined by their diffusivity.

The difficulties in the implementation of BD simulation procedures arise from the complexity of handling properly the mobility or diffusivity of a collection of beads due to the so-called hydrodynamic interaction effect (HI), which is mandatory for the quantitative prediction of dynamic properties. Also, although the internal potentials may be simple, as described above, the associate force vectors, $\mathbf{F} = -\nabla V$, are given by far more complex equations. Therefore, at the time of writing this chapter, we have not developed as yet a general BD package (although we are working on an in-house prototype, BROWFLEX[25]).

Model-specific BD algorithms can be constructed, and we have done so for the broken dumb-bell illustrated in this chapter in order to show some examples of BD results. From the BD trajectories, time-dependent functions are obtained; for instance:

- Mean square displacement of the particle's center of mass in a time interval t, $<R^2(t)>$, where the brackets mean average over the choices of the initial instant. According to the famous Einstein law, $<R^2(t)>=6D_t$, where D is the translational diffusion coefficient.
- Reorientation of some characteristic vector, v, expressed by the time-dependent function $<P_2(\cos\beta)>=[3<\cos^2\beta>-1]/2$, where b is the angle subtended by two orientations of vector v separated by a time lag t, again averaged over all the initial instants. This is a monotonically decaying function from which the relaxation or reorientation times of the particle (we stress that flexible particles do not rotate as rigid bodies) can be extracted.

In Figure 7, we provide examples of BD results for the hinged dumb-bell. The Einstein law plot gives yet another value (indeed, the most rigorous one) for D that confirms the RBMC results. The conformational variability can be grasped from a trace of the end-to-end distance (between the centers of the two terminal globules) that shows the oscillation between an upper limit (straight conformation, $\theta=0$), and a lower bound (nearly touching spheres, onset of excluded volume interactions). For quantitative purposes, in the interpretation of techniques that monitor the molecular dynamics (dynamic light scattering, NMR relaxation, transient electric birefringence, fluorescence anisotropy decay, *etc.*), $<P_2(t)>$ functions such as those shown in Figure 7 are of primary importance.

Finally, we mention the possibility of using BD without hydrodynamic interaction (BD-noHI) for the simulation of conformations. If HI is switched off, predicted overall, and internal dynamics are wrong, because of the great importance of the HI effect. However, it is known that the sample of conformations generated in the BD-noHI trajectory is in accord with the Maxwell–Boltzman statistics, and therefore can be used for predictions, as in the Monte Carlo method, of equilibrium properties as well as translational diffusion and intrinsic viscosity in the context of the RBMC treatment. Some authors say the BD-noHI is a "smart Monte Carlo" method.[35] However, it has the disadvantage of a slower and more complex computer code, and the pure Monte Carlo procedures described in the previous section can be a better choice.

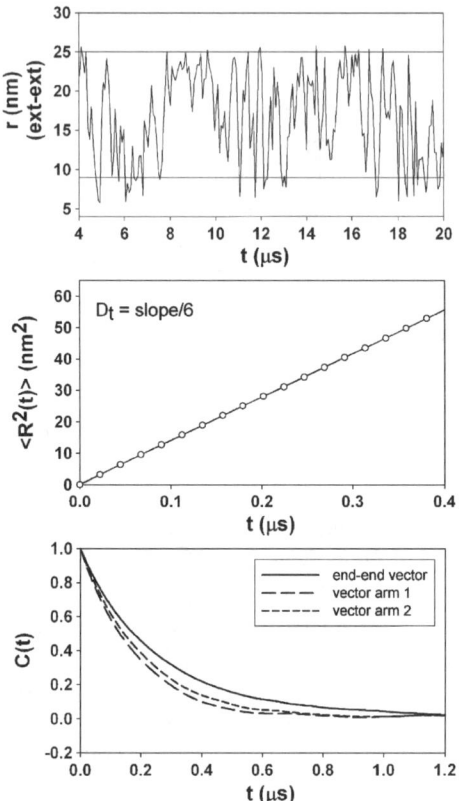

Figure 7 *(a) Einstein law plot yielding the translational diffusion coefficient. (b) Time evolution of the end-to-end distance (c) <$P_2(t)$> function for the reorientation of the end-to-end vector*

4 Concluding Remarks

In this chapter we have described some theoretical aspects of the flexibility of molecular chains represented by simple modes, and computational procedures applicable to bead (or bead-and-connector) models for molecules with arbitrary extent and type of flexibility, based on the Monte Carlo model. We describe computational tools useful for carrying out predictions of solution properties of rigid molecules based on bead models, and provide a simple example of their utilization. We have also included a brief but significant mention of the BD technique.

The computer programs mentioned in this chapter will be available from our web site (present URL is http://leonardo.fcu.um.es/macromol).

Acknowledgements

This work was supported by a grant from Ministerio de Ciencia y Tecnología to J.G.T, and A.O and H.E.P.S are the recipients of predoctoral fellowships from the FPI and FPU programs, respectively, of Ministerio de Educación y Ciencia.

References

1. O. Kratky and G. Porod, *Monatsh. Chem.*, 1949, **68**, 1106.
2. J. Yguerabide, H. F. Epstein and L. Stryer, *J. Mol. Biol.*, 1970, **51**, 573.
3. G. Barbato, M. Ikura, E. L. Kay, R. W. Pastor and A. Bax, *Biochemistry*, 1992, **31**, 759.
4. A. Perry, W. Tambyrajah, J. G. Grossman, L. Y. Lian and N. S. Scrutton, *Biochemistry*, 2004, **43**, 3167.
5. H. Brenner, *Chem. Eng. Sci.*, 1972, **27**, 1069.
6. L. D. Favro, *Phys. Rev.*, 1960, **119**, 53.
7. P. E. Rouse, *J. Chem. Phys.*, 1953, **21**, 1272.
8. B. H. Zimm, *J. Chem. Phys.*, 1956, **24**, 269.
9. P. Hagerman and B. H. Zimm, *Biopolymers*, 1981, **20**, 1481.
10. J. Garcia de la Torre, S. Navarro, M. C. Lopez Martinez, F. G. Diaz and J. J. Lopez Cascales, *Biophys. J.*, 1994, **67**, 530.
11. H. E. Perez Sanchez, J. Garcia de la Torre and F. G. Diaz Baños, *J. Phys. Chem.*, 2003, **107**, 13192.
12. J. Kirkwood and J. Riseman, *J. Chem. Phys.*, 1948, **16**, 565.
13. V. A. Bloomfield, W. O. Dalton and K. E. Van Holde, *Biopolymers*, 1967, **5**, 135.
14. J. Garcia de la Torre and V.A. Bloomfield, *Q. Rev. Biophys.*, 1981, **14**, 81.
15. P. J. Flory, *Statistical Mechanics of Chain Molecules*, Interscience, New York, 1969.
16. P. J. Flory and T. G. Fox, Jr., *J. Am. Chem. Soc.*, 1951, **73**, 1904.
17. P. J. Flory, *Principles of Polymer Chemistry*, Cornell University Press, Ithaca, NY, 1953.
18. J. Garcia de la Torre, A. Jimenez and J. J. Freire, *Macromolecules*, 1982, **15**, 148.
19. J. M. Garcia Bernal, M. M. Tirado and J. Garcia de la Torre, *Macromolecules*, 1991, **24**, 593.
20. H. Benoît and P. M. Doty, *J. Phys. Chem.*, 1953, **57**, 958.
21. H. Yamakawa and M. Fujii, *Macromolecules*, 1974, **7**, 128; *Macromolecules*, 1974, **7**, 649.
22. D. Bohdanecky, *Macromolecules*, 1983, **16**, 1483.
23. J. J. Garcia Molina, M. C. Lopez Martinez and J. Garcia de la Torre, *Biopolymers*, 1990, **29**, 883.
24. V. A. Bloomfield, D. M. Crothers and I. Tinoco, Jr., *Nucleic Acids: Structures, Properties and Functions*, Chap. 9, University Science Books, Sausalito, CA, 2000.
25. J. Garcia de la Torre, H. E. Perez Sanchez, A. Ortega, J. G. Hernandez Cifre, M. X Fernandes, F. G. Diaz Baños and M. C. Lopez Martinez, *Eur. Biophys. J.*, 2003, **32**, 477.
26. N. Metropolis, A. W. Metropolis, M. N. Rosenbluth, A. H. Teller and E. Teller, *J. Chem. Phys.*, 1953, **21**, 1087.
27. B. H. Zimm, *Macromolecules*, 1980, **13**, 592.
28. M. Fixman, *J. Chem. Phys.*, 1983, **78**, 1588.
29. J. Garcia de la Torre, S. Navarro, M. C. Lopez Martinez , F. G. Diaz and J. J. Lopez Cascales, *Biophys. J.*, 1994, **67**, 530.
30. J. García de la Torre, A. Ortega, H. E. Pérez Sánchez and J. G. Hernández Cifre, *Biophys. Chem.*, 2005, **116**, 121.
31. S. A. Allison, *Macromolecules*, 1986, **19**, 118.
32. J. A. Schellman and S. C. Harvey, *Biophys. Chem.*, 1995, **55**, 95.
33. D. L. Ermak and J. A. McCammon, *J. Chem. Phys.*, 1978, **69**, 1352.
34. A. Iniesta and J. Garcia de la Torre, *J. Chem. Phys.*, 1990, **92**, 2015.
35. P. J. Rossky, J. D. Doll and H. L. Friedman, *J. Chem. Phys.*, 1978, **69**, 4628.

CHAPTER 21

The ELLIPS Suite of Whole-Body Protein Conformation Algorithms for Microsoft WINDOWS

STEPHEN E. HARDING, HELMUT CÖLFEN AND ZAHID AZIZ

1 Introduction: ELLIPS1, 2, 3, 4

ELLIPS1, 2, 3 and 4 are simple to use algorithms for the representation of the overall hydrodynamic shape of proteins in solution in terms of tri-axial ellipsoids (three semi-axes $a>b>c$ and shape characterised by two axial ratios a/b, b/c) and bi-axial ellipsoids or "ellipsoids of revolution" where two of the semi-axes are approximated as equal: a prolate ellipsoid has semi-axes a, b, b an oblate ellipsoid has a, a, b, with $a>b$ in both cases and the asymmetry defined by the axial ratio a/b. These algorithms have previously been available only from mainframe or MSDOS platforms.[1] This short chapter briefly reviews the purpose of these FORTRAN/QUICK-BASIC algorithms and describes their fresh implementation onto a Microsoft WINDOWS platform.

Hydrodynamic methods provide a useful approach to the study of macromolecular conformation in solution. In the study of the conformation and flexibility of linear types of macromolecules – for example synthetic polymers and polysaccharides, consideration of how hydrodynamic parameters such as the intrinsic viscosity, the sedimentation coefficient or the radius of gyration vary with molecular weight of a homologous polymer series has provided the means of estimating particle dimensions and flexibility (via the persistence length) in solution.

For the representation of quasi-rigid types of macromolecule – many proteins for example (except at very short timescales) can be thought of as such – hydrodynamic approaches have also proved useful.

There are two approaches to representing the conformation of fairly rigid proteins in solution. The first approach is the bead modelling whereby a macromolecule or macromolecular assembly is approximated as an array of spherical beads. Using computer programs that are currently available (based on how these spheres interact) such as HYDRO, SOLPRO, HYDROPRO, HYDROSUB,[2-4] it is possible for a given Bead model to predict its hydrodynamic properties. One can model quite sophisticated structures by this approach, and recent efforts have focussed on its application to flexible structures.[5] There are uniqueness problems, for example, one can predict the sedimentation coefficient for a particular complicated model, but there will be many other equally complicated models which give the same sedimentation coefficient. This type of modelling is therefore best for choosing between plausible models for a structure, or for refining a close starting estimate for a structure from, say, X-ray crystallography.

A complementary approach to bead modelling is to make no assumptions concerning starting estimates and to calculate the shape directly from hydrodynamic measurements. This is called the "ellipsoid" or "whole body" approach,[6] so called because the investigator instead of approximating the macromolecule as an array of spheres approximates the macromolecule instead as a smooth whole regular structure – an ellipsoid, or "three-dimensional ellipse" characterised by three perpendicular semi-axes $a \geq b \geq c$. Of course, only simple representations are possible but by combining shape parameters together there are no hydration or uniqueness problems. This approach is best for giving a relatively quick idea of the overall dimensions or shape of a macromolecule in solution. There are two types of ellipsoid approach: the *ellipsoid of revolution* and the *general triaxial ellipsoid*. The simplest of these (which has been used in one form or another for over half a century[7]) is the ellipsoid of revolution in which two of the three semi-axes are equal ($c=b$). Ellipsoids of revolution are so-called because they are the shapes formed by rotating an ellipse of semi-axes, a, b either about the major (a) axis to give a prolate ellipsoid (semi-axes a, b, b) or about the minor (b) axis to give an oblate ellipsoid (a, a, b), both defined by the axial ratio (a/b) (where $a \geq b$). One hydrodynamic measurement can uniquely define (a/b), after assuming a value for the (time averaged) hydration of the molecule; two hydrodynamic measurements are normally sufficient to define (a/b) without assumptions concerning hydration; a third is occasionally necessary to distinguish whether an oblate ellipsoid or prolate ellipsoid is the more appropriate (usually the latter for proteins). In the extremes $a \gg b$ the prolate→rod and the oblate→disc and the other extreme of $a=b$ is of course a sphere.

The most sophisticated of the whole body approaches is the general tri-axial ellipsoid where the restriction of two equal axes $b=c$ is removed. This allows a much greater variety of conformations ranging from rods ($a \gg b=c$), discs ($a=b \gg c$) and tapes ($a \gg b \gg c$) as well as the prolate ($a>b=c$), oblate ($a=b>c$) ellipsoids of revolution and the sphere ($a=b=c$). All the necessary theoretical developments for applying either of these "ellipsoid" strategies are in place. What had been lacking is a coherent set of easy-to-use algorithms – available on PC as opposed to computer mainframes – which the general user has access to. The launch of a suite of four ELLIPS algorithms in 1997[1] for an MSDOS platform helped addressing this: Table 1 gives a summary of what these do. ELLIPS1 is written in QUICKBASIC, the others are in FORTRAN. Obviously for some classes of molecule – antibodies are a good example – this type of whole-body modelling is not applicable and bead- approaches need to be employed. Even here, however, ellipsoidal representations of the major domains (Fab, Fc) have helped in the bead modelling of the intact assembly.[8–13]

Table 1   *The ELLIPS routines*

Routine	Language	Model	Purpose
ELLIPS1	QUICKBASIC	Ellipsoid of revolution	Prediction of axial ratio (a/b) (equivalent prolate or oblate ellipsoid of revolution) from user-specified shape function
ELLIPS2	FORTRAN	General triaxial ellipsoid	Evaluates the values of all the hydrodynamic shape functions from user-specified (a, b, c) or (a/b, b/c)[a]
ELLIPS3	FORTRAN	General triaxial ellipsoid	Evaluates (a/b, b/c) from combinations of hydration-independent shape functions
ELLIPS4	FORTRAN	General triaxial ellipsoid	Evaluates (a/b, b/c) from electro-optic decay combined with other hydrodynamic data

[a] Equivalent to SOLPRO[3] for bead models.

2   Universal Shape Functions: Hydration-Dependent and Hydration-Independent

In common with the bead modelling program SOLPRO[3] the ELLIPS algorithms all use *Universal shape functions*. By this we mean each is specifically a function of shape alone (and not volume). It makes no odds what the size is: a Universal shape function will have the same value, and will only depend on the shape. All these universal shape functions have been worked out in terms of the axial ratio (a/b) for ellipsoids of revolution and now the two axial ratios (a/b, b/c) for general ellipsoids. The relations of all these to (a/b) or (a/b, b/c) are given in Harding[14] and will not be repeated here: all of these *exact* formulae are inbuilt into the ELLIPS routines.

To measure these Universal shape functions experimentally, many require knowledge of the hydration δ (mass in g of H_2O bound per g of dry macromolecule) or hydrated volume V (mL) of the particle, the others do not. Hydration is a dynamic process, and so δ and V represent time-averaged values. The particle volume V is often presented in two equivalent forms:

$$V = v_s M/N_A \qquad (1)$$

where M is the molecular weight or molar mass (g mol^{-1}), N_A is Avogadro's number (6.02205×10^{23} mol^{-1}), and v_s is the specific volume (mL g^{-1}) of the hydrated macromolecule (volume occupied by the hydrated macromolecule per unit mass of dry macromolecule) or

$$V = (\bar{v} + \delta/\rho_o)M/N_A \qquad (2)$$

where $\bar{v}$ is the partial specific volume (mL g^{-1}).

3   Hydration-Dependent Universal Shape Functions

Harding *et al.*[1] give a complete list of those Universal shape functions requiring knowledge of δ or V for their experimental measurement. We give here only the most useful ones:

- *Viscosity increment:*[15,16]

$$v = [\eta]M/(N_A V) \tag{3}$$

in which $v = 2.5$ for a sphere.[17,18]

- *Perrin*[19] *function*:

$$P = (f/f_0)\{1 + \delta/(\bar{v}\rho_o)\}^{-1/3} \tag{4}$$

where (f/f_0), the frictional ratio,[7] is related to the sedimentation coefficient $s_{20,w}^o$ by

$$(f/f_0) = M(1 - \bar{v}\rho_o)/(N_A 6\pi\eta_o s_{20,w}^o)(4\pi N_A/3\bar{v}M)^{1/3} \tag{5}$$

or the translational diffusion coefficient $D_{20,w}^o$ by

$$(f/f_0) = \frac{k_B T}{6\pi \eta_o}\left(\frac{4\pi N_A}{3\bar{v}M}\right)^{1/3}\frac{1}{D_{20,w}^o} \tag{6}$$

where $T=293.15$ K, η_o is the viscosity of water at 293.15 K (0.010 P), ρ_o is the density of water at 293.15 K (0.99823 g mL^{-1}) and k_B is the Boltzmann's constant (1.3807×10^{-16} erg K^{-1}). $P=1$ for a sphere.[19]

- *Reduced excluded volume:*[20]

$$u_{red} = u/V = \{2BM^2 - Z^2/2I\}/(N_A V) \tag{7}$$

where u is the excluded volume (mL), B the second thermodynamic (or "osmotic pressure") virial coefficient (mL mol g^{-2}) from osmotic pressure, light scattering or sedimentation equilibrium measurements, Z is the valency of the macromolecule, measurable by titration[21] and I is the ionic strength of electrolyte in the solvent (mol/mL). At sufficient ionic strengths, the $Z^2/2I$ term becomes negligible compared with $2BM^2$. Of course, for uncharged macromolecules and proteins at the isoelectric point $Z=0$. $u_{red}=8$ for a sphere.[7]

- *Harmonic mean rotation relaxation time ratio:*

$$\tau_h/\tau_o = \{k_B T/\eta_o V\}\tau_h \tag{8}$$

where τ_h (s) is the harmonic mean rotational relaxation time, traditionally measured using steady-state fluorescence depolarisation methods,[22,23] and τ_o the corresponding value for a spherical particle of the same volume:

$$\tau_o = \eta_o V/k_B T \tag{9}$$

In earlier representations a factor of 3 was introduced because the rotational relaxation time was referred to on a dielectric dispersion basis (compensated for in the equations for steady-state anisotropy depolarisation) although this is no longer necessary – compare ref. 22 with ref. 23. This is further discussed in Garcia de la Torre *et al.*[3] $\tau_h/\tau_o = 1$ for a sphere.[24]

- *Reduced electro-optic decay constants*:

$$\theta_i^{red} = (\eta_o V/k_B T)\theta_i \tag{10}$$

where θ^i are the electric birefringence or electric dichroism decay constants. For ellipsoids of revolution that are homogeneous, *i.e.* where the geometric axis of symmetry coincides with the electrical axis, $i=1$. For general ellipsoids that are homogeneous, *i.e.* where the geometric axes coincide with the electrical axes, $i=2$, termed "+" and "−" (refs. 25 and 26); for general particles $i=1–5$ (ref. 27). For a sphere, $\theta_i^{red}=0.66667$.

To assist with the calculation of the salient Universal parameters P, v and u_{red} from the sedimentation coefficient, intrinsic viscosity and second virial coefficient, respectively, a spreadsheet algorithm has been set up called ELLIPSPRIME, which can be downloaded along with the other algorithms described here from the NCMH web site http://www.nottingham.ac.uk/ncmh.

4 Hydration-Independent Universal Shape Functions

Harding *et al.*[1] also give a complete list of those Universal shape functions *NOT* requiring knowledge of δ or V for their experimental measurement. Again we give here only the most popular or useful ones:

- *The Scheraga–Mandelkern*[28] *parameter*

$$\beta = \{[\eta]^{1/3}\eta_o\}/\{M^{2/3}(1 - \bar{v}\rho_o)100^{1/3}\} = \{N_A^{1/3}/(16200\pi^2)^{1/3}\}\{\bar{v}^{1/3}/P\} \tag{11}$$

The β parameter is unfortunately very insensitive to shape, and Equation (11) is used more as an equation of consistency, or for measuring M from sedimentation velocity and viscosity measurements. $\beta=2.1115\times10^6$ for a sphere.

- *The Pi function*[29]

$$\Pi = \{2BM/[\eta]\} - \{Z^2/2IM[\eta]\} = u_{red}/v \tag{12}$$

with the second term in the parantheses (an approximation of the charge contribution for polyelectrolytes)$\rightarrow 0$ at sufficient values of I, and of course $=0$ for uncharged macromolecules or proteins at the isoelectric point ($Z=0$). $\Pi=3.2$ for a sphere

- *The Wales–van Holde*[30,31] *parameter*

$$R = k_s/[\eta] = 2(1 + P^3)/v \tag{13}$$

where k_s (mL g^{-1}) is the concentration dependence parameter of the sedimentation coefficient in the limiting relation

$$s_{20,w} = s_{20,w}^o(1 - k_s c)$$

or

$$1/s_{20,w} = \{1/s^0_{20,w}\}(1 + k_s c)$$

Although the theory behind Equation (13) is less rigorous than that for Π (because of the greater complexity of "hydrodynamic" as opposed to "thermodynamic equilibrium"-based non-ideality), it does have a strong experimental basis.[31-33] To apply k_s in this way it is important that charge contributions to k_s are absent or if the macromolecule is a polyelectrolyte, charge contributions are suppressed by working with a solvent of sufficient ionic strength. $R = 1.6$ for a sphere.

- *The reduced radius of gyration function G (ref. 34)*

$$G = R_g^2 \{4\pi N_A/(3vM)^{2/3} \tag{14}$$

where R_g is the radius of gyration (cm), determined from light scattering, X-ray scattering or neutron scattering measurements. v is the specific volume of the macromolecule. Originally it was thought that this was closest to the specific volume of the anhydrous macromolecule, *i.e.* there is no difference in scattering density of the surface-bound solvent compared with free solvent although current thinking seems to be in favour of something between the anhydrous and (time-averaged) hydrated macromolecule: strictly speaking G should be regarded as a hydration-dependent parameter. $G = 0.6$ for a sphere.

- *The Lambda function*[35]

$$\Lambda = (\eta_o[\eta]M)/(N_A k_B T \tau_h) = v/(\tau_h/\tau_o) \tag{15}$$

For spheres, $\Lambda = 2.5$.

- *Electro-optic delta functions*[36]

$$\delta_i = (6\eta_o/N_A k_B T)[\eta]M\theta_i = 6\theta_i^{red}v \tag{16}$$

(for homogeneous ellipsoids of revolution $i = 1$ and for homogeneous triaxial ellipsoids, $i = +$ and $-$). For spheres $\delta_i = 2.5$.

5 ELLIPS1

Aim: Prediction of axial ratio (*a/b*) (equivalent prolate or oblate ellipsoid of revolution) from a user specified value for a shape function.

Description: ELLIPS1 is based on simple ellipsoid of revolution models (where two of the three axes of the ellipsoid are fixed equal to each other); if the user types in a value for a shape function from sedimentation or other types of hydrodynamic measurement, it will return a value for the axial ratio of the ellipsoid. The question an experimenter wishes to address usually is not "what is the shape function for a specified value of the axial ratio *a/b* ?" but rather "what is the axial ratio *a/b* for my macromolecule specified by my (Universal) shape function which I have experimentally measured?". Although there are exact analytical formulae linking each shape function with *a/b* (ref. 37), the reverse is not true: inversion is analytically impossible. The QUICKBASIC algorithm ELLIPS1 uses the polynomial-based inversion procedure of Harding and Cölfen[37] to give a/b vs. the various Universal shape functions to an acceptable degree of accuracy (*i.e.* to better than the precision of the measurement, which is normally no better than a few per cent).

Example: Figure 1 and Table 2 show an example for part of the complement receptor CR1 (modules 16/17). In a study by Kirkitadze and co-workers[38] they showed that the axial ratio of this protein from ELLIPS1 was ~5 from P values obtained from

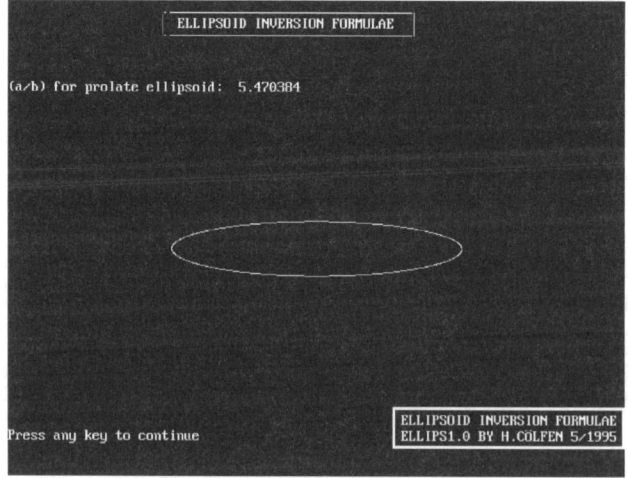

(a)

(b)

Figure 1   *ELLIPS1 screens for the determination of the axial ratio (a/b) for complement receptor CR1 (domains 16/17) using the Universal shape function P obtained from the sedimentation coefficient, $s^o_{20,w} = (1.48 \pm 0.04)$ S, molecular weight, $M = 14513$ g mol^{-1} and partial specific volume $\bar{v} = 0.725$ mL g^{-1}. (a) Input screen. (b) Output screen (result for prolate ellipsoid only shown). (Data from ref. 38)*

Table 2 *ELLIPS1 estimates of the axial ratio a/b for the complement receptor CR1 (domains 16/17) using the Universal shape function P for three different values of the time-averaged hydration parameter δ. (Adapted from ref. 38).*

δ	P	a/b
0.30	1.28	5.5
0.35	1.26	5.2
0.4	1.23	4.6

measurement of $s^o_{20,w}$ and various possible values of the hydration parameter δ. Since the individual domains (16 and 17) each had an axial ratio $\sim$3 they could conclude the domains were arranged approximately in an end-to-end fashion.

6 ELLIPS2

Aim: Evaluates the values of *all* the Universal hydrodynamic shape functions from user-specified axial dimensions (*a*, *b*, *c*) or axial ratios (*a/b*, *b/c*) for the macromolecule as modelled by a general triaxial ellipsoid.

Description: ELLIPS2 is essentially analogous to SOLPRO[3] in that from a given structure (as represented by an array of beads in SOLPRO or as a general triaxial ellipsoid in ELLIPS2) the *complete set* of Universal shape functions is returned. Most of the universal shape functions involve one or more of 10 different elliptic integrals (called alpha 1…alpha 10 – see Harding[14]). These are solved by quadrature using the NAG[39] routine D01AMF for a one-dimensional integration with an infinite upper limit and routine D01DAF for a two-dimensional integration with finite limits.

Example: Figure 2 gives the output data for an (*a/b*, *b/c*) = (1.23, 1.52), based on the crystallographic axial dimensions of $43\times35\times23$ Å for Fab′ of a chimaeric antibody B72.3.[40,41] These were then used as the basis for the construction of a surface shell bead model. This procedure was repeated for Fc and a model for the intact immunologically active antibody was then constructed.[9,10]

7 ELLIPS3

Aim: Performs the reverse of ELLIPS2 by evaluating the tri-axial shape of a macromolecule (*a/b*, *b/c*) using two possible combinations of universal shape functions.

Description: Whereas an (*a/b*, *b/c*) specifies unique values for all the hydrodynamic shape functions, the reverse is unfortunately not true: measurement of *P*, *v*, *R*, Λ, … does not uniquely fix (*a/b*, *b/c*) but rather gives a line solution of possible values. A graphical combination of the line solutions for two of these shape functions will in principle provide a unique solution for (*a/b*, *b/c*). The main criteria for selection are (i) their ease of measurement, (ii) their sensitivity to shape and insensitivity to experimental error, (iii) the two give an intersection as orthogonal as possible and (iv) the lack of requirement of an estimate for the hydration for their experimental measurement. ELLIPS3 currently offers two such combinations: the

```
**********************************************************
                     ELLIPS2 Output Data
**********************************************************

(a/b, b/c) =       1.6000     1.4200
Viscosity increment, nu =       2.9120
Perrin function, P =       1.0458
Corresponding hydration independent functions:
Wales-van Holde, R =      1.4724
Scheraga-Mandelkern, 10**-6xbeta =       2.1245
Functions based on the 2nd virial coefficient:
Reduced excluded volume, u_red        9.1442
Corresponding hydration independent function:
Pi function =       3.1401
G function (from radius of gyration) =       0.7491
Reduced electro-optic decay constants:
        Theta+ =       0.1654
        Theta- =       0.1134
Corresponding hydration independent functions:
Delta+ =       2.8907
Delta- =       1.9806
Gamma+ =       1.6965
Gamma- =       1.1624
Harmonic mean rotational relaxational time ratio:
tau_h/tau_0 =       1.1956
Corresponding hydration-independent functions:
PSI (Squire-Himmel) function =       0.9854
LAMBDA =       2.4356
Fluorescence anisotropy relaxation time ratios:
tau_1/tau_0 =       1.0075
tau_2/tau_0 =       1.3083
tau_3/tau_0 =       1.3292
tau_4/tau_0 =       1.4703
tau_5/tau_0 =       1.0074
Corresponding hydration independent functions:
lambda_1 =       2.8903
lambda_2 =       2.2259
lambda_3 =       2.1908
lambda_4 =       1.9806
lambda_5 =       2.8907
psi_1 =       1.0432
psi_2 =       0.9562
psi_3 =       0.9512
psi_4 =       0.9197
psi_5 =       1.0433
**********************************************************
```

Figure 2   *ELLIPS2 output for B72.3c Fab. The axial ratios of (a/b, b/c)=1.60, 1.42) used in this example were obtained from the crystal structure of Brady et al.[40] using the algorithm of Taylor et al.[41]*

Λ function combined with the R function and the Π function combined with the G function. The former combination satisfies all four criteria: Λ requires the experimental measurement of the harmonic mean rotational relaxation time (from *e.g.* steady-state fluorescence measurements: no complicated resolution of exponentials is required), with the intrinsic viscosity. With regard to the latter parameter, the traditional U-tube viscometers require relatively large quantities of material – the new-generation pressure imbalance methods now make this attractive for those materials available in low quantities.[13,42,43] The Wales–van Holde parameter R is measured from the ratio of the concentration-dependent sedimentation term k_s to the intrinsic viscosity measurements. The latter combination also involves a hydration-independent function, namely Π (from measurement of the intrinsic viscosity and the second-thermodynamic virial coefficient B). G can also be measured without experimental measurement of hydration provided that the radius of gyration R_g is measured using neutron scattering. If X-ray or light scattering is used, evidence now suggests the v in Equation (14) is the hydrated volume. ELLIPS3 uses as its basis the function calculation routine of ELLIPS2 except that a whole array of such values are evaluated in the (a/b, b/c) plane (a matrix of 40×40 values). A Contour plotting routine (RGCNTS from the Simpleplot Library) interpolates between these matrix points and can plot the Π, G, Λ and R functions (or any other of the universal shape functions if the programer so decides) in the (a/b, b/c) plane.

Example: Figure 3 shows an example of the determination of the triaxial shape of neurophysin monomers and dimers in solution based on data of Nicolas *et al.*[44] and Harding and Rowe.[45]

8   ELLIPS4

Aim: Evaluates the tri-axial shape of a macromolecule (a/b, b/c) from electro-optic decay based Universal shape functions combined with other hydrodynamic data.

Description: Rotational hydrodynamic shape functions, based on rotational diffusion measurements, are attractive for determining the shapes of macromolecules in solution since they are generally more sensitive functions of shape compared to other shape functions. This sensitivity comes however at a price because they are generally more difficult to measure, due to problems surrounding the resolution of multiexponential decay functions. Electro-optic measurements are more attractive than time-resolved fluorescence depolarisation anisotropy measurements in the sense that for homogeneous triaxial ellipsoids at least, there are only two exponential terms to resolve (the decay constants or reciprocal relaxation times θ_+ and θ_-) as opposed to five ($\tau_1 - \tau_5$):

$$\Delta n = A'_+ \exp(-6\theta_+ t) + A'_- \exp(-6\theta_- t) \qquad (17)$$

(see refs. 26 and 36) where Δn is the birefringence or dichroism (often expressed as "optical retardation" in degrees) at time t after the aligning electric field has been switched off. A practical problem with electro-optic decay methods is the potential local heating effects from the high electric fields used, especially if the experiments are conducted in solutions of high ionic strength: the investigator is advised to consult an article by Pörschke and Obst[46] describing how these effects can be minimised. After eliminating hydration (via *e.g.* combination with [η]) to give the Universal hydration-independent shape functions $\delta+$ and $\delta-$ and graphical combination with another Universal hydration-independent shape functions such as R[36] or Π,[47] the triaxial shape as represented by the two axial ratios (a/b, b/c) can be evaluated.

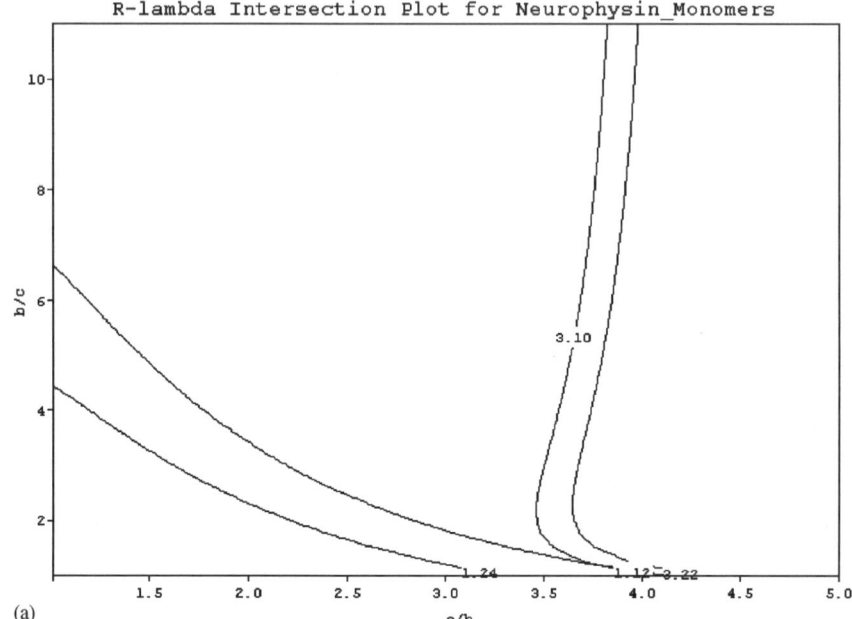

(a)

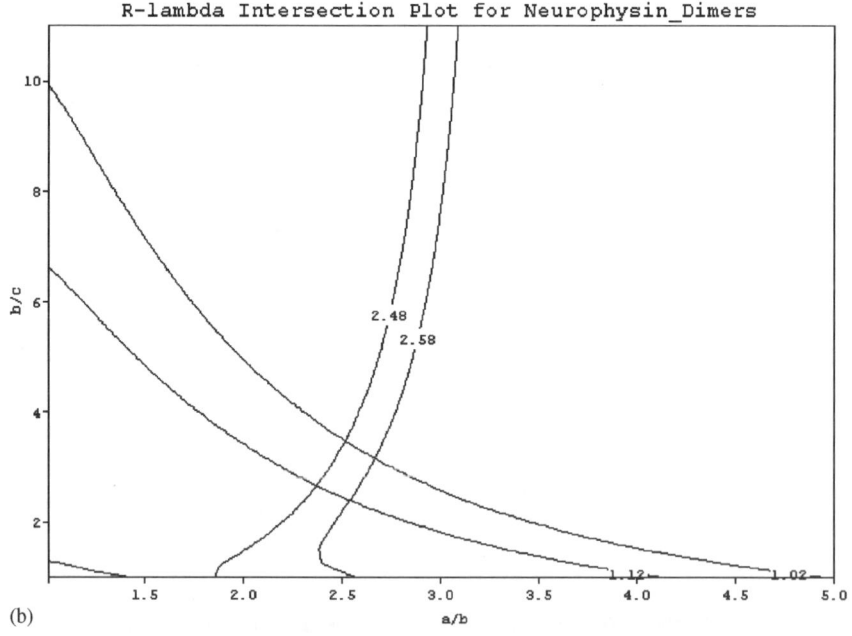

(b)

Figure 3 *ELLIPS3 output. Λ–R plot applied to neurophysin (a) monomers, Λ=3.16, R=1.18 (b) dimers, Λ=2.53 , R=1.07. The lines allow for experimental error of ±2% in Λ and ±5% in R. The intersections indicate an (a/b, b/c)=(4.0, 1.0) for neurophysin monomer (i.e. a prolate ellipsoid of axial ratio 4.0) and an (a/b, b/c)=(2.5, 2.9) for neurophysin dimers, suggesting the dimerisation is a side by side as opposed to end-to-end process*

Resolution however of even two exponential terms is not easy, particularly for globular macromolecules where θ_+ and θ_- are similar,[48] irrespective of the form of mathematical deconvolution applied, whether it be non-linear least squares or more refined types of analysis;[49-51] see Johnsen and Brown[52] for the analagous problem in dynamic light scattering analysis of polydisperse systems. In our hands[36,49] we have found a more reliable method of extraction is to use another hydrodynamic function as a constraining parameter in the analysis of the electro-optic decay data: in this way the problem is reduced from one of four variables (A'_+, θ_+, A'_-, θ_-) to one of three (A'_+, A'_-, a/b) since a/b will specify, by the constraining function a unique value for b/c (and hence θ_+, θ_-). ELLIPS4 had been written to facilitate this procedure for PC[1] based on an earlier non-interactive version of the program written for mainframe computer[49,53] and is now in WINDOWS. Its use is best illustrated by application to synthetic data (with error) generated for a macromolecule "Boningtein," which includes the following characteristics: $(a/b, b/c)=(1.5, 1.5)$; $M=71744$ Da; $[\eta]=2.74$ mL g^{-1}, and the following electro-optic decay parameters: $A'_+=0.07$, $A'_-=0.05$, $\theta_+=5.81538 \times 10^6$ s^{-1}, $\theta_-=4.15646\times10^6$ s^{-1}, $T=293.15$ K, $\eta_0=0.01$ P. Figure 4(a) shows the electro-optic decay for this based on expected error (standard deviation) of $\pm0.1°$ (optical retardation) or ±0.0017 rad random normal error on the decay data. With ELLIPS4 the user puts his electro-optic decay data (Δn *vs. t*) into a data file which is read in. The user also has to specify values for $[\eta]$ (mL g^{-1}), the molecular weight M (Da), the solvent viscosity (Poise) and temperature (K) at which the electro-optic measurements were made. The user also needs to specify the coordinates of a line of (a/b, b/c) values (based on measurement of R, Π or some other hydration-independent Universal shape function) in a second data file: Figure 4(b) shows such a constraining line of allowed (a/b, b/c) values for "Boningtein" which has an R function value of 1.479. This constrains each iteration of (a/b, b/c) and hence θ_+, θ_-, to work along the line specified by the constraining function, since each value of (a/b, b/c) specifies a value for δ_+ and δ_- (worked out using the the NAG routine D01AJF) which, combined with the user entered values for T, η_0, $[\eta]$ and M gives the θ_+, θ_- for each iteration. This reduces the risk of the fitting routine falling into subsidiary minima. When the minimum of the least-squares procedure has been formed ELLIPS4 successfully returns a/b, its corresponding value of b/c and the preexponential factors A'_+ and A'_-. The program runs automatically four times using successively the four different values of a/b entered from the constraining function (excluding the first and last data points) as starting estimates for a/b: this provides a further check against the dangers of subsidiary minima. For the starting estimates for A'_+ and A'_- the routine automatically takes these as $\Delta n_{max}/2$. The routine has various inbuilt error warnings concerning the reliability of each estimation. If no error warning is returned the result for the evaluation from a particular starting point should be reliable.

Example: Figure 4(c) shows the output for a run on the data of Figure 4(a) for Boningtein, which returns a value for (a/b, b/c)$\sim$(1.62, 1.34) – *i.e.* to within two-tenths of an axial ratio unit of the true axial ratios. In practical terms, however, the user is advised to (i) repeat the whole operation several times with various cut-off times for the decay data (at longer times the signal/noise data gets progressively worse; on the other hand more information concerning the slower relaxation time, or larger decay constant, θ_+ is contained in this region) to be certain of no subsidiary-minima

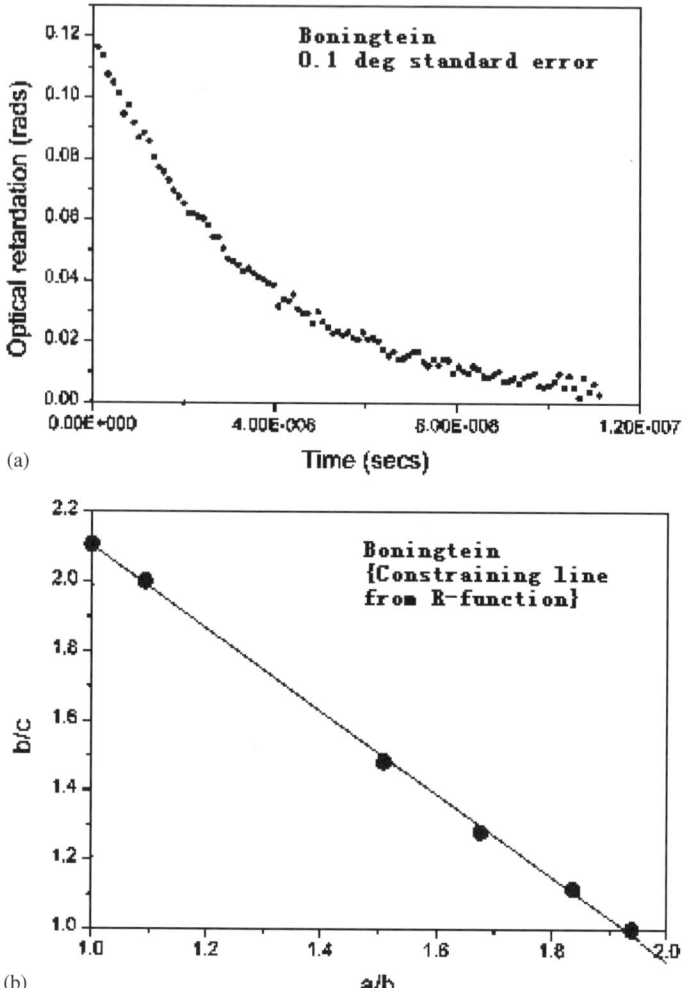

Figure 4 *ELLIPS4. (a) Electro-optic decay (expressed as the decay of optical retardation with time, t (s)). Synthetic data shown corresponding to a protein of true (a/b, b/c)=(1.5, 1.5) ("Boningtein"). The electro-optic data are fed in as a data-file into ELLIPS4. (b) R-constraining data for assisting with the resolution of the exponential terms. The user takes his experimental value of R or other suitable constraining function (Π, G, Λ ...) and plots the line of corresponding values of (a/b, b/c) using ELLIPS3 to do this for him. The user then reads off six (a/b, b/c) coordinates from this line, which can be either entered at run-time or as a data file into ELLIPS4. (c) Output giving the values ELLIPS4 returns for (a/b, b/c) and the preexponential factors for each of the four starting estimates for (a/b, b/c) and the final "best" result. A list of potential error warning estimates is also given*

problems; (ii) repeat the operation allowing for experimental error in the constraining function; (iii) check for any concentration dependence of the returned parameters: and extrapolate if necessary to zero concentration.[54] These and other features have been

```
**********************************************
              ELLIPS4 Output Data
**********************************************


Protein: Boningtein (True a/b, b/c) = (1.5, 1.5)

Constraining co-ordinates:
    a/b        b/c
( 1.000,     2.106)
( 1.092,     2.000)
( 1.509,     1.483)
( 1.676,     1.280)
( 1.836,     1.116)
( 1.937,     1.000)


Temperature  293.000 K
Solvent viscosity  0.01000 Poise
Intrinsic viscosity    2.700 ml/g
Molecular weight  0.72000E+05 Da

From a starting estimate of a/b =  1.09200
Best least squares value =  0.000258733747
a/b =  1.62197 A+ =  0.058319856203 A- =  0.061743342703

From a starting estimate of a/b =  1.50900
Best least squares value =  0.000258733747
a/b =  1.62199 A+ =  0.058319708277 A- =  0.061743502486

From a starting estimate of a/b =  1.67600
Best least squares value =  0.000258733747
a/b =  1.62198 A+ =  0.058319757156 A- =  0.061743449026

From a starting estimate of a/b =  1.83600
Best least squares value =  0.000258733747
a/b =  1.62199 A+ =  0.058319656141 A- =  0.061743559654


*****************************************************
Optimum best least squares value =  0.000258733747
          (a/b, b/c) =  1.622,  1.343
*****************************************************

In case of error output:
TYPE 2 or 3: No convergence - discard this result
TYPE 4: (unlikely) Overflow - discard this result
TYPE 5-8: some doubt about this result. The higher the
error no. the greater the doubt.
TYPE 5 means this value is almost certainly reliable;
TYPE 8 means this result is very doubtful
TYPE 9: There is probably a mistake in your constraining
data or your experimental values for mol. wt, temperature,
intrinsic viscosity or solvent viscosity
```

(c)

Figure 4   (*Continued*)

extensively explored with the earlier mainframe version of the program.[36,49,53] A possible area of further improvement includes the additional constraint that $A'_+ + A'_- = \Delta n_{max}$ although this may cause problems if the data are noisy and the $t=0$ position is not precisely defined.

9 Concluding Comment

The routines are downloadable form the NCMH web site http://www.nottingham.ac. uk/ncmh complete with full instructions. In any publication users are requested to acknowledge Salford Software and the Numerical Algorithms Group, Oxford for ELLIPS2–4 and BUSS Limited for use of SimplePlot Library routines in ELLIPS3.

References

1. S. E. Harding, H. Cölfen, and J. C. Horton, *Eur. Biophys. J.*, 1987, **25**, 347.
2. J. Garcia de la Torre, S. Navarro, M. C. Lopez Martinez, F. G. Diaz and J. J. Lopez Cascales, *Biophys. J.*, 1994, **67**, 530.
3. J. Garcia de la Torre, B. Carrasco and S. E. Harding, *Eur Biophys J.*, 1997, **25**, 361.
4. J. Garcia de la Torre and B. Carrasco, *Biopolymers*, 2002, **63**, 163.
5. J. Garcia de la Torre, this volume.
6. S. E. Harding, in *Dynamic Properties of Biomolecular Assemblies*, S. E. Harding and A. J. Rowe (eds), Royal Society of Chemistry, Cambridge, UK, 1989, 32–56.
7. C. Tanford, *Physical Chemistry of Macromolecules*, chap. 4, Wiley New York, 1961.
8. B. Carrasco, J. Garcia de la Torre, O. Byron, D. King, C. Walters, S. Jones and S. E. Harding, *Biophysical J.*, 1999, **77**, 2902.
9. B. Carrasco, J. Garcia de la Torre, K. G. Davis, S. Jones, D. Athwal, C. Walters, D. R. Burton and S. E. Harding, *Biophys. Chem.*, 2001, **93**, 181.
10. E. Longman, K. Kreusel, S. B. Tendler, I. Fiebrig, K. King, J. Adair, P. O'Shea, A. Ortega, J. Garcia de la Torre and S. E. Harding, *Eur. Biophys. J.*, 2003, **32**, 503.
11. S. E. Harding, E. Longman, A. Ortega, K. Kreusel, S. B. Tendler, K. King and J. Garcia de la Torre, *Prog. Coll. Polym. Sci.*, 2004, **127**, 113.
12. S. E. Harding, E. Longman, B. Carrasco, A. Ortega and J. Garcia de la Torre, *Method. Mol. Biol.*, 2004, **248**, 93.
13. E. Longman, Ph.D. Dissertation, University of Nottingham, 2005.
14. S. E. Harding, *Biophys. Chem.*, 1995, **55**, 69.
15. R. Simha, *J. Phys. Chem.*, 1940, **44**, 25.
16. N. Saito, *J. Phys. Soc. (Japan)*, 1951, **6**, 297.
17. A. Einstein, *Ann. Physik*, 1906, **19**, 289.
18. A. Einstein, *Ann. Physik*, 1911, **34**, 591.
19. F. Perrin, *J. Phys. Radium*, 1936, **7**, 1.
20. J. M. Rallison and S. E. Harding, *J. Coll. Int. Sci.*, 1985, **103**, 284.
21. P. D. Jeffrey, L. W., Nichol, D. R. Turner and D. J. Winzor, *J. Phys. Chem.*, 1977, **81**, 776.
22. K. E. Van Holde, *Physical Biochemistry*, 1st edn, Prentice-Hall, Englewood Cliffs, NJ, 1985, 171–172.
23. K. E. Van Holde, *Physical Biochemistry*, 2nd edn, Prentice-Hall, Englewood Cliffs NJ, 1985, 198–199.
24. F. Perrin, *J. Phys. Radium*, 1934, **5**, 497.
25. D. Ridgeway, *J. Am. Chem. Soc.*, 1966, **88**, 1104.

26. D. Ridgeway, *J. Am. Chem. Soc.*, 1968, **90**, 18.
27. W. A. Wegener, R. M. Dowben and V. J. Koester, *J. Chem. Phys.*, 1979, **70**, 622.
28. H. Scheraga and L. Mandelkern, *J. Am. Chem. Soc.*, 1953, **75**, 179.
29. S. E. Harding, *Int. J. Biol. Macromol.*, 1981, **3**, 340.
30. M. Wales and K. E. Van Holde, *J. Polym. Sci.*, 1954, **14**, 81.
31. A. J. Rowe, *Biopolymers*, 1977, **16**, 2595.
32. J. M. Creeth and C. G. Knight, *Biochim. Biophys. Acta*, 1965, **102**, 549.
33. A. J. Rowe, in *Analytical Ultracentrifugation in Biochemistry and Polymer Science*, S. E. Harding, A. J. Rowe and J. C. Horton (eds), Royal Society of Chemistry, Cambridge, UK, 1992, 394–406.
34. S. E. Harding, *Biophys. J.*, 1987, **51**, 673.
35. S. E. Harding, *Biochem. J.*, 1980, **189**, 359; *vol.* **189** corrigenda (correction in the formula for τ_h).
36. S. E. Harding and A. J. Rowe, *Biopolymers*, 1983, **22**, 1813 and **23**, 843.
37. S. E. Harding and H. Cölfen, *Anal. Biochem.*, 1995, **228**, 131.
38. M. Kirkitadze, K. Jumel, S. E. Harding, D. T. F. Dryden, M. Krych, J. P. Atkinson and P. N. Barlow, *Prog. Coll. Polym. Sci.*, 1999, **113**, 164.
39. NAG, *Workstation Library Manual*, Numerical Algorithms Group, Jordan Hill, Oxford, UK, 1991.
40. R. L. Brady, D. J. Edwards, R. E. Hubbard, J. -S. Juang, S. M. Lange, S.M. Roberts, R. J. Todd, J. R. Adair, J. S. Emtage, D. J. King and D. C. Low, *J. Mol. Biol.*, 1992, **227**, 253.
41. W. R. Taylor, J. M. Thornton and W. G. Turnell, *J. Mol. Graphics*, 1983, **1**, 30.
42. S. E. Harding, *Prog. Biophys. Mol. Biol.*, 1997, **68**, 207.
43. E. Longman, N. Marheineke and S. E. Harding, *GIT Separation*, 2005, **2**, in press.
44. P. Nicolas, G. Batelier, M. Rholam and P. Cohen, *Biochemistry*, 1981, **19**, 3563.
45. S. E. Harding and A. J. Rowe, *Int. J. Biol. Macromol.*, 1982, **4**, 357.
46. D. Pörschke and A. Obst, *Rev. Sci. Instrum.*, 1991, **62**, 818.
47. S. E. Harding, *Biochem. Soc. Trans.*, 1986, **14**, 857.
48. E. W. Small and I. Isenberg, *Biopolymers*, 1977, **16**, 1907.
49. S. E. Harding, Ph.D. Dissertation, University of Leicester, UK, 1980.
50. J. W. Jost and C. T. O'Konski, *Mol. Electro-Optics*, 1978, **2**, 529.
51. D. V. O'Connor, W. R. Ware and J. C. Andre, *J. Phys. Chem.*, 1979, **83**, 1333.
52. R. M. Johnsen and W. Brown, in *Laser Light Scattering in Biochemistry*, S. E. Harding, D. B. Sattelle, V. A. L. Bloomfield (eds), Royal Society of Chemistry, Cambridge, UK, 1992, 77–91.
53. S. E. Harding, *Comput. Biol. Med.*, 1983, **13**, 89.
54. C. L. Riddiford and B. Jennings, *Biopolymers*, 1967, **5**, 757.

CHAPTER 22

Weak Interactions: Optimal Algorithms for Their Study in the AUC

A. J. ROWE

1 Introduction

The study of solutions of macromolecular solutes at sedimentation equilibrium in the analytical ultracentrifuge (AUC) has a long history, and has lead to many insights into the properties and nature of these systems. Historically, the sedimentation equilibrium (SE) method has been of importance in that it has provided a method based upon sound thermodynamic principles for the definition of the molecular weight of solutes. No 'calibration standards' are needed when SE is employed, nor are there any significant theoretical approximations to consider. Equally, the need for knowledge of ancillary parameters is minimal: the absolute solute concentration does not need to be determined, and matters such as refractive index increment need not be considered, even though refractometric optics may be used for data acquisition. In these latter respects SE is superior to an obvious alternative fundamental method (light scattering).

The basis of the SE method derives – as do all sedimentation methods – from the Lamm equation. A centrifugal field applied to a solution of macromolecular solute(s) results in the development of a sedimentation potential which tends to displace solute towards higher radial positions in the cell: this in turn results in the development of a gradient of concentration (and hence of chemical potential) which opposes the gradient of sedimentation potential. Eventually, an equilibrium state is attained in which the opposing gradients are exactly equal in amplitude at every point in the cell. The general theory has been reviewed by many authors, and is summarised elsewhere in this volume (Chapter 4).

For present purposes, we simply note that for a single solute in the absence of thermodynamic or other non-ideality the equality of the two opposed potentials is given by

$$(\mathrm{d}c/\mathrm{d}r)_r = \sigma c_r r \qquad (1)$$

where c_r is the concentration of solute, at r the radial position in the cell, $(dc/dr)_r$ the gradient of c with respect to r at radius $= r$, and σ specifies the reduced floational mass of the solute,

$$\sigma = M(1 - \bar{v}\rho)\omega^2/RT$$

where M is the molecular weight of the solute, $\bar{v}$ the partial specific volume of the solute, ω the angular velocity of the rotor (rad s^{-1}), ρ the density of the solvent, R the gas constant, and T the rotor temperature (K).

Equation (1) is of limited practical use for data analysis. Usually the exponential form of the integrated Equation (1) is employed:

$$c_r = c_i \exp[0.5\sigma(r^2 - r_i^2)] \tag{2}$$

where c_i denotes the concentration of the solute at a defined radial position r_i. In general, a 'baseline offset' term (E) is added to the RHS of Equation (2), to allow for uncertainties in the absolute values of c_r. Curiously, the radial position r_i is often considered as being an 'initial' r value at which the data set used for analysis commences: it is even thought to be – for choice – the radial position of the meniscus. There is no algebraic basis for this belief. The definite integration of Equation (1) is valid between *any* two radial positions, and as we will consider later the choice of a value for r_i is wholly at user's discretion. It is in fact preferably *not* a near-meniscus value. Non-linear regression analysis of a data set of (r, c_r) values yields an estimate for the reduced molecular weight σ, from which the actual molecular weight is readily computed, often by use of the utility program SEDNTERP,[1] which in most cases – at least for simple proteins – provides estimates for necessary parameters without ancillary measurements.

However, the need for simple molecular weight determinations of macromolecular solutes is currently much less than it has been in the past. For proteins and polypeptides the formula weight of the chemical unit is normally known (from the gene or mass spectrometry), and rapid advances in the theory and practice of the analysis of sedimentation velocity (SV) data has made it possible to determine the order of assembly for at least the lower (*e.g.* trimer/tetramer) oligomeric states. Where SE remains the method of choice is now principally in characterisation of solute–solute interactions. Our development of the INVEQ algorithm has focussed on this area.

2 Non-ideal Systems – Thermodynamic and Specific Interactions

All systems are thermodynamically 'non-ideal'. The simple Equations (1) and (2), given above, are valid only for the case of 'infinite dilution' of the solute, where the volume occupancy of the solute particles, their charge and specific interactions (if any) can be neglected. At real concentrations all these factors must be considered. A strength of the SE method is that its theory derives simply from osmotic pressure theory, where it is long established that the effects of non-ideality can be summed up in terms of a 'virial expansion'

$$\Pi/RTc = (1/M) + Bc + Cc^2 + \cdots \tag{3}$$

where Π is the osmotic pressure.

The terms (B, C, …) are known as the second, third, … 'virial coefficients', the quantity $(1/M)$ being of course the first coefficient of the series. For many systems, especially globular proteins under conditions where charge interactions can be neglected, only the second coefficient (B) is significant, even up to around 100 mg mL^{-1}. It is convenient under these conditions to rewrite Equation (3) as

$$\Pi/RTc = (1/M)(1 + BMc) \tag{4}$$

The LHS of the equation is often equated to and written as $(1/M_{app})$, where M_{app} is the so-called 'apparent molecular weight' – *i.e.* the estimated value for M which results if the contribution of the second and higher virial coefficients is neglected under conditions where this contribution is not actually negligible. The term BMc is seen to be dimensionless, while BM has units of reciprocal solute concentration. For SE and for light scattering the term is actually 2BM, from simple theory. It follows that at sedimentation equilibrium the reduced molecular weight, σ, will have the value $\sigma/(1+2BMc)$ for finite concentrations of the macromolecular solute. The effect of thermodynamic non-ideality is always to lower the value of the apparent molecular weight.

The theory of thermodynamic non-ideality has been well reviewed[2] and need not be discussed in the present context. As noted above, where any charge interactions have been suppressed by the use of appropriate (usually around 100 mM) ionic strength in the solvent, only the second virial coefficient is significant in most cases. Where charge effects are present, then higher order terms may need to be taken into account.

So far, we have not considered specific interactions between solute particles. The simplest case here is monomer–dimer interaction, characterised by an interaction (association) constant, K_a. The presence of such interaction must lead to an *increase* in apparent molecular mass, whatever method is employed for estimation. It is customary in light scattering work to subsume this effect into the second virial coefficient: this means that in light scattering work BM can be either positive or negative in algebraic sign. We will use here the SE convention, under which BM is always positive and the presence of specific interaction (K_a) effects is considered separately and treated as additive to BM effects and of opposite algebraic sign. To first order of expansion we can thus modify Equation (2) and write

$$c_r = c_i \exp[0.5\sigma(1-\alpha c_r)(r^2 - r_i^2)] \tag{5}$$

as the basic equation for SE data fitting, where

$$\alpha = 2BM - K_a \tag{6}$$

There is an obvious problem with the use of Equation (5) in non-linear fitting routines. It is recursive in form (*i.e.* the LHS, the dependent variable c_r appears in the RHS of the equation), and no transformation is possible to make c_r explicit in r (or in r^2).

3 Inversion of the Basic Equation (the INVEQ Approach)

The problem considered above is readily demonstrated (and has been confirmed by us) on both simulated and real data sets. Not only is it incorrect in numerico–analytical terms to apply fitting algorithms (Marquadt–Levenberg or any other type) in this context: in practice equally the consequences are serious. Fits attempted either 'crash' or terminate yielding parameter sets which are demonstrably flawed, inasmuch as their values depend very much on the initial guesses used.

Historically, the problem of the interaction of non-ideal solutes has been tackled by means of graphical solutions, often based around point mass averages of various types, usually obtained via numerical differentiation.[3] Such approaches are little employed today: non-linear fitting of the data set is universally considered to be more objective and less labour-intensive. However, this latter method is not necessarily without its problems.

In widely used software, such as Beckman Origin[4] or NonLIN,[5] the problems arising from recursion are partially circumvented by segregation of terms in re-arrangements and approximations which at least (mostly) avoid 'crashes' and poor convergence. However, this is only reliably the case where 2BM and K_a are not of a similar order of magnitude. In other words, one can study *either* thermodynamic (2BM) effects *or* specific interaction effects; but not both simultaneously.

Will 2BM and K_a actually be of a similar order of magnitude under conditions of biological interest? As the majority of biochemists refer to specific interaction in terms of *dissociation* constants (K_d) rather than *association* constants (K_a), and treat the latter as the simple reciprocal of the former, we will follow that custom here. Simple calculation shows that a typical 2BM value of 10 mL g^{-1} is equivalent to a value of $K_d = 2$ mM for a protein of $M = 50\,000$. Hence there a problem must arise in the definition of weak interaction at a level of K_d around 50 μM or weaker for such a protein. Yet weak interactions are of major current interest in modern protein chemistry: to quote a recent statement on this topic:

"not all protein–protein interactions can be characterized structurally since many complexes prove next to impossible to crystallize, often because the interactions are inherently weak. The description of weak interactions represents a considerable challenge for theoretical methods, precisely because there is so little useable experimental data with which to test the models".[6]

We have aimed to replace currently unsatisfactory procedures for estimating weak interactions in the (inevitable) presence of thermodynamic non-ideality. To achieve this we have made use of a very simple fact. The *inversion* of Equation (5) yields an equation of r as a function of c, and this inverted equation *is not recursive*:

$$r = \left\{ \frac{\ln(c_r/c_i 0 = 0.5 \ (\sigma_w/(1+2BMc_r))r_i^2)}{(0.5(\sigma_w/(1+2BMc_r)))} \right\}^{0.5} \tag{7}$$

where σ_w is the mass-weighted average of the monomer and dimer values for σ. The basis for the weighting is the fraction of monomer (and hence also dimer) mass present, computed from assumed K_a values as shown below.

Equation (7) is not a 'new equation' *per se*: it has certainly been an intermediate equation in algebraic manipulation over the years. What appears to have been overlooked however is its potential for use in non-linear fitting routines. As shown below (next section) its employment leads to routinely stable fits over the range of plausible conditions. There is however an obvious objection to its use, and we will consider briefly it here.

Experimentally, the radial values constitute the independent variable, while the concentration (c) values are 'what is measured', *i.e.*, c is the dependent variable. Normal – perhaps naïve – fitting assumes that error is present in the dependent but not to any significant extent in the independent variable. However, a good modern fitting engine is not constrained in this way. We use pro Fit™ (Quantum Soft), and specify the errors in both r (normally ±0.002 cm) and c_r (normally ±0.02 fringes). Having employed the Preview Window to perform user-interactive estimates for initial parameter values, a fit is performed using the Robust fitting algorithm. This is considerably slower than the usual Levenberg–Marquadt (L–M) approach, but there are both general[6] and specific reasons (based upon current experience) for preferring it. Robust is also much more stable than L–M, and only occasionally crashes. However it does not yield estimates for standard errors of parameter estimates, and for this purpose only we often use the Robust parameters as first guesses for a subsequent L–M fit, whose stability is much improved by being provided with excellent first guesses. Such estimated error values need to be treated with a degree of caution: but they can at least indicate the order of magnitude of the uncertainty in estimates for retrieved parameters.

4 Practical Implementation of the INVEQ Algorithm

Equation (7) has been coded both directly and in modified form (see Section 4.3 below) into a set of pro Fit™ functions using the built-in programming language (essentially Pascal). The functions used in this chapter are inveq5 and inveq5a (for monomer–dimer associations) and inveq6 (for heterologous interactions, A+B). These functions serve for both generation of simulated data and for analysis. Utilities have also been programmed as needed, especially a function ADDERROR (for adding normally randomly distributed error of user-supplied variance to a data set) and a program INVERT (which takes a data set of r vs. c_r values and transforms it into a set of c_r values equi-spaced in r). In general, simulation procedures include the generation of 10 data streams each with independent (Gaussian) error. This enables a 'typical' outcome of an applied procedure to be defined.

For reasons referred to earlier, we defined the parameter c_i and the variable c_r in a manner a little different from what is conventional. The parameter c_i is defined as a reference value, at a radial position r_i given by the (estimated) harmonic mid-point of the solution column – *i.e.* at the presumed hinge-point of the solute distribution. This is sound numerical practice on general grounds, keeping the reference position in the middle of the data set, rather than at one extreme end of it. It does also, however, enable the cell loading concentration to be used as a high quality first guess for this parameter. To make this possible, and also to achieve consistency of treatments of parameter c_i and the variable c_r values, both these latter are defined as incorporating the floated additive baseline offset parameter E. inveq5a differs from inveq5 only in as much as the parameter c_i is here *fixed* rather than floated. This is possible because the data point at the chosen r value is obtained via removal of noise/entropy in the central region of the data set using a two-stage (function Fcve + orthogonal polynomial) procedure.[7] This approach reduces by one the number of floated parameters, favouring stability in fitting. The value of σ (monomer) and of r_i are fixed parameters in all fits (an alternative procedure involves the floating of as defined above within very narrow assigned limits, a practice simple to carry out in pro Fit™).

It is important to understand that it does not matter whether the *exact* hinge point has been located in radial space or not. The practice merely serves to facilitate an initial parameter estimation. Even a rather poor estimation of the hinge point radius will seldom have serious consequences, given the generally robust nature of the Robust algorithm used for fitting.

In all cases, the current value for K_a is used in the function to define a current estimate for σ_w. Here we follow a simple, classical formulation, used without approximation.

From the law of mass action the equilibrium constant K_a is given by

$$K_a = [\text{dimer}] / [\text{monomer}] [\text{monomer}] \qquad (8)$$

then if α is the fraction of moles of monomer in dimer form, and if N is the molar concentration with respect to monomer it follows that

$$K_a = N(\alpha/2)/N2(1-\alpha) = (\alpha/2)/N(1-\alpha)(1-\alpha) \qquad (9)$$

And noting that the product $K_a c$ is dimensionless (*i.e.* has the same value irrespective of whether the constituent terms are in molar or mass units), we can write in mass concentrations units

$$2K_a c(1-2\alpha+\alpha^2)-\alpha=0 \qquad (10)$$

and α can be derived simply as the solution to this quadratic equation. Then in the Robust fitting each assumed value for K_a is made to yield an estimate for α. This makes it simple to assign relative weighting in the parameter σ_w in Equation (7) leading to stable estimates for the floated parameters.

Studies on both simulated and real data sets have shown the potential of this methodology. Normally, for simulation we set up a 2–3 mm height column with some 200 data points of assumed error ±0.02 fringes. This is conservative for interference optics, but a little optimistic so far as absorption optics would be concerned. Results have been encouraging.

4.1   Simple Virial Coefficient (BM) and K_a Effects in Single Solutes

Detailed simulation (not shown) confirms that retrieval of the BM parameter from simulated data (K_a fixed at a low value of 1E-8 fringes mL^{-1}) is simple and stable. The practical limit using the INVEQ approach extends down to cell-loading concentrations <1 mg mL^{-1}. Here it is possible to retrieve an estimate for BM, with precision of the order of 20–30%, but this is improved to the level of ±3–4% when inveq5a is employed (Figure 1). Attempts to use the equations defined by McRorie and Voelker[4] (used in the Beckman Origin software) gave very poor precision in this concentration range, even with 'perfect' (*i.e.* rounding error only, or REO) data (Figure 1).

The weakest K_a which it is possible to detect at 'NMR' levels of solute concentration has been determined by simulation at 20 mg mL^{-1} solute concentration. The limiting K_a corresponds to a K_d of the order of 100 mM for a 20 000-Da protein.

Hen egg-white lysozyme has been among proteins used for evaluation of real data (Figure 2). The experiments were performed under conditions (100 mM salt, pH=7) where little significant specific interaction to form reversible dimers was expected[8]. At the low protein concentrations used (down to 0.4 mg mL^{-1}) the resolution expected required BM to be fixed rather than floated: a value for 2BM=8 mL g^{-1}, typical for globular proteins, was used. Estimates for K_d were yielded (Figure 2) which appeared to be significant in terms of our simulation criterion (<100 mM). Clearly though these were correlated with cell-loading concentration, and independent evidence showed the presence of 2–3% of irreversible dimer in the preparation used. Extrapolation to infinite concentration must logically remove any contribution from the presence of irreversibly aggregates to the K_d value. Such an extrapolation yields an estimate for the 'true' K_d=17.5 mM. This is clearly less than the method's detectability limit of 100 mM; and suggests that a very weak self-interaction is present in lysozyme under these conditions, at a level where other methods would not have detected it.

4.2   Evaluation of both BM and K_a in a Monomer–Dimer System

We have explored by detailed simulation the extent to which estimation of BM and K_a values concurrently in a weakly self-associating monomer system is feasible within likely solute concentration limits. We have found that at a loading concentrations of 5+ mg mL^{-1}, very precise estimates for BM and K_a are retrieved (<1% error, both BM and K_a floated). An important point to note is that the baseline offset (E) is a floated parameter. The problems with estimating a value for E, especially when interference optics are employed, have been a major source of concern and study:[8] work which we have now carried out that for the systems under study (which do not – it must be noted – include actually polydisperse solute systems) indicate that estimates for E yielded by the INVEQ approach are stable and precise. This contrasts starkly with the situation when the conventional 'linear' equation is employed (Figure 1).

How well would our approach cope, though, with real data, obtained not by the use of interference optics but by absorption optics? In a recent study, we have looked at the level of self-association in a series of constructs obtained from CbpA (choline-binding

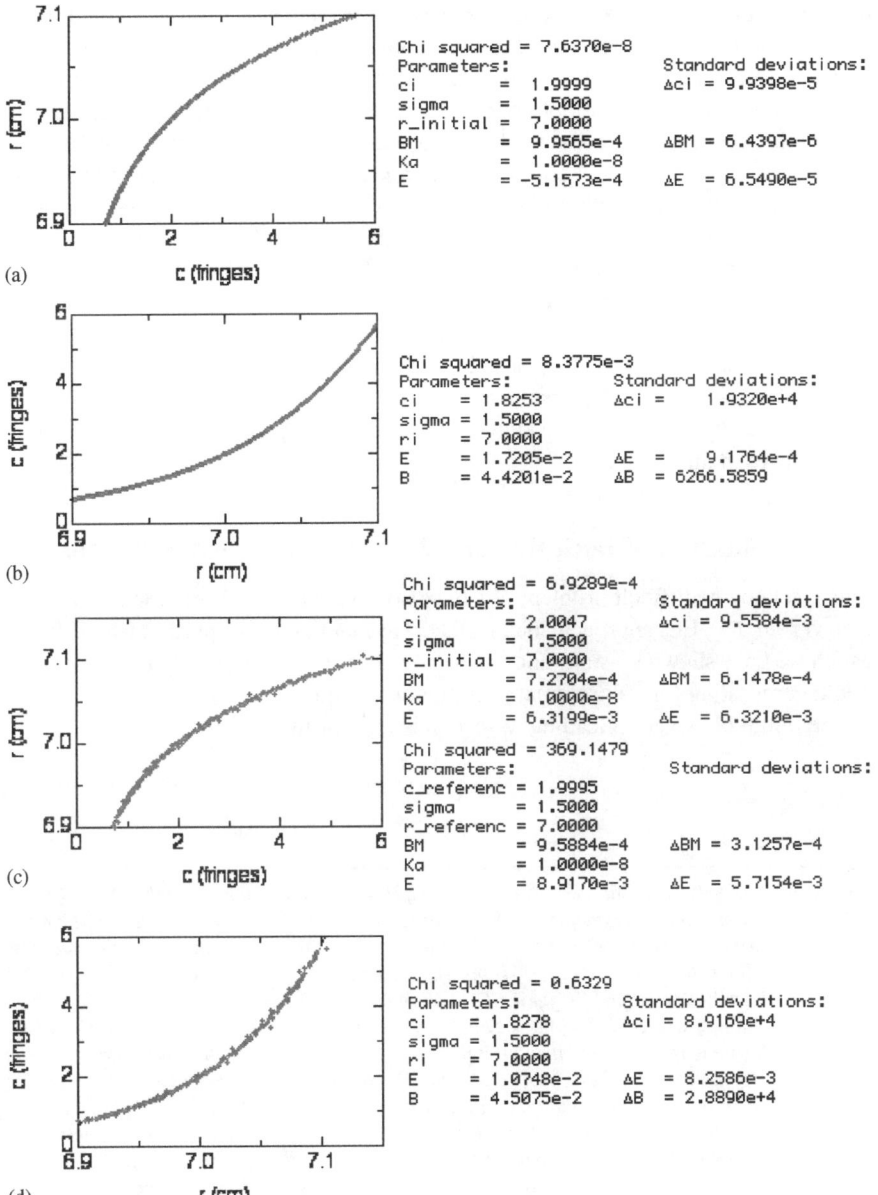

(a)

```
Chi squared = 7.6370e-8
Parameters:                    Standard deviations:
ci        =  1.9999           Δci = 9.9398e-5
sigma     =  1.5000
r_initial =  7.0000
BM        =  9.9565e-4        ΔBM = 6.4397e-6
Ka        =  1.0000e-8
E         = -5.1573e-4        ΔE  = 6.5490e-5
```

(b)

```
Chi squared = 8.3775e-3
Parameters:                   Standard deviations:
ci    = 1.8253               Δci =    1.9320e+4
sigma = 1.5000
ri    = 7.0000
E     = 1.7205e-2            ΔE  =    9.1764e-4
B     = 4.4201e-2            ΔB  = 6266.5859
```

(c)

```
Chi squared = 6.9289e-4
Parameters:                    Standard deviations:
ci        = 2.0047            Δci = 9.5584e-3
sigma     = 1.5000
r_initial = 7.0000
BM        = 7.2704e-4        ΔBM = 6.1478e-4
Ka        = 1.0000e-8
E         = 6.3199e-3        ΔE  = 6.3210e-3

Chi squared = 369.1479
Parameters:                    Standard deviations:
c_referenc = 1.9995
sigma      = 1.5000
r_referenc = 7.0000
BM         = 9.5884e-4       ΔBM = 3.1257e-4
Ka         = 1.0000e-8
E          = 8.9170e-3       ΔE  = 5.7154e-3
```

(d)

```
Chi squared = 0.6329
Parameters:                   Standard deviations:
ci    = 1.8278               Δci = 8.9169e+4
sigma = 1.5000
ri    = 7.0000
E     = 1.0748e-2            ΔE  = 8.2586e-3
B     = 4.5075e-2            ΔB  = 2.8890e+4
```

protein A), the principal adherin of *Streptococcus pneumoniae.*[9] It was found difficult with data at this more modest level of precision to estimate both BM and K_d values at protein concentrations below ~ 10 mg mL^{-1}. But as the main focus of the interaction was on the K_d values of the various constructs, it was acceptable to fix BM in lower concentration work at a level comparable to that found at higher concentrations. Figure 3 shows a typical fit from this series of experiments, and a summary of the BM and K_d values computed. It is worth noting that these constructs possess a modest degree of asymmetry in their physical conformation, and this is reflected in an average 2BM value a little above that typical of globular proteins: and for certain of the K_d values there is independent evidence from NMR measurements which agrees with the values reported from SE.[9] Thus, as with lysozyme (above) the INVEQ approach returns values for parameters which are physically plausible, and which correlate with other methodology. It has to be pointed out, of course, that K_d values as weak as (say) 10–20 mM cannot be independently checked, as we know of no other method which can return meaningful estimates in this range.

4.3 Evaluation of both BM and K_a in a Heterologous Interaction

This is a known, difficult problem. We have not found an exact solution to the problem via an INVEQ approach, but instead have defined an approximate solution which we have shown by simulation to be adequate for practical purposes. The function incorporating this approximate solution is inveq6.

The initial observation leading to inveq6 is that for the system

$$A+B \Leftrightarrow AB$$

Figure 1 *Analysis of simulated sedimentation equilibrium distributions by the INVEQ algorithm (function inveq5) and by a 'direct' fitting method (c as f(r)) as employed in NonLIN and similar. Original data set generated using the following values: r (meniscus)=6.9 cm; r (cell base)=7.1 cm; c_i=2 fringes, at r_i=7.0 cm; σ=1.5; K_a=0; BM=0.001 fringes mL^{-1}. Normally distributed random error (SD=0.02 fringes) was added to the c values by means of the function ADDERROR. Initial fitting was performed using the Robust fitting procedure, followed by the use of the Marquadt–Levenberg procedure with the Robust parameters as first guesses. No parameter estimate changed in this second fit (this is not always the case). For fitting data with error, the fit was performed with constant point errors specified as ±0.001 cm in r, ±0.02 fringes in c. A print-out of the results of the fit is shown to the right of each graph. B values should be divided by 6 to be comparable to the BM values in the inveq5/5a fits.*

It is clear that even with 'perfect' data the direct method fails to return good approximate values any of the values floated (c_i, B and E: K_a has been fixed at a low value). In the presence of noise, the linear fit returns very poor estimates, whereas the INVEQ fit, especially when the approach of fixing a c_i value is used, gives an excellent estimate for both the baseline offset (0.0089 as compared to true value of 0) and BM (0.00096 as compared to 0.001). (a) Fit to REO data using inveq5; (b) fit to REO data using the direct fit (r – f(c)) equation; (c) fit to data with error using inveq5 (upper print-out) and inveq5a (lower); and (d) fit to data with error using the direct fit (r – f(c)) equation.

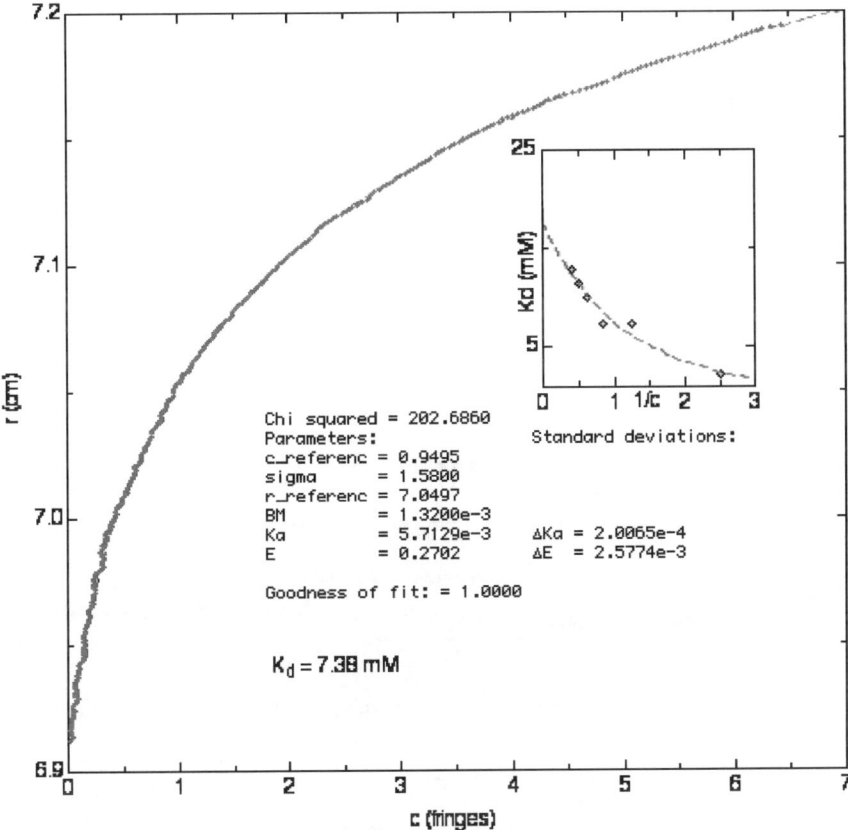

Figure 2 *Plot of interference fringe displacement vs. radial position for a solution of lysozyme at a cell-loading concentration of 0.8 mg mL⁻¹. The parameters yielded by an inveq5a fit (see text) are shown, and the fitted function (dashed line) is plotted onto the data. Inset: plot of the estimated K_d value as a function of the reciprocal of the loading concentration (0.4, 0.8, 1.2, 1.6, 2.0 and 2.5 mg mL⁻¹) for six experiments. Data not shown, but goodness of fit similar over all loading concentrations. The K_d values have been fitted by the function Fcve and extrapolated thereby to $1/c = 0$ (i.e., to infinite concentration).*

provided that A and B are present in exactly equimolar amount, the mass action equation [Equation (8)] holds and the molar fraction associated can be computed *exactly as for a monomer–dimer system.* With this knowledge, then the adaptation of inveq5 to cope with heterologous interaction is trivial. However – the above assumption only holds strictly for a homogenous system. In the AUC cell in an SE experiment there is a mass-related displacement of the components, and if the mass of B exceeds the mass of A, then there must be at least a small effect of this displacement to be considered (other than at the hinge point, where all components are conserved). Again, this issue is susceptible to analysis by simulation. We find that errors in K_d arise only for stronger levels of interaction, and at the ~1M level. It follows that in

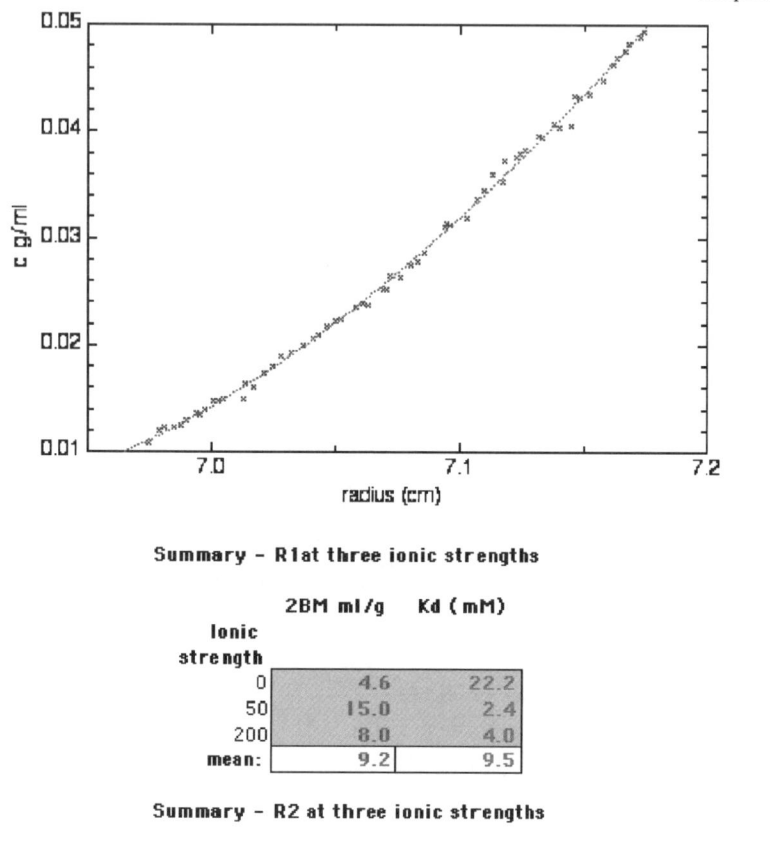

Summary – R1 at three ionic strengths

Ionic strength	2BM ml/g	Kd (mM)
0	4.6	22.2
50	15.0	2.4
200	8.0	4.0
mean:	9.2	9.5

Summary – R2 at three ionic strengths

Ionic strength	2BM ml/g	Kd (mM)
0	13.4	75.8
50	9.5	106.0
200	15.0	67.3
mean:	12.6	83.0

Figure 3   *Plot of radial value against absorbance value for the construct R1 from the CbpA of S. pneumoniae (see text); together with a tabulated summary of 2BM and K_d values estimated for this and other related constructs. The figure is re-drawn from Luo et al.[9]*

a low speed SE experiment – and looking for weak association calls for high cell loading level and hence low speeds and only modest distribution levels to avoid excessive gradients[10] – the error in inveq,[6] though real, is unlikely to perturb parameter estimations in any significant way.

On general grounds, one would expect the usable mass ratio B/A to have to be in the range 1–3, but not more. In both limits of interaction (very strong and very weak) the system reduces to a single-component system and the small error noted above is non-existent. The function inveq6 has of course been tested on 'perfect' simulated data under such conditions, and duly returns the simulated parameters within machine error.

5 The Search for Total Stability in Non-Linear Data Fitting

Non-linear data fitting is an exercise notorious for the making of repeated trials, with many sets of initial parameter guesses. The L–M algorithm is famously fast, but notoriously prone to crashes. We have defined in INVEQ a better basic algorithm for the analysis of SE data and implemented this algorithm in a set of functions which employ Robust fitting, yeast starting from interactively defined parameter guesses. The result is much improved stability. Most times a 'first attempt' is also the final attempt.

Nonetheless, a major focus of interest in the study of weak interactions lies in 'interactomics', the discipline which seeks to explore the totality of actual and potential interactions within the proteome. The existing methods used to define this 'interactome' are presently high-throughput, but extremely crude (Y2H – yeast 2-hybrid, mass spectroscopy employing a single 'probe'). It is far from clear as to what level of interaction can be detected, but it is at least plausible to think that the whole 'weak end' of the interactome is missing. Could SE in the AUC plug this gap in our knowledge base? Obviously a 'historic' approach, in which a toiling worker takes a week accumulating enough estimates of M_{app} from single determinations to get a decent '$1/M_{app}$ vs. c' plot is simply not feasible – probably even less so if light scattering is used. But we have shown that columns of 2 mm height or less suffice in SE, and with most fragments of interest being of quite low mass, then 12 h is enough to attain equilibrium. In a short path length centrepiece then around 50 μg of sample suffices. There is no need for pre-dialysis, as an approximate salt balance in the reference channel is good enough, given that E can be floated. Using eight-sector cells in an eight-hole rotor, having twin sets of rotor + cells in use, one concludes that in a 320-day working year (leaving plenty of time for machine servicing) a grand total of not less than 20 480 K_d values can be returned, using a single AUC. If this is hardly 'high throughput' at the Y2H level, it surely beats the current ability of a single workstation to express/synthesise/purify the samples.

Clearly though, the data would have to be analysed 'on line', with little or no human intervention, if this sort of high throughput work were to become a reality. Giving some consideration to this possibility, it is obvious that just one real issue stands in the way of progress. Current software – even our new INVEQ algorithm implemented using Robust fitting – is simply too liable to crash. One would require only a fraction of 1% of output to be lost via software failure, and we are nowhere near that level.

We have therefore given thought to the nature of the non-linear fitting process as applied to SE data, in the hope of seeing possible approaches to greater stability. For conceptual analysis and for visual portrayal of the issues we use the term '*criterion*' to denote the test by which a given parameter set (BM, K_a, E) is deemed to be optimal or otherwise (*e.g.* rmsd, χ^2, $\sum$|residuals|). And rather than refer to the 'solution set' as being located at the deepest well of the criterion we define the reciprocal of the criterion as being the amplitude of an 'attractor'. This usage, deriving from fractal geometry, enables four-dimensional plots to be constructed in a format which the eye readily interprets.

5.1   Development of a Brute Force Fitting Routine (BFR)

It is simple to construct a routine which computes the numerical value of the *criterion* for each point in parameter space. As noted earlier, only three parameters are needed (BM, K_a, E), routines such as NonLin being over-parameterised. It is feasible to explore a $50 \times 50 \times 50$ parameter matrix without compute time demands becoming excessive (*i.e.* in excess of several minutes). Using the Robust *criterion* the amplitude of the attractor at any point in parameter space is given by (Σ|residuals|)$^{-1}$, the summation being over all c_r, r values. The program BFR returns the parameter set corresponding to the strongest attractor, together with tabulated, sorted values of the parameter sets in order of amplitude of the attractor. BFR can be set to automatically cycle 2 or 3 times, each time reducing the range of each parameter within the set, this being a more efficient mode of convergence towards an optimal solution than simply increasing the resolution of the matrix for a single pass. There is little point in 'chasing significant figures'. It is acceptable for intervals in the final ranges in the (BM, K_a, E) matrix to be of the order of <1% of parameter value set (BM, K_a) or <0.01 fringes (E).

   As is the nature of Brute Force routines, BFR virtually never crashes. With a large number of computations it is always possible that a given number will fall outside machine precision, but the chances of this happening are small. Examples of the application of BFR to real and simulated data are given below.

5.2   The Nature of the Attractor Specified by BFR Using the Robust Criterion

We have explored some properties of the attractor for the parameter set (BM, K_a, E). While conventional use of Levenberg–Marquadt, Simplex, Robust or other fitting methods yields an optimal parameter set, all such algorithms use a 'search pathway' to arrive at an estimate for the strongest attractor. In contrast, BFR yields a four-dimensional mapping of the attractor, and thus enables us to explore the nature of the space through which the search algorithms have to progress. As a way of representing a four-dimensional surface we represent each point in (BM, K_a, E) space by a sphere, of diameter equal to the value of the attractor at that point. Some interesting and perhaps surprising facts emerge. The most striking observation is that the attractor can be cyclic (Figure 4). We see this effect in both simulated and real data. This is not unexpected on general grounds, but it does engender concern that search methods may terminate on occasion in a 'false attractor' (*i.e.* in what is a 'local' rather than a 'global' minimum in criterion-defined error).

   BFR is much less likely to return a parameter set corresponding to a false attractor. Indeed if the resolution in parameter space is adequate, then such an event is not possible.

   An equally interesting observation is that a false attractor can result from stochastic effects (Figure 5). A location on the surface of attraction may have an intense but very localised maximum, usually located very near to the main region of attraction. BFR always finds this intense, local maximum of attraction: fitting routines usually converge to the main region of attraction (Figure 5). Interestingly, if a fitting

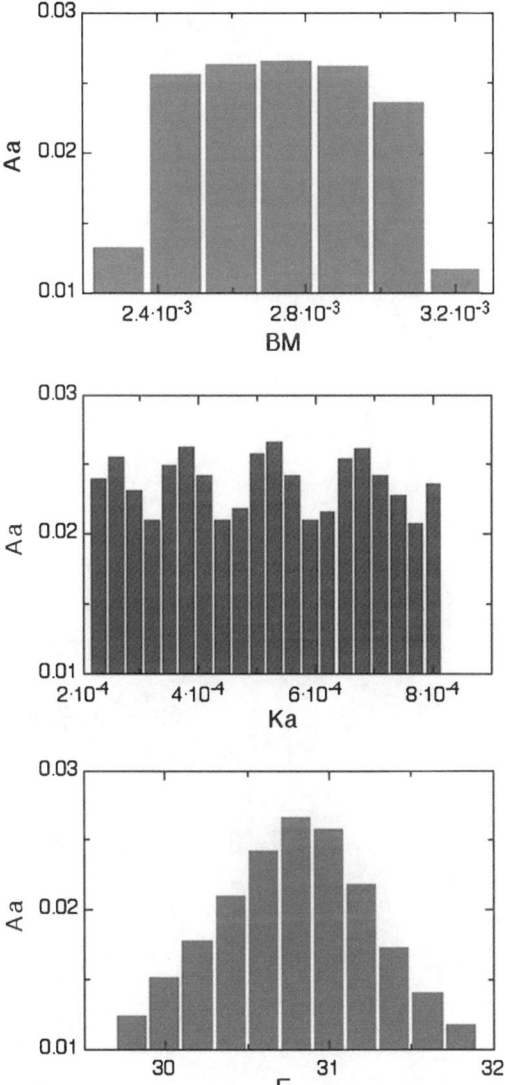

Figure 4   Plots of the amplitude of the attractor, defined as Aa (amplitude of attractor) $=\{\sum|residuals| \ (BM, \ Ka, \ E)\},^{-1}$ as a function of value of BM, Ka and E for a set of simulated data. In each case the value of the other two parameters has been fixed at the 'best estimate' yielded by running the BFR program.

routine is (re-) started with initial guesses very close to the intense local maximum, it converges towards the same. So – which is the 'real' attractor?

This fascinating – and perhaps unanswerable – question should not detain us too long. This stochastic effect is only likely to arise when we are close to an acceptable solution: to use an analogy, a large pile of stones sitting in a valley does not affect our estimate for the height and location of a nearby mountain.

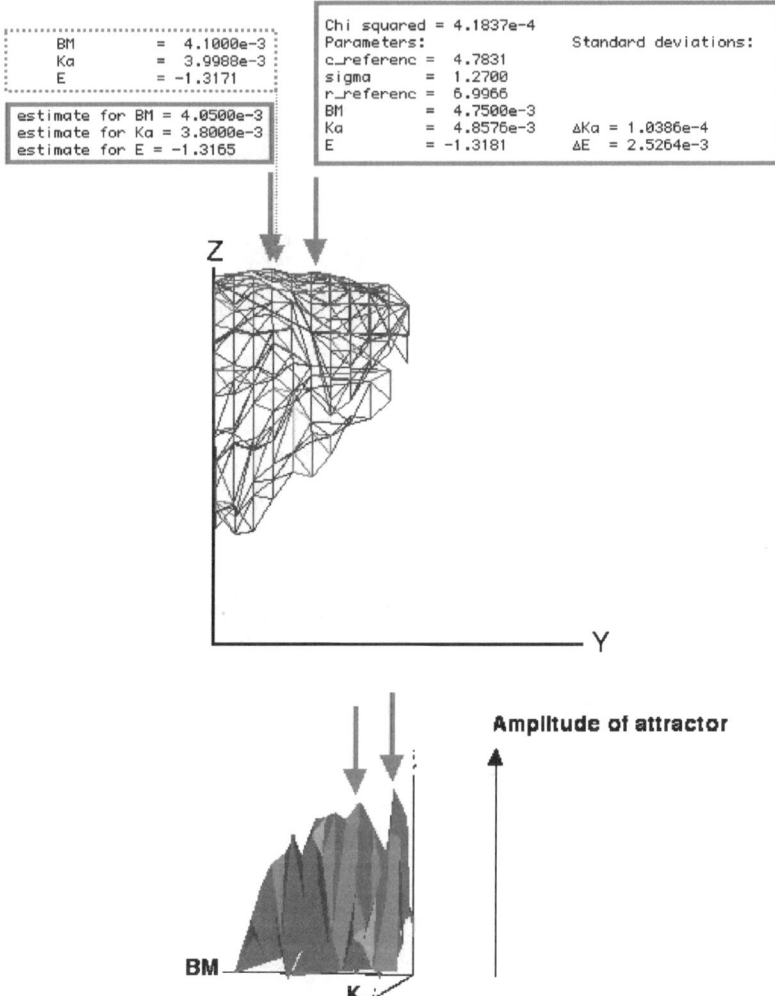

BM   = 4.1000e-3
Ka   = 3.9988e-3
E   = -1.3171

estimate for BM = 4.0500e-3
estimate for Ka = 3.8000e-3
estimate for E = -1.3165

Chi squared = 4.1837e-4
Parameters: Standard deviations:
c_referenc = 4.7831
sigma   = 1.2700
r_referenc = 6.9966
BM   = 4.7500e-3
Ka   = 4.8576e-3   ΔKa = 1.0386e-4
E   = -1.3181   ΔE = 2.5264e-3

Amplitude of attractor

Figure 5   *Surface plots of the amplitude of the attractor, defined as Aa (amplitude of attrac-
tor) =$\{\sum |residuals| (BM, Ka, E)\}$, in the plane of BM, Ka. The smaller plot is
inverted about the vertical direction, and the magnification the direction of Aa
increased by ~30 x. The results of a fit to the data using the Levenberg–Marquadt
algorithm after first using the Robust algorithm to give initial estimates is shown
top right. The program BFR returns estimates slightly different estimates (top left,
lower data) but if the starting values for fitting are set close to the BFR values,
then a Robust fit converges to yield parameter estimates (top left, upper data)
essentially identical to the BFR values. Sample: protein of mass 107 kDa.*

5.3   Preliminary Results from the Application of the BFR Routine

As would be expected, the estimate for the optimal parameter set (BM, K_a, E) is
within expected error identical to that yielded by fitting routines. This is true for
simulated and real data alike. So is the BFR routine merely a slow way of carrying

out a fitting operation? We think not. It is actually not as slow as one might suppose. For a two-stage BFR 'fit' with a 40×40×40 parameter matrix only some 5–10 min is needed, and for a given processor (single tasked) this time is known and repeatable. But on occasion a Robust fit can approach this: Simplex and (worst of all) Monte Carlo can equal or exceed it. So the time handicap of using BFR is modest, and has to be balanced against (a) the near certainty that false attractors have been avoided, and (b) the almost 'unbreakable' stability of the BFR procedure.

6 Definition of a High-throughput Methodology

As noted above, providing one has a stable compute facility for on-line data analysis, there is no reason as to why a single AUC should not return a total of 20 000+ K_d values per annum. A Brute Force Routine such as the BFR program which we have written provides such stability. Of course this assumes that one does not have to cope with the presence of irreversible dimer or other aggregates, but good modern practice with columns (*e.g.* GPC) can avoid or at least minimise this effect. At worst, one has to use the results from several channels rather than from one channel to circumvent this problem (Figure 2). Finally, the fact that we can be confident that the baseline offset E can be treated as a variable parameter is of major methodological importance: no dialysis of sample *vs.* solvent is called for, one merely has to ensure that salt concentration in the solvent channel matches that in the solution channel to an approximation good enough to avoid loss of signal from the optics of the instrument.

There are two obvious areas of application for a high throughput approach of the type which we have now described. The first is in interactomics, where the potential benefits of a knowledge of the total interactions within the proteome are very large indeed, in the areas of drug or immune therapy: the second is in the methodology of protein crystal formation, where weak interactions are considered to be of importance in defining optimal conditions.[11] In the latter case, it is thought that K_a effects must outweigh BM effects by a significant but not vast amount[11]. Interestingly, the 2BM term is expressed in most literature in this field in 'light scattering' terms, *i.e.* a composite virial coefficient is used. The INVEQ algorithm is easily set up to return a 'light scattering' second virial coefficient, merely by setting the K_a parameter to a very low, fixed value.

7 Future Developments of the INVEQ Approach

The basic equation used in the INVEQ approach [Equation (7)] can easily be expanded to incorporate, for example, higher order terms in c. This is needed to cope with systems having serious charge effects, or at very high solute concentrations. Naturally, the more parameters one uses, the higher the precision needed in the data if one is to retrieve meaningful estimates for the parameters: but the algebra needed is elementary and the danger of recursion non-existent.

Currently, the INVEQ functions also only deal with monomer–dimer or A+B interactions. Since many systems which dimerise do also form at least a small amount of, for example, trimer or tetramer, an enlargement of the algorithm to encompass further oligomers would be welcome. These and related matters are objects of current study.

References

1. T. M. Laue, B. Shah, T. Ridgeway and S. Pelletier, in *Analytical Ultracentrifugation in Biochemistry and Polymer Science*, S. E. Harding, A. J. Rowe and J. C. Horton (eds), Royal Society of Chemistry, Cambridge, 1992, 90–125.
2. P. R. Wills and D. J. Winzor, in *Analytical Ultracentrifugation in Biochemistry and Polymer Science*, S. E. Harding, A. J. Rowe and J. C. Horton (eds), Royal Society of Chemistry, Cambridge, 1992, 311–332.
3. Annals of the New York Academcy of Sciences, *Advances in Ultracentrifugal Analysis*, New York, 1969.
4. D. K. McRorie and P. J. Voelker, *Self-Associating Systems in the Analytical Ultracentrifuge*, Beckman Instruments, Palo Alto, CA, USA, 1993.
5. M. L. Johnson, J. J. Correia, D. A. Yphantis and H. R. Halvorsen, *Biophys. J.*, 1981, **36**, 575–588.
6. Robust Fitting: http://www.clecom.co.uk/science/autosignal/help/Robust_Fitting.htm
7. N. Errington, P. Mistry and A. J. Rowe, *Progr. Colloid Polym. Sci.*, 2002, 119, 58–63.
8. A. J. Rowe, *Progr. Colloid Polym. Sci.*, 2004, **127**, 1–8.
9. R. Luo, B. Mann, W. S. Lewis, A. J. Rowe, R. Heath, M. L. Stewart, A. E. Hamburger, S. Sivakolundu, E. Lacy, P. J. Bjorkman, E. Tuomanen and R. W. Kriwacki, *EMBO J.*, 2004, **24**, 1–10.
10. S. P. Spragg, *The Physical Behaviour of Macromolecules with Biological Functions*, Wiley, New York, 1980.
11. L. J. Wilson, L. Adcock-Downey and M. L. Pusey, *Biophys. J.*, 1996, **71**, 2123–2129.

CHAPTER 23

Analytical Ultracentrifugation of Colloids

HELMUT CÖLFEN

This chapter is dedicated to my new born son Manuel Sender to whom I owe many hours of postponed time for writing this manuscript.

Abstract

The application of analytical ultracentrifugation (AUC) to solve problems of colloid chemistry seems to have been almost forgotten ever since the pioneering experiments of Svedberg and co-workers in the 1920s. Although AUC was initially developed as a technique to study particle size distributions of gold colloids, it rapidly showed its versatile character with the experimental proof that macromolecules exist. Since then, AUC has mainly found application in the field of biophysics, and this dominance still exists. Nevertheless, with the current advent of nanoscience and nanotechnology – highlighted as the future technology of this century – the focus turns backwards in history, and identifies AUC as one of the most powerful colloid solution analysis techniques with a solid and mature theoretical basis. This chapter will outline how the various AUC techniques can be applied to the classical analysis problems of colloid chemistry, and will highlight potentially useful but so far less applied techniques such as sedimentation equilibrium as well as sedimentation velocity applications, the most important AUC technique for colloid analysis. Developments in this field, starting with the pioneering work of Svedberg and covering more recent work as well will be outlined, but the focus will be more on the various possible AUC techniques for colloid analysis rather than an exhaustive overview of the tremendous amount of literature from the last 70 years. This chapter is meant as a general guide to the modern applications of AUC to current problems of colloid analysis, and briefly references the methodological approaches useful for colloid analysis. A clear distinction to the so-far well-established analysis of dissolved macromolecules is maintained, showing the powerful implications of AUC for modern nanoscience.

1 Introduction

Analytical ultracentrifugation (AUC) is a powerful fractionating technique for polymer and particle characterization, and has played a significant role in facilitating the understanding of colloidal systems, especially macromolecular systems, starting with the pioneering work of Svedberg,[1,2] who initially invented this technique for the characterization of particle sizes.[1,3] However, it has emerged that this technique is also very well suited for the study of macromolecules, especially biopolymers, with the result that interest has shifted almost exclusively to the study of macromolecules. One important result of the early work was the proof that macromolecules truly exist and that they are not aggregates of small molecules, as was once heavily debated. Consequently, the focus of interest shifted from colloidal to polymeric systems, and the analysis of colloidal systems by AUC was only carried out by very few laboratories thereafter. Since then, AUC appeared almost forgotten as a powerful colloid analysis technique, and even the recent renaissance of colloid chemistry and the advent of nanoscience did not yet change this situation much, in contrast to the increased application of AUC in biophysics as a result of rapid developments in the field of biotechnology. This is understandable since AUC has predominantly found applications in biopolymer characterization rather than in colloid or synthetic polymer characterization (see Figure 1), so the experimental and methodological basis is very well developed for polymer analysis.

In addition, the Beckman XL-I AUC, launched as a modern instrument in the 1990s, was optimized for biopolymer characterization. The Schlieren optical system useful for synthetic polymers and colloids and the possibility of applying speed profiles, which are very important for the investigation of polydisperse systems did not exist for the modern instrument. A catalog of requirements for an AUC that would be

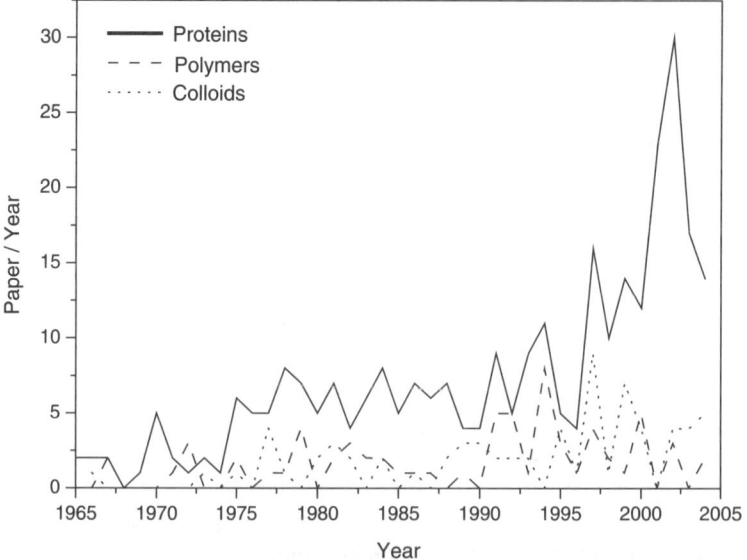

Figure 1 *Result of a literature search 12/2004 with the search phrase "Analytical Ultracentrifugation of Proteins/Polymers/Colloids. The polymer hits were corrected to include only synthetic polymers. Source: SciFinder. This figure can just qualitatively visualize the trend*

useful for both colloidal and polymeric systems was formulated by Mächtle[4] in 1991, but few of these requirements were realized in the commercial machines, and modifications have had to be made by the user.

Nevertheless, as a fractionating technique, AUC is especially well suited for the usually polydisperse and non-ideal colloidal systems, and covers a very broad application range of molar masses between 200 and 10^{14} g mol^{-1} and particle sizes between <1 and 5000 nm. The power of the technique lies in the fractionation of the sample into its components either according to its molar mass/particle size or its structure/density without the need of a stationary phase as required in many chromatographic methods. This is a big advantage over non-fractionating scattering or microscopy techniques for particle size characterization, as was manifested in a comparative study of various colloidal samples using different colloid analysis techniques by the Bayer group.[5]

In addition, the range of samples that can be investigated by AUC is extremely wide and includes all systems that consist of a solvent and a dispersed or dissolved substance, spanning from gels to microgels, dispersions, emulsions, to solutions, so that the predominantly investigated biopolymer solutions are just a special (but the most important) application example. Also, there exist no limitations to the choice of solvents, even at extreme pH. This results in a whole variety of applications, and although the technique has been in existence for about 80 years, new methods and applications continue to emerge . A noteworthy summary of the applications of AUC for polymers and nanoparticles is currently in press.[6] However, there are many special applications, such as AUC of gels, emulsions, or crystallizing systems, that cannot be covered in textbooks aiming to provide an overview of important applications. As the present book is meant to document recent advances in the field, this chapter intends to provide a general overview of the application of classical AUC methods to colloidal systems as well as an introduction to various special methods, which can otherwise only be found in the specialized primary literature.

2 Common Problems Associated with Colloid Analytics

Colloidal systems are systems characterized by mesoscopic size dimensions as visualized in Figure 2.

Colloids have special properties owing to their sizes between the atomic and the macroscopic range, and the most important colloid property is the size. As proteins, polymers, DNA, *etc.* can be considered special cases of colloids, it becomes clear that the analysis of colloids, with all of its associated questions, is even more complex than that of macromolecules. Consequently, AUC of colloids differs in many respects from that of polymeric systems; therefore special emphasis is placed on the following problems and their solutions in this chapter:

1. Colloids may exhibit extremely broad particle size distributions and associated sedimentation coefficient distributions. The danger of this is that big aggregates or small impurities are not detected.
2. Colloids can aggregate or grow during centrifugation (concentration-dependent aggregation).

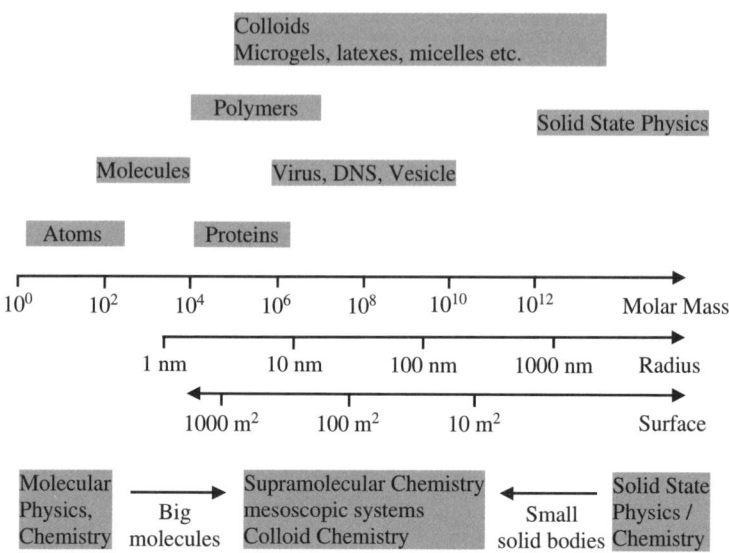

Figure 2   *The colloidal range*

3. Density of hybrid colloids is often unknown and hinders calculation of particle size.
4. Electrostatic stabilization complicates analysis owing to charge contributions.
5. Particle polydispersity in size, shape, density, and hydration.
6. High particle density often makes density gradient or density variation methods impossible.
7. Colloids are often multicomponent mixtures.

Some of these problems can be effectively addressed by AUC, whereas others are more difficult or even impossible to solve by AUC or any other currently available method alone. For example, one property of many colloidal systems (especially those of inorganic nature) is their very small particle size <10 nm. Whereas light scattering commonly fails owing to the small particle size and the associated low scattering intensity combined with the often disturbing light absorbance of these particles and electron microscopy lacks statistical significance owing to the limited number of particles detected,[5] AUC can be favorably used as it detects all particles, down to the smallest sizes. For example, the Au_{55} cluster was originally reported to be a defined and monodisperse species of special interest owing to its size at the transition from metal to molecule, which ought to give it special physical properties.[7] However, an AUC investigation of this cluster reveals that it is not monodisperse,[8] and that, at the very least, a bimodal particle size distribution is indicated (see Section 5.1.1 for further details).

Although the AUC distribution is not diffusion corrected, meaning that the distribution is broader than the real one owing to significant boundary broadening caused by the diffusion of such small particles,[9] the resolution of the particle size distribution is in the Ångström range. If this information is compared with Figure 3 (bottom) which shows the corresponding transmission electron microscopy (TEM) results, the

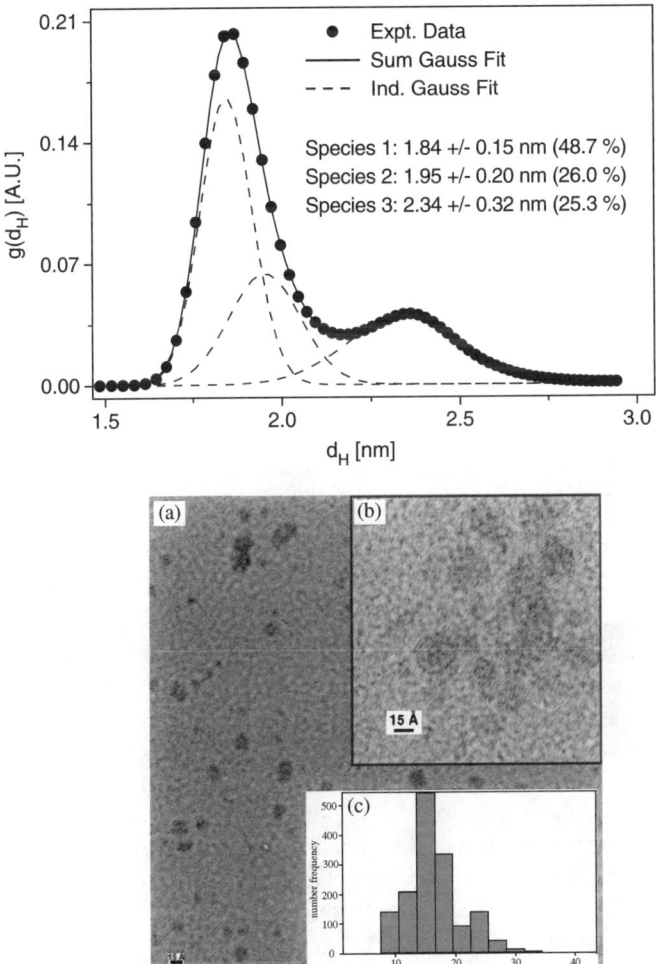

Figure 3 *Top: Apparent particle size distribution of $Au_{55}[P(Phe_3)]_{12}Cl_6$ clusters as investigated by AUC. Re-evaluated data from ref. 8. Bottom: TEM investigation taken from ref. 8. Reproduced with permission of the American Chemical Society*

higher accuracy of the particle size distribution from AUC becomes obvious although both distributions qualitatively agree. However, TEM provides important information about the particle shape, and consequently, the hydrodynamic equivalent diameters in Figure 3(a) are the real diameters.

Another common feature of colloids is their polydispersity, which is due to their inherent tendency to aggregate. This has the consequence that the s distribution can be extremely broad spanning over decades of sedimentation coefficients. An example is given in Figure 4 for aggregated $BaSO_4$. This has the consequence that a single speed is not sufficient for the characterization of such colloids. Instead, speed profiles that have proved successful for the characterization of latexes with broad

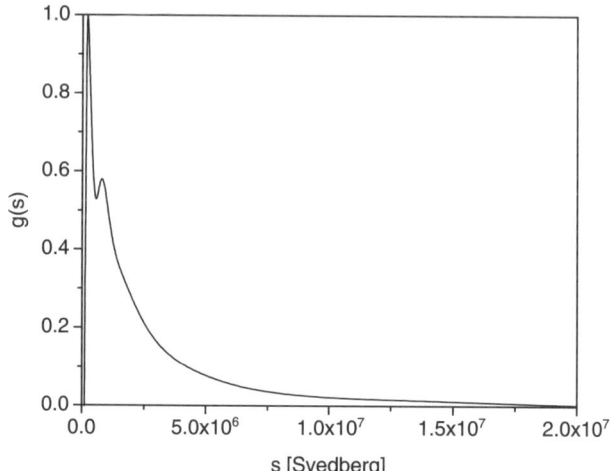

Figure 4 *Sedimentation coefficient distribution of aggregated BaSO$_4$ colloids. The distribution is not corrected for diffusion broadening and is thus apparent. Owing to the big particle size, the effect is negligible*

particle size distributions have to be applied. With such speed profiles, the entire colloidal range can be accessed in a single experiment or, if increased accuracy is needed in two experiments. This was proven for model latex mixtures.[10,11]

Another problem is the light scattered by the colloids when absorption or turbidity detection is used. Here, an MIE correction has to be applied, which is potentially problematic since this correction can amplify uncertainties or artifacts in the small particle size range. Thus, it would be much more advantageous to apply refractive index detection, but this is not yet possible on the commercial instruments as they do not allow for the necessary speed profiles. An elegant workaround was very recently developed by Müller, who overlays parts of the *s* distribution obtained at several speeds to a master curve, thus allowing one to apply refractive index detection even for very broad distributions, while simultaneously compensating for the lack of speed profiles for the XL-I AUC.[12] Refractive index detection circumvents the necessity of MIE correction and its associated potential errors so that the quantitative concentration determination, especially of small colloids, can be improved. The significance of the MIE correction is demonstrated in Figure 5, clearly indicating the benefit of a fast Rayleigh interference optical system and an automatic procedure for extracting time-dependent concentration changes.

A serious problem for AUC investigations, especially of small colloids, is density. Whereas for common polymer samples, this parameter can be measured or even predicted from monomer data, as is the case for proteins, for small colloids, the density does not correspond to the bulk density. Here, the stabilizing surface layer contributes significantly to the particle density as a result of the high surface-to-volume ratio. Either every stable colloid contains a polymer stabilization layer, which can lead to the described density alterations, or the colloid is stabilized by charge, which does not result in density variations but in non-ideal sedimentation due to charge effects. Charge

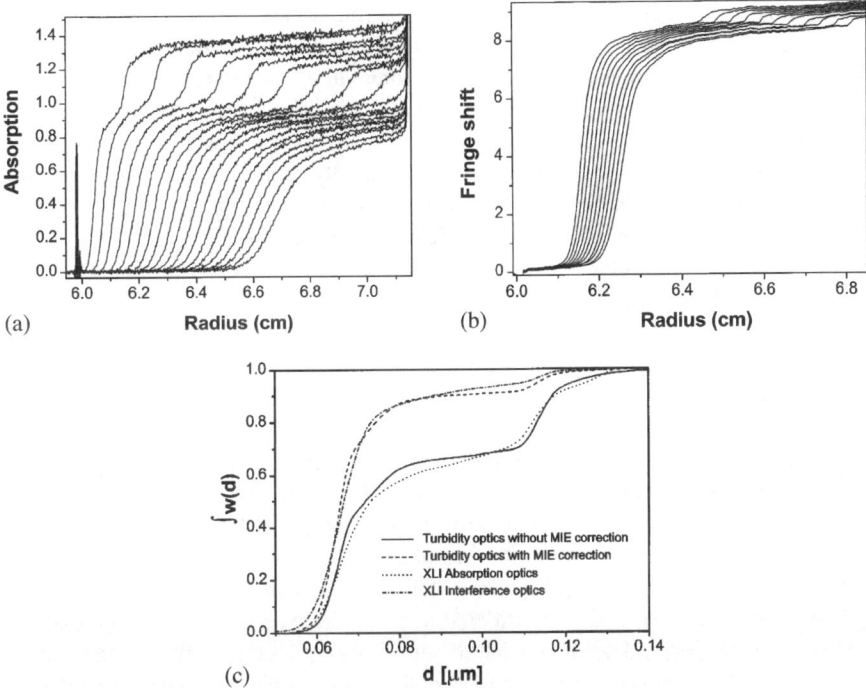

Figure 5 *Illustration of the MIE scattering effect and its impact on quantitative detection for a binary 9:1 by wt. Polystyrene latex mixture (66 and 119 nm). (a) Sedimentation velocity experiment using absorption optics at 360 nm, scan internal 5 min, (b) interference optics, scan internal 100 s. Speed 10 000 rpm, 25 °C. (c) Comparison of the integral particle size distributions with those obtained on an AUC with turbidity optics, 25 °C, linear speed profile 0–40 000 rpm*

effects can have a drastic influence on the sedimentation behavior (as known for polyelectrolytes). The usual screening of charge by electrolyte addition alters the stability of charged colloids, thus triggering aggregation, with the result that buffer application is no longer an option for electrostatically stabilized colloids.[13] As colloids are often polydisperse, a coupled polydispersity in size, density and charge is observed, which makes it impossible to characterize such samples by AUC (or any other colloid analysis technique) alone. Only the combination of various analytical techniques, or the combination of various detectors in the AUC simultaneously accessing the sedimentation and diffusion coefficient or the molar mass distributions, opens up the chance to unravel these problems. This highlights the need for the development of new detectors and a global analysis methodology.

3 Instrumentation and Methods for Colloid Analysis

AUC equipment is well known and does not need to be described here. For the design of the Optima XL-A AUC, as well as the UV–Vis absorption optical system, the reader is referred to ref. 14. The principal design of the online Rayleigh interferometer of the Optima XL-I AUC can be found in ref. 15. Besides these optical systems in modern AUCs, there is also the Schlieren optical system, which detects the refractive index/concentration gradient and belongs to the classical AUC detection systems, but is not used commercially anymore. It delivers the first derivative of the radial concentration gradient and is thus well suited to observe sedimenting boundaries during sedimentation velocity experiments. Probably, the typical Schlieren peak is the most well-known experimental output of an analytical ultracentrifuge. All three optical systems have their special advantages. The UV–Vis absorption optics combines sensitivity with selectivity owing to its variable detection wavelength, whereas the Rayleigh interference optics yields very accurate experimental data owing to the acquisition of a number of interference fringes, which are then evaluated via a fast Fourier transformation. However, the interference optics can only determine relative concentration changes with respect to a fixed point, which is usually the air/solution meniscus. The Schlieren optics is well suited for high-concentration or density gradient work and all kinds of experiments that require a derivative of the concentration gradient for evaluation (*e.g.* the determination of the z-average molar mass from sedimentation equilibrium). The Schlieren optics is similar to the setup of the Rayleigh interferometer, but has a phase plate or knife edge in the focus of the condenser lens as an additional element.

Schlieren optics was widely considered to be the least sensitive of the three detection systems, but it could be shown that its sensitivity equals that of the Rayleigh interferometer.[16] An ultra-sensitive Schlieren optical system has also been described.[17] However, even though many ultracentrifuges were equipped with video cameras to save the photographic evaluation of the experimental traces, fully automated evaluation of the images was very difficult and only reported in one case[18] until online Schlieren optical systems could be introduced.[19,20] The system adapted from the Beckman model E to the modern XL platform proved to be of good reliability producing good Schlieren pictures with an online camera system.[20] This system was further improved, soon acquiring the status of routine industrial equipment with fast, digital, data acquisition.[21] A good general description of these optical detection systems can be found in the classical study of this subject by Lloyd.[22]

Another selective optical detection system is fluorescence optics,[23] which is extremely sensitive and allows the selective investigation of components with concentrations as low as 10 ng mL^{-1}, even in mixtures with a much larger amount of other components. A prototype fluorescence detector for the Optima XL-I ultracentrifuge has been constructed,[24] and is commercially available as a third detector, which can be simultaneously used in a modern analytical ultracentrifuge. But as the fluorescence detector usually relies on labeling techniques, it is less suitable for most colloidal systems.

Besides the Rayleigh interference and Schlieren optics, the turbidity optics[10,11,25,26] is the most useful detection system for nanoparticles, especially since it is a fast online system, and can be applied with speed profiles. The optical setup is the simplest of all AUC detection optics, and when combined with multihole rotors, it allows the fast and efficient characterization of a large number of nanoparticles. Its special application is the characterization of latexes with sizes spanning the entire colloidal range. The detector is fixed at a certain radial position, thereby allowing the application of speed profiles, as in the sedimentation velocity technique reported by Runge *et al.*[27]

However, it has to be noted that with the exception of the fluorescence optics, the latter optical detection systems are only available in specialized laboratories. In addition, a general requirement for detection optics for colloidal systems is detection speed, as enough data points need to be acquired even for very rapidly sedimenting particles, to correctly detect their time-dependent sedimentation behavior. The commercially available detection systems are not ideal in that respect. These present obstacles may contribute to the rare application of AUC in colloid analysis. However, even with the UV–Vis absorption and Rayleigh interference optics of the XL-I ultracentrifuge, many colloid analysis problems can be addressed, as will be shown later in this chapter.

It is generally advantageous to combine several optical systems. The combination of the Rayleigh interference optics and the UV–Vis absorption optics can yield important information about complex systems when, for example, only some absorbing components are detected with the absorption optics, whereas the Rayleigh interferometer detects all components. One example is given in ref. 28.

Besides the simultaneous application of different optical detection systems, the application of multiplace rotors and speed profiles is often essential. Speed profiles are especially useful for unknown colloid samples, as they allow for the detection of all particles in the sample, regardless of their size. It has to be emphasized that many colloidal samples cannot be measured at all without speed profile owing to their broad particle size distribution (see Figure 4). Such speed profiles also greatly contribute to efficient measurements with multiplace rotors, as all kinds of samples can be combined in a single experiment. However, speed profiles require a fast online optical detection system (see above) and highly automated evaluation software, which has so far been fully realized only for the turbidity optics.[10,11] The workaround to apply multiple constant speeds and to combine the data after the experiment to construct the full particle size distribution from the fragments for various scans at different speeds was introduced for the Rayleigh interference optics.[12] Stafford and Braswell also reported a similar approach using a step-like speed profile called wide distribution analysis.[29] The data from absorption as well as interference optics could be evaluated on the basis of their time dependence at multiple radial positions by the time derivative method (see Section 4.1.3), eliminating both the time and radially independent noise, and permitting the determination of sedimentation coefficient distributions over orders of magnitudes. The effective upper limit is just given by the detector speed to collect enough data for proper analysis.

Methodological improvements have been reported, that enable the analysis of polydisperse systems, such as the improved van Holde–Weischet method (see Section 4.1.2, van Holde–Weischet method, for more details), which includes early and late experimental scans in the evaluation so that a broader range of sedimentation coefficients can be covered in the experiment.[30] However, this method is limited to a single speed and is therefore not applicable to polydisperse systems.

The obstacle that the commercial XL-I does not permit the execution of an experiment with a continuous speed profile appears to be partly overcome by the combination of multiple speed data, so the commercial XL-I is now increasingly useable for the analysis of colloidal systems.

4 Basic Experiment Types

Although the basic ultracentrifuge theory is well known and documented, some of the basic equations shall be repeated here wherever necessary to understand special evaluation methods, even though some of them were developed for the analysis of polymer solutions. The basic equation for ultracentrifugation experiments is the Lammequation (1), which describes the local concentration variations of the sample in the ultracentrifugal field with time owing to the sedimentation and diffusion transport processes:[31]

$$\frac{\partial c}{\partial t} = \frac{1}{r}\frac{\mathrm{d}}{\mathrm{d}r}\left(\underbrace{r\,D\,\frac{\mathrm{d}c}{\mathrm{d}r}}_{\text{Diffusion term}} - \underbrace{s\,\omega^2 r^2 c}_{\text{Sedimentation term}}\right) \quad (1)$$

Based on this equation, there are four basic types of experiments that can be performed with an analytical ultracentrifuge. Each of them can deliver its own range of physicochemical information on the sample. The different experimental approaches with a focus on colloids will be treated in the following subsections. Table 1 outlines the characteristics of each experiment type and the most important accessible parameters. However, although most experiments for the analysis of colloids are covered by the four basic experiments below, special experiments can also be described by the Lamm equation, together with additional parameters. Examples of this are the swelling pressure equilibrium experiments – a special case of sedimentation equilibrium experiments for gels – or synthetic boundary crystallization experiments, which are a special case of synthetic boundary experiments for the *in situ* formation of crystals.

Table 1 *Basic experiment types in analytical ultracentrifugation and their characteristics and basic accessible parameters*

Experiment	*Operative term in the Lamm equation (1)*	*Characteristics of experiment*	*Main accessible physicochemical parameter*
Sedimentation velocity (Section 4.1)	Sedimentation term prevails diffusion term	High rotational speed	Sedimentation coefficient
Synthetic boundary experiment (Section 4.4)	Only diffusion term operative	Low rotational speed, synthetic boundary cell	Diffusion coefficient
Sedimentation equilibrium (Section 4.2)	Both terms operative, equilibrium between sedimentation and diffusion	Low to moderate rotational speed	Molar mass, equilibrium constants, and stoichiometries of interacting systems
Density gradient (Section 4.3)	Both terms operative, equilibrium between sedimentation and diffusion	Moderate to high rotational speed, establishment of a radial solvent density gradient	Density

4.1 Sedimentation Velocity Experiment

The sedimentation velocity experiment carried out at high centrifugal fields is the most important AUC technique for nanoparticle characterization, as it allows for the determination of the sedimentation coefficient resp. particle size distribution. It is also a classical experiment for polymer characterization.[32–34] Here, the molecules/particles sediment according to their mass/size, density, and shape without significant back diffusion according to the generated concentration gradient. Under such conditions, a sample mixture is separated, and one can detect a step-like concentration profile in the ultracentrifuge cell usually exhibiting an upper and a lower plateau [see Figure 6(a)]. Each step corresponds to one species. However, colloids are often so polydisperse that no plateaus are formed. If the radial concentration gradient is detected in certain time intervals, the sedimentation of the molecules/particles can be monitored and sedimentation coefficients resp. their distributions can be calculated. This is demonstrated in Figure 6(a).

The classical calculation of the weight-average sedimentation coefficient via

$$s_w = \ln(r/r_m)/(\omega^2 t) \tag{2}$$

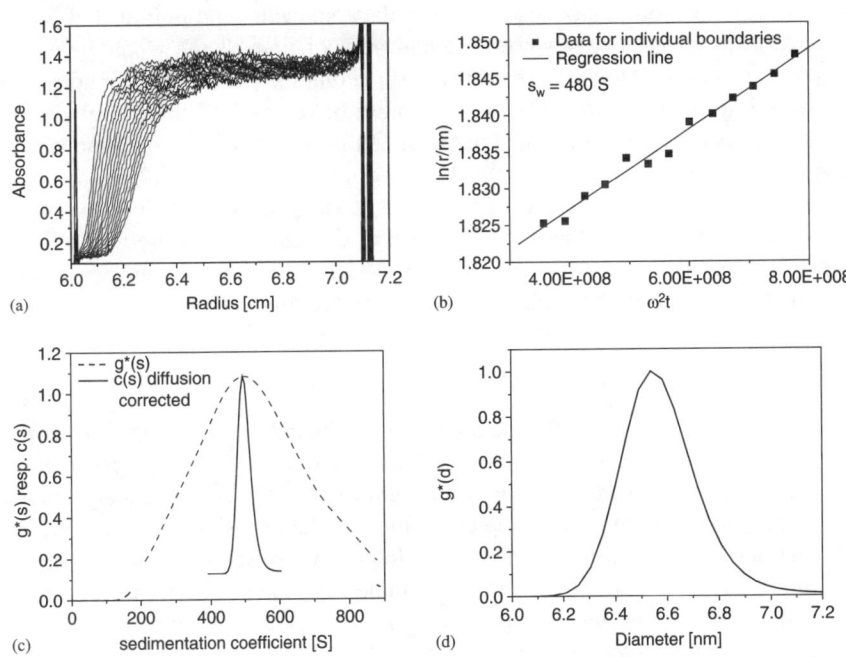

Figure 6 *Sedimentation velocity experiment on gold colloids in water at 5000 rpm and 25°C illustrating various evaluation methods. (a) Experimental raw data acquired with scanning absorption optics at 575 nm. Scan interval 2 min. (b) Sedimentation coefficient calculated from equation (2). (c) Apparent sedimentation coefficient distribution g*(s) from the time derivative method equation (3) and ref. 203 as well as diffusion-corrected sedimentation coefficient distribution c(s).[60] (d) Resulting diffusion-corrected particle size distribution. Reproduced from ref. 204 with kind permission of the American Chemical Society*

where r is the position of the midpoint or second moment point of the moving boundary, r_m the radial distance of the meniscus, t the time, and ω the angular velocity of the rotor, is not very useful for colloids [Figure 6(b)]. Fine details in the particle size distributions can remarkably alter the colloid properties so that an average quantity, which may make sense for the often monodisperse biopolymer samples, is no meaningful parameter for a colloid. However, this simple equation proves useful for the determination of sedimentation coefficient distributions if time-dependent changes are detected at a fixed radius, as in the turbidity detector.[10,11] In the following, the main evaluation techniques for sedimentation velocity experiments with colloids are briefly mentioned and the relevant primary literature cited.

4.1.1   Moving Boundary Method

Many procedures for the determination of s from sedimentation velocity data exist, but all of them cannot be named here. Some of them, like the moving boundary method,[35] allow an evaluation of the diffusion coefficient from the spreading of the boundary during the experiment[36] and thus a calculation of M via the Svedberg equation. However, the calculation of D from the boundary spreading suffers from the requirement of reasonably monodisperse or at least reasonably separated samples, because polydispersity leads to boundary spreading proportional to time, whereas diffusion goes only with the square root of time. If the whole boundary spreading is only attributed to the diffusion spreading for polydisperse samples, the determined diffusion coefficient can be found to be too high. Whereas polydispersity is often not an issue for biopolymers, it can be a serious issue in nanoparticle analysis, and erroneous results may be obtained if the diffusion coefficient is determined by this method. Therefore, in nanoparticle analysis, the diffusion coefficient should be determined by other methods, such as dynamic light scattering or Field-Flow-Fractionation, and diffusion coefficient determination from boundary spreading should generally not be applied in colloid analysis by AUC.

4.1.2   van Holde–Weischet Method

A problem arising from the diffusion broadening of boundaries is that the boundaries of several components can be smeared such that the overall boundary appears to be that of a single component. An approach to remove the effects of boundary spreading by diffusion and thus enable the calculation of the diffusion-corrected integral s distribution $G(s)$ was introduced by van Holde and Weischet.[37] This is done by selecting a fixed number of data points from one experimental scan that are evenly spaced between the baseline and the plateau. Then, an apparent sedimentation coefficient s^* is calculated for each of the data points and plotted vs. the inverse root of the runtime, yielding the typical van Holde–Weischet plot (see Figure 7).

If a linear fit of the corresponding s^* (one slice) is performed, the integral diffusion-corrected sedimentation coefficient distribution $G(s)$ can be obtained from the y values at infinite time in the van Holde–Weischet plot. In the case of a single monodisperse component, the lines intersect at one point (see Figure 7). For multiple components, the corresponding number of intersects is obtained, whereas the

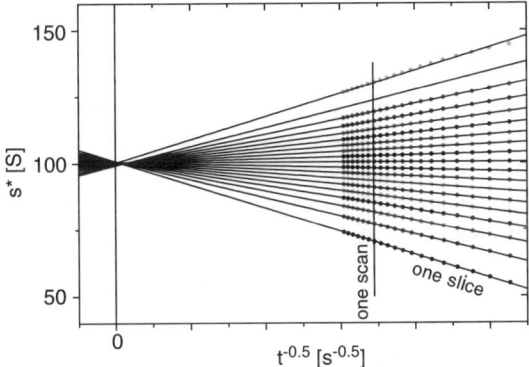

Figure 7 *Typical van Holde–Weischet plot of a sedimentation velocity experiment with a monodisperse system. Reproduced from ref. 205, with kind permission of Dr. K. Schilling*

intersection point is shifted to times less than infinity in the case of non-ideality. Therefore, the van Holde–Weischet analysis is a rigorous test for sample homogeneity or non-ideality[37–42] and does not require any assumptions about the sample. Thus, it is potentially well suited for the analysis of colloids but suffers from the requirement of an upper and lower plateau as well as the consideration of only a single-speed experiment. In fact, the van Holde–Weischet method is a useful diagnostic tool for colloids and for monodisperse colloids, and is a tool for the determination of true diffusion-corrected particle size distributions. The restriction of necessary plateaus for the van Holde–Weischet evaluation was recently overcome,[30] and distributions spanning a wide range of s values are now accessible by this method. However, the restriction of a constant speed for the evaluation still applies, so very polydisperse colloids with sedimentation coefficients spanning orders of magnitudes (Figure 4) and which require speed profiles, cannot be evaluated. However, it must be mentioned that for speed profiles, diffusion correction is so far not possible, so all of these sedimentation coefficient distributions have apparent character with respect to diffusion spreading. Nevertheless, for the usually investigated particle sizes >50 nm, the boundary spreading effect is quite small.

4.1.3 Time Derivative Method

In many cases, particles are polydisperse or one detects a multimodal distribution. In such cases, it is of interest to determine the sedimentation coefficient distribution $G(s)$ or the differential form $g(s)$. Although this is in principle possible by the van Holde–Weischet method, another method suited for the determination of $g(s)$ is the time derivative method,[203,43,44] which determines the time derivative of the radial scans acquired at different times according to

$$g(s^*)_t = \left(\frac{\partial \{ c(r,t)/c_0 \}}{\partial t} \right) \left(\frac{\omega^2 t^2}{\ln(r_m/r)} \right) \left(\frac{r}{r_m} \right)^2 \tag{3}$$

where $g(s^*)$ equals the true distribution $g(s)$ in cases where diffusion can be ignored. An example for a gold colloid is given in Figure 6(c). An important advantage of this procedure is a significant improvement of the signal-to-noise ratio of the experimental data because two scans are subtracted from each other such that systematic errors in the optical patterns cancel out and the random noise decreases. This approach especially takes advantage of the rapid data acquisition of modern analytical ultracentrifuges (Rayleigh interference optics) where 100 or more scans per velocity experiment pose no experimental problem anymore. Hence, even scans for much diluted solutions where the sedimenting boundary can hardly be seen anymore in the raw scans can be evaluated with the time derivative method. By that means, concentrations as low as 10 μg mL^{-1} can be investigated so that interacting macromolecules and particles can be addressed in a concentration range previously not accessible with the analytical ultracentrifuge. However, a drawback of the time derivative method is that only scans from a relatively narrow time interval can be used for a single evaluation, so that in fact, no full advantage is taken of the possibility to scan several hundreds of experimental scans throughout an experiment.

If diffusion is significant, extrapolation of $g(s^*)$ curves calculated for different times to infinite time yield the true distribution. The diffusion coefficient can as well be derived from the $g(s^*)$ distribution using the maximum of the $g(s^*)$ curve:[203]

$$g(s^*)_{max} = \left(\frac{s}{2\pi D}\right)^{1/2}(\omega^3 r_m)\left(\frac{t}{\sqrt{1-e^{-2\omega^2 st}}}\right) \tag{4}$$

and thus a plot of $g(s^*)_{max}$ vs. $t/\sqrt{1-e^{-2\omega^2 st}}$ yields a line with a slope proportional to the square root of s/D so that the Svedberg equation can be applied to derive the molar mass as follows:

$$M = \frac{2\pi RT}{(1-\bar{v}\rho)\omega^6 r_m^2}(\text{slope})^2 \tag{5}$$

However, for heterogeneous systems, the molar mass can be underestimated up to 10–20%, and for very polydisperse colloids, this error is certainly much bigger. Nevertheless, a former limitation that $g(s^*)$ could only be evaluated for a single-speed experiment limiting the application of the method to reasonably narrow particle size distributions could be overcome by application of speed profiles for polydisperse systems.[29] Now, the time derivative method can even be applied to colloids with sedimentation coefficients spanning over magnitudes.

The determination of $g(s^*)$ can yield a lot of important information besides the sample homogeneity and number of components. In case of interacting systems, for example, interaction constants or stoichiometries can be derived.[203,45–47] If multiple Gauss curves with a maximum at fixed s are fitted to $g(s)$, each of them corresponds to one component.[48] By that means, aggregation state as well as the corresponding concentration of the different aggregates can be determined for aggregating systems so that the equilibrium constant and thus the free enthalpy of the association steps is accessible. An example of the advantageous application of this technique is the precrystallization aggregation of lysozyme, which yielded the smallest oligomer able to form a crystal.[49]

4.1.4 Fitting to approximate or finite element solutions of the Lamm equation

One new and recently much investigated approach has proved useful for the determination of s and D and thus M as well as the concentrations of individual components from sedimentation velocity data, shown in Figure 3(a). This is the fit of a series of radial concentration profiles to approximate solutions of the Lamm differential equation (1) of the ultracentrifuge[50-55] or finite element solutions of the Lamm equation.[56-59]

As the Lamm equation (1) is the fundamental equation in AUC capable of describing all types of ultracentrifuge experiments, fitting of experimental data to this equation is potentially a very powerful approach. One drawback is that this method is clearly model dependent. Nevertheless, it is widely applicable and can quite accurately determine the sedimentation and diffusion coefficients even in mixtures of up to three components as long as the proper fitting function is used, the individual components are reasonably monodisperse (see discussion about boundary broadening in 4.1.2), and the individual components differ with respect to s by a factor of 1.5 and more. For smaller s differences, the van Holde–Weischet method is better suited. In the case of unknown samples, the van Holde–Weischet method can be used first to determine the type of system under investigation so that the correct model can be used for the fitting of the Lamm equation.

If mixtures of more than three components have to be analyzed, s and D values of some components have to be known in order to get reliable fitting results. The fitting of experimental data to solutions of the Lamm equation has its special merits for small macromolecules or nanoparticles that sediment so slowly ($s<1$S) that they do not clear the meniscus. Furthermore, such approaches permit the rapid determination of molar masses with an accuracy of 10% within 15–30 min after the start of a sedimentation equilibrium experiment using a modified Archibald approach.[59] This is especially important for unstable samples, which have to be characterized rapidly. However, for the majority of colloids, this approach is not useful, as the requirement of sufficient monodispersity is not fulfilled.

Very recently, the requirement of monodisperse samples for fitting of the Lamm equation was overcome so that now, even sedimentation coefficients and molar mass distributions of polydisperse samples can be investigated.[60] However, this approach suffers from the necessary prior knowledge of $\bar{v}$ (no serious problem) and the frictional ratio f/f_0 of the sample. If the frictional ratio is not known, it can also be fitted, which in turn allows one to draw conclusions about the particle shape.

The most significant merit of this approach for nanoparticle analysis is the possibility of correcting for the effects of diffusion on the broadening of the sedimenting boundary so that diffusion-corrected sedimentation coefficient distributions can be obtained reflecting the true polydispersity of the sample [Figure 6(c)]. The method was shown to yield reliable results for different model systems.[61] It was recently extended to account for solvent compressibility effects, which are significant for organic solvents and also non-negligible for water at very high speeds, which have to be applied for small colloids.[62] In addition, the sedimentation in a dynamic density gradient formed by the sedimentation of co-solutes was treated, which is a relevant case for the AUC analysis of colloids.[63] Therefore, fitting to approximate solutions of the Lamm equation with constrained algorithms for the determination of diffusion coefficients associated with a given sedimentation coefficient distribution is a very appropriate method for the analysis of the sedimentation coefficient distribution of colloids.

Meanwhile, equilibrium constants can also be obtained from sedimentation velocity data by combining multiple runs and concentrations, and mixing ratios of the interacting compounds by fitting time-difference data, and applying finite element solutions of the Lamm equation.[64]

4.1.5   Comparison of Sedimentation Velocity Evaluation Methods

The above-presented methods have their special merits and are thus briefly compared to allow an assessment of the methods in view of colloid analysis. The moving boundary method and with it all methods, which calculate only average sedimentation coefficients are of very limited use for the analysis of colloids as their particle size distribution is the most important property.

The van Holde–Weischet method is a rigorous test for sample homogeneity and non-ideality, and allows correcting for diffusion broadening of sedimenting boundaries. However, as only an experiment at a single speed can be evaluated, it is of limited value for the analysis of nanoparticles, although the accessible range was extended to evaluate broader distributions.[30] The time derivative method is valuable, if distributions of a good quality are desired, but it does not allow for diffusion correction of the distribution. Another drawback is that only scans of a limited time interval can be analyzed, with the result that this method does not make full use of the possibility of modern ultracentrifuges to acquire hundreds of experimental scans in a single experiment, unless scans are evaluated over several time intervals. In the latter case, the calculation of the diffusion coefficient is possible. However, very recently, a step-speed profile became applicable for evaluation of experimental data by the time derivative method, so colloids with sedimentation coefficients spanning over magnitudes can now be evaluated by this method.[29]

Fitting to approximate solutions of the Lamm equation can make full use of the possibility of acquiring large experimental data sets and thus can yield very accurate results. In addition, time and radially invariant noise can be removed and it is possible to obtain diffusion-corrected sedimentation coefficient distributions – even for polydisperse particles. However, in case of polydispersity, artificial peaks can be generated by the diffusion correction, so it is good to compare the diffusion corrected with the uncorrected distribution.

4.2   Sedimentation Equilibrium Experiment

In contrast to sedimentation velocity, which is the most important technique for colloid analysis by AUC reflecting the major analytical problem of determining distributions, sedimentation equilibrium techniques can also be applied to a limited extent. These techniques concern very small colloidal precursor particles and supramolecular systems, and focus mainly on the determination of their molar mass as well as their interactions. A solid basis for this method has already been laid by the numerous AUC studies on biopolymer systems. Nevertheless, it has to be stated that the wealth of sedimentation equilibrium analysis has not yet been applied for colloidal systems although the potential amount of precise information concerning interaction constants and stoichiometries is remarkable. Interaction analysis is a problem most relevant in colloid chemistry, but often, the resulting aggregates quickly grow to a size beyond the possibilities of a sedimentation experiment, or the species are so polydisperse, that an interaction analysis is simply not possible anymore. Nevertheless, future potential is seen in sedimentation equilibrium experiments on colloidal systems as they will help to address colloid stabilization issues, thermodynamics of particle interaction and address the important class of gels, microgels, and lyotropic phases. Especially for the latter systems, little knowledge exists compared to the well-established sedimentation equilibrium analysis of soluble macromolecules. But even for the classical questions of particle size determination, sedimentation equilibrium experiments have much to offer. Molar mass determinations are a precise measure of particle size as the molar mass is proportional to the third power of the particle size. If the error of a conventional molar mass determination can be stated to be around 5%, the corresponding error in the colloid particle size would only be 1.7%. Thus, sedimentation equilibrium experiments should in principle be a very accurate method to determine particle size averages – especially of very small particles, which sediment very slowly.

Therefore, a short overview will be given on the most relevant sedimentation equilibrium techniques for soluble or dispersed systems. Besides, this subsection also focuses on the available analysis methods for gel and microgel systems, which should be directly applicable to the analysis of lyotropic phases as well, since sedimentation equilibrium experiments are based on solid thermodynamics.

It is well established that the concentration gradient contains information about the molar mass of the sample, the second osmotic virial coefficient, or interaction constants in the case of interacting systems, independently of the particle shape. An advantage is that the detection of the concentration gradient is possible without disturbing the chemical equilibrium even of weak interactions. Sedimentation equilibrium is of more value for polymeric systems, because the molar mass of particles can be too high to allow a significant diffusion counteracting the particle sedimentation. Nevertheless, for smaller particles down to the smallest nanoparticles or even sub-critical clusters and complexes, it can be applied successfully, as shown for the example of complexes of Zr(IV) and Hf(IV) formed by hydrolytic polymerization in acidic medium.[65,66] In addition, sedimentation equilibrium analysis was successfully applied in supramolecular systems by Schubert *et al.*[67,68] and Tziafzios *et al.*[69]

Various procedures exist for the evaluation of sedimentation equilibrium experiments of dissolved macromolecules, which are very briefly mentioned in the following sections and are critically considered concerning their use for the analysis of colloidal systems.

4.2.1   Classical Approaches

The classical approach to the evaluation of sedimentation equilibrium concentration gradients is to plot ln (c) *vs.* r^2 to obtain the weight-average molar mass M_w from the slope according to Equation (5).

$$M_{w,app} = \frac{2RT}{(1-\bar{v}\rho)\omega^2} \frac{d(\ln c)}{dr^2} \qquad (6)$$

It must be noted that these M_w values have apparent character because they are calculated for finite sample concentrations. Consequently, they are named $M_{w,app.}$. The true M_w can be obtained by an extrapolation of a concentration series to infinite dilution. The evaluation according to Equation (6) works well for monodisperse ideal samples because in such cases, the ln (c) *vs.* r^2 plot is linear. However, in most cases in colloid analysis, this plot is curved owing to multiple overlay effects, which normally yield erroneous $M_{w,app.}$ values from linear regression. Nevertheless, the curvature indicates whether the sample is non-ideal (downward curvature), heterogeneous, or self-associating/aggregating (upward curvature). One advantage of Equation (6) is that it allows the computation of local $M_{w,app.}(r)$ values for every acquired data point. $1/M_{w,app.}(r)$ can be plotted against the corresponding concentrations to yield M_w and the second osmotic virial coefficient A_2 by using Equation (7):

$$1/M_{w,app.}(r) = 1/M_w + 2A_2 c(r) \qquad (7)$$

From the radial concentration gradient and the local $M_{w,app.}(r)$ values, the apparent z-average molar mass $M_{z,app.}(r)$ can also be obtained according to Equation (8), if the data set is of good quality (double differentiation of the experimental raw data).[70] But this case is very rare for colloidal samples.

$$M_{z,app.}(r) = M_{w,app.}(r) + \frac{2RT}{(1-\bar{v}\rho)\omega^2}\left[\frac{d\ln M_{w,app.}(r)}{d(r)^2}\right]$$

$$= \frac{2RT}{(1-\bar{v}\rho)\omega^2}\frac{d}{d(r)^2}\ln\left(\frac{1}{r}\frac{dc(r)}{dr}\right) \qquad (8)$$

An alternative evaluation procedure was described by Lansing and Kraemer:[71]

$$M_{w,app.} = \frac{2RT}{(1-\bar{v}\rho)\omega^2}\frac{c_b - c_m}{c_0(r_b^2 - r_m^2)} \qquad (9)$$

with the indices b=bottom and m=meniscus. This procedure allows the calculation of $M_{z,app.}$[72,73]

$$M_{z,app.} = \frac{2RT}{(1-\bar{v}\rho)\omega^2}\frac{1}{(c_b - c_m)}\left[\frac{1}{r_b}\left(\frac{dc(r)}{dr}\right)_b - \frac{1}{r_m}\left(\frac{dc(r)}{dr}\right)_m\right] \qquad (10)$$

which yields average molar masses for the whole cell and no local values. A significant disadvantage of this evaluation is that it completely relies upon the correct determination of the concentration at the meniscus and the cell bottom where there might be considerable uncertainty. Furthermore, the obtained molar masses can be strongly influenced by impurities being much smaller (c_m influenced) or larger (c_b influenced) than the sample (more common for colloids) owing to the evaluation of a whole-cell average molar mass. In addition, it has to be stated that the average molecular mass from such approaches is by far more inaccurate for colloids than it is for dissolved macromolecules. The reason for this is the sample polydispersity, which is small or negligible for most biomacromolecules, moderate for synthetic macromolecules, but very large for colloids – often spanning several magnitudes in the sedimentation coefficient (see Figure 4). This becomes clear from the consideration that the molar mass of a colloid scales with its radius to the power of 3, so that even slight particle size variations result in a large molecular mass polydispersity. Polydispersity was up to now only rarely treated in the analysis of sedimentation equilibrium experiments as the main focus of this analysis technique was and still is the analysis of defined biomacromolecules and their interactions. The so-far available approaches will be briefly outlined in the following sections and can be found in much greater detail in the current literature (for most recent advances, see other chapters of this book).

4.2.2 Fitting Concentration Gradients to a Model

An approach for the analysis of sedimentation equilibrium data greatly facilitated by cheap and commonly available computers is the fit of the radial equilibrium concentration gradient to a theoretical model using non-linear least-squares analysis.[74,75] This approach is highly popular today and a whole variety of programs is available for this purpose, both commercial as well as public domain software.[74,76] Furthermore, such fitting routines can easily be set up on commercial software platforms, offering freely definable functions for the curve-fitting process. In general, the radial equilibrium concentration gradient in the ultracentrifuge is described as the exponential form of Equation (6) for a single ideal component, as a modified exponential equation in case of non-ideality

$$c(r) = c(r_{\text{ref.}}) \exp\left(\frac{\omega^2}{2RT} M(1-\bar{v}\rho)(r^2 - r_{\text{ref.}}^2) - A_2 M(c(r) - c(r_{\text{ref.}})) \right) \qquad (11)$$

or as a sum of exponentials for different components in a mixture or in an interacting system

$$c(r) = c_{\text{Mon.}}(r_{\text{ref.}}) \exp\left(\frac{\omega^2}{2RT} M_{\text{Mon.}}(1-\bar{v}\rho)(r^2 - r_{\text{ref.}}^2) \right)$$

$$+ c_{\text{Mon.}}(r_{\text{ref.}})^{n_2} K_2 \exp\left(\frac{\omega^2}{2RT} n_2 M_{\text{Mon.}}(1-\bar{v}\rho)(r^2 - r_{\text{ref.}}^2) \right) + \cdots \qquad (12)$$

$$+ c_{\text{Mon.}}(r_{\text{ref.}})^{n_i} K_i \exp\left(\frac{\omega^2}{2RT} n_i M_{\text{Mon.}}(1-\bar{v}\rho)(r^2 - r_{\text{ref.}}^2) \right)$$

For an ideal self-associating system of the type [n monomer $\rightarrow$ n-mer with $K=c(n\text{-}mer)/c(monomer)^n$], the first two exponentials can be used as a fitting function, whereas for a more complicated self-associating system with several n-mers, the sum of exponentials in the form of Equation (12) can be applied. With such an approach, the quantitative determination of the association constant(s), stoichiometries, and second virial coefficients for self-associating systems is possible.[77–82] However, for polydisperse colloids, this has not yet been demonstrated although it should be possible in principle. One can think of many more models for a system that can be used to fit the experimentally detected radial concentration gradient. However, the case of a polydisperse and often highly non-ideal system – which colloid chemists are very often confronted with – is not yet covered by a model in the commonly available programs, as the experimentally obtained radial exponential function has to be deconvoluted into a sum of n exponential functions for the individual components. As a consequence, in the above simple fitting, the models for a monodisperse system must be applied knowing that the obtained values for the polydisperse system are only averages as fitting to equations such as Equation (12) only works for up to four components.

Very often, a baseline absorbance which originates from non-sedimenting but light-absorbing material is used as a fitting parameter for data derived from absorption optics just by adding a variable to the right-hand side of Equations (11) and (12). This can have an enormous influence on the results, and extreme care has to be taken when allowing the baseline absorbance to be a fitting variable. Wherever possible, one should determine the baseline absorbance using the so-called over-speeding technique, where the sample is sedimented to the cell base, leaving the non-sedimenting material behind. If the Rayleigh interference optics, which yields interference fringe shifts relative to the meniscus is applied , it is recommended to apply synthetic boundary experiments in order to determine the fringe shifts at the meniscus for low-speed equilibrium experiments.[83] Another possibility is to use the so-called high-speed equilibrium technique, which applies such high speeds that the meniscus is cleared of the sample, leaving behind an exponential radial concentration gradient. Whereas this technique works well for relatively monodisperse samples, in highly polydisperse systems, the high molecular weight material has sedimented to the cell bottom and is thus lost for detection, resulting in a molar mass that is found to be too low, so that over-speeding techniques cannot be commonly applied for colloidal systems.

The advantage of fitting the radial equilibrium concentration gradient to a model is that the experimental data can be directly analyzed without differentiation, an operation that amplifies experimental noise. Furthermore, multiple data sets can be fitted to one model, allowing an analysis with better statistical significance. On the other hand, a possible danger of every fitting process that the least squares fit just corresponds to a side minimum can become serious even if the fitted curve well describes the experimental data. The fact that sums of exponentials have to be used for the analysis of the equilibrium concentration gradient limits the analysis of multicomponent systems considerably to three or four components[84] because the fitting parameters of the exponentials are badly determined by the data. A further serious disadvantage is that a model has to be known prior to the fitting process, *i.e.* one must already have information about the system from other sources. However, very often, one does not know anything about the sample under investigation despite the density of the colloid. In such cases, several models can lead to answers of similar accuracy from the statistical viewpoint. This either requires other independent techniques to seek more information, applying model independent approaches like the $M*$ function (see below), or an evaluation using a classical approach to learn something about the system of interest prior to fitting.

4.2.3 Model Independent Approaches

There are two main model independent approaches to evaluate sedimentation equilibrium data. The first one is the so-called $M*$ function[85] defined as

$$M*(r) = \frac{c(r) - c_m}{\Phi c_m (r^2 - r_m^2) + 2\Phi \int_{r_m}^{r} r[c(r) - c_m] dr} \quad \text{with} \quad \Phi = \frac{(1 - \bar{v}\rho)\omega^2}{2RT} \quad (13)$$

The most useful property of the $M*$ function is that it equals $M_{w,app.}$ at the cell bottom. This means that one can get a cell-average molar mass by extrapolating $M*(r)$ to r_b. The $M*$ function has been combined with the classical evaluation according to Equation (6) in evaluation algorithms[86,87] to yield whole-cell average $M_{w,app.}$ via $M*(r)$ as well as local $M_{w,app.}(r)$ via Equation (6) to allow detailed insights into the kind of system under investigation besides the evaluation of $M_{w,app.}$. Afterward, the selection of a model for fitting the radial concentration gradient should lead to realistic results in coincidence with those already derived for $M_{w,app.}$, allowing more detailed insights into the system under consideration (*e.g.* concentration of different components in a mixture, equilibrium constants of interacting systems, *etc.*). The $M*$ function has proven to be of special use for polydisperse systems in a model study[87] so that it seems to be well suited for the analysis of colloid sedimentation equilibrium data, although its application for colloids is not yet reported in the literature.

The second model independent approach – the Ω function[88,89] or in a later derived form ψ function[90] – is particularly well suited for interacting systems. The Ω function is a transformed type of the concentration distribution at sedimentation equilibrium:

$$\Omega_i(r) = \frac{c(r)}{c(r_{ref.})} \exp\left[\frac{(1 - \bar{v}_i\rho)\omega^2}{2RT} M_i(r_{ref.}^2 - r^2)\right] = \frac{a_i(r_{ref.})c(r)}{a_i(r)c(r_{ref.})} \quad (14)$$

where M_i is the molar mass of the smallest sedimenting species i (1 at the beginning of the evaluation). M_i usually has to be determined in a separate experiment using routines like MSTAR.[87] This can be a severe problem in colloid analysis since often, the primary species are not stable for long enough on the timescale of an AUC experiment. a_i is the thermodynamic activity of the smallest sedimenting species i equal to the concentration in the ideal case. The detailed analysis procedure of further unraveling the detected concentration gradient into those for the individual components can be found in the literature.[88,89]

From the concentration gradients of the individual species, one can calculate the equilibrium constants in case of interacting systems. The elegance of this model independent approach is that it avoids any differentiation/integration, the former being especially sensitive to the amplification of experimental noise. Furthermore, it allows for the calculation of activity distributions, which allows a rigorous thermodynamic treatment of non-ideal systems. However, one difficulty is still the curvature in the $\Omega(r)$ vs. $c(r)$ plots especially for low $c(r)$, which are important for the extrapolation to infinite dilution to give the correct intercept. Therefore, the Ω function was modified, giving the ψ function[90] which shows a much better linearity when plotted as $c(r)/\psi_i(r)$ (y-axis) vs. ψ_{i+1}/ψ_i on the x-axis and thus allows a safer determination of the intercept and $c_i(r_{\text{ref.}})$, where $\psi_i(r)$ is defined as

$$\psi_i(r) = \exp\left[\frac{(1 - \bar{v}_i \rho)\omega^2}{2RT} M_i(r^2 - r_{\text{ref.}}^2)\right] \tag{15}$$

Then, the calculation of the concentration distribution of i can be performed using $c_i(r) = c_i(r_{\text{ref.}})\psi_i(r)$, and the analysis is repeated for the next smaller sedimenting species. This evaluation was implemented into a computer program.[91] Such an analysis is especially useful if nothing is known about the kind of interaction taking place in a complicated mixture (homogeneous or heterogeneous interaction), and when several components are involved.[92] If the interaction is traced to be homogeneous, the total concentration gradient $c(r)$ can be fitted to simple polynomials of $c_1(r)$ of the form:

$$c(r) + c_1(r) + K_2[c_1(r)]^2 + K_3[c_1(r)]^3 + \cdots$$

or

$$c(r) = c_1(r) + K_n[c_1(r)]^n \text{ for monomer} \rightarrow n\text{-mer} \tag{16}$$

This allows the analysis of complicated self-associations with a general polynomial fit or of monomer $\rightarrow$ n-mer interactions using the second part of Equation (16).

The Ω resp. ψ functions have several other useful features. They are a sensitive diagnostic test for the discrimination between polydispersity and self-association. If $\Omega(r)$ is plotted for different cell-loading concentrations but a common $c(r_{\text{ref.}})$, the plots will superimpose in the case of self-associating systems, whereas they do not in the case of a polydisperse or impure system. This test is more sensitive than the commonly applied plot of $M_{\text{w,app.}}(r)$ vs. $c(r)$ for different cell-loading concentrations.[93] However, the Ω resp. ψ functions have not yet been applied to colloidal systems, although they are of potential use. This appears to be a reflection of the up-to-now predominantly applied sedimentation velocity technique for colloid analysis in favor of the sophisticated sedimentation equilibrium analyses and is also a reflection of the often very polydisperse colloids.

In summary, the M^* approach seems to be best suited for the analysis of polydisperse colloidal systems, and the Ω resp. ψ functions could be useful for interacting systems, whereas classical approaches or curve fitting are less useful as they cannot cope very well with polydispersity.[87] Therefore, approaches that allow for the determination of molar mass distributions from sedimentation equilibrium experiments are better suited for the analysis of polydisperse colloidal systems.

4.2.4   Molar Mass Distributions from Sedimentation Equilibrium

One method for the evaluation of molar mass distributions from sedimentation equilibrium experiments was suggested by Scholte.[94] He defines a value $U(\lambda,\xi) = - (1/c_0)$ $(dc/d\xi)$, which can be determined from the sedimentation equilibrium concentration gradient for every data point where $\xi = (r^2 - r_m^2)/(r_b^2 - r_m^2)$ is a normalized radial position and $\lambda = (1 - \bar{v}\rho)\omega^2(r_b^2 - r_m^2)/2RT$.

$$U(\lambda,\xi) = \sum_i w_i \underbrace{\frac{\lambda^2 M_i^2 \exp(-\xi\lambda M_i)}{1 - \exp(-M_i)}}_{k_i} \tag{17}$$

Equation (17) relates the experimentally determined $U(\lambda,\xi)$ to molar masses M_i and their corresponding weight fractions w_i. Now, four series of M_i are generated where $M_i/M_{i-1} = 2$ and M_i covers the whole range of suspected molar masses contained in the distribution. Variation of w_i in Equation (17), $U(\lambda,\xi)_j = \sum_i w_i k_{i,j} + \rho_j$, is carried out until a set of w_i is found that has minimized $\sum_j |\rho_j|$. From the M_i/w_i data pairs, the molar mass distribution can be plotted. This calculation-intensive approach should be highly supported by modern computer facilities.

Another approach for the determination of the molar mass distribution relies on the knowledge of the type of molar mass distribution itself (Poisson, Schulz–Flory, etc.).[73,95,96] For polymers, this is a useful approach, as the type of molar mass distribution is often known from the type of the applied polymerization. For example, a Schulz–Flory distribution is expected for polymers from radical polymerization whereas a polymer from anionic polymerization should give a Poisson distribution. Once the model of the molar mass distribution is defined, the radial equilibrium concentration profiles for several initial concentrations are used and converted into the so-called reduced concentration profile $U_w(r)$. $U_w(r)$ can then be converted into the multimodal molar mass distribution function $W(M)_i$, which is given by the selected model distribution using a weighing factor. This approach can only be applied for a monomodal distribution. Monomodality can be checked by a sedimentation velocity experiment. This is a severe restriction for colloidal systems, which are often multicomponent systems.

It has also been reported that the exponential concentration gradient in sedimentation equilibrium can be fitted to sums of exponentials for ideal monodisperse components.[97] From the concentrations of these components and their molar mass, the molar mass distribution can be constructed. Although such an approach is potentially liable to yield wrong results due to too many unknown variables, it was reported that M_w and M_z determined from the distribution were reliable. Nevertheless, a highly resolved molar mass distribution cannot be expected from such an approach.

4.2.5   Sedimentation Equilibrium of Gels

The analytical ultracentrifuge has proved to be very useful for the characterization of thermodynamic and elastic properties of gels and microgels, as exhaustively reviewed in refs. 98 and 99. Whereas sedimentation velocity techniques are useful for the characterization of microgels (see Section 5.4), they are not useful for the characterization of bulk gel systems, although speaking, they were applied first, starting with the first experiments by Svedberg and later on by Johnson.[2,100–105] Nevertheless, the difficulties of interpreting nonlinear ln r *vs.* $\omega^2 t$ plots [see Figure 6(b)] with steadily decreasing slopes resulting in a horizontal equilibrium position showed that a sedimentation coefficient of a gel is not defined as for the classical polymer or particle case. In fact, a meaningful definition of the sedimentation coefficient of a gel was only published decades later as the movement of the center of mass and not the movement of the gel phase boundary is considered in analogy to the solution case before. Nowadays, even the Lamm equation can be generalized for gels so that sedimentation velocity experiments with gels can be carried out.[106] This enables the determination of sedimentation and diffusion coefficients of a gel.[107]

If bulk gels are investigated in an ultracentrifuge, the centrifugal field acts as a pressure generator, which leads to a radially dependent deswelling of the gel.[108] Thus, the continuous dependence of the swelling pressure Π_S of a gel on the radial polymer concentration gradient (here expressed as local gel density $\rho_2(r)$) inside the gel phase at sedimentation equilibrium can be determined in only one experiment using

$$\Pi_S = \omega^2 \int_{r_m^{g/s}}^{r} \left(\rho_2(r) - \frac{1}{\overline{V}_{01}} \right) r \, dr \tag{18}$$

where $\overline{V}_{01}$ is the specific volume of the solvent and $r_m^{g/s}$ the position of the meniscus gel/sol. Classical techniques for the determination of swelling pressures only allow the measurement of single swelling pressure/concentration data points.

If a bulk gel is placed in an ultracentrifugal field in the sector-shaped ultracentrifuge cell, two cases can be distinguished, which are schematically presented in Figure 8. The first case (a) is the beginning of the experiment or an experiment at low rotational speed where no sedimentation of the macroscopic gel phase occurs (as sedimentation of the gel phase, the sedimentation of the gel meniscus is understood). Nevertheless, a concentration gradient of the polymer in the gel phase will occur at these lower speeds owing to the sedimentation of the cross-linked polymer. The gradient indicates the locally dependent deswelling of the gel, which is caused by the swelling pressure generated by the centrifugal field. The concentration gradient changes until a final equilibrium gradient is established.

The second case (b) is observed at higher rotational speeds, *i.e.* above 10 000 rpm for the system gelatine/water. Again, the polymer concentration is increased at the cell bottom, whereas it is decreased at the meniscus gel/vapor. These processes are illustrated in Figure 9.

At the beginning of the experiment ($\omega=0$), the polymer concentration in the gel is constant. It is increased at ω_1, which leads to a decrease of c_2 at the meniscus gel/vapor. At a critical angular velocity ω_2, the polymer concentration has dropped to the value of the maximum swollen gel $c_{2,s}$. As the polymer concentration in the gel cannot be lower than $\rho_{2,s}$, a sol phase is introduced as soon as the polymer concentration has reached this lower limit, and the meniscus gel/sol begins to sediment (ω_3). This corresponds to case (b) in Figure 8. The sol phase might consist of pure solvent as well as a solution of non-gelling material. The gel phase sediments in certain cases until an equilibrium is reached. From the equilibrium states of both case (a) and case (b) in Figure 8, information about thermodynamic, elastic, structural, and molecular parameters of the gel can be obtained.

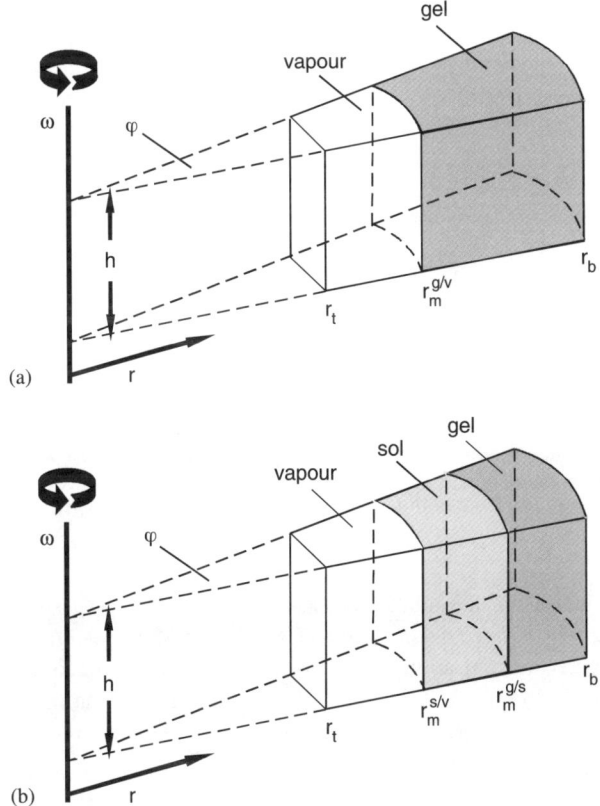

Figure 8 *Gel in an ultracentrifuge cell. (a) At the beginning of the experiment or at low speeds, where no sedimentation of the gel phase occurs. (b) At high speeds, where the gel phase has sedimented establishing a sol phase. ω is the angular velocity, h the height of the ultracentrifuge cell, φ the sector angle of the ultracentrifuge cell and r the distance to the axis of rotation with the indices t=top, m=meniscus, b=bottom, g/v=boundary gel/vapor, s/v=boundary solvent/vapor and g/s= gel/solvent. Redrawn from ref. 206 with kind permission of Dr. Köster, Berlin, FRG*

The concentration dependence of the swelling pressure can be related to changes in the chemical potential of the solvent in the gel. It can be used for the determination of thermodynamic and elastic properties of gels by applying a modified Flory–Huggins equation with a concentration-dependent interaction parameter and an elastic term,[109] which semi-empirically relates the swelling pressure of the gel at a known concentration to the molecular parameters of the gel:

$$\frac{\Pi_S V_1}{RT} = \underbrace{\ln(1-w_2) + w_2 + \chi_{w,0}\, w_2^2 + \chi_{w,1}\, w_2^3}_{\text{Mixing Term}} + \underbrace{C_w w_2^{1/3}}_{\text{Network Term}} \qquad (19)$$

The important point about Equation (19) is that a whole swelling pressure–concentration curve can be determined in a single sedimentation equilibrium experiment

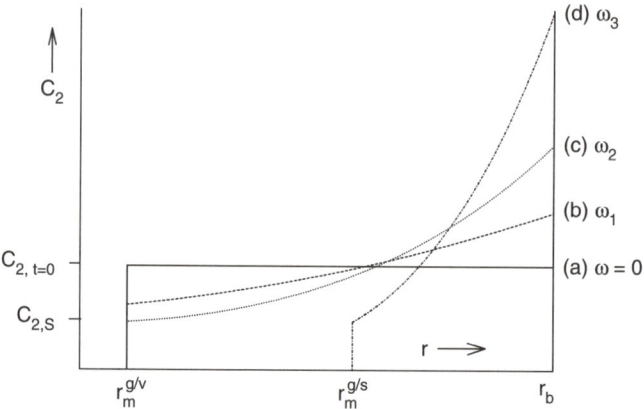

Figure 9 *Radial dependence of the local polymer concentration in the gel c_2 at different angular velocities ω_i. $c_{2,s}$=concentration of the maximum swollen gel. Redrawn from ref. 206 with kind permission of Dr. Köster, Berlin, FRG*

with a gel in contrast to the multiple time-consuming static experiments at a single concentration, which were necessary to determine these curves in the past. Typical swelling pressure–concentration curves are shown in Figure 10 for the system gelatin/water. The good reproducibility of identical experiments can be seen, and it can also be observed that experiments for various speeds yield a common master curve, clearly proving the swelling pressure equilibrium.

The change in the chemical potential could even be expressed for an n-component system:[206]

$$[\Delta\tilde{\mu}_i(c_1,c_2,c_3,\ldots,c_{N-1})]_{T,P} = \omega^2\int_{r=r_m^{g/s}}^{r}(1-\tilde{V}_i\rho)r\,dr; \quad i = 1, 2, 3, \ldots, N \quad (20)$$

Also, extensions were made for polyelectrolyte gels in terms of an additional ionic term, which has to be added to the right-hand side of Equation (19):[110]

$$\Delta\mu_{1,\text{Ion}} = 2V_{01}\left[c_s - \sqrt{\left\{c_s^2 + \left(\frac{\rho\phi_2}{2M_2}\right)^2\right\}}\right] \quad (21)$$

where ϕ=volume fraction, with index 1=solvent, 2=polymer, c_s=salt concentration, ρ=density of the dry polymer and M_2=molar mass of the polymer per free monovalent counter ion. It must be noted that the volume fractions have to be converted to the mass fraction concentration unit in Equation (19). Therefore, sedimentation equilibrium experiments can now be applied to multicomponent polyelectrolyte gels so that all kinds of gel systems become accessible.

However, despite the advantages outlined above, sedimentation equilibrium experiments with gels have the disadvantage of a very long experimental duration

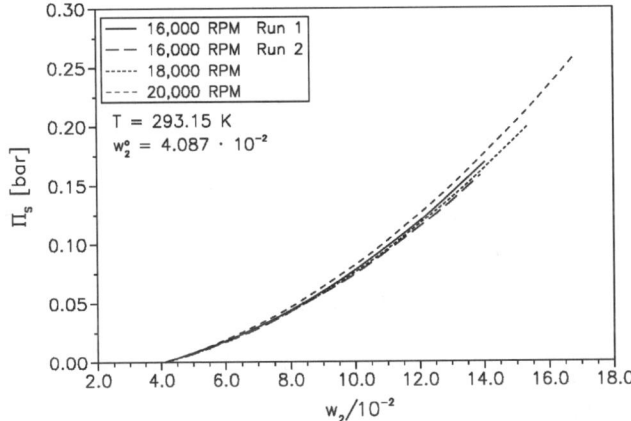

Figure 10 *Swelling pressure Π_S of the system gelatin/water vs. concentration of gelatin for different rotational speeds. w_2 is the weight fraction of the polymer. Reproduced from ref. 109 with kind permission of Springer-Verlag, Berlin*

lasting up to weeks or even months if the AUC cell is filled with the usual 1 cm column. Additionally, adhesion to the cell walls and ultracentrifuge windows is a problem. Therefore, short column techniques are especially interesting for gels as well as multisample measurements in a single experiment.

It could be shown that it is even possible to investigate up to 70 samples simultaneously – *e.g.* to characterize a complete system – by thermodynamic and elastic parameters in a single experiment.[206] Since the sedimentation of a bulk gel phase caused several problems (polymer concentration detection, possible adhesion), the sedimentation equilibria of gels are now investigated at much lower speeds where deswelling of the gel occurs without sedimentation of the whole gel phase.[111] Also, gelation itself can be monitored via the turbidity changes upon network formation.[112]

4.2.6 Sedimentation Equilibrium of Emulsions

Sedimentation equilibrium is a fast and convenient technique to test emulsion stability. Under the action of a centrifugal field, an emulsion will demix and form not only an oil and water layer, but also a transparent and non-transparent emulsion layer (see Figure 11; top). The equilibrium extension of these layers is correlated to the centrifugal force and the emulsion stability (see Figure 11; bottom).

From Figure 11 (bottom), it can be seen that the centrifugation of emulsions leads to constant values of the layer thickness but equilibrium was not yet proven. Nevertheless, it is possible to calculate a coalescence pressure $P_{\text{coal.}}$, which is a measure of the emulsion stability.

$$P_{\text{coal.}} = 0.5\omega^2 \Delta\rho\phi_2(r_1^2 - r_2^2) \tag{22}$$

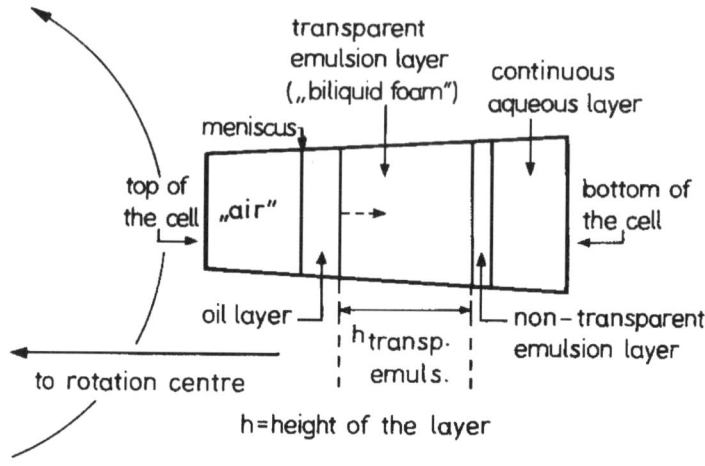

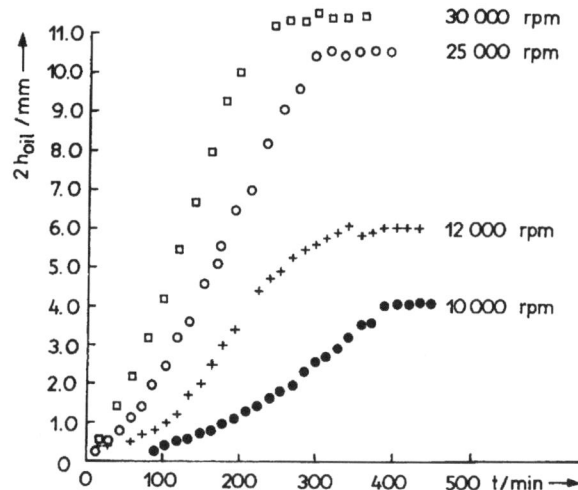

Figure 11   *Top: Layers formed during demulsification of n-decane-in-water emulsions stabilized by 0.1 mol L^{-1} aqueous sodium dodecylsulfate (SDS) solution in an AUC.
Bottom: Analytical Ultracentrifugation of n-decane-in-water emulsions stabilized by 0.1 mol L^{-1} aqueous SDS solution. Dependence of the height of separated oil layers on the time of centrifugation at various rotor speeds. Taken from
ref. 207 with permission of Springer-Verlag*

where $\Delta\rho$=density difference between dispersed and continuous phase, ϕ_2=volume
fraction of the dispersed phase in the emulsion layers, and r_1, r_2=distances from the
lower (1) and upper (2) boundary of the emulsion layers to the rotation center. It has
to be noted that Equation (22) has the character of an approximation, as the particle
size of the emulsion droplets is of influence but not yet considered, which may also

be a reason for the observed speed dependence of coalescence pressures. Therefore, a deeper understanding of the processes taking place during the ultracentrifugation of emulsions is required to develop an appropriate theoretical treatment of coalescence pressures. With this in mind, ultracentrifugation of emulsions can only be used for qualitative assessment of emulsion stability at the moment.

4.2.7 pH Gradients

pH gradients are a new and exciting sedimentation equilibrium experiment method. They rely on the sedimentation of a dense anion/polyanion so that a pH gradient is established by electrostatic attraction of protons.[113] These pH gradients exhibit high Donnan potential, which can effectively be suppressed by salt addition. pH gradients can span a notable pH range, and in the so-far investigated examples, a range of almost 3 units could be established in an analytical cell (Figure 12; top). This is a remarkable range; and if one considers the large amount of pH-dependent reactions, such as crystallization, polymer complex formation, *etc.*, pH gradients could be a versatile new tool for the study of pH-dependent sample properties with the advantage of a continuous pH gradient. However, if the gradient material itself is a component that can potentially interact with the sample, like poly(styrene sulfonate), care has to be applied for the interpretation of the experimental results.

Nevertheless, the first experiments with pH gradients applied to crystallization reactions were encouraging. For the model system $BaCrO_4/BaCr_2O_7$, which has the advantage that the two pH-dependent species transform reversibly and can be identified by different colors resp. UV–Vis spectra, the dissolution of the sparingly soluble $BaCrO_4$ could clearly be demonstrated upon pH decrease via its transformation into the more soluble $BaCr_2O_7$.[113] It is an advantage that pH gradients are true sedimentation equilibria proven by the path independence of reaching the equilibrium (Figure 12). Such equilibrium experiments can be a valuable tool to study crystallization reactions, as early species could be traceable in an equilibrium situation between dissolution and recrystallization, where the smallest just-metastable species is called the critical crystal nucleus. This is still a matter of debate even after decades of intense crystallization research, mainly caused by the extreme experimental difficulties in observing the smallest crystallization precursor particles directly in solution.

It is remarkable, that the sudden decrease of the nanoparticle size ending in dissolution could be effectively demonstrated for the system CdS with quantum size effect dimensions, where the decrease of the particle size is reflected in a blue shift of the so-called band gap, which is the onset of UV–Vis absorption (Figure 12; bottom).[113,114] In Figure 12 (bottom), a typical profile of a dissolving crystal in a pH gradient is demonstrated. Up to a certain point, the concentration profile is similar to the normal exponential concentration profile in sedimentation equilibrium experiments. However, deviations soon become obvious, coupled with a particle size decrease as evidenced by UV–Vis spectra, until a sharp decrease in the particle concentration can be observed, hinting at the dissolution of the crystal (Figure 12; bottom). These experiments can also be performed in a dynamic fashion with the same qualitative results.[113] Another application of pH gradients was reported as an

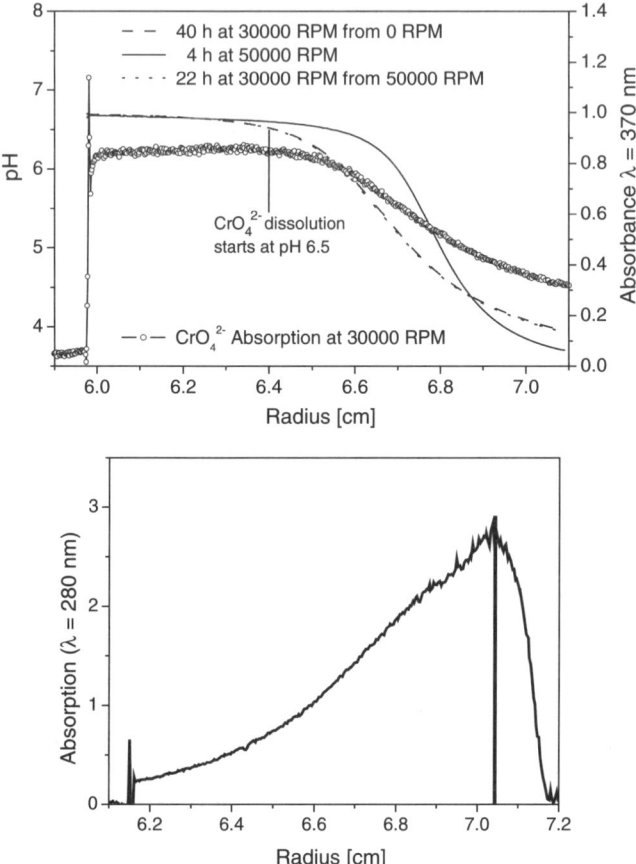

Figure 12 *Top: proof of Poly(styrene sulfonate) pH gradient, $M_w = 140\ 000\ g\ mol^{-1}$, pH gradient equilibrium and equilibrium concentration gradient of dissolving $BaCrO_4$. The pH was determined via the PSS concentration. Bottom: CdS in a HI pH gradient detected at 280 nm and 60 000 rpm. The vertical line indicates the point of CdS nanoparticle dissolution. Reproduced from ref. 113 with kind permission of Springer-Verlag*

attempt toward the determination of the charge distribution of particles upon their sedimentation along the pH gradient.[115]

Neither is any theory yet available for any of these experiments, nor is any quantitative evaluation possible. Nevertheless, these experiments indicate that even 70 years after the invention of AUC, new methods can be developed.

4.3   Density Gradient Experiment

The second principal possibility of separation in an analytical ultracentrifuge despite the separation according to the sample mass is the separation due to the chemical structure expressed in different solute densities in a density gradient. The range of densities that can be covered using density gradients is limited (0.8–2.0 g mL^{-1})[25] but sufficient for the separation of purely polymeric substances. However, inorganic or organic–inorganic hybrid colloids have a density that is in most cases too high for the successful application of a density gradient. Also, the high salt concentrations in aqueous density gradients may be a problem for the analysis of electrostatically stabilized colloids. Furthermore, these so-called static density gradient experiments take a long time, usually in the order of several days (Figure 13), owing to the slow banding of the sample. Also, potential solvent binding of the sample in a density gradient may be a problem. Nevertheless, density gradients are an excellent tool for the investigation of structural differences in mixtures and have proved to be an invaluable tool for the analysis of polymer latexes.[25]

The separation capabilities of a static density gradient are demonstrated in Figure 13 for the separation of a mixture of an acrylonitrile–vinylacetate copolymer in a DMF/CHBr$_3$ density gradient. The sample consists of three fractions, which could be successfully separated: (1) linear polymer (0.5×10^6 g mol^{-1}, $\rho=1.0347$ g mL^{-1}), (2) highly branched polymer (25×10^6 g moL^{-1}, $\rho=1.0365$ g mL^{-1}), and (3) weakly cross-linked polymer (75×10^6 g mol^{-1}, $\rho=1.0371$ g mL^{-1}). If one looks at the

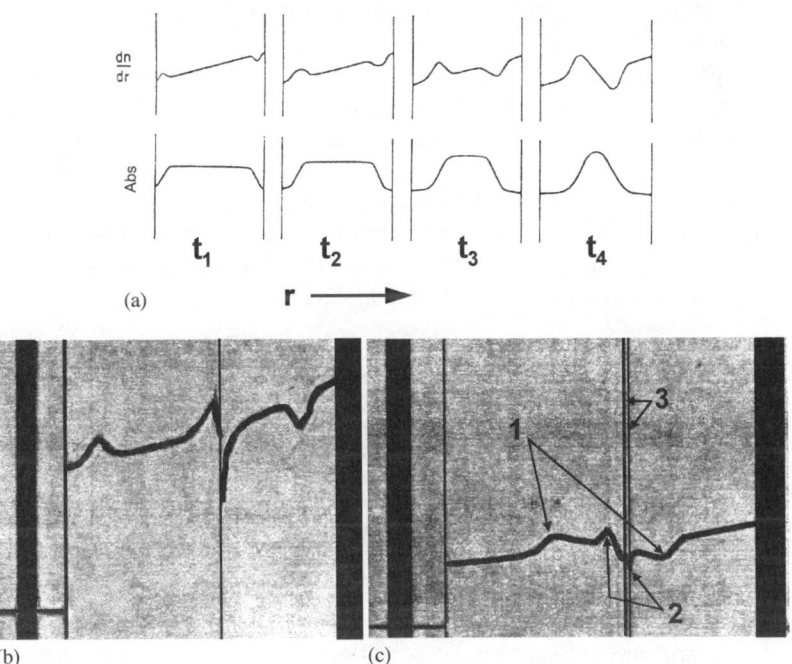

Figure 13 *Static analytical density gradient. (a) Banding of a sample in a density gradient observed with Schlieren optics (upper) or absorption/Rayleigh interference optics (lower) at different times. (b) Acrylonitrile–vinylacetate copolymer in a DMF/CHBr$_3$ (136 g l^{-1}) density gradient after 35 h at 33 450 rpm, fractions 1 and 2 are separated already. (c) After 108 h at 33 450 rpm (equilibrium). Redrawn from ref. 208 with kind permission of John Wiley & Sons*

densities of the separated components, one sees that the resolution of the separation is in the fourth digit of the density.

Much of the theory for the analysis of analytical density gradient experiments was established in the classical papers of Meselson,[116] Hermans,[117,118] and Ende.[119] However, the Hermans–Ende equation was restricted to ideal systems, which can result in serious errors in the calculation of the radial density profile. Therefore it was recently improved for real systems.[120–122] The basic equations that allow the calculation of the density of the gradient at every point in the ultracentrifuge cell can be found in these references. From the width σ of the Gaussian concentration profile of the sample, the molar mass can be calculated.[116] The accuracy is much smaller, however, than that of the molar mass derived by sedimentation equilibrium experiments, so it can realistically only serve as an estimate.

The disadvantage of the prolonged experimental times for static density gradient experiments can be overcome by using so-called dynamic density gradients.[123,124] Here, a layer of H_2O is usually layered upon D_2O in a synthetic boundary experiment (see the corresponding chapter for details of this technique), establishing the fast formation of a dynamic H_2O/D_2O density gradient within a few minutes. Although the density range of this type of gradient is rather limited $(1.0–1.1 \text{ g mL}^{-1})$, it is very well suited for the fast characterization of latexes – especially polystyrene – and can be extended to $0.85–1.25 \text{ g mL}^{-1}$ under certain circumstances.[123]

An example is given in Figure 14 where an 11-component latex mixture with different densities is subjected to an H_2O/D_2O density gradient. As five latexes have a density in the density range between H_2O and D_2O, the gradient selectively separates these components, whereas the other six components are not detected although their density is partly quite close to the solvent. This example illustrates very well how fast information on the structural composition of a mixture can be derived. After only 2 min of centrifugation, 4 of the 5 latexes are separated, after 4 min, all 5 components are separated, and the separation is further improved within

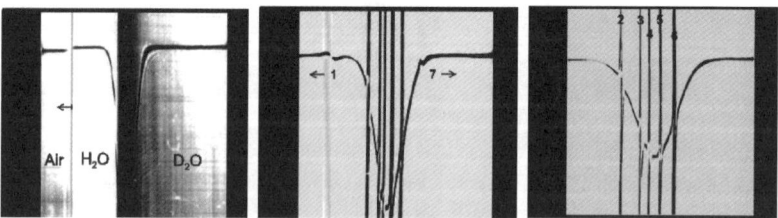

Figure 14   *Formation of a dynamic H_2O/D_2O density gradient with a mixture of 11 different ethyl hexylacrylate/methylacrylate (EHA/MA) copolymer latexes which have been polymerized separately [w[MA] = 0/ 10/ 20/ 30/ 40/ 50/ 60/ 70/ 80/90/ 100 wt%). All particles had approximately the same diameter 200 nm and exhibit the following densities: 0.980/1.000/ 1.021/1.043/ 1.066/1.089/ 1.114/ 1.167/1.196/ 1.225 / 1.140 g mL⁻¹. Just five of these are in the density range between 0.997 (pure H_2O) and 1.095 g mL⁻¹ (pure D_2O). Run conditions: 40 000 rpm, 25 °C. Left = 4 min, center = 6 min and right = 10 min. Reproduced from ref. 20 with permission of Springer-Verlag*

the next 12 min. Therefore, dynamic density gradients offer the fastest way of learning about the structural heterogeneity of mixtures as long as they have densities in the limited range for this type of density gradient. This condition is fulfilled in the case of polymer latexes.

4.4 Synthetic Boundary Experiment

In a synthetic boundary experiment, changes of a boundary between solution and solvent with time are observed at low centrifugal fields where no sedimentation of the sample occurs. Such experiments require special synthetic boundary cells where the solvent is layered upon the solution column under the action of the centrifugal field. A pseudo-synthetic boundary experiment can be constructed from a normal sedimentation velocity experiment if $G(s)$ is obtained by the van Holde–Weischet method and subtracted from the experimental scans at various times.[125] This has the advantage that any layering imperfections, which can occur in a conventional synthetic boundary experiment, are avoided and that furthermore, even samples that sediment at the low speeds of a conventional synthetic boundary experiment can be investigated, which is valid for some colloidal systems. However, compared with dynamic light scattering, a synthetic boundary experiment is more tedious and time-consuming for the determination of the particle diffusion coefficient, so that here, dynamic light scattering is the method of choice. Nevertheless, if aggregates are present in a sample, the light scattering results can be seriously obstructed, whereas the synthetic boundary experiment will be unaffected as the aggregates will sediment even at the low applied speeds. Also, light absorption by the sample can be a problem for light scattering analysis.

Whereas the classical determination of diffusion coefficients by synthetic boundary experiments is not important anymore, performance of physical or chemical reactions in a synthetic boundary cell are attractive in colloid chemistry. These experiments, which are a special derivative of synthetic boundary experiments, are described below (Section 5.5).

5 Analysis of Colloids

There are numerous examples of information derived by one of the four basic experiments or combinations thereof using an ultracentrifuge. They cannot be treated completely here, but the selected applications will show that analytical ultracentrifugation is a universal absolute technique for the characterization of polymers or colloids, especially in mixtures. The given examples have been chosen in a way that the range of information which was obtained by AUC is broad. At the end of the chapter, an overview table that will give the reader a guide to the primary literature for specific systems is provided, although this table is by no means exhaustive.

5.1 Particle Size Distributions

The application of AUC for the determination of particle sizes and their distributions to address problems of colloid analysis was already realized by the pioneers of this technique because sedimentation velocity experiments provide a sensitive fractionation according to particle sizes/molar masses.[3,126–129] Nevertheless, it appears that the potential of this application is still not yet commonly recognized. It is relatively straightforward to convert a sedimentation coefficient distribution, which can be calculated using Equation (2) for every data point (a) r_i (if a radial scan has been acquired at a specified time) or (b) t_i (if concentration detection at a specified radius has been performed in dependence of time), to a particle size distribution. Assuming the validity of Stoke's law (*e.g.* the sample is spherical), the following derivative of the Svedberg equation is obtained:

$$d_i = \sqrt{\frac{18\eta s_i}{\rho_2 - \rho}} \qquad (23)$$

where d_i is the particle diameter corresponding to s_i, ρ_2 the density of the *sedimenting* particle (including solvent/polymer, *etc.* adhering to the sample), and η the solvent viscosity. If the particles are not spherical, only the hydrodynamically equivalent diameter is obtained unless form factors are applied if the axial ratio of the particles is known from other sources such as electron microscopy.

The conversion of sedimentation coefficient distributions to a particle size distribution relies highly on the knowledge of the density of the sedimenting particle. For hybrid particles or very small nanoparticles <5 nm, this issue can be a severe problem, especially in the case of mixtures, as the density of the particles is usually not known. Measurements of the average particle density in the mixture can lead to erroneous results so that in such cases, the correlation of the sedimentation coefficient distribution with a distribution obtained from a density-insensitive method like Fl-FFF (Flow-Field-Flow Fractionation) or dynamic light scattering is meaningful. This can in turn yield the particle density, which can give information about the relative amount of the materials building up the hybrid particle[130] (Section 5.1.4). But even an apparent particle size distribution calculated within the limits of reasonable particle densities can yield very valuable information[131] (Section 5.1.2).

However, in the case of the industrially important latexes, the particle density is usually exactly known from the chemistry of particle formation/polymerization. Thus, the determination of particle size distributions with the analytical ultracentrifuge is a rapid technique providing a high statistical accuracy (*e.g.* every sedimenting particle is detected) in contrast to electron microscopy, which delivers information about the particle shape but often suffers from drying artifacts. The determination of a particle size distribution from microscopy images requires counting of hundreds and thousands of particles. This problem has only partly been diminished by the advent of commercially available picture evaluation algorithms. A recent test of the worldwide operating Bayer group in 17 laboratories dedicated to particle size analysis confirmed the view that TEM and AUC are the best techniques for the determination of particle size distributions[5] as discussed above. A combination of TEM, AUC, and X-ray diffraction techniques can provide a complete insight into a colloidal system.[8] AUC in combination with electron microscopy in its various forms can be considered the most powerful characterization approach for particle size distributions and particle morphologies known to date. The following examples illustrate the fractionation power of the analytical ultracentrifuge for latexes[25,132,133] [see Figure 15(a)] and especially for inorganic colloids.[134]

In the case of latexes, the accessible size range is between 10 and 5000 nm, and the baseline resolution for monodisperse components differs just by 10% in diameter,[135] whereas for inorganic colloids, AUC separates dispersions with an almost atomic resolution. The most striking example is shown in Figure 15(b), where Pt was quenched during the growth process and all species in an eight-component mixture were successfully analyzed with Ångström baseline resolution, the smallest compound being just a cluster of 21 Pt atoms. The shown particle size distribution allows the conclusion that the particles grow via coalescence of primary particles.[134] HRTEM only yielded an overall bimodal distribution with the same sizes as in Figure 15(b) but suffered from the lack of statistical significance, in contrast to AUC where all particles are detected (data not shown).

The determination of particle size distributions is nowadays a routine task – even for very small colloids. The smallest reported species that could be detected was for

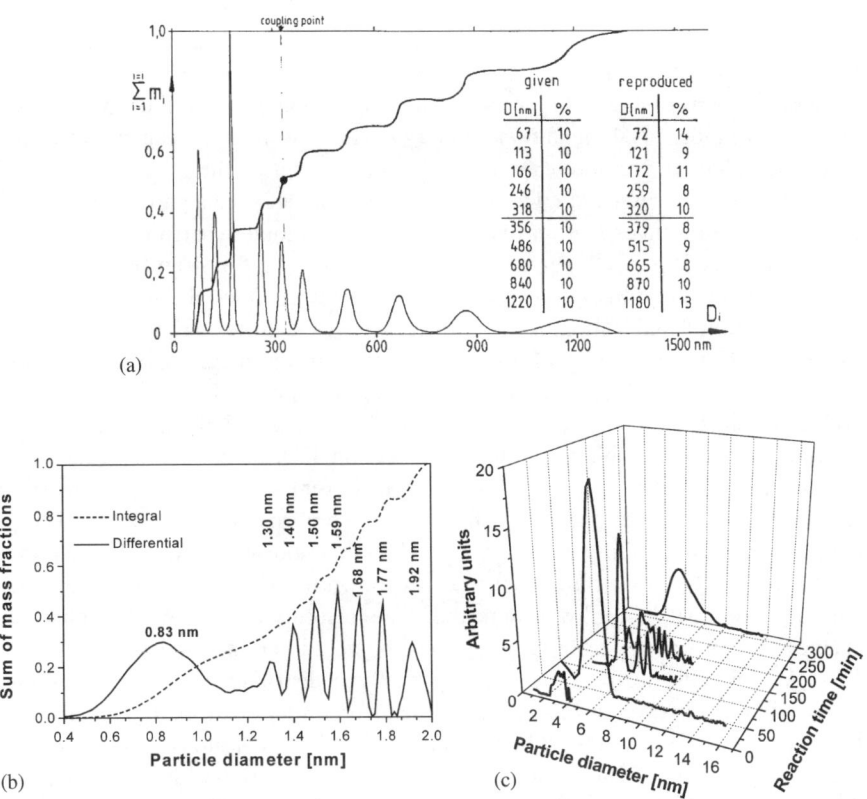

(a)

given		reproduced	
D[nm]	%	D[nm]	%
67	10	72	14
113	10	121	9
166	10	172	11
246	10	259	8
318	10	320	10
356	10	379	8
486	10	515	9
680	10	665	8
840	10	870	10
1220	10	1180	13

(b)

(c)

Figure 15 *Examples for particle size distributions determined with the analytical ultracentrifuge. (a) 10 component mixture of polystyrene standard latexes corrected for MIE scattering,[25] (b) Pt colloid during particle growth[134] and (c) Particle growth of ZnO[209] Figure 15(a) reproduced from ref. 11 with kind permission of the Royal Society of Chemistry, (b) and (c) from refs. 134 and 209 with kind permission of Springer-Verlag*

a ZrO_2 precursor and just 0.4 nm in diameter[131] (see also Section 5.1.2), which shows the potential of the analytical ultracentrifuge for the analysis of smallest nanoparticles or sub-critical complexes. This means that particle size distributions derived by ultracentrifugation can well be used to investigate particle growth mechanisms[131,134] from its smallest species, especially if the growth is slow enough to detect the time-dependent particle growth as in the example of ZnO [see Figure 15(c)].[134,209] However, this is a rare case, and if the samples cannot be successfully quenched as shown in Figures 15(b) and (c), the ultracentrifuge is usually too slow to pick up kinetic information.

5.1.1 High Resolution Particle Size Distributions

It is interesting to note that in the very first reported AUC experiments on colloidal systems, particles of only a few nm size could be successfully analyzed. The classical Svedberg and Rinde work from 1924 characterized gold colloids with a size of only 1.5 nm,[3] suggesting that AUC is a precise technique for the determination of particle size distributions.[136–139] If it is taken into account that the optical detection systems of those days were by far not as sophisticated as today's, this is most remarkable. Nevertheless, it is not astonishing that modern detection systems combined with modern sedimentation velocity evaluation methods can reveal details that were hidden in previous examinations. One example is the Au_{55} cluster, which is of interest as it is considered to be at the transition between a particle and a molecule. The initial investigation by AUC revealed a monodisperse population.[7] In the method used, the sedimentation of the clusters was followed by a Schlieren optical system (gray/black transition on a photographic record due to the deeply colored sample) and a homogeneous sample was deduced from the boundary movement. Concentration profiles, which represent the integral particle size distribution, were not given even though the authors noted that there was a pronounced transition area. The determination of a precise sedimentation coefficient distribution is not a problem anymore (see Section 4.1), as radially dependent concentration values can be precisely determined from the UV–Vis absorption or Rayleigh interference optical system of modern analytical ultracentrifuges. Consequently, the initially assumed defined and monodisperse Au_{55} cluster turned out to be a mixture of different species,[8] as shown in Figure 3. At least a bimodal particle size distribution was detected. The resolution between the observed species was in the Ångström range. It must be noted that the modern optics of an XL-I AUC is able to trace such highly resolved distributions [see also Figures 15(b), (c) and 17] even in routine experiments. For very small particles, diffusion becomes significant, so it must be corrected. Therefore, it has to be maintained that either the particle diffusion can be suppressed by application of high speeds,[134] or that the diffusion boundary broadening effect can be calculated out of the distribution by extrapolating to infinite time,[37] or by applying regularization methods for the determination of diffusion coefficients to correct for boundary broadening.[60] Another method was suggested, which is based on the calculation of the diffusion coefficient from the particle size.[9] From this, the diffusion broadening of the sedimenting boundary can be calculated and subtracted from the measured concentration distribution, giving the diffusion-corrected particle size distribution. This procedure works best for monomodal and relatively narrow distributions, whereas it is less accurate for broad and multimodal distributions.

As with the exception of the van Holde–Weischet method,[37] the diffusion correction methods have been reported only quite recently, it is not surprising that high-resolution particle size distributions have also been reported rather recently, despite the long time for which particles have been investigated by AUC. The other reasons are the sophisticated modern detection systems, which allow the determination of high-resolution particle size distributions even without diffusion correction (Figure 3 Section 5.1.2). One of the first examples for such a high-resolution particle size distribution is that in Figure 15(b).[115]

This figure underlines two important points: (a) AUC can detect particle size distributions with almost atomar resolution for particles consisting only of a few atoms/molecules; and (b) The resolution is extremely high with only 0.1 nm baseline resolution between the detected species. With the modern machines, the experiments for high-resolution particle size distributions are no problem but the evaluation of the results can be complicated, as in almost every case, the boundary broadening by polydispersity has to be deconvoluted from that of diffusion.

5.1.2 A Practical Example: ZrO_2 Catalyst Film Deposition

To demonstrate the role AUC plays in the solution of colloid chemistry problems, a few examples will be treated in the following sections, which rely on the information from AUC. They have been selected from the routine service work in our laboratory. The first example is thin films from ZrO_2.

ZrO_2 has important applications as a strongly acidic catalyst. Therefore, it is desirable to design coatings with a high surface area, which means homogeneous coatings of very small nanoparticles. One way of achieving this is the application of self-assembled monolayers, which on the one hand homogeneously assemble on a substrate, and on the other hand provide sticking functional groups for ZrO_2. When ZrO_2 nanoparticles were prepared via a hydrolysis reaction of $Zr(SO_4)_2$ at 70 °C, the resulting film showed cracks; furthermore, the solution particle growth was not controlled, so particles in the μm range formed by aggregation and led to an inhomogeneous film, as shown in Figure 16.[131]

AUC, which was performed on samples where the reaction was quenched by cooling the reaction mixture with liquid nitrogen and subsequent re-thawing, revealed a fast nanoparticle growth at the early growth stages (Figure 17).

It is remarkable how much the particle size can change during minutes (Figure 17). This indicates fast aggregation processes caused by insufficient particle stabilization. It is evident that four different early growth stages can be observed by multiple Gauss fits to the sedimentation coefficient distributions [Figure 17(a)]:

Species 1: $r_H = 0.30 \{0.43\}$ nm
Species 2: $r_H = 0.55 \{0.80\} - 0.60 \{0.87\}$ nm
Species 3: $r_H = 1.04 \{1.50\} - 1.10 \{1.61\}$ nm
Species 4: $r_H = 1.58 \{2.29\}$ nm

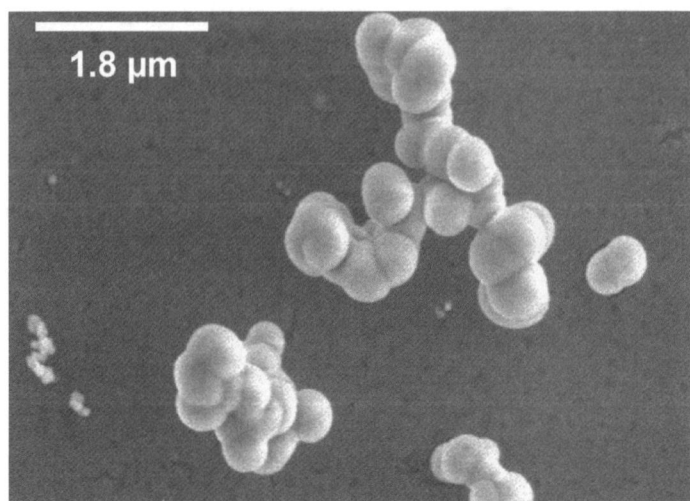

Figure 16   *ZrO$_2$ film deposited on a self-assembled monolayer via Zr(SO$_4$)$_2$ hydrolysis at 70 °C. The cracks and large particles are clearly visible. Reproduced from ref. 131 with permission of the American Chemical Society*

However, their exact particle size cannot be determined according to Equation (23), as the particle density is unknown, ranging between 3.22 g mL^{-1} for Zr(SO$_4$)$_2$ (particle sizes in brackets) and 5.66 g mL^{-1} for ZrO$_2$. As numerous complexes can be formed in the hydrolysis reaction before the first crystalline particle is formed, only a density range can be used as limit for the particle size calculation. This is a limitation, which often occurs in colloid analysis with AUC, especially for hybrid colloids and very small particles. For the very small particles, the bulk density cannot be used anymore owing to the dominance of the particle surface with its solvation effects. On the other hand, hybrid colloids consist of several components so that they can even exhibit a density distribution (see Section 5.1.4). Nevertheless, the particle size range is often rather small, as in the above example, so comparisons with literature-reported sizes of precursor species can be made that provide an insight into the reaction mechanism. In the above example, it can be seen that the resolution of AUC is again very high and in the Ångström range. The smallest resolved species is just 0.3–0.4 nm in diameter, which corresponds to the literature-described [Zr$_4$(OH)$_8$(H$_2$O)$_{16}$]$^{8+}$ complex with a radius of gyration found to be between 3.8 and 5 Å by small-angle X-ray scattering (SAXS).[140–142] This tetrameric complex further oligomerizes to the octameric species [Zr$_8$(OH)$_{20}$(H$_2$O)$_{24}$Cl$_{12}$] with a literature-reported gyration radius of 6 Å,[140] which is in very good agreement with species 2 in Figure 17. The octamer then oligomerizes to higher oligomers (species 3), which then precipitate (species 4).

This shows that particle size distributions determined by AUC show almost atomic resolution even if the densities of the inorganic species are only moderate. This is important to point out as it shows that AUC is very well able to investigate complex particle growth mechanisms with highest particle size resolution in solution, which is not possible by any other known analytical solution technique to date. The drawback of AUC is the low time resolution, so the growth reaction has to be quenched for a successful AUC characterization. Nevertheless, alternative techniques like the

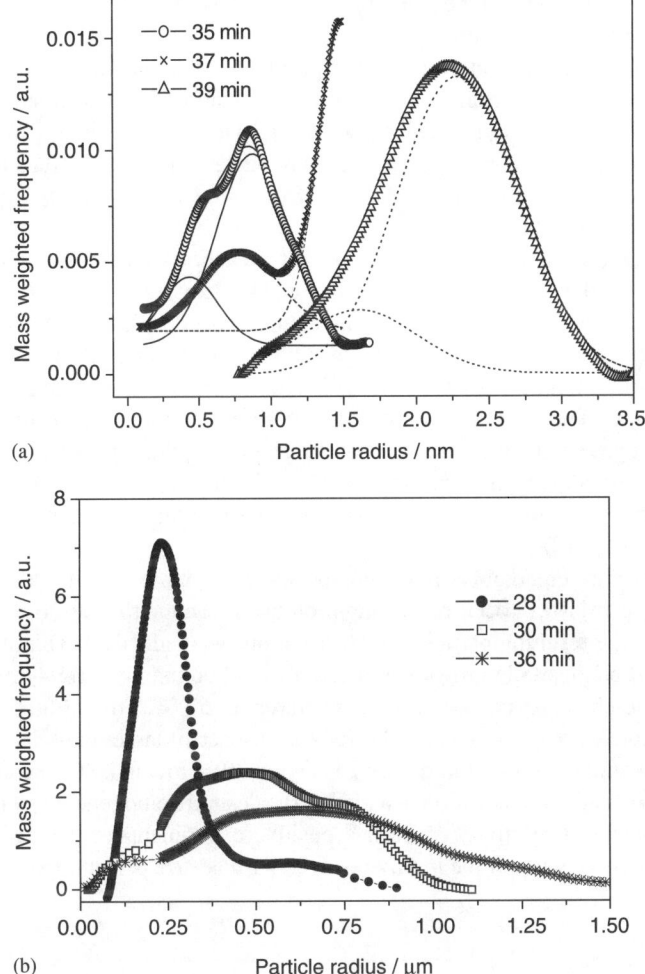

Figure 17 *Particle size distributions of ZrO₂ colloids at 70 °C at different reaction times. Particle size calculated with ρ=3.22 g cm⁻³. (a) after quenching in liquid nitrogen and re-thawing and (b) evolution of particle size distribution with time, representative of the transition from clear to cloudy medium. The distributions were not corrected for diffusion broadening. Please note that the different reaction times in (a) and (b) are due to the different heating up rates in these experiments. Reproduced from ref. 131 with permission of the American Chemical Society*

experimentally extremely demanding SAXS using synchrotron radiation can on the one hand give a very high time resolution in the observation of the particle growth processes well into the ms range,[143] but lacks the fractionation capability which can be problematic, as in the above case of multiple growth species where only average values can be determined.

After the growth process on the nm scale was elucidated, further growth was followed up to the μm scale (Figure 17). Here, it could be shown that the particles grow/aggregate very fast to the μm range within a few minutes, and the particle size distribution gets extremely broad.

However, AUC alone cannot solve the problem of the elucidation of an experimental procedure for controlled particle growth owing to the lack of sufficient time resolution. Alternatively, dynamic light scattering (DLS) was applied, which shows a time resolution in the second range but could not detect the very small nanoparticles. However, the particle growth could be well observed under various experimental conditions, as shown in Figure 18.

From the DLS curves, it became obvious that a reaction temperature 'inhibition' period was observed where no particle growth could be detected as the particles were too small for DLS detection as well as a temperature-dependent linear growth rate. This experiment revealed fast growth resp. aggregation rates of 34 nm s^{-1} at 80 °C reaction temperature, 11 nm min^{-1} at 70 °C, 0.7 nm/s at 60 °C and no detectable growth/aggregation at 50 °C. Thus, 50 °C is either suitable to produce very small nanoparticles, which do not grow but which are also not detectable by DLS, or no particles are produced at all. To address this question, AUC was performed on the reaction mixture with 50 °C reaction temperature. The result clearly shows that *ca.* 1.6 nm big particles are grown (Figure 19).

This species is remarkably similar to species 4 in Figure 17. In addition, it was checked whether growth occurred during prolonged reaction times of 6, 12, and 24 h by AUC, and the resulting particle size distributions were identical. This showed that the undesired nanoparticle growth/aggregation could be inhibited right after the particle nucleation by applying a reaction temperature of 50 °C. Thin films deposited on a self-assembled monolayer clearly showed the absence of larger particles and cracks such as those initially present in Figure 16 (Figure 19), proving that the control of the solution-based particle nucleation can lead to the desired homogeneous, thin, catalyst film with enhanced reactivity.[131] Such a positive experimental outcome would not have been possible without the highly resolved particle size distributions from AUC.

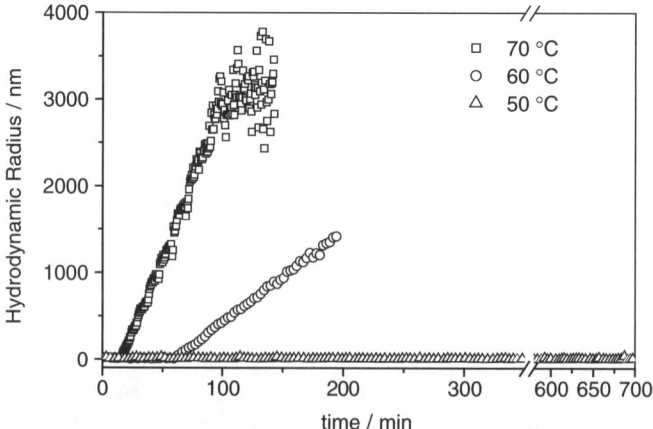

Figure 18   *Time resolved DLS of growing ZrO$_2$ colloids under varying reaction conditions. Reproduced from ref. 131 with permission of the American Chemical Society*

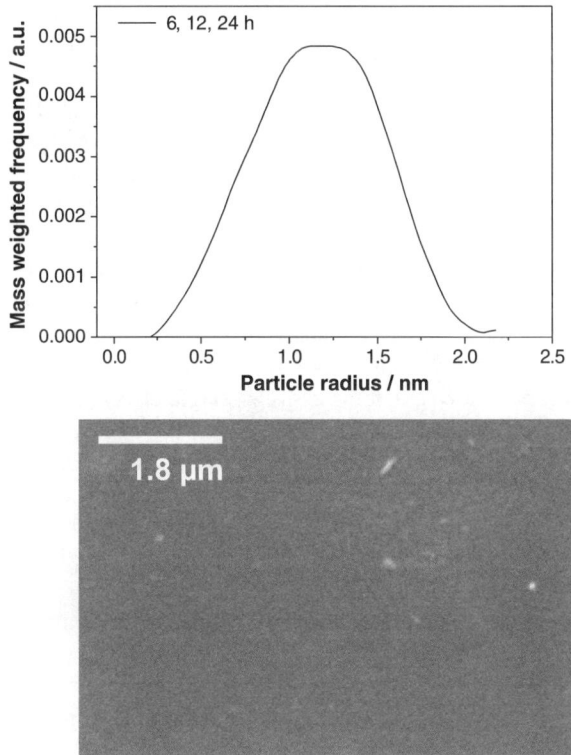

Figure 19 *Top: AUC-derived ZrO$_2$ particle size distribution obtained for the reaction temperature 50 °C. Bottom: ZrO$_2$ thin film deposited on a self-assembled monolayer at 50 °C, 24 h deposition time. Reproduced from ref. 131 with permission of the American Chemical Society*

5.1.3 Aggregating Colloids

A common feature of colloidal systems is their inherent tendency to aggregate unless sufficient particle stabilization is provided. In fact, a major application of AUC in colloid chemistry concerns the question of whether a nanoparticle sample is defined in particle size or aggregating. To demonstrate that AUC is very well suited for the investigation of such systems by comparison with other commonly applied techniques for particle size measurement such as TEM or light scattering, we investigated a model system for a polydisperse aggregating system. Here, a mixture of two monodisperse latexes (127 and 221 nm) was set up with 92 wt% of the small lattices.[144]

The latexes were coated with antibodies for C-reactive protein (CRP): the smaller particles were coated with an antibody of low reactivity and the bigger ones with a highly reactive antibody. Such coated latex mixtures are applied as turbidimetric assays with an enhanced dynamic range for the detection of antigen concentrations via simple turbidity measurements.[145] (see Figure 20).

To achieve a linear turbidity response with increasing antigen concentrations, the large latexes with the highly reactive antibody have to aggregate first at low antigen concentrations, giving an enhanced sensitivity in the low antigen concentration range owing to the higher turbidity of the larger particles. At increasing antigen concentrations, the smaller particles then have to aggregate in addition to detect high antigen concentrations. The measurement problem was to detect in a quantitative manner, if the above concept works, so that an aggregating polydisperse system was characterized with respect to its particle size distribution. TEM investigations indicated that at low CRP concentrations, the larger latexes aggregate first and that at higher CRP concentrations, the smaller ones follow, but besides the visual information on the aggregate morphologies, no quantitative information could be derived from this investigation (see Figure 21). In addition, the samples had to be dried for the TEM measurement so that drying artifacts cannot be excluded.

Therefore, static light scattering (SLS) was applied as a solution technique with a rather good time resolution so that the aggregation kinetics could be monitored. The results are shown in Figure 22. It can be seen that the aggregation kinetics can be effectively followed, and indicate that the larger particles indeed aggregate first. However, the resolution of the particle size distribution is low as can be seen for the distributions at 0 min where 92 wt% 127 nm latexes should be detected next to 8 wt% 221 nm latexes (Figure 22). This is a consequence of the lack of fractionation in the SLS measurement.

Highly resolved particle size distributions could be obtained by AUC where the 92 wt% of the 127 nm latexes were effectively resolved after the particle size distributions were corrected for MIE scattering. Also, it could be shown that the larger particles aggregate first (Figure 23). Overall, these results were in agreement with the model of a diffusion-controlled aggregation process.[144]

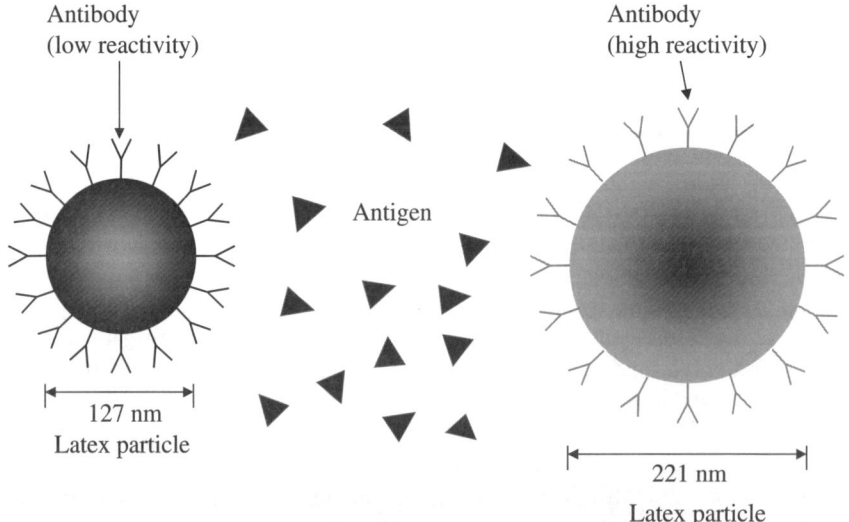

Figure 20   *Schematic representation of a turbidimetric assay based on two latexes, which are coated with antibodies of different reactivity (see text). Reproduced from ref. 212 with permission of Springer-Verlag*

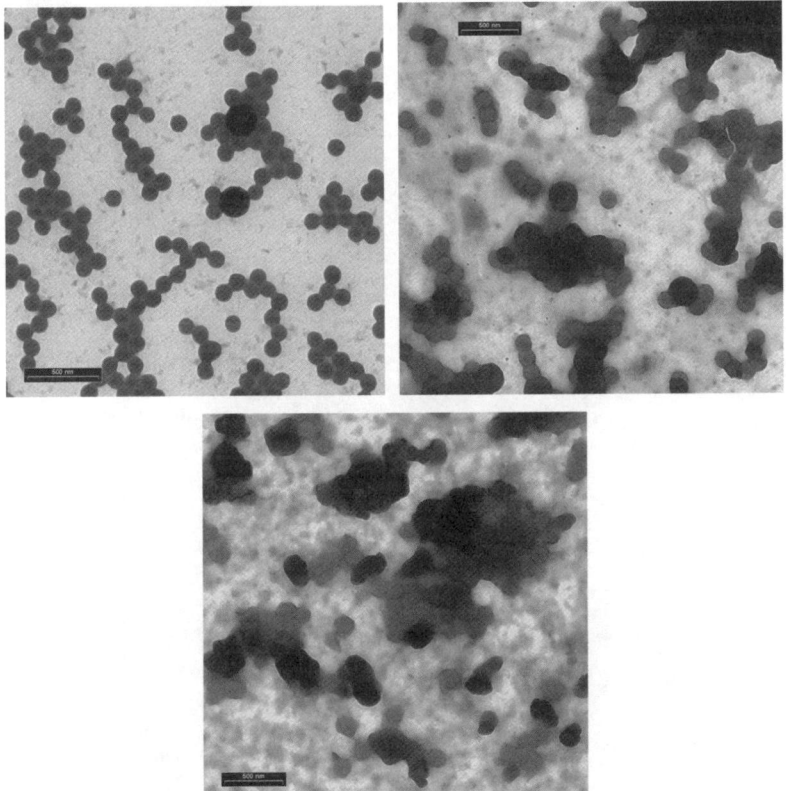

Figure 21 *Representative TEM pictures of anti-CRP coated latex particles after agglutina-tion with 0 (upper left), 4 (upper right), 25 (lower left), and 156 mg L^{-1} (lower right) C-reactive protein. Scale bar=500 nm. Reproduced from ref. 144 with per-mission of the American Chemical Society*

This demonstrates that AUC is able to characterize aggregating systems with a high resolution in the particle size, but with the drawback of a low kinetic resolution as the temperature equilibration time for an AUC experiment is >10 min even with a pre-tempered rotor (Figure 23). SLS (or alternatively DLS) on the other hand can yield a good kinetic resolution but lack the capability of a high particle size resolution. Nevertheless, both techniques complement each other in a favorable way (see also Section 5.1.4). However, if very broad particle size distributions with particle sizes spanning the whole colloidal range from 1 nm to a few μm result from aggregation processes, scattering techniques are likely to fail and a fractionating technique like AUC is the only realistic option to obtain quantitative particle size distributions.

5.1.4 Hybrid Colloids

Another colloid substance class rapidly gaining importance is hybrid colloids between polymers and inorganic matter. Here, AUC shows all its merits for the

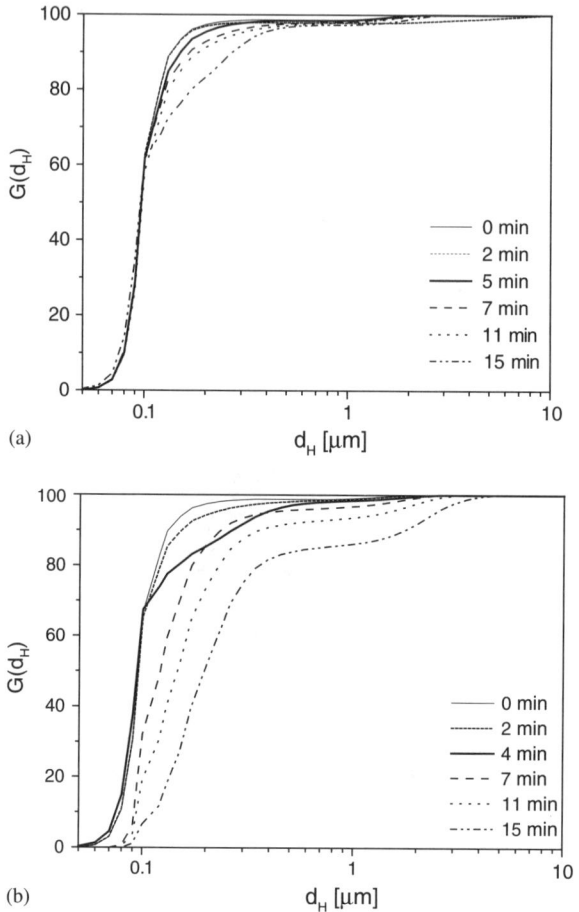

Figure 22 *Integral particle size distribution $G(d_H)$ from a SLS kinetic measurement with (a) 50 mg L^{-1} and (b) 171 mg L^{-1} C-reactive protein. Reproduced from ref.* ***144*** *with permission of the American Chemical Society*

investigation of transformations, aggregation processes, *etc.* For example, the encapsulation of a molybdenum cluster with a surfactant could be characterized as well as the aggregation of the primary clusters in different solvents with high resolution.[146] An example for monitoring a restructuration process of a complex hybrid colloid in solution is given in Figure 24. Here, calcium phosphate was synthesized within functional polymeric aggregates, and a very unusual neuron-like crystal morphology was obtained within the polymeric aggregates. This neuron-like morphology slowly transforms into a more compact spherical structure (see photos from left to right), which can be monitored by AUC to determine the relative proportions of each species.[147] As the particle density is not known, only the consideration of the sedimentation coefficient distributions makes sense.

From the sedimentation coefficient distributions, the whisker structure identifies the species with the lower sedimentation coefficient, whereas the dense spherical

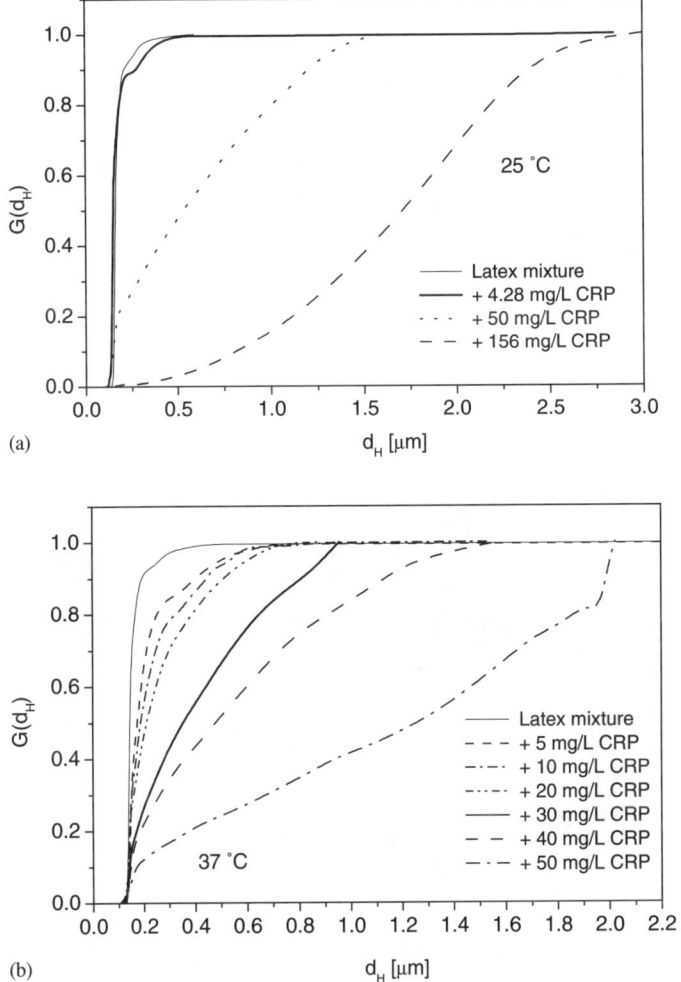

Figure 23 *Mass weighted integral AUC particle size distributions of an agglutinated latex mixture at different CRP concentrations (a) after 10 min at 25 °C and (b) incubated for 50 min at 37 °C. (d_H = diameter of the particles, $G(d_H)$ = integral particle size distribution). Reproduced from ref. 144 with permission of the American Chemical Society*

structure can be identified by its high sedimentation coefficients. The formation of the dense species has a clear maximum before its amount decreases again owing to macroscopic precipitation. The observation of such transformation in solution adds to the visual information from TEM in a favorable manner and shows the potential of ultracentrifuge experiments combined with a TEM investigation.

Often, however, the simple information on the various species in hybrid colloid mixtures via the sedimentation coefficient distribution is not sufficient to characterize a hybrid colloid system. For complex hybrid colloid mixtures, a combined analysis of AUC data with those from a different independent technique is clearly

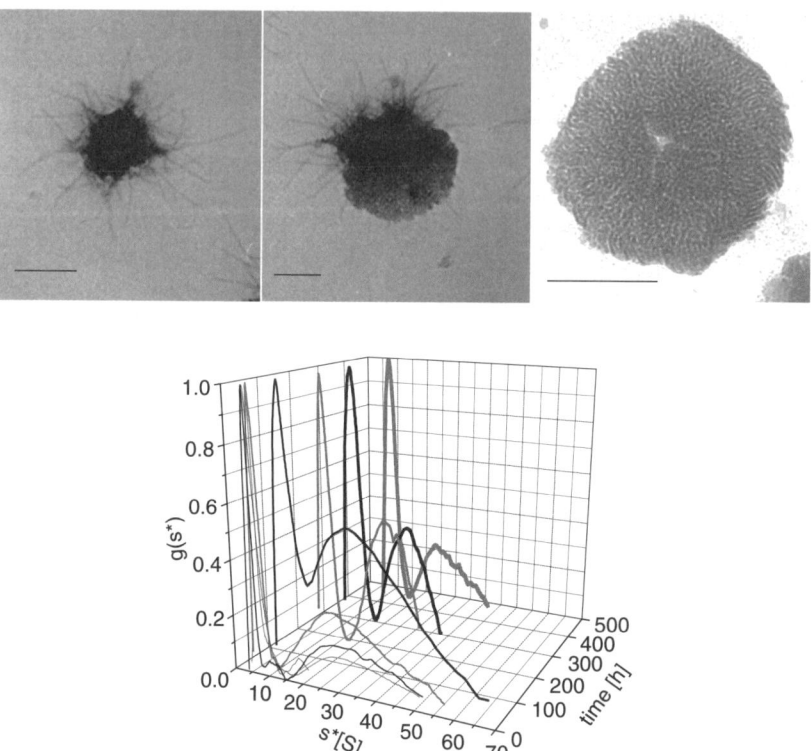

Figure 24 *Transformation of 'neuron-like' calcium phosphate into a more compact spherical form and the time-resolved observation of this process by analytical ultracentrifugation. The ultracentrifuge data was reproduced from ref. 147 with kind
permission of Wiley-VCH Verlag*

advantageous. As discussed above, the density of hybrid colloids is often unknown,
and their particle size cannot be accessed. An example for such hybrid colloid with
superimposed particle size and density distribution is the iron storage protein ferritin, which can show different iron loadings and thus different particle densities.
Considering the sedimentation coefficient and diffusion coefficient distributions
from AUC and Flow-Field-Flow Fractionation (Fl-FFF), respectively, one can see
differences on the first sight, which reflect the above facts (Figure 25). The particle
density-independent D distribution (proportional to the particle size distribution via
the Stokes–Einstein equation) shows the defined oligomer species (Figure 24; right),
where the ferritin monomer size of 11.9 nm is found to be in good agreement to that
of 12.5 nm given in the literature.[148] In contrast, the density-dependent s distribution
(Figure 25; top) just indicates different oligomers, and only peak deconvolution can
lead to a picture resolving the different oligomers.[149]

This clearly shows that the distributions in Figure 25 are not identical, and that the
only difference between AUC and Fl-FFF for the same sample is that AUC depends on
the particle density whereas Fl-FFF does not. Consequently, the results in Figure 25

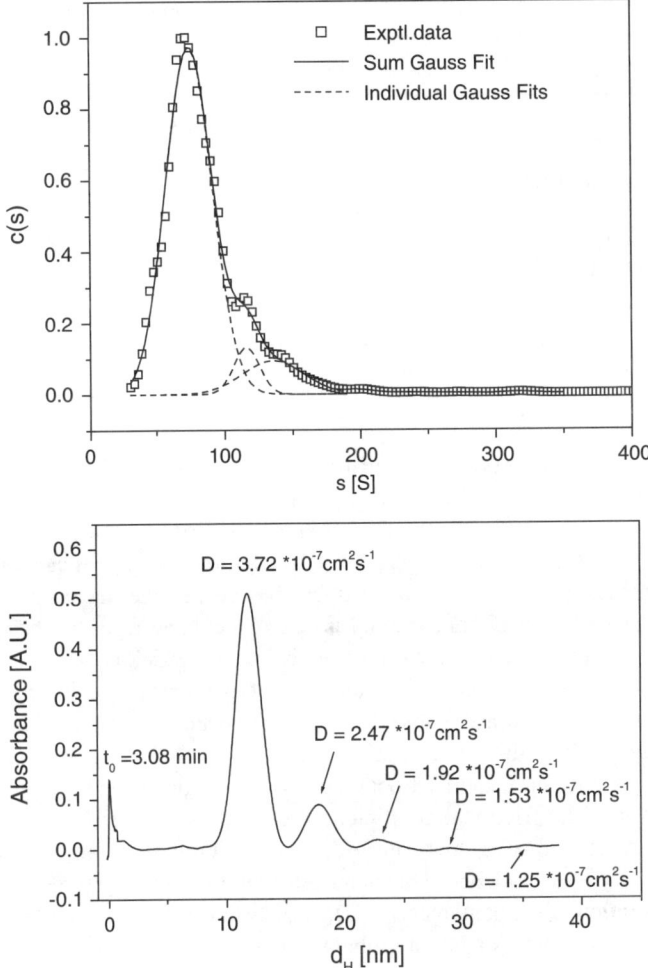

Figure 25 *Ferritin: Top: Diffusion-corrected s distribution from AUC. Bottom: Fl-FFF elu-gram after conversion to the particle size distribution. Reproduced from ref. 149 with kind permission of Springer-Verlag*

reflect a particle density and particle size distribution. In view of this, conventional sedimentation analysis is likely to fail, as it is usually based on constant density resp. partial-specific volume $\bar{v}$; therefore, a global analysis has clear advantages.

From the given s and D values in Table 2, the particle density-independent buoyant molar mass M_b can be calculated. It can be seen that it is relatively constant between 500 and 580 kDa for monomer–trimer. Deviations arise for the higher oligomers but as their amounts are very small (about 2 wt% as detected by Fl-FFF, which shows more confidence owing to the better resolved distribution), they have to be treated with care in the further analysis, at least with respect to the sedimentation analysis, which is based on a fitting procedure of the experimental sedimentation profiles. In

Table 2 *Sedimentation coefficients s, diffusion coefficients D, hydrodynamic diameters d_H, and buoyant molar mass M_b of each of the ferritin oligomers from AUC & Fl-FFF measurements*

	Monomer	Dimer	Trimer	Tetramer	Pentamer
s (S) at 25 °C (AUC)	74.6	116.3	135.7	195.3	317.0
$D \times 10^7$ (cm^2 s$^-$) at 25 °C (Fl-FFF)	3.72	2.47	1.92	1.53	1.25
d_H (nm) (Fl-FFF)	11.9	17.7	22.7	28.9	35.1
M_b per monomer (g mol^{-1})	496 900	583 300	583 700	790 700	1 256 700
Oligomer amount (wt%) from AUC[a]	82.50	6.17	10.46	0.64	0.23
Oligomer amount (wt%) from Fl-FFF[a]	80.70	14.63	2.59	0.47	1.61

[a]The relative oligomer amounts were calculated on the basis of proportionality between absorption signal and ferritin concentration implying constant ferrihydrite content for every capsule.

view of this, it is indicated that the ferritin oligomers have a rather constant $\bar{v}$, indicating a constant average iron loading per capsule of the different oligomers. Nevertheless, the smeared s distribution in Figure 25 (top) compared with the D distribution (Figure 25; bottom) indicates the $\bar{v}$ distribution, although it cannot be too big as the different oligomers can still be distinguished despite the smearing. In principle, the molar mass of each oligomer could now be calculated from the buoyant molar mass, if the density of each of the oligomers is known, which is, for example, possible by density gradient ultracentrifugation. From this molar mass, the iron content of each oligomer would be accessible. Nevertheless, even if $\bar{v}$ remains unknown, the present data sets allow further conclusions with respect to the oligomer shape.[149]

Assuming a constant $\bar{v}$ and frictional ratio, which neglects the small contributions of asymmetry to the frictional coefficient, the s_n/s_1 resp. the D_1/D_n ratio of a hard sphere aggregate with oligomerization degree n should show an $n^{2/3}$ dependence reflecting the $M^{2/3}$ dependence. This is indeed found in good agreement for s and D with the exception of the pentamer s value, which is unsafely detected.[149] This means that the ferritin oligomers adapt a globular solution structure and can be hydrodynamically treated as a sphere.

5.1.5 Combined Particle Size and Density Analysis

A combined particle size and density gradient analysis is a powerful tool for the analysis of complex mixtures where the particle size as well as the density of the components is different. This is illustrated in Figure 26. Here, a mixture of four different polybutylacrylate latexes (PBA) was grafted with a styrene–acrylonitrile copolymer (SAN), which has a higher density than PBA.

In the density gradient, the four grafted particles are clearly resolved, which indicates that the degree of grafting is different for every particle size. From the particle size distributions of the ungrafted and grafted PBA particles, it becomes obvious that the mass fraction of the smaller particles increases after grafting. This leads to the conclusion that the amount of grafting is proportional to the particle surface.

Another possibility to obtain the density of unknown particles is to run two velocity experiments in chemically similar solvents with different density (*e.g.* H$_2$O/D$_2$O).[132,150]

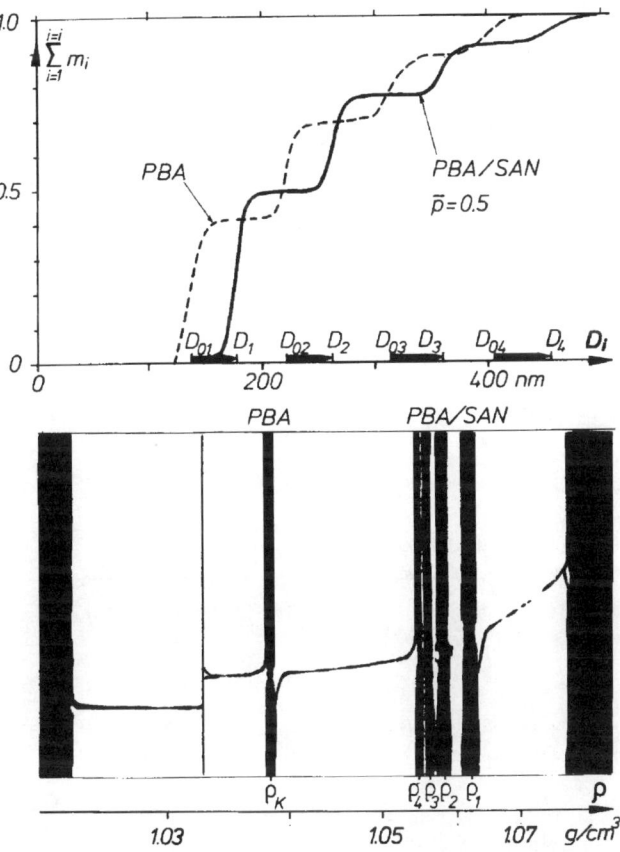

Figure 26 *Particle size distribution and density gradient of a four-modal ungrafted and a SAN-grafted PBA dispersion (40:30:20:10 wt% mixture). D refers to the particle diameter, the index 0i to ungrafted and i to the grafted latex. Reproduced from ref. 11 with permission of the Royal Society of Chemistry*

This allows to simultaneously determine particle size and density distributions (Figure 27) according to

$$\rho_P = \frac{s_1 \eta_{01} \rho_{02} - s_2 \eta_{02} \rho_{01}}{s_1 \eta_{01} - s_2 \eta_{02}} \tag{24}$$

$$d = \sqrt{\frac{18(s_2 \eta_{02} - s_1 \eta_{01})}{\rho_{01} - \rho_{02}}} \tag{25}$$

with d=particle diameter, ρ=density, η=viscosity with the index, p=particle, 01=dispersion medium 1, and 02=dispersion medium 2.

The method works best for particles with densities <1.5 g mL^{-1} but requires rather good data quality. Recently, a diffusion correction has been implemented into

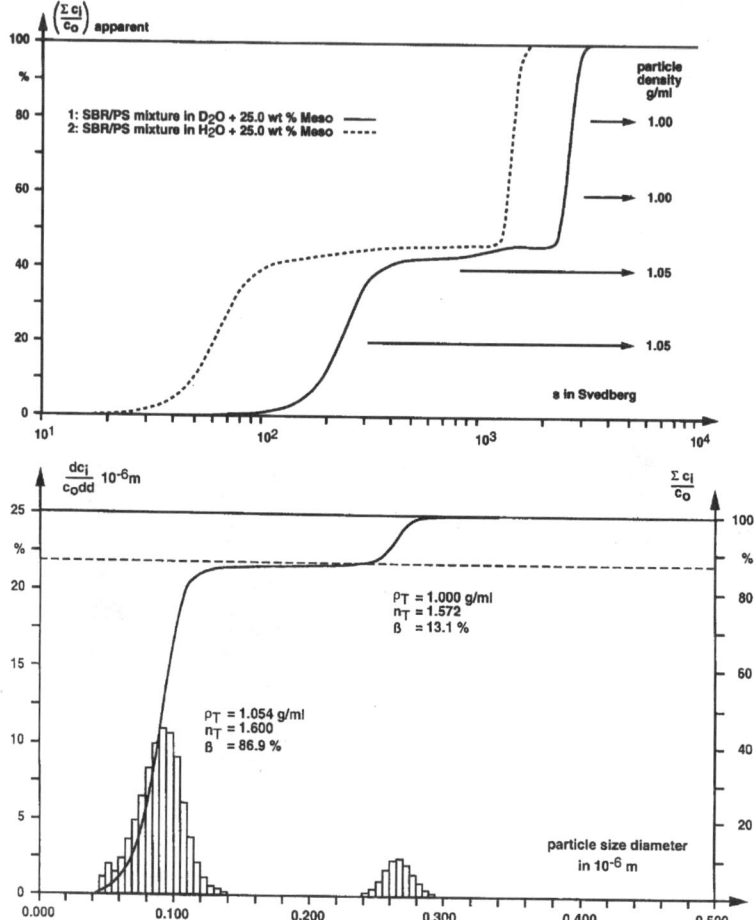

Figure 27   *Top: Sedimentation coefficient distribution of a polystyrene (PS) and poly-
styrene/butadiene copolymer (SBR) latex mixture in H_2O and D_2O.[132] Bottom:
Integral and differential particle size distribution evaluated from the sedimentation coefficient distributions in the left figure by application of the MIE scattering theory. Reprinted from ref. 132 with permission of Springer-Verlag*

this method, especially important for small particles/molecules, so that even complex mixtures become accessible by this method.[151]

5.2   Particle charge determination

The influence of charge on the sedimentation behavior has been established since the
very early days of AUC.[2] Although the term primary charge effect is well-known in
the ultracentrifugation literature, no full theoretical treatment of charge effects is
available yet, which would allow for the routine consideration of particle charge,
although the presence of an electric field was clearly identified to be of importance to

the sedimentation of charged species.[152,153] The problem for the analysis of charged species in the case of polydisperse polymers or colloids is that the particle size/molar mass distribution is superimposed by a particle charge distribution. With respect to the important role of particle charge in particle stabilization issues in colloid chemistry, the determination of particle charge distributions would be of major importance for colloid analytics. Nevertheless, the obstacle is that the particle charge distribution information is folded with the particle size and particle density distributions for colloids and cannot be unraveled, even if parallel information on the particle size distribution by techniques such as Fl-FFF is obtained. One workaround is to work with particles with constant density and size. However, a much better way would be the application of multiple detectors to simultaneously determine the sedimentation coefficient distribution next to the particle size distribution by scattering techniques. If such experiments are performed in various electrolyte solutions, particle charge distributions should become accessible through a global analysis approach. However, this analysis approach is not yet available, which is understandable if the difficulties in analyzing polyelectrolytes[154] or hybrid colloids[149] (see Section 5.1.4) are taken into account.The predominant approach in the literature of the last decades was to avoid charge contributions by buffer application, increasing the solvent ionic strength. Although this is certainly a good approach for the usually investigated (bio)polymer samples with a more or less expressed polyelectrolyte character, the increase of the ionic strength in a colloid dispersion unavoidably leads to a particle destabilization and the associated more or less expressed aggregation tendency. The same effect as ionic strength variations for the variation of colloid stabilization can be achieved by pH changes as input parameters for a global analysis of the sedimentation coefficient distribution of charged particles at various pH or ionic strengths.

A first step toward the determination of particle charge distributions was recently reported.[13] Here, two TiO$_2$ anatase nanoparticle samples with similar particle size and shape, almost identical average particle charge as determined by ζ-potential measurements, but different exposed nanoparticle faces, were subjected to AUC experiments at various pH values, and the sedimentation coefficient distributions were reported (Figure 28).[13] The two sedimentation coefficient distributions at various pH values of the two samples with different surface structure are different, (Figure 28) revealing the high sensitivity of sedimentation velocity experiments toward differences in nanoparticle surface structures, even when just the different exposed faces of a nanoparticle vary.[13] Although the charge influence could be qualitatively demonstrated, the actual particle charge could not be determined, as the particle size distribution would have to be known to unravel the particle size and charge contributions toward particle sedimentation.

This situation could be changed in a very recent report, where sedimentation equilibrium profiles on charged silica particles in ethanol were recorded.[155] Here, the sedimentation equilibrium equation was extended by an electrical term assuming ideal species and charge neutrality, neglecting colloid–colloid and ion–ion interactions as well as electrical charge distribution or potential variations on the scale of colloids and ions. Nevertheless, the charge contribution could be extracted from the sedimentation equilibrium profiles, allowing for the determination of an average particle charge (see Figure 29).

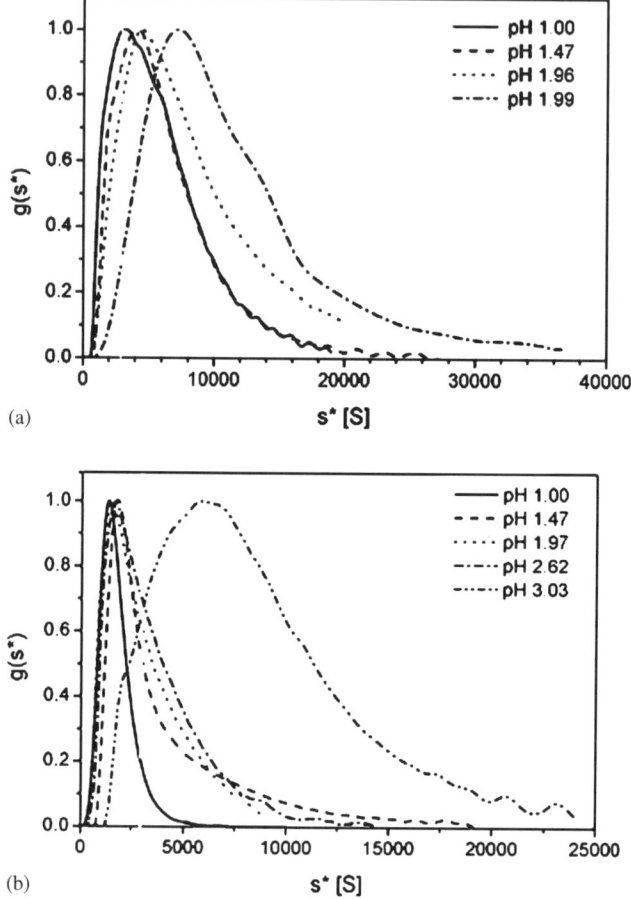

Figure 28 *pH-dependent sedimentation coefficient distributions of TiO₂ in the anatase modification with different exposed nanoparticle faces and thus different charge distributions. The overall particle charge is almost identical according to ζ-potential measurements. Reprinted from ref. 13 with permission of the American Chemical Society*

Three regions (I–III) could be identified in the equilibrium gradients of the charged species [Figure 29(a)]. Region I at low concentrations, following the sedimentation equilibrium theory, has the highest electrical potential [Figure 29(b)], flattening out toward the electrical potential of the background electrolyte in the meniscus region. In region I, the molar mass of the sample can be determined using conventional sedimentation equilibrium theory. The electrical potential decreases with the radial distance r. Two regions were identified where charge influence comes into play. Whereas in region II at moderate colloid concentrations, the sedimentation equilibrium theory extended by the electrical term yields particle charges in coincidence with results from electrophoresis, in region III, the charges were determined to be too low, indicating that the extension of the sedimentation equilibrium theory

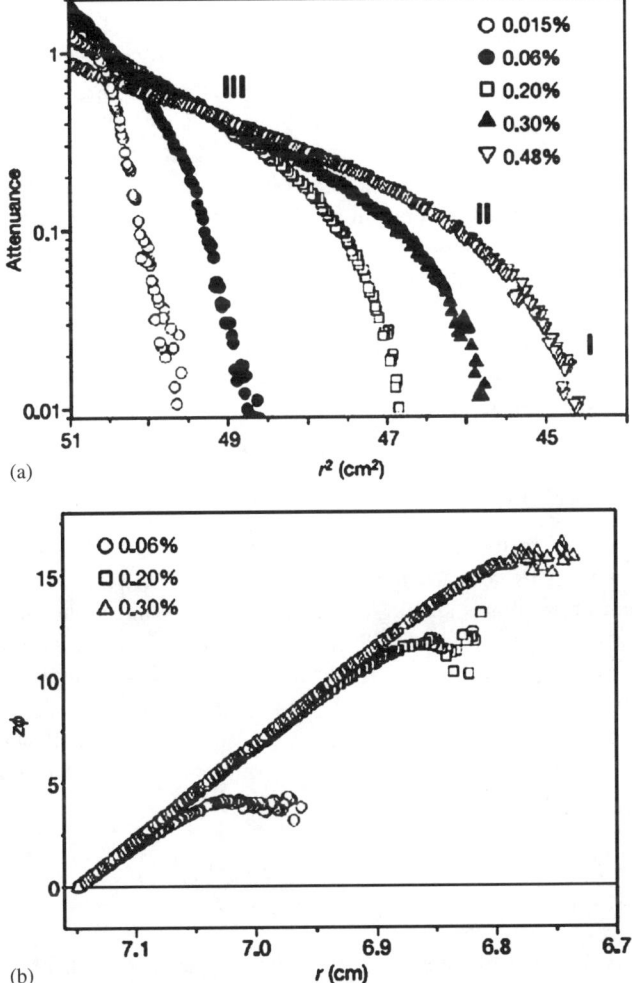

Figure 29 *(a) Experimental sedimentation–diffusion equilibrium profiles for dispersions of charged silica spheres in ethanol plotted as attenuance against the square of the radial distance from the center of rotation. (a) Representative profiles for dispersions having different initial volume fractions, with no added salt. Increase in the (very low) initial silica volume fraction significantly expands the exponential decay III as a consequence of the colloidal charge. (b) Non-dimensional electrical potential times charge z plotted against the radial distance for three initial volume fractions of particles. Reproduced from ref. 155 with permission of Nature Publishing Group*

by a simple electrical term is not valid at low concentrations of the charged species. Although an extension of the theory is certainly needed, this remarkable study points out that the molar mass can be determined (from region I) simultaneously with particle charge information (from region II). Therefore, the first steps are undertaken to determine particle charge distributions from AUC experiments.

5.3   Characterization of Emulsions

In analogy to dispersions, emulsions resp. mini or micro emulsions can be characterized by the particle size distribution of the dispersed phase if flat centerpieces
are used. These permit the passage of light to the detector through the turbid emulsion as the emulsion cannot be diluted, in contrast to a dispersion. One such example
was reported for water/AOT/heptane water in oil micro–emulsions.[156] It is possible to
force the coalescence of the emulsion by the ultracentrifugal field, which can easily
be detected as the formation of a new phase by any optical detection system of the
ultracentrifuge. This is a fast and effective means of evaluating the stability of emulsions in a qualitative way by determining either the centrifugal field necessary for
spontaneous coalescence or the time until coalescence occurs at a given centrifugal
field.[207,157,158] From the experimental parameters, such as the rotational speed, densities, *etc.*, a coalescence pressure can be calculated as a more quantitative measure.[207]

5.4   Characterization of Microgels

AUC has proved to be very useful for the characterization of thermodynamic and
elastic properties of gels and microgels, as reviewed in refs. 98 and 99, revealing
information on attributes such as whole swelling pressure–concentration curves in
the case of gels. In the case of microgels in a thermodynamically good solvent, the
degree of swelling Q can be directly determined from sedimentation velocity experiments.[159] If the microgel contains uncross-linked polymer, two components are
resolved in the sedimentation coefficient distribution calculated via Equation (2).
From $G(s)$ (Figure 30), the amount of each component can be derived. The swelling
degree Q can then be calculated using:

$$Q = \frac{bd^2}{s} \frac{\rho_2 - \rho}{18\eta} \tag{26}$$

where d=diameter of the compact, unswollen particle, ρ_2=density of the compact,
unswollen particle, ρ=density of the dispersion medium, and η=viscosity of the
diluted dispersion. b is a factor according to $m_r = bm$, where the mass of the particle
m_r reduced by the soluble part is related to the mass m of the particle consisting of
soluble and insoluble components. It can be derived from interference optical traces.
   The particle diameter of the unswollen sample has to be determined in a non-solvent in a separate experiment, which may be a problem if no solvent can be found,
where the microgel completely deswells, or if the stabilization in the non-solvent is
insufficient and aggregation occurs. The degree of swelling can be related to the
molar mass of the elastically effective network chains and thus to elastic properties
of the microgel, applying the Flory–Rehner theory. Such experiments allow not only
the characterization of microgels but can also be used to investigate the efficiency of
cross-linking reactions by specifying the amount and physicochemical properties of
the uncross-linked polymer.[160] Figure 30 shows the sedimentation coefficient distributions, which are derived for different cross-linking degrees of latexes. One can
now easily compare samples from different cross-linking reactions.

particles in water	particles in an organic solvent	s-distribution of these particles in the organic solvent

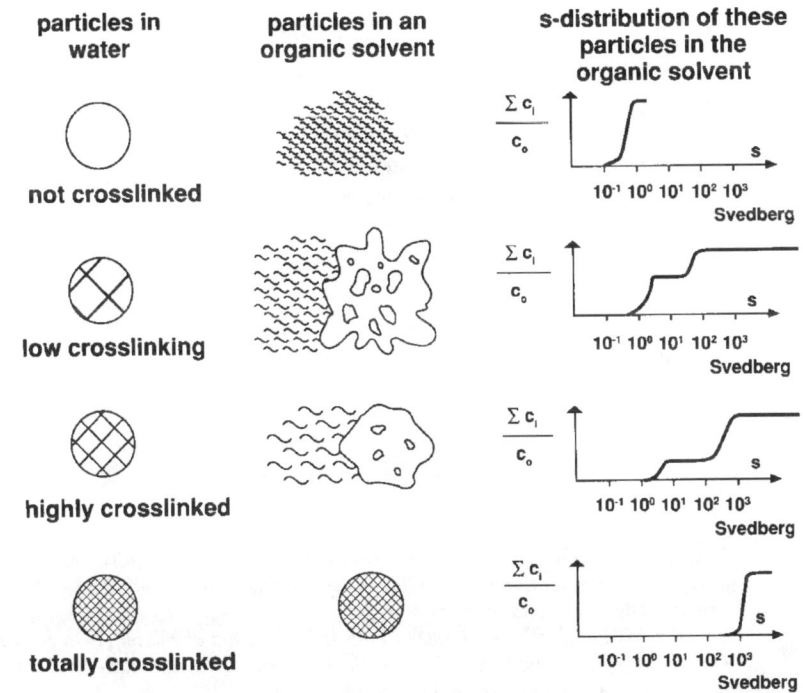

Figure 30 *Sedimentation coefficient distributions for latexes with different degrees of cross linking. Reproduced from ref. 159 with kind permission of Springer-Verlag*

Using the above procedure, it is possible to determine not only average swelling degrees, but also whole swelling degree distributions, by combining the whole sedimentation coefficient distributions for the swollen and deswollen particles in such a way that each data point in the sedimentation coefficient distributions is measured for the swollen and deswollen sample (Figure 31). This enables a much more detailed view into the efficiency of cross-linking reactions.

5.5 Observation of Chemical Reactions

It is possible to perform chemical reactions in the analytical ultracentrifuge with a synthetic boundary experiment using a special cell [see Figure 32(a)]. In biochemistry, this kind of centrifugation is called 'active enzyme centrifugation,'[161] but the principle can be applied to every chemical reaction. In the synthetic boundary cell [Figure 32(a)], a small amount of a reactant (usually 10–15 μL) is layered onto a column of the second reactant while the centrifuge is speeded up to about 3000 rpm. Thus, a reaction boundary is established. It is important that the density of the solution being layered is smaller than that of the solution in the sample sector to prevent convection problems. The detection wavelength of the absorption optics is set to a wavelength where the reaction product (here, the enzyme–substrate complex) absorbs. One can then observe the product formation and sedimentation as visualized in Figure 32(b).

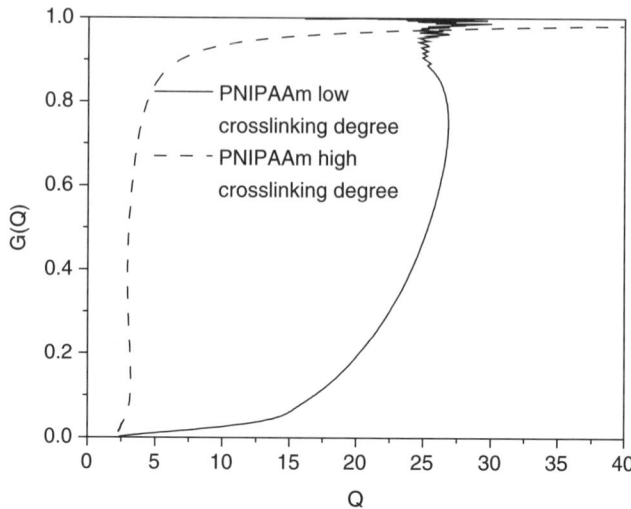

Figure 31   *Integral distribution of the volume swelling degree for two fully cross-linked poly(N-isopropylacrylamide) (PNIPAAm) microgels with a different cross-linking density. The non-monotonic increase of the distribution for the higher cross-linked gel is an artifact from noise in the sedimentation coefficient distributions. Deswelling of the sample was achieved by temperature variation. Reprinted from ref. 210 with permission of Taylor and Francis, Inc*

The absorbance increases with the time of sedimentation indicating that the sedimenting enzyme is reacting continuously. Such experiments can yield sedimentation and diffusion coefficients of the enzyme in its catalytically active state, which enables the detection of differences in polymerization state, hydration, or conformation with the non-reacting enzyme. If the extinction coefficient of the reaction product is known, the product concentration dependent on time can be calculated, allowing the extraction of kinetic information from such experiments. Figure 32(c) shows that a slight change in pH can turn the enzyme to its inactive state. An advantage of the application of synthetic boundary cells of the type shown in Figure 32(a) is that only a few microliters corresponding to a few micro/nanograms of reactant are required.

This type of synthetic boundary experiment was recently adapted to carry out crystallization reactions inside the spinning cell of the ultracentrifuge, and was named "Synthetic Boundary Crystallization Ultracentrifugation.[162,163] This method was first tested for CdS nanoparticles with sizes in the quantum confinement region where the UV–Vis absorption spectra depend on the particle size. Here, a small amount of Na_2S was layered upon a solution of $CdCl_2$ containing a stabilizer. A fast reaction to CdS with subsequent stabilization of the very small nanoparticles formed takes place as shown in Figure 33.[163]

The advantage of this technique lies is the high speed of the fast chemical reaction within the very small reaction zone and the quenching of further particle growth due to the lack of a second reaction partner as soon as the particles move out of the reaction zone by sedimentation or diffusion processes. These particles are then fractionated

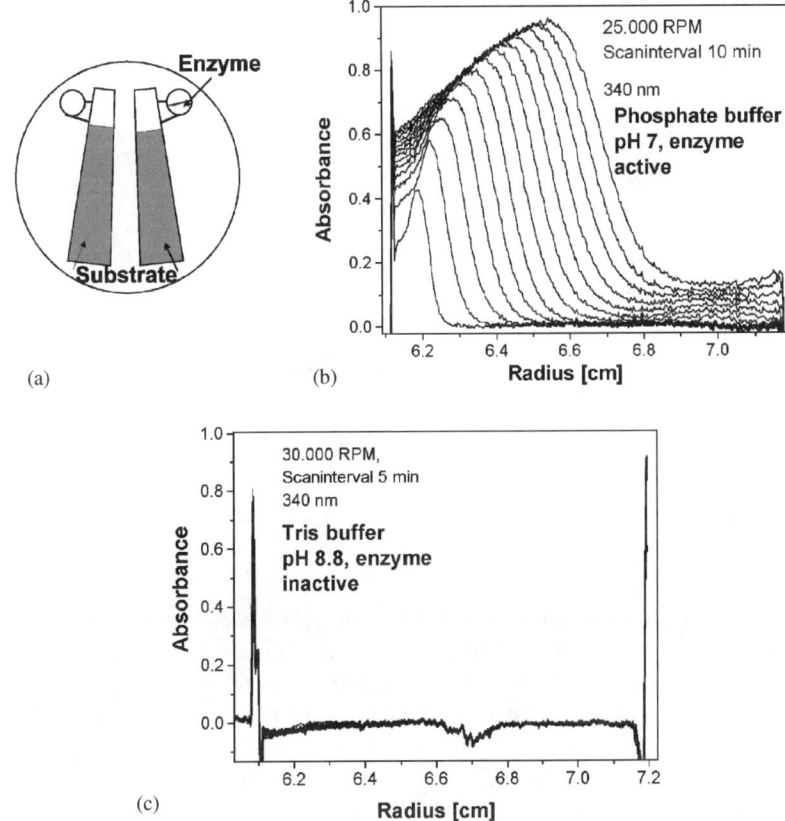

Figure 32 *(a) Synthetic boundary cell with small sample compartments, b) sedimentation velocity profile of a glutamate dehydrogenase mutant enzyme which reacts with its substrate, and (c) sedimentation velocity profile of the same enzyme at a different pH. The enzyme is now inactive and no chemical reaction takes place. Reprinted from ref. 211 with permission of American Scientific Publishers*

according to their size and density and a particle size distribution can be determined as discussed above. However, as the formed nanoparticles are very small, they show significant diffusion, and diffuse back into the reaction boundary and grow further, thus forming the second growth generation of particles. This process will be repeated until the reaction partner in the reaction zone is used up (which usually occurs within a few minutes), so that the particles only sediment owing to their size/density anymore with the usual diffusion broadening of the boundary. By this technique, different growth stages of nanoparticles can be investigated. However, owing to the extensive diffusion of the particles, the individual particle size distributions become extensively smeared, and they are only detected as a continuous distribution if no diffusion correction is applied. Software has recently become available to simultaneously evaluate several experimental scans so that only one experimental scan was taken for the determination of the apparent particle size distribution, as shown in Figure 34 (top). Nevertheless,

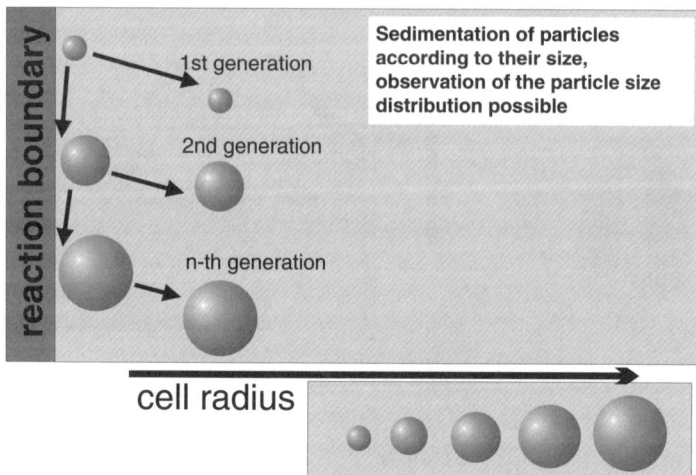

Figure 33 *Schematic representation of Synthetic Boundary Crystallization Ultracentrifugation. Reproduced from ref. 163 with kind permission of Elsevier Science Publishers*

even such data can show different stabilizer capabilities of chemically very similar stabilizer molecules with a resolution in the Ångström range, as shown for CdS in Figure 34.[162,163] Note that the particle size distributions only have apparent character as the density of the very small particles being investigated is lower than the bulk density owing to the importance of stabilizer and solvent surface layers.

However, application of the SEDFIT software allows the simultaneous evaluation of several experimental scans, increasing the accuracy of the determined particle size distribution. The difference between the evaluation of a single scan (Figure 34; top) and that of multiple scans is tremendous (Figure 34; bottom). The particle size distribution for thioglycerine-stabilized CdS (Figure 34; bottom) shows multiple peaks. If the particle sizes of the individually detected species in the growth process are compared with literature-reported species for thioglycerine-stabilized CdS, a remarkable coincidence can be observed. The particle sizes for six detected species coincide with the literature values with an accuracy of only 1 Å, although all particle sizes were calculated under the assumption of an average constant density, which is certainly not fulfilled for the different species. A smallest species with a size of only 3 Å is also indicated. Again, it is shown that AUC can precisely determine particle size distributions with an almost atomic resolution. The particle size distribution in Figure 34 (bottom) shows, in an impressive way, that synthetic boundary crystallization ultracentrifugation is a very powerful technique to study crystal growth processes.

The power of this technique is the transformation of a fast kinetics into a radial distribution, which can be detected by AUC. The faster the crystallization kinetics, the better the accuracy of the results. For the given example of CdS, the growth sequence in the presence of a thioglycerine stabilizer can be precisely revealed. It starts with sub-critical complexes (SR = thioglycerin ligand) as small as one complexed Cd ion and continues in stabilized crystalline clusters: $[CdSR_4]^{2-}$ – $[Cd_4SR_{10}]^{2-}$ – $[Cd_{10}S_4SR_{16}]^{4-}$ – unknown species – $Cd_{17}S_4SR_{26}$ – unknown species

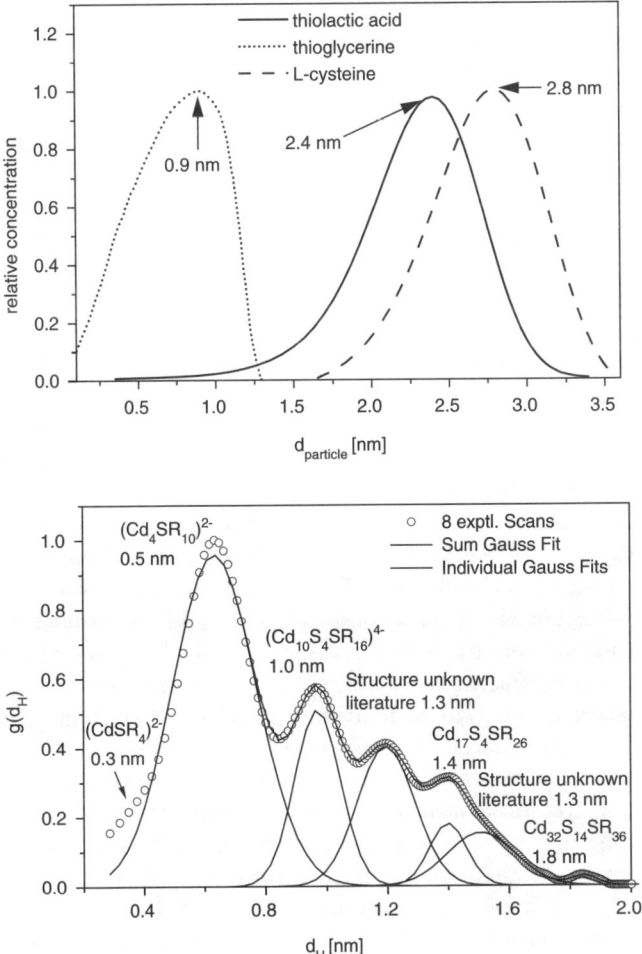

Figure 34 *Top: Apparent particle size distributions of CdS in the presence of different sta-bilizer molecules as detected by Synthetic Boundary Crystallization Ultracentrifugation and evaluated from one experimental scan with the density of bulk CdS.[162,163] Reproduced from ref. 163 with kind permission of Elsevier Science Publishers. Bottom: Particle size distribution evaluated from several experimental scans for the thioglycerine-stabilized CdS experiment in the left figure using the SEDFIT software by P. Schuck clearly showing a much higher resolution. This distribution was evaluated with an average particle density determined by combination of dynamic light scattering and sedimentation veloc-ity ultracentrifugation data*

– $Cd_{32}S_{14}SR_{36}$. Such a particle size distribution can also help to detect the smallest crystalline species – the so-called critical crystal nucleus. In the given CdS example, it would be either the unknown species with a size of 1.3 nm or $Cd_{17}S_4SR_{26}$. The crit-ical crystal nucleus is a matter of great fundamental interest and has been studied for decades owing to its importance for the understanding of crystallization processes.

The experimental detection of these species in the relevant solution state is extremely difficult but of great importance to test crystallization theories. Although synthetic boundary crystallization ultracentrifugation is a very promising technique, it also has some drawbacks. One is that this technique is designed for the detection of the smallest species, with the associated problem that the density of the particles is not known and can be expected to significantly deviate from the density of the bulk mineral. The suitable workaround is the determination of the particle density for one species out of the growth sequence as in the above example, or the estimation of a density range as applied for the example of ZrO_2.[131] Another drawback is that the suitable systems for synthetic boundary crystallization ultracentrifugation are presently limited to light-absorbing reaction products. This problem is associated with the linearity of the detector response to the reaction product concentration. Whereas this can be easily fulfilled for a light-absorbing reaction product, application of Rayleigh interference detection of refractive index difference yields results that are not interpretable at the moment. Although the detected sedimentation profiles have the form of the typical sigmoidal sedimentation velocity traces, they cannot yet be interpreted.[114] This is because the signals of several processes are overlaid during a crystallization reaction, all of them being detected via refractive index changes. For example, the nucleation of a crystal decreases the refractive index at the location of nucleation owing to ion consumption, although this would be a desirable detector signal. However, overlaying two salt solutions leads to rapid ion diffusion with the associated refractive index changes. Up to now, it was not yet possible to unravel the detected signals to the individual processes although this might become possible if the ion diffusion coefficients can be taken into account. Application of the Rayleigh interference optics would be the detector system of choice for synthetic boundary crystallization ultracentrifugation and would cover all crystallizing systems, including many of the technologically interesting crystalline systems. Synthetic boundary crystallization ultracentrifugation is the technique of choice to study early crystallization events and the influence of different stabilizers on the growth or aggregation processes. This is of great importance for scientific as well as industrial research.

Another elegant way to observe reactions with a synthetic boundary technique was introduced by Wandrey and Bartkowiak,[164] and Wandrey *et al.*[165] They were able to observe the formation of polyelectrolyte complex membranes between two oppositely charged polyelectrolytes, and the influence of parameters like pH, ionic strength, component ratio, and temperature. The membrane formation kinetics could also be detected, as the overlaying of the two polymer solutions resulted in the formation of a thin membrane at the solution interface that could be precisely detected and its thickness measured. The schematic experiment is shown in Figure 35.

The example shown is for a non-absorbing solution of alginate polyanions and a light-absorbing chitosan polycation solution.[164] Therefore, the membrane formation can be studied with the absorption optics and the layering velocity by simultaneous application of the Rayleigh interference optics. The absorption scans show the five regions in Figure 35. The negative absorption in ranges 2, 3, and 5 results from the subtraction of the absorption signal of the chitosan solution in the reference sector from the signal in the sample sector, namely, air (region 2), diluted chitosan after a

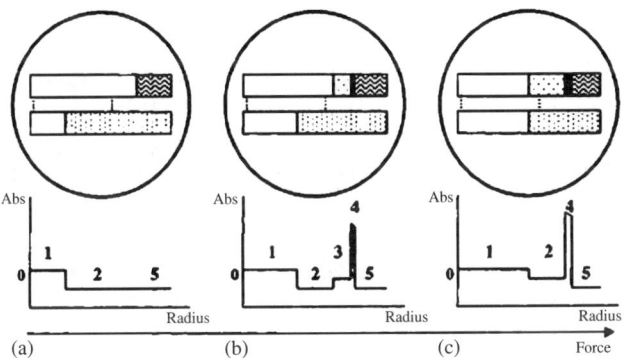

Figure 35 *Principle of the membrane formation experiments in a synthetic boundary cell for the example of a chitosan–alginate membrane. Component positions and absorption scans (a) before layering; the upper sector contains only alginate, (b) during layering of the chitosan solution onto the alginate solution, a membrane (black line) is formed and (c) after layering is finished; though, chitosan is not yet completely complexed. Ranges: 1 =air–air, 2 =air–chitosan, 3 =diluted chitosan–chitosan, 4 =Membrane–chitosan, 5 =alginate–chitosan. The trapezoidal cross section of the two sectors is not shown. Reproduced from ref. 164 with permission of Elsevier Science Publishers*

part of the chitosan was used up for the membrane formation (region 3), and non-absorbing alginate solution (region 5). The interesting region 4 marks the forming membrane, which creates an apparent absorption profile owing to its turbidity. Therefore, the thickness of the forming membrane can be precisely measured with the absorption optics. This is illustrated in Figure 36 (top), where the kinetics of membrane formation can be followed. In addition, details such as the symmetry of membrane formation can also be visualized. The membrane is visible in the ultracentrifuge cell (Figure 36; bottom) and can be isolated for subsequent characterization.

In a subsequent study, the membrane formation with alginate was studied for chitosan, polyvinylamine hydrochloride, and poly-L-lysine,[165] also with a focus on the elucidation of the membrane-formation process. The formation of gradient membranes could be revealed. Very recently, several characteristics of a forming membrane in an AUC cell were defined as indicators for a quantitative comparison of membranes formed under different conditions (polymer concentration, pH, ionic strength).[166] With this toolbox, a fast characterization of membrane-formation processes between polyelectrolytes, and potentially also between surfactants and polyelectrolytes, becomes quantitatively accessible in a rapid way for the first time.

5.6 Determination of Sample Homogeneity, Efficiency of Chemical Reactions, and Determination of Extinction Coefficients

In many fields of polymer or colloid chemistry, complicated complexes or hybrid colloids are synthesized. It is often desirable to have a quick and convenient check for the efficiency of the reaction as well as a check of sample homogeneity.

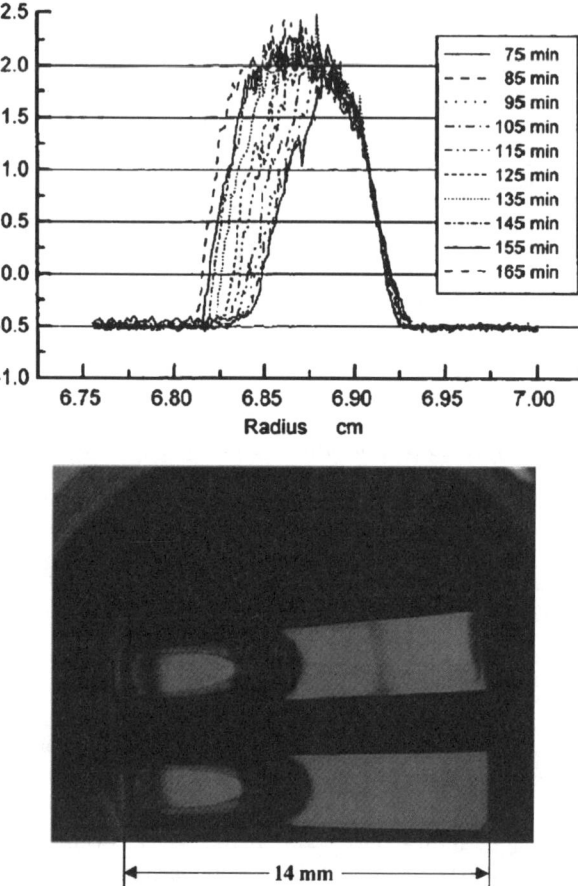

Figure 36 *Top: Chitosan/alginate membrane equilibration after the complete polycation consumption at 5000 rpm, 370 nm, and 20 °C. Bottom: Membrane location in the right (here upper) centerpiece sector of a synthetic boundary cell after completing the experiment. Reproduced from ref. 164 with permission of Elsevier Science Publishers*

Sedimentation velocity experiments can be used very advantageously here owing to fractionation without any stationary phase, as is demonstrated for the case of iron oxyhydroxide particles partly stabilized in a self-assembling microgel of κ-carrageenan (Figure 37).[167] If one component absorbs light (FeOOH) and the second does not (κ-carrageenan), one can selectively detect the absorption of the iron oxide in the polymeric superstructures with absorption optics, whereas all components together are detected by the simultaneously used Rayleigh interference optics. Here, it must be stated that most colored inorganic compounds have a high extinction coefficient and are thus almost exclusively detected with the UV–Vis absorption optics, whereas the simultaneously applied interference optics almost exclusively represents the local polymer concentrations, which can be converted to the real concentrations via the refractive index increment and the known refractive index of the solvent.

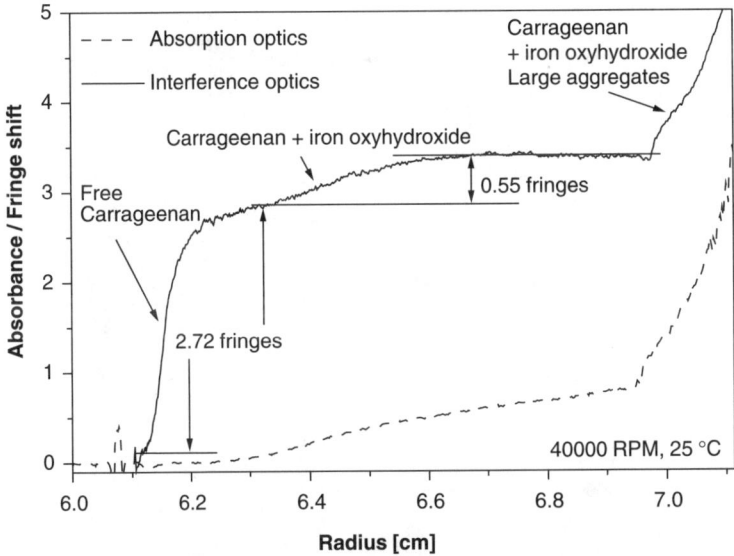

Figure 37 *Iron oxyhydroxide in κ-carrageenan microgels. Simultaneous detection of different species by combination of UV–Vis and refractive index detectors. Reproduced from ref. 167 with kind permission of Springer–Verlag*

From Figure 37, three species are quantitatively detected: free unbound κ-carrageenan, κ-carrageenan+ iron oxyhydroxide microgels, and larger cross-linked aggregates. The efficiency control of chemical reactions was also reported for complicated complexes between oppositely charged polyelectrolytes.[168]

If the change in the refractive index of a sample (Rayleigh interference optics) is simultaneously detected with its absorption at a specified wavelength, extinction coefficients can be determined.[169] This is not particularly interesting if a single component is considered, but for mixtures, the analytical ultracentrifuge is often the only instrument capable of determining the extinction coefficient of a specified component in a complicated mixture. The extinction coefficient can be calculated from a concentration reading in the Rayleigh interference scan, which must be converted to the desired concentration unit using the refractive index increment and the corresponding absorption. Figure 38 demonstrates this for ZnO prepared in polystyrene–polymethacrylic acid (PS–PMAA) micelles.

An extinction coefficient based on an absorption reading taken in a spectrophotometer under the assumption that all Zn^{2+} have reacted to ZnO delivers a low extinction coefficient of less than 1000 L mol^{-1} cm^{-1} owing to the majority of UV–V is inactive species. Therefore, it is of interest to detect which amount of the sample is the spectroscopically active ZnO. Figure 38 shows the results and confirms that only a minor amount of the mixture is responsible for the detected light absorption. From the interference scans in Figure 38(a), it can be seen that the component with a sedimentation velocity similar to that of the ZnO-filled micelles detected by the corresponding UV-absorption scans has a concentration of only 0.18 fringes. The major part is a considerably slower sedimenting component (micelles with unreacted Zn^{2+}).

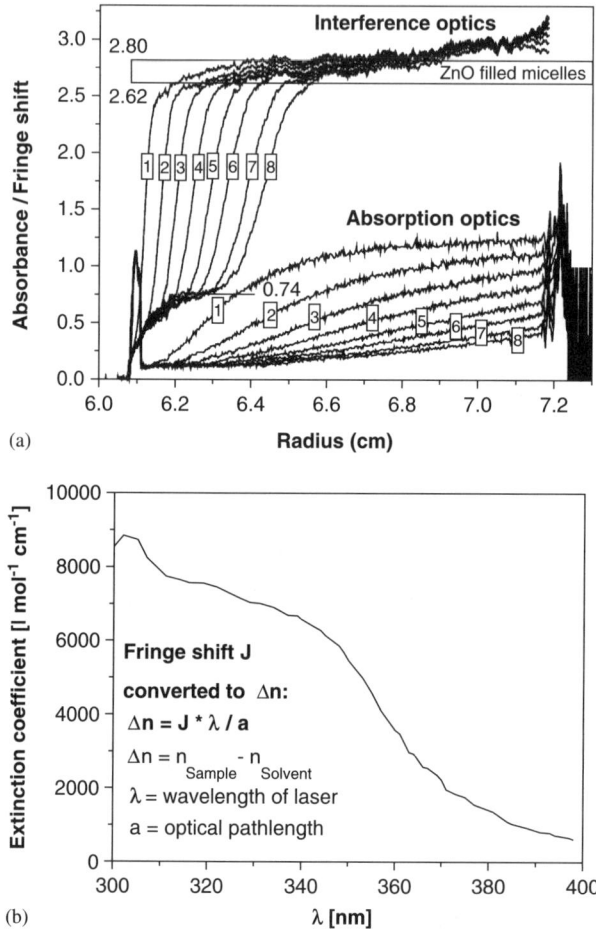

Figure 38 *Determination of the extinction coefficient of ZnO in PS-PMAA micelles.[209] (a) Sedimentation velocity profiles acquired simultaneously with Rayleigh interference and absorption optics (315 nm). Run parameters: 20 000 rpm, scan interval 2 min. The numbers represent the corresponding interference and absorption scans. (b) Wavelength dependence of the extinction coefficient for ZnO prepared in PS-PMAA micelles. The particle diameter of the ZnO (from TEM) is 3.2 – 6.7 nm. The diameter of the micelles (from light scattering) is 64.4 +/−33.9 nm. Reproduced from ref. 209 with kind permission of Springer-Verlag*

A third component (polymer or empty micelles) is also detected. The interference fringe shift of the ZnO-filled micelles can be converted to a concentration, and this can be related to a spectrum of the ZnO in the ultracentrifuge cell, giving the wavelength dependence of the ZnO extinction coefficient [Figure 38(b)]. The extinction coefficient is found to reach about 9000 L mol^{-1} cm^{-1}. This illustrates how important the separation is to determine the spectral properties of the optically active component in a complicated mixture. Many further variations of such experiments are possible, showing that the separation capability of the AUC can be used to determine not only the concentrations of the individual compounds but also their spectral properties if they absorb light.

An *s* distribution can yield insights into the success of chemical reactions forming hybrid colloids. An example is 11-mercaptoundecanoic acid-capped gold colloids, which were coated with bovine serum albumin (BSA).[170] The thiol is used as phase compatibilizer as the thiol groups chemically bind to the gold surface, whereas the carboxy groups on the other end of the molecule electrostatically interact with cationic amino acid residues of the BSA molecule (Figure 39).

It is of primary interest, if the coating reaction was successful. For this, AUC can be advantageously applied and the raw experimental data can reveal the amount of soluble BSA, which is not bound to the particles. However, the success of the coating is also visible in the *s* distribution, which shifts to higher *s* values (Figure 40).

From the *s* distributions in Figure 40, it can be concluded that the BSA coating reaction was successful. The *s* distribution of the coated gold colloid clearly shifts to higher *s* values, although the BSA coating results in a decrease in the density of the overall particle, which is compensated for by the particle molar mass increase. It is noteworthy that the *s* distribution of the coated colloids is only slightly broadened by diffusion effects but no shoulders can be traced in the distribution, which indicates a rather homogeneous particle coating by BSA.

Another example is particle coatings for a phase transfer from hydrophobic solvents to water. Often, stable colloids can be prepared in organic solvents, but their preparation in aqueous solutions results in undesired aggregation. Therefore, it is advantageous to synthesize the colloids in organic solvents, equipping them with the appropriate stabilizer for a subsequent phase transfer to aqueous solutions. The above gold colloids can serve as example for this. They were prepared in toluene, and then capped with mercaptoundecanoic acid, which makes it possible to dissolve the coated particles in water.[170] The same is possible with other metal nanoparticles (Figure 41).

For the proof of successful phase transfer, it is of crucial importance to show that the particles do not aggregate upon re-dispersion in another phase. This can again be easily monitored by the investigation of the colloid *s* distributions in both aqueous and organic solvent (see Figure 40; solid and dashed line). Despite the slightly different peak widths as a result of the different diffusion broadening at the different

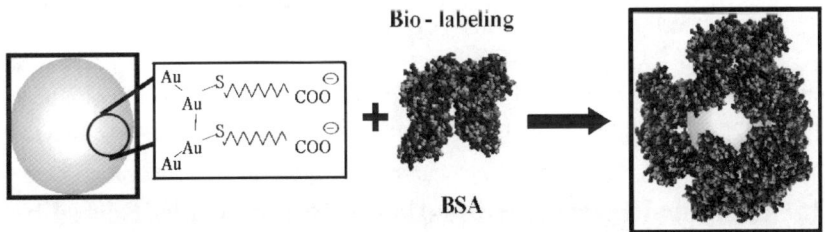

Figure 39 *Schematic presentation of the BSA coating of gold nanoparticles using 11-mer-captoundecanoic acid as phase compatilizer. Reproduced from ref. 170 with permission of Wiley-VCH Weinheim*

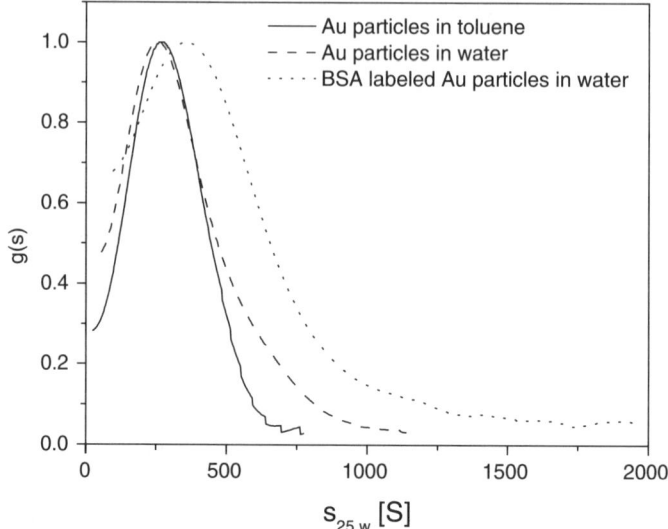

Figure 40 *Sedimentation coefficient distributions (not diffusion corrected) of Au particles with different coatings corrected to water at 25 °C.[170] Solid line: Hydrophobic Au in toluene, dashed line: Hydrophilic mercaptoundecanoic acid capped Au parti- cles in water, dotted line: BSA-labeled Au particles in water. The peak width extending the 0 Svedberg limit is a result of diffusion broadening. Reproduced from ref. 212 with permission of Springer-Verlag, Berlin*

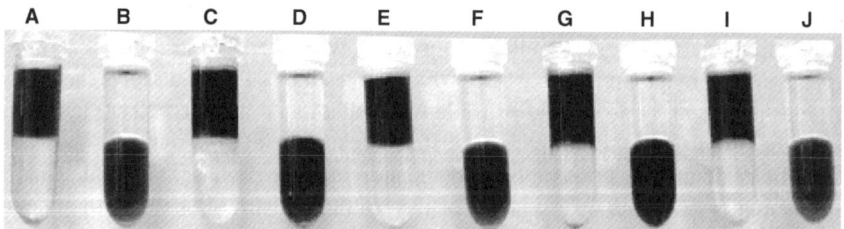

Figure 41 *Photograph of five diluted nanoparticle samples before and after phase transfer into water from their reaction mixtures (toluene) using mercaptoundecanoic acid. A, B: silver; C, D: gold, prepared according to ref. 213; E, F: platinum; G, H: gold, prepared according to ref. 214; I, J: palladium. Reproduced from ref. 170 with permission of Wiley-VCH Weinheim*

applied speeds in the two experiments, the distributions are very similar, indicating that the capped Au particles can be re-dispersed in water without aggregation.

5.7 Analytical Ultracentrifugation of Nanoparticles Sorted by Systems

In Table 3, some examples are given for colloidal systems that have been investi- gated by AUC, with reference to the original literature. The table is by no means exhaustive or representative and is just meant to guide the reader to some literature if solutions for a particular colloidal system are sought.

6 Some Current Trends in AUC of Colloids

The current trends in AUC of colloids can be divided into three sub categories. These are: (a) Improvement of data-evaluation methodologies, (b) new detection systems, and (c) application of AUC to novel systems. Whereas point (a) is mainly brought forward by the community interested in biophysical questions, new detection systems are (and have been) developed by very few groups worldwide. This could be a reflection that most AUC users are satisfied with the currently commercially available UV–Vis spectrophotometer and Rayleigh interference optics of the Beckman XL-I, as these optical systems are sufficient to address the large majority of questions related to the ultracentrifugation of biopolymers. Point (c) in the above list is very relevant to the investigation of colloidal systems, as most AUC methodology has been developed for biopolymer problems and is not always directly adaptable to problems of colloid analysis. For example, for a polydisperse colloid, a simple sedimentation velocity experiment may make the application of a speed profile necessary. Therefore, colloid analysis by AUC is somehow still in its infancy with respect to methodology and data evaluation. The good thing about colloid analysis by AUC is the versatility of the technique, which is displayed in this chapter. Therefore, the potential of AUC for the analysis of colloids is by no means exhausted yet, and many future methods are expected to be developed in the future, likely catalyzed by development of new detection systems. In addition, the growing interest in nanotechnology will also catalyze the application of AUC, as this is one of the most rigorous and versatile techniques for the direct analysis of colloids in the dispersed state with high statistical relevance.

6.1 Improvement of Data Evaluation Methodologies

Progress in computer and electronic technology has greatly facilitated the analysis capabilities of AUC and catalyzed the rebirth of this technique. Nowadays, it is relatively simple to apply fitting functions for all kinds of experimental output, and it is also possible to acquire data quickly and conveniently even when simultaneously using several different methods of detection. Therefore, a major current trend in AUC is the development of better evaluation methodologies and algorithms. One example is the simultaneous radial and wavelength analysis of sedimentation equilibrium experiments with the Optima XL-I ultracentrifuge.[171] A complicated interacting system exhibiting three different chromophores and consisting of the Band3 membrane protein, a small ligand, oxyhemoglobin, and the complexes between these components – an 11-component system overall – could be successfully analyzed. These investigations are also applicable to complicated colloid mixtures. This amounts to seven more components than possible before when analyzing a concentration profile acquired at a single wavelength. The increase in the number of components that can be resolved is achieved by the acquisition of as much experimental information as possible – in this case, wavelength scans at different radii as well as radial scans at different wavelengths. From this information it is possible to create a two–dimensional

Table 3 *Application of AUC to different colloidal systems*

System	Experiment Type / Quantity	References
Latex dispersions	Sedimentation velocity/particle size distribution; static and dynamic density gradient/particle density	Polystyrene [25, 215], polystyrene, polybutylacrylatestyrene, polybutadiene, & polybutylacrylate [26], polystyrene [10], polystyrene, polybutylacrylate (also styrene & acrylonitrile grafted), polybutadiene, acrylic homopolymer, and copolymer dispersions [11], polychloropropene [128], polystyrene, polystyrene/polybutadiene, polystyrene-co-butadiene, polychloropropene [5], polystyrene, polystyrene-co-butadiene, styrene, and acrylonitrile grafted polybutadiene [132], polystyrene [133], polyurethane [135], polystyrene [9] polystyrene, polybutadiene, polybutylacrylate-co-butadiene [150], poly(butyl cyanoacrylate) [216], protein grafted polystyrene [144], polystyrene core shell, and inverted core shell particles [217]
Inorganic nanoparticles	Sedimentation velocity/particle size distribution, spectral properties of quantum sized semiconductors, particle surface properties	Au [3, 126], Au_{55}-cluster [8], polyelectrolyte coated Au [218], $ZrO2$ [131] Pt & ZnO [134], ZnO, Au, CdS, Pt [209], various semiconductors [209, 219, 220], silica [9], proteins adsorbed on silica [221], β-FeOOH [150], CdS [162], CdS[163], iron oxide [222], Au, Pd [223], surfactant encapsulated, $(NH4)[H_3Mo_{57}V_6(NO)(6)O_{183}·H2O(18)]$ [146], CeO_2–ZrO_2 mixed nanoparticles [224], TiO_2 [13, 225], ZrO_2 [131].
Inorganic complexes	Sedimentation equilibrium/molar mass	Zr-complexes [65, 66]
Magnetic fluids	Sedimentation velocity	Fe_3O_4 [226, 227]
Microgels	Sedimentation velocity/composition and density, cross-linking degree, particle size distribution	Styrene-butadiene [159], acrylic acid [160], polystyrene-poly-4-vinylpyridine microgels [228], polystyrene-polyethyleneglycol latexes [229], human serum albumin microgels [230, 231]
Gels	Sedimentation velocity and equilibrium /swelling degree and pressure, amount of soluble compounds	Gelatin, κ-carrageenan, agar, casein [99], gelatin, κ-carrageenan, agar, casein [98], and other systems (reviews)
Emulsions	Sedimentation equilibrium, sedimentation velocity	SDS stabilized n-decane in water [207], food emulsions [157], legumin stabilized n-decane in water [158, 232], nujol in glycerol [233]

Supramolecular assemblies and polymers	Sedimentation equilibrium/molar mass, sedimentation velocity / particle size	Co-coordination arrays [67-69], Fe-coordination polymers [187], Ru-coordination polymers [234, 235]
Polyelectrolyte complexes	Sedimentation velocity/particle size and composition, sedimentation equilibrium/molar mass and interaction and membrane characteristics; synthetic boundary/membrane formation	Alginate/chitosan [164], alginate/poly(vinylamine) hydrochloride, poly(1-lysine) hydrochloride, and chitosan [165], polystyrenesulfonate/polydiallyldimethylaminchloride – acrylamide [168], poly(isobutylene)-block-poly(sodium methacrylate)/poly(N-ethyl-4-vinylpyridinium bromide) [236], polyelectrolyte membranes
Micelles	Sedimentation velocity/composition; dynamic density gradient/composition	Polyampholyte/fluorinated & hydrogenated dodecanoic acid [237], lipid and detergent micelles [238], SDS–liposome micelles [239], review on different systems [193], polystyrene-b-polyisoprene, polystyrene-b-poly(ethylene-co-propylene) [194], polylactide-co-polyethyleneglycol [195], enzymes in reverse micelles [196], various surfactant micelles [240], polyethyleneoxide-peptide micelles [241]
Dendrimers	Sedimentation velocity and sedimentation equilibrium/hydrodynamics	Carbohydrate-coated polypropyleneimine dendrimers [197], lactosylated polyamidoamine dendrimers [198]
Hybrid colloids	Sedimentation velocity/composition /particle size, global analysis by combination with other methods	Au/polystyrenesulfonate microgels, CdS or Au in polystyrene-b-poly-4-vinylpyridin micelles, Pt in polyethyleneoxide-Polymethacrylic acid [209], calcium phosphates in alkylated polyethyleneoxide-b-polymethacrylic acid [147], FeOOH, NiOOH & CoOOH in κ-carrageenan microgels [167], review of various systems [130], CdS in reverse micelles [242], Pt and Pd in polystyrene-b-polyethyleneoxide/cetylpyridiniumchloride mixed micelles [243], BSA-coated Au [244] Au, Pt, & Pd in poly-2-vinylpyridine-b-polyethyleneoxide micelles [199], Pt, Pd, Rh, & Cu in polyethyleneoxide-b-polyethyl-eneimine [200], ferritin [149].
Organic colloids	Sedimentation equilibrium; sedimentation velocity /composition, molar mass	Cu-phtalocyanine [133], κ-casein particles [245, 246], humic substances cross-linked by metal ions [247].
Nanocapsules	Sedimentation velocity, density gradient/composition	Oil-filled polybutylcyanoacrylate nanocapsules [248]

data surface in a radius–wavelength–absorption space, which allows a proper analysis by fitting procedures. Therefore, it is not surprising that more and more scientists opt for scanning at multiple wavelengths,[172–174] and that a fast, multiwavelength detector is currently being developed.[175] A further enhancement of the data space is achieved by the application of several rotational speeds in the experiments. As this increases the experimental time, so-called short-column techniques,[176,177] which considerably decrease the time taken to reach the sedimentation equilibrium by the application of a shorter solution column in the ultracentrifuge cell, are becoming popular, as long as the resolution of the optical system is still sufficient for the short solution column. However, for the mainly polydisperse nanoparticles, such an approach is restricted.

Significant effort has also been directed toward other efficient methods of analyzing ultracentrifuge data and increasing the sensitivity of ultracentrifuge experiments. Nowadays, components that are barely visible in the experimental raw data and virtually unrecognizable in the old days of AUC can be identified. As the Rayleigh interference system of the XL-I ultracentrifuge allows a rapid acquisition of large data sets, the confidence intervals of fitting procedures in general can be significantly improved. However, as the interference data contain time invariant and radial invariant noise components in addition to the random noise, an efficient algebraic method was developed for the elimination of the time and radial invariant noise.[178] This leads to a high sensitivity of Rayleigh interference data even for such experiments where the sedimentation boundary can hardly be traced owing to the noise components. These improved data sets can then be used for fitting to approximate analytical and numerical solutions of the Lamm equation. It is also possible to eliminate systematic noise components in interference data of sedimentation equilibrium profiles by extraction of the time-invariant noise during the approach to equilibrium.[179] It turned out by this approach that the sensitivity of the Rayleigh interference optics for sedimentation equilibrium experiments can be improved by an order of magnitude such that equilibrium experiments with concentrations as low as $50~\mu\text{g mL}^{-1}$ are no problem for Rayleigh interference detection anymore. Recently, the described systematic noise decomposition method[178] has been applied for linear least squares modeling of the sedimenting boundary by a superposition of $g(s)$ of ideal non-diffusing components.[180] This approach turned out to be especially advantageous for data acquired during a large time interval; in such cases, an improved resolution could be achieved, especially for heterogeneous mixtures. A very recent advance is the determination of molar mass distributions from fitting of finite-element solutions of the Lamm equation (1) paired with regularization algorithms, as discussed before.[60] However, the full range of successful applications for colloids still needs to be explored. In the author's experience, the latter approach works very well for defined samples but can fail for polydisperse samples, resulting in the generation of artificial ghost peaks and making it hard to determine which data are reliable.

Another development that promises to influence the analysis of colloidal systems, is the so-called global analysis approach (see also Section 5.1.4). This approach is supported by the cheap computer resources available nowadays. A global analysis approach that is already available is the combination of sedimentation velocity, equilibrium, and DLS data.[181] These techniques constrain each other in the evaluation as they contain supplementary information in the Svedberg equation (M, s and D). Therefore, an easy determination of absolute molar mass distributions appears possible, as the sedimentation coefficient distribution is constrained by the average molar mass and diffusion coefficient, leading to a molar mass distribution based on experiments with three different techniques. For colloids with the often problematic folded particle size and density distribution as shown above for ferritin hybrid colloids (Section 5.1.4), the application of global analysis approaches coupled with the application of multiple optical detectors will be the only chance to analyze the complex distributions of particle size, density, charge, shape, *etc.*

In general, data-evaluation methods seem to be much more developed than the experimental methodologies or detection systems. Therefore, the limiting step for the future development of AUC in general but especially in the field of colloid analysis is seen in the development of new optical detection systems. Only the multidetection of many particle properties can satisfy the high demands for the analysis of complex colloid mixtures. In addition, simultaneous application of multiple detectors supports the global analysis approach.

6.2 New Detection Systems for the XL-I Ultracentrifuge

The development of AUC depends on the availability of new optical detection systems that extend the commercially available UV–Vis and Rayleigh interference optics, and the recently developed and commercially available fluorescence optics for the XL-I ultracentrifuge.[24] Especially important will be the adaptation of detectors that can reveal new sample properties such as the molar mass or diffusion coefficient (light scattering), or provide spectral information on the sample (IR or Raman detection), thereby opening up new possibilities for selective sample detection.

Schlieren optics is one of the classical detection systems for colloids as it allows for the detection of high concentrations and has a sensitivity similar to that of the Rayleigh interference optical system. It is the technique of choice for all density-gradient work as well as studies on emulsions or gels. Unfortunately, this system is no longer commercially available on the XL-I platform, although it was a traditional detection system in the model E ultracentrifuge.

Some improvements of Schlieren optical systems have been reported. Computer-based picture evaluation allows the online capture of Schlieren patterns, enabling the fast and efficient evaluation of these records, which earlier had to be photographed.[18,182–184] Whereas digital detection is one issue, automatic picture evaluation is another one. The on-line digitization and picture evaluation of Schlieren optical records is improved by proper detection of the zerothorder Fresnel fringe next to the Schlieren curve. Either by means of picture manipulation performed in order to facilitate a proper detection of the Schlieren curve, even with a non-uniformly illuminated picture, or by additional consideration of higher order Fresnel fringes, the concentration gradient curve can be properly detected automatically. Especially with the system built up at the University of Leicester/Nottingham,[182] high rates of data acquisition can be achieved with good data quality, proving the virtues of the online Schlieren system. This system is currently being adapted to the XL-I ultracentrifuge by the Nottingham group. The Schlieren optics has also been set up on a preparative ultracentrifuge,[20] providing for the well-accepted quality of the Beckman Model E Schlieren optics on the platform of the new computer-controllable XL preparative machines with subsequent digital image processing.

A potentially interesting optical system that should also be adaptable to XL-ultra-centrifuges is the Lebedev interferometer, which gives an interference pattern with a Schlieren peak for a sedimenting boundary.[185] Therefore, the fast evaluation algorithms developed for the Rayleigh interferometer should be applicable here as well, resulting in high-quality Schlieren images.

Another system that is now commercially available on the XL-I ultracentrifuge is the fluorescence detector developed by Schmidt and Riesner[23] and adapted to the XL-I by Macgregor and Laue.[24] These systems are designed to be simultaneously applicable with the present absorption and Rayleigh interference optics but not with the Schlieren optics, which uses the optical path of the interference optics. Further current approaches towards the faster detection of UV–Vis data are on the way in the laboratories of Laue and Cölfen.[175,186] However, all of these efforts are not yet considered to be sufficient to launch AUC as a universal colloid analysis technique in the coming time of the nanotechnology. Therefore, further detector development is absolutely essential to reveal more sample properties besides the so-far accessible ones.

6.3 Application of Analytical Ultracentrifugation to Novel Systems

Apart from biological systems, AUC is being more frequently applied to complicated synthetic systems, such as supramolecular assemblies and polymers,[67–69,187] colloidal clusters,[146] smallest colloids and their interactions with stabilizer,[8, 146, 162, 163] synthetic polyelectrolytes[188–192] and complexes thereof,[168] micelles,[193–195] components inside micelles,[196] dendrimers,[197,198] or hybrid colloids between organic and inorganic matter.[147,199,200] In most of these applications, advantage is taken of the simultaneous acquisition of Rayleigh interference and UV–Vis absorption data. It is possible to speculate how much a simultaneous application of further detectors could yield further information and thus open up further applications. A visionary paper about future requirements in ultracentrifuge methodology and detection systems was published by Mächtle[4] more than 10 years ago, and it can be stated that, at the very least, the simultaneous application of as many detectors as possible as realized in size exclusion chromatography is advantageous and will open new applications for AUC.

7 Future Perspectives

To cut a long story short, modern AUC is more versatile than ever, and despite its major applications in biopolymer analysis, it returned to its roots in colloid analytics. AUC is one of the most powerful techniques known to date for the characterization of very small colloids <10 nm. The power of AUC lies in the fractionation of the sample and the possibility of measuring distributions without any interactions with the stationary phase or solvent flows. as occurrs in the commonly applied chromatographic techniques nowadays. Wherever information is sought for the individual components in a mixture, AUC is one of the first techniques of choice and, especially for nanoparticles <10 nm, the only realistic choice to determine a particle size distribution in solution. Therefore, it is expected that AUC will play an important role in colloid chemistry if researchers recognize the tremendous potential of this method.

Although the very early colloid work by AUC was of most remarkable precision,[3] it cannot compare with the present possibilities, especially in the data evaluation sector. The last six decades of AUC research were mainly dedicated to biopolymers and put forward many useful methodological developments, reflected in the large pool of freely available software (for example, *http://www.cauma.uthscsa.edu/software*). However, the commercial hardware development could not catch up with the demand for colloid analysis. The most significant instrumental drawback for colloid analysis was the loss of the Schlieren optics in the modern XL-I instrument, which was designed to meet the needs of the low-concentration work for the characterization of biopolymers, but not the high-concentration work often necessary to study stabilized highly concentrated colloid systems by Schlieren optics. Although a model E Schlieren optics adaptation onto the XL-AUC was reported,[20] this is not generally feasible as the average AUC user cannot invest time in instrument development, which is the responsibility of commercial manufacturers. Therefore, hardware detector development did not effectively take place despite the development of the fluorescence optics, which is of limited value for colloid research. A patent was meanwhile filed to adapt light scattering as a detector for AUC.[201] However, the light scattering detector investigations in the authors' laboratory revealed many practical obstacles to such experiments. Since only a limited experimental range is available for light scattering experiments, this patent appears to be only of theoretical use, especially if one depends on the lenses in an ultracentrifuge rotor. Also the experimental proof of the promise of this detector has not been reported, so alternative methods are still being sought in the authors' laboratory to allow for light scattering detection in the ultracentrifuge cell.

A sophisticated methodological and theoretical AUC base is now available as a consequence of the research on biopolymers over the last decades. Many of the biopolymer AUC evaluation methods can be adapted to colloids. However, polydispersity will still be an obstacle in the standard evaluation techniques.

If the actual trends are considered, the popularity of biotechnology has already reached its uppermost limit, whereas nanotechnology is just at its beginning, Therefore, it can be expected that one of the main future AUC applications will be colloid analysis. However, so far, the AUC community is not yet prepared for this and the colloid community is also not yet aware of the power of AUC for colloid analysis. Only a very small number of students are currently qualified to analyze colloids with the associated polydispersity problems, linear detector response problems due to turbidity, *etc.* Whether the wealth of present knowledge can be transferred and adapted to the fast-growing field of nanosystems will depend on the community of AUC users.

8 Conclusion

AUC is one of the most universally applicable and variable physicochemical techniques for the characterization of colloidal or polymeric systems in solution, dispersion, or emulsion. The power of AUC lies in the fractionation of the sample due to either the molar mass/size (sedimentation velocity/sedimentation equilibrium) or due to the chemical structure (density gradient) in the solvent and, therefore, its capacity to measure distributions without any interaction with stationary phase or solvent flows, as occurs in the applied chromatographic techniques commonly nowadays. Wherever information is sought for the individual components in a mixture, AUC is among the first techniques of choice. On the other hand, sedimentation equilibrium is an equilibrium well described by thermodynamics. Thus, it is possible to derive information even on complicated or interacting systems on an absolute basis without disturbing this equilibrium. Many different physicochemical quantities can be determined. Very often, even simple investigations such as those of sample homogeneity can provide much information on a system. But a complete physicochemical characterization is also possible, often with information that cannot be obtained by means of other techniques. As for other analytical techniques, the rapid computer and electronics development of the past years was of benefit for AUC. This development provides many new applications, improved sensitivity of the experiments and even the realization of evaluation approaches not possible so far. Not only has the amount of experimental data that can be obtained from an ultracentrifuge experiment increased, the amount of information it yields has also increased. This has led to a resurgence of interest in this old technique. If multidetection systems can be realized on ultracentrifuge platforms, new dimensions of analytical information will be available from a single sedimentation velocity experiment. Therefore, a synergy between ultracentrifuge hardware and methodology development can be expected, which will lead to a still increasing number of applications.

In biochemistry or biophysics, AUC could gain some importance again, even though the pace had slowed considerably in the 1980s. The same trend is not unlikely for the field of polymer or colloid chemistry although, so far, many synthetically working polymer or colloid chemists are not yet aware of the potential of AUC.

AUC is a versatile technique well equipped to handle large number of future experimental demands for colloid analysis. As already shown by Svedberg in his very first experiments, AUC is a sophisticated high-resolution technique for colloid analysis. A gold colloid, not much bigger than 1 nm, could be characterized about 80 years ago with most remarkable resolution.[3] Current standard centrifugation protocols allow for the Ångström-resolved particle size distribution investigations of dense metal colloids[134] or inorganic colloids/hybrid particles with densities down to 4–5 g mL^{-1}.[131]

More than 10 years ago, W. Mächtle wrote a visionary paper about the future requirements for modern analytical ultracentrifuges.[4] It is amazing that only parts of the outlined requirements were so far realized in the commercial XL-I, including the multihole rotors. Other requirements have been realized only in a few laboratories (multiple detection optics, automatic online data analysis, application of speed profiles), but these requirements are important and essential for the analysis of complex colloids. It is important to obtain as much analytical information as possible simultaneously; therefore, multiple detection systems have to be applied for colloid analysis (Schlieren/interference, UV–Vis, SLS, turbidity, and others) in order to unravel the complex folded distributions of size, shape, density and charge. Even existing optical systems such as the UV–Vis optics should be improved so that they can give further information. For example, a fast fiber based UV–Vis optics with a charged coupled device (CCD) array spectrophotometer, which we currently design[175] can yield a whole spectrum consisting of 800 wavelengths in a few ms.

This means that such a detector is not only an excellent, new, potential turbidity detector with varying sensitivity due to the wavelength dependence of light scattering, it could also be used to determine the wavelength dependence of turbidity, which can yield the particle size. As the same spectrum could yield the particle size via the wavelength dependence of turbidity and the sedimentation coefficient from the time when the spectrum was taken, the application of a hard sphere model would yield the particle density via a modified Svedberg equation. This just illustrates, with an example, what can be expected from efforts in detector development.

As colloids are often very polydisperse, the application of speed profiles is essential. If the detectors can be designed in such a way that the data acquisition rate is in the ms–s rate, which is possible with modern techniques, the speed profiles can be designed to be very fast to suppress diffusion broadening. So a future sedimentation velocity experiment could be designed so that a fast acceleration profile from 0–60 000 rpm is used and the experiment is finished after 5–15 min, provided that the sample is big enough to have fully sedimented.

The increasing amount of information from multiple AUC detection systems, as well as its useful combination with other analytical techniques to increase the information content for complex systems, makes global analysis essential. Owing to the increasing complexity of colloidal systems, the increasing importance of hybrid systems (organic–organic, inorganic–inorganic, organic–inorganic), and the importance of functional colloids with programmed self-assembly properties – potential candidates for size, shape, density, and charge polydispersity – global analysis will be the only option to characterize such systems in terms of whole distributions of the above quantities.

Despite the existence of better detection systems and analysis approaches, the future could show the establishment of new ultracentrifuge methods. AUC is so versatile that it is still possible, after 80 years of application of this technique, to set up new methods such as, for example, a synthetic boundary technique for the kinetic study of membrane growth processes[164] or crystallization.[162,163] Further actual examples include pH gradients,[113] which can be potentially used for particle charge determination or observation of the smallest crystallizing species. The number of quantities made accessible by AUC are expected to increase further, especially after the expected nanotechnology hype brought further complicated samples into the analysis labs. This may help to lead to a broader application of AUC than at present with a focus on biopolymers, and could extend AUC to a general application in the industrially and scientifically important fields of synthetic polymers, latexes, pigments, functional colloids, nucleation, crystal growth studies, and so on.

Acknowledgements

The author thanks numerous people for their contributions to this work: Prof. Dr. Dr. h.c. M. Antonietti for the working environment and useful discussions. B. Zilske, A. Völkel, Drs. G. Lucas L. Börger, T. Pauck and K. Schilling for their skilful AUC measurements. Dr. H. Schnablegger is acknowledged for the DLS studies on ZrO_2 and Dr. H. Wachernig for the SLS on antibody coated latexes. We also thank Dr. C. Göltner for the TEM micrographs of the antibody coated latexes. Drs. G. A. Braun,

F. Caruso, S. Eda, A. Fischer, D. I. Gittins, J. Kaufmann, U. Kobold, and D. Rapoport are acknowledged for the supply of interesting colloidal samples. Dr. D. I. Gittins is also thanked for making graphical material available to the authors. Also, big thanks go to my wife Stefanie Sender for her patience. Finally, I thank the Max Planck Society for the financial support of this work and Erich C. for stimulating discussions.

References

1. T. Svedberg and J. B. Nichols, *J. Am. Chem. Soc.*, 1923, **45**, 2910.
2. T. Svedberg and K. O. Pedersen, *The Ultracentrifuge*, Clarendon Press, Oxford 1940.
3. T. Svedberg and H. Rinde, *J. Am. Chem. Soc.*, 1924, **46**, 2677.
4. W. Mächtle, *Progr. Colloid Polym. Sci.*, 1991, **86**, 111.
5. H. Lange, *Part. Part. Syst. Charact.*, 1995, **12**, 148.
6. W. Mächtle and L. Börger, *Analytical Ultracentrifugation of Polymers and Nanoparticles*, Springer, Berlin 2005.
7. G. Schmid, R. Pfeil, R. Boese, F. Bandermann, S. Meyer, G. H. M. Calis and J. W. A. van der Velden, *Chem. Ber.*, 1981, **114**, 3634.
8. D. H. Rapoport, W. Vogel, H. Cölfen and R. Schlögl, *J. Phys. Chem. B*, 1997, **101**, 4175.
9. M. D. Lechner and W. Mächtle, *Progr. Colloid Polym. Sci.*, 1999, **113**, 37.
10. H. G. Müller, *Colloid Polym. Sci.*, 1989, **267**, 1113.
11. W. Mächtle, "Analysis of polymer dispersions with an eight cell AUC multiplexer: High resolution particle size distribution and density gradient techniques" in *Analytical Ultracentrifugation in Biochemistry and Polymer Science*, S. E. Harding, A. J. Rowe and J. C. Horton (eds), Royal Society of Chemistry, Cambridge, UK, 1992, 147.
12. H. G. Müller, *Progr. Colloid Polym. Sci.*, 2004, **127**, 9.
13. H. Cölfen, S. Tirosh and A. Zaban, *Langmuir*, 2003, **19**, 10654.
14. R. Giebeler "The Optima XL-A: A new analytical ultracentrifuge with a novel precision absorption optical system" in *Analytical Ultracentrifugation in Biochemistry and Polymer Science*, S. E. Harding, A. J. Rowe and J. C. Horton (eds), Royal Society of Chemistry, Cambridge, UK, 1992, 16.
15. T. M. Laue, "On-line data aquisition and analysis from the Rayleigh interferometer" in *Analytical Ultracentrifugation in Biochemistry and Polymer Science*, S. E. Harding, A. J. Rowe and J. C. Horton (eds), Royal Society of Chemistry, Cambridge, UK, 1992, 63.
16. A. Rowe, S. Wynne Jones, D. G. Thomas and S. E. Harding, "Methods for off-line analysis of sedimentation velocity and sedimentation equilibrium patterns" in *Analytical Ultracentrifugation in Biochemistry and Polymer Science*, S. E. Harding, A. J. Rowe and J. C. Horton (eds), Royal Society of Chemistry, Cambridge, UK, 1992, 49.
17. H. Cölfen and W. Borchard, "Ultrasensitive Schlieren optical system" in *Biochemical Diagnostic Instrumentation*, R. F. Bonner, G. E. Cohn, T. M. Laue, A. Priezzhev (eds), Proc. SPIE 1994, **2136**, 307.
18. U. Klodwig and W. Mächtle, *Colloid Polym. Sci.*, 1989, **267**, 1117.
19. A. C. Clewlow, N. Errington and A. J. Rowe, *Eur. Biophys. J.*, 1997, **25**, 311.
20. W. Mächtle, *Progr. Colloid Polym. Sci.*, 1999, **113**, 1.
21. L. Börger, M. D. Lechner and M. Stadler, *Progr. Colloid Polym. Sci.*, 2004, **127**, 19.
22. P. H. Lloyd, *Optical Methods in Ultracentrifugation, Electrophoresis and Diffusion*, Oxford University Press, Oxford, 1974.
23. B. Schmidt and D. Riesner "A fluorescence detection system for the analytical ultracentrifuge and its application to proteins, nucleic acids, viroids and viruses" in *Analytical*

Ultracentrifugation in Biochemistry and Polymer Science, S. E. Harding, A. J. Rowe and J. C. Horton (eds), Royal Society of Chemistry, Cambridge, UK, 1992, 176.

24. L. K. MacGregor and T. M. Laue, *Biophys. J.*, 1999 **76**, A357.
25. H. J. Cantow, *Makromol. Chem.*, 1964, **70**, 130.
26. W. Scholtan and H. Lange, *Kolloid Z. Z. Polym.*, 1972, **250**, 782.
27. M. S. Runge, T. M. Laue, D. A. Yphantis, M. R. Lifsics, A. Saito, M. Altin,, K. Reinke and R. C. Williams, Jr., *Proc. Natl. Acad. Sci. USA*, 1981, **78**, 1431.
28. A. Böhm, S. Kielhorn Bayer and P. Rossmanith, *Progr. Colloid Polym. Sci.*, 1999, **113**, 121.
29. W. E. Stafford and E. H. Braswell, *Biophys. Chem.* 2004, **108**, 273.
30. B. Demeler and K. E. van Holde, *Anal. Biochem.*, 2004, **335**, 279.
31. O. Lamm, *Ark. Math. Astron. Fysik*, 1929, **21B**, Nr. 2, 1.
32. J. Kubal, *Chemie (Prague)*, 1949, **5**, 21.
33. P. O. Kinell, *Acta Chem. Scand.*, 1947, **1**, 335.
34. R. Signer and H. Gross, *Helv. Chim. Acta*, 1934, **17**, 726.
35. R. J. Goldberg, *J. Phys. Chem.*, 1953, **57**, 194.
36. N. Muramatsu and A. P. Minton, *Anal. Biochem.*, 1988, **168**, 345.
37. K. E. van Holde and W. O. Weischet, *Biopolymers*, 1978, **17**, 1387.
38. B. Demeler, H. Saber and J. C. Hansen, *Biophys. J.*, 1997, **72**, 397.
39. J. Geiselmann, T. D. Yager, S. C. Gill, P. Camettes and P. H. von Hippel, *Biochemistry*, 1992, **31**, 111.
40. S. C. Gill, T. D. Yager and P. H. von Hippel, *J. Mol. Biol.*, 1991, **220**, 325.
41. J. C. Hansen and D. Lohr, *J. Biol. Chem.*, 1993, **268**, 5840.
42. J. C. Hansen, J. Ausio, V. H. Stanik and K. E. van Holde, *Biochemistry*, 1989, **28**, 9129.
43. D. A. Yphantis, *Biophys. J.*, 1984, **45**, 324a.
44. W. F. Stafford, *Anal. Biochem.*, 1992, **203**, 295.
45. W. F. Stafford, *Method. Enzymol.*, 2000, **323**, 302.
46. W. F. Stafford, *Biophys. J.*, 1994, **66**, A280.
47. W. F. Stafford, *Biophys. J.*, 1998, **74**, A301.
48. J. Behlke and O. Ristau, *Eur. Biophys. J.*, 1997, **25**, 325.
49. J. Behlke and A. Knespel, *J. Cryst. Growth*, 1996, **158**, 388.
50. L. A. Holladay, *Biophys. Chem.*, 1979, **10**, 187.
51. J. Behlke and O. Ristau, *Progr. Colloid Polym. Sci.*, 1997, **107**, 27.
52. L. A. Holladay, *Biophys. Chem.*, 1980, **11**, 303.
53. J. Philo, "Measuring sedimentation, diffusion and molecular weights of small mmolecules by direct fitting of sedimentation velocity profiles" in *Modern Analytical Ultracentrifugation*, T. M. Schuster and T. M. Laue (eds), Birkhäuser, Boston, Basel, Berlin, 1994, 156.
54. J. Behlke and O. Ristau, *Biophys. J.*, 1997, **72**, 428.
55. J. Philo, *Biophys. J.*, 1997, **72**, 435.
56. B. Demeler and H. Saber, *Biophys. J.*, 1998, **74**, 444.
57. P. Schuck, C. E. McPhee and G. J. Howlett, *Biophys. J.*, 1998, **74**, 466.
58. P. Schuck, *Biophys. J.*, 1998, **75**, 1503.
59. P. Schuck and D. B. Millar, *Anal. Biochem.*, 1998, **259**, 48.
60. P. Schuck, *Biophys. J.*, 2000, **78**, 1606.
61. P. Schuck, M. A. Perugini, N. R. Gonzales, G. J. Howlett and D. Schubert, *Biophys. J.*, 2002, **82**, 1096.
62. P. Schuck, *Biophys. Chem.*, 2004, **108**, 201.
63. P. Schuck, *Biophys. Chem.*, 2004, **108**, 187.
64. W. F Stafford and P. J. Sherwood, *Biophys. Chem.*, 2004, **108**, 231.
65. K. A. Kraus and J. S. Johnson, *J. Am. Chem. Soc.*, 1953, **75**, 5769.

66. J. S. Johnson and K. A. Kraus, *J. Am. Chem. Soc.*, 1956, **78**, 3937.
67. D. Schubert, J. A. van den Broek, B. Sell, H. Durchschlag, W. Mächtle, U. S. Schubert and J. M. Lehn, *Progr. Colloid Polym. Sci.*, 1997, **107**, 166.
68. D. Schubert, C. Tziatzios, P. Schuck and U. S. Schubert, *Chem. Eur. J.*, 1999, **5**, 1377.
69. C. Tziatzios, H. Durchschlag, B. Sell, J. A. van den Broek, W. Mächtle, W. Haase, J. M. Lehn, C. H. Weidl, C. Eschbaumer, D. Schubert and U. S. Schubert, *Progr. Colloid Polym. Sci.*, 1999, **113**, 114.
70. J. M. Creeth and R. H. Pain, *Progr. Biophys. Mol. Biol.*, 1967, **17**, 217.
71. W. D. Lansing and E. O. Kraemer, *J. Am. Chem. Soc.*, 1935, **57**, 1369.
72. H. Fujita, *Foundations of Ultracentrifugal Analysis*, John Wiley, New York, London, Sydney, Toronto, 1975.
73. M. D. Lechner, "Determination of molecular weight averages and molecular weight distributions from sedimentation equilibrium" in *Analytical Ultracentrifugation in Biochemistry and Polymer Science*, S. E. Harding, A. J. Rowe and J. C. Horton (eds), Royal Society of Chemistry, Cambridge, UK, 1992, 295.
74. M. L. Johnson, J. J. Correia, D. A. Yphantis and H. R. Halvorson, *Biophys. J.*, 1981, **36**, 575.
75. M. L. Johnson and S. G. Frazier "Nonlinear least squares analysis" in *Methods in Enzymology*, Vol 117, C. H. W. Hirs and S. N. Timasheff (eds), Academic Press, San Diego, 1985.
76. D. K. McRorie and P. J. Voelker, *Self-Associating Systems in the Analytical Ultracentrifuge*, Beckman Instruments, Fullerton, California, 1993.
77. D. A. Yphantis and T. Arakawa, *Biochemistry*, 1987, **26**, 5422.
78. N. G. Dolinnaya, E. H. Braswell, J. A. Fossella, H. Klump and J.R. Fresco, *Biochemistry*, 1993, **32**, 10263.
79. P. B. Harbury, T. Zhang, P. S. Kim and T. Alber, *Science*, 1993, **262**, 1401.
80. J. Liu, T. M. Laue, H. U. Choi, L. H. Tang and L. Rosenberg, *J. Biol. Chem.*, 1994, **269**, 28366.
81. T. M. Laue, M. A. Starovasnick, R. E. Klevit and H. Weintraub, *Proc. Natl. Acad. Sci.*, 1995, **92**, 11824.
82. C. G. Lon, E. H. Braswell, D. Zhu, J. Apigo, J. Baum and B. Bodsky, *Biochemistry*, 1993, **32**, 11688.
83. D. R. Hall, S. E. Harding and D. J. Winzor, *Progr. Colloid Polym. Sci.*, 1999, **113**, 62.
84. D. Schubert and P. Schuck, *Progr. Colloid Polym. Sci.*, 1991, **86**, 12.
85. J. M. Creeth and S. E. Harding, *J. Biochem. Biophys. Methods*, 1982, **7**, 25.
86. S. E. Harding, J. C. Horton and P. J. Morgan, "MSTAR: A Fortran program for the model independent molecular weight analysis of macromolecules using low speed or high speed sedimentation equilibrium" in *Analytical Ultracentrifugation in Biochemistry and Polymer Science*, S. E. Harding, A. J. Rowe and J. C. Horton (eds), Royal Society of Chemistry, Cambridge, UK, 1992, 275.
87. H. Cölfen and S. E. Harding, *Eur. Biophys. J.*, 1997, **25**, 333.
88. B. K. Milthorpe, P. D. Jeffrey and L. W. Nichol, *Biophys. Chem.*, 1975, **3**, 169.
89. D. J. Winzor and P. R. Wills, "The omega analysis and the characterization of solute self-association by sedimentation equilibrium" in *Modern Analytical Ultracentrifugation*, T. M. Schuster and T. M. Laue (eds), Birkhäuser, Boston, Basel, Berlin, 1994, 66.
90. P. R. Wills, M. P. Jacobsen and D. J. Winzor, *Biopolymers*, 1996, **38**, 119.
91. H. Cölfen and D. J. Winzor, *Progr. Colloid Polym. Sci.*, 1997, **107**, 36.
92. P. R. Wills, M. P. Jacobsen and D. J, Winzor, *Progr. Colloid Polym. Sci.*, 1997, **107**, 1.
93. G. B. Ralston and M. B. Morris, "The use of the Omega function for sedimentation equilibrium analysis" in *Analytical Ultracentrifugation in Biochemistry and Polymer*

Science, S. E. Harding, A. J. Rowe and J. C. Horton (eds), Royal Society of Chemistry, Cambridge, UK, 1992, 253.

94. T. G. Scholte, *J. Polym. Sci.*, 1968, **6** Part A-2, 111.

95. M. D. Lechner and W. Mächtle, *Makromol. Chem. Rapid Commun.*, 1992, **13**, 555.

96. M. D. Lechner , "Determination of molecular weight averages and molecular weight distributions from sedimentation equilibrium" in *Analytical Ultracentrifugation in Biochemistry and Polymer Science*, S. E. Harding, A. J. Rowe and J. C. Horton (eds), Royal Society of Chemistry, Cambridge, UK, 1992, 295.

97. J. Mlynar and S. Sarkanen, "Renaissance in ultracentrifugal sedimentation equilibrium calibrations of size exclusion chromatographic elution profiles" *ACS Sym. Ser.*, 1996, **635**, 379.

98. H. Cölfen, *Colloid Polym. Sci.*, 1995, **273**, 1101.

99. H. Cölfen, *Biotechnol. Genet. Eng.*, 1999, **16**, 87,

100. P. Johnson, *Proc. Royal Soc.*, 1964, **A278**, 527.

101. P. Johnson and J. C. Metcalfe, *J. Photogr. Sci.*, 1963, **11**, 214.

102. J. C, Metcalfe, *Ph.D. Thesis*, Cambr idge, 1965.

103. P. Johnson and J. C. Metcalfe, *Eur. Polym. J.*, 1967, **3**, 423.

104. P. Johnson and R. W. King, *J. Photograph. Sci.*, 1968, **16**, 82.

105. P. Johnson, Chem. Soc. Special Publication No. 23, Burlington House, London, 1968, 243.

106. D. Kisters, A. Straatmann and W. Borchard, *Progr. Colloid Polym. Sci.*, 2002, **119**, 92.

107. W. Borchard and A. Straatmann, *Progr. Colloid Polym. Sci.*, 2004, **127**, 14.

108. W. Borchard and H. Cölfen, *Macromol. Chem., Macromol. Symp.*, 1992, **61**, 143.

109. G. Holtus, H. Cölfen and W. Borchard, *Progr. Colloid Polym. Sci.*, 1991, **86**, 92.

110. U. P. Schröder and W. Oppermann, "Properties of polyelectrolyte gels" in *Physical Properties of Polymer Gels*, J. P. Cohen Addad, John Wiley & Sons, 1996, 19.

111. H. Hinsken, E. Selic and W. Borchard, *Progr. Colloid Polym. Sci.*, 1995, **99**, 154.

112. J. M. Messer, T. M. Laue and T. P. Moody, *Biophys. J.*, 1997, **72**, TU 469.

113. G. Lucas, L. Börger and H. Cölfen, *Progr. Colloid Polym. Sci.*, 2002, **119**, 11.

114. L. Börger, Ph.D. Thesis, Potsdam, 2000.

115. G. Lucas, Ph.D. Thesis, Potsdam, 2003.

116. M. Meselson, F. W. Stahl and J. Vinograd, *Proc. Nat. Acad. Sci.* (USA), 1957, **43**, 581.

117. J. J. Hermans and H. A. Ende, *J. Polym. Sci.*, 1963, **1** Part. C, 161.

118. J. J. Hermans, *J. Polym. Sci.*, 1963, **1** Part. C, 179.

119. H. A. Ende, *Makromol. Chem.*, 1965, **88**, 159.

120. M. D. Lechner, *Macromol. Rapid Commun.*, 1997, **18**, 781.

121. M. D. Lechner, W. Mächtle and U. Sedlack, *Progr. Colloid Polym. Sci.*, 1997, **107**, 148.

122. M. D. Lechner and W. Borchard, *Eur. Polym. J.*, 1999, **35**, 371.

123. H. Lange, *Colloid Polym. Sci.*, 1980, **258**, 1077.

124. W. Mächtle, *Colloid Polym. Sci.*, 1984, **262**, 270.

125. H. Cölfen and K. Schilling, *Progr. Collod Polym. Sci.*, 1999, **113**, 44.

126. H. Rinde, Ph.D. Thesis, Upsala. 1928.

127. J. B. Nichols, *Physics*, 1931, **1**, 254.

128. J. B. Nichols, E. O. Kramer and E. D. Bailey, *J. Phys. Chem.*, 1932, **36**, 326.

129. J. B. Nichols, E. O. Kraemer and E. D. Bailey, *J. Phys. Chem.*, 1932, **36**, 505.

130. H. Cölfen, Habilitation Thesis, Potsdam. 2001.

131. H. Cölfen, H. Schnablegger, A. Fischer, F. C. Jentoft, G. Weinberg and R. Schlögl, *Langmuir*, 2002, **18**, 3500.

132. H. G. Müller and F. Herrmann, *Progr. Colloid Polym. Sci.*, 1995, **99**, 114.

133. W. Mächtle, *Biophys. J.*, 1999, **76**, 1080.

134. H. Cölfen and T. Pauck, *Colloid Polym. Sci.*, 1997, **275**, 175.
135. H. G. Müller, *Progr. Colloid Polym. Sci.*, 1997, **107**, 180.
136. E. Suito, N. Ueda and M. Katsumi, *Bull. Inst. Chem. Res. Kyoto University*, 1959, **36**, 197.
137. E. A. Hauser and H. K. Schachmann, *J. Phys. Chem.*, 1940, **44**, 584.
138. E. G. Pickels, *Colloid Chem.*, 1944, **5**, 411.
139. P. von Mutzenbecher, *Angew. Chem.*, 1938, **51**, 633.
140. A. Singhal, L. M. Toth, J. S. Lin and K. Affholter, *J. Am. Chem. Soc.*, 1996, **118**, 11529.
141. A. Singhal, L. M. Toth, G. Beaucage, J. S. Lin and J. Peterson, *J. Coll. Interf. Sci.*, 1997, **194**, 470.
142. L. M. Toth, J. S. Lin and L. K. Felker, *J. Phys. Chem.*, 1991, **95**, 3106.
143. J. Bolze, B. Peng, N. Dingenouts, P. Panine, T. Narayanan and M. Ballauff *Langmuir*, 2002, **18**, 8364.
144. H. Cölfen, A. Völkel, S. Eda, U. Kobold, J. Kaufmann, A. C. G. Puhlmann and H. Wachernig, *Langmuir*, 2002, **18**, 7623.
145. S. Eda, J. Kaufmann, W. Roos and S. Pohl, *J. Clin. Lab. Anal.*, 1998, **12**, 137.
146. D. G. Kurth, P. Lehmann, D. Volkmer, H. Cölfen, M. J. Koop, A. Müller and A. DuChesne, *Chem. Eur. J.*, 2000, **6**, 385.
147. M. Antonietti, M. Breulmann, C. G. Göltner, H. Cölfen, K. K. W. Wong, D. Walsh and S. Mann, *Chem. Eur. J.*, 1998, **4**, 2493.
148. S. Mann, *Chem. in unserer Zeit*, 1986, **20**, 69.
149. H. Cölfen and A. Völkel, *Eur. Biophys. J.*, 2003, **32**, 432.
150. W. Mächtle, *Makromol. Chem.*, 1984, **185**, 1025.
151. K. Schilling and H. Cölfen, *Progr. Colloid Polym. Sci.*, 1999, **113**, 50.
152. R. van Roij, *J. Condens. Matter*, 2003, **15**, S3569.
153. A. P. Hynninen, R. van Roij and M. Dijkstra, *Europhys. Lett.*, 2004, **65**, 719.
154. P. Budd, "Sedimentation analysis of synthetic polyelectrolytes" in *Analytical Ultracentrifugation in Biochemistry and Polymer Science*, S. E. Harding, A. J. Rowe and J. C. Horton (eds), Royal Society of Chemistry, Cambridge, UK, 1992, 593.
155. M. Rasa and A. P. Philipse, *Nature*, 2004, **429**, 857.
156. P. M. Budd, R. K. Pinfield and C. Price, *Progr. Colloid Polym. Sci.*, 1997, **107**, 189.
157. A. Seifert, K. Strenge, M. Schultz and H. Schmandtke, *Die Nahrung*, 1991, **9**, 989.
158. A. Seifert and K. D. Schwenke, *Progr. Colloid Polym. Sci.*, 1995, **99**, 31.
159. H. G. Müller, A. Schmidt and D. Kranz, *Progr. Colloid Polym. Sci.*, 1991, **86**, 70.
160. W. Mächtle, G. Ley and J. Streib, *Progr. Colloid Polym. Sci.*, 1995, **99**, 144.
161. R. Cohen, *C. R. Acad. Sci. (Paris)*, 1963, **256**, 3513.
162. L. Börger and H. Cölfen, *Progr. Colloid Polym. Sci.*, 1999, **113**, 23.
163. L. Börger, H. Cölfen and M. Antonietti, *Colloid. Surface. A*, 2000, **163**, 29.
164. C. Wandrey and A. Bartkowiak, *Colloid. Surface. A*, 2001, **180**, 141.
165. C. Wandrey, G. Grigorescu and D. Hunkeler, *Progr. Colloid Polym. Sci.*, 2002, **119**, 84.
166. L. Bourdillon and C. Wandrey, *Colloid. Polym. Sci.*, 2004, **282**, 1247.
167. F. Jones, H. Cölfen and M. Antonietti, *Colloid Polym. Sci.* 2000, **278**, 491.
168. N. Karibyants, H. Dautzenberg and H. Cölfen, *Macromolecules*, 1997, **30**, 7803.
169. P. Voelker, *Progr. Colloid Polym. Sci.*, 1995, **99**, 162.
170. D. I. Gittins and F. Caruso, *Chem. Phys. Chem.*, 2002, **3**, 110.
171. P. Schuck, *Progr. Colloid Polym. Sci.* 1994, **94**, 1.
172. M. S. Lewis, R. I. Shrager and S. J. Kim, "Analysis of protein–nucleic acid and protein–protein interactions using multi-wavelength scans from the XL-I analytical ultracentrifuge" in *Modern Analytical Ultracentrifugation*, T. M. Schuster and T. M. Laue (eds), Birkhäuser, Boston, Basel, Berlin, 1994, 94.
173. A. P. Minton, *Progr. Colloid Polym. Sci.*, 1997, **107**, 11.

174. A. Balbo, K. H. Minor, C. A. Velikovsky, R. A. Mariuzza, C. B. Peterson and P. Schuck, *Proc. Natl. Acad. Sci. USA.*, 2005, **102**, 81.

175. S. Bhattacharyya, P. Maciejewska, L. Börger, A. M. Gülsün, H. B. Cicek and H. Cölfen, *Progr. Colloid Polym. Sci.*, 2005, submitted.

176. K. E. van Holde and R. L. Baldwin, *J. Phys. Chem.*, 1958, **62**, 734.

177. J. J. Correia and D. A. Yphantis, "Equilibrium sedimentation in short columns" in *Analytical Ultracentrifugation in Biochemistry and Polymer Science*, S. E. Harding, A. J. Rowe and J. C. Horton (eds), Royal Society of Chemistry, Cambridge, UK, 1992, 231.

178. P. Schuck and B. Demeler, *Biophys. J.*, 1999, **76**, 2288.

179. P. Schuck, *Anal. Biochem.*, 1999, **272**, 199.

180. P. Schuck and P. Rossmanith, *Biopolymers*, 2000, **54**, 328.

181. P. Schuck, *Presentation on Adv. in AUC and Hydro 2002 Meeting*, Autrans, France 8–11.6, 2002.

182. A. C. Clewelow, N. Errington and A. J. Rowe, *Eur. Biophys. J.*, 1997, **25**, 311.

183. H. Cölfen and W. Borchard, *Progr. Colloid. Polym. Sci.*, 1994, **94**, 90.

184. D. Kisters and W. Borchard, *Progr. Colloid. Polym. Sci.*, 1999, **113**, 10.

185. P. Lavrenko, V. Lavrenko and V. Tsvetkov, *Progr. Colloid Polym. Sci.*, 1999, **113**, 14.

186. T. M. Laue and H. Cölfen, *Presentations on Adv. in AUC and Hydro 2002 Meeting*, Autrans, France 8 – 11.6., 2002.

187. M. Schütte, D. G. Kurth, M. R. Linford, H. Cölfen and H. Möhwald, *Angew. Chem. Int. Ed.*, 1998, **37**, 2891.

188. P. M. Budd, "Sedimentation analysis of synthetic polymers" in *Analytical Ultracentrifugation in Biochemistry and Polymer Science*, S. E. Harding, A. J. Rowe and J. C. Horton (eds), Royal Society of Chemistry, Cambridge, UK, 1992, 593.

189. E. Görnitz, M. Hahn, W. Jaeger and H. Dautzenberg, *Progr. Colloid Polym. Sci.*, 1997, **107**, 127.

190. G. J. Timofejeva, S. A. Pavlova, C. Wandrey, W. Jaeger, M. Hahn. K. J. Linow and E. Görnitz, *Acta Polymer.*, 1990, **41**, 479.

191. C. Wandrey and E. Görnitz, *Acta Polymer*, 1992, **43**, 320.

192. M. Hahn, E. Görnitz and H. Dautzenberg, *Macromolecules*, 1998, **31**, 5616.

193. R.W. Roxby, "Sedimentation analysis of micelle forming systems" in *Analytical Ultracentrifugation in Biochemistry and Polymer Science*, S. E. Harding, A. J. Rowe and J. C. Horton (eds), Royal Society of Chemistry, Cambridge, UK, 1992, 609.

194. M. Pacovska, K. Prochazka, Z. Tuzar and P. Munk, *Polymer*, 1993, **34**, 4585.

195. S. A. Hagan, A. G. A. Coombes, M. C. Garnett, S. E. Dunn, M. C. Davies, L. Illum and S. S. Davis, *Langmuir*, 1996, **12**, 2153.

196. N. A. Chebotareva, B. I. Kurganov and A. A. Burlakova, *Progr. Colloid Polym. Sci.*, 1999, **113**, 129

197. G. M. Pavlov, E. V. Korneeva, K. Jumel, S. E. Harding, E. W. Meijer, H. W. I. Peerlings, J. Fraser Stoddart and S. A. Nepogodiev, *Carbohydr. Polym.*, 1999, **38**, 195.

198. G. M. Pavlov, E. V. Korneeva, R Roy, N. A. Michailova, P. C. Ortega and M. A. Perez, *Progr. Colloid Polym. Sci.*, 1999, **113**, 150.

199. L. H. Bronstein, S. N. Sidorov, P. M. Valetsky, J. Hartmann, H. Cölfen and M. Antonietti, *Langmuir*, 1999, **15**, 6256.

200. S. N. Sidorov, L. M. Bronstein, P. M. Valetsky, J. Hartmann, H. Cölfen, H. Schnablegger and M. Antonietti, *J. Colloid Interf. Sci.*, 1999, **212**, 197.

201. P. Wyatt, US Patent 0017568 A1, 2004.

203. W. F. Stafford, "Methods for obtaining sedimentation coefficient distributions" in *Analytical Ultracentrifugation in Biochemistry and Polymer Science*, S. E. Harding, A. J. Rowe and J. C. Horton (eds), Royal Society of Chemistry, Cambridge, UK, 1992, 359.

204. H. Cölfen, "Analysis of nanoparticles < 10 nm by analytical ultracentrifugation" in T. Provder and J. Texter (eds), *Particle Sizing and Characterization*, American Chemical Society, Washington, DC, 2004, 119.
205. K. Schilling, Ph.D. Thesis, Potsdam, 1999.
206. H. Cölfen, Ph.D. Thesis, Duisburg, 1993.
207. K. Strenge and A. Seifert, *Progr. Colloid Polym. Sci.*, 1991, **86**, 76.
208. R. Buchdahl, H. A. Ende and L. H. Peebles, *J. Polym. Sci.*, 1963, **1** Part C, 143.
209. H. Cölfen, T. Pauck and M. Antonietti, *Progr. Colloid Polym. Sci.*, 1997, **107**, 136.
210. H. Cölfen, *Polymer News*, 2004, **29**, 101.
211. H. Cölfen, "Analytical ultracentrifugation of nanoparticles", in *Encyclopedia of Nanoscience and Nanotechnology*, H.S. Nalwa (ed), American Scientific Publishers, Los Angeles, 2004, 67.
212. H. Cölfen and A. Völkel, *Progr. Colloid Polym. Sci.*, 2004, **127**, 31.
213. M. Brust, D. Bethell, D. J. Schiffrin and C. J. Kiely, *Adv. Mater.*, 1995, **7**, 795.
214. J. P. Wilcoxon, US Patent 5147841, 1992.
215. W. Mächtle and M. D. Lechner, *Progr. Colloid Polym. Sci.*, 2002, **119**, 1.
216. A. Bootz, V. Vogel, D. Schubert and J. Kreuter, *Eur. J. Pharma. Biopharma*, 2004, **57**, 369.
217. S. Kirsch, A. Dörk, E. Bartsch, H. Sillescu, K. Landfester, H. W. Spiess and W. Mächtle, *Macromolecules*, 1999, **32**, 4508.
218. K. S. Mayya, B. Schöler and F. Caruso, *Adv. Funct. Mater.*, 2003, **13**, 183.
219. H. Döllefeld, K. Hoppe, J. Kolny, K. Schilling, H. Weller and A. Eychmüller, *Phys. Chem. Chem. Phys.*, 2002, **4**, 4747.
220. T. Kuntzsch, M. Hollatz, M. Stintz and S. Ripperger, *Chem. Ingenieur Tech.*, 2002, **74**, 1151.
221. M. Lundqvist, I. Sethson and B. H. Johnson, *Langmuir*, 2004, **20**, 10639.
222. J. B. Nichols, E. O. Kraemer and E. D. Bailey, *J. Phys. Chem.*, 1932, **36**, 326.
223. D. I. Gittins and F. Caruso, *Angew. Chem. Int. Ed.*, 2001, **40**, 3001.
224. A. S. Deshpande, N. Pinna, P. Beato, M. Antonietti and M. Niederberger, *Chem. Mater.*, 2004, **16**, 2599.
225. M. Niederberger, G. Garnweitner, F. Krummeich, R. Nesper, H. Cölfen and M. Antonietti, *Chem. Mater.*, 2004, **16**, 1202.
226. A. Seifert, N. Buske and K. Strenge, *Colloids. Surface*, 1991, **57**, 267.
227. A. Seifert and N. Buske, *J. Magnetism Magn. Mater.*, 1993, **122**, 115.
228. E. E. Remsen, K. B. Thurmond and K. L. Wooley, *Macromolecules*, 1999, **32**, 3685.
229. K. Tauer and U. Yildiz, *Macromolecules*, 2003, **36**, 8638.
230. V. Vogel, K. Langer, S. Balthasar, P. Schuck, W. Mächtle, W. Haase, J. A. van den Broek, C. Tziatzios and D. Schubert, *Progr. Colloid Polym. Sci.*, 2002, **119**, 31.
231. K. Langer, S. Balthasar, V. Vogel, N. Dinauer, H. von Briesen and D. Schubert., *Int. J. Pharmaceut.*, 2003, **257**, 169.
232. J. P. Krause, R. Wustneck, A. Seifert and K. D. Schwenke, *Coll. Surf. B.*, 1998, **10**, 119.
233. E. D. Bailey, J. B. Burton and E. O. Krämer, *J. Phys. Chem.*, 1936, **40**, 1149.
234. V. Vogel, J. F. Gohy, B. G. G. Lohmeijer, J. A. van den Broek, W. Haase, U. S. Schubert and D. Schubert, *J. Polym. Sci. A1*, 2003, **41**, 3159.
235. G. Mayer, V. Vogel, B. G. G. Lohmeijer, J. F. Gohy, J. A. van den Broek, W. Haase, U. S. Schubert and D. Schubert, *J. Polym. Sci. A1*, 2004, **42**, 4458.
236. D. V. Pergushov, E. V. Remizova, J. Feldthusen, A. B. Zezin, A. H. E. Müller and V. A. Kabanov, *J. Phys. Chem. B*, 2003, **107**, 8093.
237. A. F. Thünemann, K. Sander, W. Jaeger and R. Dimova, *Langmuir*, 2002, **18**, 5099.
238. A. Lustig, A. Engel and M. Zulauf, *Biochim. Biophys. Acta*, 1991, **1115**, 89.
239. H. D. Lee and P. Somasundaran, *J. Disper. Sci. Technol.*, 2002, **23**, 483.

240. R. Zhang and P. Somasundaran, *Langmuir*, 2004, **20**, 8552.
241. G. W. M. Vandermeulen, C. Tziatzios and H. A. Klok, *Macromolecules*, 2003, **36**, 4107.
242. B. H. Robinson, T. F. Towey, S. Zourab, A. J. W. G. Visser and A. Vanhoek, *Colloid. Surface.*, 1991, **61**, 175.
243. L. M. Bronstein, D. M. Chernychov, G. I. Timofeeva, L. V. Dubrovina, P. M. Valetsky, E. S. Obolonkova and A. R. Khokhlov, *Langmuir*, 2000, **16**, 3626.
244. D. I. Gittins and F. Caruso, *Chem. Phys. Chem.*, 2002, **1**, 110.
245. H. M. Farrell, E. D. Wickham, H. J. Dower, Piotrowski, P. D. Hoagland, P. H. Cooke and M. L. Groves, *J. Protein Chem.*, 1999, **18**, 637.
246. H. M. Farrell, T. F. Kumosinski, P. H. Cooke, P. D. Hoagland, E. D. Wickham, J. J. Unruh and M. L. Groves, *Int. Dairy J.*, 1999, **9**, 193.
247. N. D. Bryan, M. N. Jones, J. Birkett and F. R. Livens, *Anal. Chim. Acta*, 2001, **437**, 291.
248. M. Wohlgemuth, W. Mächtle and C. Mayer, *J. Microencapsul.*, 2000, **17**, 437.

Subject Index